Instructor's Complete Solu[...]

for

Differential Equations

Third Edition

Paul Blanchard
Boston University

Robert L. Devaney
Boston University

Glen R. Hall
Boston University

THOMSON
BROOKS/COLE

Australia • Brazil • Canada • Mexico • Singapore • Spain • United Kingdom • United States

MW01680222

© 2006 Thomson Brooks/Cole, a part of The Thomson Corporation. Thomson, the Star logo, and Brooks/Cole are trademarks used herein under license.

ALL RIGHTS RESERVED. No part of this work covered by the copyright hereon may be reproduced or used in any form or by any means—graphic, electronic, or mechanical, including photocopying, recording, taping, Web distribution, information storage and retrieval systems, or in any other manner—except as may be permitted by the license terms herein.

Printed in the United States of America

1 2 3 4 5 6 7 09 08 07 06

Printer: Thomson/West

0-495-01404-4

Thomson Higher Education
10 Davis Drive
Belmont, CA 94002-3098
USA

For more information about our products, contact us at:
Thomson Learning Academic Resource Ce[illegible]
1-800-423-0563

For permission to use material from this text or product, submit a request online at **http://www.thomsonrights.com.**
Any additional questions about permissions can be submitted by email to **thomsonrights@thomson.com.**

CONTENTS

3 Linear Systems 203

4 Forcing and Resonance 353

5 Nonlinear Systems 421

6 Laplace Transforms 501

SECTION BY SECTION SUMMARIES

CHAPTER 1: FIRST-ORDER DIFFERENTIAL EQUATIONS

Two of our major goals throughout the text are

- to give a balanced presentation of analytic, qualitative, and numerical techniques for studying solutions of ordinary differential equations and
- to make the applications of the subject as transparent as possible.

To accomplish these goals, we continually introduce new models and modify old ones using all three types of techniques whenever possible. In Chapter 1 we introduce, by example, all of these approaches in the context of first-order equations. Most of the topics in this chapter can be found in many other texts. The difference lies in the order and emphasis we place on the topics.

Chapter 1 divides naturally into two parts. Sections 1.1–1.5 constitute an introduction to modeling and to examples of analytic, qualitative, and numerical techniques along with the Existence and the Uniqueness Theorems (and their applications). These sections are essential both in material and in outlook to the rest of the text.

Sections 1.6–1.9 discuss more specialized topics, and not every section is essential. Section 1.6 (phase lines) anticipates the discussion of the phase plane in Chapter 2, and Section 1.8 (linear equations) is also very important. Students tend to enjoy the models discussed in Section 1.7 (bifurcations), but this section can be skipped if necessary. Also, integrating factors for linear equations (Section 1.9) are not used in subsequent chapters.

At some schools with an engineering-oriented course, an early introduction to Laplace transforms is desirable. Consequently, we have written Sections 6.1 and 6.2 so that they can be covered immediately after Chapter 1. However, in our experience, it is best to wait until Chapter 4 has been covered before discussing Laplace transforms.

1.1 Modeling via Differential Equations

Our main goal in this section is to demonstrate that differential equations arise naturally in "real world" systems and that solutions should be interpreted in terms of the original system. In building models, we emphasize that there is a connection between the assumptions made regarding the physical system and the individual terms in the differential equation. In studying solutions, we stress that the formula and/or the graph of a solution provides a prediction about the future behavior of the physical system under consideration. We also emphasize that models are imperfect and should be criticized and improved if their predictions do not correspond to reality.

DETools

Using `HPGSolver` to graph solutions to the logistic equation is a good way to introduce students to the use of technology in this course, but keep the slope field turned off until Section 1.3.

Comments on selected exercises

Exercises 1 and 2 are straightforward exercises involving exponential growth.

Exercises 3–5 begin the process of thinking qualitatively about differential equations.

In Exercise 4, students are asked to model data with an exponential model. However, exponential growth gives a poor fit to the data as well as ridiculous predictions of the future population. If pressed, students will suggest a logistic model as an alternative.

Exercises 7–9 involve a simple learning model (see also Lab 1.1).

Exercises 10–14 deal with radioactive decay. Many students already know the formula for the solutions and some wonder why bother with the differential equation. Exercise 13 points out the importance of the differential equation.

Exercises 15 and 16 are "do-able" modeling problems. In general, students find modifying existing models easier than creating new ones from scratch. A standard mistake in part (a) of Exercise 15 is to subtract $100t$ instead of 100.

Exercise 17 is very challenging for students, perhaps because there are multiple "correct" answers, i.e., reasonable models that match the assumptions.

Exercises 19–22 deal with systems of differential equations (population models with two populations), but only in terms of understanding the relationship between terms in the equations and assumptions about the physical system.

1.2 Analytic Technique: Separation of Variables

In this section we give the first examples of the "analytic" approach (i.e., finding closed-form formulas for solutions) to differential equations. We begin by emphasizing that solutions of differential equations can be checked by substitution into the differential equation. Understanding this fact is fundamental to learning the relationship between a differential equation and its solutions, but it is a fact that students often forget later in the course. Repeated reference to this fact throughout the course is important.

The only equations that we consider at this point are separable, and we include the standard mixing and compound interest examples. The only nonstandard aspect of this section is the observation that, even if an equation is separable, there is no guarantee that the solution is obtainable in closed-form since either the integrals or the algebra may be impossible. The mixing example returns in Section 1.3 where it is analyzed qualitatively.

Students find this section fairly straightforward and familiar. Although the students may need some review of their techniques of integration, the only real difficulty is in dealing with absolute value signs. Most textbooks ignore this technical detail, and there is merit to that approach.

DETools

`HPGSolver` can be used to illustrate the solution to an initial-value problem as well as the general solution of a differential equation. `FirstOrderExamples` also does a good job of illustrating the concept of a general solution, and five of its equations are separable.

Comments on selected exercises

Knowing what it means to check a solution by "plugging it back into" the differential equation is an excellent way to reinforce what a solution is. Exercise 1 illustrates this fact, and this check is requested in many other problems throughout the book.

In Exercises 2 and 3, students are asked to construct a differential equation given a solution. This change in point of view is difficult for many students.

In Exercise 4, the students solve $dP/dt = kP$ by separating variables.

Exercises 5–32 are standard examples of separable differential equations. However, the algebra necessary to solve for y as a function of t in Exercises 19 and 22 is impossible. The answers to these problems must be left in implicit form. All of the integrals in Exercises 5–32 have closed-form expressions. However, Exercise 22 is challenging.

Exercises 33–39 are fairly standard (and difficult) "word problems." However, Exercise 36 is difficult because setting up a mixing problem in terms of concentration is tricky. In Exercise 35, the first step is to determine the proportionality constant for Newton's law of cooling. Wherever possible, we ask for a description of the long-term behavior of a solution to promote qualitative thinking.

1.3 Qualitative Technique: Slope Fields

In this section we introduce our first geometric and qualitative technique—slope fields. We emphasize that the slope field is a tool to help in sketching the graphs of solutions. This point helps drive home the idea that solutions are functions. We also encourage describing the behavior of solutions qualitatively. The discussion of slope fields leads naturally to Euler's method as well as issues of uniqueness of solutions.

We introduce RC circuits as an example in this section, but we do not develop the physics or the circuit theory. Many, but not all, of our students take electric circuit theory concurrently. For these students, developing the theory of RC circuits is review (or preview). For students who are not studying engineering, the topic is entirely new. Hence, we omit the theory and start with the differential equation.

DETools

It is extremely helpful at this point to have some sort of technology available for drawing slope fields, e.g., `HPGSolver`. Start with the "Draw Slopes" box checked and place individual slope marks at random places in the ty-plane. Then turn on the slope field, and illustrate the relationship between the slope field and graphs of solutions. `FirstOrderExamples` and `TargetPractice` are also useful for illustrating slope fields.

Comments on selected exercises

Exercises 1–6 are very easy with technology, but it is worthwhile to encourage students to do them to gain intuition about what slope fields look like and to practice sketching solutions.

Exercises 7–10 involve practice going from slope field pictures to graphs of solutions.

In Exercise 15, since technology can be used to determine which slope field is which and that is not the point of the exercise, students should be encouraged (required) to write a paragraph that describes what features of the slope fields they used to determine which is which.

Exercises 11, 13, 14, 16, and 17 are somewhat more challenging theoretical problems relating slope fields and solutions.

Exercise 18(c) is a good excuse to start worrying about the uniqueness of solutions.

Exercises 19–23 involve the RC circuit model. They serve as a good advertisement for qualitative and numeric methods.

Exercise 24 involves a discontinuous differential equation (a population model where a disease is introduced at time $t = 5$). Again, qualitative and numeric methods are best. `HPGSolver` has a `step` function to handle equations of this sort. Assigning only parts (a), (b), and (d), avoiding finding solutions in closed form, is a reasonable modification of this problem.

1.4 Numerical Technique: Euler's Method

In this section we introduce Euler's method. This topic naturally follows slope fields (as a way to have the computer sketch accurately what we can sketch roughly by hand from the slope field). Introducing a numerical method early, even one as elementary as Euler's method, is important to legitimize the use of the technology which is essential for the labs and for the development of qualitative techniques in our approach.

A careful analysis of error in Euler's method is given in Section 7.1, and that section can be covered at this point in the course. Improved Euler's method and Runge-Kutta are also presented in Chapter 7 (see the section-by-section comments for Chapter 7).

We have tried to promote a healthy skepticism—a middle ground between blind faith and paranoia—for numerical methods. Numerical methods usually do work, and in many cases they are the technique of choice even when more theoretical approaches are available. The importance of interpretation must be stressed.

DETools

`EulersMethod` is a tool that illustrates Euler's method for a variety of equations. It can be used in class when Euler's method is first introduced, and Exercises 1–4 refer to it specifically.

Comments on selected exercises

Exercises 1–10 are computational practice with Euler's method. The first four refer to `EulersMethod` but could be done without using the tool. The computations in Exercises 5–10 must be done some other way. We have had very good luck using a spreadsheet for these computations. The step sizes in Exercises 1–8 are very large so that only a few computations must be done.

In Exercises 11–13, the inaccuracies of "large step" Euler's method are discussed. This is another good opportunity to begin speculation on the Uniqueness Theorem. (How do we know solutions don't jump over equilibrium solutions?)

Exercises 14 and 15 discuss the geometry of Euler's method. Comparing approximate solutions with the slope field is encouraged.

In Exercises 16–19, the appropriate step size must be determined via experimentation.

Exercises 20 and 21 are applications of Euler's method to root finding. They preview the discussion of phase lines (see Section 1.6).

1.5 Existence and Uniqueness of Solutions

In this section we present the existence and uniqueness theory for first-order equations. We also consider the issue of the domain of definition of a solution—a theoretical point that has been ignored in Sections 1.1–1.4.

Existence is stated and then taken for granted. Uniqueness, however, is emphasized as a very useful tool for the qualitative study of solutions. This is a good opportunity to show that "abstract theorems" are actually quite useful in applied mathematics.

We also deal with questions of the domain of definition for solutions. The exceptional student will have wondered about domains of definition in Section 1.2 where restricted domains are the norm. We take the dynamical systems point of view that a solution that escapes to infinity or that encounters a singular point of the differential equation at a finite time cannot be extended beyond that time. For example, the function $y(t) = 1/t$ for $t > 0$ is a completely different solution to the equation $dy/dt = -y^2$ than the solution $y(t) = 1/t$ for $t < 0$.

DETools

`HPGSolver` is useful both during class and on the homework. Also, the equation $dy/dt = y/t + t\cos t$ in `FirstOrderExamples` is a good example for the discussion of uniqueness.

Comments on selected exercises

Exercises 1–9 encourage the use of the Uniqueness Theorem to take a small amount of information about solutions of a differential equation and derive information about other solutions. The information thus gained is qualitative.

Exercises 10 and 18 both deal with the situation where the Uniqueness Theorem does not hold. In general we try not to make too big a deal of lack of uniqueness. However, for some of our students, this is the first time they have used a theoretical result in a practical way and they need to remember that theorems have hypotheses.

Exercise 11 (showing that solutions of autonomous equations cannot have local maxima and minima) is difficult.

Exercises 12–17 are exercises that address the domain of definition issue. They are also a good review of Section 1.2.

1.6 Equilibria and the Phase Line

This section deals exclusively with autonomous equations. It introduces the notion of a phase line. The goal is to promote qualitative analysis. The section also sets the stage for the concept of the phase plane, which will be introduced in Chapter 2.

At this point, students sometimes feel overwhelmed with graphs. For a single equation of the form $dy/dt = f(y)$, they have the graph of $f(y)$, the slope field, the graphs of solutions, and the phase line. We try to emphasize that this is good—the more ways of representing a differential equation we have, the more ways we have of obtaining information about its solutions.

The section is long and difficult to cover in a single one-hour lecture. The material at the end of the section on linearization (pp. 87–89) can be skipped with no serious ramifications. This material is included as a preview of the topic of linearization as it is discussed in Chapter 5, but that discussion is a long way off.

DETools

`PhaseLines` can be used to illustrate the relationships among the graphs of solutions, the phase line, and the graph of the right-hand side of the differential equation. At this point you probably want to keep the bifurcation plane hidden. The two autonomous equations in `FirstOrderExamples` are also useful early in the discussion of this material.

Comments on selected exercises

Exercises 1–28 involve drawing phase lines and using them to obtain sketches and other qualitative information about solutions. The role of the Uniqueness Theorem can be emphasized.

Exercises 29–36 concern the drawing of phase lines using only qualitative information regarding the right-hand side of the differential equation. In Exercises 33–36, the graph of the right-hand side of the equation is sketched from the phase line.

In Exercise 37, students match phase lines with equations from a list. We always insist on valid justifications for their choices.

Exercise 38 concerns possible types of phase lines. Even though this exercise is quite abstract, students usually understand the ideas easily.

Exercises 39 and 40 also involve the drawing of phase lines using only qualitative information regarding the right-hand side of the differential equation.

Exercises 41 and 42 use the `PhaseLines` tool to anticipate the discussion of one-parameter families of differential equations in Section 1.7.

Exercise 43 concerns the behavior of solutions near equilbria in cases where linearization is not conclusive.

Exercise 44 compares solutions for an equation with singularities on the phase line.

Exercises 45–48 can be applied to any transit system where busses and trains run frequently without a rigid schedule. The interpretation is difficult but, from painful experiment, fairly accurate.

1.7 Bifurcations

Highlighting the role and importance of parameters in the study of differential equations is one of our major goals. Using phase lines, parameters can be made quite geometric and accessible. This section also reinforces the notion of a phase line.

Simply having some familarity with the term "bifurcation" is very useful in subsequent sections. For example, an understanding of node equilibrium points and bifurcations on phase lines helps a great deal when discussing critically damped oscillators and linear systems that have 0 as an eigenvalue. If time permits, the bifurcation diagram can also be discussed.

Some of us were initially hesitant to tackle this subject at this level. However, after doing so, we became true believers. Exercise 20 illustrates the power of the ideas introduced in this section.

Some of us view this section as central to our approach while others consider it to be an interesting optional section. In the latter case, this section is skipped during "short" semesters.

DETools

This section is the appropriate place to turn on the bifurcation plane in `PhaseLines`.

Comments on selected exercises

Exercises 1–8 consider one-parameter families. The students are asked to find the bifurcation values.

Exercises 9 and 10 ask for bifurcation values based on graphs for the right-hand side of the differential equation.

In Exercises 11 and 12, students must provide graphs that are consistent with the bifurcations described.

Exercises 13 and 14 are more abstract. They ask if it is possible to find a one-parameter family of a certain type with specified bifurcations.

Exercises 15–21 return to population and harvesting examples. Exercise 20 requires interpretation of bifurcation diagram. The conclusions are quite striking.

Exercises 22 and 23 use the bifurcation plane feature in `PhaseLines`.

1.8 Linear Equations

In this section, we introduce linear differential equations, and we discuss the special structure of their solution sets as given by the Linearity Principle and the Extended Linearity Principle. This presentation anticipates the treatment of linear systems in Chapter 3 and second-order linear equations in Chapter 4. We also discuss the Method of Undetermined Coefficients for constant-coefficient equations rather than integrating factors. (Integrating factors are discussed in the next section.) This approach pays dividends later and is also consistent with the what our colleagues in engineering do in their courses.

DETools

`HPGSolver` can be used to illustrate the Linearity Principle and the Extended Linearity Principle.

Comments on selected exercises

Exercises 1–12 provide routine practice finding general solutions and solutions to initial-value problems.

In Exercises 13 and 14, students are asked to explain why the method works the way that it does.

Exercises 15 and 16 involve graphical illustrations of the two linearity principles.

Exercises 17 and 18 involve theoretical aspects of the Linearity Principle.

Exercises 19–24 illustrate how we deal with forcing functions that are sums of the elementary examples.

Exercises 25–28 are qualitative in nature. In fact, they are a little tricky. If you assign them, you may want to hint to your students that they should visualize the slope fields.

Exercises 29–31 are investment problems.

In Exercise 32, students are asked to verify an assertion made in the section. This exercise is not difficult if they remember the easy way to check a possible solution.

Exercise 33 leads the students through a verification of the Extended Linearity Principle. This verification was left to the reader in the section.

Exercise 34 is a tricky theoretical exercise involving the Linearity Principle.

1.9 Integrating Factors for Linear Equations

For many instructors, using integrating factors to solve first-order linear equations is a necessity. However, others feel that the approach to linear equations that we take in Section 1.8 is sufficient. Consequently, we end this chapter with a discussion of integrating factors and leave the decision to cover this technique to the instructor. Some prefer to start Chapter 2 as quickly as possible while others think that it is good for the students to see another analytic technique at this point in the course. Integrating factors are mentioned two or three times throughout the rest of the text, but they are never used to solve an equation. Therefore, this section can be omitted without serious consequences.

Our presentation in this section is fairly standard, although we have made an attempt to motivate the ideas behind integrating factors as much as possible. We have also tried to be honest about the difficulties that frequently arise in the integrals that come from even relatively simple linear equations.

Comments on selected exercises

Exercises 1–12 are standard. In some cases, the integration is challenging.

Exercises 13–20 consider the possibility of encountering impossible integrals when solving linear equations. They point out how often this occurs.

Exercise 21 asks the student to compare the method of integrating factors with the guessing technique described in Section 1.8.

Exercise 22 is a theoretical exercise that relates the abstract formula for the general solution obtained in this section with the Extended Linearity Principle from Section 1.8.

Exercise 23 uses integrating factors to justify the "second" guess that we introduced in Section 1.8.

Exercises 24–27 are mixing problems. Exercise 27 considers the extreme case when the initial volume is zero (division by zero is a danger).

Review Exercises

Exercises 1–10 are "short answer" exercises. The answers are (usually) one or two sentences. Most (but not all) are relatively straightforward.

Exercises 11–20 are true/false problems. We always expect our students to justify their answers.

Exercises 21–40 are routine exercises where students must derive general solutions or solve initial-value problems. Exercises 26, 32, and 39 involve integrating factors.

Exercise 41 involves Euler's method and a solution that blows up in finite time.

Exercises 42–45 involve qualitative analysis in one form or another.

Exercise 46 investigates an equation that is both linear and separable.

Exercise 47 concerns the Uniqueness Theorem.

Exercise 48 involves Euler's method and phase lines.

Exercise 49 uses Newton's law of cooling

Exercises 50–51 are bifurcation problems.

Exercise 52 involves compound interest.

Exercise 53 investigates the phase line of a simple linear equation.

Exercise 54 is a modeling exercise based on the number of cell phone subscribers in the U.S.

Exercise 55 examines solutions that are defined for all time for a particular equation.

Exercises 56 and 57 are mixing problems.

Comments on the Labs

The labs are written with the expectation that the students will use a numerical solver such as `HPGSolver`. It is surprising how reluctant some students are to use technology.

Lab 1.1: Rate of Memorization Model
Because the model equation can be solved easily, this lab can be done without technology. (In fact, most students chose to solve the equation analytically and fit the data to the solution rather than take advantage of numerical techniques.) There is considerable variability in the results of the experiment. This lab can be started immediately after Section 1.1.

Lab 1.2: Growth of a Population of Mold
While this lab is a lot of fun, it is also subject to many difficulties. Some pieces of bread just dry out, leaving the student with no data. Also, be prepared for all sorts of strange stories why the report is not done on time—my dog ate my homework might actually be true. It seems that, on most pieces of bread, the mold stops growing before the entire piece is covered, i.e., the carrying capacity is less than 1. This lab can be started immediately after Section 1.1, and it takes at least two weeks to collect a reasonable amount of mold.

Lab 1.3: Logistic Population Models with Harvesting
The equations in this lab are difficult to analyze theoretically. Hence, the use of a solver is a must. This lab relates most closely with Section 1.7, but it can be done even if the material in Section 1.7 is not discussed in class. It has proven to be one of our most successful.

Lab 1.4: Exponential and Logistic Population Models
This lab points out the difficulties in making generalizations while modeling. There is considerable variability in the behavior of the populations of the states. The process of accepting or rejecting and adapting a model is new to many, if not most, students. Also, determining parameter values from the data is an important concept that many have not seen before. This lab can be started immediately after Section 1.1 is covered.

Lab 1.5: Modeling Oil Production
This lab involves U.S. and world oil production from 1920–2000. The students are asked to use the logistic to model oil production and then analyze their results.

CHAPTER 2: FIRST-ORDER SYSTEMS

Chapter 2 marks a radical departure from the traditional ordering of topics. Rather than following first-order equations with second-order, linear equations, we go directly to autonomous first-order systems. The emphasis in this chapter is on interpreting first-order systems and their solutions and on qualitative techniques. One reason for going directly to systems is that it leads to a better qualitative understanding of autonomous second-order equations (achieving a major goal of the standard approach). It opens up the course to a wealth of topics, both theoretical and applied, associated with the modern theory of dynamical systems. A second, more technical reason is that the numerical treatment of second-order equations requires the use of first-order systems. Consequently, if numerical techniques are to have an equal role in the course, we must be able to discuss systems and vector fields at this point.

This chapter is organized in much the same way as the first half of Chapter 1. It begins with a section on modeling in which we return to the predator-prey models mentioned in Section 1.1. We also discuss the undamped harmonic oscillator. The next three sections are devoted to introducing our three-pronged approach in terms of the study of systems of differential equations. Section 2.2 introduces geometric and qualitative concepts, Section 2.3 presents some analytic techniques that apply to special systems, and Section 2.4 describes Euler's method as it is applied to systems. Unlike Chapter 1, we do not feel the need for a separate section on existence and uniqueness theory, so the Existence and Uniqueness Theorem is presented at the end of Section 2.4. Section 2.5 begins the qualitative discussion of the Lorenz system as an introduction to three-dimensional systems. This last section is the first part of a discussion of three-dimensional systems that reappears in Sections 3.8, 5.5, and 8.5. These four sections can be safely omitted if necessary. (At Boston University, we often omit these sections due to time constraints.)

Although all of Sections 2.1–2.4 contain material that is consistently used throughout the remainder of the text, it is conceivable that one could skip to Chapter 3 immediately after Section 2.2.

The use of technology to draw vector fields, direction fields, and phase portraits is essential. In order to begin to get a feel for what pictures to expect, students must see many examples accurately drawn. `HPGSystemSolver` often fits the bill. The CD also contains other special purpose tools that are useful at various times (see the comments below).

2.1 Modeling via Systems

This modeling section is similar in purpose to Section 1.1. Our goal is to introduce the relationship between various graphical representations of a system of differential equations, its solutions, and the interpretation of the solutions in terms of the model. The graphical representations include the phase portrait and graphs of the component functions of the solutions.

The transition to systems is natural for students. Relating the solution curve in the phase plane to the graphs of the components is considered difficult by some but is generally mastered after sufficient effort. (More practice to help develop this skill is provided in Section 2.2.) It is important for students to realize that both types of graphs are necessary because neither graph alone contains all of the information about a solution. A good analogy is trying to understand a slinky from its shadows (see Figure 2.16). Bring a slinky to class. It can be used both to demonstrate a mass-spring system and to illustrate Figure 2.16.

The section is based on predator-prey models and the undamped harmonic oscillator. We focus on these two models because we feel that these are two of the best to use at this point. In the case of the predator-prey equations, we delay an explanation of how we obtain the graphs of solutions until Section 2.4. For the undamped harmonic oscillator we simply guess solutions. At this point, we do not have enough theory to discuss the general solution. Many of the exercises focus on interpretation—both of the equations in a system and of the behavior of the solutions.

DETools

In addition to the general purpose `HPGSystemSolver`, the special purpose tools `PredatorPrey` and `MassSpring` provide easy-to-run classroom demonstrations that are very relevant to this section.

Comments on selected exercises

Exercises 1 and 15 involve the interpretation of the parameters in a system while Exercises 9–14 involve the interpretation of the equations.

Exercises 2–6 require an analysis of a predator-prey system similar to the one carried out in the section.

Exercises 7, 8, 16, and 17 give practice in going from the phase plane to the graphs of component functions and in the interpretation of solutions. Exercise 17 is particularly good for assigning an essay. (This predator-prey phenomenon really does occur.)

In Exercises 9–14 and 18, modifications are made to a predator-prey model. Modifying an existing model is easier than developing models from scratch, but it is still challenging. In Exercise 18, there is more than one reasonable answer.

Exercises 19 and 20 provide practice with solutions of equations for undamped harmonic oscillators. Both of these exercises prepare the students for ideas that are discussed in more detail in Sections 2.2 and 2.3.

Exercises 21–24 are modeling problems that involve Hooke's law applied to mass-spring systems.

In Exercises 25–30, models for concentrations of reactants in simple chemical reactions are developed. These models reappear in Section 5.2 (Exercises 16–20).

2.2 The Geometry of Systems

This section introduces geometric and qualitative ideas that we use throughout the remainder of the text. Vector fields, direction fields, solution curves in the phase plane, equilibrium solutions, and phase portraits are discussed. Vector fields for autonomous systems play the role of slope fields for first-order equations. They are the fundamental geometric representation of a system of differential equations.

DETools

In addition to the general purpose `HPGSystemSolver`, the special purpose tools `DESketchPad` and `GraphingSolutionsQuiz` help students understand the relationship between a solution curve in the phase plane and the corresponding $x(t)$- and $y(t)$-graphs.

Comments on selected exercises

Exercises 1–6 involve vector fields and phase portraits for very simple systems.

Exercises 7 and 8 involve vector fields and phase portraits for simple second-order equations.

In Exercises 9 and 10, we provide the direction field, and the student sketches the phase portrait. A description of the long-term behavior of a particular solution is also requested.

Exercise 11 is a "matching" exercise reminiscent of the matching exercise in Section 1.3. This type of exercise is an excellent way to test understanding of geometric concepts such as slope fields and direction fields. If brief explanations are required to justify why a given system corresponds to a particular direction field, students must examine the fields closely.

In Exercise 12, students find equilibrium solutions for a predator-prey system discussed in Section 2.1.

In Exercises 13–18, students are expected to solve for the equilibrium points by hand, but then they produce the phase portraits with the aid of a computer. One important goal is to give them practice using paper-and-pencil techniques and technology in the same exercise. Exercise 19 is a similar exercise for a second-order equation (an example of Duffing's equation).

In Exercise 21, students match solution curves with their corresponding $x(t)$- and $y(t)$-graphs.

Exercise 22 is one of our favorite "animated" homework problems.

In Exercises 23–26, students must produce $x(t)$- and $y(t)$-graphs from solution curves in the xy-phase plane. Their answers should include a scale on the x, y-axis, but they cannot provide a scale on the t-axis.

Exercise 27 goes the opposite direction as Exercises 23–26. The $x(t)$- and $y(t)$-graphs are given, and the student must draw the corresponding solution curve. In theory, doing this exercise is fairly mechanical, but try it before you assign it.

Exercises 28 and 29 are meant to motivate the study of uniqueness. These two exercises are repeated (exactly) as Exercises 12 and 13 in Section 2.4. Logically they make more sense in Section 2.4, but they are also stated here in case you want to get your students thinking about these issues before uniqueness is discussed.

2.3 Analytic Methods for Special Systems

Many students are uncomfortable with the heavy emphasis on geometry in Sections 2.1 and 2.2. Consequently, this section is devoted to certain special systems for which closed-form solutions can be obtained. This is also a good time to repeat the observation that any given function can always be checked to see if it is a solution, and we begin this section with that observation.

Using the method of undetermined coefficients from Section 1.8 or integrating factors from Section 1.9, we also show how one can derive the general solution of a "partially" decoupled linear system. Knowing how to solve this class of systems is useful when we study repeated eigenvalues in Chapter 3, and it provides the students with a systematic method for deriving the solutions to a system.

We also take this opportunity to introduce damped harmonic oscillators in this section. This model is used throughout the text, so the terms and equations presented on pages 193 and 194 are extremely important. Since this section is devoted to techniques that produce closed-form solutions, we also introduce the technique of guessing solutions of the form $y(t) = e^{st}$. The Linearity Principle does not appear until Section 3.1, so we stop short of finding general solutions of the form $y(t) = k_1e^{s_1t} + k_2e^{s_2t}$. However, we do make the connection between solutions of the form $y(t) = e^{st}$ and the corresponding vector-valued solution $\mathbf{Y}(t) = (y(t), v(t)) = (e^{st}, se^{st})$.

DETools

Some of the solutions obtained in this section can be illustrated with `MassSpring`, and `HPGSystem-Solver` can be used to plot phase portraits and $x(t)$- and $y(t)$-graphs.

Comments on selected exercises

Exercises 1–4 provide routine practice at checking functions to see if they are solutions.

Exercises 5–12 involve a given partially decoupled system. They illustrate the technique introduced in this section, and they provide additional practice with the geometric ideas introduced in Section 2.2.

Exercises 13–14 involve the analysis of a mass-spring system that involves two springs.

Exercises 15–18 provide the same type of practice for second-order equations. These equations reappear in Exercises 21–24 of Section 3.2.

Exercise 19 is a nonlinear partially decoupled system.

2.4 Euler's Method for Autonomous Systems

Section 2.4 generalizes Euler's method to systems, and we are now able to explain how the graphs and solution curves for the predator-prey equations in Sections 2.1 and 2.2 are obtained. Students usually have little difficulty understanding Euler's method for systems if they have a solid understanding of Section 1.4 and the vector field ideas introduced in Section 2.2.

If Section 1.4 was supplemented with material from Chapter 7, then it is also possible to supplement this section with material from Section 7.3.

As has already been mentioned, it is also convenient to discuss existence and uniqueness theory for systems at this point.

The example of a swaying building is presented in this section because quantitative information determines which model is more appropriate.

DETools

`EulersMethodForSystems` can be used in class to demonstrate the method, and the students are encouraged to use this tool when they answer Exercises 3–6.

Comments on selected exercises

Exercises 1 and 2 involve computing Euler's method solutions with fairly large step sizes and comparing the results with the direction field and/or actual solutions. The computations are tedious but manageable if done by hand, but the expectation is that the students will use some type of technology such as a spreadsheet to help with the calculations.

Exercises 3–6 were designed to be completed using `EulersMethodForSystems`. The phase portraits in Exercises 5 and 6 are entertaining.

Exercises 7–9 provide some practice at checking solutions, but their main intent is to illustrate the implications of the Uniqueness Theorem. Exercises 12–15 are more abstract versions of the same question.

Exercises 10–11 illustrate how Euler's method for systems applies to second-order equations.

2.5 The Lorenz Equations

We introduce the Lorenz system here mainly because it is possible to do so. Almost none of our students have seen any modern (i.e., post 1800) mathematics, and they are surprised to learn that there are unanswered questions and that there is active research in mathematics. At this point, we can only describe the Lorenz system and display some numerical solutions. Consequently, this is something of a "golly-gee-whiz" section. Three-dimensional linear systems are discussed in Section 3.8, and the Lorenz system is studied more carefully in Sections 5.5 and 8.5.

If you cover this section, we recommend your mentioning James Gleick's book *Chaos*. There are also a number of interesting videos that have been produced. They usually do a better job of illustrating the solution curves than we can do with our solvers. One of our favorite videos is *Fractals: An Animated Discussion*, by Peitgen, Jürgens, Saupe, and Zahlten, published by W. H. Freeman, New York. At this writing, there is also a Java applet (`http://www.geom.uiuc.edu/java/Lorenz/`) written by Patrick Worfolk available at the Geometry Center.

DETools

There are two tools, `LorenzEquations` and `ButterflyEffect`, that illustrate the Lorenz attractor.

Comments on selected exercises

Exercises 1–3 and 5 cover aspects of the Lorenz system that can be verified by hand.

Exercise 4 uses `ButterflyEffect` to illustrate sensitive dependence on initial conditions for the Lorenz attractor.

Review Exercises

Exercises 1–12 are "short answer" exercises. The answers are (usually) one or two sentences. Most (but not all) are relatively straightforward.

Exercises 13–24 are true/false problems. We always expect our students to justify their answers. Exercises 19-24 all refer to the same phase portrait.

Exercise 25 involves solving for equilibrium points and interpreting solutions.

Exercises 26–28 involve decoupled or partially decoupled systems.

In Exercises 29–32, students sketch $x(t)$- and $y(t)$-graphs from solution curves in the phase plane.

Exercise 33 corresponds to the simple model of a glider that is given in `HMSGlider`.

Exercise 34 relates to Euler's method for systems.

Comments on the Labs

All of these labs require technology that is capable of plotting phase portraits. The ability to draw graphs of the coordinate functions is also very useful. `HPGSystemSolver` is one option that works well.

Lab 2.1: Two Magnets and a Spring
This lab is a computer exploration that involves a bifurcation. It can be started as soon as Section 2.2 is covered, and it pairs up nicely with Lab 4.1.

Lab 2.2: Cooperative and Competitive Species Population Models
This lab can be started as soon as Section 2.2 is covered. It is mostly a computer exploration, but item 2 anticipates certain qualitative ideas that are discussed more specifically in later chapters. Particular attention should be paid to the interpretation of the solutions in physical terms.

Lab 2.3: The Harmonic Oscillator with Modified Damping
Section 2.3 must be covered before this lab can be assigned. This lab is designed to give students practice applying the ideas of this chapter to various second-order equations, both linear and nonlinear ones.

Lab 2.4: A Mass-Spring System with a Rubber Band
This lab is a modification of the standard mass-spring model. It involves both a spring and rubber band. It anticipates the discussion of the Tacoma Narrows Bridge in Chapter 4 and pairs up nicely with Lab 4.3.

Lab 2.5: Active Shock Absorbers
Recently developed magnetorheological fluids allow the damping coefficient of a harmonic oscillator to be adjusted in real time, and this lab examines various models where the damping coefficient is a function of the velocity of the system.

CHAPTER 3: LINEAR SYSTEMS

This chapter covers linear systems (mostly two-dimensional systems) as well as homogeneous, constant-coefficient, second-order linear equations. Linear algebra is not a prerequisite for our course, and we introduce the required concepts and notation as needed. We find that even those students who have had some linear algebra generally do not have a good geometric understanding of eigenvalues and eigenvectors, so most students learn some linear algebra while studying this chapter.

There are two examples that appear frequently throughout the chapter. They are the harmonic oscillator and Paul's and Bob's CD stores. The purpose of the CD store model is to provide an example for understanding the meaning of the coefficients of a linear system and for practicing interpretation of solutions in everyday language. Hopefully the readers will not be annoyed by our poking fun at each other.

Section 3.1 covers matrix notation and the basic properties of linear systems. In Sections 3.2 and 3.3, we study linear systems with two real, distinct, nonzero eigenvalues. Section 3.4 generalizes the discussion to the case of nonreal eigenvalues, and Section 3.5 covers certain special cases—repeated eigenvalues and zero eigenvalues. Section 3.6 is a summary of everything presented in Chapter 3 as it applies to second-order, linear *equations*. It completes the discussion started in Section 2.3. In Section 3.7 we summarize the results of the chapter in terms of the trace-determinant plane. Finally, in preparation for a more detailed discussion of the Lorenz equations, we briefly discuss three-dimensional linear systems in Section 3.8. Sections 3.1–3.6 are fundamental and should be covered completely. Sections 3.7 and 3.8 are less central and may be safely skipped.

DETools

We use `HPGSystemSolver` constantly throughout the chapter along with the special purpose tools mentioned below.

3.1 Properties of Linear Systems and The Linearity Principle

In this section we introduce matrix notation for linear systems. Most of our students have seen 2×2 matrices previously, often in high school. The determinant of a 2×2 matrix arises naturally when checking for equilibrium points of a linear system. We also mention that that a linear system of algebraic equations has nontrivial solutions if and only if the determinant of the matrix vanishes. This property is used in Section 3.2.

The main goal of the section is to present the Linearity Principle (or Principle of Superposition) and the idea that the general solution of a two-dimensional, linear system can be obtained from two independent solutions. Linearly independent solutions are defined as those with linearly independent initial conditions. The Wronskian appears in Exercise 35 in this section.

Comments on selected exercises

Exercises 1–4 consider the meaning of the parameters in the CD store model. Students are asked to describe the underlying assumptions built into these choices of coefficients.

Exercises 5–9 are straightforward practice with matrix notation for systems.

Exercises 10–13 use the geometric methods of Chapter 2 to view the direction field and the $x(t)$- and $y(t)$-graphs. Students should use technology to help with the direction field, but then they should be able to produce rough sketches of the phase portrait and the $x(t)$- and $y(t)$-graphs without technology.

Exercise 14 deals with the equilibrium points of a matrix whose determinant is nonzero (completing the discussion in the text). Exercise 15 discusses the case where $\det \mathbf{A} = 0$.

There are a number of exercises throughout the chapter on second-order equations, and it is important that the students have a lot of practice with the relationship between second-order equations and first-order systems. Exercises 16–19 are their first opportunity for this practice.

Exercises 20–23 consider a simple model of a housing market. These questions are similar to those of Exercises 1–4 (the CD stores).

Exercises 24–30, 33, and 34 concern the Linearity Principle. Exercises 26 and 27 involve the same initial-value problem, so the answers to part (c) are indentical. Making this observation in class after the students have done both exercises can lead to a valuable discussion.

Exercises 31, 32, and 35 are theoretical. They concern linear independence of vectors and solutions as well as the Wronskian.

3.2 Straight-Line Solutions

Sections 3.2 and 3.3 are closely related. In Section 3.2, we find solutions using eigenvalues and eigenvectors. In Section 3.3, we show how this technique leads to an understanding of the phase portrait if the eigenvalues of the linear system are nonzero, real, and distinct.

In Section 3.2, we begin with the geometric observation that certain solution curves for linear systems form straight lines (actually rays) in the phase plane. Using this geometric observation, we derive algebraic tools for finding solutions. Eigenvectors and eigenvalues are introduced geometrically ("Where does the vector field point directly toward or directly away from the origin?"). Then this geometric condition is converted into an algebraic condition. In turn, the algebra gives rise to the characteristic polynomial. Once eigenvalues and eigenvectors have been discussed, the formula for the corresponding particular solution is obtained and justified using a guess-and-test technique.

Although these solutions are exponentials, we call them "straight-line solutions" because they manifest themselves as lines in the phase plane. It is important to emphasize that these lines really correspond to functions $x(t)$ and $y(t)$ that are exponentials.

DETools

The `MatrixFields` tool helps with the geometric interpretation of eigenvectors. The vector **AY** is attached to the end of the position vector **Y**. Start the tool with both the direction field and eigenvalue plane turned off. After playing with the tool for a few minutes, turn on the direction field. Ultimately, you should turn on the eigenvalue plane to illustrate the eigendirections. The discussion of why the eigenvalues are plotted in the complex plane is delayed until Section 3.4.

Comments on selected exercises

Exercises 1–10 provide routine practice with the computation of eigenvalues, eigenvectors, straight-line solutions, and general solutions. The results of Exercises 1–3 and 6–10 can be used in Exercises 1–8 in Section 3.3.

Exercises 11–14 use eigenvalues and eigenvectors to solve initial-value problems. These initial-value problems reappear in the Exercises 9–12 of Section 3.3.

Exercises 15–18 are more theoretical. The students are asked to make general observations regarding eigenvalues of matrices of certain special forms.

Exercises 19–25 apply the techniques developed in this section to the important special case of second-order, homogeneous linear equations. Exercises 21–24 involve the same equations as those in Exercises 13–16 in Section 2.3. They also reappear in Exercises 13–16 in Section 3.3.

3.3 Phase Planes for Linear Systems with Real Eigenvalues

The qualitative analysis of linear systems with real, distinct, nonzero eigenvalues is completed in this section. We emphasize that a good qualitative picture of the system can be obtained from the eigenvalues and eigenvectors alone and that one can often get a good idea of the behavior of the solutions without having to compute them in complete detail.

DETools

In addition to `HPGSystemSolver`, one can use `LinearPhasePortraits` to illustrate the various types of phase portraits. We recommend keeping the trace-determinant plane hidden until Section 3.7.

Comments on selected exercises

Exercises 1–8 involve phase portraits corresponding to some of the systems that arose in the exercise set for Section 3.2. Similarly, Exercises 9–12 and 13–16 are related to Exercises 11–14 and 21–24 in Section 3.2, respectively.

The CD store returns in Exercises 17 and 18. Now the students have more tools at their disposal.

Exercises 19 and 20 have been very successful examination questions to test to see if students understand the relationship between the phase portrait and $x(t)$- and $y(t)$-graphs. Although one is tempted to use technology to help with these problems, they are most instructive if they are done without the use of technology.

Exercises 21 and 22 apply the techniques of this section to a particular harmonic oscillators.

Exercises 23–26 consider simple linear models of a pond inhabited by two species of fish. We ask the students to make predictions based on the phase portraits. These are good examples to revisit during the discussion of linearization in Section 5.1.

3.4 Complex Eigenvalues

The careful student will have already asked if all linear systems have straight-line solutions. In this section we deal with the case of complex eigenvalues. The idea is to use the algebra developed in the previous two sections to obtain a formula for the general solution and then to use the formula to obtain the qualitative analysis. Of particular importance is the idea of a natural period for a system. This important quantity is qualitative information that cannot be seen in the phase plane.

Most students have seen complex numbers before but need a little review in the basics, i.e., multiplication and division of complex numbers. Also, a review of (or an introduction to) Euler's formula is necessary. We have provided a very brief summary of some of this material in Appendix B.

Again, it is stressed that a great deal of qualitative (and quantitative) information can be obtained from the eigenvalues alone. When sketching phase portraits, the easiest way to determine whether solutions spiral clockwise or counter-clockwise is by looking at any nonzero vector in the vector field.

DETools

Both `MatrixFields` and `LinearPhasePortraits` illustrate the geometric significance of complex eigenvalues. When you use these tools, you should turn on the eigenvalue plane, but we still recommend that you keep the trace-determinant plane turned off in `LinearPhasePortraits` until Section 3.7.

Comments on selected exercises

Exercises 1 and 2 ask the student to produce the general solution from one complex eigenvalue and one eigenvector.

Exercises 3–8 and 9–14 are parallel groups of problems. In the first group, the students are asked to determine qualitative information about the solutions without actually finding closed-form expressions for them. In the second group, the students are asked to compute closed-form expressions for solutions and compare their results to those they obtained in the first group. At this point, the arithmetic of finding solutions starts to become fairly involved, and it is easy to assign more problems than the students can reasonably do.

Exercises 15 and 24 give the students practice working with the graphs that arise if the eigenvalues are complex.

Exercises 16–20 consider certain theoretical issues that arise in the case of complex eigenvalues.

Exercises 21 and 22 connect our approach with the approach that is more more common in core engineering courses (amplitude, phase, ...). We expand on these connections in Sections 4.2–4.4.

Exercise 23 is a theoretical exercise that relates the material of this section to the behavior of harmonic oscillators.

Exercise 25 is an essay on why there are no spiral saddles in two dimensions.

Exercise 26 involves an elliptical center.

3.5 The Special Cases: Repeated and Zero Eigenvalues

In this section we study the atypical cases. The amount of time spent on this section varies widely depending on how much detail you want the students to master. If you are so inclined, the details of how to find the general solution can be safely skipped. Lab 3.1 can be used as a way for students to discover much of the qualitative material from this section on their own.

We derive the general solution for a 2×2 system with repeated eigenvalues by taking advantage of the fact that every vector is a generalized eigenvector. This approach leads to general solutions that are expressed in terms of their initial conditions, and it should be contrasted with the approach that was taken in Sections 3.2–3.4.

DETools

`LinearPhasePortraits` can be used to illustrate the fact that the repeated eigenvalue case separates the case of real and distinct eigenvalues from complex eigenvalues. This is a good point to illustrate here as well as when the trace-determinant plane is discussed.

Comments on selected exercises

Exercises 1–4 and 5–8 is another parallel grouping of exercises. Once again, qualitative information is determined initially, and closed-form solutions are obtained later.

Exercises 17–19 and 21–22 are also computational, involving double roots and zero eigenvalues. Exercises 21 and 22 are a pair similar to Exercises 1–8.

Exercises 9–16, 20, and 23 are theoretical. In particular, Exercises 15 and 16 verify two claims that are made in the section.

3.6 Second-Order Linear Equations

When we mentioned the harmonic oscillator as an example in Sections 3.1–3.5, the students were expected to write the second-order equation, convert it to a first-order system, and then complete the analysis described in this chapter. In this section, we classify the harmonic oscillators and explain why it is not necessary to go through all of these steps. In other words, we complete the connection between the methods of this chapter and the "guessing" method described in Section 2.3. It is important that the students completely understand this simplification since it saves a great deal of labor in Chapter 4.

DETools

This is a good place to bring back the `MassSpring` tool. `RLCCircuits` with $A = 0$ is another tool that is appropriate for this section.

Comments on selected exercises

Exercises 1–12 are routine computational exercises. They ensure that the students have a basic understanding of the algebraic techniques described in this section.

Exercises 13–20 and 21–28 are parallel groups of exercises involving harmonic oscillators. By the end of these pairings, we hope that students have a good idea of what they can determine easily from the eigenvalues and eigenvectors and what they learn from calculating the closed-form solution.

Exercise 29 asks for a table of all possible harmonic oscillators. It is a precursor to the discussion included in Section 3.7.

Exercises 30 and 31 revisit the Linearity Principle for second-order equations.

Exercises 32 and 33 involve the relationship between the method described in this section and the eigenvalue/eigenvector methods of Sections 3.2–3.5.

Exercises 34 and 35 involve the behavior of solutions for a particular one-parameter family of second-order equations.

Exercises 36–40 describe applications that can be modeled by a second-order equation. We have always enjoyed discussing Exercise 37 with our students. Exercises 37–40 are designed to see if the students can make qualitative predictions if certain parameters are varied slightly. You may want to think about the answers to these exercises before the students appear at your office hours.

3.7 The Trace-Determinant Plane

In this section we present a classification of the behavior of linear systems using the trace-determinant plane. This section is essentially a compilation of everything that has been discussed earlier in the chapter, and therefore it can be omitted. However, the resulting diagram is quite appealing, and students seem to enjoy this method of classification (as opposed to the construction of a table as is suggested in Exercise 1 at the end of this section).

DETools

Now turn on the trace-determinant plane in the `LinearPhasePortraits` tool. Also, note the comment about Exercise 14 below.

Comments on selected exercises

A word of warning regarding the exercises for this section: A detailed analysis of the qualitative behavior of any of the one- and two-parameter families that are given in the exercises is not a task for the faint of heart. Do not assign many problems from this exercise set.

Exercise 14 is an "animated" exercise involving one-parameter families of phase portraits and paths in the trace-determinant plane.

3.8 Linear Systems in Three Dimensions

The algebra and geometry of three-dimensional linear systems is presented quickly and in a very matter-of-fact fashion in this section, and students without a background in linear algebra may find this section a bit terse. Our goals are to show that the ideas involved in the classification of two-dimensional linear systems generalize to higher dimensions and that, while the fundamental ideas remain the same in higher dimensions, the calculations become more complicated. This section is used only in the analysis of the Lorenz system in Section 5.5.

Comments on selected exercises

Exercises 1, 4–7, and 10–18 are computational. They take advantage of simple matrices or given information (either solutions or eigenvectors) to make the computations reasonable. However, even for these exercises, it is often difficult to visualize the solution curves in phase space.

Exercises 2 and 3 concern linear independence in three dimensions.

Exercises 8 and 9 recall some facts about cubic polynomials that help explain certain facts about eigenvalues.

Exercises 19–22 consider the CD store one last time (in three dimensions).

Review Exercises

Exercises 1–10 are "short answer" exercises. The answers are (usually) one or two sentences. Most (but not all) are relatively straightforward.

Exercises 11–18 are true/false problems. We always expect our students to justify their answers.

Exercises 19 and 21 are matching problems involving phase portraits and $x(t)$- and $y(t)$-graphs respectively.

Exercise 20 involves analyzing a one-parameter family of linear systems using the trace-determinant plane.

Exercise 22 involves systems that have a given line of equilibrium points and systems that have a given straight-line solution.

Exercises 23–26 are routine initial-value problems for second-order equations.

Exercises 27–32 involve general solutions, phase portraits, initial-value problems, and $x(t)$- and $y(t)$-graphs for an assorted collection of linear systems.

Comments on the Labs

Lab 3.1: Bifurcations in Linear Systems
Students find this lab difficult but very useful in understanding linear systems. Some students attempt to discover the diagram necessary for part 1 by using technology and trying (many) different values of a and b. Others realize immediately that this approach is hopelessly tedious and have no idea how to start. Compare this lab to some of the exercises in Section 3.7.

Lab 3.2: RLC Circuits
The equations modeling RLC circuits, which are frequently given as a main motivation why second-order equations must be studied prior to first-order systems, are actually first expressed as a first-order system in most circuit theory courses. The dependent variables are voltage across the capacitor and the current. In most circuit theory texts, this system is immediately converted into a second-order equation. This lab is meant to emphasize this point and to help our engineering students make the connection between our approach and what they are learning in their circuit theory class. This lab can be paired with Lab 4.2.

Lab 3.3: Measuring Mass in Space
A "harmonic-oscillator-like" device is actually used to measure the mass of astronauts during long space flights. ("How would you measure mass in space?" is a good question for discussion.)

Lab 3.4: Exploring a Parameter Space
This investigation of a three-parameter family of linear systems has resulted in some of our most spectacular lab "reports" ever. A number of our students who are artistically gifted have come up with especially creative ways of presenting their results.

Lab 3.5: Find Your Own Harmonic Oscillator
Part 1 of this lab can also be assigned as a "mini-lab" or homework essay. Some students have difficulty distinguishing external forces from restoring forces. Also, some students identify anything that oscillates as a harmonic oscillator. On the other hand, other students will produce some very clever models.

Lab 3.6: A Baby Bottle Harmonic Oscillator
Baby bottles are fairly foreign to the everyday life of most students in this course. The point of including the lab is that systems that can be modeled using the harmonic oscillator really do occur in unlikely places. (It also gave us something to think about when we were feeding Gib.) Students who have tried this lab have had difficulty obtaining useful data. If you assign this lab, it may be the only time that you ever hear "my baby ate my homework."

CHAPTER 4: FORCING AND RESONANCE

In this chapter we study a few important second-order equations with nonautonomous terms due to external forcing. Sections 4.1–4.4 cover nonhomogeneous linear equations, and our treatment is relatively traditional. For the most part we rely on the Method of Undetermined Coefficients to obtain particular solutions. We discuss steady-state solutions and resonance. We augment the usual discussion with a little more qualitative analysis than is often the case, but on the whole, the methods discussed in these sections are versions of those used in most texts. Sections 4.1–4.3 include the standard material, and Section 4.4 is a brief and more technical section designed for courses that are oriented toward engineering applications.

In Section 4.5, we discuss the collapse of the Tacoma Narrows Bridge. The work of A. C. Lazer and P. J. McKenna indicates that resonance due to the forcing caused by vortex shedding was *not* the cause of the collapse. (It is extremely unlikely that vortex shedding could have occurred at sufficiently regular intervals for traditional resonance to have been the cause.) Their work indicates that a more likely explanation is that the collapse was due to nonlinear effects occurring during large oscillations. Like the Lorenz equations, we include this section to give our students a glimpse of current mathematical research. The results discussed there do not reappear in subsequent sections and can safely be skipped if necessary.

Chapters 4 and 5 are relatively independent of each other, and some instructors prefer to discuss linearization near an equilibrium point for a nonlinear system (Section 5.1) immediately after Chapter 3. We do not adopt that approach mainly because our engineering students take electric circuit theory concurrently, and therefore it is important that we study forced, second-order linear equations as soon as possible.

DETools

The main tool for this chapter is `ForcedMassSpring`. However, we also use the `Pendulums` tool with $F \neq 0$ to illustrate the concept of external forcing in our introduction to the material.

4.1 Forced Harmonic Oscillators

In this section we introduce the nonhomogeneous linear equation for a forced harmonic oscillator, and we discuss the Extended Linearity Principle and the Method of Undetermined Coefficients. Only exponential forcing terms are used in the section, but other forcing functions are covered in the exercises. Trigonometric forcing is covered in Section 4.2. Our approach here is traditional except that we are honest that the "method" is really just a a guess-and-test technique.

The section ends with an RLC circuit example. The coefficients of this example are realistic values, so the time scales in Figures 4.7 and 4.8 are typical.

Comments on selected exercises

Exercises 1–19 involve the use of the Method of Undetermined Coefficients for standard exponential forcing functions.

The remaining exercises involve other forms of forcing—both for damped and undamped systems.

4.2 Sinusoidal Forcing

Harmonic oscillators with sinusoidal forcing is covered in this section. The approach is again fairly standard except for the fact that complex exponentials are used in the guess-and-test step of the Method of Undetermined Coefficients. The use of complex exponentials simplifies the guessing procedure and reinforces the use of complex numbers begun in Chapter 3. This technique is also consistent with the notion of *phasors*, which are standard descriptors used in the theory of electric circuits.

Unlike many discussions of this technique, we continue to employ the use of the phase plane, which provides a nice geometric representation for the steady-state solution. In this section, we concentrate on the damped case leaving the undamped phenomena for Section 4.3.

DETools

`ForcedMassSpring` and and `HPGSystemSolver` are the two main programs for this section.

Comments on selected exercises

Exercises 1–14 are typical exercises that use the techniques described in this section.

Exercise 15 presents an alternate approach that does not use complex numbers.

Exercise 16 illustrates the linearity of the equation.

Exercise 17 involves matching the graph of a solution with its corresponding differential equation. You may want to try this exercise before you assign it.

Exercises 18–20 involve the theory and practice for solving equations where the forcing function includes both exponential and trigonometric functions.

Exercises 21–23 make the connection between the phasor approach that engineering students often see in their circuit theory courses and the approach discussed in this section.

4.3 Undamped Forcing and Resonance

In this section we continue the discussion initiated in Section 4.2, and we study the effects of periodic forcing in the undamped case. In particular, we study beats and resonance using the techniques introduced in Section 4.2.

DETools

`ForcedMassSpring`, `BeatsAndResonance`, and `HPGSystemSolver` are the main programs for this section.

Comments on selected exercises

Exercises 1–14 provide practice with the Method of Undetermined Coefficients.

Exercises 15–18 give practice determining the frequency of beats and the frequency of their rapid oscillations.

Exercise 21 is a matching problem similar to the one in Section 4.2.

Exercises 20 and 22 are essay questions involving resonance. We believe that the story in Exercise 22 is true, but we cannot remember the name of the university.

4.4 Amplitude and Phase of the Steady State

This brief section is intended for courses that are significantly oriented toward engineering applications. It uses the complex arithmetic developed in Section 4.2 to discuss the amplitude and phase of the steady-state solutions derived in Section 4.2. The exercises have a distinct engineering orientation.

DETools

`AmplitudeAndPhase` illustrates the results of this section.

4.5 The Tacoma Narrows Bridge

The official investigation into the collapse of the Tacoma Narrows bridge concluded that resonant forcing was **not** the cause of the collapse. Periodic forcing caused by the shedding of vortices during high winds could not reasonably be expected to maintain a sufficiently precise period for a sufficient time to cause the collapse.

In this section, we discuss the work of A.C. Lazer and P.J. McKenna (see The American Mathematical Monthly, Vol. 106, No. 1, 1999, pp. 1–18, SIAM Reviews, Vol. 32, No. 4, 1990, pp. 537–578, and the references cited in those two articles). They model the motion of light flexible suspension bridges and observe that there is a significant nonlinearity that occurs when the bridge oscillates with moderately high amplitude. When the bridge is below equilibrium, the stretched cables act as springs pulling the bridge up and a linear harmonic oscillator model is reasonable. On the other hand, when the bridge is significantly above its equilibrium position, the cables are slack and *do not* push down. Only the constant force of gravity pulls down on the bridge. They show that a simple nonlinear system with this sort of nonlinearity can exhibit stable, large amplitude oscillations in the presence periodic forcing over a range of frequencies. Based on this idea, they also develop more accurate models for the full motion of the bridge. These models employ partial differential equations rather than ordinary differential equations.

This section develops the Lazer-McKenna system and reports on the results of their numerical and analytical investigations. While the numerical work is fairly delicate and the theorems moderately difficult, the results are within the reach of the typical student.

There are many good examples where resonance leads to dramatic behavior in physical systems. However, this is not the case for the Tacoma Narrows Bridge. The real culprit was the use of a linear model when a nonlinear model was required. This is an important lesson that we certainly want our students to understand and appreciate.

Comments on selected exercises

Exercises 1–4 are modifications of the Lazer-McKenna system for suspension bridges.

Exercises 5–8 discuss another application given by Lazer and McKenna to objects bobbing up and down in water but which can also lift out of the water.

Review Exercises

Exercises 1–10 are “short answer” exercises. The answers are (usually) one or two sentences. Most (but not all) are relatively straightforward.

Exercises 11–14 are true/false problems. We always expect our students to justify their answers.

Exercises 15–22 are routine exercises in finding the general solution of a forced second-order equation.

Exercise 23 is a matching exercise that includes both homogeneous and forced equations.

In Exercises 24–27, the student must find the general solution of sinusoidally forced equations as well as compute the amplitude and phase of the steady-state solution. Exercise 27 is somewhat tricky because the forcing is expressed in terms of sine rather than cosine.

Comments on the Labs

Lab 4.1: Two Magnets and a Spring Revisited
In this lab, the students continue their study of the system involving two magnets and a spring from Lab 2.1. In this case, the system is subject to periodic forcing.

Lab 4.2: A Periodically Forced RLC Circuit
The equations modeling RLC circuits, which are frequently given as a main motivation why second-order equations must be studied instead of first-order systems, are actually first developed as a first-order system. The dependent variables are voltage over the capacitor and the current. This system is then converted into a second-order equation. This lab is meant to emphasize this point. It can be assigned after Section 4.2 and goes nicely with Lab 3.2.

Lab 4.3: The Tacoma Narrows Bridge
This lab illustrates the ideas discussed in Section 4.5. It is designed to follow Lab 2.4. The forced harmonic oscillator parts of the lab are relatively straightforward. However, the students will probably need help with the computation and meaning of the large-amplitude periodic solution. (See the article by Lisa Humphreys and Ray Shammas, College Mathematics Journal, Vol. 31, No. 5, 2000, pp. 338–346.)

CHAPTER 5: NONLINEAR SYSTEMS

In Chapter 5, we study nonlinear systems. In many ways, this material is a continuation of the discussion that began in Chapter 2. However, we are now able to take advantage of the results of Chapter 3 to discuss linearization near an equilibrium point.

Section 5.1 introduces linearization. We feel that this section is one of the most important, as it relates the more traditional material on linear systems to nonlinear systems. In most semesters, this is the only section in Chapter 5 that we have time to cover.

Section 5.2 uses nullclines as a tool for qualitative analysis. This section is a nice application of qualitative techniques. Even though the techniques seem straightforward, it is surprising to see our students struggle with this material. Nevertheless, this material is appropriate for all students at this level.

The rest of the chapter covers special types of systems, and the level of sophistication is somewhat higher. Hamiltonian systems and systems with an integral are discussed in Section 5.3, and Section 5.4 considers systems with a Lyapunov function and gradient systems. Both sections are of particular interest to physics majors, and we teach these sections in our honors class. Section 5.5 discusses two examples of systems in three dimensions, a food chain model and the Lorenz system. In Section 5.6, we discuss nonlinear, periodically-forced systems and introduce return maps. This section is considerably more challenging than most other sections in the book. It is intended for advanced students.

DETools

In addition to `HPGSystemSolver`, Hu Hohn has written many special purpose demonstrations that go nicely with the examples in this chapter. Make sure that you take a look at:

- `ChemicalOscillator`
- `CompetingSpecies`
- `Duffing`
- `HMSGlider`
- `Pendulums`
- `PredatorPrey`
- `VanderPol`

5.1 Equilibrium Point Analysis

The technique of linearization of an equilibrium point of a system is introduced in this section. It can be viewed as a generalization of the linearization technique introduced at the end of Section 1.6. Using the classification of linear systems obtained in Chapter 3, equilibrium points of nonlinear systems are classified. The major difficulty in this section is to keep clear the distinction between nonlinear and linear systems. (Some students try to compute eigenvalues for equilibrium points of nonlinear systems.)

DETools

The `HPGLinearizer` tool is designed to help illustrate the material in this section. It is based on `HPGSystemSolver`, but it can also find equilibria numerically and determine their linearizations. In addition, the "Zoom In" feature of `HPGSystemSolver` is also useful here.

Comments on selected exercises

Exercises 1–4 practice linearization at the origin (dropping higher order terms).

In Exercise 5, solutions of a linear and nonlinear system at a saddle are compared in an example for which separatrices can be computed explicitly.

Exercises 7–16 ask for a classification of all equilibria. These systems also appear in Exercises 5–14 of Section 5.2.

Exercise 17 compares the linearization to the nonlinear system at a degenerate equilibrium point.

Exercises 18–26 use linearization to study bifurcation of equilibrium points in one-parameter families.

Exercises 27–30 are modeling problems where information regarding the linearization of the vector field at the origin is the only information given.

5.2 Qualitative Analysis

In this section we use the direction field, along with some numerics when necessary, to study the long-term behavior of solutions of nonlinear systems. The only new technique introduced is the location of nullclines in the phase plane. Unfortunately, many students are confused initially about the difference between nullclines and straight-line solutions.

Geometric analysis of this sort is particularly hard for students because it involves many steps and many different ideas and techniques. (They keep hoping you will just give them the magic bullet for understanding systems and are skeptical when you say there isn't one.) Extended projects are particularly helpful in making students realize that there is no template that leads to a complete description of a phase portrait.

DETools

All of the tools listed in the introduction can be used in this section, but `CompetingSpecies` and `ChemicalOscillator` are particularly well suited for analysis involving nullclines.

Comments on selected exercises

In Exercises 1–3, 5–14, and 16–20, we request a qualitative analysis of the given system. This analysis should go beyond what a student can print out from a good numerical solver. Note that Exercises 5–14 are the same systems as in Exercises 7–16 of Section 5.1 and Exercises 16–20 relate to the chemical reaction models created in Exercises 25–30 of Section 2.1.

Exercises 4 and 15 concern the general Volterra-Lotka models of a pair of species.

Exercises 21–23 study a nonlinear saddle.

5.3 Hamiltonian Systems

This section discusses systems with a conserved quantity (an integral of motion). Although it is an optional section, it may be of particular interest to students in physics and mechanics. Hamiltonian systems are introduced as a type of system for which the existence of a conserved quantity is automatic.

The nonlinear pendulum is studied in this section. This example can be covered in conjunction with Section 5.1, obtaining the period of oscillation from the linearization around the origin. For the ideal pendulum, the linearization at the origin is a case where linearization alone fails to describe the long-term behavior of the system.

DETools

The following five tools are especially useful in this section:

- `Duffing` with $b = 0$ and $F = 0$
- `HMSGlider` with $D = 0$
- `MassSpring` with $b = 0$
- `Pendulums` with $b = 0$ and $F = 0$
- `PredatorPrey`

Comments on selected exercises

Exercises 1–3 involve checking that a given system is Hamiltonian given H. Then H is used to sketch the phase portrait.

Exercises 4–8 study the ideal pendulum. (For the damped pendulum, see Exercises 4–11 of Section 5.4.)

Exercises 9–14 involve determining if a given system is Hamiltonian and determination of the Hamiltonian function.

Exercises 15 and 16 introduce the technique of rescaling a vector field so that the system obtained is Hamiltonian. This rescaling can be thought of as a position-dependent change of time scale.

Exercises 18–20 study examples of bifurcations of Hamiltonian systems.

5.4 Dissipative Systems

Adding damping to the pendulum or the harmonic oscillator turns the energy function into a Lyapunov function. In this section, we discuss Lyapunov functions. We also discuss gradient systems as a type of system that automatically has a Lyapunov function. Gradient systems lead naturally to applications where the motion is in the direction of steepest ascent.

DETools

Four of the tools cited in Section 5.3 are also helpful here with different parameter values:

- `Duffing` with $b \neq 0$ and $F = 0$
- `HMSGlider` with $D \neq 0$
- `MassSpring` with $b \neq 0$
- `Pendulums` with $b \neq 0$ and $F = 0$

Comments on selected exercises

Exercises 1–3 involve checking that a given function is Lyapunov and using this function to sketch the phase portrait.

Exercises 4–11 study the damped pendulum using the energy as a Lyapunov function.

Exercises 12–22 study gradient systems with Exercises 21 and 22 comparing gradient and Hamiltonian systems.

Exercises 23 and 24 involve the tuned-mass damper discussed in the section.

5.5 Nonlinear Systems in Three Dimensions

Two examples of three-dimensional systems are discussed in this section. The first is a food chain model with three species. The system has one equilibrium point where the three species coexist, and this point is a sink. The dependence of this point on parameters is considered.

The second example is the Lorenz system, first introduced in Section 2.5. (Section 2.5 can be covered quickly now if it was skipped earlier.) An analysis of the equilibrium points yields a qualitative description of solutions which agrees with numerically-generated pictures. The Lorenz system is studied again at the end of Chapter 8.

DETools

The `LorenzEquations` tool helps students visualize the Lorenz attractor in three dimensions.

Comments on selected exercises

All of the exercises involve the equilibrium points of the food chain model.

5.6 Periodic Forcing of Nonlinear Systems and Chaos

This section considers periodically forced nonlinear systems and Poincaré return maps. This is a difficult section that requires considerable mathematical maturity. In Chapter 8, we provide a more accessible introduction to recent results in the study of dynamical systems, e.g., chaos.

DETools

The `Duffing` and `Pendulums` tools illustrate some of the mathematical issues involved.

Comments on selected exercises

One problem with presenting this material at this level is the lack of reasonable exercises that can be done without fancy technology. Exercises 1–9 reinforce the relationship between the graphs of solutions and the return map pictures.

Review Exercises

Exercises 1–10 are "short answer" exercises. The answers are (usually) one or two sentences. Most (but not all) are relatively straightforward.

Exercises 11–14 are true/false problems. We always expect our students to justify their answers.

Exercises 15–18 involve linearization, qualitative analysis via nullclines, and phase portraits for a collection of nonlinear systems.

Exercises 19–24 result in a complete analysis of a two-parameter family of systems.

In Exercise 25, most of the techniques of this chapter are applied to a system.

Exercise 26 involves linearization of equilibrium points that come from a nonlinear second-order equation.

Exercises 27 and 28 involve the manner in which some of the notions introduced in this chapter relate to 2×2 linear systems.

Exercise 29 involves linearization for the simple model of the glider that was introduced in Chapter 2.

Comments on the Labs

Lab 5.1: Hard and Soft Springs
This lab studies the harmonic oscillator with a modified restoring force (either hard and soft springs). The soft spring is identical to the model of a swaying building considered in Section 2.4. This lab can be assigned immediately after Section 5.1, but if Sections 5.3 and 5.4 are covered, then the students should be expected to use techniques from those sections as well.

Lab 5.2: Higher Order Approximations of the Pendulum
This is the same system as the soft spring in Lab 5.1 and the swaying building of Section 2.4 but couched in different language.

Lab 5.3: A Family of Predator-Prey Equations
Students are asked to analyze a one-parameter family of predator-prey equations. In fact, they can complete their analysis without the use of a computer, and this lab does a good job of distinguishing those students who can use both a computer *and* a pencil. This lab has been one of our most successful.

Lab 5.4: The Glider
This lab is based on a simple model of a glider. Students are asked to analyze the motion of the glider with or without drag. Hu Hohn based his `HMSGlider` tool on this model, but we recommend that you assign the lab without mentioning that tool to the students. Show them the tool after they have submitted their reports. It is interesting to note that this model has been traced back to 1908.

CHAPTER 6: LAPLACE TRANSFORMS

Laplace transforms are widely used in engineering, particularly electrical engineering, but there seems to be considerable variation in when they are first encountered in the engineering curriculum. Most of our engineering students take electric circuit theory concurrently with our course. However, Laplace transforms are not used until the signals and systems course the subsequent semester, and that course is not taken by all engineering students at Boston University.

Using Laplace transforms to find formulas for solutions frequently involves a lot of algebra. This aspect of Laplace transforms will disappear as symbolic software becomes cheaper and better. However, Laplace transforms can be used as a tool for qualitative analysis of equations. The poles of the Laplace transform fill the same role as the eigenvalues of a linear system (of which they are a generalization). We have attempted to present at least the beginning of this theory.

Sections 6.1 and 6.2 form a self-contained introduction to Laplace transforms. In fact, since some schools would prefer an early introduction to the topic, we have written these sections so that they can be covered immediately after Chapter 1. In these two sections, we confine our discussion to first-order equations with or without discontinuities. Section 6.3 extends the discussion to second-order equations. Delta functions are covered in Section 6.4, and convolution is discussed in Section 6.5. Section 6.6 is an introduction to the qualitative use of the poles of the Laplace transform.

This chapter is independent of Chapter 5. You can easily skip from Section 4.3 directly to Chapter 6.

6.1 Laplace Transforms

The basics of Laplace transforms are discussed in this section. As mentioned earlier, we confine our discussion to first-order equations.

Comments on selected exercises

Exercises 1–6 provide practice with the definition of $\mathcal{L}$, and Exercises 7–14 involve $\mathcal{L}^{-1}$.

Exercises 15–24 are initial-value problems that can be solved using the techniques of Sections 1.8 or 1.9, but the problems are stated so that they should be solved using Laplace transforms.

Exercise 25 requests that the students use Laplace transforms to derive the general solution of a first-order linear equation.

Exercise 27 illustrates why the equations considered in this chapter are always linear.

6.2 Discontinuous Functions

This section is a standard presentation of Laplace transforms applied to first-order equations with discontinuous terms. Even though the techniques discussed here are mainly algebraic, the students benefit from slope fields just as they did in Chapter 1.

DETools

This is a good place to resurrect `HPGSolver`. It really helps with the explanation of the terms in the solution that involve the Heaviside function. Use of the `step` function in the solver is the key to entering discontinuous differential equations.

Comments on selected exercises

In Exercises 1–3, students compute Laplace transforms of piecewise-defined functions. They get practice working with the Heaviside function.

Exercises 4–7 provide practice inverting the transform when it includes terms of the form e^{-sa}.

In Exercises 8–13, the Laplace transform is used to solve first-order discontinuous initial-value problems.

Exercises 14 and 15 are also initial-value problems, but the computations are more difficult.

Exercises 16–20 involve Laplace transforms of periodic forcing functions such as the square wave and sawtooth wave function. Exercises 17 and 18 can be done without using Exercise 16, but Exercise 16 simplifies the calculation considerably.

6.3 Second-Order Equations

This section is a relatively standard discussion of the Laplace transform method applied to second-order linear equations. The most difficult equations considered are those with discontinuous forcing and resonance. The Laplace transform of $t \sin \omega t$ is derived in two different ways in the exercise set.

This section depends on Sections 4.2 and 4.3 since no motivation is given here for considering second-order forced equations.

Comments on selected exercises

Exercises 1–4 involve computing the Laplace transform directly from the definition whereas Exercise 5 involves computing $\mathcal{L}[\cos \omega t]$ using the fact that $\cos \omega t$ satisfies the equation for a simple harmonic oscillator.

Exercises 6–10 also involve computing Laplace transforms, but the instructions suggest a clever way of avoiding the integration.

Exercises 11–14 and 15–18 go together. The first group simply involves completing the square while the second group uses the results to compute inverse Laplace transforms.

Exercises 19–26 show how to use complex arithmetic in place of completing the square. Exercises 23–26 are the same as Exercises 15–18.

Exercises 27–34 apply the methods of this section to various initial-value problems. Exercises 27–29 and 31 could easily be done using the methods of Chapter 4, but they are here to give the students some practice before they tackle the more complicated initial-value problems in Exercises 30 and 32–34.

Exercise 35 suggests another way to do the Laplace transforms in Exercises 6–10.

6.4 Delta Functions and Impulse Forcing

This is a relatively standard section on the Dirac delta function. The "limit" approach is used. Thinking of the delta function as the "derivative" of the Heaviside function is discussed in Exercise 7.

Our students seem to enjoy our bringing a hammer to class and pounding it on the desk. It also seems to improve their attention spans.

Comments on selected exercises

In Exercise 1, the limit required to compute the Laplace transform of the delta function is computed (by L'Hôpital's rule).

Exercises 2–6 are standard second-order initial-value problems with delta function forcing.

Exercise 7 considers the relationship between the delta function and the Heaviside function.

Exercises 8–10 consider periodic delta function forcing (using Section 6.2, Exercise 16). These problems are considerably more difficult than Exercises 2–6.

6.5 Convolution

This is a typical section on convolution. However, it ends with a discussion of how one can find the solution of an initial-value problem without ever knowing the differential equation.

Comments on selected exercises

Exercises 1–5 involve computing convolutions from the definition. Exercise 5 requires a number of trigonometric identities.

Exercise 6 is a verification of the commutativity of convolution. It involves the definition of convolution.

Exercises 7–11 reinforce the points made at the end of the section.

6.6 The Qualitative Theory of Laplace Transforms

This section is not a standard one. It is a brief introduction to how Laplace transforms can be used to obtain qualitative information. The emphasis is on the idea that the poles of a Laplace transform of a solution for a forced harmonic oscillator play the same role as the eigenvalues for an unforced harmonic oscillator. This point of view is standard in electrical engineering, and Figure 6.26 can be found in circuit theory textbooks.

Comments on selected exercises

All of these exercises involve familiar equations that model forced harmonic oscillators. The goal here is the use the poles of the Laplace transform to obtain qualitative information about solutions without computing the inverse Laplace transform. Particularly in Exercises 3 and 4, analysis of the poles must be combined with common sense, since the forcing term turns off at some value of t. Exercises 9 and 10 refer to square wave and sawtooth forcing (see Section 6.2, Exercises 17–20).

Review Exercises

Exercises 1–10 are "short answer" exercises. The answers are (usually) one or two sentences. Most (but not all) are relatively straightforward.

Exercise 11 is a "multiple choice" matching exercise. The students match functions with their transforms from a list of transforms that is larger than the list of functions.

Exercises 12–17 are true/false problems. We always expect our students to justify their answers.

Exercises 18–23 give students practice computing inverse Laplace transforms.

In Exercises 24 and 25, the student is asked to solve the equation in two ways—using the methods of Chapter 4 and using the Laplace transform. Then they are asked to comment on which method they prefer.

Exercises 26–30 provide practice solving initial-value problems that involve Heaviside functions and delta functions.

Comments on the Labs

Lab 6.1: Poles
In this lab, the students are asked to formulate a conjecture regarding the relationship between multiple poles and growth rates of solutions.

Lab 6.2: Convolutions
In this lab, the students are given "experimental data" and are asked to use convolutions to compute the underlying differential equation.

CHAPTER 7: NUMERICAL METHODS

The goals of this brief chapter are to give an introduction to numerical methods other than Euler's method and to discuss error analysis in more detail. In particular, we introduce the concept of the order of a numerical method. Section 7.1 considers the errors involved in Euler's method. Sections 7.2 and 7.3 present improved Euler's method and Runge-Kutta, and Section 7.4 deals with the problem of roundoff errors arising from finite arithmetic.

Since the entire chapter deals almost exclusively with first-order equations, most of this material can be covered immediately after Section 1.4 (Euler's method). However, Sections 7.1 and 7.4 are somewhat more sophisticated than the rest of the text, so it may be appropriate to skim Section 7.1 and jump directly to Section 7.2.

We expect that the students will use some type of computer technology as they solve most of the exercises in this chapter.

DETools

Both `EulersMethod` and `EulersMethodForSystems` provide four different step sizes for Euler's method as well as Runge-Kutta with one step size. These tools may help illustrate some of the ideas discussed in this chapter, but alone they are probably not sufficient. Actually it is surprisingly easy to do most of the necessary computation using a spreadsheet.

7.1 Numerical Error in Euler's Method

This section is a detailed discussion of the errors involved in Euler's method including an explanation of the fact that Euler's method is a first-order method. Precise error bounds are derived in Exercises 11 and 12, but the text contains a discussion of the reasons why estimates of this type are often far too conservative.

Comments on selected exercises

Exercises 1–5 study the errors involved in Euler's method as a function of step size.

Exercise 6 is a good exercise to see if the students have a practical understanding of the fact that Euler's method is a first-order method.

Exercises 7–9 use a convenient method to estimate the error in an approximation.

Exercises 10–12 are theoretical. In particular, Exercise 11 is a long step-by-step derivation of precise error bounds.

7.2 Improving Euler's Method

Improved Euler's method is introduced as an example of a second-order method. One goal of the section is to develop the analogy between numerical methods to approximate solutions to differential equations and numerical methods to approximate integrals. The other goal is to contrast the accuracy of a second-order method with that of a first-order method (Euler's method).

Comments on selected exercises

Exercises 1–4 of this section involve the initial-value problems in Exercises 1–4 of Section 1.4. Similarly, Exercises 5–8 involve the initial-value problems in Exercises 6–9 of Section 1.4. These exercises compare the accuracy of improved Euler's method with that of Euler's method. The computations are possible by hand, but they would be tedious.

Exercises 9–13 give students practical experience with the fact that improved Euler's method is a second-order method.

Exercises 14–17 compare the accuracy of Euler's method versus improved Euler's method.

7.3 The Runge-Kutta Method

In this section the (fourth-order) Runge-Kutta algorithm is derived. As this is a method of choice for many applications, implementations of the method are given for the TI calculator, in *Mathematica*, and in the C programming language. We also use vector notation to discuss Runge-Kutta for first-order systems.

Comments on the exercises

In Exercises 1–5, we revisit initial-value problems that were studied in Sections 1.4 and 7.2. Using the results, we can compare the accuracy of Runge-Kutta with that of Euler's and improved Euler's methods.

Exercise 6 gives students practice with the fact that Runge-Kutta is a fourth-order method.

In Exercises 7 and 8, we illustrate how Runge-Kutta can be used to approximate solutions to first-order systems as well as second-order equations.

7.4 The Effects of Finite Arithmetic

The goal of this short section is to point out that, in practice, using a ridiculously small step size may not yield more accurate numerical approximations. The exercises study this phenomenon for particular examples.

Review Exercises

Exercises 1–5 are true/false problems. We always expect our students to justify their answers.

Exercise 6 compares the results of Euler's method, improved Euler's method, and Runge-Kutta for a first-order equation.

Exercise 7 compares the results of Euler's method, improved Euler's method, and Runge-Kutta for a competing species system.

Comments on the Labs

Lab 7.1: Errors of Numerical Approximations
This lab relates the geometry of an equation with the accuracy of the numerical approximation. It addresses the question of why some methods are more accurate than expected on certain equations.

Lab 7.2: Lost in Space
In this lab, numerical methods are used to approximate solutions of Newton's equations. Parts of the lab could be used as detailed exercises or projects much earlier in the text.

CHAPTER 8: DISCRETE DYNAMICAL SYSTEMS

This chapter covers some basic topics in discrete dynamical systems in one dimension. The student is introduced to the idea of iteration together with an overview of what happens when simple functions are iterated. As in previous sections, a particular model is highlighted throughout. In this chapter, it is the discrete logistic model

$$P_{n+1} = kP_n(1 - P_n).$$

The pace of this chapter is brisk and the level of sophistication assumed is somewhat higher than previous chapters, especially in Sections 8.4 (Chaos) and 8.5 (Chaos in the Lorenz System).

One goal of the chapter is to introduce mathematical topics that are both accessible to students at this level and quite modern. Very few of our students have ever seen a theorem of a living mathematician and have no idea that mathematical research is ongoing. Topics such as chaos and the bifurcation diagram are easy ways to overcome this myth. (See also Sections 2.5 and 5.5 on the Lorenz system and Section 4.5 on recent work on the Tacoma Narrows Bridge.)

Section 8.1 gives definitions and develops graphical techniques for plotting orbits. Sections 8.2 and 8.3 deal with the calculus of fixed and periodic points and bifurcations respectively. A discussion of chaos is given in Section 8.4. In Section 8.5, we discuss the behavior of the Lorenz equations using a one-dimensional model. The discrete logistic map appears repeatedly in Sections 8.1–8.4.

Available Technology

There are many programs available on the web that help illustrate this material. For the PC, one of our favorites is Rick Parris's *Winfeed*. The web site for the Dynamical Systems and Technology Project at Boston University (`http://math.bu.edu/DYSYS`) has java applets that compute orbits and histograms, perform graphical iteration, and compute the bifurcation diagram.

8.1 The Discrete Logistic Equation

This section introduces the idea of discrete time systems and the relation to iteration in one dimension. We discuss graphical techniques for displaying the iterates, including time series, histograms, and the graphs of higher iterates of the function. We also define and discuss fixed points.

Comments on selected exercises

Many of the exercises in this section can be done by hand, but others (iteration of $\cos x$ and $\sin x$, for example) require the use of a calculator or computer.

Exercises 1–8 involve computation of the orbit of 0 for various functions. Students observe orbits tending to fixed points, to cycles, to infinity, and so forth.

In Exercises 9–21, students are asked to find fixed points and periodic points of period 2. Many of these exercises can be done by hand. Occasionally the student must resort to looking at the graph of the function or its second iterate to answer this question (as in the case of e^x, $\sin x$, or $-e^x$).

Exercises 24–35 involve experimentation with the tent map.

Exercises 36–39 consider the role of a parameter by studying the one-parameter family $F(x) = x^2 + c$. These exercises foreshadow some of our work with bifurcations in Section 8.3.

8.2 Fixed and Periodic Points

In this section, we study fixed and periodic points. We classify them as sources, sinks, or nodes. We also introduce the process of graphical iteration (the "web diagram"). Graphing calculators that possess the ability to produce web diagrams are very useful.

Comments on selected exercises

In Exercises 1–11, the students find and classify fixed points.

In Exercises 12–17, the number 0 lies on a cycle. The student must determine the period of the cycle and then classify it as attracting, repelling, or neutral.

In Exercises 18–25, students determine whether certain fixed points whose derivative is ± 1 are attracting, repelling, or neutral. The students generally use the graphs of the function to classify the cycle.

In Exercises 26 and 28, the fixed points of certain one-parameter families are considered.

Newton's method is studied as a discrete dynamical system in Exercise 30. Newton's method also appears as Lab 8.1.

8.3 Bifurcations

In this section, we consider one-parameter families of maps and bifurcations of fixed and periodic points. The logistic map and the period doubling route to chaos are discussed in some detail. Toward the end of the section, the level of sophistication demanded of students is higher than the preceding sections.

Comments on selected exercises

Exercises 1–7 involve the analysis of basic bifurcations. Students find these problems difficult because they must explain what happens to orbits "at, before, and after" the bifurcation. Students at first do not know how close to take nearby parameter values in order to be "before" or "after" the bifurcation. They often simply choose the nearest integer which may be too far away.

Exercises 8 and 9 involve other families of functions where the students must determine the specific bifurcation value.

8.4 Chaos

The phenomenon of chaos is observed in the logistic map and studied for the shift map on decimal expansion (multiplication by 10 modulo 1). The second example gives an introduction to symbolic dynamics, dense orbits, and sensitive dependence on initial conditions. Again, the level of mathematical sophistication required is somewhat higher than in previous sections.

Comments on selected exercises

Exercise 1 studies chaos in the logistic map numerically.

Exercises 2–14 consider the tent map, including an introduction to symbolic dynamics for this map. This series of exercises could be a good project for students in an honors class.

8.5 Chaos in the Lorenz System

In this section we return to the Lorenz equations and, using the template construction, develop the one-dimensional model for the dynamics on the attractor. If the Lorenz equations have been studied earlier in the course (Sections 2.5 and 5.5), this section can be covered after Section 8.1.

Review Exercises

Exercises 1–10 are "short answer" exercises. The answers are (usually) one or two sentences. Most (but not all) are relatively straightforward.

Exercises 11–16 are true/false problems. We always expect our students to justify their answers.

In Exercise 17, the student is asked to compute cycles of period two for the chopping function.

Exercise 18 requests a bifurcation diagram for a two-parameter family of linear maps.

Exercise 19 asks for the number of points that lie on cycles for $F(x) = x^2 - 2$.

Exercise 20 involves a one-parameter family of tent functions.

Comments on the Labs

Lab 8.1: Newton's Method as a Difference Equation
Newton's method on a cubic is studied using the terminology of discrete dynamical systems. A programmable calculator is sufficient technology for this lab, and it can be assigned after Section 8.1. It provides another approach to complicated behavior of one-dimensional maps.

Lab 8.2: The Delayed Logistic and Iteration in Two Dimensions
This lab requires the ability to plot phase planes of functions of two variables. While this idea is natural given our previous work, students will need some guidance in viewing orbits in this manner.

Lab 8.3: The Bifurcation Diagram
This lab necessitates that students either write a simple program or use some available software to sketch the bifurcation diagram(s) requested.

APPENDICES

There are three appendices. Appendix A discusses the general technique of changing variables. Appendix B is a very brief description of the use of power series, and Appendix C is a short summary of some of the basic properties of complex arithmetic.

Appendix A: Changing Variables

Many of the standard analytic techniques and many of the special cases for analytically solving differential equations can be viewed as a "change of variables." This appendix illustrates this fact by discussing a number of examples. Included is a short presentation of Bernoulli and Riccati equations. Only the dependent variable is changed in the text and exercises (with the exception of Exercise 22).

Comments on selected exercises

Exercises 1–13 relate solutions of equations in different variables, both analytically and geometrically.

Exercises 14 and 15 apply this technique to mixing problems (changing the dependent variable to concentration).

Exercise 16 introduces homogeneous equations, i.e., equations of the form $dy/dt = g(y/t)$.

Exercises 17–21 illustrate changes of variables to help with linearization near equilibrium points.

Exercise 22 involves changing the independent variable.

Exercises 23–33 provide practice with Bernoulli and Riccati equations.

Appendix B: The Ultimate Guess

This appendix is an ultra-lite presentation of the idea of a power series solution to a differential equation. We illustrate the technique with a first-order example and a second-order example (Hermite's equation). Legendre's equation appears in Exercise 15.

Comments on selected exercises

Exercises 1–13 illustrate the power series technique for various first- and second-order equations. Closed form solutions are not requested.

Exercises 14 and 15 involve Hermite's equation and Legendre's equation respectively.

Exercises 17 and 18 provide motivation for the second guessing technique that is discussed in Sections 1.8, 3.5, and 3.6.

Appendix C: Complex Numbers and Euler's Formula

This appendix is not intended to teach students the arithmetic of complex numbers, but it does summarize the basic properties that we use, especially in Chapters 3 and 4. There is a power series derivation of Euler's formula and a brief mention of the polar representation of a complex number. Both of these ideas play important roles in Section 3.4 and Sections 4.2–4.4.

CHAPTER 1

First-Order Differential Equations

EXERCISES FOR SECTION 1.1

1. We know the solution of the differential equation is $P(t) = P_0 e^{kt}$. It is a little easier to use the year 1800 as $t = 0$, so 1900 corresponds to $t = 100$ and 2000 corresponds to $t = 200$. To compute k, we use

$$5.3 = P(0) = P_0 e^{k \cdot 0} = P_0$$

and

$$75 = P(100) = P_0 e^{k \cdot 100} = 5.3 e^{100k}$$

from Table 1.1. Therefore, $75 = 5.3 e^{100k}$, which yields

$$k = \frac{\ln(75/5.3)}{100} \approx 0.026498.$$

Hence, the prediction for the year 2000 is

$$P(200) = 5.3 e^{(0.026498)(200)} \approx 1061.$$

Since the data is given in millions, this "prediction" is more than one billion people. It is too high by a factor of almost 4.

2. Let $B(t)$ be the number of bacteria at time t (measured in hours). We assume the exponential growth model $dB/dt = kB$. Consequently, $B(t) = B_0 e^{kt}$. Since $B(0) = 50$ and $B(12) = 1000$, we have

$$1000 = B(12) = 50 e^{k \cdot 12},$$

so

$$k = \frac{\ln(20)}{12} \approx 0.25.$$

(a) After 18 hours there will be $B(18) = 50 e^{(0.25)(18)} \approx 4500$ bacteria.

(b) Let T be the time required for the number of bacteria to double. In other words,

$$B(t + T) = 2B(t).$$

Then

$$B_0 e^{0.25(t+T)} = 2 B_0 e^{0.25t}.$$

Canceling B_0 and $e^{0.25t}$ from both sides, we get $e^{0.25T} = 2$, which implies that

$$T = \frac{\ln(2)}{0.25} \approx 2.77.$$

So the time necessary for the number of bacteria to double is approximately 2.77 hours.

3. **(a)** The equilibrium solutions correspond to the values of P for which $dP/dt = 0$ for all t. For this equation, $dP/dt = 0$ for all t if $P = 0$ or $P = 230$.

(b) The population is increasing if $dP/dt > 0$. That is, $P(1 - P/230) > 0$. Hence, $0 < P < 230$.

(c) The population is decreasing if $dP/dt < 0$. That is, $P(1 - P/230) < 0$. Hence, $P > 230$ or $P < 0$. Since this is a population model, $P < 0$ might be considered "nonphysical."

4. **(a)** The equilibrium solutions correspond to the values of P for which $dP/dt = 0$ for all t. For this equation, $dP/dt = 0$ for all t if $P = 0$, $P = 50$, or $P = 200$.

(b) The population is increasing if $dP/dt > 0$. That is, $P < 0$ or $50 < P < 200$. Note, $P < 0$ might be considered "nonphysical" for a population model.

(c) The population is decreasing if $dP/dt < 0$. That is, $0 < P < 50$ or $P > 200$.

5. In order to answer the question, we first need to analyze the sign of the polynomial $y^3 - y^2 - 12y$. Factoring, we obtain

$$y^3 - y^2 - 12y = y(y^2 - y - 12) = y(y - 4)(y + 3).$$

(a) The equilibrium solutions correspond to the values of y for which $dy/dt = 0$ for all t. For this equation, $dy/dt = 0$ for all t if $y = -3$, $y = 0$, or $y = 4$.

(b) The solution $y(t)$ is increasing if $dy/dt > 0$. That is, $-3 < y < 0$ or $y > 4$.

(c) The solution $y(t)$ is decreasing if $dy/dt < 0$. That is, $y < -3$ or $0 < y < 4$.

6. Let t represent time measured in years with $t = 0$ corresponding to 1939.

(a) The general solution of the differential equation is

$$A(t) = A_0 e^{kt}$$

(see the "guess-and-test" method described in this section). Here A_0 is the initial value 32, 800, so the solution of the initial-value problem is

$$A(t) = 32,800e^{kt}.$$

(b) One reasonable way to determine k is to proceed as in the text and choose k so that the solution agrees exactly with the first data point. That is, choose k so that

$$55,800 = A(5) = 32,800e^{5k}.$$

Solving for k we have

$$e^{5k} = \frac{55,800}{32,800}$$

hence,

$$5k = \ln \frac{55,800}{32,800},$$

which yields

$$k = \frac{1}{5} \ln \frac{55,800}{32,800} \approx 0.10627.$$

Other reasonable methods include using $A(10)$, $A(15)$, ... in place of $A(5)$.

(c) The years 2010, 2050, and 2100 correspond to $t = 71$, $t = 111$, and $t = 161$, respectively. Hence, using $k = 0.10627$, the model predicts

$$A(71) \approx 6.2 \times 10^7$$
$$A(111) \approx 4.4 \times 10^9$$
$$A(161) \approx 8.8 \times 10^{11}.$$

(d) The following graph compares the data with the solution of the model (assuming $k = 0.10627$).

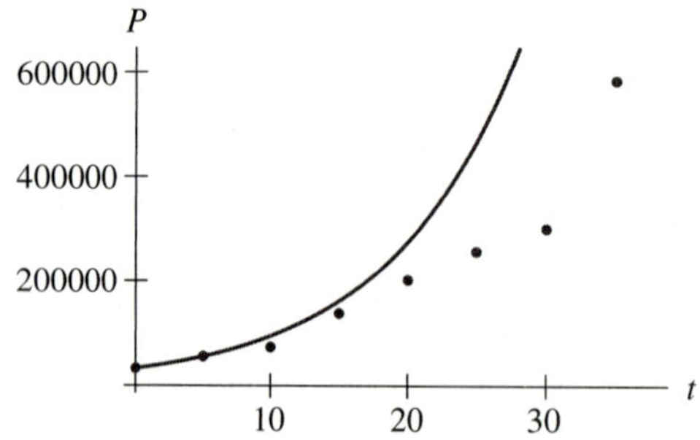

Since the predictions in part (c) are much too large, the model should not be believed. A logistic model might be more appropriate.

7. The rate of learning is dL/dt. Thus, we want to know the values of L between 0 and 1 for which dL/dt is a maximum. As $k > 0$ and $dL/dt = k(1 - L)$, dL/dt attains it maximum value at $L = 0$.

8. **(a)** Let $L_1(t)$ be the solution of the model with $L_1(0) = 1/2$ (the student who starts out knowing one-half of the list) and $L_2(t)$ be the solution of the model with $L_2(0) = 0$ (the student who starts out knowing none of the list). At time $t = 0$,

$$\frac{dL_1}{dt} = 2\,(1 - L_1(0)) = 2\left(1 - \tfrac{1}{2}\right) = 1,$$

and

$$\frac{dL_2}{dt} = 2\,(1 - L_2(0)) = 2.$$

Hence, the student who starts out knowing none of the list learns faster at time $t = 0$.

(b) The solution $L_2(t)$ with $L_2(0) = 0$ will learn one-half the list in some amount of time $t_* > 0$. For $t > t_*$, $L_2(t)$ will increase at exactly the same rate that $L_1(t)$ increases for $t > 0$. In other words, $L_2(t)$ increases at the same rate as $L_1(t)$ at t_* time units later. Hence, $L_2(t)$ will never catch up to $L_1(t)$ (although they both approach 1 as t increases). In other words, after a very long time $L_2(t) \approx L_1(t)$, but $L_2(t) < L_1(t)$.

9. **(a)** We have $L_B(0) = L_J(0) = 0$. So Jillian's rate of learning at $t = 0$ is dL_J/dt evaluated at $t = 0$. At $t = 0$, we have

$$\frac{dL_J}{dt} = 2(1 - L_J) = 2.$$

Beth's rate of learning at $t = 0$ is

$$\frac{dL_B}{dt} = 3(1 - L_B)^2 = 3.$$

Hence Beth's rate is larger.

(b) In this case, $L_B(0) = L_J(0) = 1/2$. So Jillian's rate of learning at $t = 0$ is

$$\frac{dL_J}{dt} = 2(1 - L_J) = 1$$

because $L_J = 1/2$ at $t = 0$. Beth's rate of learning at $t = 0$ is

$$\frac{dL_B}{dt} = 3(1 - L_B)^2 = \frac{3}{4}$$

because $L_B = 1/2$ at $t = 0$. Hence Jillian's rate is larger.

(c) In this case, $L_B(0) = L_J(0) = 1/3$. So Jillian's rate of learning at $t = 0$ is

$$\frac{dL_J}{dt} = 2(1 - L_J) = \frac{4}{3}.$$

Beth's rate of learning at $t = 0$ is

$$\frac{dL_B}{dt} = 3(1 - L_B)^2 = \frac{4}{3}.$$

They are both learning at the same rate when $t = 0$.

10. **(a)** The rate of change of the amount of radioactive material is dr/dt. This rate is proportional to the amount r of material present at time t. With $-\lambda$ as the proportionality constant, we obtain the differential equation

$$\frac{dr}{dt} = -\lambda r.$$

Note that the minus sign (along with the assumption that λ is positive) means that the material decays.

(b) The only additional assumption is the initial condition $r(0) = r_0$. Consequently, the corresponding initial-value problem is

$$\frac{dr}{dt} = -\lambda r, \quad r(0) = r_0.$$

11. The general solution of the differential equation $dr/dt = -\lambda r$ is $r(t) = r_0 e^{-\lambda t}$ where $r(0) = r_0$ is the initial amount.

(a) We have $r(t) = r_0 e^{-\lambda t}$ and $r(5230) = r_0/2$. Thus

$$\frac{r_0}{2} = r_0 e^{-\lambda \cdot 5230}$$

$$\frac{1}{2} = e^{-\lambda \cdot 5230}$$

$$\ln \frac{1}{2} = -\lambda \cdot 5230$$

$$-\ln 2 = -\lambda \cdot 5230$$

because $\ln 1/2 = -\ln 2$. Thus,

$$\lambda = \frac{\ln 2}{5230} \approx 0.000132533.$$

(b) We have $r(t) = r_0 e^{-\lambda t}$ and $r(8) = r_0/2$. By a computation similar to the one in part (a), we have

$$\lambda = \frac{\ln 2}{8} \approx 0.0866434.$$

(c) If $r(t)$ is the number of atoms of C-14, then the units for dr/dt is number of atoms per year. Since $dr/dt = -\lambda r$, λ is "per year." Similarly, for I-131, λ is "per day." The unit of measurement of r does not matter.

(d) We get the same answer because the original quantity, r_0, cancels from each side of the equation. We are only concerned with the proportion remaining (one-half of the original amount).

12. We will solve for k percent. In other words, we want to find t such that $r(t) = (k/100)r_0$, and we know that $r(t) = r_0 e^{-\lambda t}$, where $\lambda = (\ln 2)/5230$ from Exercise 11. Thus we have

$$\begin{aligned} r_0 e^{-\lambda t} &= \frac{k}{100} r_0 \\ e^{-\lambda t} &= \frac{k}{100} \\ -\lambda t &= \ln\left(\frac{k}{100}\right) \\ t &= \frac{-\ln\left(\frac{k}{100}\right)}{\lambda} \\ t &= \frac{\ln 100 - \ln k}{\lambda} \\ t &= \frac{5230(\ln 100 - \ln k)}{\ln 2}. \end{aligned}$$

Thus, there is 88% left when $t \approx 964.54$ years; there is 12% left when $t \approx 15{,}998$ years; 2% left when $t \approx 29{,}517$ years; and 98% left when $t \approx 152.44$ years.

13. The most important difference between Carbon-14 and Carbon-12 is that C-14 is radioactive. The models we constructed above are based on the assumption that the number of C-14 atoms that decay in a particular time period is proportional to the number of C-14 atoms present where the constant of proportionality λ is determined from the half-life. Hence we can determine the amount of C-14 in a sample by counting the number of atoms that decay in a given time period. Divide this number by the time elapsed to obtain an approximation for dr/dt. Then, using $dr/dt = -\lambda r$, divide by λ to get r (making sure the units agree).

14. We use $\lambda \approx 0.0866434$ from part (b) of Exercise 11.

(a) Since 72 hours is 3 days, we have $r(3) = r_0 e^{-\lambda \cdot 3} = r_0 e^{-.2598} \approx 0.77 r_0$. Approximately 77% of the original amount arrives at the hospital.

(b) Similarly, $r(5) = r_0 e^{-\lambda \cdot 5} = r_0 e^{-.4330} \approx 0.65 r_0$. Approximately 65% of the original amount is left when it is used.

(c) It will never *completely* decay since $e^{-\lambda t}$ is never zero. However, after one year, the proportion of the original amount left will be $e^{-\lambda \cdot 365} \approx 1.85 \times 10^{-14}$. Unless you start with a very large amount I-131, the amount left after one year should be safe to throw away. In practice, samples are stored for ten half-lives (80 days for I-131) and then disposed.

15. Let $P(t)$ be the population at time t, k be the growth-rate parameter, and N be the carrying capacity. The modified models are

(a) $dP/dt = k(1 - P/N)P - 100$

(b) $dP/dt = k(1 - P/N)P - P/3$

(c) $dP/dt = k(1 - P/N)P - a\sqrt{P}$, where a is a positive parameter.

16. **(a)** The differential equation is $dP/dt = 0.3P(1 - P/2500) - 100$. The equilibrium solutions of this equation correspond to the values of P for which $dP/dt = 0$ for all t. Using the quadratic formula, we obtain two such values, $P_1 \approx 396$ and $P_2 \approx 2104$. If $P > P_2$, $dP/dt < 0$, so $P(t)$ is decreasing. If $P_1 < P < P_2$, $dP/dt > 0$, so $P(t)$ is increasing. Hence the solution that satisfies the initial condition $P(0) = 2500$ decreases toward the equilibrium $P_2 \approx 2104$.

(b) The differential equation is $dP/dt = 0.3P(1 - P/2500) - P/3$. The equilibrium solutions of this equation are $P_1 \approx -277$ and $P_2 = 0$. If $P > 0$, $dP/dt < 0$, so $P(t)$ is decreasing. Hence, for $P(0) = 2500$, the population decreases toward $P = 0$ (extinction).

17. Several different models are possible. Let $R(t)$ denote the rhinoceros population at time t. The basic assumption is that there is a minimum threshold that the population must exceed if it is to survive. In terms of the differential equation, this assumption means that dR/dt must be negative if R is close to zero. Three models that satisfy this assumption are:

- If k is a growth-rate parameter and M is a parameter measuring when the population is "too small", then
$$\frac{dR}{dt} = kR\left(\frac{R}{M} - 1\right).$$
- If k is a growth-rate parameter and b is a parameter that determines the level the population will start to decrease ($R < b/k$), then
$$\frac{dR}{dt} = kR - b.$$
- If k is a growth-rate parameter and b is a parameter that determines the extinction threshold, then
$$\frac{dR}{dt} = kR - \frac{b}{R}.$$

In each case, if R is below a certain threshold, dR/dt is negative. Thus, the rhinos will eventually die out. The choice of which model to use depends on other assumptions. There are other equations that are also consistent with the basic assumption.

18. **(a)** Since the population seems to level off at 64, we try a logistic model. Let t be the time measured in years and $P(t)$ be the population. The model is then
$$\frac{dP}{dt} = kP\left(1 - \frac{P}{N}\right)$$
where k is the growth-rate parameter and N is the carrying capacity.

(b) Take $N = 64$ as the carrying capacity. This is reasonable since the population has "leveled off" for the four years 1956–1959. To compute k we could use the same technique as we did for the U.S. population, but since we already know the carrying capacity and since $P(0) = 34$ is a significant fraction of the carrying capacity, it is better to use the differential equation to determine k. To approximate dP/dt at time $t = 0$ (which we take to be 1947), we calculate the slope of the line connecting the first two data points $(0, 34)$ and $(1, 40)$. The slope of this line gives us an approximation of dP/dt at $t = 0$. That is,

$$\frac{40 - 34}{1 - 0} \approx \left.\frac{dP}{dt}\right|_{t=0} = kP(0)\left(1 - \frac{P(0)}{64}\right),$$

so

$$6 = k \cdot 34\left(1 - \frac{34}{64}\right),$$

which yields $k \approx 0.38$.

(c) Since the population is at the carrying capacity in 1959, it will remain there, so we predict that the population today is 64.

19. **(a)** The term governing the effect of the interaction of x and y on the rate of change of x is $+\beta xy$. Since this term is positive, the presence of y's helps the x population grow. Hence, x is the predator. Similarly, the term $-\delta xy$ in the dy/dt equation implies that when $x > 0$, y's grow more slowly, so y is the prey. If $y = 0$, then $dx/dt < 0$, so the predators will die out; thus, they must have insufficient alternative food sources. The prey has no limits on its growth other than the predator since, if $x = 0$, then $dy/dt > 0$ and the population increases exponentially.

(b) Since $-\beta xy$ is negative and $+\delta xy$ is positive, x suffers due to its interaction with y and y benefits from its interaction with x. Hence, x is the prey and y is the predator. The predator has other sources of food than the prey since $dy/dt > 0$ even if $x = 0$. Also, the prey has a limit on its growth due to the $-\alpha x^2/N$ term.

20. **(a)** We consider dx/dt in each system. Setting $y = 0$ yields $dx/dt = 5x$ in system (i) and $dx/dt = x$ in system (ii). If the number x of prey is equal for both systems, dx/dt is larger in system (i). Therefore, the prey in system (i) reproduce faster if there are no predators.

(b) We must see what effect the predators (represented by the y-terms) have on dx/dt in each system. Since the magnitude of the coefficient of the xy-term is larger in system (ii) than in system (i), y has a greater effect on dx/dt in system (ii). Hence the predators have a greater effect on the rate of change of the prey in system (ii).

(c) We must see what effect the prey (represented by the x-terms) have on dy/dt in each system. Since x and y are both nonnegative, it follows that

$$-2y + \tfrac{1}{2}xy < -2y + 6xy,$$

and therefore, if the number of predators is equal for both systems, dy/dt is smaller in system (i). Hence more prey are required in system (i) than in system (ii) to achieve a certain growth rate.

21. **(a)** The independent variable is t, and x and y are dependent variables. Since each xy-term is positive, the presence of either species increases the rate of change of the other. Hence, these species cooperate. The parameter α is the growth-rate parameter for x, and γ is the growth-rate parameter for y. The parameter N represents the carrying capacity for x, but y has no carrying capacity. The parameter β measures the benefit to x of the interaction of the two species, and δ measures the benefit to y of the interaction.

(b) The independent variable is t, and x and y are the dependent variables. Since both xy-terms are negative, these species compete. The parameter γ is the growth-rate coefficient for x, and α is the growth-rate parameter for y. Neither population has a carrying capacity. The parameter δ measures the harm to x caused by the interaction of the two species, and β measures the harm to y caused by the interaction.

22. **(a)** The effect of y on dx/dt is through the term $-by\sqrt{x}$. Because this term is negative (x and y are positive), a larger y population decreases dx/dt, so x is the prey. Similarly, the term $cy\sqrt{x}$ gives the effect of x on dy/dt and a larger x gives a larger dy/dt. So y is the predator.

(b) If $x = 0$, $dy/dt = 0$, so the predator population is constant. The predator must have alternate sources of food or be able to hibernate without need for prey.

EXERCISES FOR SECTION 1.2

1. **(a)** Let's check Bob's solution first. Since $dy/dt = 1$ and

$$\frac{y(t)+1}{t+1} = \frac{t+1}{t+1} = 1,$$

Bob's answer is correct.

Now let's check Glen's solution. Since $dy/dt = 2$ and

$$\frac{y(t)+1}{t+1} = \frac{2t+2}{t+1} = 2,$$

Glen's solution is also correct.

Finally let's check Paul's solution. We have $dy/dt = 2t$ on one hand and

$$\frac{y(t)+1}{t+1} = \frac{t^2-1}{t+1} = t-1$$

on the other. Paul is wrong.

(b) At first glance, they should have seen the equilibrium solution $y(t) = -1$ for all t because $dy/dt = 0$ for any constant function and $y = -1$ implies that

$$\frac{y+1}{t+1} = 0$$

independent of t.

Strictly speaking the differential equation is not defined for $t = -1$, and hence the solutions are not defined for $t = -1$.

2. We note that $dy/dt = 2e^{2t}$ for $y(t) = e^{2t}$. If $y(t) = e^{2t}$ is a solution to the differential equation, then we must have

$$\begin{aligned} 2e^{2t} &= 2y(t) - t + g(y(t)) \\ &= 2e^{2t} - t + g(e^{2t}). \end{aligned}$$

Hence, we need

$$g(e^{2t}) = t.$$

This equation is satisfied if we let $g(y) = (\ln y)/2$. In other words, $y(t) = e^{2t}$ is a solution of the differential equation

$$\frac{dy}{dt} = 2y - t + \frac{\ln y}{2}.$$

3. In order to find one such $f(t, y)$, we compute the derivative of $y(t)$. We obtain

$$\frac{dy}{dt} = \frac{de^{t^3}}{dt} = 3t^2 e^{t^3}.$$

Now we replace e^{t^3} in the last expression by y and get the differential equation

$$\frac{dy}{dt} = 3t^2 y.$$

4. Starting with $dP/dt = kP$, we divide both sides by P to obtain

$$\frac{1}{P}\frac{dP}{dt} = k.$$

Then integrating both sides with respect to t, we have

$$\int \frac{1}{P}\frac{dP}{dt}\,dt = \int k\,dt,$$

and changing variables on the left-hand side, we obtain

$$\int \frac{1}{P}\,dP = \int k\,dt.$$

(Typically, we jump to the equation above by "informally" multiplying both sides by dt.) Integrating, we get

$$\ln|P| = kt + c,$$

where c is an arbitrary constant. Exponentiating both sides gives

$$|P| = e^{kt+c} = e^c e^{kt}.$$

For population models we consider only $P \geq 0$, and the absolute value sign is unnecessary. Letting $P_0 = e^c$, we have

$$P(t) = P_0 e^{kt}.$$

In general, it is possible for $P(0)$ to be negative. In that case, $e^c = -P_0$, and $|P| = -P$. Once again we obtain

$$P(t) = P_0 e^{kt}.$$

5. The constant function $y(t) = 0$ is an equilibrium solution.

For $y \neq 0$ we separate the variables and integrate

$$\int \frac{dy}{y} = \int t\,dt$$
$$\ln|y| = \frac{t^2}{2} + c$$
$$|y| = c_1 e^{t^2/2}$$

where $c_1 = e^c$ is an arbitrary positive constant.

If $y > 0$, then $|y| = y$ and we can just drop the absolute value signs in this calculation. If $y < 0$, then $|y| = -y$, so $-y = c_1 e^{t^2/2}$. Hence, $y = -c_1 e^{t^2/2}$. Therefore,

$$y = ke^{t^2/2}$$

where $k = \pm c_1$. Moreover, if $k = 0$, we get the equilibrium solution. Thus, $y = ke^{t^2/2}$ yields all solutions to the differential equation if we let k be any real number. (Strickly speaking we need a theorem from Section 1.5 to justify the assertion that this formula provides all solutions.)

6. Separating variables and integrating, we obtain

$$\int \frac{1}{y}\,dy = \int t^4\,dt$$
$$\ln|y| = \frac{t^5}{5} + c$$
$$|y| = c_1 e^{t^5/5},$$

where $c_1 = e^c$. As in Exercise 5, we can eliminate the absolute values by replacing the positive constant c_1 with $k = \pm c_1$. Hence, the general solution is

$$y(t) = ke^{t^5/5},$$

where k is any real number. Note that $k = 0$ gives the equilibrium solution.

7. We separate variables and integrate to obtain

$$\int \frac{dy}{2y+1} = \int dt.$$

We get

$$\frac{1}{2}\ln|2y+1| = t + c$$
$$|2y+1| = c_1 e^{2t},$$

where $c_1 = e^{2c}$. As in Exercise 5, we can drop the absolute value signs by replacing $\pm c_1$ with a new constant k_1. Hence, we have

$$2y + 1 = k_1 e^{2t}$$
$$y = \frac{1}{2}\left(k_1 e^{2t} - 1\right),$$

and letting $k = k_1/2$, $y(t) = ke^{2t} - 1/2$. Note that, for $k = 0$, we get the equilibrium solution.

8. Separating variables and integrating, we obtain

$$\int \frac{1}{2-y}\,dy = \int dt$$

$$-\ln|2-y| = t + c$$

$$\ln|2-y| = -t + c_1,$$

where we have replaced $-c$ with c_1. Then

$$|2-y| = k_1 e^{-t},$$

where $k_1 = e_1^c$. We can drop the absolute value signs if we replace $\pm k_1$ with k_2, that is, if we allow k_2 to be either positive or negative. Then we have

$$2 - y = k_2 e^{-t}$$

$$y = 2 - k_2 e^{-t}.$$

This could also be written as $y(t) = ke^{-t} + 2$, where we replace $-k_2$ with k. Note that $k = 0$ gives the equilibrium solution.

9. We separate variables and integrate to obtain

$$\int e^y\,dy = \int dt$$

$$e^y = t + c,$$

where c is any constant. We obtain $y(t) = \ln(t + c)$.

10. We separate variables and obtain

$$\int \frac{dx}{1+x^2} = \int 1\,dt.$$

Integrating both sides, we get

$$\arctan x = t + c,$$

where c is a constant. Hence, the general solution is

$$x(t) = \tan(t + c).$$

11. First note that the differential equation is not defined if $y = 0$.

In order to separate the variables, we write the equation as

$$\frac{dy}{dt} = \frac{t}{y(t^2+1)}$$

to obtain

$$\int y\,dy = \int \frac{t}{t^2+1}\,dt$$

$$\frac{y^2}{2} = \frac{1}{2}\ln(t^2+1) + c,$$

where c is any constant. So we get

$$y^2 = \ln\left(k(t^2+1)\right),$$

where $k = e^{2c}$ (hence any positive constant). We have

$$y(t) = \pm\sqrt{\ln\left(k(t^2+1)\right)},$$

where k is any positive constant and the sign is determined by the initial condition.

12. Separating variables and integrating, we obtain

$$\int y^{-1/3}\,dy = \int t\,dt$$

$$\frac{3}{2}y^{2/3} = \frac{t^2}{2} + k$$

$$y^{2/3} = \frac{t^2}{3} + c,$$

where $c = 2k/3$. Hence,

$$y(t) = \pm\left(\frac{t^2}{3} + c\right)^{3/2}.$$

Note that this form does not include the equilibrium solution $y = 0$.

13. First note that the differential equation is not defined for $y = -1/2$. We separate variables and integrate to obtain

$$\int (2y+1)\,dy = \int dt$$

$$y^2 + y = t + k,$$

where k is any constant. So

$$y(t) = \frac{-1 \pm \sqrt{4t+4k+1}}{2} = \frac{-1 \pm \sqrt{4t+c}}{2},$$

where c is any constant and the $\pm$ sign is determined by the initial condition.

We can rewrite the answer in the more simple form

$$y(t) = -\frac{1}{2} \pm \sqrt{t + c_1}$$

where $c_1 = k + 1/4$. If k can be any possible constant, then c_1 can be as well.

14. First we write the equation as $dy/dt = y^2(3-4t)$. Then we separate variables and integrate to obtain

$$\int y^{-2}\,dy = \int (3-4t)\,dt,$$

which yields $-y^{-1} = 3t - 2t^2 + c_1$, where c_1 is a constant. Consequently,

$$y(t) = \frac{1}{2t^2 - 3t + c}$$

where $c = -c_1$. This formula represents all solutions to the differential equation except the equilibrium solution $y = 0$.

15. First of all, the equilibrium solutions are $y = 0$ and $y = 1$. Now suppose $y \neq 0$ and $y \neq 1$. We separate variables to obtain

$$\int \frac{1}{y(1-y)}\,dy = \int dt = t + c,$$

where c is any constant. To integrate, we use partial fractions. Write

$$\frac{1}{y(1-y)} = \frac{A}{y} + \frac{B}{1-y}.$$

We must have $A = 1$ and $-A + B = 0$. Hence, $A = B = 1$ and

$$\frac{1}{y(1-y)} = \frac{1}{y} + \frac{1}{1-y}.$$

Consequently,

$$\int \frac{1}{y(1-y)}\,dy = \ln|y| - \ln|1-y| = \ln\left|\frac{y}{1-y}\right|.$$

After integration, we have

$$\ln\left|\frac{y}{1-y}\right| = t + c$$

$$\left|\frac{y}{1-y}\right| = c_1 e^t,$$

where $c_1 = e^c$ is any positive constant. To remove the absolute value signs, we replace the positive constant c_1 with a constant k that can be any real number and get

$$y(t) = \frac{ke^t}{1+ke^t},$$

where $k = \pm c_1$. If $k = 0$, we get the first equilibrium solution. The formula $y(t) = ke^t/(1+ke^t)$ yields all the solutions to the differential equation except for the equilibrium solution $y(t) = 1$.

16. Separating variables and integrating, we obtain

$$\int y\,dy = \int t\,dt$$

$$\frac{y^2}{2} = \frac{t^2}{2} + k$$

$$y^2 = t^2 + c,$$

where $c = 2k$. Hence,

$$y(t) = \pm\sqrt{t^2 + c},$$

where the initial condition determines the choice of sign.

17. The equation can be written in the form

$$\frac{dv}{dt} = (v+1)(t^2 - 2),$$

and we note that $v(t) = -1$ for all t is an equilibrium solution. Separating variables and integrating, we obtain

$$\int \frac{dv}{v+1} = \int t^2 - 2\,dt$$

$$\ln|v+1| = \frac{t^3}{3} - 2t + c,$$

where c is any constant. Thus,

$$|v+1| = c_1 e^{-2t+t^3/3},$$

where $c_1 = e^c$. We can dispose of the absolute value signs by allowing the constant c_1 to be any real number. In other words,

$$v(t) = -1 + ke^{-2t+t^3/3},$$

where $k = \pm c_1$. Note that, if $k = 0$, we get the equilibrium solution.

18. Rewriting the equation as

$$\frac{dy}{dt} = \frac{1}{(t+1)(y+1)}$$

we separate variables and obtain

$$\int (y+1)\,dy = \int \frac{1}{t+1}\,dt.$$

Hence,

$$\frac{y^2}{2} + y = \ln|t+1| + k.$$

We can solve using the quadratic formula. We have

$$y(t) = -1 \pm \sqrt{1 + 2\ln|t+1| + 2k}$$

$$= -1 \pm \sqrt{2\ln|t+1| + c},$$

where $c = 1 + 2k$ is any constant and the choice of sign is determined by the initial condition.

19. The function $y(t) = 0$ for all t is an equilibrium solution.

Suppose $y \neq 0$ and separate variables. We get

$$\int y + \frac{1}{y}\, dy = \int e^t\, dt$$

$$\frac{y^2}{2} + \ln|y| = e^t + c,$$

where c is any real constant. We cannot solve this equation for y, so we leave the expression for y in this implicit form. Note that the equilibrium solution $y = 0$ cannot be obtained from this implicit equation.

20. Since $y^2 - 4 = (y+2)(y-2)$, there are two equilibrium solutions, $y_1(t) = -2$ for all t and $y_2(t) = 2$ for all t. If $y \neq \pm 2$, we separate variables and obtain

$$\int \frac{dy}{y^2 - 4} = \int dt.$$

To integrate the left-hand side, we use partial fractions. If

$$\frac{1}{y^2 - 4} = \frac{A}{y+2} + \frac{B}{y-2},$$

then $A + B = 0$ and $2(B - A) = 1$. Hence, $A = -1/4$ and $B = 1/4$, and

$$\frac{1}{(y+2)(y-2)} = \frac{-1/4}{y+2} + \frac{1/4}{y-2}.$$

Consequently,

$$\int \frac{dy}{y^2 - 4} = -\frac{1}{4}\ln|y+2| + \frac{1}{4}\ln|y-2|.$$

Using this integral on the separated equation above, we get

$$\frac{1}{4}\ln\left|\frac{y-2}{y+2}\right| = t + c,$$

which yields

$$\left|\frac{y-2}{y+2}\right| = c_1 e^{4t},$$

where $c_1 = e^{4c}$. As in Exercise 5, we can drop the absolute value signs by replacing $\pm c_1$ with a new constant k. Hence, we have

$$\frac{y-2}{y+2} = ke^{4t}.$$

Solving for y, we obtain

$$y(t) = \frac{2(1 + ke^{4t})}{1 - ke^{4t}}.$$

Note that, if $k = 0$, we get the equilibrium solution $y_2(t)$. The formula $y(t) = 2(1 + ke^{4t})/(1 - ke^{4t})$ provides all of the solutions to the differential equation except the equilibrium solution $y_1(t)$.

21. The constant function $w(t) = 0$ is an equilibrium solution. Suppose $w \neq 0$ and separate variables. We get

$$\int \frac{dw}{w} = \int \frac{dt}{t}$$

$$\begin{aligned}\ln|w| &= \ln|t| + c \\ &= \ln c_1|t|,\end{aligned}$$

where c is any constant and $c_1 = e^c$. Therefore,

$$|w| = c_1|t|.$$

We can eliminate the absolute value signs by allowing the constant to assume positive or negative values. We have

$$w = kt,$$

where $k = \pm c_1$. Moreover, if $k = 0$ we get the equilibrium solution.

22. Write the equation as

$$\frac{dy}{dt} = \frac{y^2 + 1}{y^2}.$$

Separating variables and integrating, we obtain

$$\begin{aligned}\int \frac{y^2}{y^2 + 1}\, dy &= \int dt \\ \int \left(1 - \frac{1}{y^2 + 1}\right) dy &= t + c \\ y - \arctan y &= t + c.\end{aligned}$$

We cannot solve for y as a function of t.

23. From Exercise 7, we already know that the general solution is

$$y(t) = ke^{2t} - \tfrac{1}{2},$$

so we need only find the constant k for which $y(0) = 3$. We solve

$$3 = ke^0 - \tfrac{1}{2}$$

for k and obtain $k = 7/2$. The solution of the initial-value problem is

$$y(t) = \tfrac{7}{2}e^{2t} - \tfrac{1}{2}.$$

24. First we find the general solution by writing the differential equation as

$$\frac{dy}{dt} = (t+2)y^2,$$

separating variables, and integrating. We have

$$\int \frac{1}{y^2}\,dy = \int (t+2)\,dt$$

$$-\frac{1}{y} = \frac{t^2}{2} + 2t + c$$

$$= \frac{t^2+4t+c_1}{2},$$

where $c_1 = 2c$. Inverting and multiplying by -1 produces

$$y(t) = \frac{-2}{t^2+4t+c_1}.$$

Setting

$$1 = y(0) = \frac{-2}{c_1}$$

and solving for c_1, we obtain $c_1 = -2$. So

$$y(t) = \frac{-2}{t^2+4t-2}.$$

25. Separating variables and integrating, we obtain

$$\int \frac{dy}{y^2} = -\int dt$$

$$-\frac{1}{y} = -t + c.$$

So we get

$$y = \frac{1}{t-c}.$$

Now we need to find the constant c so that $y(0) = 1/2$. To do this we solve

$$\frac{1}{2} = \frac{1}{0-c}$$

and get $c = -2$. The solution of the initial-value problem is

$$y(t) = \frac{1}{t+2}.$$

26. First we separate variables and integrate to obtain

$$\int y^{-3}\,dy = \int t^2\,dt,$$

which yields

$$-\frac{y^{-2}}{2} = \frac{t^3}{3} + c.$$

Solving for y gives

$$y^2 = \frac{1}{c_1 - 2t^3/3},$$

where $c_1 = -2c$. So

$$y(t) = \pm\frac{1}{\sqrt{c_1 - 2t^3/3}}.$$

The initial value $y(0)$ is negative, so we choose the negative square root and obtain

$$y(t) = -\frac{1}{\sqrt{c_1 - 2t^3/3}}.$$

Using $-1 = y(0) = -1/\sqrt{c_1}$, we see that $c_1 = 1$ and the solution of the initial-value problem is

$$y(t) = -\frac{1}{\sqrt{1 - 2t^3/3}}.$$

27. We do not need to do any computations to solve this initial-value problem. We know that the constant function $y(t) = 0$ for all t is an equilibrium solution, and it satisfies the initial condition.

28. Rewriting the equation as

$$\frac{dy}{dt} = \frac{t}{(1 - t^2)y},$$

we separate variables and integrate obtaining

$$\int y\,dy - \int \frac{t}{1 - t^2}\,dt$$

$$\frac{y^2}{2} = -\frac{1}{2}\ln|1 - t^2| + c$$

$$y = \pm\sqrt{-\ln|1 - t^2| + k}.$$

Since $y(0) = 4$ is positive, we use the positive square root and solve

$$4 = y(0) = \sqrt{-\ln|1| + k} = \sqrt{k}$$

for k. We obtain $k = 16$. Hence,

$$y(t) = \sqrt{16 - \ln(1 - t^2)}.$$

We may replace $|1 - t^2|$ with $(1 - t^2)$ because the solution is only defined for $-1 < t < 1$.

29. We write the equation in the form

$$\frac{dx}{dt} = \frac{t^2}{x(t^3+1)}$$

and separate variables to obtain

$$\int x\,dx = \int \frac{t^2}{t^3+1}\,dt$$

$$\frac{x^2}{2} = \frac{1}{3}\ln|t^3+1| + c,$$

where c is a constant. Hence,

$$x^2 = \frac{2}{3}\ln|t^3+1| + 2c.$$

The initial condition $x(0) = -2$ implies

$$4 = (-2)^2 = \frac{2}{3}\ln|1| + 2c.$$

Thus, $c = 2$. Solving for $x(t)$, we choose the negative square root because $x(0)$ is negative, and we drop the absolute value sign because $t^3 + 1 > 0$ for t near 0. The result is

$$x(t) = -\sqrt{\tfrac{2}{3}\ln(t^3+1) + 4}.$$

30. Separating variables, we have

$$\int \frac{y\,dy}{1-y^2} = \int dt$$

$$= t + c,$$

where c is any constant. To integrate the left-hand side, we substitute $u = 1 - y^2$. Then $du = -2y\,dy$. We get

$$\int \frac{y\,dy}{1-y^2} = -\frac{1}{2}\int \frac{du}{u} = -\frac{1}{2}\ln|u| = -\frac{1}{2}\ln|1-y^2|.$$

Using this integral, we have

$$-\frac{1}{2}\ln|1-y^2| = t + c$$

$$|1-y^2| = c_1 e^{-2t},$$

where $c_1 = e^{-2c}$. As in Exercise 5, we can drop the absolute value signs by replacing $\pm c_1$ with a new constant k. Hence, we have

$$y(t) = \pm\sqrt{1 - ke^{-2t}}$$

Because $y(0)$ is negative, we use the negative square root and solve

$$-2 = y(0) = -\sqrt{1 - ke^0} = -\sqrt{1-k}$$

for k. We obtain $k = -3$. Hence, $y(t) = -\sqrt{1 + 3e^{-2t}}$.

31. We separate variables to obtain

$$\int \frac{dy}{1+y^2} = \int t\,dt$$

$$\arctan y = \frac{t^2}{2} + c,$$

where c is a constant. Hence the general solution is

$$y(t) = \tan\left(\frac{t^2}{2} + c\right).$$

Next we find c so that $y(0) = 1$. Solving

$$1 = \tan\left(\frac{0^2}{2} + c\right)$$

yields $c = \pi/4$, and the solution to the initial-value problem is

$$y(t) = \tan\left(\frac{t^2}{2} + \frac{\pi}{4}\right).$$

32. Separating variables and integrating, we obtain

$$\int (2y+3)\,dy = \int dt$$

$$y^2 + 3y = t + c$$

$$y^2 + 3y - (t + c) = 0.$$

We can use the quadratic formula to obtain

$$y = -\tfrac{3}{2} \pm \sqrt{t + c_1},$$

where $c_1 = c + 9/4$. Since $y(0) = 1 > -3/2$ we take the positive square root and solve

$$1 = y(0) = -\tfrac{3}{2} + \sqrt{c_1},$$

so $c_1 = 25/4$. The solution to the initial-value problem is

$$y(t) = -\tfrac{3}{2} + \sqrt{t + \tfrac{25}{4}}.$$

33. Let $S(t)$ denote the amount of salt (in pounds) in the bucket at time t (in minutes). We derive a differential equation for S by considering the difference between the rate that salt is entering the bucket and the rate that salt is leaving the bucket. Salt is entering the bucket at the rate of 1/4 pounds per minute. The rate that salt is leaving the bucket is the product of the concentration of salt in the mixture and the rate that the mixture is leaving the bucket. The concentration is $S/5$, and the mixture is leaving the bucket at the rate of $1/2$ gallons per minute. We obtain the differential equation

$$\frac{dS}{dt} = \frac{1}{4} - \frac{S}{5} \cdot \frac{1}{2},$$

which can be rewritten as

$$\frac{dS}{dt} = \frac{5 - 2S}{20}.$$

This differential equation is separable, and we can find the general solution by integrating

$$\int \frac{1}{5 - 2S}\, dS = \int \frac{1}{20}\, dt.$$

We have

$$-\frac{\ln|5 - 2S|}{2} = \frac{t}{20} + c$$

$$\ln|5 - 2S| = -\frac{t}{10} + c_1$$

$$|5 - 2S| = c_2 e^{-t/10}.$$

We can eliminate the absolute value signs and determine c_2 using the initial condition $S(0) = 0$ (the water is initially free of salt). We have $c_2 = 5$, and the solution is

$$S(t) = 2.5 - 2.5e^{-t/10} = 2.5(1 - e^{-t/10}).$$

(a) When $t = 1$, we have $S(1) = 2.5(1 - e^{-0.1}) \approx 0.238$ lbs.
(b) When $t = 10$, we have $S(10) = 2.5(1 - e^{-1}) \approx 1.58$ lbs.
(c) When $t = 60$, we have $S(60) = 2.5(1 - e^{-6}) \approx 2.49$ lbs.
(d) When $t = 1000$, we have $S(1000) = 2.5(1 - e^{-100}) \approx 2.50$ lbs.
(e) When t is very large, the $e^{-t/10}$ term is close to zero, so $S(t)$ is very close to 2.5 lbs. In this case, we can also reach the same conclusion by doing a qualitative analysis of the solutions of the equation. The constant solution $S(t) = 2.5$ is the only equilibrium solution for this equation, and by examining the sign of dS/dt, we see that all solutions approach $S = 2.5$ as t increases.

34. Rewrite the equation as

$$\frac{dC}{dt} = -k_1 C + (k_1 C_0 + k_2 E),$$

separate variables, and integrate to obtain

$$\int \frac{1}{-k_1 C + (k_1 C_0 + k_2 E)}\, dC = \int dt$$

$$-\frac{1}{k_1} \ln|-k_1 C + k_1 C_0 + k_2 E| = t + c$$

$$-k_1 C + k_1 C_0 + k_2 E = c_1 e^{-k_1 t},$$

where c_1 is a constant determined by the initial condition. Hence,

$$C(t) = C_0 + \frac{k_2}{k_1}E - c_2 e^{-k_1 t},$$

where c_2 is a constant.

(a) Substituting the given values for the parameters, we obtain

$$C(t) = 600 - c_2 e^{-0.1t},$$

and the initial condition $C(0) = 150$ gives $c_2 = 450$, which implies that

$$C(t) = 600 - 450e^{-0.1t}.$$

Hence, $C(2) \approx 232$.

(b) Using part (a), $C(5) \approx 328$.

(c) When t is very large, $e^{-0.1t}$ is very close to zero, so $C(t) \approx 600$. (We could also obtain this conclusion by doing a qualitative analysis of the solutions.)

(d) Using the new parameter values and $C(0) = 600$ yields

$$C(t) = 300 + 300e^{-0.1t},$$

so $C(1) \approx 571$, $C(5) \approx 482$, and as $t \to \infty$, $C \to 300$.

(e) Again changing the parameter values and using $C(0) = 600$, we have

$$C(t) = 500 + 100e^{-0.1t},$$

so $C(1) \approx 590$, $C(5) \approx 560$, and as $t \to \infty$, $C \to 500$.

35. **(a)** If we let k denote the proportionality constant in Newton's law of cooling, the differential equation satisfied by the temperature T of the chocolate is

$$\frac{dT}{dt} = k(T - 70).$$

We also know that $T(0) = 170$ and that $dT/dt = -20$ at $t = 0$. Therefore, we obtain k by evaluating the differential equation at $t = 0$. We have

$$-20 = k(170 - 70),$$

so $k = -0.2$. The initial-value problem is

$$\frac{dT}{dt} = -0.2(T - 70), \quad T(0) = 170.$$

(b) We can solve the initial-value problem in part (a) by separating variables. We have

$$\int \frac{dT}{T - 70} = \int -0.2\, dt$$

$$\ln|T - 70| = -0.2t + k$$

$$|T - 70| = ce^{-0.2t}.$$

Since the temperature of the chocolate cannot become lower than the temperature of the room, we can ignore the absolute value and conclude

$$T(t) = 70 + ce^{-0.2t}.$$

Now we use the initial condition $T(0) = 170$ to find the constant c because

$$170 = T(0) = 70 + ce^{-0.2(0)},$$

which implies that $c = 100$. The solution is

$$T = 70 + 100e^{-0.2t}.$$

In order to find t so that the temperature is 110° F, we solve

$$110 = 70 + 100e^{-0.2t}$$

for t obtaining

$$\tfrac{2}{5} = e^{-0.2t}$$

$$\ln \tfrac{2}{5} = -0.2t$$

so that

$$t = \frac{\ln(2/5)}{-0.2} \approx 4.6.$$

36. The easiest way to do this problem is to use the fact that $C = S/100$, so

$$\frac{dC}{dt} = \frac{1}{100}\frac{dS}{dt}.$$

Hence,

$$\frac{dC}{dt} = \frac{1}{100}\frac{dS}{dt} = \frac{1}{100}\left(20 - \frac{3S}{100}\right) = \frac{1}{100}(20 - 3C)$$

We obtain

$$\frac{dC}{dt} = \frac{20 - 3C}{100} = \frac{1}{5} - \frac{3C}{100}.$$

We can also derive this differential equation directly, but in order to do so, we need to determine how the concentrations of the incoming mixtures affect the total concentration in the vat. Since the total concentration $C(t) = S(t)/V(t)$ where $V(t)$ is the total volume,

$$\begin{aligned}\frac{dC}{dt} &= \frac{1}{V}\frac{dS}{dt} - \frac{1}{V^2}\frac{dV}{dt}S \\ &= \frac{1}{V}\frac{dS}{dt} - \frac{1}{V}\frac{S}{V}\frac{dV}{dt} \\ &= \frac{\dfrac{dS}{dt} - C\dfrac{dV}{dt}}{V}.\end{aligned}$$

If we want to consider only the contribution to dC/dt from one pipe, we let C_{new} be the concentration of the sugar entering from that pipe alone. Then

$$\frac{dS}{dt} = C_{\text{new}} \frac{dV}{dt},$$

where both derivatives represent the rates associated to the pipe in question. Considering just the one pipe, the formula for dC/dt is

$$\frac{dC}{dt} = \frac{(C_{\text{new}} - C)\dfrac{dV}{dt}}{V}.$$

(When using this formula, it is important to realize that both dC/dt and dV/dt refer only to the pipe in question, but V refers to the total volume.)

For Pipe A, we have $C_{\text{new}} = 5$ and $dV/dt = 2$. For Pipe B, we have $C_{\text{new}} = 10$ and $dV/dt = 1$. For Pipe C, we have $C_{\text{new}} = C$ and $dV/dt = -3$. In this case, $V(t) = 100$ for all t. Totaling the contributions to dC/dt for all three pipes gives us

$$\frac{dC}{dt} = \frac{(5 - C)2 + (10 - C)1 + (C - C)3}{100}.$$

We obtain

$$\frac{dC}{dt} = \frac{20 - 3C}{100}.$$

37. Let $C(t)$ denote the concentration. From Exercise 36, the differential equation is

$$\frac{dC}{dt} = \frac{20 - 3C}{100}$$

Note that $dC/dt = 0$ for all t if $C = 20/3$. Hence, the constant function $C(t) = 20/3$ is an equilibrium solution.

Separating variables and integrating, we obtain

$$\int \frac{dC}{20 - 3C} = \int \frac{1}{100}\, dt$$

$$-\frac{1}{3} \ln |20 - 3C| = \frac{t}{100} + c$$

$$|20 - 3C| = c_1 e^{-3t/100}.$$

To remove the absolute value signs we note that $e^{-3t/100}$ is never zero, so $20 - 3C(t)$ is either positive for all t, negative for all t, or zero for all t. Hence the general solution is

$$20 - 3C = k_1 e^{-3t/100},$$

where k_1 can be any constant. Finally, we have

$$C(t) = \tfrac{20}{3} + ke^{-3t/100}$$

where k can be any constant.

38. Let t be time measured in minutes and let $H(t)$ represent the hot sauce in the chili measured in teaspoons at time t. Then $H(0) = 12$.

The pot contains 32 cups of chili, and chili is removed from the pot at the rate of 1 cup per minute. Since each cup of chili contains $H/32$ teaspoons of hot sauce, the differential equation is

$$\frac{dH}{dt} = -\frac{H}{32}.$$

The general solution of this equation is

$$H(t) = ke^{-t/32}.$$

(We could solve this differential equation by separation of variables, but this is also the equation for which we guessed solutions in Section 1.1.) Since $H(0) = 12$, we get the solution

$$H(t) = 12e^{-t/32}.$$

We wish to find t such that $H(t) = 4$ (two teaspoons per gallon in two gallons). We have

$$\begin{aligned} 12e^{-t/32} &= 4 \\ -\frac{t}{32} &= \ln\frac{1}{3} \\ t &= 32\ln 3. \end{aligned}$$

So, $t \approx 35.16$ minutes. A reasonable approximation is 35 minutes and in that time 35 cups will have been eaten.

39. First we find the general solution without using the specific values of the parameters. The equation

$$\frac{dM}{dt} = rM - p$$

is separable, so we separate variables and integrate to obtain

$$\begin{aligned} \int \frac{dM}{rM - p} &= \int dt \\ \frac{1}{r}\ln|rM - p| &= t + c_1 \\ |rM - p| &= c_2e^{rt} \end{aligned}$$

where $c_2 = e^{rc_1}$. Hence,

$$rM - p = ke^{rt}.$$

There is an equilibrium solution if $dM/dt = rM - p = 0$ for all t. Consequently there is a constant solution if $p = rM$ (the amount of the payments is exactly equal to the interest).

Since e^{rt} is never zero, the sign of $rM - p$ never changes. Hence $rM(t) - p > 0$ for all t, $rM(t) - p < 0$ for all t, or $rM(t) - p = 0$ for all t. The general solution is

$$M = \frac{p + ke^{rt}}{r},$$

where k is a constant which is determined by the initial condition. (Typically k is negative if $M(t)$ is eventually zero.)

We know that $M(0) = 150{,}000$ and $M(30) = 0$. Thus, we obtain the system of equations

$$\begin{cases} \dfrac{p+k}{r} = 150{,}000 \\ \dfrac{p+ke^{30r}}{r} = 0. \end{cases}$$

Subtracting the second equation from the first yields

$$\frac{k(1-e^{30r})}{r} = 150{,}000,$$

and consequently,

$$k = \frac{150{,}000\,r}{1-e^{30r}}.$$

Since

$$p + ke^{30r} = 0,$$

we have

$$p = -ke^{30r} = -\frac{150{,}000\,r}{1-e^{30r}}e^{30r} = \frac{150{,}000\,r}{1-e^{-30r}}.$$

Note that the total amount paid over the thirty year life of the loan is $30p$.

(a) Substituting $r = 0.07$, we get a total amount paid of $30p = \$358{,}957$. At $r = 0.065$, we get $30p = \$341{,}018$, and adding the \$4,500 extra due to points, we get a total cost of \$345,518.

(b) The 6.5% option is better.

(c) If the \$4,500 is invested at 5%, then after thirty years the \$4,500 has grown to \$20,168 (the solution of $dy/dt = 0.05y$, $y(0) = 4{,}500$.) The 7% option is the better deal.

EXERCISES FOR SECTION 1.3

1.

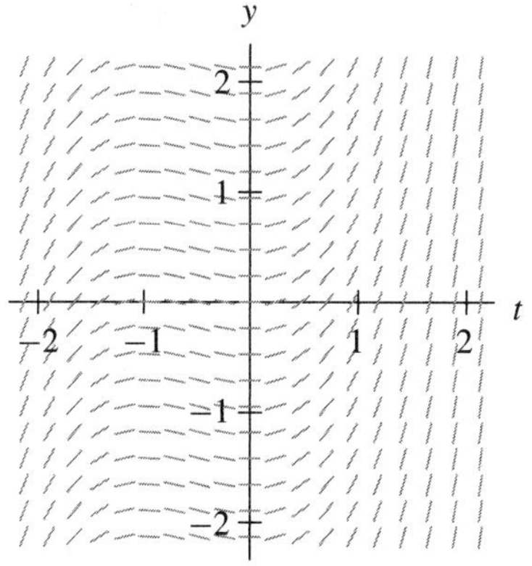

2.

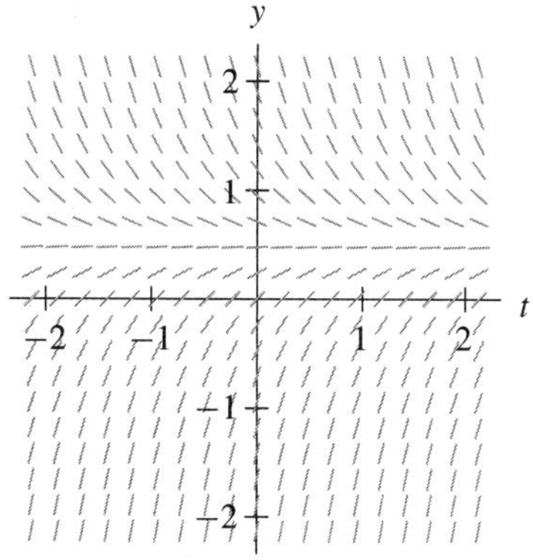

3.

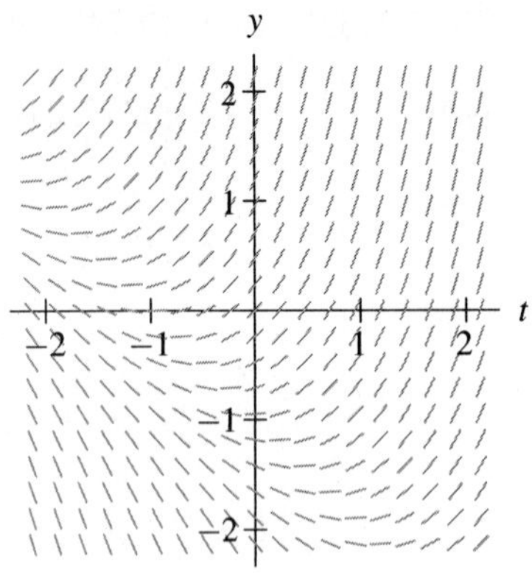

4.

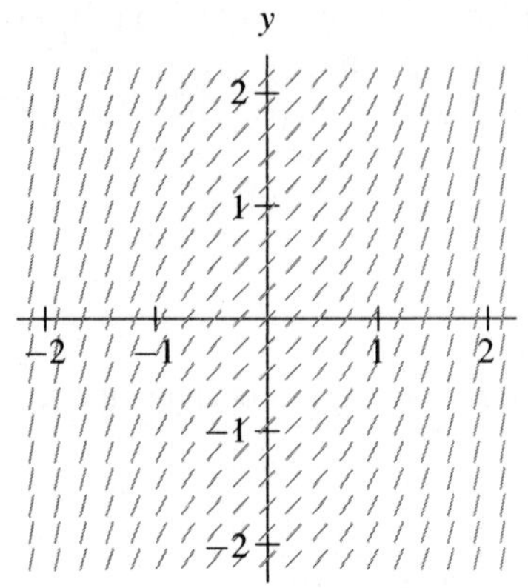

5.

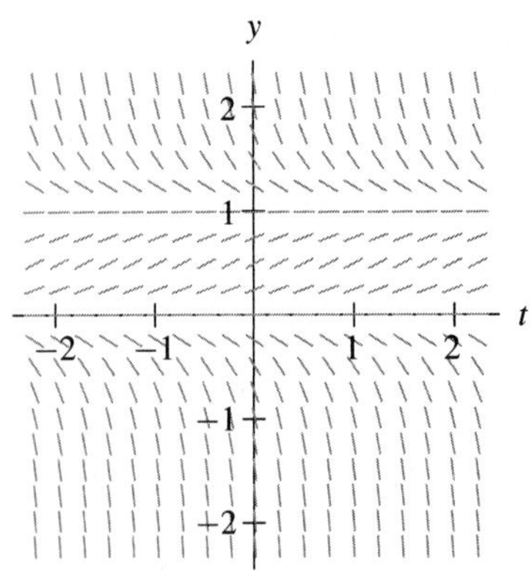

6.

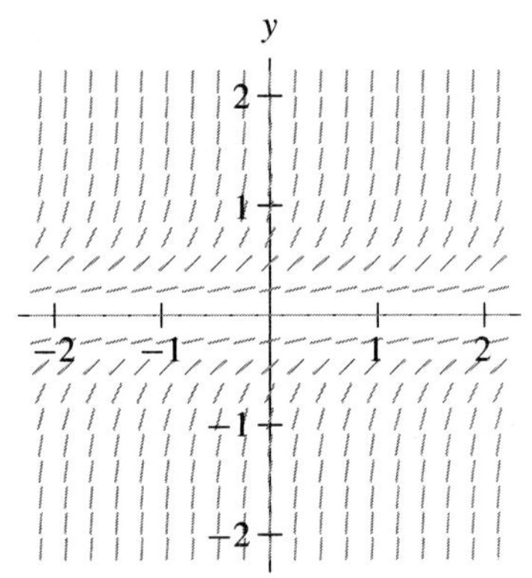

7. **(a)**

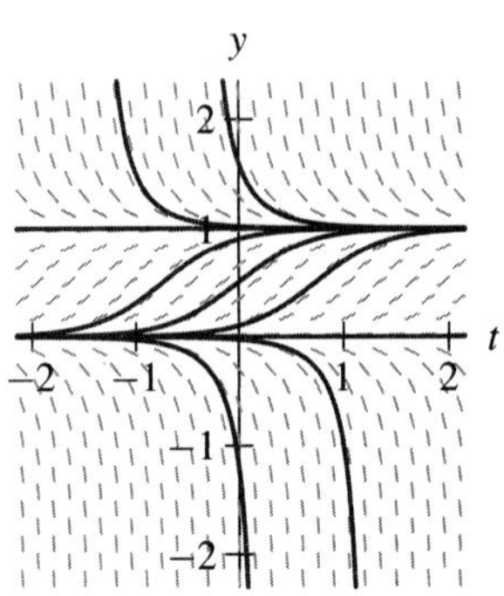

(b) The solution with $y(0) = 1/2$ approaches the equilibrium value $y = 1$ from below as t increases. It decreases toward $y = 0$ as t decreases.

8. **(a)**

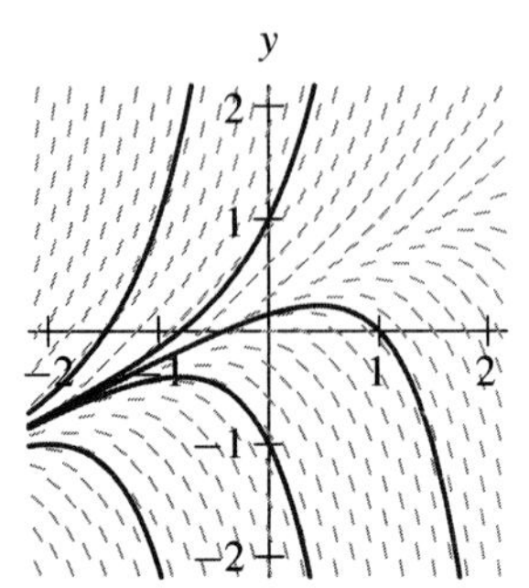

(b) The solution $y(t)$ with $y(0) = 1/2$ increases with $y(t) \to \infty$ as t increases. As t decreases, $y(t) \to -\infty$.

9. **(a)**

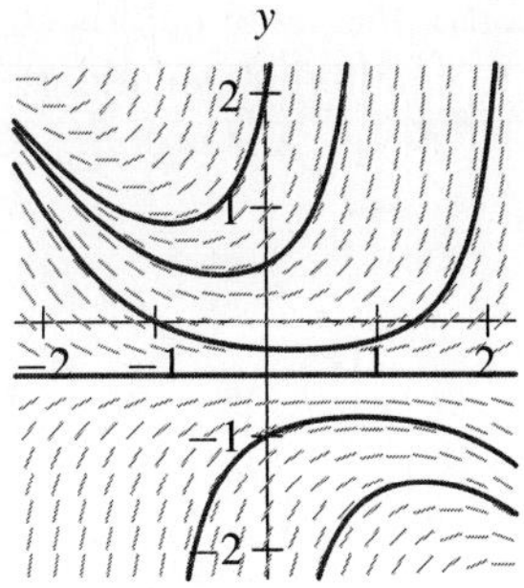

(b) The solution $y(t)$ with $y(0) = 1/2$ has $y(t) \to \infty$ both as t increases and as t decreases.

10. **(a)**

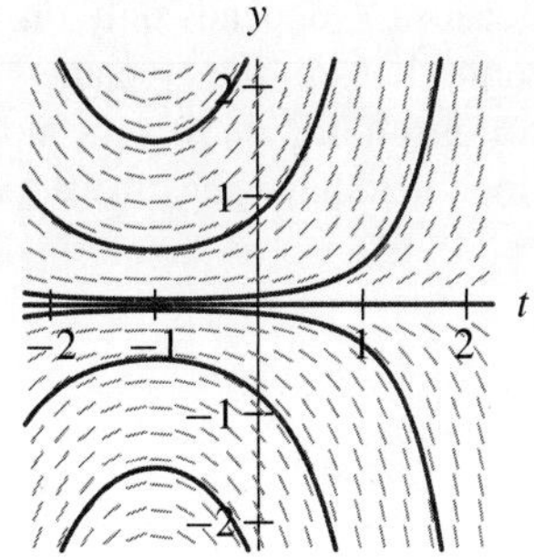

(b) The solution $y(t)$ with $y(0) = 1/2$ has $y(t) \to \infty$ both as t increases and as t decreases.

11. **(a)** On the line $y = 3$ in the ty-plane, all of the slope marks have slope -1.

(b) Because f is continuous, if y is close to 3, then $f(t, y) < 0$. So any solution close to $y = 3$ must be decreasing. Therefore, solutions $y(t)$ that satisfy $y(0) < 3$ can never be larger than 3 for $t > 0$, and consequently $y(t) < 3$ for all t.

12.

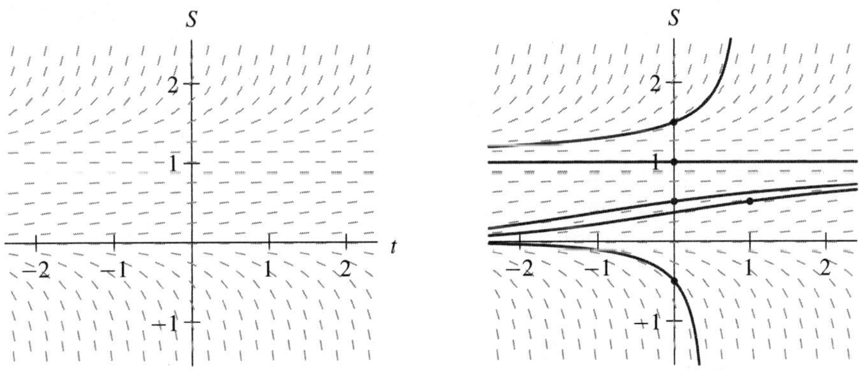

13. The slope field in the ty-plane is constant along vertical lines.

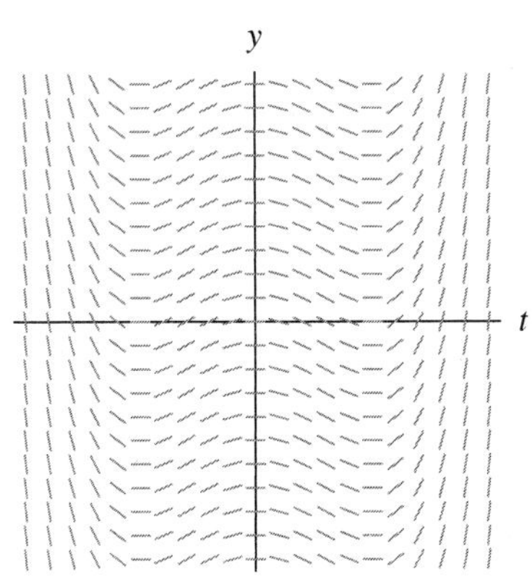

14. Because f depends only on y (the equation is autonomous), the slope field is constant along horizontal lines in the ty-plane. The roots of f correspond to equilibrium solutions. If $f(y) > 0$, the corresponding lines in the slope field have positive slope. If $f(y) < 0$, the corresponding lines in the slope field have negative slope.

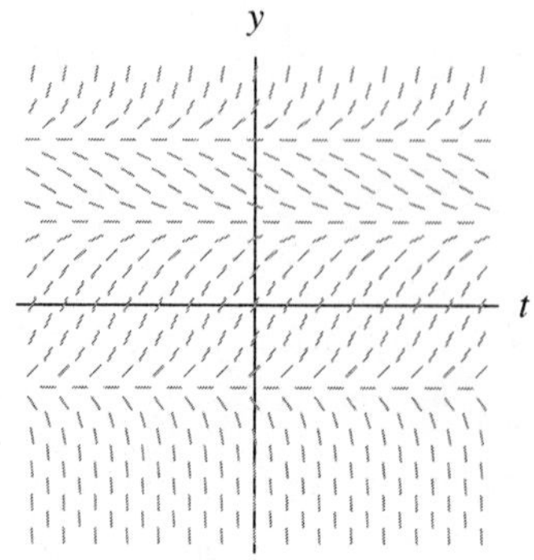

15. **(a)** Note that the slopes are constant along vertical lines—lines along which t is constant, so the right-hand side of the corresponding equation depends only on t. The only choices are equations (i) and (iv). Because the slopes are negative for $t > 1$ and positive for $t < 1$, this slope field corresponds to equation (iv).

(b) This slope field has an equilibrium solution corresponding to the line $y = 1$, as does equations (ii), (v), (vii), and (viii). Equations (ii), (v), and (viii) are autonomous, and this slope field is not constant along horizontal lines. Consequently, it corresponds to equation (vii).

(c) This slope field is constant along horizontal lines, so it corresponds to an autonomous equation. The autonomous equations are (ii), (v), and (viii). This field does not correspond to equation (v) because it has the equilibrium solution $y = -1$. The slopes are negative between $y = -1$ and $y = 1$. Consequently, this field corresponds to equation (viii).

(d) This slope field depends both on y and on t, so it can only correspond to equations (iii), (vi), or (vii). It does not correspond to (vii) because it does not have an equilibrium solution at $y = 1$. Also, the slopes are positive if $y > 0$. Therefore, it must correspond to equation (vi).

16. **(a)** Because the slope field is constant on vertical lines, the given information is enough to draw the entire slope field.

(b) The solution with initial condition $y(0) = 2$ is a vertical translation of the given solution. We only need change the "constant of integration" so that $y(0) = 2$.

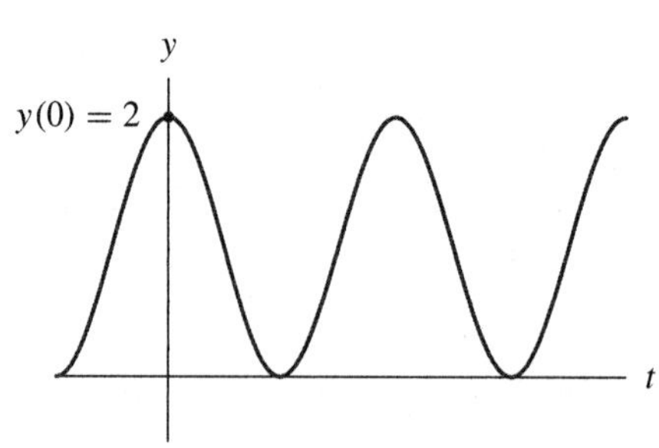

17. **(a)** Because the equation is autonomous, the slope field is constant on horizontal lines, so this solution provides enough information to sketch the slope field on the entire upper half plane. Also, if we assume that f is continuous, then the slope field on the line $y = 0$ must be horizontal.

(b) The solution with initial condition $y(0) = 2$ is a translate to the left of the given solution.

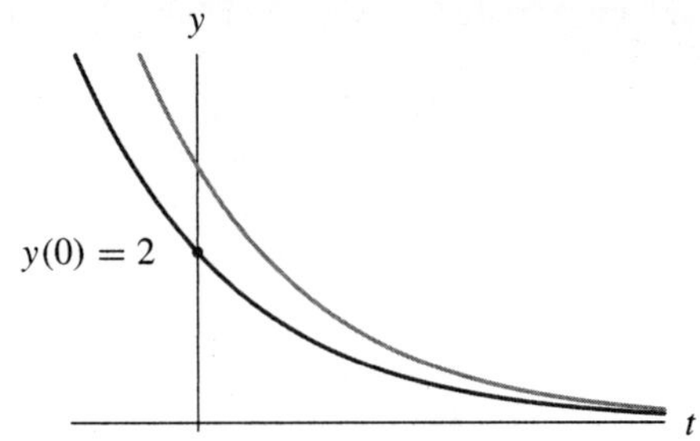

18. **(a)** Since $y(t) = 2$ for all t is a solution and $dy/dt = 0$ for all t, $f(t, y(t)) = f(t, 2) = 0$ for all t.

(b) Therefore, the slope marks all have zero slope along the horizontal line $y = 2$.

(c) If the graphs of solutions cannot cross in the ty-plane, then the graph of a solution must stay on the same side of the line $y = 2$ as it is at time $t = 0$. In Section 1.5, we discuss conditions that guarantee that graphs of solutions do not cross.

19. With $V(t) = K$, the differential equation is

$$\frac{dv_c}{dt} = \frac{K - v_c}{RC}.$$

Separating variables, we obtain

$$\frac{dv_c}{K - v_c} = \frac{dt}{RC}.$$

Integrating both sides yields

$$-\ln|K - v_c| = \frac{t}{RC} + k_1,$$

where k_1 is the constant of integration. Thus,

$$|K - v_c| = k_2 e^{-t/RC}$$

where $k_2 = e^{-k_1}$. We can eliminate the absolute values by allowing k_2 to assume either positive or negative values. We obtain the general solution

$$v_c = K + ke^{-t/RC}$$

where k (replacing $-k_2$) is the constant depending on the initial condition.

20. From Exercise 19, we have

$$v_c(t) = K + ke^{-t/(RC)}$$

for $t < 3$. Using the initial condition $1 = v_c(0) = K + k$, we obtain $k = 1 - K$, so

$$v_c(t) = K + (1 - K)e^{-t/(RC)}.$$

For $t \geq 3$, the differential equation is

$$\frac{dv_c}{dt} = -\frac{v_c}{RC},$$

which we can solve by separating variables or by the technique discussed in Section 1.1. We have

$$\ln |v_c| = -\frac{t}{RC} + c$$

$$v_c(t) = k_1 e^{-t/(RC)}.$$

Using the formula for $v_c(t)$ for $t < 3$, we obtain $v_c(3) = K + (1 - K)e^{-3/(RC)}$, and we use this value as the initial condition for $t \geq 3$. That is,

$$k_1 e^{-3/(RC)} = K + (1 - K)e^{-3/(RC)}$$

so

$$k_1 = Ke^{3/(RC)} + (1 - K)$$

and

$$v_c(t) = (Ke^{3/(RC)} + (1 - K))e^{-t/(RC)}.$$

Hence,

$$v_c(t) = \begin{cases} K + (1 - K)e^{-t/(RC)} & \text{for } t < 3; \\ e^{(3-t)/(RC)} K + (1 - K)e^{-t/(RC)} & \text{for } t \geq 3. \end{cases}$$

21. **(a)**

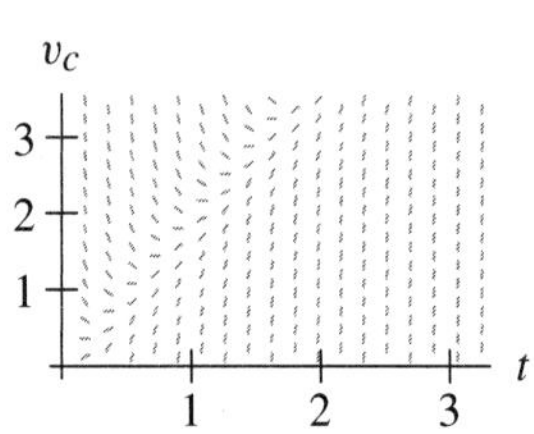

(b)

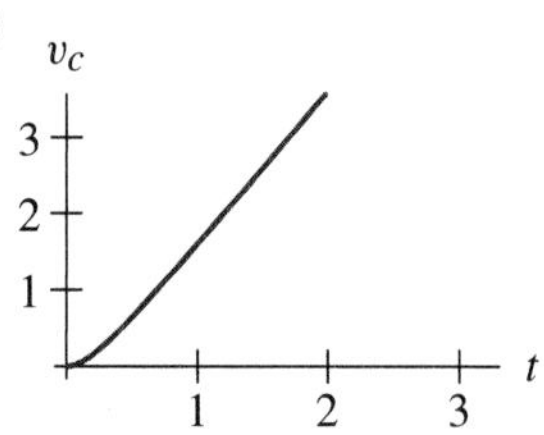

(c)

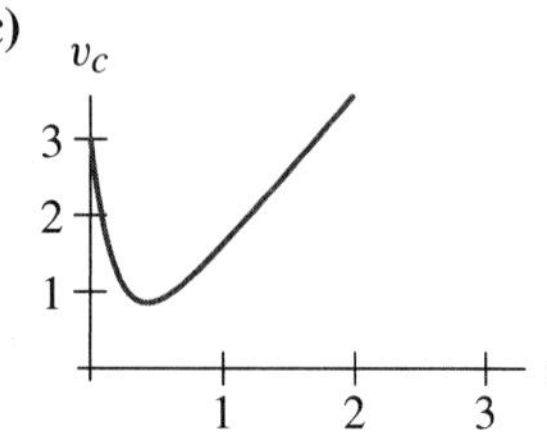

22. **(a)**

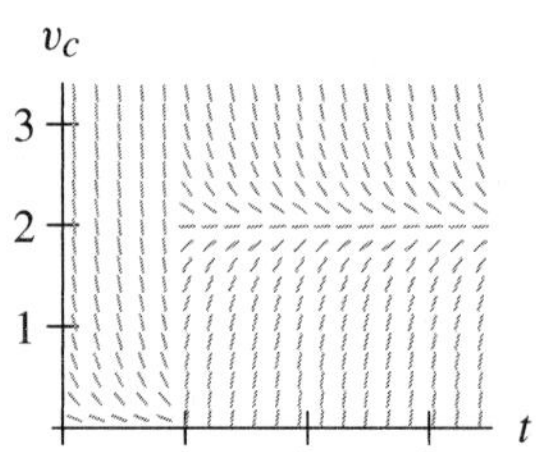

(b)

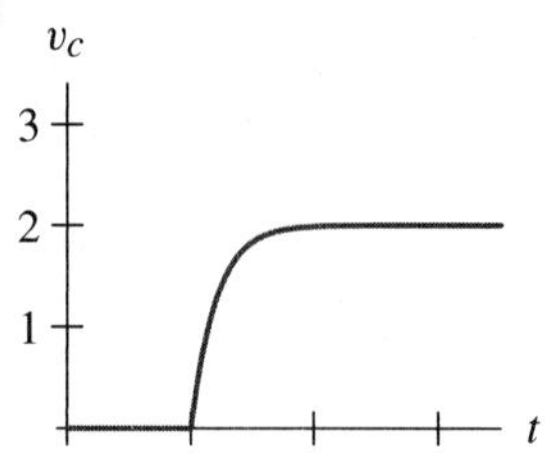

(c)

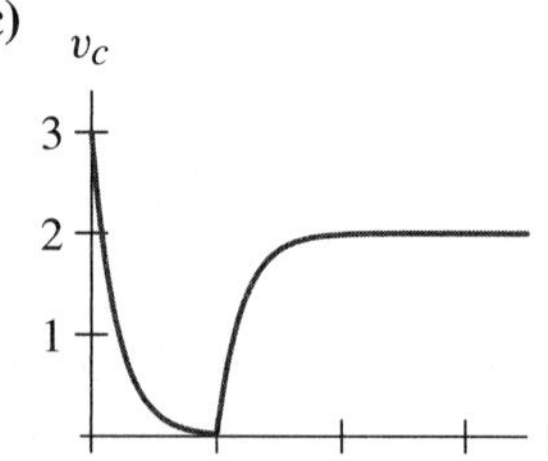

23. Our differential equation has two different expressions depending on whether $t < 1$ or $t \geq 1$. Substituting $2t$ or 2 for $V(t)$ depending on the value of t, we have

$$\frac{dv_c}{dt} = \begin{cases} (2t - v_c)/0.2 = 5(2t - v_c) & \text{for } t < 1; \\ (2 - v_c)/0.2 = 10 - 5v_c & \text{for } t \geq 1. \end{cases}$$

(a)

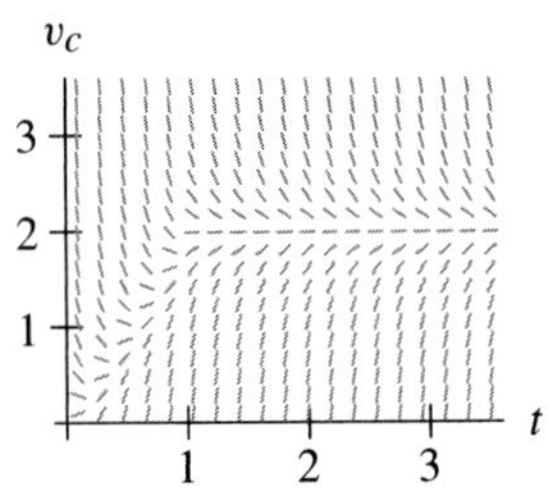

(b)

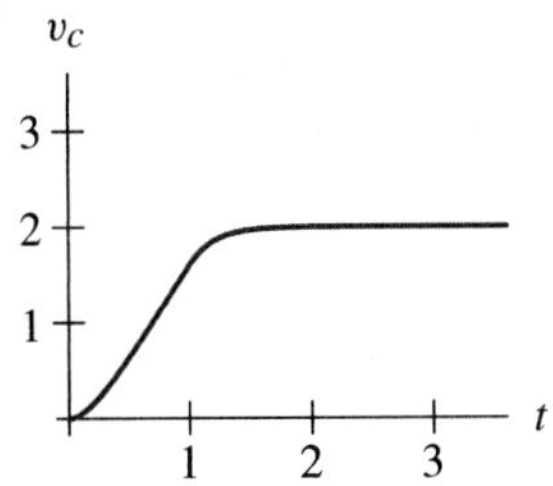

(c)

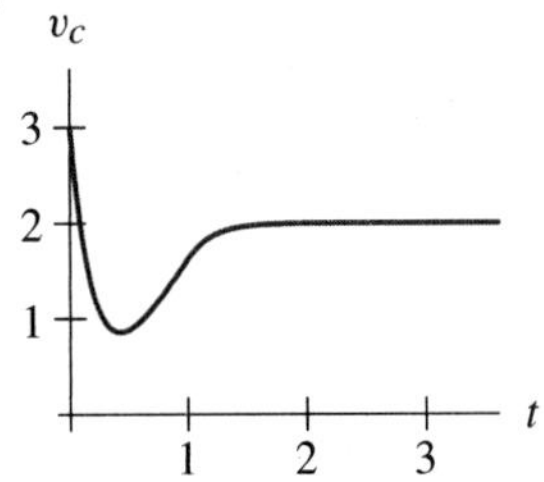

(e) For $t < 1$, the graph of the solution is identical to the one for $t < 1$ in Exercise 21. For $t \geq 1$, the graph of the solution is qualitatively similar to the one for $t \geq 1$ in Exercise 22. The complete graph is the "combination" of the graphs from Exercises 21 and 22.

24. **(a)**

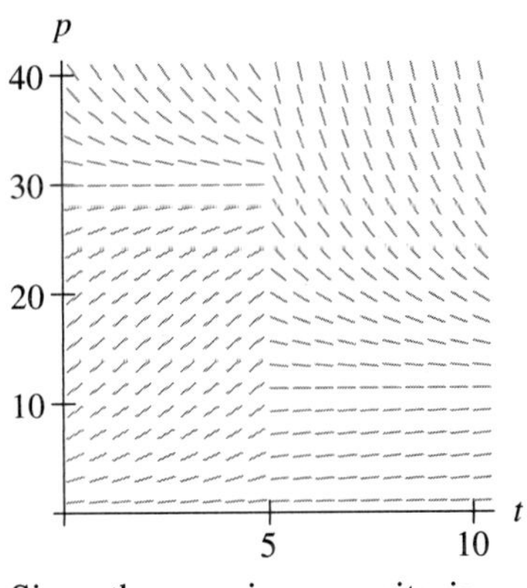

(b)

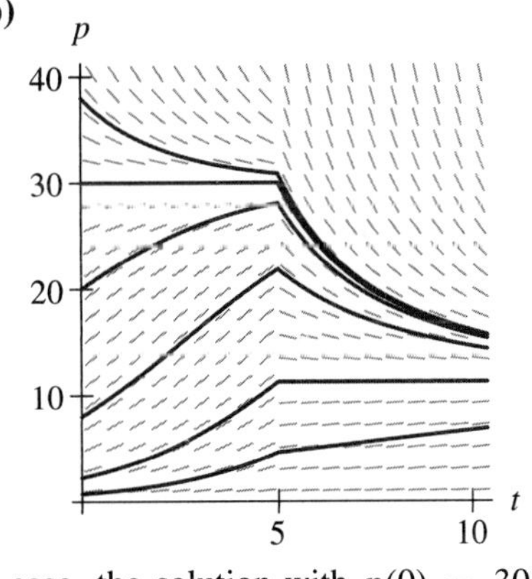

(c) Since the carrying capacity is $p = 30$ in this case, the solution with $p(0) = 30$ is just the equilibrium solution $p(t) = 30$ for $t < 5$.

For $t < 5$ and $p(0) = 20$, we separate variables and integrate to obtain the solution

$$p(t) = \frac{30ke^{0.4t}}{1 + ke^{0.4t}},$$

where k is any constant. To obtain the desired solution, we solve for k using $p(0) = 20$. We get $k = 2$. Hence, the solution is

$$p(t) = \frac{60e^{0.4t}}{1 + 2e^{0.4t}},$$

which gives $p(5) \approx 28.1$.

For $t \geq 5$ we have the equation

$$\frac{dp}{dt} = 0.15p - \frac{0.4p^2}{30},$$

which is also separable. Separating variables, integrating, and solving for p yields

$$p(t) = \frac{-0.15\,ke^{3t/20}}{1 - 0.0133\,ke^{3t/20}}.$$

The solution with $p(0) = 30$ has $p(5) = 30$. Using $p(5) = 30$ as the initial condition, we get

$$p(t) = \frac{-0.15 \cdot 59\,e^{3t/20}}{1 - 0.0133 \cdot 59\,e^{3t/20}}$$

for $t \geq 5$. Using $p(5) = 28.1$ as the initial condition, we obtain

$$p(t) = \frac{0.15 \cdot 61.6\,e^{3t/20}}{1 - 0.0133 \cdot 61.6\,e^{3t/20}}.$$

(d) The solution with $p(0) = 30$ is constant for $0 \leq t \leq 5$ because 30 is the carrying capacity. For $t > 5$, the solution decays toward a new, smaller equilibrium at $p = 11.25$. The solution with initial condition $p(0) = 20$ grows toward the carrying capacity until $t = 5$. Then it also decreases toward the new, smaller equilibrium at $p = 11.25$.

EXERCISES FOR SECTION 1.4

1.

Table 1.1
Results of Euler's method

k	t_k	y_k	m_k
0	0	3	7
1	0.5	6.5	14
2	1.0	13.5	28
3	1.5	27.5	56
4	2.0	55.5	

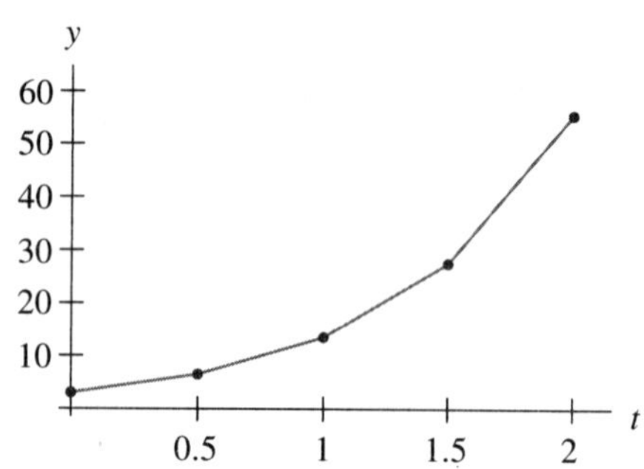

2.

Table 1.2
Results of Euler's method (y_k rounded to two decimal places)

k	t_k	y_k	m_k
0	0	1	-1
1	0.25	0.75	-0.3125
2	0.5	0.67	0.0485
3	0.75	0.68	0.282
4	1.0	0.75	

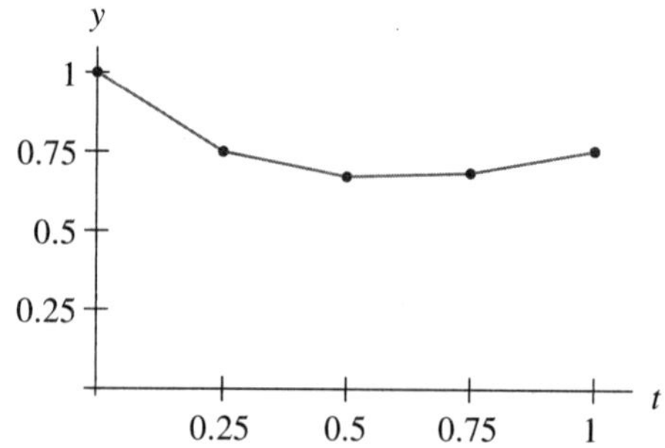

3.

Table 1.3
Results of Euler's method (shown rounded to two decimal places)

k	t_k	y_k	m_k
0	0	0.5	0.25
1	0.25	0.56	−0.68
2	0.50	0.39	−1.85
3	0.75	−0.07	−2.99
4	1.00	−0.82	−3.33
5	1.25	−1.65	−2.27
6	1.50	−2.22	−1.07
7	1.75	−2.49	−0.81
8	2.00	−2.69	

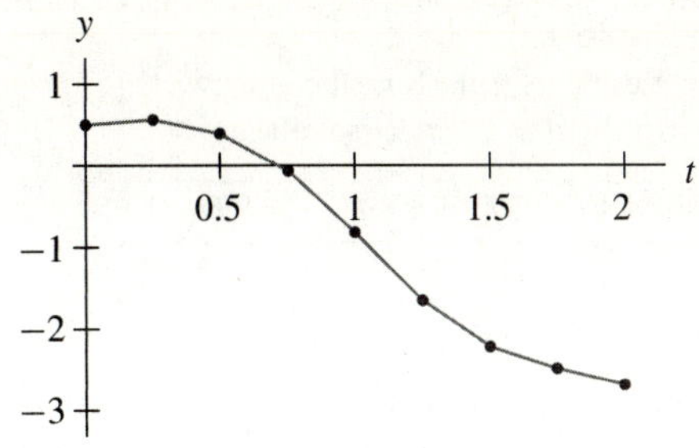

4.

Table 1.4
Results of Euler's method (to two decimal places)

k	t_k	y_k	m_k
0	0	1	0.84
1	0.5	1.42	0.99
2	1.0	1.91	0.94
3	1.5	2.38	0.68
4	2.0	2.73	0.40
5	2.5	2.93	0.21
6	3.0	3.03	

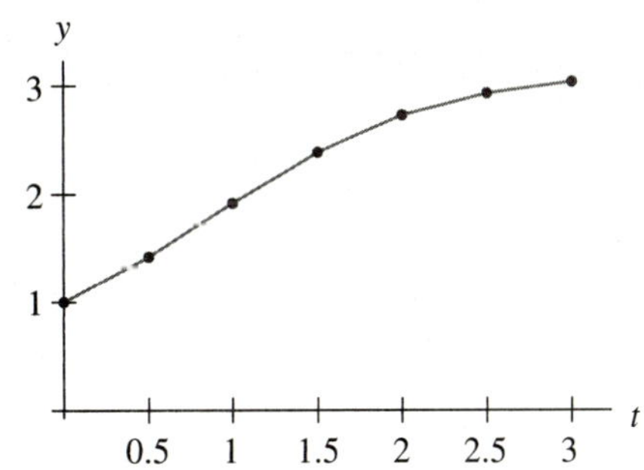

5.

Table 1.5
Results of Euler's method

k	t_k	w_k	m_k
0	0	4	−5
1	1	−1	0
2	2	−1	0
3	3	−1	0
4	4	−1	0
5	5	−1	

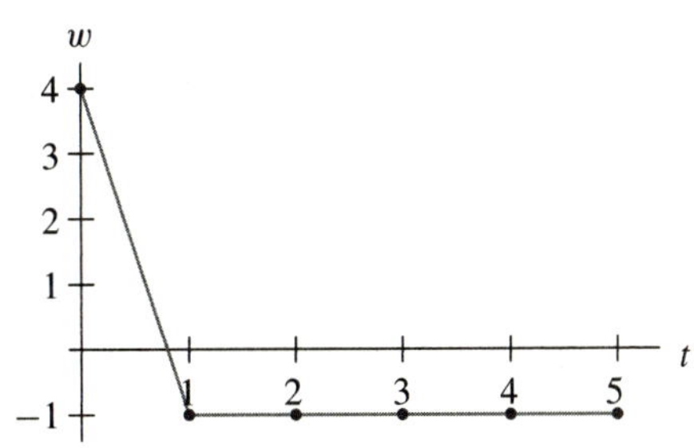

6.

Table 1.6
Results of Euler's method (shown rounded to two decimal places)

k	t_k	w_k	m_k
0	0	0	3
1	0.5	1.5	3.75
2	1.0	3.38	−1.64
3	1.5	2.55	1.58
4	2.0	3.35	−1.50
5	2.5	2.59	1.46
6	3.0	3.32	−1.40
7	3.5	2.62	1.36
8	4.0	3.31	−1.31
9	4.5	2.65	1.28
10	5.0	3.29	

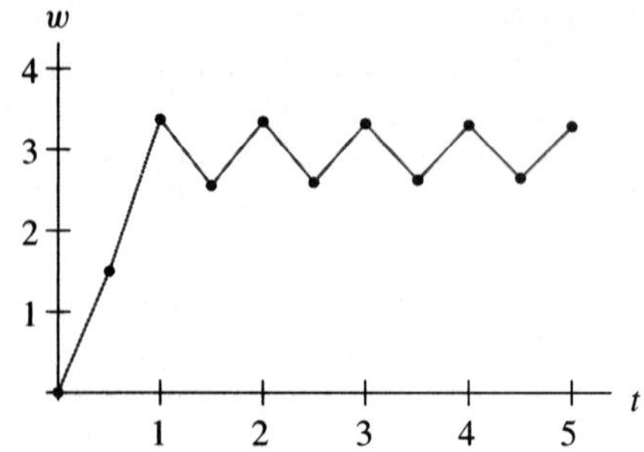

7.

Table 1.7
Results of Euler's method (shown rounded to two decimal places)

k	t_k	y_k	m_k
0	0	2	2.72
1	0.5	3.36	1.81
2	1.0	4.27	1.60
3	1.5	5.06	1.48
4	2.0	5.81	

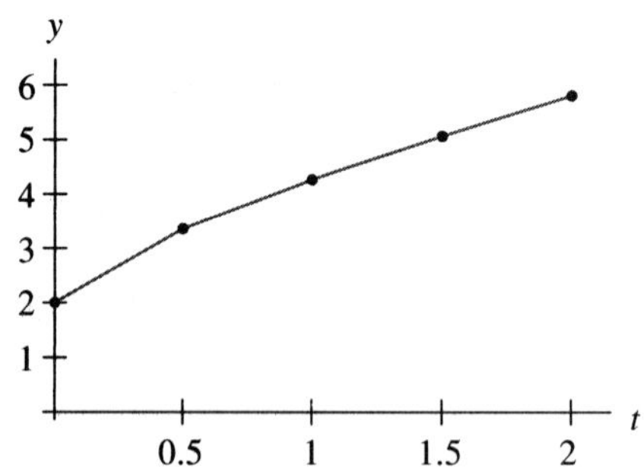

8.

Table 1.8
Results of Euler's method (shown rounded to two decimal places)

k	t_k	y_k	m_k
0	1.0	2	2.72
1	1.5	3.36	1.81
2	2.0	4.27	1.60
3	2.5	5.06	1.48
4	3.0	5.81	

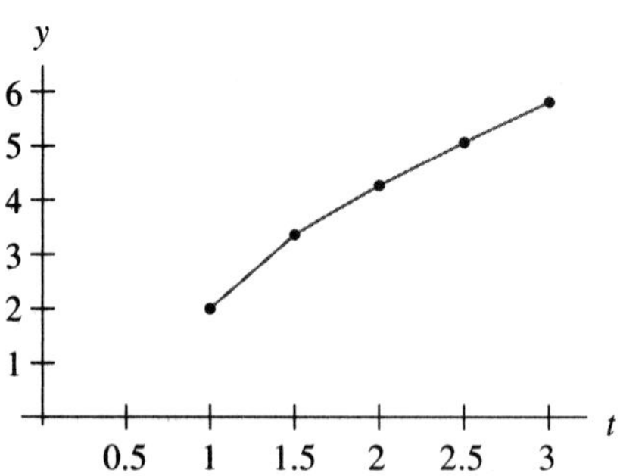

9.

Table 1.9
Results of Euler's method (shown rounded to three decimal places)

k	t_k	y_k	m_k
0	0.0	0.2	0.032
1	0.1	0.203	0.033
2	0.2	0.206	0.034
3	0.3	0.210	0.035
⋮	⋮	⋮	⋮
99	9.9	0.990	0.010
100	10.0	0.991	

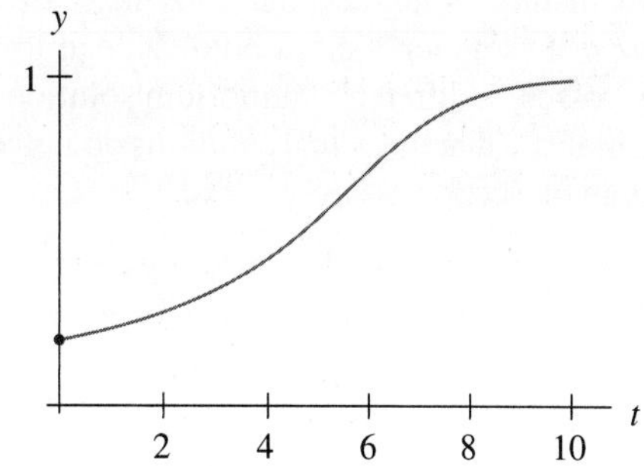

10.

Table 1.10
Results of Euler's method with Δt negative (shown rounded to three decimal places)

k	t_k	y_k	m_k
0	0	−0.5	−0.25
1	−0.1	−0.475	−0.204
2	−0.2	−0.455	−0.147
3	−0.3	−0.440	−0.080
⋮	⋮	⋮	⋮
19	−1.9	−1.160	0.488
20	−2.0	−1.209	0.467

Table 1.11
Results of Euler's method with Δt positive (shown rounded to three decimal places)

k	t_k	y_k	m_k
0	0	−0.5	−0.25
1	0.1	−0.525	−0.279
2	0.2	−0.553	−0.298
3	0.3	−0.583	−0.306
⋮	⋮	⋮	⋮
19	1.9	0.898	5.058
20	2.0	1.404	9.532

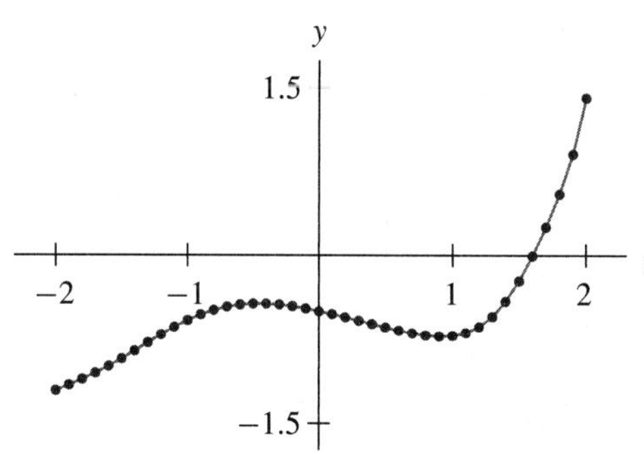

11. Because the differential equation is autonomous, the computation that determines y_{k+1} from y_k depends only on y_k and Δt and not on the actual value of t_k. Hence the approximate y-values that are obtained in both exercises are the same. It is useful to think about this fact in terms of the slope field of an autonomous equation.

12. Euler's method is not accurate in either case because the step size is too large. In Exercise 5, the approximate solution "jumps onto" an equilibrium solution. In Exercise 6, the approximate solution "crisscrosses" a different equilibrium solution. Approximate solutions generated with smaller values of Δt indicate that the actual solutions do not exhibit this behavior (see the Existence and Uniqueness Theorem of Section 1.5).

13. As the solution approaches the equilibrium solution corresponding to $w = 3$, its slope decreases. We do not expect the solution to "jump over" an equilibrium solution (see the Existence and Uniqueness Theorem in Section 1.5).

14.

Table 1.12
Results of Euler's method with $\Delta t = 1.0$ (shown to two decimal places)

k	t_k	y_k	m_k
0	0	1	1
1	1	2	1.41
2	2	3.41	1.85
3	3	5.26	2.29
4	4	7.56	

Table 1.13
Results of Euler's method with $\Delta t = 0.5$ (shown to two decimal places)

k	t_k	y_k	m_k
0	0	1	1
1	0.5	1.5	1.22
2	1.0	2.11	1.45
3	1.5	2.84	1.68
4	2.0	3.68	1.92
5	2.5	4.64	2.15
6	3.0	5.72	2.39
7	3.5	6.91	2.63
8	4.0	8.23	

Table 1.14
Results of Euler's method with $\Delta t = 0.25$ (shown to two decimal places)

k	t_k	y_k	m_k	k	t_k	y_k	m_k
0	0	1	1	9	2.25	4.32	2.08
1	0.25	1.25	1.12	10	2.50	4.84	2.20
2	0.50	1.53	1.24	11	2.75	5.39	2.32
3	0.75	1.84	1.36	12	3.0	5.97	2.44
4	1.0	2.18	1.48	13	3.25	6.58	2.56
5	1.25	2.55	1.60	14	3.50	7.23	2.69
6	1.50	2.94	1.72	15	3.75	7.90	2.81
7	1.75	3.37	1.84	16	4.0	8.60	
8	2.0	3.83	1.96				

The slopes in the slope field are positive and increasing. Hence, the graphs of all solutions are concave up. Since Euler's method uses line segments to approximate the graph of the actual solution, the approximate solutions will always be less than the actual solution. This error decreases as the step size decreases.

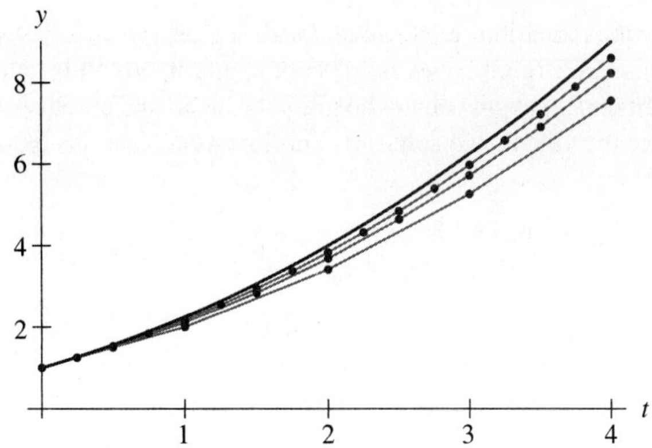

15.

Table 1.15
Results of Euler's method with $\Delta t = 1.0$ (shown to two decimal places)

k	t_k	y_k	m_k
0	0	1	1
1	1	2	0
2	2	2	0
3	3	2	0
4	4	2	

Table 1.16
Results of Euler's method with $\Delta t = 0.5$ (shown to two decimal places)

k	t_k	y_k	m_k
0	0	1	1
1	0.5	1.5	0.5
2	1.0	1.75	0.26
3	1.5	1.88	0.12
4	2.0	1.94	0.06
5	2.5	1.97	0.02
6	3.0	1.98	0.02
7	3.5	1.99	0.02
8	4.0	2.0	

Table 1.17
Results of Euler's method with $\Delta t = 0.25$ (shown to two decimal places)

k	t_k	y_k	m_k	k	t_k	y_k	m_k
0	0	1	1	9	2.25	1.92	0.08
1	0.25	1.25	0.76	10	2.50	1.94	0.06
2	0.50	1.44	0.56	11	2.75	1.96	0.04
3	0.75	1.58	0.40	12	3.0	1.97	0.03
4	1.0	1.68	0.32	13	3.25	1.98	0.02
5	1.25	1.76	0.24	14	3.50	1.98	0.02
6	1.50	1.82	0.18	15	3.75	1.99	0.01
7	1.75	1.87	0.13	16	4.0	1.99	
8	2.0	1.90	0.10				

From the differential equation, we see that dy/dt is positive and decreasing as long as $y(0) = 1$ and $y(t) < 2$ for $t > 0$. Therefore, $y(t)$ is increasing, and its graph is concave down. Since Euler's method uses line segments to approximate the graph of the actual solution, the approximate solutions will always be greater than the actual solution. This error decreases as the step size decreases.

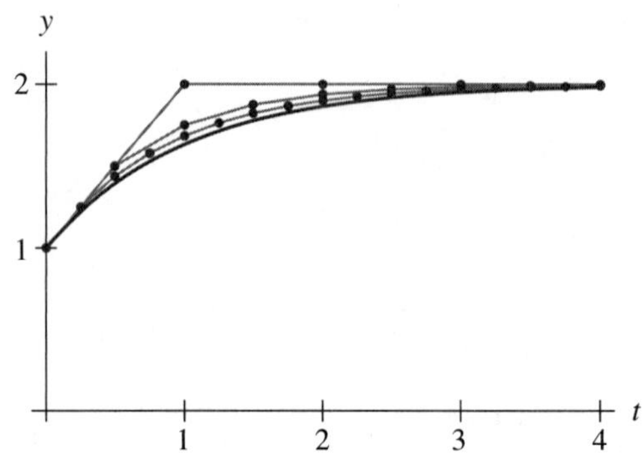

16.

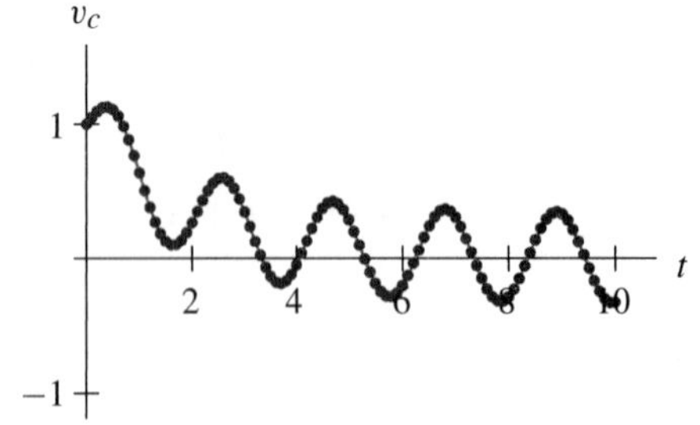

Graph of approximate solution obtained using Euler's method with $\Delta t = 0.1$.

17.

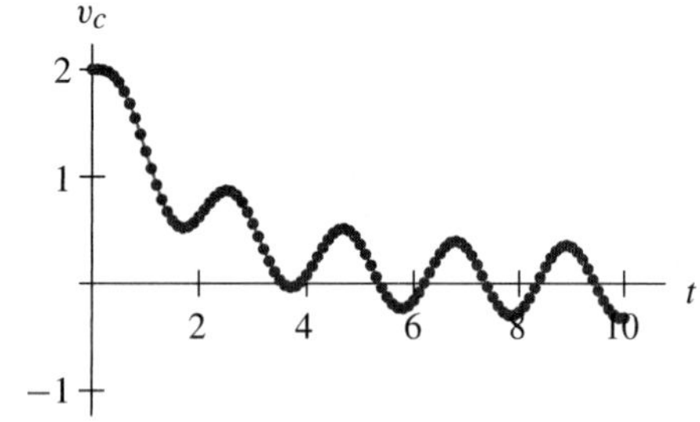

Graph of approximate solution obtained using Euler's method with $\Delta t = 0.1$.

18.

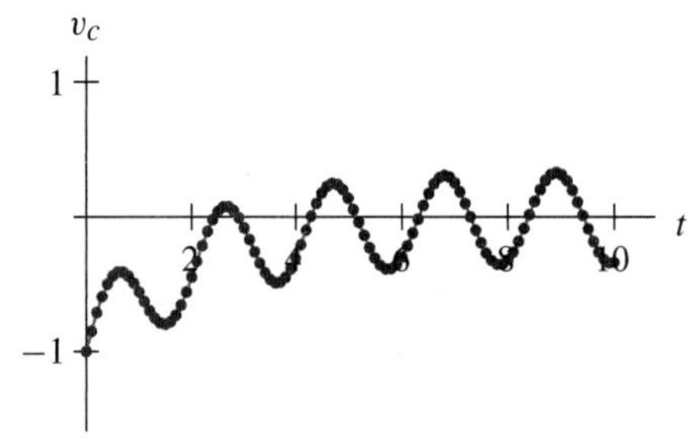

Graph of approximate solution obtained using Euler's method with $\Delta t = 0.1$.

19.

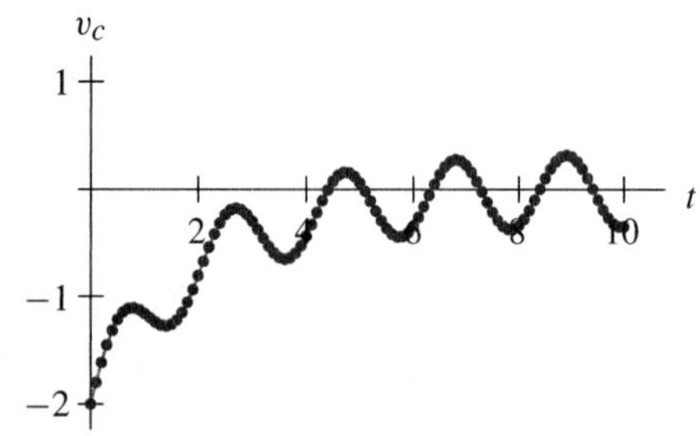

Graph of approximate solution obtained using Euler's method with $\Delta t = 0.1$.

20. **(a)**

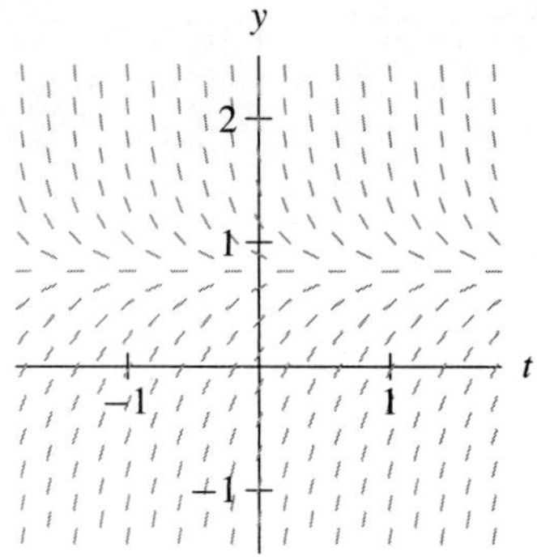

(b)

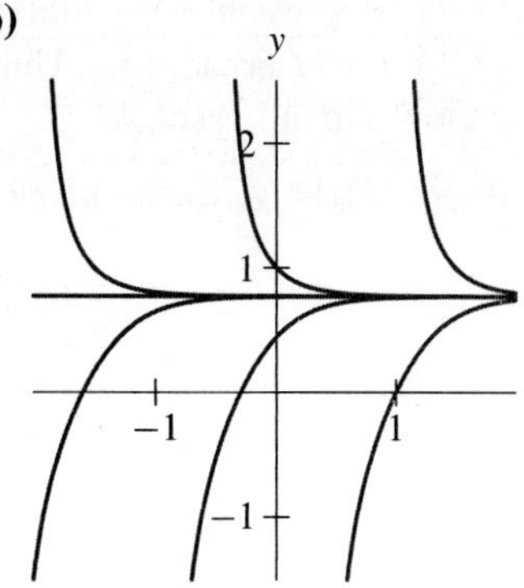

(c) The roots of $p(y)$ correspond to equilibrium solutions.

(d) From the graphs of the solutions, we see that there is one equilibrium solution, hence only one real root. Solutions tend toward it as t increases. Hence, approximate solutions using Euler's method with a relatively small step size will also tend toward this value. The approximate root is $y \approx 0.771$. Note $p(0.770) \approx 0.0035$ and $p(0.772) \approx -0.0041$.

21. **(a)**

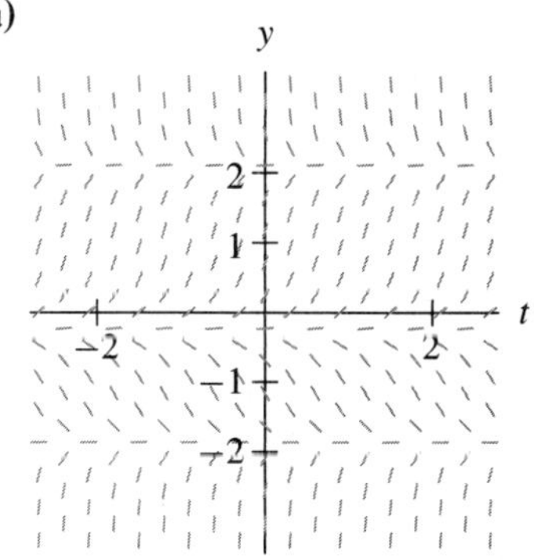

(b)

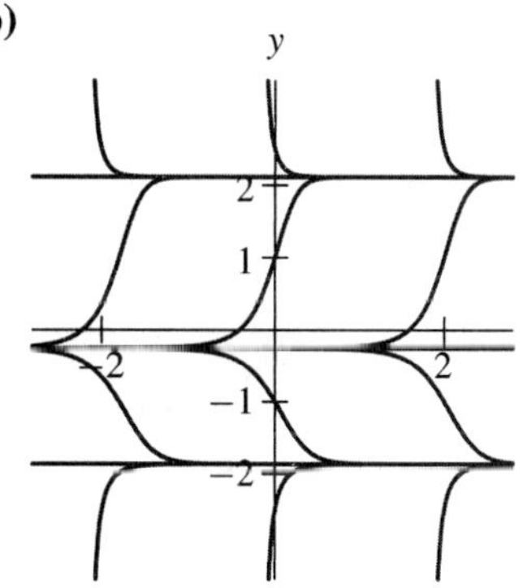

(c) The roots of $p(y)$ correspond to the equilibrium solutions.

(d) From the graphs of the solutions, we see that there are three equilibrium points, hence three real roots. We can find two of the roots by using Euler's method with a positive value of Δt. The other root can be found using Euler's method with a negative Δt. Since solutions tend toward the root, we can use a fairly large step size to obtain the required accuracy, but since this is only an approximate solution, it is best to double check the location of the root by substituting the value into $p(y)$. The approximate roots are $y \approx 2.115$, $y \approx -1.861$ and $y \approx -0.254$.

EXERCISES FOR SECTION 1.5

1. Since the constant function $y_1(t) = 3$ for all t is a solution, then the graph of any other solution $y(t)$ with $y(0) < 3$ cannot cross the line $y = 3$ by the Uniqueness Theorem. So $y(t) < 3$ for all t in the domain of $y(t)$.

2. Since $y(0) = 1$ is between the equilibrium solutions $y_2(t) = 0$ and $y_3(t) = 2$, we must have $0 < y(t) < 2$ for all t because the Uniqueness Theorem implies that graphs of solutions cannot cross (or even touch in this case).

3. Because $y_2(0) < y(0) < y_1(0)$, we know that

$$-t^2 = y_2(t) < y(t) < y_1(t) = t + 2$$

for all t. This restricts how large positive or negative $y(t)$ can be for a given value of t (that is, between $-t^2$ and $t + 2$). As $t \to -\infty$, $y(t) \to -\infty$ between $-t^2$ and $t + 2$ ($y(t) \to -\infty$ as $t \to -\infty$ at least linearly, but no faster than quadratically).

4. Because $y_1(0) < y(0) < y_2(0)$, the solution $y(t)$ must satisfy $y_1(t) < y(t) < y_2(t)$ for all t by the Uniqueness Theorem. Hence $-1 < y(t) < 1 + t^2$ for all t.

5. The Existence Theorem implies that a solution with this initial condition exists, at least for a small t-interval about $t = 0$. This differential equation has equilibrium solutions $y_1(t) = 0$, $y_2(t) = 2$, and $y_3(t) = 3$. Since $y(0) = 4$, the Uniqueness Theorem implies that $y(t) > 3$ for all t in the domain of $y(t)$. Also, $dy/dt > 0$ for all $y > 3$, so the solution $y(t)$ is increasing for all t in its domain.

6. Note that $dy/dt = 0$ if $y = 3$. Hence, $y_1(t) = 3$ for all t is an equilibrium solution. By the Uniqueness Theorem, this is the only solution that is 3 at $t = 0$. Therefore, $y(t) = 3$ for all t.

7. Because $0 < y(0) < 2$ and $y_1(t) = 0$ and $y_2(t) = 2$ are equilibrium solutions of the differential equation, we know that $0 < y(t) < 2$ for all t by the Uniqueness Theorem. Also, $dy/dt > 0$ for $0 < y < 2$, so dy/dt is always positive for this solution. Hence, $y(t) \to 2$ as $t \to \infty$, and $y(t) \to 0$ as $t \to -\infty$.

8. Note that $y(0) < 0$. Since $y_1(t) = 0$ is an equilibrium solution, the Uniqueness Theorem implies that $y(t) < 0$ for all t. Also, $dy/dt < 0$ if $y < 0$, so $y(t)$ is decreasing for all t, and $y(t) \to -\infty$ as t increases. As $t \to -\infty$, $y(t) \to 0$.

9. **(a)** To check that $y_1(t) = t^2$ is a solution, we compute

$$\frac{dy_1}{dt} = 2t$$

and

$$\begin{aligned} -y_1^2 + y_1 + 2y_1 t^2 + 2t - t^2 - t^4 &= -(t^2)^2 + (t^2) + 2(t^2)t^2 + 2t - t^2 - t^4 \\ &= 2t. \end{aligned}$$

To check that $y_2(t) = t^2 + 1$ is a solution, we compute

$$\frac{dy_2}{dt} = 2t$$

and

$$\begin{aligned} -y_2^2 + y_2 + 2y_2 t^2 + 2t - t^2 - t^4 &= -(t^2+1)^2 + (t^2+1) + 2(t^2+1)t^2 \\ &\quad + 2t - t^2 - t^4 \\ &= 2t. \end{aligned}$$

(b) The initial values of the two solutions are $y_1(0) = 0$ and $y_2(0) = 1$. Thus if $y(t)$ is a solution and $y_1(0) = 0 < y(0) < 1 = y_2(0)$, then we can apply the Uniqueness Theorem to obtain

$$y_1(t) = t^2 < y(t) < t^2 + 1 = y_2(t)$$

for all t. Note that since the differential equation satisfies the hypothesis of the Existence and Uniqueness Theorem over the entire ty-plane, we can continue to extend the solution as long as it does not escape to $\pm\infty$ in finite time. Since it is bounded above and below by solutions that exist for all time, $y(t)$ is defined for all time also.

(c)

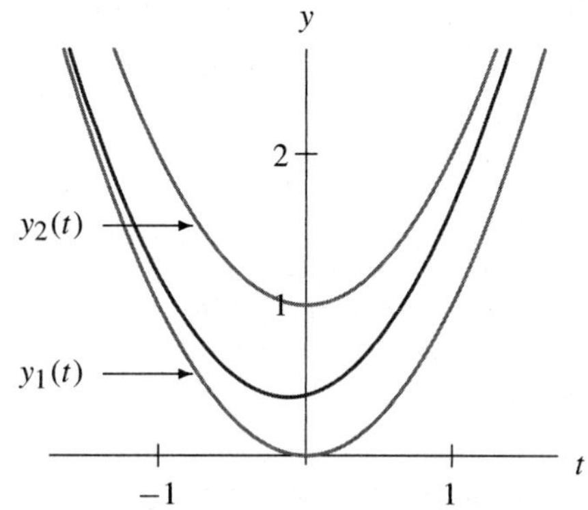

10. **(a)** If $y(t) = 0$ for all t, then $dy/dt = 0$ and $2\sqrt{|y(t)|} = 0$ for all t. Hence, the function that is constantly zero satisfies the differential equation.

(b) First, consider the case where $y > 0$. The differential equation reduces to $dy/dt = 2\sqrt{y}$. If we separate variables and integrate, we obtain

$$\sqrt{y} = t - c,$$

where c is any constant. The graph of this equation is the half of the parabola $y = (t - c)^2$ where $t \geq c$.

Next, consider the case where $y < 0$. The differential equation reduces to $dy/dt = 2\sqrt{-y}$. If we separate variables and integrate, we obtain

$$\sqrt{-y} = d - t,$$

where d is any constant. The graph of this equation is the half of the parabola $y = -(d - t)^2$ where $t \leq d$.

To obtain all solutions, we observe that any choice of constants c and d where $c \geq d$ leads to a solution of the form

$$y(t) = \begin{cases} -(d-t)^2, & \text{if } t \leq d; \\ 0, & \text{if } d \leq t \leq c; \\ (t-c)^2, & \text{if } t \geq c. \end{cases}$$

(See the following figure for the case where $d = -2$ and $c = 1$.)

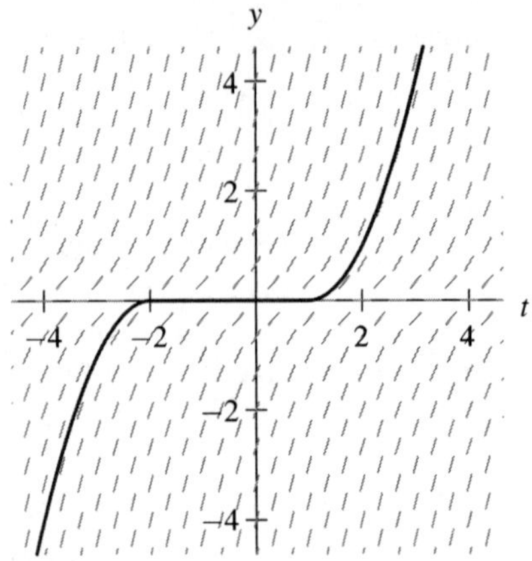

(c) The partial derivative $\partial f/\partial y$ of $f(t, y) = \sqrt{|y|}$ does not exist along the t-axis.

(d) If $y_0 = 0$, `HPGSolver` plots the equilibrium solution that is constantly zero. If $y_0 \neq 0$, it plots a solution whose graph crosses the t-axis. This is a solution where $c = d$ in the formula given above.

11. This exercise shows that solutions of autonomous equations cannot have local maximums or minimums. Hence they must be either constant or monotonically increasing or monotonically decreasing. A useful corollary is that a function $y(t)$ that oscillates cannot be the solution of an autonomous differential equation.

(a) Note $dy_1/dt = 0$ at $t = t_0$ because $y_1(t)$ has a local maximum. Because $y_1(t)$ is a solution, we know that $dy_1/dt = f(y_1(t))$ for all t in the domain of $y_1(t)$. In particular,

$$0 = \left.\frac{dy_1}{dt}\right|_{t=t_0} = f(y_1(t_0)) = f(y_0),$$

so $f(y_0) = 0$.

(b) This differential equation is autonomous, so the slope marks along any given horizontal line are parallel. Hence, the slope marks along the line $y = y_0$ must all have zero slope.

(c) For all t,

$$\frac{dy_2}{dt} = \frac{d(y_0)}{dt} = 0$$

because the derivative of a constant function is zero, and for all t

$$f(y_2(t)) = f(y_0) = 0.$$

So $y_2(t)$ is a solution.

(d) By the Uniqueness Theorem, we know that two solutions that are in the same place at the same time are the same solution. We have $y_1(t_0) = y_0 = y_2(t_0)$. Moreover, $y_1(t)$ is assumed to be a solution, and we showed that $y_2(t)$ is a solution in parts (a) and (b) of this exercise. So $y_1(t) = y_2(t)$ for all t. In other words, $y_1(t) = y_0$ for all t.

(e) Follow the same four steps as before. We still have $dy_1/dt = 0$ at $t = t_0$ because y_1 has a local minimum at $t = t_0$.

12. **(a)** Note that

$$\frac{dy_1}{dt} = \frac{d}{dt}\left(\frac{1}{t-1}\right) = -\frac{1}{(t-1)^2} = -(y_1(t))^2$$

and

$$\frac{dy_2}{dt} = \frac{d}{dt}\left(\frac{1}{t-2}\right) = -\frac{1}{(t-2)^2} = -(y_2(t))^2,$$

so both $y_1(t)$ and $y_2(t)$ are solutions.

(b) Note that $y_1(0) = -1$ and $y_2(0) = -1/2$. If $y(t)$ is another solution whose initial condition satisfies $-1 < y(0) < -1/2$, then $y_1(t) < y(t) < y_2(t)$ for all t by the Uniqueness Theorem. Also, since $dy/dt < 0$, $y(t)$ is decreasing for all t in its domain. Therefore, $y(t) \to 0$ as $t \to -\infty$, and the graph of $y(t)$ has a vertical asymptote between $t = 1$ and $t = 2$.

13. The key observation is that the differential equation is not defined when $t = 0$.

(a) Note that $dy_1/dt = 0$ and $y_1/t^2 = 0$, so $y_1(t)$ is a solution.

(b) Separating variables, we have

$$\int \frac{dy}{y} = \int \frac{dt}{t^2}.$$

Solving for y we obtain $y(t) = ce^{-1/t}$, where c is any constant. Thus, for any real number c, define the function $y_c(t)$ by

$$y_c(t) = \begin{cases} 0 & \text{for } t \le 0; \\ ce^{-1/t} & \text{for } t > 0. \end{cases}$$

For each c, $y_c(t)$ satisfies the differential equation for all $t \neq 0$.

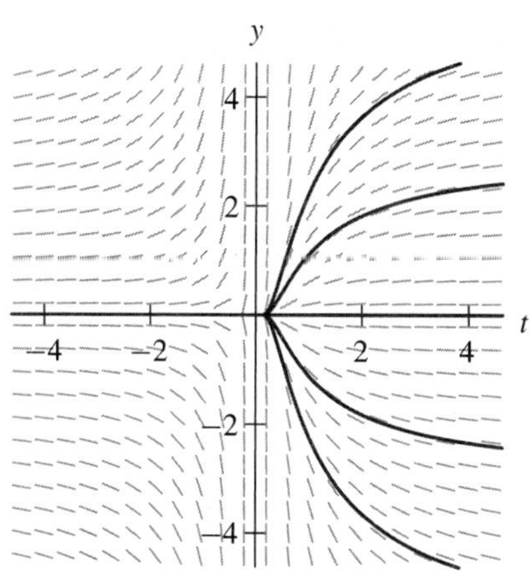

There are infinitely many solutions of the form $y_c(t)$ that agree with $y_1(t)$ for $t < 0$.

(c) Note that $f(t, y) = y/t^2$ is not defined at $t = 0$. Therefore, we *cannot* apply the Uniqueness Theorem for the initial condition $y(0) = 0$. The "solution" $y_c(t)$ given in part (b) actually represents two solutions, one for $t < 0$ and one for $t > 0$.

14. **(a)** The equation is separable. We separate the variables and compute

$$\int y^{-3}\,dy = \int dt.$$

Solving for y, we obtain

$$y(t) = \frac{1}{\sqrt{c - 2t}}$$

for any constant c. To find the desired solution, we use the initial condition $y(0) = 1$ and obtain $c = 1$. So the solution to the initial-value problem is

$$y(t) = \frac{1}{\sqrt{1 - 2t}}.$$

(b) This solution is defined when $-2t + 1 > 0$, which is equivalent to $t < 1/2$.

(c) As $t \to 1/2^-$, the denominator of $y(t)$ becomes a small positive number, so $y(t) \to \infty$. We only consider $t \to 1/2^-$ because the solution is defined only for $t < 1/2$. (The other "branch" of the function is also a solution, but the solution that includes $t = 0$ in its domain is not defined for $t \geq 1/2$.) As $t \to -\infty$, $y(t) \to 0$.

15. **(a)** The equation is separable, so we obtain

$$\int (y + 1)\,dy = \int \frac{dt}{t - 2}.$$

Solving for y with help from the quadratic formula yields the general solution

$$y(t) = -1 \pm \sqrt{1 + \ln(c(t - 2)^2)}$$

where c is a constant. Substituting the initial condition $y(0) = 0$ and solving for c, we have

$$0 = -1 \pm \sqrt{1 + \ln(4c)},$$

and thus $c = 1/4$. The desired solution is therefore

$$y(t) = -1 + \sqrt{1 + \ln((1 - t/2)^2)}$$

(b) The solution is defined only when $1 + \ln((1 - t/2)^2) \geq 0$, that is, when $|t - 2| \geq 2/\sqrt{e}$. Therefore, the domain of the solution is

$$t \leq 2(1 - 1/\sqrt{e}).$$

(c) As $t \to 2(1 - 1/\sqrt{e})$, then $1 + \ln((1 - t/2)^2) \to 0$. Thus

$$\lim_{t \to 2(1 - 1/\sqrt{e})} y(t) = -1.$$

Note that the differential equation is not defined at $y = -1$. Also, note that

$$\lim_{t \to -\infty} y(t) = \infty.$$

16. **(a)** The equation is separable. We separate, integrate

$$\int (y+2)^2\,dy = \int dt,$$

and solve for y to obtain the general solution

$$y(t) = (3t+c)^{1/3} - 2,$$

where c is any constant. To obtain the desired solution, we use the initial condition $y(0) = 1$ and solve

$$1 = (3 \cdot 0 + c)^{1/3} - 2$$

for c to obtain $c = 27$. So the solution to the given initial-value problem is

$$y(t) = (3t+27)^{1/3} - 2.$$

(b) This function is defined for all t. However, $y(-9) = -2$, and the differential equation is not defined at $y = -2$. Strictly speaking, the solution exists only for $t > -9$.

(c) As $t \to \infty$, $y(t) \to \infty$. As $t \to -9^+$, $y(t) \to -2$.

17. **(a)** The equation is separable. Separating variables we obtain

$$\int (y-2)\,dy = \int t\,dt.$$

Solving for y with help from the quadratic formula yields the general solution

$$y(t) = 2 \pm \sqrt{t^2+c}\,.$$

To find c, we let $t = -1$ and $y = 0$, and we obtain $c = 3$. The desired solution is therefore $y(t) = 2 - \sqrt{t^2+3}$

(b) Since t^2+2 is always positive and $y(t) < 2$ for all t, the solution $y(t)$ is defined for all real numbers.

(c) As $t \to \pm\infty$, $t^2 + 3 \to \infty$. Therefore,

$$\lim_{t\to\pm\infty} y(t) = -\infty.$$

18. **(a)** The partial derivative with respect to v of dv/dt does not exist at $v = 0$. Hence the Uniqueness Theorem tells us nothing about the uniqueness of solutions that involve $v = 0$. In fact, if we use the techniques described in the section related to the uniqueness of solutions for $dy/dt = 3y^{2/3}$, we can find infinitely many solutions with this initial condition.

(b) Since it does not make sense to talk about rain drops with negative volume, we always have $v \geq 0$. Once $v > 0$, the evolution of the drop is completely determined by the differential equation.

What is the physical significance of a drop with $v = 0$? It is tempting to interpret the fact that solutions can have $v = 0$ for an arbitrary amount of time before beginning to grow as a statement that the rain drops can spontaneously begin to grow at any time. Since the model gives no information about when a solution with $v = 0$ starts to grow, it is not very useful for the understanding the initial formation of rain drops. The safest assertion is to say is the model breaks down if $v = 0$.

EXERCISES FOR SECTION 1.6

1. The equilibrium points of $dy/dt = f(y)$ are the numbers y where $f(y) = 0$. For

$$f(y) = 3y(y - 1),$$

the equilibrium points are $y = 0$ and $y = 1$. Since $f(y)$ is positive for $y < 0$, negative for $0 < y < 1$, and positive for $y > 1$, the equilibrium point $y = 0$ is a sink and the equilibrium point $y = 1$ is a source.

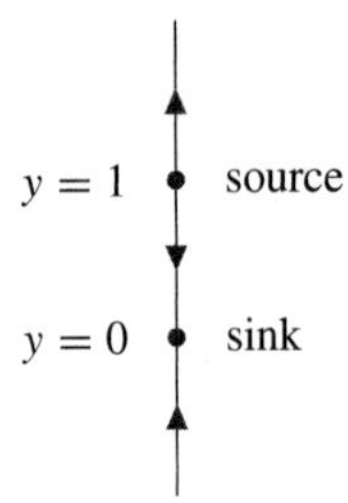

2. The equilibrium points of $dy/dt = f(y)$ are the numbers y where $f(y) = 0$. For

$$f(y) = y^2 - 6y - 7 = (y - 7)(y + 1),$$

the equilibrium points are $y = -1$ and $y = 7$. Since $f(y)$ is positive for $y < -1$, negative for $-1 < y < 7$, and positive for $y > 7$, the equilibrium point $y = -1$ is a sink and the equilibrium point $y = 7$ is a source.

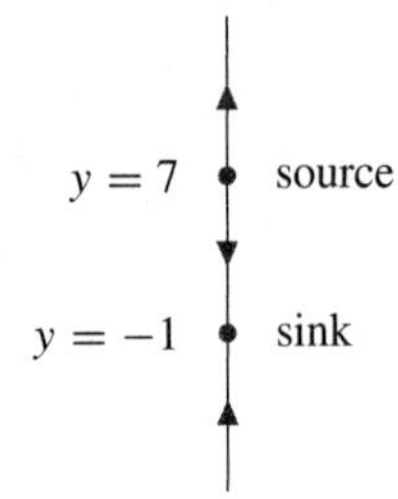

3. The equilibrium points of $dy/dt = f(y)$ are the numbers y where $f(y) = 0$. For $f(y) = \cos y$, the equilibrium points are $y = \pi/2 + n\pi$, where $n = 0, \pm 1, \pm 2, \ldots$. Since $\cos y > 0$ for $-\pi/2 < y < \pi/2$ and $\cos y < 0$ for $\pi/2 < y < 3\pi/2$, we see that the equilibrium point at $y = \pi/2$ is a sink. Since the sign of $\cos y$ alternates between positive and negative in a period fashion, we see that the equilibrium points at $y = \pi/2 + 2n\pi$ are sinks and the equilibrium points at $y = 3\pi/2 + 2n\pi$ are sources.

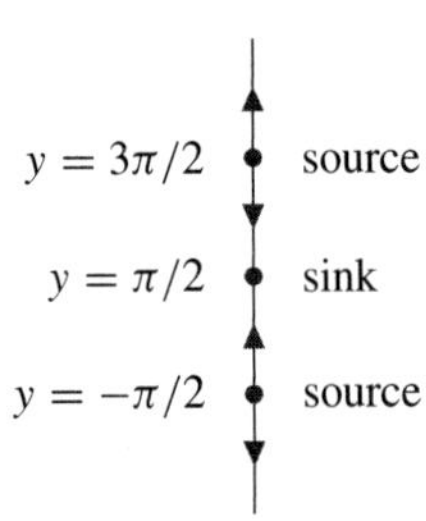

4. The equilibrium points of $dw/dt = f(w)$ are the numbers w where $f(w) = 0$. For $f(w) = w \cos w$, the equilibrium points are $w = 0$ and $w = \pi/2 + n\pi$, where $n = 0, \pm 1, \pm 2, \ldots$. The sign of $w \cos w$ alternates positive and negative at successive zeros. It is negative for $-\pi/2 < w < 0$ and positive for $0 < w < \pi/2$. Therefore, $w = 0$ is a source, and the equilibrium points alternate back and forth between sources and sinks.

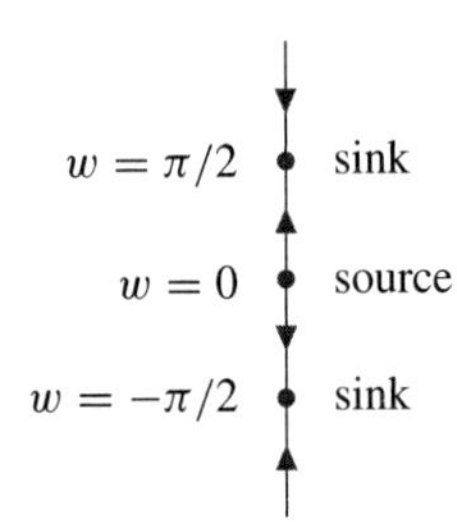

5. The equilibrium points of $dw/dt = f(w)$ are the numbers w where $f(w) = 0$. For $f(w) = (w-1)\sin w$, the equilibrium points are $w = 1$ and $w = n\pi$, where $n = 0, \pm 1, \pm 2, \ldots$. The sign of $(w - 1)\sin w$ alternates between positive and negative at successive zeros. It is positive for $-\pi < w < 0$ and negative for $0 < w < 1$. Therefore, $w = 0$ is a sink, and the equilibrium points alternate between sources and sinks.

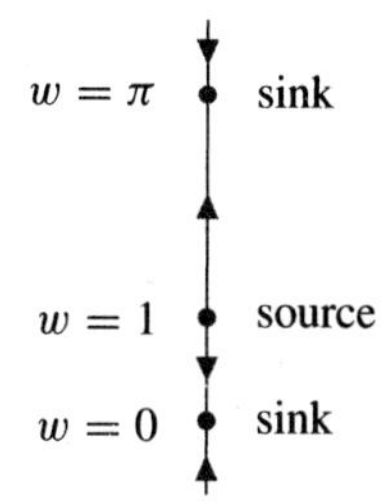

6. This equation has no equilibrium points, but the equation is not defined at $y = 2$. For $y > 2$, $dy/dt > 0$, so solutions increase. If $y < 2$, $dy/dt < 0$, so solutions decrease. The solutions approach the point $y = 2$ as time decreases and actually arrive there in finite time.

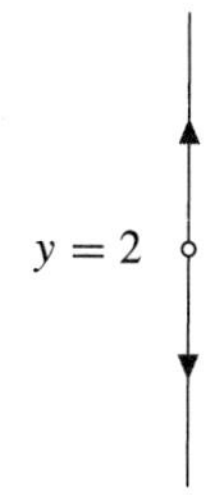

7. The derivative dv/dt is always positive, so there are no equilibrium points, and all solutions are increasing.

8. The equilibrium points of $dw/dt = f(w)$ are the numbers w where $f(w) = 0$. For $f(w) = 3w^3 - 12w^2$, the equilibrium points are $w = 0$ and $w = 4$. Since $f(w) < 0$ for $w < 0$ and $0 < w < 4$, and $f(w) > 0$ for $w > 4$, the equilibrium point at $w = 0$ is a node and the equilibrium point at $w = 4$ is a source.

9. The equilibrium points of $dy/dt = f(y)$ are the numbers y where $f(y) = 0$. For $f(y) = -1 + \cos y$, the equilibrium points are $y = 2n\pi$, where $n = 0, \pm 1, \pm 2, \ldots$. Since $f(y)$ is nonpositive for all values of y, all of the equilibrium points are nodes.

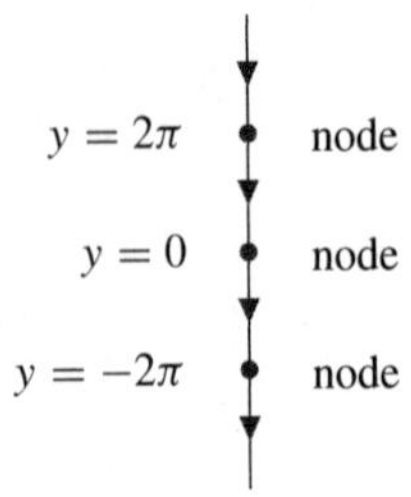

10. The equilibrium points of $dy/dt = f(y)$ are the numbers y where $f(y) = 0$. For $f(y) = \tan y$, the equilibrium points are $y = n\pi$ for $n = 0, \pm 1, \pm 2, \ldots$. Since $\tan y$ changes from negative to positive at each of its zeros, all of these equilibria are sources.
The differential equation is not defined at $y = \pi/2 + n\pi$ for $n = 0, \pm 1, \pm 2, \ldots$. Solutions increase or decrease toward one of these points as t increases and reach it in finite time.

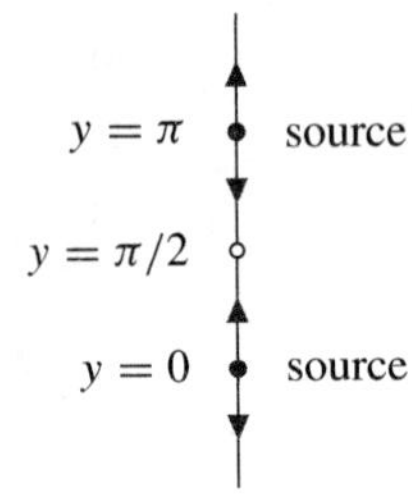

11. The equilibrium points of $dy/dt = f(y)$ are the numbers y where $f(y) = 0$. For $f(y) = y \ln |y|$, there are equilibrium points at $y = \pm 1$. In addition, although the function $f(y)$ is technically undefined at $y = 0$, the limit of $f(y)$ as $y \to 0$ is 0. Thus we can treat $y = 0$ as another equilibrium point. Since $f(y) < 0$ for $y < -1$ and $0 < y < 1$, and $f(y) > 0$ for $y > 1$ and $-1 < y < 0$, $y = -1$ is a source, $y = 0$ is a sink, and $y = 1$ is a source.

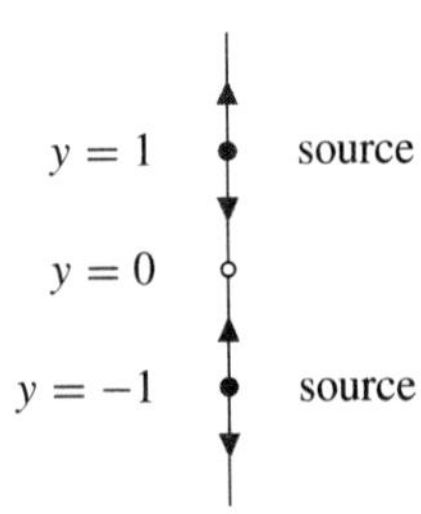

12. The equilibrium points of $dw/dt = f(w)$ are the numbers w where $f(w) = 0$. For $f(w) = (w^2 - 1)\arctan w$, there are equilibrium points at $w = \pm 1$ and $w = 0$. Since $f(w) > 0$ for $w > 1$ and $-1 < w < 0$, and $f(w) < 0$ for $w < -1$ and $0 < w < 1$, the equilibrium points at $w = \pm 1$ are sources, and the equilibrium point at $w = 0$ is a sink.

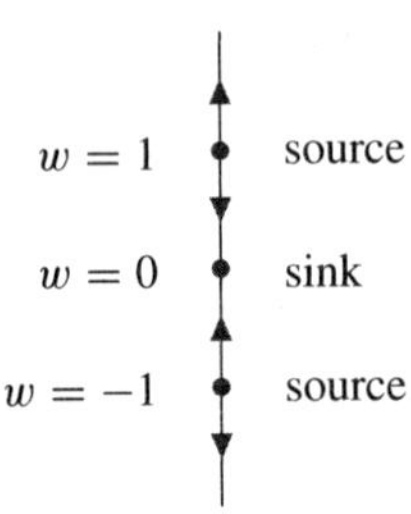

13.

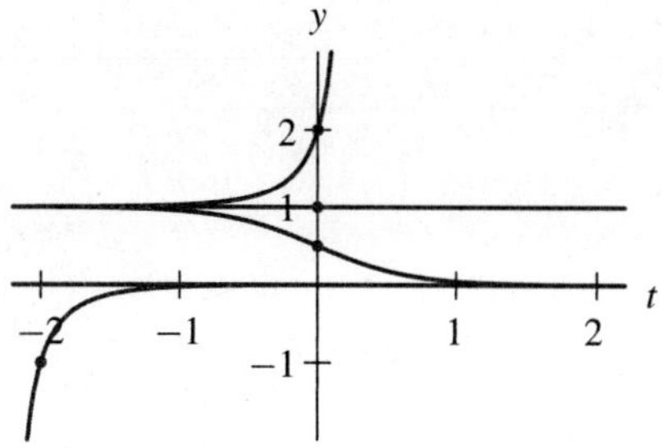

14.

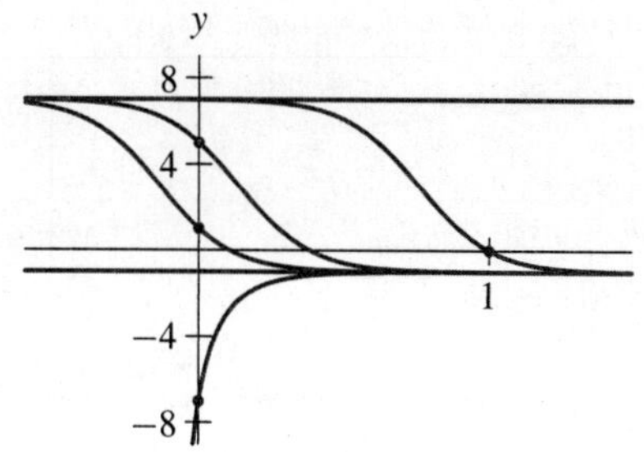

15.

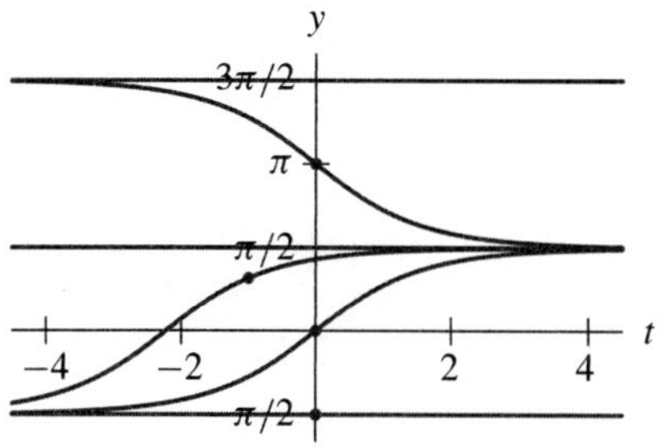

16.

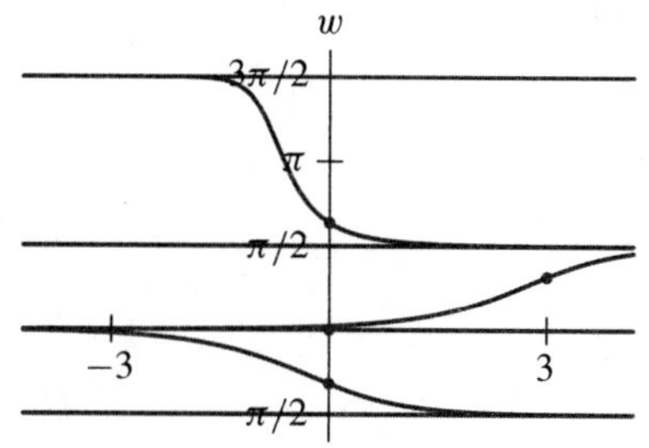

17.

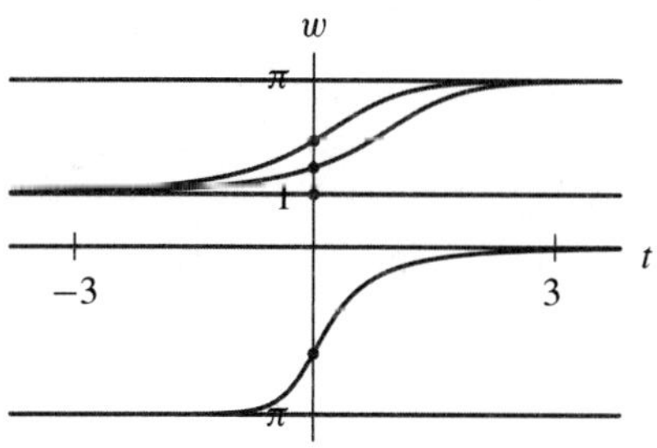

18.

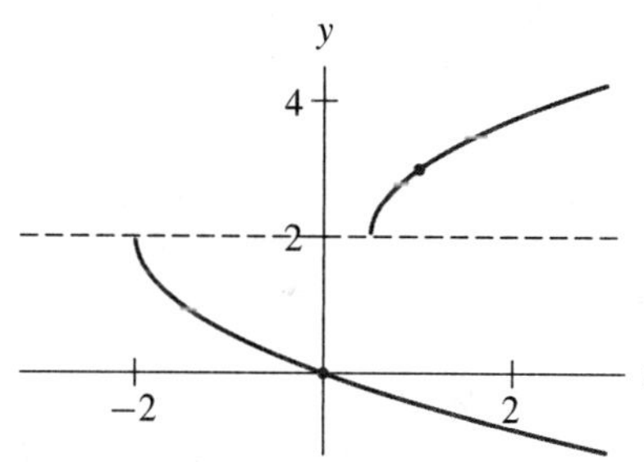

The equation is undefined at $y = 2$.

19.

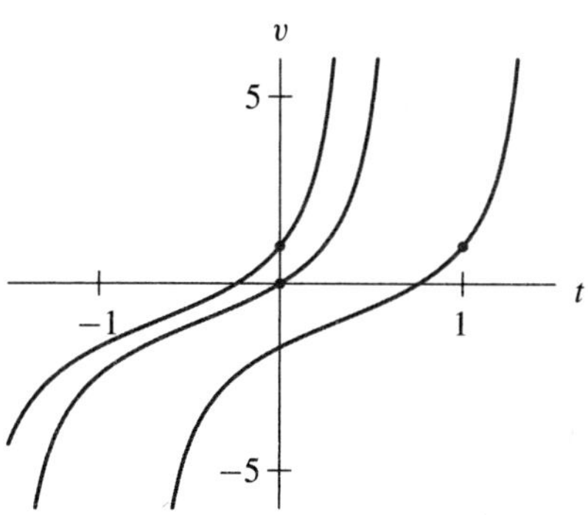

20.

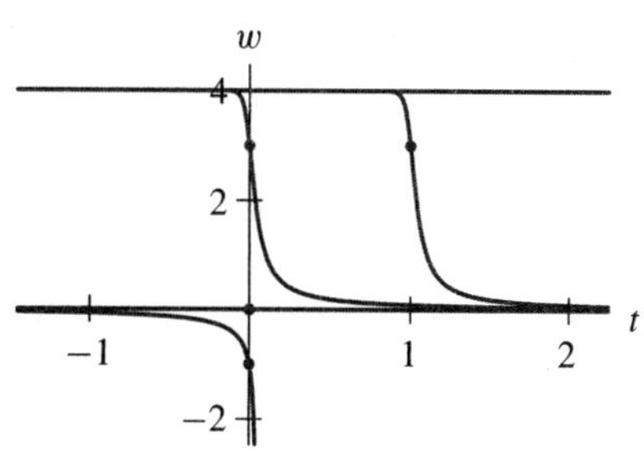

21.

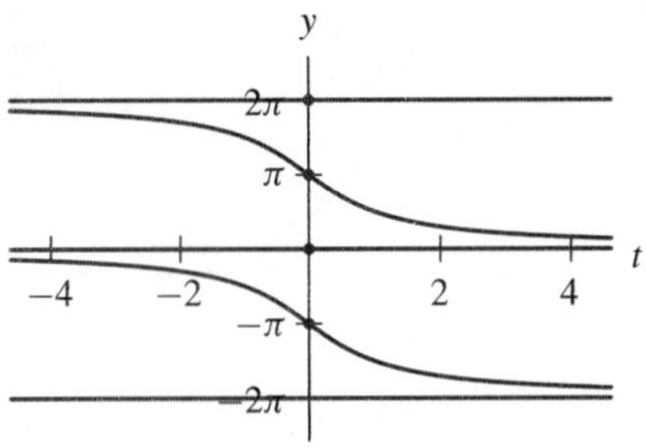

22. The initial value $y(0) = 0$ is below the two equilibrium points $y = 2 \pm \sqrt{2}$. Since $dy/dt > 0$ for $y < 2 - \sqrt{2}$, this solution is increasing, and $y(t) \to 2 - \sqrt{2}$ as $t \to \infty$. As t decreases, it becomes unbounded in the negative direction in finite time.

23. The initial value $y(0) = 1$ is between the equilibrium points $y = 2 - \sqrt{2}$ and $y = 2 + \sqrt{2}$. Also, $dy/dt < 0$ for $2 - \sqrt{2} < y < 2 + \sqrt{2}$. Hence the solution is decreasing and tends toward the smaller equilibrium point $y = 2 - \sqrt{2}$ as $t \to \infty$. It tends toward the larger equilibrium point $y = 2 + \sqrt{2}$ as $t \to -\infty$.

24. Because $y(0) = -1 < 2 - \sqrt{2}$, this solution increases toward $2 - \sqrt{2}$ as t increases and decreases as t decreases. In fact, because $y(0) = -1 < 0$, this solution is always below the solution in Exercise 22.

25. The initial value $y(0) = -5$ is below both equilibrium points. Since dy/dt is positive for $y < 2 - \sqrt{2}$, the solution is increasing for all t and tends to the equilibrium point $y = 2 - \sqrt{2}$ as $t \to \infty$. As t decreases, $y(t) \to -\infty$ in finite time.

26. The initial value $y(0) = 5$ is greater than the largest equilibrium point $2 + \sqrt{2}$, and $dy/dt > 0$ if $y > 2 + \sqrt{2}$. Hence, this solution increases without bound as t increases. (In fact, it blows up in finite time). As $t \to -\infty$, $y(t) \to 2 + \sqrt{2}$.

27. The initial value $y(3) = 1$ is between the equilibrium points $y = 2 - \sqrt{2}$ and $y = 2 + \sqrt{2}$. Also, $dy/dt < 0$ for $2 - \sqrt{2} < y < 2 + \sqrt{2}$. Hence the solution is decreasing and tends toward the smaller equilibrium point $y = 2 - \sqrt{2}$ as $t \to \infty$. It tends toward the larger equilibrium point $y = 2 + \sqrt{2}$ as $t \to -\infty$.

28. **(a)** Any solution that has an initial value between the equilibrium points at $y = -1$ and $y = 2$ must remain between these values for all t, so $-1 < y(t) < 2$ for all t.

(b) The extra assumption implies that the solution is increasing for all t such that $-1 < y(t) < 2$. Again assuming that the Uniqueness Theorem applies, we conclude that $y(t) \to 2$ as $t \to \infty$ and $y(t) \to -1$ as $t \to -\infty$.

29. The function $f(y)$ has two zeros $\pm y_0$, where y_0 is some positive number. So the differential equation $dy/dt = f(y)$ has two equilibrium solutions, one for each zero. Also, $f(y) < 0$ if $-y_0 < y < y_0$ and $f(y) > 0$ if $y < -y_0$ or if $y > y_0$. Hence y_0 is a source and $-y_0$ is a sink.

30. The function $f(y)$ has three zeros. We denote them as y_1, y_2, and y_3, where $y_1 < 0 < y_2 < y_3$. So the differential equation $dy/dt = f(y)$ has three equilibrium solutions, one for each zero. Also, $f(y) > 0$ if $y < y_1$, $f(y) < 0$ if $y_1 < y < y_2$, and $f(y) > 0$ if $y_2 < y < y_3$ or if $y > y_3$. Hence y_1 is a sink, y_2 is a source, and y_3 is a node.

31. The function $f(y)$ has two zeros, one positive and one negative. We denote them as y_1 and y_2, where $y_1 < y_2$. So the differential equation $dy/dt = f(y)$ has two equilibrium solutions, one for each zero. Also, $f(y) > 0$ if $y_1 < y < y_2$ and $f(y) < 0$ if $y < y_1$ or if $y > y_2$. Hence y_1 is a source and y_2 is a sink.

32. The function $f(y)$ has four zeros, which we denote $y_1, \ldots, y_4$ where $y_1 < 0 < y_2 < y_3 < y_4$. So the differential equation $dy/dt = f(y)$ has four equilibrium solutions, one for each zero. Also, $f(y) > 0$ if $y < y_1$, if $y_2 < y < y_3$, or if $y_3 < y < y_4$; and $f(y) < 0$ if $y_1 < y < y_2$ or if $y > y_4$. Hence y_1 is a sink, y_2 is a source, y_3 is a node, and y_4 is a sink.

33. Since there are two equilibrium points, the graph of $f(y)$ must touch the y-axis at two distinct numbers y_1 and y_2. Assume that $y_1 < y_2$. Since the arrows point up if $y < y_1$ and if $y > y_2$, we must have $f(y) > 0$ for $y < y_1$ and for $y > y_2$. Similarly, $f(y) < 0$ for $y_1 < y < y_2$.

The precise location of the equilibrium points is not given, and the direction of the arrows on the phase line is determined only by the sign (and not the magnitude) of $f(y)$. So the following graph is one of many possible answers.

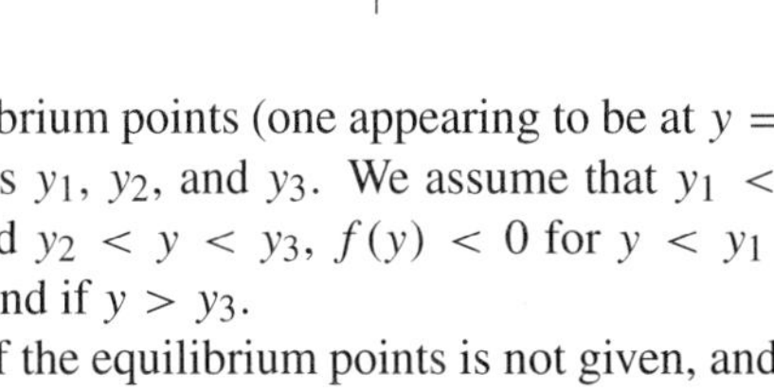

34. Since there are three equilibrium points (one appearing to be at $y = 0$), the graph of $f(y)$ must touch the y-axis at three numbers y_1, y_2, and y_3. We assume that $y_1 < y_2 = 0 < y_3$. Since the arrows point down for $y < y_1$ and $y_2 < y < y_3$, $f(y) < 0$ for $y < y_1$ and for $y_2 < y < y_3$. Similarly, $f(y) > 0$ if $y_1 < y < y_2$ and if $y > y_3$.

The precise location of the equilibrium points is not given, and the direction of the arrows on the phase line is determined only by the sign (and not the magnitude) of $f(y)$. So the following graph is one of many possible answers.

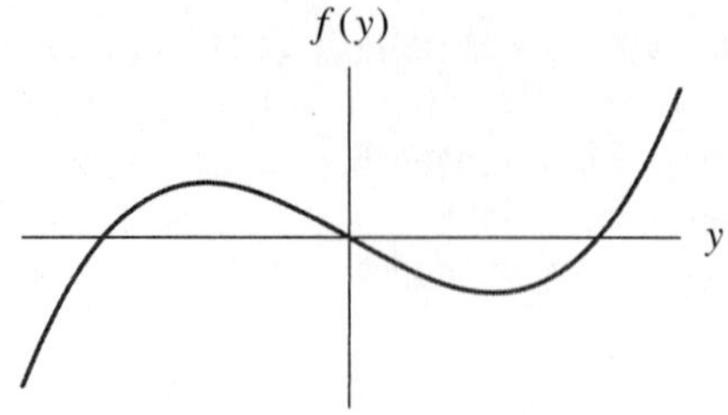

35. Since there are four equilibrium points, the graph of $f(y)$ must touch the y-axis at four distinct numbers y_1, y_2, y_3, and y_4. We assume that $y_1 < y_2 < y_3 < y_4$. Since the arrows point up only if $y_1 < y < y_2$ or if $y_2 < y < y_3$, we must have $f(y) > 0$ for $y_1 < y < y_2$ and for $y_2 < y < y_3$. Moreover, $f(y) < 0$ if $y < y_1$, if $y_3 < y < y_4$, or if $y > y_4$. Therefore, the graph of f crosses the y-axis at y_1 and y_3, but it is tangent to the y-axis at y_2 and y_4.

The precise location of the equilibrium points is not given, and the direction of the arrows on the phase line is determined only by the sign (and not the magnitude) of $f(y)$. So the following graph is one of many possible answers.

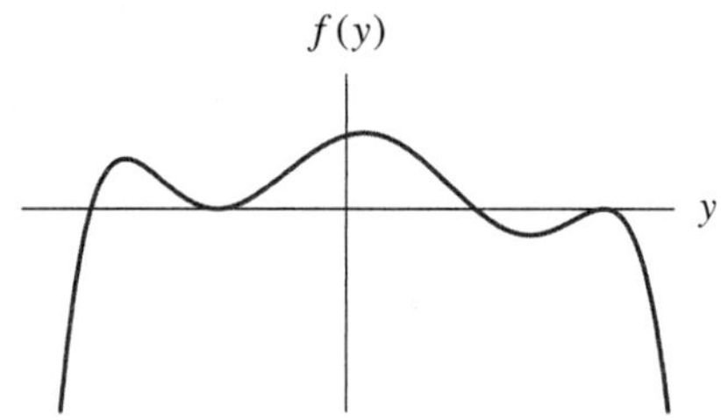

36. Since there are three equilibrium points (one appearing to be at $y = 0$), the graph of $f(y)$ must touch the y-axis at three numbers y_1, y_2, and y_3. We assume that $y_1 < y_2 = 0 < y_3$. Since the arrows point up only for $y < y_1$, $f(y) > 0$ only if $y < y_1$. Otherwise, $f(y) \leq 0$.

The precise location of the equilibrium points is not given, and the direction of the arrows on the phase line is determined only by the sign (and not the magnitude) of $f(y)$. So the following graph is one of many possible answers.

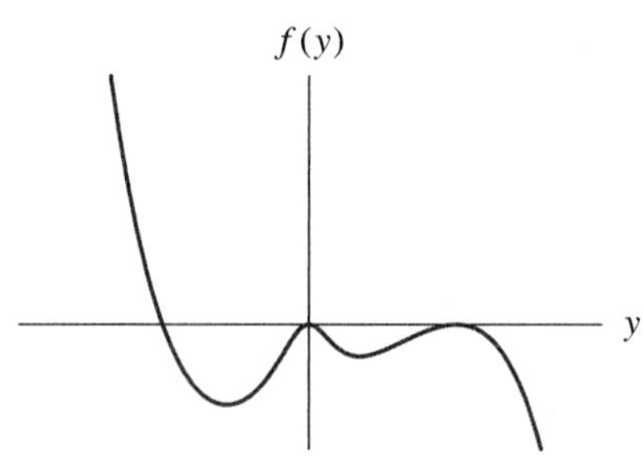

37. **(a)** This phase line has three equilibrium points, $y = 0, \pm 1$. Only equations (vii) and (viii) have three equilibria. For this phase line, $dy/dt < 0$ for $0 < y < 1$. Only equation (vii) satisfies this property. Consequently, the phase line corresponds to equation (vii).

(b) This phase line has two equilibrium points, $y = 0$ and $y = 1$. Equations (i), (ii), (iii), and (iv) have exactly these equilibria. For this phase line, $dy/dt \geq 0$ for $y > 0$. Only equations (i) and (ii) satisfy this property. Moreover, for this phase line, $dy/dt < 0$ for $y < 0$. Only equation (ii) satisfies this property. Consequently, the phase line corresponds to equation (ii).

(c) This phase line has two equilibrium points, $y = 0$ and $y = 2$. Equations (v) and (vi) have exactly these equilibria. For this phase line, $dy/dt > 0$ for $0 < y < 2$. Only equation (vi) satisfies this property. Consequently, the phase line corresponds to equation (vi).

(d) This phase line has two equilibrium points, $y = 0$ and $y = 1$. Equations (i), (ii), (iii), and (iv) have exactly these equilibria. For this phase line, $dy/dt < 0$ for $0 < y < 1$. Only equation (iii) satisfies this property. Consequently, the phase line corresponds to equation (iii).

38. **(a)** Because $f(y)$ is continuous we can use the Intermediate Value Theorem to say that there must be a zero of $f(y)$ between -10 and 10. This value of y is an equilibrium point of the differential equation. In fact, $f(y)$ must cross from positive to negative, so if there is a single equilibrium point, it must be a sink (see part (b)).

(b) We know that $f(y)$ must cross the y-axis between -10 and 10. Moreover, it must cross from positive to negative because $f(-10)$ is positive and $f(10)$ is negative. Where $f(y)$ crosses the y-axis from positive to negative, we have a sink. If $y = 1$ is a source, then crosses the y-axis from negative to positive at $y = 1$. Hence, $f(y)$ must cross the y-axis from positive to negative at least once between $y = -10$ and $y = 1$ and at least once between $y = 1$ and $y = 10$. There must be at least one sink in each of these intervals. (We need the assumption that the number of equilibrium points is finite to prevent cases where $f(y) = 0$ along an entire interval.)

39. **(a)** In terms of the phase line with $P \geq 0$, there are three equilibrium points. If we assume that $f(P)$ is differentiable, then a decreasing population at $P = 100$ implies that $f(P) < 0$ for $P > 50$. An increasing population at $P = 25$ implies that $f(P) > 0$ for $10 < P < 50$. These assumptions leave two possible phase lines since the arrow between $P = 0$ and $P = 10$ is undetermined.

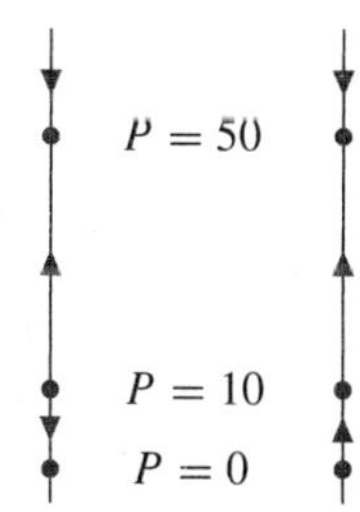

(b) Given the observations in part (a), we see that there are two basic types of graphs that go with the assumptions. However, there are many graphs that correspond to each possibility. The following two graphs are representative.

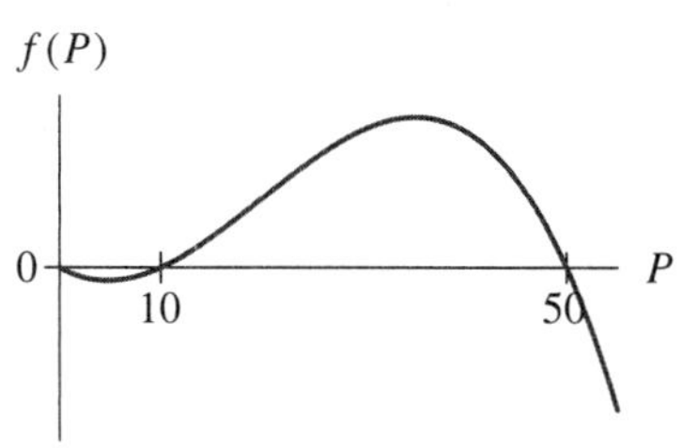

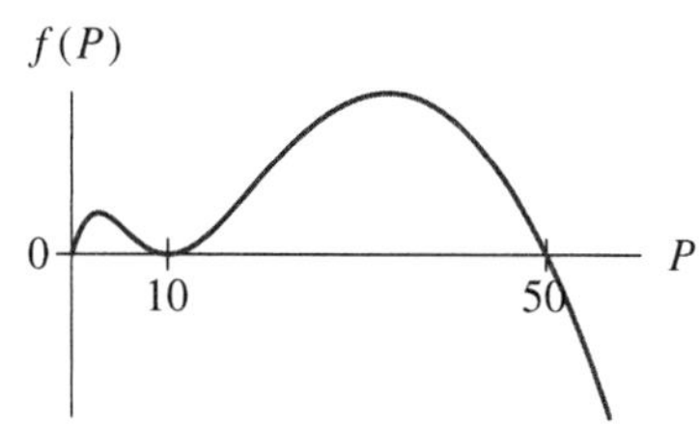

(c) The functions $f(P) = P(P-10)(50-P)$ and $f(P) = P(P-10)^2(50-P)$ respectively are two examples but there are many others.

40. **(a)** From the information given, there is an equilibrium point at $P = 0$ and at least two more equilibrium points, $P = A$ and $P = B$, where $0 < A < 20$ and $20 < B < 100$. The following phase line is the simplest one that is consistent with the assumptions.

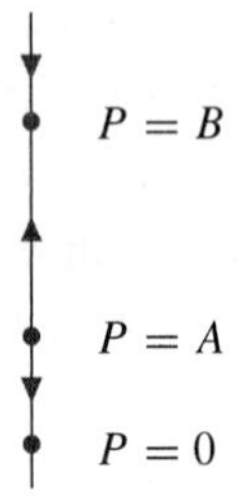

(b) The following graph is one of many possible answers.

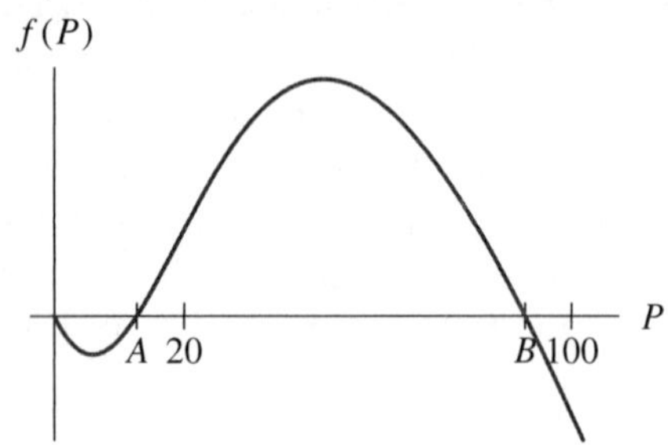

(c) There could be many more equilibrium points between $P = 20$ and $P = 100$ and between $P = 0$ and $P = 20$.

41. The equilibrium points occur at solutions of $dy/dt = y^2 + a = 0$. For $a > 0$, there are no equilibrium points. For $a = 0$, there is one equilibrium point, $y = 0$. For $a < 0$, there are two equilibrium points, $y = \pm\sqrt{-a}$.

To draw the phase lines, note that:

- If $a > 0$, $dy/dt = y^2 + a > 0$, so the solutions are always increasing.
- If $a = 0$, $dy/dt > 0$ unless $y = 0$. Thus, $y = 0$ is a node.
- For $a < 0$, $dy/dt < 0$ for $-\sqrt{-a} < y < \sqrt{-a}$, and $dy/dt > 0$ for $y < -\sqrt{-a}$ and for $y > \sqrt{-a}$.

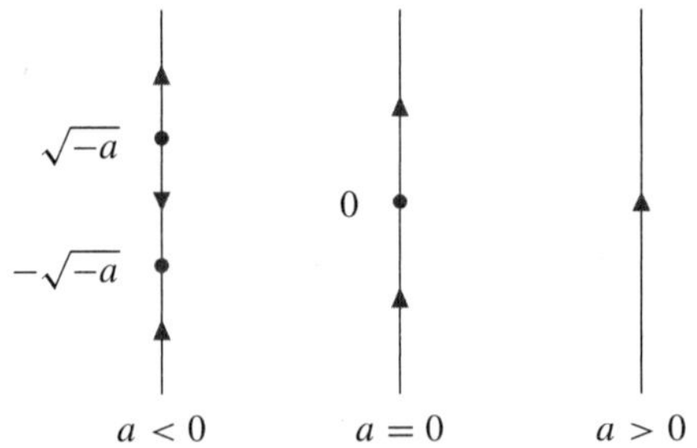

(a) The phase lines for $a < 0$ are qualitatively the same, and the phase lines for $a > 0$ are qualitatively the same.

(b) The phase line undergoes a qualitative change at $a = 0$.

42. The equilibrium points occur at solutions of $dy/dt = ay - y^3 = 0$. For $a \le 0$, there is one equilibrium point, $y = 0$. For $a > 0$, there are three equilibrium points, $y = 0$ and $y = \pm\sqrt{a}$.

To draw the phase lines, note that:

- For $a \leq 0$, $dy/dt > 0$ if $y < 0$, and $dy/dt < 0$ if $y > 0$. Consequently, the equilibrium point $y = 0$ is a sink.
- For $a > 0$, $dy/dt > 0$ if $y < -\sqrt{a}$ or $0 < y < \sqrt{a}$. Similarly, $dy/dt < 0$ if $-\sqrt{a} < y < 0$ or $y > \sqrt{a}$. Consequently, the equilibrium point $y = 0$ is a source, and the equilibria $y = \pm\sqrt{a}$ are sinks.

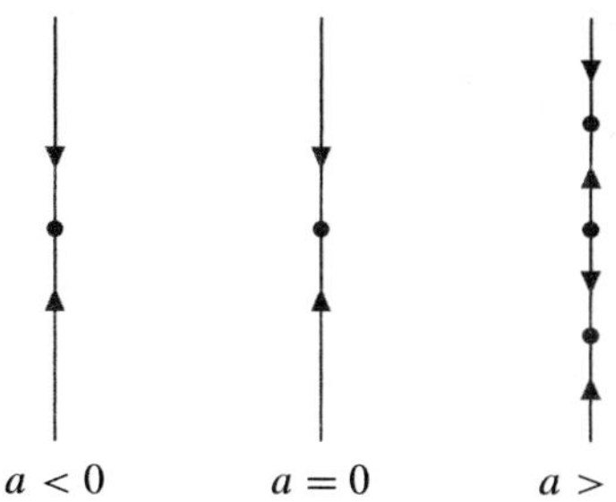

(a) The phase lines for $a \leq 0$ are qualitatively the same, and the phase lines for $a > 0$ are qualitatively the same.

(b) The phase line undergoes a qualitative change at $a = 0$.

43. **(a)** Because the first and second derivative are zero at y_0 and the third derivative is positive, Taylor's Theorem implies that the function $f(y)$ is approximately equal to

$$\frac{f'''(y_0)}{3!}(y - y_0)^3$$

for y near y_0. Since $f'''(y_0) > 0, f(y)$ is increasing near y_0. Hence, y_0 is a source.

(b) Just as in part (a), we see that $f(y)$ is decreasing near y_0, so y_0 is a sink.

(c) In this case, we can approximate $f(y)$ near y_0 by

$$\frac{f''(y_0)}{2!}(y - y_0)^2.$$

Since the second derivative of $f(y)$ at y_0 is assumed to be positive, $f(y)$ is positive on both sides of y_0 for y near y_0. Hence y_0 is a node.

44. **(a)** The differential equation is not defined for $y = -1$ and $y = 2$ and has no equilibria. So the phase line has holes at $y = -1$ and $y = 2$. The function $f(y) = 1/((y - 2)(y + 1))$ is positive for $y > 2$ and for $y < -1$. It is negative for $-1 < y < 2$. Thus, the phase line to the right corresponds to this differential equation.

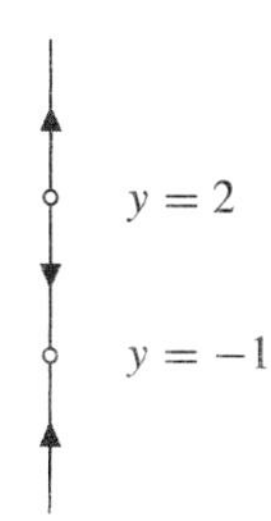

Since the value, $1/2$, of the initial condition $y(0) = 1/2$ is in the interval where the function $f(y)$ is negative, the solution is decreasing. It reaches $y = -1$ in finite time. As t decreases, the solution reaches $y = 2$ in finite time. Strictly speaking, the solution does not continue beyond the values $y = -1$ and $y = 2$ because the differential equation is not defined for $y = -1$ and $y = 2$.

(b) We can solve the differential equation analytically. We separate variables and integrate. We get

$$\int (y-2)(y+1)\,dy = \int dt$$

$$\frac{y^3}{3} - \frac{y^2}{2} - 2y = t + c,$$

where c is a constant. Using $y(0) = 1/2$, we get $c = 13/12$. Therefore the solution to the initial-value problem is the unique solution $y(t)$ that satisfies the equation

$$4y^3 - 6y^2 - 24y - 24t + 13 = 0$$

with $-1 < y(t) < 2$. It is not easy to solve this equation explicitly. However, in order to obtain the domain of this solution, we substitute $y = -1$ and $y = 2$ into the equation, and we get $t = -9/8$ and $t = 9/8$ respectively.

45. One assumption of the model is that, if no people are present, then the time between trains decreases at a constant rate. Hence the term $-\alpha$ represents this assumption. The parameter α should be positive, so that $-\alpha$ makes a negative contribution to dx/dt.

The term βx represents the effect of the passengers. The parameter β should be positive so that βx contributes positively to dx/dt.

46. **(a)** Solving $\beta x - \alpha = 0$, we see that the equilibrium point is $x = \alpha/\beta$.

(b) Since $f(x) = \beta x - \alpha$ is positive for $x > \alpha/\beta$ and negative for $x < \alpha/\beta$, the equilibrium point is a source.

(c) and **(d)**

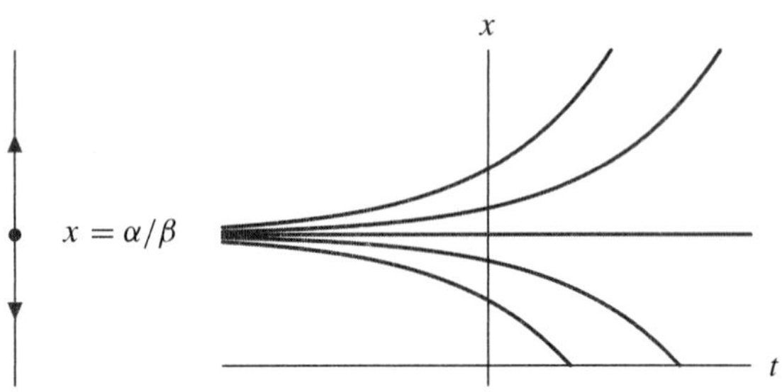

(e) We separate the variables and integrate to obtain

$$\int \frac{dx}{\beta x - \alpha} = \int dt$$

$$\tfrac{1}{\beta} \ln|\beta x - \alpha| = t + c,$$

which yields the general solution $x(t) = \alpha/\beta + ke^{\beta t}$, where k is any constant.

47. Note that the only equilibrium point is a source. If the initial gap between trains is too large, then x will increase without bound. If it is too small, x will decrease to zero. When $x = 0$, the two trains are next to each other, and they will stay together since $x < 0$ is not physically possible in this problem.

If the time between trains is exactly the equilibrium value ($x = \alpha/\beta$), then theoretically $x(t)$ is constant. However, any disruption to x causes the solution to tend away from the source. Since it is very likely that some stops will have fewer than the expected number of passengers and some stops will have more, it is unlikely that the time between trains will remain constant for long.

48. If the trains are spaced too close together, then each train will catch up with the one in front of it. This phenomenon will continue until there is a very large time gap between two successive trains. When this happens, the time between these two trains will grow, and a second cluster of trains will form.

For the "B branch of the Green Line," the clusters seem to contain three or four trains during rush hour. For the "D branch of the Green Line," clusters seem to contain only two trains or three trains.

It is tempting to say that the trains should be spaced at time intervals of exactly α/β, and nothing else needs to be changed. In theory, this choice will result in equal spacing between trains, but we must remember that the equilibrium point, $x = \alpha/\beta$, is a source. Hence, anything that perturbs x will cause x to increase or decrease in an exponential fashion.

The only solution that is consistent with this model is to have the trains run to a schedule that allows for sufficient time for the loading of passengers. The trains will occasionally have to wait if they get ahead of schedule, but this plan avoids the phenomenon of one tremendously crowded train followed by two or three relatively empty ones.

EXERCISES FOR SECTION 1.7

1. The equilibrium points occur at solutions of $dy/dt = y^2 + a = 0$. For $a > 0$, there are no equilibrium points. For $a = 0$, there is one equilibrium point, $y = 0$. For $a < 0$, there are two equilibrium points, $y = \pm\sqrt{-a}$. Thus, $a = 0$ is a bifurcation value.

To draw the phase lines, note that:

- If $a > 0$, $dy/dt = y^2 + a > 0$, so the solutions are always increasing.
- If $a = 0$, $dy/dt > 0$ unless $y = 0$. Thus, $y = 0$ is a node.
- For $a < 0$, $dy/dt < 0$ for $-\sqrt{-a} < y < \sqrt{-a}$, and $dy/dt > 0$ for $y < -\sqrt{-a}$ and for $y > \sqrt{-a}$.

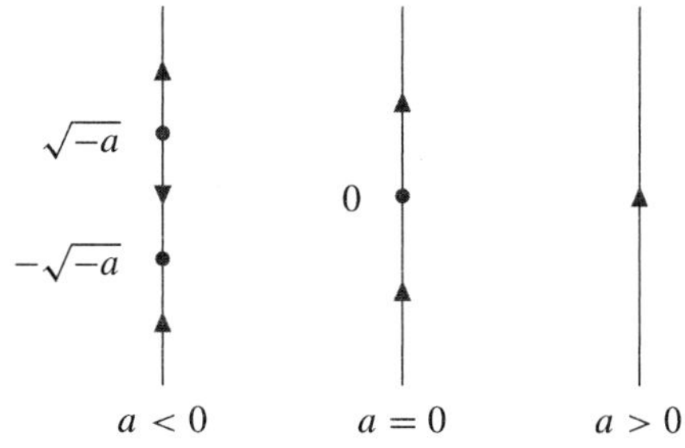

Phase lines for $a < 0$, $a = 0$, and $a > 0$.

2. The equilibrium points occur at solutions of $dy/dt = y^2 + 3y + a = 0$. From the quadratic formula, we have

$$y = \frac{-3 \pm \sqrt{9 - 4a}}{2}.$$

Hence, the bifurcation value of a is $9/4$. For $a < 9/4$, there are two equilibria, one source and one sink. For $a = 9/4$, there is one equilibrium which is a node, and for $a > 9/4$, there are no equilibria.

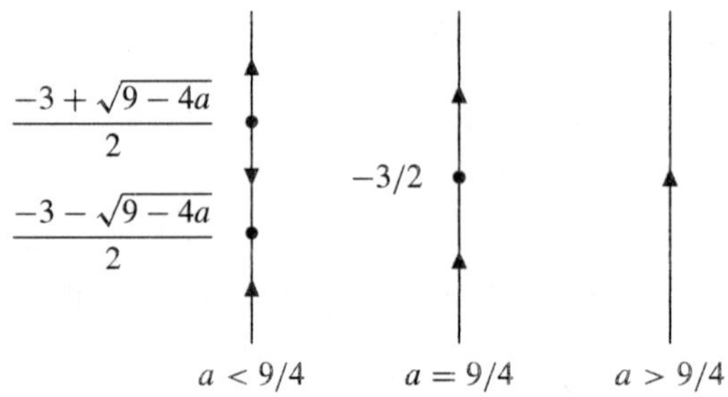

Phase lines for $a < 9/4$, $a = 9/4$, and $a > 9/4$.

3. The equilibrium points occur at solutions of $dy/dt = y^2 - ay + 1 = 0$. From the quadratic formula, we have

$$y = \frac{a \pm \sqrt{a^2 - 4}}{2}.$$

If $-2 < a < 2$, then $a^2 - 4 < 0$, and there are no equilibrium points. If $a > 2$ or $a < -2$, there are two equilibrium points. For $a = \pm 2$, there is one equilibrium point at $y = a/2$. The bifurcations occur at $a = \pm 2$.

To draw the phase lines, note that:

- For $-2 < a < 2$, $dy/dt = y^2 - ay + 1 > 0$, so the solutions are always increasing.
- For $a = 2$, $dy/dt = (y - 1)^2 \geq 0$, and $y = 1$ is a node.
- For $a = -2$, $dy/dt = (y + 1)^2 \geq 0$, and $y = -1$ is a node.
- For $a < -2$ or $a > 2$, let

$$y_1 = \frac{a - \sqrt{a^2 - 4}}{2} \quad \text{and} \quad y_2 = \frac{a + \sqrt{a^2 - 4}}{2}.$$

Then $dy/dt < 0$ if $y_1 < y < y_2$, and $dy/dt > 0$ if $y < y_1$ or $y > y_2$.

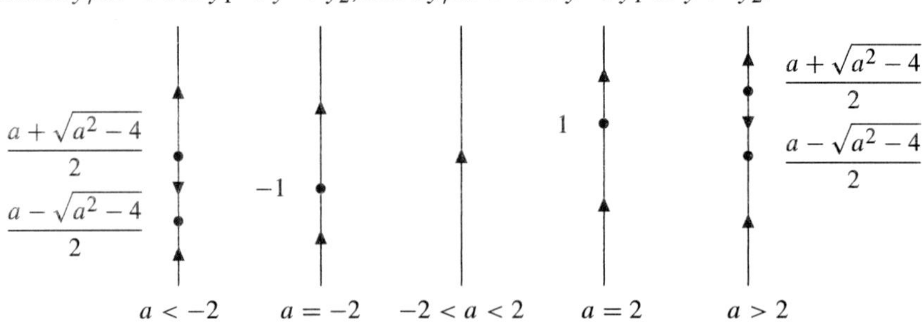

The five possible phase lines.

4. The equilibrium points occur at solutions of $dy/dt = y^3 + \alpha y^2 = 0$. For $\alpha = 0$, there is one equilibrium point, $y = 0$. For $\alpha \neq 0$, there are two equilibrium points, $y = 0$ and $y = -\alpha$. Thus, $\alpha = 0$ is a bifurcation value.

To draw the phase lines, note that:

- If $\alpha < 0$, $dy/dt > 0$ only if $y > -\alpha$.
- If $\alpha = 0$, $dy/dt > 0$ if $y > 0$, and $dy/dt < 0$ if $y < 0$.
- If $\alpha > 0$, $dy/dt < 0$ only if $y < -\alpha$.

Hence, as α increases from negative to positive, the source at $y = -\alpha$ moves from positive to negative as it "passes through" the node at $y = 0$.

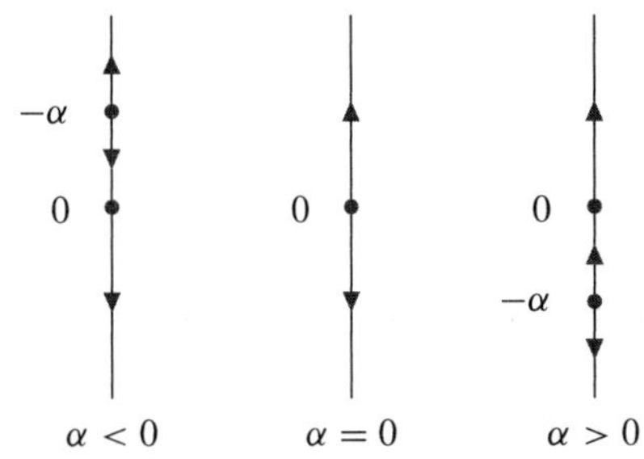

5. To find the equilibria we solve

$$(y^2 - \alpha)(y^2 - 4) = 0,$$

obtaining $y = \pm 2$ and $y = \pm\sqrt{\alpha}$ if $\alpha \geq 0$. Hence, there are two bifurcation values of α, $\alpha = 0$ and $\alpha = 4$.

For $\alpha < 0$, there are only two equilibria. The point $y = -2$ is a sink and $y = 2$ is a source. At $\alpha = 0$, there are three equilibria. There is a sink at $y = -2$, a source at $y = 2$, and a node at $y = 0$. For $0 < \alpha < 4$, there are four equilibria. The point $y = -2$ is still a sink, $y = -\sqrt{\alpha}$ is a source, $y = \sqrt{\alpha}$ is a sink, and $y = 2$ is still a source.

For $\alpha = 4$, there are only two equilibria, $y = \pm 2$. Both are nodes. For $\alpha > 4$, there are four equilibria again. The point $y = -\sqrt{\alpha}$ is a sink, $y = -2$ is now a source, $y = 2$ is now a sink, and $y = \sqrt{\alpha}$ is a source.

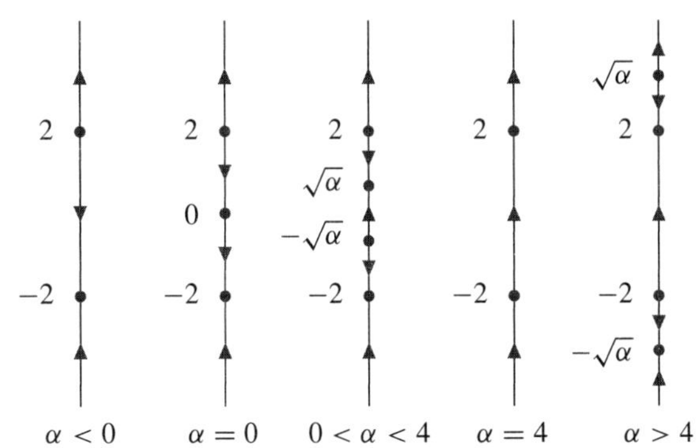

6. The equilibrium points occur at solutions of $dy/dt = \alpha - |y| = 0$. For $\alpha < 0$, there are no equilibrium points. For $\alpha = 0$, there is one equilibrium point, $y = 0$. For $\alpha > 0$, there are two equilibrium points, $y = \pm\alpha$. Therefore, $\alpha = 0$ is a bifurcation value.

To draw the phase lines, note that:

- If $\alpha < 0$, $dy/dt = \alpha - |y| < 0$, so the solutions are always decreasing.
- If $\alpha = 0$, $dy/dt < 0$ unless $y = 0$. Thus, $y = 0$ is a node.
- For $\alpha > 0$, $dy/dt > 0$ for $-\alpha < y < \alpha$, and $dy/dt < 0$ for $y < -\alpha$ and for $y > \alpha$.

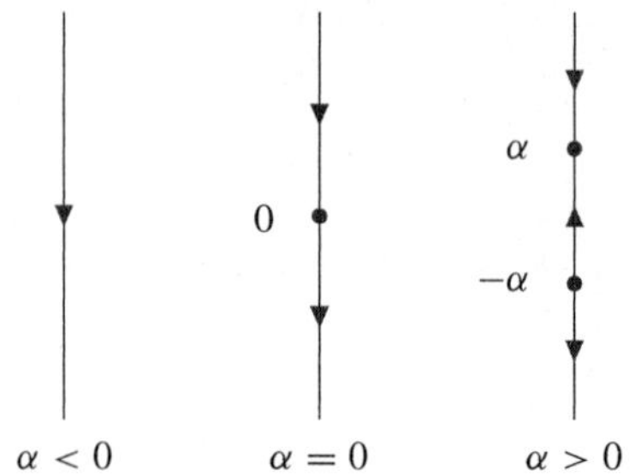

7. The bifurcations occur at values of α for which the graph of $\sin y + \alpha$ is tangent to the y-axis. That is, $\alpha = -1$ and $\alpha = 1$.

For $\alpha < -1$, there are no equilibria, and all solutions become unbounded in the negative direction as t increases.

If $\alpha = -1$, there are equilibrium points at $y = \pi/2 \pm 2n\pi$ for every integer n. All equilibria are nodes, and as $t \to \infty$, all other solutions decrease toward the nearest equilibrium solution below the given initial condition.

For $-1 < \alpha < 1$, there are infinitely many sinks and infinitely many sources, and they alternate along the phase line. Successive sinks differ by 2π. Similarly, successive sources are separated by 2π.

As α increases from -1 to $+1$, nearby sink and source pairs move apart. This separation continues until α is close to 1 where each source is close to the next sink with larger value of y.

At $\alpha = 1$, there are infinitely many nodes, and they are located at $y = 3\pi/2 \pm 2n\pi$ for every integer n. For $\alpha > 1$, there are no equilibria, and all solutions become unbounded in the positive direction as t increases.

8. Note that $0 < e^{-y^2} \leq 1$ for all y, and its maximum value occurs at $y = 0$. Therefore, for $\alpha < -1$, dy/dt is always negative, and the solutions are always decreasing.

If $\alpha = -1$, $dy/dt = 0$ if and only if $y = 0$. For $y \neq 0$, $dy/dt < 0$, and the equilibrium point at $y = 0$ is a node.

If $-1 < \alpha < 0$, then there are two equilibrium points which we compute by solving

$$e^{-y^2} + \alpha = 0.$$

We get $-y^2 = \ln(-\alpha)$. Consequently, $y = \pm\sqrt{\ln(-1/\alpha)}$. As $\alpha \to 0$ from below, $\ln(-1/\alpha) \to \infty$, and the two equilibria tend to $\pm\infty$.

If $\alpha \geq 0$, dy/dt is always positive, and the solutions are always increasing.

9. For $\alpha = 0$, there are three equilibria. There is a sink to the left of $y = 0$, a source at $y = 0$, and a sink to the right of $y = 0$.

As α decreases, the source and sink on the right move together. A bifurcation occurs at $\alpha \approx -2$. At this bifurcation value, there is a sink to the left of $y = 0$ and a node to the right of $y = 0$. For α below this bifurcation value, there is only the sink to the left of $y = 0$.

As α increases from zero, the sink to the left of $y = 0$ and the source move together. There is a bifurcation at $\alpha \approx 2$ with a node to the left of $y = 0$ and a sink to the right of $y = 0$. For α above this bifurcation value, there is only the sink to the right of $y = 0$.

10. Note that if α is very negative, then the equation $g(y) = -\alpha y$ has only one solution. It is $y = 0$. Furthermore, $dy/dt > 0$ for $y < 0$, and $dy/dt < 0$ for $y > 0$. Consequently, the equilibrium point at $y = 0$ is a sink.

In the figure, it appears that the tangent line to the graph of g at the origin has slope 1 and does not intersect the graph of g other than at the origin. If so, $\alpha = -1$ is a bifurcation value. For $\alpha \leq -1$, the differential equation has one equilibrium, which is a sink. For $\alpha > -1$, the equation has three equilibria, $y = 0$ and two others, one on each side of $y = 0$. The equilibrium point at the origin is a source, and the other two equilibria are sinks.

11. The graph of f needs to cross the y-axis exactly four times so that there are exactly four equilibria if $\alpha = 0$. The function must be greater than -3 everywhere so that there are no equilibria if $\alpha \geq 3$. Finally, the graph of f must cross horizontal lines three or more units above the y-axis exactly twice so that there are exactly two equilibria for $\alpha \leq -3$. The following graph is an example of the graph of such a function.

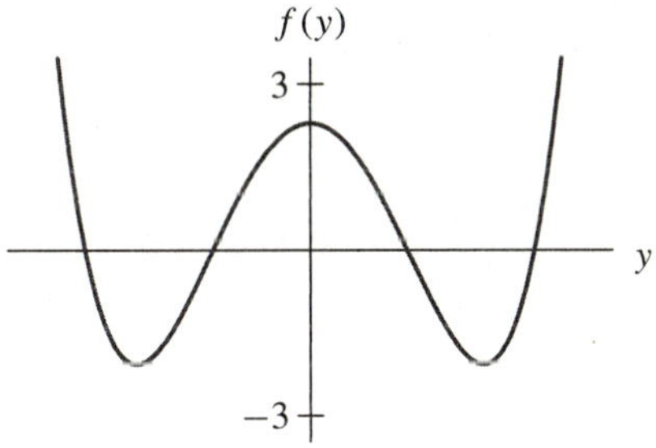

12. The graph of g can only intersect horizontal lines above 4 once, and it must go from above to below as y increases. Then there is exactly one sink for $\alpha \leq -4$.

Similarly, the graph of g can only intersect horizontal lines below -4 once, and it must go from above to below as y increases. Then there is exactly one sink for $\alpha \geq 4$.

Finally, the graph of g must touch the y-axis at exactly six points so that there are exactly six equilibria for $\alpha = 0$.

The following graph is the graph of one such function.

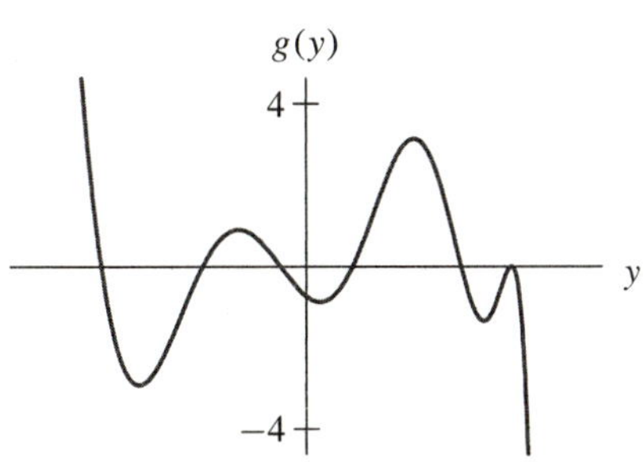

13. No such $f(y)$ exists. To see why, suppose that there is exactly one sink y_0 for $\alpha = 0$. Then, $f(y) > 0$ for $y < y_0$, and $f(y) < 0$ for $y > y_0$. Now consider the system $dy/dt = f(y) + 1$. Then $dy/dt \geq 1$ for $y < y_0$. If this system has an equilibrium point y_1 that is a source, then $y_1 > y_0$ and $dy/dt < 0$ for y slightly less than y_1. Since $f(y)$ is continuous and $dy/dt \geq 1$ for $y \leq y_0$, then dy/dt must have another zero between y_0 and y_1.

14. No, it is not possible to find such a function $g(y)$.

To see why, consider the two cases $\alpha = -1$ and $\alpha = +1$. If $\alpha = -1$ and the equation has exactly one sink y_0 and no other equilibrium points, then $g(y) > 1$ for $y < y_0$. If $\alpha = 1$ and the equation has exactly three equilibria—one sink and two sources, then the sink must be between the two sources. Let y_1 be the lower source. Since $dy/dt = g(y) + 1$, we see that $dy/dt > 0$ for all y less than the minimum of y_0 and y_1. Therefore, y_1 cannot be a source with no other equilibrium points below it.

15. **(a)** For all $C \geq 0$, the equation has a source at $P = C/k$, and this is the only equilibrium point. Hence all of the phase lines are qualitatively the same, and there are no bifurcation values for C.

(b) If $P(0) > C/k$, the corresponding solution $P(t) \to \infty$ at an exponential rate as $t \to \infty$, and if $P(0) < C/k$, $P(t) \to -\infty$, passing through "extinction" ($P = 0$) after a finite time.

16. First we set $C = 0$ so that the population can grow. Once the population reaches the desired level ($P = 100{,}000$), then we set $C = 100{,}000k = 200{,}000$. With this value of C, the population $P = 100,000$ is an equilibrium point.

We must be diligent in our management of the population since the equilibrium point is a source. A small change in P due to random fluctuations will eventually cause extinction or explosion of the population.

17. **(a)** A model of the fish population that includes fishing is

$$\frac{dP}{dt} = 2P - \frac{P^2}{50} - 3L,$$

where L is the number of licenses issued. The coefficient of 3 represents the average catch of 3 fish per year. As L is increased, the two equilibrium points for $L = 0$ (at $P = 0$ and $P = 100$) will move together. If L is sufficiently large, there are no equilibrium points. Hence we wish to pick L as large as possible so that there is still an equilibrium point present. In other words, we want the bifurcation value of L. The bifurcation value of L occurs if the equation

$$\frac{dP}{dt} = 2P - \frac{P^2}{50} - 3L = 0$$

has just one solution for P in terms of L. Using the quadratic formula, we see that there is exactly one equilibrium point if $L = 50/3$. Since this value of L is not an integer, the largest number of licenses that should be allowed is 16.

(b) If we allow the fish population to come to equilibrium then the population will be at the carrying capacity, which is $P = 100$ if $L = 0$. If we then allow 16 licenses to be issued, we expect that the population is a solution to the new model with $L = 16$ and initial population $P = 100$. The model becomes

$$\frac{dP}{dt} = 2P - \frac{P^2}{50} - 48,$$

which has a source at $P = 40$ and a sink at $P = 60$.

Thus, any initial population greater than 40 when fishing begins tends to the equilibrium level $P = 60$. If the initial population of fish was less than 40 when fishing begins, then the model predicts that the population will decrease to zero in a finite amount of time.

(c) The maximum "number" of licenses is $16\frac{2}{3}$. With $L = 16\frac{2}{3}$, there is an equilibrium at $P = 50$. This equilibrium is a node, and if $P(0) > 50$, the population will approach 50 as t increases. However, it is dangerous to allow this many licenses since an unforeseen event might cause the death of a few extra fish. That event would push the number of fish below the equilibrium value of $P = 50$. In this case, $dP/dt < 0$, and the population decreases to extinction.

If, however, we restrict to $L = 16$ licenses, then there are two equilibria, a sink at $P = 60$ and source at $P = 40$. As long as $P(0) > 40$, the population will tend to 60 as t increases. In this case, we have a small margin of safety. If $P \approx 60$, then it would have to drop to less than 40 before the fish are in danger of extinction.

18. **(a)**

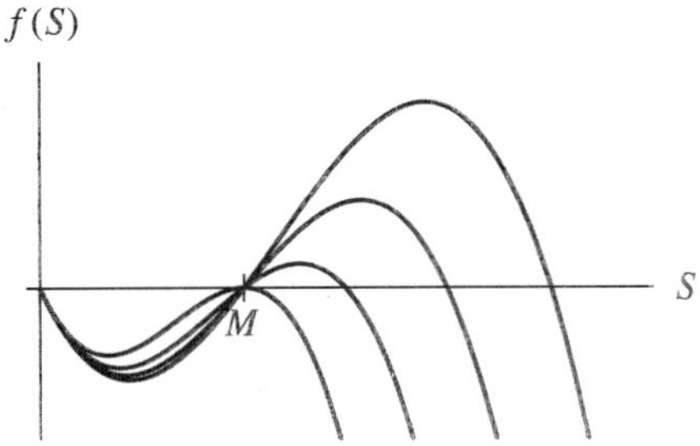

(b) The bifurcation occurs at $N = M$. The sink at $S = N$ coincides with the source at $S = M$ and becomes a node.

(c) Assuming that the population $S(t)$ is approximately N, the population adjusts to stay near the sink at $S = N$ as N slowly decreases. If $N < M$, the model is no longer consistent with the underlying assumptions.

19. **(a)**

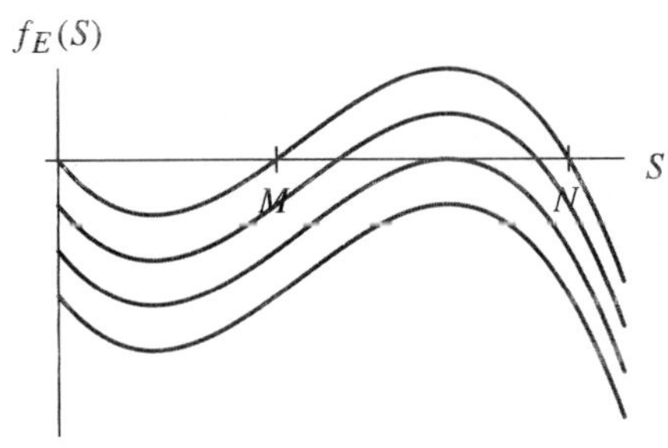

(b) As E increases, the graph of $f_E(S)$ is translated down until it has a double root between M and N. This value of E, which we denote by E_b, is the bifurcation value for this model. If $0 < E < E_b$, $f_E(S)$ has three real roots S_1, S_2, and S_3, where $S_1 < 0$ and $M < S_2 < S_3 < N$.

Note that $f_E(S) > 0$ if $S_2 < S < S_3$ and $f_E(S) < 0$ if $S_1 < S < S_2$ or $S > S_3$. Thus, S_2 is a source and S_3 is a sink. Consequently, if the fox squirrel population starts above S_2, it tends to the carrying capacity S_3.

If $E > E_b$, $f_E(S)$ no longer has any positive roots. Hence, there are no positive equilibria. Moreover, $f_E(S) < 0$ for all $S > 0$, and thus the fox squirrel population will plummet toward extinction.

(c) We must find the value of E for which $f_E(S)$ has a double root. Differentiation with respect to S yields

$$f'(S) = -\frac{3k}{MN}\left(S^2 - \frac{2}{3}(M+N)S + \frac{MN}{3}\right).$$

Setting $f'(S) = 0$ gives the two critical points

$$S = \frac{M + N \pm \sqrt{M^2 + N^2 - MN}}{3}.$$

Let S_c denote the largest of these two numbers.

As E increases, the graph of $f_E(S)$ shifts down. To find the bifurcation value E_b, we need to find the value of E such that $f_E(S_c) = 0$. Therefore, we set

$$f_E\left(\frac{M + N + \sqrt{M^2 + N^2 - MN}}{3}\right) = 0$$

and solve for E. After much computation, we find that

$$E_b = \frac{k}{27MN}\left((N - 2M)(M + N)(2N - M) + 2(M^2 - MN + N^2)^{3/2}\right).$$

20. If $C < kN/4$, the differential equation has two equilibria

$$P_1 = \frac{N}{2} - \sqrt{\frac{N^2}{4} - \frac{CN}{k}} \quad \text{and} \quad P_2 = \frac{N}{2} + \sqrt{\frac{N^2}{4} - \frac{CN}{k}}.$$

The smaller one, P_1, is a source, and the larger one, P_2, is a sink. Note that they are equidistant from $N/2$. Also, note that any population below P_1 tends to extinction.

If C is near $kN/4$, then P_1 and P_2 are near $N/2$. Consequently, if the population is near zero, it will tend to extinction. As C is decreased, P_1 and P_2 move apart until they reach $P_1 = 0$ and $P_2 = N$ for $C = 0$.

Once P is near zero, the parameter C must be reset essentially to zero so that P will be greater than P_1. Simply reducing C slightly below $kN/4$ leaves P in the range where $dP/dt < 0$ and the population will still die out.

21. **(a)** To find the equilibria, we solve

$$kP\left(1 - \frac{P}{N}\right) - \alpha P = 0.$$

We obtain the two points $P = 0$ and $P = (1 - (\alpha/k))N$.

(b) These equilibria correspond to the intersection of the parabola determined by the graph of

$$q(P) = kP\left(1 - \frac{P}{N}\right)$$

and the line that is the graph of $g_\alpha(P) = \alpha P$. As α increases, the larger equilibrium moves toward $P = 0$, and these equilibria coalesce at $\alpha = k$. For $0 \leq \alpha < k$, $P = 0$ is a source, and the other equilibrium is a sink.

If α is increased slowly from zero, then the population remains close to the sink

$$P = (1 - (\alpha/k))N$$

until $k = \alpha$ and the population is extinct.

(c) This model does not have a "point of no return." The bifurcation occurs at $\alpha = k$. Hence, harvesting that is proportional to population is easier to manage.

22. **(a)** If $a = 0$, there is a single equilibrium point at $y = 0$. For $a \neq 0$, the equilibrium points occur at $y = 0$ and $y = a$. If $a < 0$, the equilibrium point at $y = 0$ is a sink and the equilibrium point at $y = a$ is a source. If $a > 0$, the equilibrium point at $y = 0$ is a source and the equilibrium point at $y = a$ is a sink.

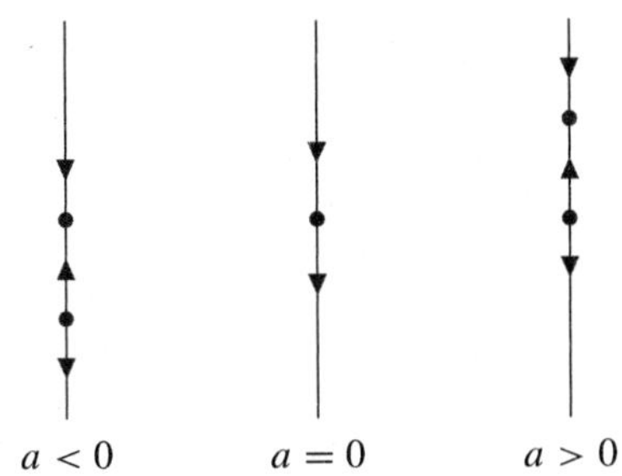

Phase lines for $dy/dt = ay - y^2$.

(b) Given the results in part (a), there is one bifurcation value, $a = 0$.

(c) The equilibrium points satisfy the equation

$$r + ay - y^2 = 0.$$

Solving it, we obtain

$$y = \frac{a \pm \sqrt{a^2 + 4r}}{2}.$$

Hence, there are no equilibrium points if $a^2 + 4r < 0$, one equilibrium point if $a^2 + 4r = 0$, and two equilibrium points if $a^2 + 4r > 0$.

If $r > 0$, we always have two equilibrium points.

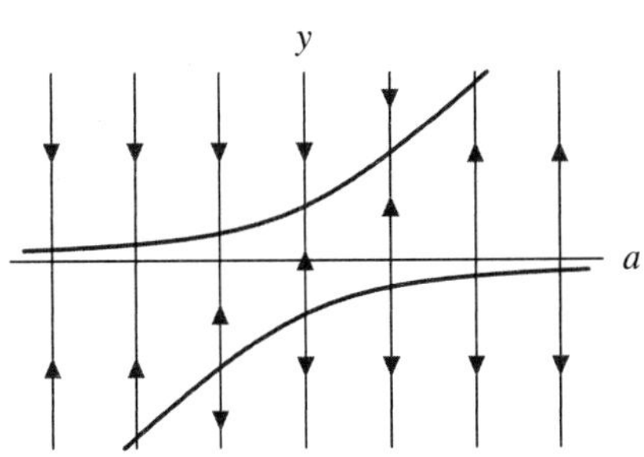

The bifurcation diagram for $r > 0$.

(d) If $r < 0$, there are no equilibrium points if $a^2+4r < 0$. In other words, there are no equilibrium points if $-2\sqrt{-r} < a < 2\sqrt{-r}$. If $a = \pm 2\sqrt{-r}$, there is a single equilibrium point, and if $|a| > 2\sqrt{-r}$, there are two equilibrium points.

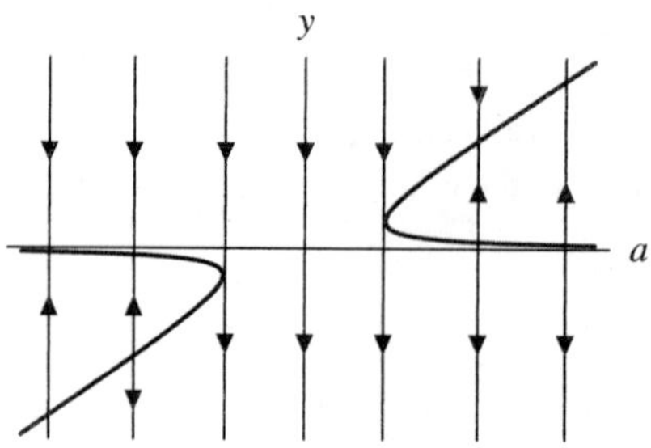

The bifurcation diagram for $r < 0$.

23. **(a)** If $a \leq 0$, there is a single equilibrium point at $y = 0$, and it is a sink. For $a > 0$, there are equilibrium points at $y = 0$ and $y = \pm\sqrt{a}$. The equilibrium point at $y = 0$ is a source, and the other two are sinks.

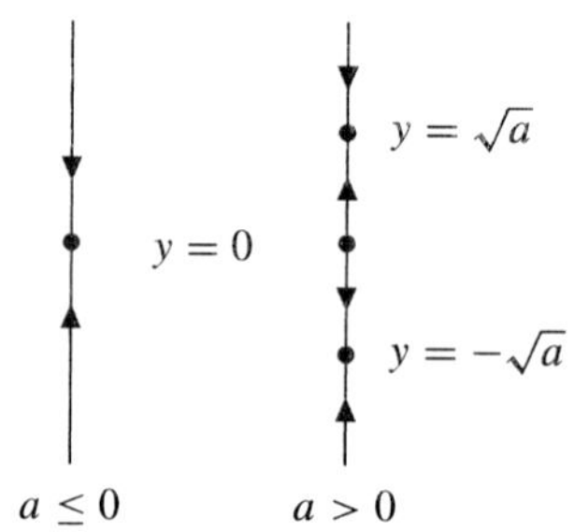

Phase lines for $dy/dt = ay - y^3$.

(b) Given the results in part (a), there is one bifurcation value, $a = 0$.

(c) The equilibrium points satisfy the cubic equation

$$r + ay - y^3 = 0.$$

Rather than solving it explicitly, we rely on `PhaseLines`.

If $r > 0$, there is a positive bifurcation value $a = a_0$. For $a < a_0$, the phase line has one equilibrium point, a positive sink. If $a > a_0$, there are two negative equilibria in addition to the positive sink. The larger of the two negative equilibria is a source and the smaller is a sink.

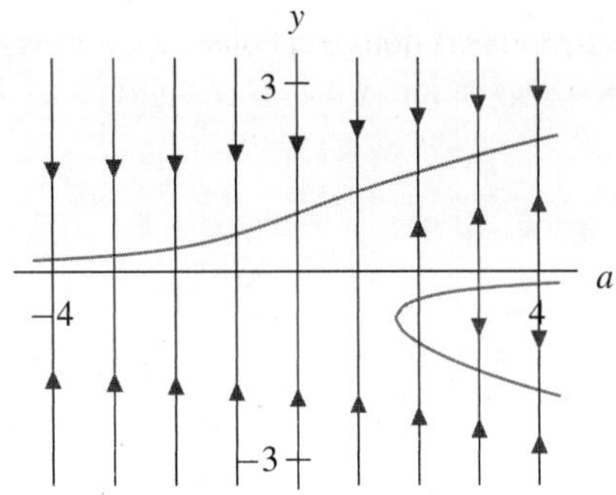

The bifurcation diagram for $r = 0.8$.

(d) If $r < 0$, there is a positive bifurcation value $a = a_0$. For $a < a_0$, the phase line has one equilibrium point, a negative sink. If $a > a_0$, there are two positive equilibria in addition to the negative sink. The larger of the two positive equilibria is a sink and the smaller is a source.

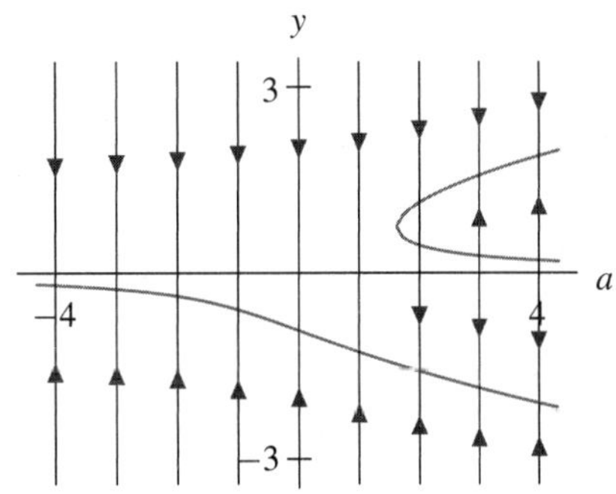

The bifurcation diagram for $r = -0.8$.

EXERCISES FOR SECTION 1.8

1. The general solution to the associated homogeneous equation is $y_h(t) = ke^t$. For a particular solution of the nonhomogeneous equation, we guess a solution of the form $y_p(t) = \alpha e^{-2t}$. Then

$$\begin{aligned}\frac{dy_p}{dt} - y_p &= -2\alpha e^{-2t} - \alpha e^{-2t} \\ &= -3\alpha e^{-2t}.\end{aligned}$$

Consequently, we must have $-3\alpha = 3$ for $y_p(t)$ to be a solution. Hence, $\alpha = -1$, and the general solution to the nonhomogeneous equation is

$$y(t) = ke^t - e^{-2t}.$$

2. The general solution to the associated homogeneous equation is $y_h(t) = ke^{-4t}$. For a particular solution of the nonhomogeneous equation, we guess a solution of the form $y_p(t) = \alpha e^{-t}$. Then

$$\begin{aligned}\frac{dy_p}{dt} + 4y_p &= -\alpha e^{-t} + 4\alpha e^{-t} \\ &= 3\alpha e^{-t}.\end{aligned}$$

Consequently, we must have $3\alpha = 3$ for $y_p(t)$ to be a solution. Hence, $\alpha = 1$, and the general solution to the nonhomogeneous equation is

$$y(t) = ke^{-4t} + e^{-t}.$$

3. The general solution to the associated homogeneous equation is $y_h(t) = ke^t$. For a particular solution of the nonhomogeneous equation, we guess a solution of the form $y_p(t) = \alpha \cos 2t + \beta \sin 2t$. Then

$$\begin{aligned}\frac{dy_p}{dt} - y_p &= -2\alpha \sin 2t + 2\beta \cos 2t - (\alpha \cos 2t + \beta \sin 2t) \\ &= (2\beta - \alpha)\cos 2t + (-2\alpha - \beta)\sin 2t\end{aligned}$$

Consequently, we must have

$$(2\beta - \alpha)\cos 2t + (-2\alpha - \beta)\sin 2t = \cos 2t$$

for $y_p(t)$ to be a solution. We must solve

$$\begin{cases} 2\beta - \alpha = 1 \\ -2\alpha - \beta = 0. \end{cases}$$

Hence, $\alpha = -1/5$ and $\beta = 2/5$. The general solution is

$$y(t) = ke^t - \tfrac{1}{5}\cos 2t + \tfrac{2}{5}\sin 2t.$$

4. The general solution to the associated homogeneous equation is $y_h(t) = ke^{2t}$. For a particular solution of the nonhomogeneous equation, we guess $y_p(t) = \alpha \cos 2t + \beta \sin 2t$. Then

$$\begin{aligned}\frac{dy_p}{dt} - 2y_p &= -2\alpha \sin 2t + 2\beta \cos 2t - 2(\alpha \cos 2t + \beta \sin 2t) \\ &= (2\beta - 2\alpha)\cos 2t + (-2\alpha - 2\beta)\sin 2t.\end{aligned}$$

Consequently, we must have

$$(2\beta - 2\alpha)\cos 2t + (-2\alpha - 2\beta)\sin 2t = \sin 2t$$

for $y_p(t)$ to be a solution, that is, we must solve

$$\begin{cases} -2\alpha - 2\beta = 1 \\ -2\alpha + 2\beta = 0. \end{cases}$$

Hence, $\alpha = -1/4$ and $\beta = -1/4$. The general solution of the nonhomogeneous equation is

$$y(t) = ke^{2t} - \tfrac{1}{4}\cos 2t - \tfrac{1}{4}\sin 2t.$$

5. The general solution to the associated homogeneous equation is $y_h(t) = ke^{4t}$. For a particular solution of the nonhomogeneous equation, we guess $y_p(t) = \alpha te^{4t}$ rather than αe^{4t} because αe^{4t} is a solution of the homogeneous equation. Then

$$\begin{aligned}\frac{dy_p}{dt} - 4y_p &= \alpha e^{4t} + 4\alpha te^{4t} - 4\alpha te^{4t}\\ &= \alpha e^{4t}.\end{aligned}$$

Consequently, we must have $\alpha = -5$ for $y_p(t)$ to be a solution. Hence, the general solution to the nonhomogeneous equation is

$$y(t) = ke^{4t} - 5te^{4t}.$$

6. The general solution of the associated homogeneous equation is $y_h(t) = ke^{t/2}$. For a particular solution of the nonhomogeneous equation, we guess $y_p(t) = \alpha te^{t/2}$ rather than $\alpha e^{t/2}$ because $\alpha e^{t/2}$ is a solution of the homogeneous equation. Then

$$\begin{aligned}\frac{dy_p}{dt} - \frac{y_p}{2} &= \alpha e^{t/2} + \frac{\alpha}{2}te^{t/2} - \frac{\alpha te^{t/2}}{2}\\ &= \alpha c^{t/2}.\end{aligned}$$

Consequently, we must have $\alpha = 4$ for $y_p(t)$ to be a solution. Hence, the general solution to the nonhomogeneous equation is

$$y(t) = ke^{t/2} + 4te^{t/2}.$$

7. The general solution to the associated homogeneous equation is $y_h(t) = ke^{-2t}$. For a particular solution of the nonhomogeneous equation, we guess a solution of the form $y_p(t) = \alpha e^{t/3}$. Then

$$\begin{aligned}\frac{dy_p}{dt} + 2y_p &= \tfrac{1}{3}\alpha e^{t/3} + 2\alpha e^{t/3}\\ &= \tfrac{7}{3}\alpha e^{t/3}.\end{aligned}$$

Consequently, we must have $\frac{7}{3}\alpha = 1$ for $y_p(t)$ to be a solution. Hence, $\alpha = 3/7$, and the general solution to the nonhomogeneous equation is

$$y(t) = ke^{-2t} + \tfrac{3}{7}e^{t/3}.$$

Since $y(0) = 1$, we have

$$1 = k + \tfrac{3}{7},$$

so $k = 4/7$. The function $y(t) = \frac{4}{7}e^{-2t} + \frac{3}{7}e^{t/3}$ is the solution of the initial-value problem.

8. The general solution to the associated homogeneous equation is $y_h(t) = ke^{2t}$. For a particular solution of the nonhomogeneous equation, we guess a solution of the form $y_p(t) = \alpha e^{-2t}$. Then

$$\begin{aligned}\frac{dy_p}{dt} - 2y_p &= -2\alpha e^{-2t} - 2\alpha e^{-2t} \\ &= -4\alpha e^{-2t}.\end{aligned}$$

Consequently, we must have $-4\alpha = 3$ for $y_p(t)$ to be a solution. Hence, $\alpha = -3/4$, and the general solution to the nonhomogeneous equation is

$$y(t) = ke^{2t} - \tfrac{3}{4}e^{-2t}.$$

Since $y(0) = 10$, we have

$$10 = k - \tfrac{3}{4},$$

so $k = 43/4$. The function

$$y(t) = \tfrac{43}{4}e^{2t} - \tfrac{3}{4}e^{-2t}$$

is the solution of the initial-value problem.

9. The general solution of the associated homogeneous equation is $y_h(t) = ke^{-t}$. For a particular solution of the nonhomogeneous equation, we guess a solution of the form $y_p(t) = \alpha\cos 2t + \beta\sin 2t$. Then

$$\begin{aligned}\frac{dy_p}{dt} + y_p &= -2\alpha\sin 2t + 2\beta\cos 2t + \alpha\cos 2t + \beta\sin 2t \\ &= (\alpha + 2\beta)\cos 2t + (-2\alpha + \beta)\sin 2t.\end{aligned}$$

Consequently, we must have

$$(\alpha + 2\beta)\cos 2t + (-2\alpha + \beta)\sin 2t = \cos 2t$$

for $y_p(t)$ to be a solution. We must solve

$$\begin{cases} \alpha + 2\beta = 1 \\ -2\alpha + \beta = 0. \end{cases}$$

Hence, $\alpha = 1/5$ and $\beta = 2/5$. The general solution to the differential equation is

$$y(t) = ke^{-t} + \tfrac{1}{5}\cos 2t + \tfrac{2}{5}\sin 2t.$$

To find the solution of the given initial-value problem, we evaluate the general solution at $t = 0$ and obtain

$$y(0) = k + \tfrac{1}{5}.$$

Since the initial condition is $y(0) = 5$, we see that $k = 24/5$. The desired solution is

$$y(t) = \tfrac{24}{5}e^{-t} + \tfrac{1}{5}\cos 2t + \tfrac{2}{5}\sin 2t.$$

10. The general solution of the associated homogeneous equation is $y_h(t) = ke^{-3t}$. For a particular solution of the nonhomogeneous equation, we guess a solution of the form $y_p(t) = \alpha \cos 2t + \beta \sin 2t$. Then

$$\begin{aligned}\frac{dy_p}{dt} + 3y_p &= -2\alpha \sin 2t + 2\beta \cos 2t + 3\alpha \cos 2t + 3\beta \sin 2t \\ &= (3\alpha + 2\beta) \cos 2t + (-2\alpha + 3\beta) \sin 2t.\end{aligned}$$

Consequently, we must have

$$(3\alpha + 2\beta) \cos 2t + (-2\alpha + 3\beta) \sin 2t = \cos 2t$$

for $y_p(t)$ to be a solution. We must solve

$$\begin{cases} 3\alpha + 2\beta = 1 \\ -2\alpha + 3\beta = 0. \end{cases}$$

Hence, $\alpha = 3/13$ and $\beta = 2/13$. The general solution to the differential equation is

$$y(t) = ke^{-3t} + \tfrac{3}{13} \cos 2t + \tfrac{2}{13} \sin 2t.$$

To find the solution of the given initial-value problem, we evaluate the general solution at $t = 0$ and obtain

$$y(0) = k + \tfrac{3}{13}.$$

Since the initial condition is $y(0) = -1$, we see that $k = -16/13$. The desired solution is

$$y(t) = -\tfrac{16}{13}e^{-3t} + \tfrac{3}{13} \cos 2t + \tfrac{2}{13} \sin 2t.$$

11. The general solution to the associated homogeneous equation is $y_h(t) = ke^{-5t}$. For a particular solution of the nonhomogeneous equation, we guess $y_p(t) = \alpha te^{-5t}$ rather than αe^{-5t} because αe^{-5t} is a solution of the homogeneous equation. Then

$$\begin{aligned}\frac{dy_p}{dt} + 5y_p &= \alpha e^{-5t} - 5\alpha te^{-5t} + 5\alpha te^{-5t} \\ &= \alpha e^{-5t}.\end{aligned}$$

Consequently, we must have $\alpha = 3$ for $y_p(t)$ to be a solution. Hence, the general solution to the nonhomogeneous equation is

$$y(t) = ke^{-5t} + 3te^{-5t}.$$

Note that $y(0) = k$, so the solution to the initial-value problem is

$$y(t) = -2e^{-5t} + 3te^{-5t} = (3t - 2)e^{-5t}.$$

12. The general solution to the associated homogeneous equation is $y_h(t) = ke^{2t}$. For a particular solution of the nonhomogeneous equation, we guess $y_p(t) = \alpha te^{2t}$ rather than αe^{2t} because αe^{2t} is a solution of the homogeneous equation. Then

$$\begin{aligned}\frac{dy_p}{dt} - 2y_p &= \alpha e^{2t} + 2\alpha te^{2t} - 2\alpha te^{2t} \\ &= \alpha e^{2t}.\end{aligned}$$

Consequently, we must have $\alpha = 7$ for $y_p(t)$ to be a solution. Hence, the general solution to the nonhomogeneous equation is

$$y(t) = ke^{2t} + 7te^{2t}.$$

Note that $y(0) = k$, so the solution to the initial-value problem is

$$y(t) = 3e^{2t} + 7te^{2t} = (7t + 3)e^{2t}.$$

13. **(a)** For the guess $y_p(t) = \alpha \cos 3t$, we have $dy_p/dt = -3\alpha \sin 3t$, and substituting this guess into the differential equation, we get

$$-3\alpha \sin 3t + 2\alpha \cos 3t = \cos 3t.$$

If we evaluate this equation at $t = \pi/6$, we get $-3\alpha = 0$. Therefore, $\alpha = 0$. However, $\alpha = 0$ does not produce a solution to the differential equation. Consequently, there is no value of α for which $y_p(t) = \alpha \cos 3t$ is a solution.

(b) If we guess $y_p(t) = \alpha \cos 3t + \beta \sin 3t$, then the derivative

$$\frac{dy_p}{dt} = -3\alpha \sin 3t + 3\beta \cos 3t$$

is also a simple combination of terms involving $\cos 3t$ and $\sin 3t$. Substitution of this guess into the equation leads to two linear algebraic equations in two unknowns, and such systems of equations usually have a unique solution.

14. Consider two different solutions $y_1(t)$ and $y_2(t)$ of the nonhomogeneous equation. We have

$$\frac{dy_1}{dt} = \lambda y_1 + \cos 2t \quad \text{and} \quad \frac{dy_2}{dt} = \lambda y_2 + \cos 2t.$$

By subtracting the first equation from the second, we see that

$$\begin{aligned}\frac{dy_2}{dt} - \frac{dy_1}{dt} &= \lambda y_2 + \cos 2t - \lambda y_1 - \cos 2t \\ &= \lambda y_2 - \lambda y_1.\end{aligned}$$

In other words,

$$\frac{d(y_2 - y_1)}{dt} = \lambda(y_2 - y_1),$$

and consequently, the difference $y_2 - y_1$ is a solution to the associated homogeneous equation.

Whether we write the general solution of the nonhomogeneous equation as

$$y(t) = y_1(t) + k_1 e^{\lambda t} \quad \text{or as} \quad y(t) = y_2(t) + k_2 e^{\lambda t},$$

we get the same set of solutions because $y_1(t) - y_2(t) = k_3 e^{\lambda t}$ for some k_3. In other words, both representations of the solutions produce the same collection of functions.

15. The Linearity Principle says that all nonzero solutions of a homogeneous linear equation are constant multiples of each other.

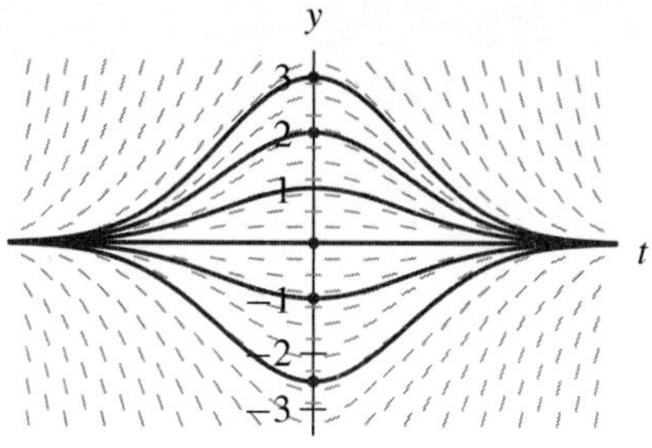

16. The Extended Linearity Principle says that any two solutions of a nonhomogeneous linear equation differ by a solution of the associated homogeneous equation.

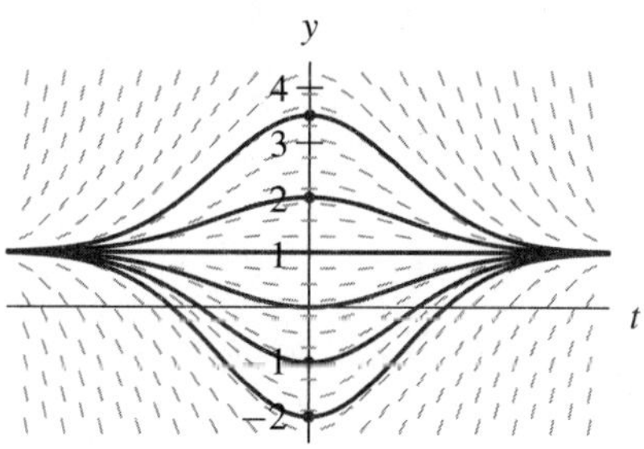

17. **(a)** We compute

$$\frac{dy_1}{dt} = \frac{1}{(1-t)^2} = (y_1(t))^2$$

to see that $y_1(t)$ is a solution.

(b) We compute

$$\frac{dy_2}{dt} = 2\frac{1}{(1-t)^2} \neq (y_2(t))^2$$

to see that $y_2(t)$ is not a solution.

(c) The equation $dy/dt = y^2$ is not linear. It contains y^2.

18. **(a)** The constant function $y(t) = 2$ for all t is an equilibrium solution.

(b) If $y(t) = 2 - e^{-t}$, then $dy/dt = e^{-t}$. Also, $-y(t) + 2 = e^{-t}$. Consequently, $y(t) = 2 - e^{-t}$ is a solution.

(c) Note that the solution $y(t) = 2 - e^{-t}$ has initial condition $y(0) = 1$. If the Linearity Principle held for this equation, then we could multiply the equilibrium solution $y(t) = 2$ by $1/2$ and obtain another solution that satisfies the initial condition $y(0) = 1$. Two solutions that satisfy the same initial condition would violate the Uniqueness Theorem.

19. Let $y(t) = y_h(t) + y_1(t) + y_2(t)$. Then

$$\begin{aligned}\frac{dy}{dt} + a(t)y &= \frac{dy_h}{dt} + \frac{dy_1}{dt} + \frac{dy_2}{dt} + a(t)y_h + a(t)y_1 + a(t)y_2 \\ &= \frac{dy_h}{dt} + a(t)y_h + \frac{dy_1}{dt} + a(t)y_1 + \frac{dy_2}{dt} + a(t)y_2 \\ &= 0 + b_1(t) + b_2(t).\end{aligned}$$

This computation shows that $y_h(t) + y_1(t) + y_2(t)$ is a solution of the original differential equation.

20. If $y_p(t) = at^2 + bt + c$, then

$$\begin{aligned}\frac{dy_p}{dt} + 2y_p &= 2at + b + 2at^2 + 2bt + 2c \\ &= 2at^2 + (2a + 2b)t + (b + 2c).\end{aligned}$$

Then $y_p(t)$ is a solution if this quadratic is equal to $3t^2 + 2t - 1$. In other words, $y_p(t)$ is a solution if

$$\begin{cases} 2a = 3 \\ 2a + 2b = 2 \\ b + 2c = -1. \end{cases}$$

From the first equation, we have $a = 3/2$. Then from the second equation, we have $b = -1/2$. Finally, from the third equation, we have $c = -1/4$. The function

$$y_p(t) = \tfrac{3}{2}t^2 - \tfrac{1}{2}t - \tfrac{1}{4}$$

is a solution of the differential equation.

21. To find the general solution, we use the technique suggested in Exercise 19. We calculate two particular solutions—one for the right-hand side $t^2 + 2t + 1$ and one for the right-hand side e^{4t}.

With the right-hand side $t^2 + 2t + 1$, we guess a solution of the form

$$y_{p_1}(t) = at^2 + bt + c.$$

Then

$$\begin{aligned}\frac{dy_{p_1}}{dt} + 2y_{p_1} &= 2at + b + 2(at^2 + bt + c) \\ &= 2at^2 + (2a + 2b)t + (b + 2c).\end{aligned}$$

Then y_{p_1} is a solution if

$$\begin{cases} 2a = 1 \\ 2a + 2b = 2 \\ b + 2c = 1. \end{cases}$$

We get $a = 1/2$, $b = 1/2$, and $c = 1/4$.

With the right-hand side e^{4t}, we guess a solution of the form

$$y_{p_2}(t) = \alpha e^{4t}.$$

Then

$$\frac{dy_{p_2}}{dt} + 2y_{p_2} = 4\alpha e^{4t} + 2\alpha e^{4t} = 6\alpha e^{4t},$$

and y_{p_2} is a solution if $\alpha = 1/6$.

The general solution of the associated homogeneous equation is $y_h(t) = ke^{-2t}$, so the general solution of the original equation is

$$ke^{-2t} + \tfrac{1}{2}t^2 + \tfrac{1}{2}t + \tfrac{1}{4} + \tfrac{1}{6}e^{4t}.$$

To find the solution that satisfies the initial condition $y(0) = 0$, we evaluate the general solution at $t = 0$ and obtain

$$k + \tfrac{1}{4} + \tfrac{1}{6} = 0.$$

Hence, $k = -5/12$.

22. To find the general solution, we use the technique suggested in Exercise 19. We calculate two particular solutions—one for the right-hand side t^3 and one for the right-hand side $\sin 3t$.

With the right-hand side t^3, we are tempted to guess that there is a solution of the form at^3, but there isn't. Instead we guess a solution of the form

$$y_{p_1}(t) = at^3 + bt^2 + ct + d.$$

Then

$$\begin{aligned}\frac{dy_{p_1}}{dt} + y_{p_1} &= 3at^2 + 2bt + c + at^3 + bt^2 + ct + d \\ &= at^3 + (3a + b)t^2 + (2b + c)t + (c + d)\end{aligned}$$

Then y_{p_1} is a solution if

$$\begin{cases} a = 1 \\ 3a + b = 0 \\ 2b + c = 0 \\ c + d = 0. \end{cases}$$

We get $a = 1$, $b = -3$, $c = 6$, and $d = -6$.

With the right-hand side $\sin 3t$, we guess a solution of the form

$$y_{p_2}(t) = \alpha \cos 3t + \beta \sin 3t.$$

Then

$$\begin{aligned}\frac{dy_{p_2}}{dt} + y_{p_1} &= -3\alpha \sin 3t + 3\beta \cos 3t + \alpha \cos 3t + \beta \sin 3t \\ &= (\alpha + 3\beta) \cos 3t + (-3\alpha + \beta) \sin 3t.\end{aligned}$$

Then y_{p_2} is a solution if

$$\begin{cases} \alpha + 3\beta = 0 \\ -3\alpha + \beta = 1. \end{cases}$$

We get $\alpha = -3/10$ and $\beta = 1/10$.

The general solution of the associated homogeneous equation is $y_h(t) = ke^{-t}$, so the general solution of the original equation is

$$ke^{-t} + t^3 - 3t^2 + 6t - 6 - \tfrac{3}{10}\cos 3t + \tfrac{1}{10}\sin 3t.$$

To find the solution that satisfies the initial condition $y(0) = 0$, we evaluate the general solution at $t = 0$ and obtain

$$k - 6 - \tfrac{3}{10} = 0.$$

Hence, $k = 63/10$.

23. To find the general solution, we use the technique suggested in Exercise 19. We calculate three particular solutions—one for the right-hand side $\cos 2t$, one for the right-hand side e^{3t}, and one for the right-hand side e^{-4t}.

With the right-hand side $\cos 2t$, we guess a solution of the form

$$y_{p_1}(t) = \alpha \cos 2t + \beta \sin 2t.$$

Then

$$\begin{aligned} \frac{dy_{p_1}}{dt} + 3y_{p_1} &= -2\alpha \sin 2t + 2\beta \cos 2t + 3(\alpha \cos 2t + \beta \sin 2t) \\ &= (3\alpha + 2\beta)\cos 2t + (-2\alpha + 3\beta)\sin 2t. \end{aligned}$$

Then y_{p_1} is a solution if

$$\begin{cases} 3\alpha + 2\beta = 1 \\ -2\alpha + 3\beta = 0. \end{cases}$$

We get $\alpha = 3/13$ and $\beta = 2/13$.

With the right-hand side e^{3t}, we guess a solution of the form

$$y_{p_2}(t) = ae^{3t}.$$

Then

$$\frac{dy_{p_2}}{dt} + 3y_{p_2} = 3ae^{3t} + 3ae^{3t} = 6ae^{3t},$$

and y_{p_2} is a solution if $a = 1/6$.

With the right-hand side e^{-4t}, we guess a solution of the form

$$y_{p_3}(t) = be^{-4t}.$$

Then

$$\frac{dy_{p_3}}{dt} + 3y_{p_3} = -4be^{-4t} + 3be^{-4t} = -be^{-4t},$$

and y_{p_3} is a solution if $b = -1$.

The general solution of the associated homogeneous equation is $y_h(t) = ke^{-3t}$, so the general solution of the original equation is

$$ke^{-3t} + \tfrac{3}{13}\cos 2t + \tfrac{2}{13}\sin 2t + \tfrac{1}{6}e^{3t} - e^{-4t}.$$

To find the solution that satisfies the initial condition $y(0) = 0$, we evaluate the general solution at $t = 0$ and obtain

$$k + \tfrac{3}{13} + \tfrac{1}{6} - 1 = 0.$$

Hence, $k = 47/78$.

24. To find the general solution, we use the technique suggested in Exercise 19. We calculate two particular solutions—one for the right-hand side $\cos 2t + 3\sin 2t$ and one for the right-hand side e^{-t}.

With the right-hand side $\cos 2t + 3\sin 2t$, we guess a solution of the form

$$y_{p_1}(t) = \alpha\cos 2t + \beta\sin 2t.$$

Then

$$\begin{aligned}\frac{dy_{p_1}}{dt} + y_{p_1} &= -2\alpha\sin 2t + 2\beta\cos 2t + \alpha\cos 2t + \beta\sin 2t \\ &= (\alpha + 2\beta)\cos 2t + (-2\alpha + \beta)\sin 2t.\end{aligned}$$

Then y_{p_1} is a solution if

$$\begin{cases} \alpha + 2\beta = 1 \\ -2\alpha + \beta = 3. \end{cases}$$

We get $\alpha = -1$ and $\beta = 1$.

With the right-hand side e^{-t}, making a guess of the form $y_{p_2}(t) = ae^{-t}$ does not lead to a solution of the nonhomogeneous equation because the general solution of the associated homogeneous equation is $y_h(t) = ke^{-t}$.

Consequently, we guess

$$y_{p_2}(t) = ate^{-t}.$$

Then

$$\frac{dy_{p_2}}{dt} + y_{p_2} = a(1-t)e^{-t} + ate^{-t} = ae^{-t},$$

and y_{p_2} is a solution if $a = 1$.

The general solution of the original equation is

$$ke^{-t} - \cos 2t + \sin 2t + te^{-t}.$$

To find the solution that satisfies the initial condition $y(0) = 0$, we evaluate the general solution at $t = 0$ and obtain

$$k - 1 = 0.$$

Hence, $k = 1$.

25. Since the general solution of the associated homogeneous equation is $y_h(t) = ke^{-2t}$ and since these $y_h(t) \to 0$ as $t \to \infty$, we only have to determine the long-term behavior of one solution to the nonhomogeneous equation. However, that is easier said than done.

Consider the slopes in the slope field for the equation. We rewrite the equation as

$$\frac{dy}{dt} = -2y + b(t).$$

Using the fact that $b(t) < 2$ for all t, we observe that $dy/dt < 0$ if $y > 1$ and, as y increases beyond $y = 1$, the slopes become more negative. Similarly, using the fact that $b(t) > -1$ for all t, we observe that $dy/dt > 0$ if $y < -1/2$ and, as y decreases below $y = -1/2$, the slopes become more positive. Thus, the graphs of all solutions must approach the strip $-1/2 \le y \le 1$ in the ty-plane as t increases. More precise information about the long-term behavior of solutions is difficult to obtain without specific knowledge of $b(t)$.

26. Since the general solution of the associated homogeneous equation is $y_h(t) = ke^{2t}$ and since these $y_h(t) \to \pm\infty$ as $t \to \infty$ if $k \neq 0$, the long-term behavior of one solution says a lot about the long-term behavior of all solutions.

Consider the slopes in the slope field for the equation. We rewrite the equation as

$$\frac{dy}{dt} = 2y + b(t).$$

Using the fact that $b(t) > -1$ for all t, we observe that $dy/dt > 0$ if $y > 1/2$, and as y increases beyond $y = 1/2$, the slopes increase. Similarly, using the fact that $b(t) < 2$ for all t, we observe that $dy/dt < 0$ if $y < -1$. and as y decreases below $y = -1$, the slopes decrease.

Thus, if a value of a solution $y(t)$ is larger than $1/2$, then $y(t) \to \infty$ as $t \to \infty$, and if a value of a solution $y(t)$ is less than -1, then $y(t) \to -\infty$ as $t \to \infty$. If one solution $y_p(t)$ satisfies $-1 \le y_p(t) \le 1/2$, then all other solutions become unbounded as $t \to \infty$. (In fact, there is exactly one solution that satisfies $-1 \le y(t) \le 1/2$ for all t, but demonstrating its existence is somewhat difficult.)

27. Since the general solution of the associated homogeneous equation is $y_h(t) = ke^{-t}$ and since these $y_h(t) \to 0$ as $t \to \infty$, we only have to determine the long-term behavior of one solution to the nonhomogeneous equation. However, that is easier said than done.

Consider the slopes in the slope field for the equation. We rewrite the equation as

$$\frac{dy}{dt} = -y + b(t).$$

For any number $T > 3$, let ϵ be a positive number less than $T - 3$, and fix t_0 such that $b(t) < T - \epsilon$ if $t > t_0$. If $t > t_0$ and $y(t) > T$, then

$$\frac{dy}{dt} < -T + T - \epsilon = -\epsilon.$$

Hence, no solution remains greater than T for all time. Since $T > 3$ is arbitrary, no solution remains greater than 3 (by a fixed amount) for all time.

The same idea works to show that no solution can remain less than 3 (by a fixed amount) for all time. Hence, every solution tends to 3 as $t \to \infty$.

28. Since the equation is linear, we can consider the two separate differential equations

$$\frac{dy_1}{dt} + ay_1 = \cos 3t \quad \text{and} \quad \frac{dy_2}{dt} + ay_2 = b$$

(see Exercise 19). One particular solution of the equation for y_1 is of the form

$$y_1(t) = \alpha \cos 3t + \beta \sin 3t,$$

and one particular solution of the equation for y_2 is the equilibrium solution $y_2(t) = b/a$. The solution $y_1(t)$ oscillates in a periodic fashion. In fact, we can use the techniques introduced in Section 4.4 to show that the amplitude of the oscillations is no larger than $1/3$.

The general solution of the associated homogeneous equation is $y_h(t) = ke^{-at}$, so the general solution of the original differential equation can be written as

$$y(t) = y_h(t) + y_1(t) + y_2(t).$$

As $t \to \infty$, $y_h(t) \to 0$, and therefore all solutions behave like the sum $y_1(t) + y_2(t)$ over the long term. In other words, they oscillate about $y = b/a$ with periodic oscillations of amplitude at most $1/3$.

29. **(a)** The differential equation modeling the problem is

$$\frac{dP}{dt} = .04P + 520$$

where 520 is the amount of money added to the account per year. (Assuming 52 weeks in a year and a "continuous deposit".)

(b) To find the general solution, we first compute the general solution of the associated homogeneous equation. It is $P_h(t) = ke^{0.04t}$.

To find a particular solution of the nonhomogeneous equation, we observe that the equation is autonomous, and we calculate its equilibrium solution. It is $P(t) = -520/.04 = -13,000$ for all t. (This equilibrium solution is what we would have calculated if we had guessed a constant.)

Hence, the general solution is

$$P(t) = -13,000 + ke^{0.04t}.$$

Since the account initially has \$500 in it, the initial condition is $P(0) = 500$. Solving

$$500 = -13,000 + ke^{.04(0)}$$

yields $k = 13,500$. Therefore, our model is

$$P(t) = -13,000 + 13,500e^{.04t}.$$

To find the amount on deposit after 5 years, we evaluate $P(5)$ and obtain

$$-13,000 + 13,500e^{.04(5)} \approx 3,488.94.$$

30. Let $M(t)$ be the amount of money left at time t. Then, we have the initial condition $M(0) = \$30{,}000$. Money is being added to the account at a rate of 6% and removed from the account at a rate of \$10,000 per year, so

$$\frac{dM}{dt} = 0.06M - 10{,}000.$$

To find the general solution, we first compute the general solution of the associated homogeneous equation. It is $M_h(t) = ke^{0.06t}$.

To find a particular solution of the nonhomogeneous equation, we observe that the equation is autonomous, and we calculate its equilibrium solution. It is $M(t) = 10{,}000/.06 \approx \$166,666.67$ for all t. (This equilibrium solution is what we would have calculated if we had guessed a constant.)

Simplifying and rounding to the nearest dollar, we have

$$M(t) = 166,667 + ke^{0.06t}.$$

Using the initial condition $M(0) = 30{,}000$, we have

$$166,667 + k = 30{,}000,$$

so $k = -136,667$ and

$$M(t) = 166,667 - 136,667e^{0.06t}.$$

Solving for the value of t when $M(t) = 0$, we have

$$166,667 - 136,667e^{0.06t} = 0,$$

which is equivalent to

$$e^{0.06t} = \frac{166,667}{136,667}.$$

In other words,

$$0.06t = \ln(1.21951),$$

which yields $t \approx 3.307$ years.

31. Step 1: *Before retirement*

First we calculate how much money will be in her retirement fund after 30 years. The differential equation modeling the situation is

$$\frac{dy}{dt} = .08y + 1200,$$

where $y(t)$ represents the fund's balance at time t.

The general solution of the homogeneous equation is $y_h(t) = ke^{0.08t}$.

To find a particular solution, we observe that the nonhomogeneous equation is autonomous and that it has an equilibrium solution at $y = -1200/0.08 = -15,000$. We can use this equilibrium solution as the particular solution. (It is the solution we would have computed if we had guessed a constant solution). We obtain

$$y(t) = ke^{0.08t} - 15,000.$$

From the initial condition, we see that $k = 15,000$, and

$$y(t) = 15,000(e^{0.08t} - 1).$$

Letting $t = 30$, we compute that the fund contains \$150, 347.65 after 30 years.

Step 2: *After retirement*
We need a new model for the remaining years since the professor is withdrawing rather than depositing. Since she withdraws at a rate of \$3, 000 per month (\$36, 000 per year), we write

$$\frac{dy}{dt} = .08y - 36,000,$$

where we continue to measure time t in years.

Again, the solution of the homogeneous equation is $y_h(t) = ke^{0.08t}$.

To find a particular solution of the nonhomogeneous equation, we note that the equation is autonomous and that it has an equilibrium at $y = 36,000/0.08 = 450,000$. Hence, we may take the particular solution to be this equilibrium solution. (Again, this solution is what we would have computed if we had guessed a constant function for y_p.)

The general solution is

$$y(t) = ke^{0.08t} + 450,000.$$

In this case, we have the initial condition $y(0) = 150,347.65$ since now $y(t)$ is the amount in the fund t years after she retires. Solving $150,347.65 = k + 450,000$, we get $k = -299,652.35$. The solution in this case is

$$y(t) = -299,652.35e^{0.08t} + 450,000.$$

Finally, we wish to know when her money runs out. That is, at what time t is $y(t) = 0$? Solving

$$y(t) = -299,652.35e^{0.08t} + 450,000 = 0$$

yields $t \approx 5.08$ years (approximately 61 months).

32. Note that $dy/dt = 1/5$ for this function. Substituting $y(t) = t/5$ in the right-hand side of the differential equation yields

$$(\cos t)\left(\frac{t}{5}\right) + \frac{1}{5}(1 - t\cos t),$$

which also equals $1/5$. Hence, $y(t) = t/5$ is a solution.

33. **(a)** We know that

$$\frac{dy_h}{dt} = a(t)y_h \quad \text{and} \quad \frac{dy_p}{dt} = a(t)y_p + b(t)$$

Then

$$\begin{aligned}\frac{d(y_h + y_p)}{dt} &= a(t)y_h + a(t)y_p + b(t)\\ &= a(t)(y_h + y_p) + b(t).\end{aligned}$$

(b) We know that

$$\frac{dy_p}{dt} = a(t)y_p + b(t) \quad \text{and} \quad \frac{dy_q}{dt} = a(t)y_q + b(t).$$

Then

$$\begin{aligned}\frac{d(y_p - y_q)}{dt} &= (a(t)y_p + b(t)) - (a(t)y_q + b(t))\\ &= a(t)(y_p - y_q).\end{aligned}$$

34. Suppose k is a constant and $y_1(t)$ is a solution. Then we know that $ky_1(t)$ is also a solution. Hence,

$$\frac{d(ky_1)}{dt} = f(t, ky_1)$$

for all t. Also,

$$\frac{d(ky_1)}{dt} = k\frac{dy_1}{dt} = kf(t, y_1)$$

because $y_1(t)$ is a solution. Therefore, we have

$$f(t, ky_1) = kf(t, y_1)$$

for all t. In particular, if $y_1(t) \neq 0$, we can pick $k = 1/y_1(t)$, and we get

$$f(t, 1) = \frac{1}{y_1(t)} f(t, y_1(t)).$$

In other words,

$$y_1(t) f(t, 1) = f(t, y_1(t))$$

for all t for which $y_1(t) \neq 0$. If we ignore the dependence on t, we have

$$yf(t, 1) = f(t, y)$$

for all $y \neq 0$ because we know that there is a solution $y_1(t)$ that solves the initial-value problem $y_1(t) = y$. By continuity, we know that the equality

$$yf(t, 1) = f(t, y)$$

holds even as y tends to zero.

If we define $a(t) = f(t, 1)$, we have

$$f(t, y) = a(t)y.$$

The differential equation is linear and homogeneous.

EXERCISES FOR SECTION 1.9

1. We rewrite the equation in the form

$$\frac{dy}{dt} + \frac{y}{t} = 2$$

and note that the integrating factor is

$$\mu(t) = e^{\int (1/t)\, dt} = e^{\ln t} = t.$$

Multiplying both sides by $\mu(t)$, we obtain

$$t\frac{dy}{dt} + y = 2t.$$

Applying the Product Rule to the left-hand side, we see that this equation is the same as

$$\frac{d(ty)}{dt} = 2t,$$

and integrating both sides with respect to t, we obtain

$$ty = t^2 + c,$$

where c is an arbitrary constant. The general solution is

$$y(t) = \frac{1}{t}(t^2 + c) = t + \frac{c}{t}.$$

2. We rewrite the equation in the form

$$\frac{dy}{dt} - \frac{3}{t}y = t^5$$

and note that the integrating factor is

$$\mu(t) = e^{\int(-3/t)\,dt} = e^{-3\ln t} = e^{\ln(t^{-3})} = t^{-3}.$$

Multiplying both sides by $\mu(t)$, we obtain

$$t^{-3}\frac{dy}{dt} - 3t^{-4}y = t^2.$$

Applying the Product Rule to the left-hand side, we see that this equation is the same as

$$\frac{d(t^{-3}y)}{dt} = t^2$$

and integrating both sides with respect to t, we obtain

$$t^{-3}y = \frac{t^3}{3} + c,$$

where c is an arbitrary constant. The general solution is

$$y(t) = \frac{t^6}{3} + ct^3.$$

3. We rewrite the equation in the form

$$\frac{dy}{dt} + \frac{y}{1+t} = t^2$$

and note that the integrating factor is

$$\mu(t) = e^{\int(1/(1+t))\,dt} = e^{\ln(1+t)} = 1 + t.$$

Multiplying both sides by $\mu(t)$, we obtain

$$(1+t)\frac{dy}{dt} + y = (1+t)t^2.$$

Applying the Product Rule to the left-hand side, we see that this equation is the same as

$$\frac{d((1+t)y)}{dt} = t^3 + t^2,$$

and integrating both sides with respect to t, we obtain

$$(1+t)y = \frac{t^4}{4} + \frac{t^3}{3} + c,$$

where c is an arbitrary constant. The general solution is

$$y(t) = \frac{3t^4 + 4t^3 + 12c}{12(t+1)}.$$

4. We rewrite the equation in the form

$$\frac{dy}{dt} + 2ty = 4e^{-t^2}$$

and note that the integrating factor is

$$\mu(t) = e^{\int 2t\,dt} = e^{t^2}.$$

Multiplying both sides by $\mu(t)$, we obtain

$$e^{t^2}\frac{dy}{dt} + 2te^{t^2}y = 4.$$

Applying the Product Rule to the left-hand side, we see that this equation is the same as

$$\frac{d(e^{t^2}y)}{dt} = 4,$$

and integrating both sides with respect to t, we obtain

$$e^{t^2}y = 4t + c,$$

where c is an arbitrary constant. The general solution is

$$y(t) = 4te^{-t^2} + ce^{-t^2}.$$

5. Note that the integrating factor is

$$\mu(t) = e^{\int(-2t/(1+t^2))\,dt} = e^{-\ln(1+t^2)} = \left(e^{\ln(1+t^2)}\right)^{-1} = \frac{1}{1+t^2}.$$

Multiplying both sides by $\mu(t)$, we obtain

$$\frac{1}{1+t^2}\frac{dy}{dt} - \frac{2t}{(1+t^2)^2}\,y = \frac{3}{1+t^2}.$$

Applying the Product Rule to the left-hand side, we see that this equation is the same as

$$\frac{d}{dt}\left(\frac{y}{1+t^2}\right) = \frac{3}{1+t^2}.$$

Integrating both sides with respect to t, we obtain

$$\frac{y}{1+t^2} = 3\arctan(t) + c,$$

where c is an arbitrary constant. The general solution is

$$y(t) = (1+t^2)(3\arctan(t) + c).$$

6. Note that the integrating factor is

$$\mu(t) = e^{\int(-2/t)\,dt} = e^{-2\ln t} = e^{\ln(t^{-2})} = t^{-2}.$$

Multiplying both sides by $\mu(t)$, we obtain

$$t^{-2}\frac{dy}{dt} - 2t^{-3}y = te^t.$$

Applying the Product Rule to the left-hand side, we see that this equation is the same as

$$\frac{d(t^{-2}y)}{dt} = te^t,$$

and integrating both sides with respect to t, we obtain

$$t^{-2}y = (t-1)e^t + c,$$

where c is an arbitrary constant. The general solution is

$$y(t) = t^2(t-1)e^t + ct^2.$$

7. We rewrite the equation in the form

$$\frac{dy}{dt} + \frac{y}{1+t} = 2$$

and note that the integrating factor is

$$\mu(t) = e^{\int (1/(1+t))\,dt} = e^{\ln(1+t)} = 1 + t.$$

Multiplying both sides by $\mu(t)$, we obtain

$$(1+t)\frac{dy}{dt} + y = 2(1+t).$$

Applying the Product Rule to the left-hand side, we see that this equation is the same as

$$\frac{d((1+t)y)}{dt} = 2(1+t),$$

and integrating both sides with respect to t, we obtain

$$(1+t)y = 2t + t^2 + c,$$

where c is an arbitrary constant. The general solution is

$$y(t) = \frac{t^2 + 2t + c}{1+t}.$$

To find the solution that satisfies the initial condition $y(0) = 3$, we evaluate the general solution at $t = 0$ and obtain

$$c = 3.$$

The desired solution is

$$y(t) = \frac{t^2 + 2t + 3}{1+t}.$$

8. We rewrite the equation in the form

$$\frac{dy}{dt} - \frac{1}{t+1}y = 4t^2 + 4t$$

and note that the integrating factor is

$$\mu(t) = e^{\int (-1/(t+1))\,dt} = e^{-\ln(t+1)} = \left(e^{\ln((t+1)^{-1})}\right) = \frac{1}{t+1}.$$

Multiplying both sides by $\mu(t)$, we obtain

$$\frac{1}{t+1}\frac{dy}{dt} - \frac{1}{(t+1)^2}y = \frac{4t^2 + 4t}{t+1}.$$

Applying the Product Rule to the left-hand side, we see that this equation is the same as

$$\frac{d}{dt}\left(\frac{y}{t+1}\right) = 4t.$$

Integrating both sides with respect to t, we obtain

$$\frac{y}{t+1} = 2t^2 + c,$$

where c is an arbitrary constant. The general solution is

$$y(t) = (2t^2 + c)(t + 1) = 2t^3 + 2t^2 + ct + c.$$

To find the solution that satisfies the initial condition $y(1) = 10$, we evaluate the general solution at $t = 1$ and obtain $c = 3$. The desired solution is

$$y(t) = 2t^3 + 2t^2 + 3t + 3.$$

9. In Exercise 1, we derived the general solution

$$y(t) = t + \frac{c}{t}.$$

To find the solution that satisfies the initial condition $y(1) = 3$, we evaluate the general solution at $t = 1$ and obtain $c = 2$. The desired solution is

$$y(t) = t + \frac{2}{t}.$$

10. In Exercise 4, we derived the general solution

$$y(t) = 4te^{-t^2} + ce^{-t^2}.$$

To find the solution that satisfies the initial condition $y(0) = 3$, we evaluate the general solution at $t = 0$ and obtain $c = 3$. The desired solution is

$$y(t) = 4te^{-t^2} + 3e^{-t^2}.$$

11. Note that the integrating factor is

$$\mu(t) = e^{\int -(2/t)\,dt} = e^{-2\int (1/t)\,dt} = e^{-2\ln t} = e^{\ln(t^{-2})} = \frac{1}{t^2}.$$

Multiplying both sides by $\mu(t)$, we obtain

$$\frac{1}{t^2}\frac{dy}{dt} - \frac{2y}{t^3} = 2.$$

Applying the Product Rule to the left-hand side, we see that this equation is the same as

$$\frac{d}{dt}\left(\frac{y}{t^2}\right) = 2,$$

and integrating both sides with respect to t, we obtain

$$\frac{y}{t^2} = 2t + c,$$

where c is an arbitrary constant. The general solution is

$$y(t) = 2t^3 + ct^2.$$

To find the solution that satisfies the initial condition $y(-2) = 4$, we evaluate the general solution at $t = -2$ and obtain

$$-16 + 4c = 4.$$

Hence, $c = 5$, and the desired solution is

$$y(t) = 2t^3 + 5t^2.$$

12. Note that the integrating factor is

$$\mu(t) = e^{\int(-3/t)\,dt} = e^{-3\ln t} = e^{\ln(t^{-3})} = t^{-3}.$$

Multiplying both sides by $\mu(t)$, we obtain

$$t^{-3}\frac{dy}{dt} - 3t^{-4}y = 2e^{2t}.$$

Applying the Product Rule to the left-hand side, we see that this equation is the same as

$$\frac{d(t^{-3}y)}{dt} = 2e^{2t},$$

and integrating both sides with respect to t, we obtain

$$t^{-3}y = e^{2t} + c,$$

where c is an arbitrary constant. The general solution is

$$y(t) = t^3(e^{2t} + c).$$

To find the solution that satisfies the initial condition $y(1) = 0$, we evaluate the general solution at $t = 1$ and obtain $c = -e^2$. The desired solution is

$$y(t) = t^3(e^{2t} - e^2).$$

13. We rewrite the equation in the form

$$\frac{dy}{dt} - (\sin t)y = 4$$

and note that the integrating factor is

$$\mu(t) = e^{\int(-\sin t)\,dt} = e^{\cos t}.$$

Multiplying both sides by $\mu(t)$, we obtain

$$e^{\cos t}\frac{dy}{dt} - e^{\cos t}(\sin t)y = 4e^{\cos t}.$$

Applying the Product Rule to the left-hand side, we see that this equation is the same as

$$\frac{d(e^{\cos t}y)}{dt} = 4e^{\cos t},$$

and integrating both sides with respect to t, we obtain

$$e^{\cos t}y = \int 4e^{\cos t}\,dt.$$

Since the integral on the right-hand side is impossible to express using elementary functions, we write the general solution as

$$y(t) = 4e^{-\cos t}\int e^{\cos t}\,dt.$$

14. We rewrite the equation in the form

$$\frac{dy}{dt} - t^2y = 4$$

and note that the integrating factor is

$$\mu(t) = e^{\int(-t^2)\,dt} = e^{-t^3/3}.$$

Multiplying both sides of the equation by $\mu(t)$, we obtain

$$e^{-t^3/3}\frac{dy}{dt} - t^2e^{-t^3/3}y = 4e^{-t^3/3}.$$

Applying the Product Rule to the left-hand side, we see that this equation is the same as

$$\frac{d(e^{-t^3/3}y)}{dt} = 4e^{-t^3/3},$$

and integrating both sides with respect to t, we obtain

$$e^{-t^3/3}y = \int 4e^{-t^3/3}\,dt.$$

Since the integral on the right-hand side is impossible to express using elementary functions, we write the general solution as

$$y(t) = 4e^{t^3/3}\int e^{-t^3/3}\,dt.$$

15. We rewrite the equation in the form

$$\frac{dy}{dt} - \frac{y}{t^2} = 4\cos t$$

and note that the integrating factor is

$$\mu(t) = e^{\int (-1/t^2)\,dt} = e^{1/t}.$$

Multiplying both sides by $\mu(t)$, we obtain

$$e^{1/t}\frac{dy}{dt} - \frac{e^{1/t}}{t^2}y = 4e^{1/t}\cos t.$$

Applying the Product Rule to the left-hand side, we see that this equation is the same as

$$\frac{d(e^{1/t}y)}{dt} = 4e^{1/t}\cos t,$$

and integrating both sides with respect to t, we obtain

$$e^{1/t}y = \int 4e^{1/t}\cos t\,dt.$$

Since the integral on the right-hand side is impossible to express using elementary functions, we write the general solution as

$$y(t) = 4e^{-1/t}\int e^{1/t}\cos t\,dt.$$

16. We rewrite the equation in the form

$$\frac{dy}{dt} - y = 4\cos t^2$$

and note that the integrating factor is

$$\mu(t) = e^{\int -1\,dt} = e^{-t}.$$

Multiplying both sides of the equation by $\mu(t)$, we obtain

$$e^{-t}\frac{dy}{dt} - e^{-t}y = 4e^{-t}\cos t^2.$$

Applying the Product Rule to the left-hand side, we see that this equation is the same as

$$\frac{d(e^{-t}y)}{dt} = 4e^{-t}\cos t^2,$$

and integrating both sides with respect to t, we obtain

$$e^{-t}y = \int 4e^{-t}\cos t^2\,dt.$$

Since the integral on the right-hand side is impossible to express using elementary functions, we write the general solution as

$$y(t) = 4e^{t}\int e^{-t}\cos t^2\,dt.$$

17. We rewrite the equation in the form

$$\frac{dy}{dt} + e^{-t^2} y = \cos t$$

and note that the integrating factor is

$$\mu(t) = e^{\int e^{-t^2}\,dt}.$$

This integral is impossible to express in terms of elementary functions. Multiplying both sides by $\mu(t)$, we obtain

$$\left(e^{\int e^{-t^2}\,dt}\right)\frac{dy}{dt} + \left(e^{\int e^{-t^2}\,dt}\right)e^{-t^2} y = \left(e^{\int e^{-t^2}\,dt}\right)\cos t.$$

Applying the Product Rule to the left-hand side, we see that this equation is the same as

$$\frac{d\left(\left(e^{\int e^{-t^2}\,dt}\right)y\right)}{dt} = \left(e^{\int e^{-t^2}\,dt}\right)\cos t,$$

and integrating both sides with respect to t, we obtain

$$\left(e^{\int e^{-t^2}\,dt}\right)y = \int\left(e^{\int e^{-t^2}\,dt}\right)\cos t\,dt.$$

These integrals are also impossible to express in terms of elementary functions, so we write the general solution in the form

$$y(t) = \left(e^{-\int e^{-t^2}\,dt}\right)\int\left(e^{\int e^{-t^2}\,dt}\right)\cos t\,dt.$$

18. We rewrite the equation in the form

$$\frac{dy}{dt} - \frac{y}{\sqrt{t^3-3}} = t$$

and note that the integrating factor is

$$\mu(t) = e^{-\int \frac{1}{\sqrt{t^3-3}}\,dt}.$$

This integral is impossible to express in terms of elementary functions. Multiplying both sides by $\mu(t)$, we obtain

$$\left(e^{-\int \frac{1}{\sqrt{t^3-3}}\,dt}\right)\frac{dy}{dt} - \left(e^{-\int \frac{1}{\sqrt{t^3-3}}\,dt}\right)\frac{y}{\sqrt{t^3-3}} = t\left(e^{-\int \frac{1}{\sqrt{t^3-3}}\,dt}\right).$$

Applying the Product Rule to the left-hand side, we see that this equation is the same as

$$\frac{d\left(\left(e^{-\int \frac{1}{\sqrt{t^3-3}}\,dt}\right)y\right)}{dt} = t\left(e^{-\int \frac{1}{\sqrt{t^3-3}}\,dt}\right),$$

and integrating both sides with respect to t, we obtain

$$\left(e^{-\int \frac{1}{\sqrt{t^3-3}}\,dt}\right) y = \int t \left(e^{-\int \frac{1}{\sqrt{t^3-3}}\,dt}\right) dt.$$

These integrals are also impossible to express in terms of elementary functions, so we write the general solution in the form

$$y(t) = \left(e^{\int \frac{1}{\sqrt{t^3-3}}\,dt}\right) \int t \left(e^{-\int \frac{1}{\sqrt{t^3-3}}\,dt}\right) dt.$$

19. We rewrite the equation in the form

$$\frac{dy}{dt} - aty = 4e^{-t^2}$$

and note that the integrating factor is

$$\mu(t) = e^{\int (-at)\,dt} = e^{-at^2/2}.$$

Multiplying both sides by $\mu(t)$, we obtain

$$e^{-at^2/2}\frac{dy}{dt} - ate^{-at^2/2}y = 4e^{-t^2}e^{-at^2/2}.$$

Applying the Product Rule to the left-hand side and simplifying the right-hand side, we see that this equation is the same as

$$\frac{d(e^{-at^2/2}y)}{dt} = 4e^{-(1+a/2)t^2}.$$

Integrating both sides with respect to t, we obtain

$$e^{-at^2/2}y = \int 4e^{-(1+a/2)t^2}\,dt.$$

The integral on the right-hand side can be expressed in terms of elementary functions only if $1 + a/2 = 0$ (that is, if the factor involving e^{t^2} really isn't there). Hence, the only value of a that yields an integral we can express in terms of elementary functions form is $a = -2$ (see Exercise 4).

20. We rewrite the equation in the form

$$\frac{dy}{dt} - t^r y = 4$$

and note that the integrating factor is

$$\mu(t) = e^{-\int t^r\,dt}.$$

There are two cases to consider.

(a) If $r \neq -1$, then

$$\mu(t) = e^{-t^{r+1}/(r+1)}.$$

Multiplying both sides of the differential equation by $\mu(t)$, we obtain

$$\left(e^{-t^{r+1}/(r+1)}\right)\frac{dy}{dt} - t^r\left(e^{-t^{r+1}/(r+1)}\right)y = 4\left(e^{-t^{r+1}/(r+1)}\right).$$

Applying the Product Rule to the left-hand side, we see that this equation is the same as

$$\frac{d\left(\left(e^{-t^{r+1}/(r+1)}\right)y\right)}{dt} = 4\left(e^{-t^{r+1}/(r+1)}\right).$$

The next step is to integrate both sides with respect to t. The integral

$$\int 4\left(e^{-t^{r+1}/(r+1)}\right)\,dt$$

on the right-hand side can only be expressed in terms of elementary functions if $r = 0$.

(b) If $r = -1$, then the integrating factor is

$$\mu(t) = e^{-\int t^{-1}\,dt} = e^{-\ln t} = \frac{1}{t}.$$

Multiplying both sides by $\mu(t)$ yields the equation

$$\frac{d\left(t^{-1}y\right)}{dt} = \frac{4}{t},$$

and since $\int (4/t)\,dt = 4\ln t$, we can express the solution without integrals in this case.

Hence, the values of r that give solutions in terms of elementary functions are $r = 0$ and $r = -1$.

21. **(a)** The integrating factor is

$$\mu(t) = e^{0.4t}.$$

Multiplying both sides of the differential equation by $\mu(t)$ and collecting terms, we obtain

$$\frac{d(e^{0.4t}v)}{dt} = 3e^{0.4t}\cos 2t.$$

Integrating both sides with respect to t yields

$$e^{0.4t}v = \int 3e^{0.4t}\cos 2t\,dt.$$

To calculate the integral on the right-hand side, we must integrate by parts twice.

For the first integration, we pick $u_1(t) = \cos 2t$ and $v_1(t) = e^{0.4t}$. Using the fact that $0.4 = 2/5$, we get

$$\int e^{0.4t}\cos 2t\,dt = \tfrac{5}{2}e^{0.4t}\cos 2t + 5\int e^{0.4t}\sin 2t\,dt.$$

For the second integration, we pick $u_2(t) = \sin 2t$ and $v_2(t) = e^{0.4t}$. We get

$$\int e^{0.4t}\sin 2t\,dt = \tfrac{5}{2}e^{0.4t}\sin 2t - 5\int e^{0.4t}\cos 2t\,dt.$$

Combining these results yields

$$\int e^{0.4t}\cos 2t\,dt = \tfrac{5}{2}e^{0.4t}\cos 2t + \tfrac{25}{2}e^{0.4t}\sin 2t - 25\int e^{0.4t}\cos 2t\,dt.$$

Solving for $\int e^{0.4t}\cos 2t\,dt$, we have

$$\int e^{0.4t}\cos 2t\,dt = \frac{5\,e^{0.4t}\cos 2t + 25\,e^{0.4t}\sin 2t}{52}.$$

To obtain the general solution, we multiply this integral by 3, add the constant of integration, and solve for v. We obtain the general solution

$$v(t) = ke^{-0.4t} + \tfrac{15}{52}\cos 2t + \tfrac{75}{52}\sin 2t.$$

(b) The solution of the associated homogeneous equation is

$$v_h(t) = e^{-0.4t}.$$

We guess

$$v_p(t) = \alpha\cos 2t + \beta\sin 2t$$

for the a solution to the nonhomogeneous equation and solve for α and β. Substituting this guess into the differential equation, we obtain

$$-2\alpha\sin 2t + 2\beta\cos 2t + 0.4\alpha\cos 2t + 0.4\beta\sin 2t = 3\cos 2t.$$

Collecting sine and cosine terms, we get the system of equations

$$\begin{cases} -2\alpha + 0.4\beta = 0 \\ 0.4\alpha + 2\beta = 3. \end{cases}$$

Using the fact that $0.4 = 2/5$, we solve this system of equations and obtain

$$\alpha = \tfrac{15}{52} \quad\text{and}\quad \beta = \tfrac{75}{52}.$$

The general solution of the original nonhomogeneous equation is

$$v(t) = ke^{-0.4t} + \tfrac{15}{52}\cos 2t + \tfrac{75}{52}\sin 2t.$$

Both methods require quite a bit of computation. If we use an integrating factor, we must do a complicated integral, and if we use the guessing technique, we have to be careful with our algebra.

22. **(a)** Note that

$$\frac{d\mu}{dt} = \mu(t)(-a(t))$$

by the Fundamental Theorem of Calculus. Therefore, if we rewrite the differential equation as

$$\frac{dy}{dt} - a(t)y = b(t)$$

and multiply the left-hand side of this equation by $\mu(t)$, the left-hand side becomes

$$\begin{aligned}\mu(t)\frac{dy}{dt} - \mu(t)a(t)y &= \mu(t)\frac{dy}{dt} + \frac{d\mu}{dt}y \\ &= \frac{d(\mu\,y)}{dt}.\end{aligned}$$

Consequently, the function $\mu(t)$ satisfies the requirements of an integrating factor.

(b) To see that $1/\mu(t)$ is a solution of the associated homogeneous equation, we calculate

$$\begin{aligned}\frac{d\left(\frac{1}{\mu(t)}\right)}{dt} &= \frac{-1}{\mu(t)^2}\frac{d\mu}{dt}\\ &= \frac{-1}{\mu(t)^2}\mu(t)(-a(t))\\ &= a(t)\frac{1}{\mu(t)}.\end{aligned}$$

Thus, $y(t) = 1/\mu(t)$ satisfies the equation $dy/dt = a(t)y$.

(c) To see that $y_p(t)$ is a solution to the nonhomogeneous equation, we compute

$$\begin{aligned}\frac{dy_p}{dt} &= \frac{d\left(\frac{1}{\mu(t)}\right)}{dt}\left(\int_0^t \mu(\tau)\,b(\tau)\,d\tau\right) + \frac{1}{\mu(t)}\mu(t)b(t)\\ &= a(t)\frac{1}{\mu(t)}\left(\int_0^t \mu(\tau)\,b(\tau)\,d\tau\right) + b(t)\\ &= a(t)y_p(t) + b(t).\end{aligned}$$

(d) Let k be an arbitrary constant. Since $k/\mu(t)$ is the general solution of the associated homogeneous equation and

$$\frac{1}{\mu(t)}\int_0^t \mu(\tau)\,b(\tau)\,d\tau$$

is a solution to the nonhomogeneous equation, the general solution of the nonhomogeneous equation is

$$\begin{aligned}y(t) &= \frac{k}{\mu(t)} + \frac{1}{\mu(t)}\int_0^t \mu(\tau)\,b(\tau)\,d\tau\\ &- \frac{1}{\mu(t)}\left(k + \int_0^t \mu(\tau)\,b(\tau)\,d\tau\right).\end{aligned}$$

(e) Since

$$\int \mu(t)\,b(t)\,dt = \int_0^t \mu(\tau)\,b(\tau)\,d\tau + k$$

by the Fundamental Theorem of Calculus, the two formulas agree.

(f) In this equation, $a(t) = -2t$ and $b(t) = 4e^{-t^2}$. Therefore,

$$\mu(t) = e^{\int_0^t 2\tau\,d\tau} = e^{t^2}.$$

Consequently, $1/\mu(t) = e^{-t^2}$. Note that,

$$\frac{d\left(\frac{1}{\mu(t)}\right)}{dt} = (-2t)e^{-t^2} = a(t)\frac{1}{\mu(t)}.$$

Also,

$$y_p(t) = e^{-t^2} \int_0^t e^{\tau^2} \left(4e^{-\tau^2}\right) d\tau = e^{-t^2} \int_0^t 4\,d\tau = 4te^{-t^2}.$$

It is easy to see that $4te^{-t^2}$ satisfies the nonhomogeneous equation.

Therefore, the general solution to the nonhomogeneous equation is

$$ke^{-t^2} + 4te^{-t^2},$$

which can also be written as $(4t + k)e^{-t^2}$. Finally, note that

$$\frac{1}{\mu(t)} \int \mu(t)\, b(t)\, dt = e^{-t^2} \int e^{t^2} \left(4e^{-t^2}\right) dt = (4t + k)e^{-t^2}.$$

23. The integrating factor is

$$\mu(t) = e^{\int 2\,dt} = e^{2t}.$$

Multiplying both sides by $\mu(t)$, we obtain

$$\begin{aligned} e^{2t}\frac{dy}{dt} + 2e^{2t}y &= 3e^{2t}e^{-2t} \\ &= 3. \end{aligned}$$

Applying the Product Rule to the left-hand side, we see that this equation is the same as

$$\frac{d(e^{2t}y)}{dt} = 3,$$

and integrating both sides with respect to t, we obtain

$$e^{2t}y = 3t + k,$$

where k is an arbitrary constant. The general solution is

$$y(t) = (3t + k)e^{-2t}.$$

We know that ke^{-2t} is the general solution of the associated homogeneous equation, so $y_p(t) = 3te^{-2t}$ is a particular solution of the nonhomogeneous equation. Note that the factor of t arose after we multiplied the right-hand side of the equation by the integrating factor and ended up with the constant 3. After integrating, the constant produces a factor of t.

24. Let $S(t)$ be the amount of salt (in pounds) in the tank at time t. Then noting the amounts of salt that enter and leave the tank per minute, we have

$$\frac{dS}{dt} = 2 - \frac{S}{V(t)},$$

where $V(t)$ is the volume of the tank at time t. We have $V(t) = 15 + t$ since the tank starts with 15 gallons and one gallon per minute more is pumped into the tank than leaves the tank. So

$$\frac{dS}{dt} = 2 - \frac{S}{15 + t}.$$

This equation is linear, and we can rewrite it as

$$\frac{dS}{dt} + \frac{S}{15+t} = 2.$$

The integrating factor is

$$\mu(t) = e^{\int 1/(15+t)\,dt} = e^{\ln(15+t)} = 15+t.$$

Multiplying both sides of the equation by $\mu(t)$, we obtain

$$(15+t)\frac{dS}{dt} + S = 2(15+t),$$

which via the Product Rule is equivalent to

$$\frac{d((15+t)S)}{dt} = 30 + 2t.$$

Integration and simplification yields

$$S(t) = \frac{t^2 + 30t + c}{15+t}.$$

Using the initial condition $S(0) = 6$, we have $c/15 = 6$, which implies that $c = 90$ and

$$S(t) = \frac{t^2 + 30t + 90}{15+t}.$$

The tank is full when $t = 15$, and the amount of salt at that time is $S(15) = 51/2$ pounds.

25. We will use the term "parts" as shorthand for the product of parts per billion of dioxin and the volume of water in the tank. Basically this product represents the total amount of dioxin in the tank. The tank initially contains 200 gallons at a concentration of 2 parts per billion, which results in 400 parts of dioxin.

Let $y(t)$ be the amount of dioxin in the tank at time t. Since water with 4 parts per billion of dioxin flows in at the rate of 5 gallons per minute, 20 parts of dioxin enter the tank each minute. Also, the volume of water in the tank at time t is $200 + 2t$, so the concentration of dioxin in the tank is $y/(200 + 2t)$. Since well-mixed water leaves the tank at the rate of 2 gallons per minute, the differential equation that represents the change in the amount of dioxin in the tank is

$$\frac{dy}{dt} = 20 - 2\left(\frac{y}{200+2t}\right),$$

which can be simplified and rewritten as

$$\frac{dy}{dt} + \left(\frac{1}{100+t}\right) y = 20.$$

The integrating factor is

$$\mu(t) = e^{\int (1/(100+t))\,dt} = e^{\ln(100+t)} = 100+t.$$

Multiplying both sides by $\mu(t)$, we obtain

$$(100+t)\frac{dy}{dt} + y = 20(100+t),$$

which is equivalent to

$$\frac{d((100+t)y)}{dt} = 20(100+t)$$

by the Product Rule. Integrating both sides with respect to t, we obtain

$$(100+t)y = 2000t + 10t^2 + c.$$

Since $y(0) = 400$, we see that $c = 40,000$. Therefore,

$$y(t) = \frac{10t^2 + 2000t + 40,000}{t + 100}.$$

The tank fills up at $t = 100$, and $y(100) = 1,700$. To express our answer in terms of concentration, we calculate $y(100)/400 = 4.25$ parts per billion.

26. Let $S(t)$ denote the amount of sugar in the tank at time t. Sugar is added to the tank at the rate of p pounds per minute. The amount of sugar that leaves the tank is the product of the concentration of the sugar in the water and the rate that the water leaves the tank. At time t, there are $100 - t$ gallons of sugar water in the tank, so the concentration of sugar is $S(t)/(100-t)$. Since sugar water leaves the tank at the rate of 1 gallon per minute, the differential equation for S is

$$\frac{dS}{dt} = p - \frac{S}{100-t}.$$

Since this equation is linear, we rewrite it as

$$\frac{dS}{dt} + \frac{S}{100-t} = p,$$

and the integrating factor is

$$\mu(t) = e^{\int (1/(100-t))\,dt} = e^{-\ln(100-t)} = \frac{1}{100-t}.$$

Multiplying both sides of the differential equation by $\mu(t)$ yields

$$\left(\frac{1}{100-t}\right)\frac{dS}{dt} + \frac{S}{(100-t)^2} = \frac{p}{100-t},$$

which is equivalent to

$$\frac{d}{dt}\left(\frac{S}{100-t}\right) = \frac{p}{100-t}$$

by the Product Rule. We integrate both sides and obtain

$$\frac{S}{100-t} = -p\ln(100-t) + c,$$

where c is some constant. Note that the left-hand side of this formula is the concentration of sugar in the tank at time t.

At $t = 0$, the concentration of sugar is 0.25 pounds per gallon, so we can determine c by evaluating at $t = 0$. We obtain

$$0.25 = -p\ln(100) + c,$$

so

$$\begin{aligned}\frac{S}{100-t} &= -p\ln(100-t) + 0.25 + p\ln(100)\\ &= 0.25 + p\ln\left(\frac{100}{100-t}\right).\end{aligned}$$

(a) To determine the value of p such that the concentration is 0.5 when there are 5 gallons left in the tank, we note that $t = 95$. We get

$$0.5 = 0.25 + p\ln 20,$$

so $p = 0.25/(\ln 20) \approx 0.08345$.

(b) We can rephrase the question: Can we find p such that

$$\lim_{t\to 100^-}\frac{S}{100-t} = 0.75?$$

Using the formula for the concentration $S/(100-t)$, we have

$$\lim_{t\to 100^-}\frac{S}{100-t} = 0.25 + p\lim_{t\to 100^-}\ln\left(\frac{100}{100-t}\right).$$

As $t \to 100^-$, $100 - t \to 0^+$, so

$$\lim_{t\to 100^-}\ln\left(\frac{100}{100-t}\right) = \infty.$$

If $p \neq 0$, then the concentration is unbounded as $t \to 100^-$. If $p = 0$, then the concentration is constant at 0.25. Hence it is impossible to choose p so that the "last" drop out of the bucket has a concentration of 0.75 pounds per gallon.

27. **(a)** Let $y(t)$ be the amount of salt in the tank at time t. Since the tank is being filled at a total rate of 1 gallon per minute, the volume at time t is $V_0 + t$ and the concentration of salt in the tank is

$$\frac{y}{V_0 + t}.$$

The amount of salt entering the tank is the product of 2 gallons per minute and 0.25 pounds of salt per minute. The amount of salt leaving the tank is the product of the concentration of salt in the tank and the rate that brine is leaving. In this case, the rate is 1 gallon per minute, so the amount of salt leaving the tank is $y/(V_0 + t)$. The differential equation for $y(t)$ is

$$\frac{dy}{dt} = \frac{1}{2} - \frac{y}{V_0 + t}.$$

Since the water is initially clean, the initial condition is $y(0) = 0$.

(b) If $V_0 = 0$, the differential equation above becomes

$$\frac{dy}{dt} = \frac{1}{2} - \frac{y}{t}.$$

Note that this differential equation is undefined at $t = 0$. Thus, we *cannot* apply the Existence and Uniqueness Theorem to guarantee a unique solution at time $t = 0$. However, we can still solve the equation using our standard techniques assuming that $t \neq 0$.

Rewriting the equation as

$$\frac{dy}{dt} + \frac{y}{t} = \frac{1}{2},$$

we see that the integrating factor is

$$\mu(t) = e^{\int (1/t)\,dt} = e^{\ln t} = t.$$

Multiplying both sides of the differential equation by $\mu(t)$, we obtain

$$t\frac{dy}{dt} + y = \frac{t}{2},$$

which is equivalent to

$$\frac{d(ty)}{dt} = \frac{t}{2}.$$

Integrating both sides with respect to t, we get

$$ty = \frac{t^2}{4} + c,$$

so that the general solution is

$$y(t) = \frac{t}{4} + \frac{c}{t}.$$

Since the above expression is undefined at $t = 0$, we cannot make use of the initial condition $y(0) = 0$ to find the desired solution.

However, if the tank is initially empty, the concentration of salt in the tank remains constant over time at 0.25 pounds of salt per gallon. Therefore, we reconsider the equation

$$\frac{y}{t} = \frac{1}{4} + \frac{c}{t^2}.$$

If $c = 0$, we have $y/t = 1/4$. Hence, $c = 0$ yields the solution $y(t) = t/4$ which *is* a valid model for this situation.

It is useful to note that, if $V_0 = 0$, then we do not really need a differential equation to model the amount of the salt in the tank as a function of time. Clearly the concentration is constant as a function of time, and therefore the amount of salt in the tank is the product of the concentration and the volume of brine in the tank.

REVIEW EXERCISES FOR CHAPTER 1

1. The simplest differential equation with $y(t) = 2t$ as a solution is $dy/dt = 2$. The initial condition $y(0) = 3$ specifies the desired solution.

2. By guessing or separating variables, we know that the general solution is $y(t) = y_0 e^{3t}$, where $y(0) = y_0$ is the initial condition.

3. There are no values of y for which dy/dt is zero for all t. Hence, there are no equilibrium solutions.

4. Since the question only asks for one solution, look for the simplest first. Note that $y(t) = 0$ for all t is an equilibrium solution. There are other equilibrium solutions as well.

5. The right-hand side is zero for all t only if $y = -1$. Consequently, the function $y(t) = -1$ for all t is the only equilibrium solution.

6. The equilibria occur at $y = \pm n\pi$ for $n = 0, 1, 2, \ldots$, and dy/dt is positive otherwise. So all of the arrows between the equilibrium points point up.

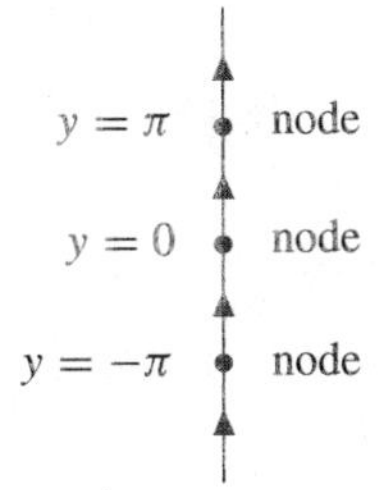

7. The equations $dy/dt = y$ and $dy/dt = 0$ are first-order, autonomous, separable, linear, and homogeneous.

8. The equation $dy/dt = y - 2$ is autonomous, linear, and nonhomogeneous. Moreover, if $y = 2$, then $dy/dt = 0$ for all t.

9. The graph of $f(y)$ must cross the y-axis from negative to positive at $y = 0$. For example, the graph of the function $f(y) = y$ produces this phase line.

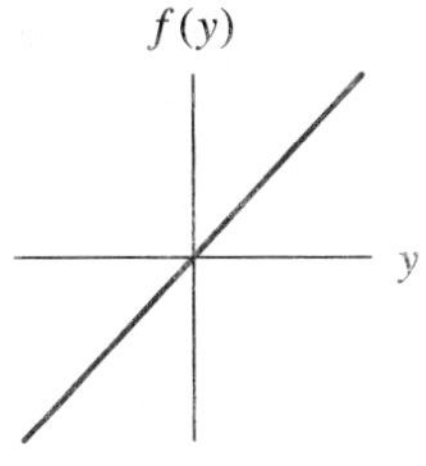

10. For $a > -4$, all solutions increase at a constant rate, and for $a < -4$, all solutions decrease at a constant rate. Consequently, a bifurcation occurs at $a = -4$, and all solutions are equilibria.

11. True. We have $dy/dt = e^{-t}$, which agrees with $|y(t)|$.

12. False. A separable equation has the form $dy/dt = g(t)h(y)$. So if $g(t)$ is not constant, then the equation is not separable. For example, $dy/dt = ty$ is separable but not autonomous.

13. True. Autonomous equations have the form $dy/dt = f(y)$. Therefore, we can separate variables by dividing by $f(y)$. That is,

$$\frac{1}{f(y)}\frac{dy}{dt} = 1.$$

14. False. For example, $dy/dt = y + t$ is linear but not separable.

15. False. For example, $dy/dt = ty^2$ is separable but not linear.

16. True. A homogeneous linear equation has the form $dy/dt = a(t)y$. We can separate variables by dividing by y. That is,

$$\frac{1}{y}\frac{dy}{dt} = a(t).$$

17. True. Note that the function $y(t) = 3$ for all t is an equilibrium solution for the equation. The Uniqueness Theorem says that graphs of different solutions cannot touch. Hence, a solution with $y(0) > 3$ must have $y(t) > 3$ for all t.

18. False. For example, $dy/dt = y$ has one source ($y = 0$) and no sinks.

19. False. By the Uniqueness Theorem, graphs of different solutions cannot touch. Hence, if one solution $y_1(t) \to \infty$ as t increases, any solution $y_2(t)$ with $y_2(0) > y_1(0)$ satisfies $y_2(t) > y_1(t)$ for all t. Therefore, $y_2(t) \to \infty$ as t increases.

20. False. The general solution of this differential equation has the form $y(t) = ke^t + \alpha e^{-t}$, where k is any constant and α is a particular constant (in fact, $\alpha = -1/2$). Choosing $k = 0$, we obtain a solution that tends to 0 as $t \to \infty$.

21. **(a)** The equation is autonomous, separable, and linear and nonhomogeneous.

(b) The general solution to the associated homogeneous equation is $y_h(t) = ke^{-2t}$. For a particular solution of the nonhomogeneous equation, we guess a solution of the form $y_p(t) = \alpha$. Then

$$\frac{dy_p}{dt} + 2y_p = 2\alpha.$$

Consequently, we must have $2\alpha = 3$ for $y_p(t)$ to be a solution. Hence, $\alpha = 3/2$, and the general solution to the nonhomogeneous equation is

$$y(t) = \tfrac{3}{2} + ke^{-2t}.$$

22. **(a)** This equation is separable. (It is nonlinear and nonautonomous as well.)

(b) We separate variables and integrate to obtain

$$\int \frac{1}{y^2}\,dy = \int t^2\,dt$$

$$-\frac{1}{y} = \frac{t^3}{3} + c$$

$$y(t) = \frac{-1}{(t^3/3) + c},$$

where c is any real number. This function can also be written in the form

$$y(t) = \frac{-3}{t^3 + k}$$

where k is any constant. The constant function $y(t) = 0$ for all t is also a solution of this equation. It is the equilibrium solution at $y = 0$.

23. **(a)** The equation is linear and nonhomogeneous. (It is nonautonomous as well.)

(b) The general solution of the associated homogeneous equation is $y_h(t) = ke^{3t}$. For a particular solution of the nonhomogeneous equation, we guess a solution of the form $y_p(t) = \alpha e^{7t}$. Then

$$\frac{dy_p}{dt} - 3y_p = 7\alpha e^{7t} - 3\alpha e^{7t} = 4\alpha e^{7t}.$$

Consequently, we must have $4\alpha = 1$ for $y_p(t)$ to be a solution. Hence, $\alpha = 1/4$, and the general solution to the nonhomogeneous equation is

$$y(t) = ke^{3t} + \tfrac{1}{4}e^{7t}.$$

24. **(a)** This equation is linear and homogeneous as well as separable.

(b) The Linearity Principle implies that

$$\begin{aligned} y(t) &= ke^{\int t/(1+t^2)\,dt} \\ &= ke^{\frac{1}{2}\ln(1+t^2)} \\ &= k\sqrt{1+t^2}, \end{aligned}$$

where k can be any real number (see page 115 in Section 1.8).

25. **(a)** This equation is separable.

(b) We separate variables and integrate to obtain

$$\begin{aligned} \int \frac{1}{y^2}\,dy &= \int (2t+3)\,dt \\ -\frac{1}{y} &= t^2 + 3t + k \\ y(t) &= \frac{-1}{t^2 + 3t + k}, \end{aligned}$$

where k is any constant. The constant function $y(t) = 0$ for all t is also a solution of this equation. It is the equilibrium solution at $y = 0$.

26. **(a)** This equation is linear and nonhomogeneous.

(b) We rewrite the equation in the form

$$\frac{dy}{dt} - \frac{2y}{1+t} = t$$

and note that the integrating factor is

$$\mu(t) = e^{\int -2/(1+t)\,dt} = e^{-2\ln(1+t)} = \frac{1}{(1+t)^2}.$$

Multiplying both sides of the differential equation by $\mu(t)$, we obtain

$$\frac{1}{(1+t)^2}\frac{dy}{dt} - \frac{2y}{(1+t)^3} = \frac{t}{(1+t)^2}.$$

Applying the Product Rule to the left-hand side, we see that this equation is the same as

$$\frac{d}{dt}\left(\frac{y}{(1+t)^2}\right) = \frac{t}{(1+t)^2}.$$

Integrating both sides with respect to t and using the substitution $u = 1 + t$ on the right-hand side, we obtain

$$\frac{y}{(1+t)^2} = \frac{1}{1+t} + \ln|1+t| + k,$$

where k can be any real number. The general solution is

$$y(t) = (1+t) + (1+t)^2 \ln|1+t| + k(1+t)^2.$$

27. **(a)** This equation is linear and nonhomogeneous.

(b) To find the general solution, we first note that $y_h(t) = ke^{-5t}$ is the general solution of the associated homogeneous equation.

To get a particular solution of the nonhomogeneous equation, we guess

$$y_p(t) = \alpha\cos 3t + \beta\sin 3t.$$

Substituting this guess into the nonhomogeneous equation gives

$$\begin{aligned}\frac{dy_p}{dt} + 5y_p &= -3\alpha\sin 3t + 3\beta\cos 3t + 5\alpha\cos 3t + 5\beta\sin 3t \\ &= (5\alpha + 3\beta)\cos 3t + (5\beta - 3\alpha)\sin 3t.\end{aligned}$$

In order for $y_p(t)$ to be a solution, we must solve the simultaneous equations

$$\begin{cases} 5\alpha + 3\beta = 0 \\ 5\beta - 3\alpha = 1. \end{cases}$$

From these equations, we get $\alpha = -3/34$ and $\beta = 5/34$. Hence, the general solution is

$$y(t) = ke^{-5t} - \tfrac{3}{34}\cos 3t + \tfrac{5}{34}\sin 3t.$$

28. **(a)** This equation is linear and nonhomogeneous.

(b) First we note that the general solution of the associated homogeneous equation is ke^{-3t}.

Next we use the technique suggested in Exercise 19 of Section 1.8. We could find particular solutions of the two nonhomogeneous equations

$$\frac{dy}{dt} = -3y + e^{-2t} \qquad \text{and} \qquad \frac{dy}{dt} = -3y + 4$$

separately and add the results to obtain a particular solution for the original equation. However, these two steps can be combined by making a more complicated guess for the particular solution.

We guess $y_p(t) = ae^{-2t} + b$, and we have

$$\begin{aligned} \frac{dy_p}{dt} + 3y_p &= -2ae^{-2t} + 3ae^{2t} + 3b \\ &= ae^{-2t} + 3b. \end{aligned}$$

Hence, for $y_p(t)$ to be a solution we must have $a = 1$ and $b = 4/3$. Therefore, a particular solution is $y_p(t) = e^{-2t} + \frac{4}{3}$, and the general solution is

$$y(t) = ke^{-3t} + e^{-2t} + \tfrac{4}{3}.$$

29. **(a)** The equation is linear and nonhomogeneous. (It is nonautonomous as well.)

(b) The general solution to the associated homogeneous equation is $y_h(t) = ke^{-t}$. For a particular solution of the nonhomogeneous equation, we guess a solution of the form $y_p(t) = at^2 + bt + c$. Then we must have

$$\frac{dy_p}{dt} + y_p = t^2.$$

In other words, we must have $(2at + b) + (at^2 + bt + c) = t^2$ for $y_p(t)$ to be a solution. Rewriting the left-hand side, we get

$$at^2 + (2a + b)t + (b + c) = t^2,$$

so we must solve

$$\begin{cases} u = 1 \\ 2a + b = 0 \\ b + c = 0. \end{cases}$$

Hence, $a = 1$, $b = -2$, and $c = 2$, and our particular solution is

$$y_p(t) = t^2 - 2t + 2.$$

The general solution is $y(t) = ke^{-t} + t^2 - 2t + 2$.

30. **(a)** This equation is separable and autonomous.

(b) First, note that $y = 0$ and $y = 2$ are the equilibrium points. Assuming that $y \neq 0$ and $y \neq 2$, we separate variables to obtain

$$\int \frac{1}{2y - y^2}\, dy = \int dt.$$

To integrate the left-hand side, we use partial fractions. We write

$$\frac{1}{y(2-y)} = \frac{A}{y} + \frac{B}{2-y},$$

which gives $2A = 1$ and $-A + B = 0$. So $A = B = 1/2$, and

$$\int \frac{1}{2y - y^2}\,dy = \frac{1}{2}\left(\int \frac{1}{y} + \frac{1}{2-y}\,dy\right) = \frac{1}{2}\ln\left|\frac{y}{y-2}\right|.$$

After integrating, we have

$$\ln\left|\frac{y}{y-2}\right| = 2t + c$$

$$\left|\frac{y}{y-2}\right| = c_1 e^{2t},$$

where $c_1 = e^c$ is any positive constant. To remove the absolute value signs, we replace the positive constant c_1 by a constant k that can be any real number and get

$$\frac{y}{y-2} = ke^{2t}.$$

After solving for y, we obtain

$$y(t) = \frac{2ke^{2t}}{ke^{2t} - 1}.$$

Note that $k = 0$ corresponds to the equilibrium solution $y = 0$. However, no value of k yields the equilibrium solution $y = 2$.

31. **(a)** The equation is autonomous and separable.

(b) When we separate variables, we obtain

$$\int \frac{1}{3 + y^2}\,dy = \int dt.$$

Integrating, we get

$$\frac{1}{\sqrt{3}}\arctan\left(\frac{y}{\sqrt{3}}\right) = t + c,$$

and solving for $y(t)$ produces

$$y(t) = \sqrt{3}\tan\left(\sqrt{3}\,t + k\right).$$

32. **(a)** This equation is linear and nonhomogeneous.

(b) We rewrite the equation in the form

$$\frac{dy}{dt} - \frac{t^3 y}{1 + t^4} = 2,$$

and note that the integrating factor is

$$\mu(t) = e^{\int -t^3/(1+t^4)\,dt} = (1+t^4)^{-1/4}.$$

Multiplying both sides of the differential equation by $\mu(t)$, we obtain

$$\frac{d}{dt}\left((1+t^4)^{-1/4}y\right) = \frac{2}{\sqrt[4]{1+t^4}}.$$

Consequently,

$$\frac{y}{\sqrt[4]{1+t^4}} = 2\int \frac{1}{\sqrt[4]{1+t^4}}\,dt.$$

Since the integral on the right-hand side is impossible to express using elementary functions, we write the general solution as

$$y(t) = 2\sqrt[4]{1+t^4}\int \frac{1}{\sqrt[4]{1+t^4}}\,dt.$$

33. **(a)** The equation is separable, linear and homogeneous.

(b) We know that the general solution of this equation has the form

$$x(t) = ke^{\int -2t\,dt},$$

where k is an arbitrary constant. We get $x(t) = ke^{-t^2}$.

To satisfy the initial condition $x(0) = e$, we note that $x(0) = k$, so $k = e$. The solution of the initial-value problem is

$$x(t) = ee^{-t^2} = e^{1-t^2}.$$

34. **(a)** This equation is separable.

(b) We separate variables and integrate to obtain

$$\int \frac{1}{y^2}\,dy = \int (2t+3t^2)\,dt$$

$$-\frac{1}{y} = t^2 + t^3 + k$$

$$y = \frac{-1}{t^2+t^3+k}.$$

To find the solution of the initial-value problem, we evaluate the general solution at $t = 1$ and obtain

$$y(1) = \frac{-1}{2+k}.$$

Since the initial condition is $y(1) = -1$, we see that $k = -1$. The solution to the initial-value problem is

$$y(t) = \frac{1}{1-t^2-t^3}.$$

35. **(a)** This equation is linear and nonhomogeneous. (It is nonautonomous as well.)

(b) The general solution of the associated homogeneous equaion is $y_h(t) = ke^{2t}$. To find a particular solution of the nonhomogeneous equation, we guess $y_p(t) = \alpha \cos 4t + \beta \sin 4t$. Then

$$\begin{aligned}\frac{dy_p}{dt} - 2y_p &= -4\alpha \sin 4t + 4\beta \cos 4t - 2(\alpha \cos 4t + \beta \sin 4t) \\ &= (-2\alpha + 4\beta) \cos 4t + (-4\alpha - 2\beta) \sin 4t.\end{aligned}$$

Consequently, we must have

$$(-2\alpha + 4\beta) \cos 4t + (-4\alpha - 2\beta) \sin 4t = \cos 4t$$

for $y_p(t)$ to be a solution. We must solve

$$\begin{cases} -2\alpha + 4\beta = 1 \\ -4\alpha - 2\beta = 0. \end{cases}$$

Hence, $\alpha = -1/10$ and $\beta = 1/5$, and the general solution of the nonhomogeneous equation is

$$y(t) = ke^{2t} - \tfrac{1}{10} \cos 4t + \tfrac{1}{5} \sin 4t.$$

To find the solution of the given initial-value problem, we evaluate the general solution at $t = 0$ and obtain

$$y(0) = k - \tfrac{1}{10}.$$

Since the initial condition is $y(0) = 1$, we see that $k = 11/10$. The desired solution is

$$y(t) = \tfrac{11}{10}e^{2t} - \tfrac{1}{10} \cos 4t + \tfrac{1}{5} \sin 4t.$$

36. **(a)** This equation is linear and nonhomogeneous.

(b) We first find the general solution. The general solution of the associated homogeneous equation is $y_h(t) = ke^{3t}$. For a particular solution of the nonhomogeneous equation, we guess $y_p(t) = \alpha te^{3t}$ rather than αe^{3t} because αe^{3t} is a solution of the homogeneous equation. Then

$$\begin{aligned}\frac{dy_p}{dt} - 3y_p &= \alpha e^{3t} + 3\alpha te^{3t} - 3\alpha te^{3t} \\ &= \alpha e^{3t}.\end{aligned}$$

Consequently, we must have $\alpha = 2$ for $y_p(t)$ to be a solution. Hence, the general solution to the nonhomogeneous equation is

$$y(t) = ke^{3t} + 2te^{3t}.$$

Note that $y(0) = k$, so the solution to the initial-value problem is

$$y(t) = -e^{3t} + 2te^{3t} = (2t - 1)e^{3t}.$$

37. **(a)** The equation is separable because

$$\frac{dy}{dt} = (t^2 + 1)y^3.$$

(b) Separating variables and integrating, we have

$$\int y^{-3}\,dy = \int (t^2 + 1)\,dt$$

$$\frac{y^{-2}}{-2} = \frac{t^3}{3} + t + c$$

$$y^{-2} = -\frac{2}{3}t^3 - 2t + k.$$

Using the initial condition $y(0) = -1/2$, we get that $k = 4$. Therefore,

$$y^2 = \frac{1}{4 - 2t - \frac{2}{3}t^3}.$$

Taking the square root of both sides yields

$$y = \frac{\pm 1}{\sqrt{4 - 2t - \frac{2}{3}t^3}}.$$

In this case, we take the negative square root because $y(0) = -1/2$. The solution to the initial value problem is

$$y(t) = \frac{-1}{\sqrt{4 - 2t - \frac{2}{3}t^3}}.$$

38. **(a)** This equation is separable.

(b) We separate variables and integrate to obtain

$$\int (y + 1)^2\,dy = \int (t + 1)^2\,dt$$

$$\tfrac{1}{3}(y + 1)^3 = \tfrac{1}{3}(t + 1)^3 + k,$$

where k is a constant.

We could solve for $y(t)$ now, but it is much easier to find k first. Using the initial condition $y(0) = 0$, we see that $k = 0$. Hence, the solution of the initial-value problem satisfies the equality

$$\tfrac{1}{3}(y + 1)^3 = \tfrac{1}{3}(t + 1)^3,$$

and therefore, $y(t) = t$.

39. **(a)** This equation is linear and nonhomogeneous. (It is nonautonomous as well.)

(b) We rewrite the equation as

$$\frac{dy}{dt} - 2ty = 3te^{t^2}$$

and note that the integrating factor is

$$\mu(t) = e^{\int -2t\,dt} = e^{-t^2}.$$

Multiplying both sides by $\mu(t)$, we obtain

$$e^{-t^2}\frac{dy}{dt} - 2te^{-t^2}y = 3t.$$

Applying the Product Rule to the left-hand side, we see that this equation is the same as

$$\frac{d}{dt}\left(e^{-t^2}y\right) = 3t,$$

and integrating both sides with respect to t, we obtain $e^{-t^2}y = \frac{3}{2}t^2 + k$, where k is an arbitrary constant. The general solution is

$$y(t) = \left(\tfrac{3}{2}t^2 + k\right)e^{t^2}.$$

To find the solution that satisfies the initial condition $y(0) = 1$, we evaluate the general solution at $t = 0$ and obtain $k = 1$. The desired solution is

$$y(t) = \left(\tfrac{3}{2}t^2 + 1\right)e^{t^2}.$$

40. **(a)** This equation is autonomous and separable.

(b) Note that the equilibrium points are $y = \pm 1$. Since the initial condition is $y(0) = 1$, we know that the solution to the initial-value problem is the equilibrium solution $y(t) = 1$ for all t.

41. **(a)** **(b)**

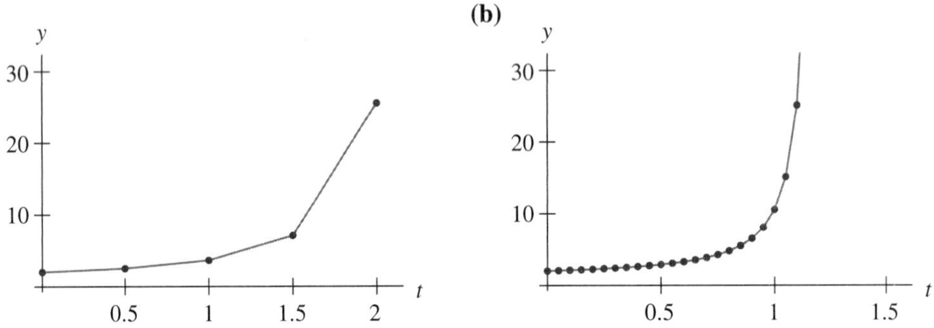

(c) Note that

$$\frac{dy}{dt} = (y-1)^2.$$

Separating variables and integrating, we get

$$\int \frac{1}{(y-1)^2}\,dy = \int 1\,dt$$

$$\frac{1}{1-y} = t + k.$$

From the intial condition, we see that $k = -1$, and we have

$$\frac{1}{1-y} = t - 1.$$

Solving for y yields

$$y(t) = \frac{t-2}{t-1},$$

which blows up as $t \to 1$ from below.

42. **(a)**

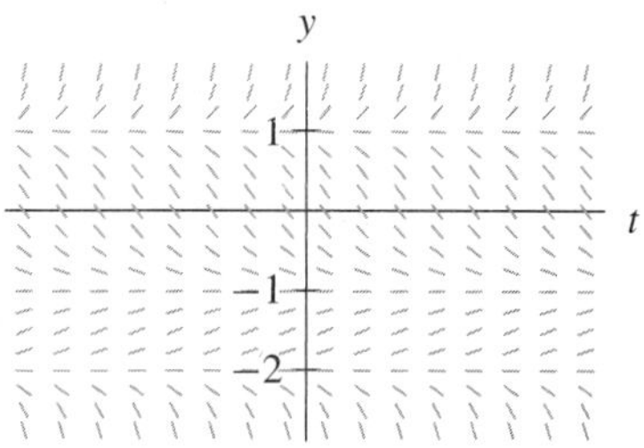

(b)

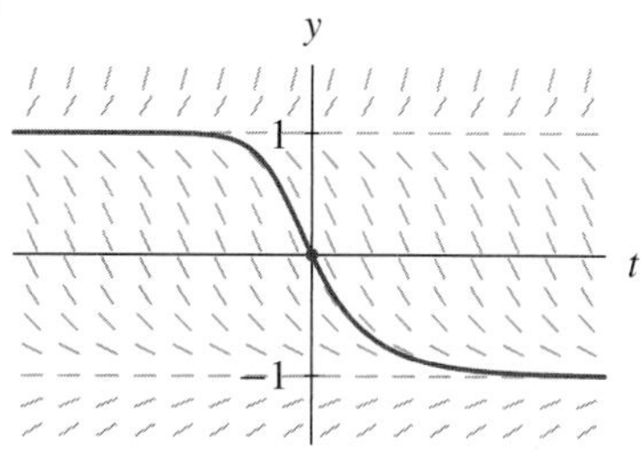

43. **(a)**

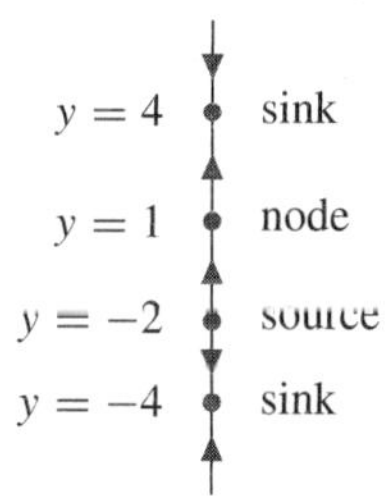

(b)

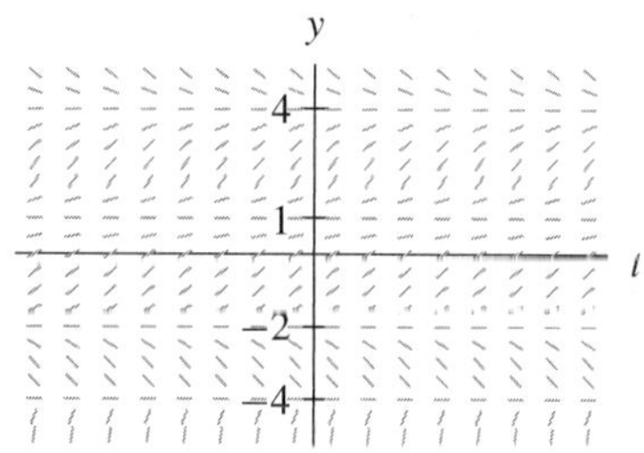

(c)

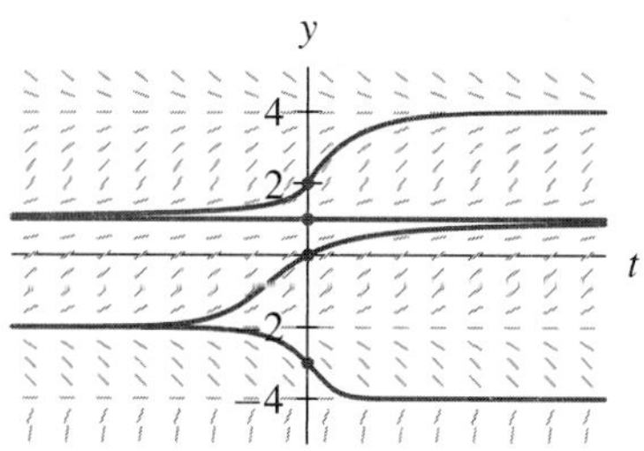

44. The constant function $y(t) = 2$ for all t is an equilibrium solution. If $y > 2$, then $dy/dt > 0$. Moreover, solutions with initial conditions above $y = 2$ satisfy $y(t) \to \infty$ as t increases and $y(t) \to 2$ as $t \to -\infty$.

If $y < -2$, then $dy/dt > 0$, so solutions with initial conditions below $y = -2$ increase until they cross the line $y = -2$. If $0 < y < 2$, then $dy/dt < 0$, and solutions in this strip decrease until they cross the t-axis.

For all initial conditions on the y-axis below $y = 2$, the solutions tend toward a periodic solution of period 2π as t increases. This periodic solution crosses the y-axis at $y_0 \approx -0.1471$. If $y(0) < y_0$, then the solution satisfies $y(t) \to -\infty$ as t decreases. If $y_0 < y(0) < 2$, the $y(t) \to 2$ as $t \to -\infty$.

45. From the equation, we can see that the functions $y_1(t) = 1$ for all t and $y_2(t) = 2$ for all t are equilibrium solutions. The Uniqueness Theorem tells us that solutions with initial conditions that satisfy $1 < y(0) < 2$ must also satisfy $1 < y(t) < 2$ for all t. An analysis of the sign of dy/dt within this strip indicates that $y(t) \to 2$ as $t \to \pm\infty$ if $1 < y(0) < 2$. All such solutions decrease until they intersect the curve $y = e^{t/2}$ and then they increase thereafter.

Solutions with $y(0)$ slightly greater than 2 increase until they intersect the curve $y = e^{t/2}$ and then they decrease and approach $y = 2$ as $t \to \infty$.

Solutions with $y(0)$ somewhat larger (approximately $y(0) > 2.1285$) increase quickly. It is difficult to determine if they eventually decrease, if they blow up in finite time, or if they increase for all time. In all cases where $y(0) > 2$, $y(t) \to 2$ as $t \to -\infty$.

Solutions with $y(0) < 1$ satisfy $y(t) \to -\infty$ as t increases, perhaps in finite time. As $t \to -\infty$, $y(t) \to 0$ for these solutions.

46. Note that

$$\frac{dy}{dt} = (1+t^2)y + 1 + t^2 = (1+t^2)(y+1).$$

(a) Separating variables and integrating, we obtain

$$\int \frac{1}{y+1}\,dy = \int (1+t^2)\,dt$$

$$\ln|y+1| = t + \frac{t^3}{3} + c,$$

where c is any constant. Thus, $|y+1| = c_1 e^{t+t^3/3}$, where $c_1 = e^c$. We can dispose of the absolute value signs by allowing the constant c_1 to be any real number. In other words,

$$y(t) = -1 + ke^{t+t^3/3},$$

where $k = \pm c_1$. Note that, if $k = 0$, we have the equilibrium solution $y(t) = -1$ for all t.

(b) The associated homogeneous equation is $dy/dt = (1+t^2)y$, and the Linearity Principle implies that

$$\begin{aligned} y(t) &= ke^{\int (1+t^2)\,dt} \\ &= ke^{t+t^3/3}. \end{aligned}$$

where k can be any real number (see page 115 in Section 1.8).

(c) When we write the differential equation as $dy/dt = (1+t^2)(y+1)$, we can immediately see that $y = -1$ corresponds to the equilibrium solution $y(t) = -1$ for all t.

(d) This equilibrium solution is a particular solution of the nonhomogeneous equation. Therefore, using the result of part (b), we get the general solution

$$y(t) = -1 + ke^{t+t^3/3}$$

of the nonhomogeneous equation using the Extended Linearity Principle. Note that this result agrees with the result of part (a).

47. **(a)** Note that there is an equilibrium solution of the form $y = -1/2$.
Separating variables and integrating, we obtain

$$\int \frac{1}{2y+1}\,dy = \int \frac{1}{t}\,dt$$

$$\tfrac{1}{2}\ln|2y+1| = \ln|t| + c$$

$$\ln|2y+1| = (\ln t^2) + c$$

$$|2y+1| = c_1 t^2,$$

where $c_1 = e^c$. We can eliminate the absolute value signs by allowing the constant to be either positive or negative. In other words, $2y + 1 = k_1 t^2$, where $k_1 = \pm c_1$. Hence

$$y(t) = kt^2 - \tfrac{1}{2},$$

where $k = k_1/2$.

(b) As t approaches zero all the solutions approach $-1/2$. In fact, $y(0) = -1/2$ for every value of k.

(c) This example does not violate the Uniqueness Theorem because the differential equation is not defined at $t = 0$. So functions $y(t)$ can only be said to be solutions for $t \neq 0$.

48. **(a)** Using Euler's method, we obtain the values $y_0 = 0$, $y_1 = 1.5$, $y_2 = 1.875$, $y_3 = 1.617$, and $y_4 = 1.810$ (rounded to three decimal places).

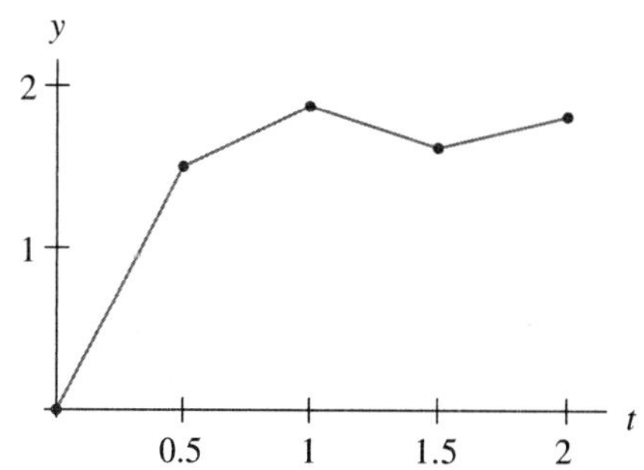

(b)

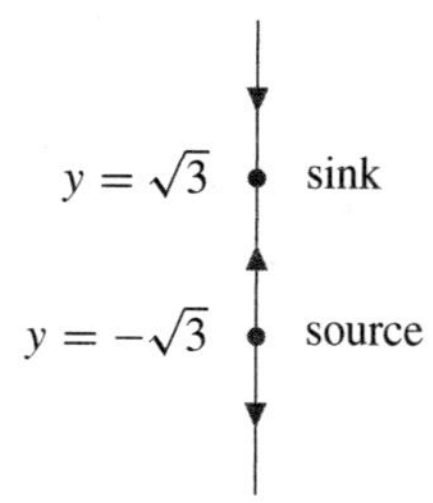

(c) The phase line tells us that the solution with initial condition $y(0) = 0$ must be increasing. Moreover, its graph is below and asymptotic to the line $y = \sqrt{3}$ as $t \to \infty$. The oscillations obtained using Euler's method come from numerical error.

49. **(a)** If we let k denote the proportionality constant in Newton's law of cooling, the initial-value problem satisfied by the temperature T of the soup is

$$\frac{dT}{dt} = k(T - 70), \quad T(0) = 150.$$

(b) We can solve the initial-value problem in part (a) using the fact that this equation is a nonhomogeneous linear equation. The function $T(t) = 70$ for all t is clearly an equilibrium solution to the equation. Therefore, the Extended Linearity Principle tells us that the general solution is

$$T(t) = 70 + ce^{kt},$$

where c is a constant determined by the initial condition. Since $T(0) = 150$, we have $c = 80$.

To determine k, we use the fact that $T(1) = 140$. We get

$$\begin{aligned} 140 &= 70 + 80e^k \\ 70 &= 80e^k \\ \tfrac{7}{8} &= e^k. \end{aligned}$$

We conclude that $k = \ln(7/8)$.

In order to find t so that the temperature is 100°, we solve

$$100 = 70 + 80e^{\ln(7/8)t}$$

for t. We get $\ln(3/8) = \ln(7/8)t$, which yields $t = \ln(3/8)/\ln(7/8) \approx 7.3$ minutes.

50. The equilibrium points occur at solutions of

$$\frac{dy}{dt} = y^6 - 2y^3 + \alpha = 0.$$

Using the quadratic formula to solve for y^3, we obtain

$$y^3 = \frac{2 \pm \sqrt{4 - 4\alpha}}{2}.$$

So the equilibrium points are at

$$y = \left(1 \pm \sqrt{1 - \alpha}\right)^{1/3}.$$

If $\alpha > 1$, there are no equilibrium points because this equation has no real solutions. If $\alpha < 1$, the differential equation has two equilibrium points. A bifurcation occurs at $\alpha = 1$ where the differential equation has one equilibrium point at $y = 1$.

51. We have

$$\frac{dy}{dt} = y^4 + \alpha y^2 = y^2(y^2 + \alpha).$$

If $\alpha > 0$, there is one equilibrium point at $y = 0$, and $dy/dt > 0$ otherwise. Hence, $y = 0$ is a node.

If $\alpha < 0$, there are equilibria at $y = 0$ and $y = \pm\sqrt{-\alpha}$. From the sign of $y^4 + \alpha y^2$, we know that $y = 0$ is a node, $y = -\sqrt{-\alpha}$ is a sink, and $y = \sqrt{-\alpha}$ is a source.

The bifurcation value of α is $\alpha = 0$. As α increases through 0, a sink and a source come together with the node at $y = 0$, leaving only the node. For $\alpha < 0$, there are three equilibria, and for $\alpha \geq 0$, there is only one equilibrium.

52. **(a)** Let t be time measured in years with $t = 0$ corresponding to the time of the first deposit, and let $M(t)$ be Beth's balance at time t. The 52 weekly deposits of \$10 are approximately the same as a continuous yearly rate of \$520. Therefore, the initial-value problem that models the growth in savings is

$$\frac{dM}{dt} = 0.03M + 520, \qquad M(0) = 400.$$

(b) The differential equation is both linear and separable, so we can solve the initial-value problem by separating variables, using an integrating factor, or using the Extended Linearity Principle. We use the Extended Linearity Principle.

The general solution of the associated homogeneous equation is $ke^{0.03t}$. We obtain one particular solution of the nonhomogeneous equation by determining its equilibrium solution. The equilibrium point is $M = -520/0.03 \approx -17,333$. Therefore, the general solution of the nonhomogeneous equation is

$$M(t) = ke^{0.03t} - 17,333.$$

Since $M(0) = 400$, we have $k = 17,733$, and after four years, Beth balance is $M(4) \approx 17,733e^{0.12} - 17,333 \approx \$2,661$.

53. **(a)**

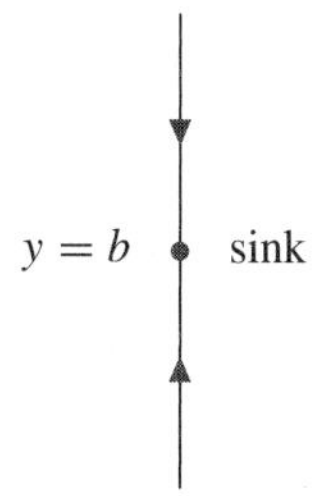

(b) As $t \to \infty$, $y(t) \to b$ for every solution $y(t)$.

(c) The equation is separable and linear. Hence, you can find the general solution by separating variables or by either of the methods for solving linear equations (undetermined coefficients or integrating factors).

(d) The associated homogeneous equation is $dy/dt = -(1/a)y$, and its general solution is $ke^{-t/a}$. One particular solution of the nonhomogeneous equation is the equilibrium solution $y(t) = b$ for all t. Therefore, the general solution of the nonhomogeneous equation is

$$y(t) = ke^{-t/a} + b.$$

(e) The authors love all the methods, just in different ways and for different reasons.

(f) Since $a > 0$, $e^{-t/a} \to 0$ as $t \to \infty$. Hence, $y(t) \to b$ as $t \to \infty$ independent of k.

54. **(a)** Let $p(t)$ be the number of cell phones in millions in the U.S. and t be time in years with 1990 corresponding to $t = 0$. An exponential model has the form $dp/dt = kp$ with $p(0) = 5.2$. If we use the 1990 and 1994 data to compute a value for k, we have

$$24.1 = 5.2e^{4k},$$

and therefore,

$$k = \frac{1}{4} \ln\left(\frac{24.1}{5.2}\right) \approx 0.383.$$

Similarly, using the 1994 and 1995 data to compute k, we get $k \approx 0.338$.

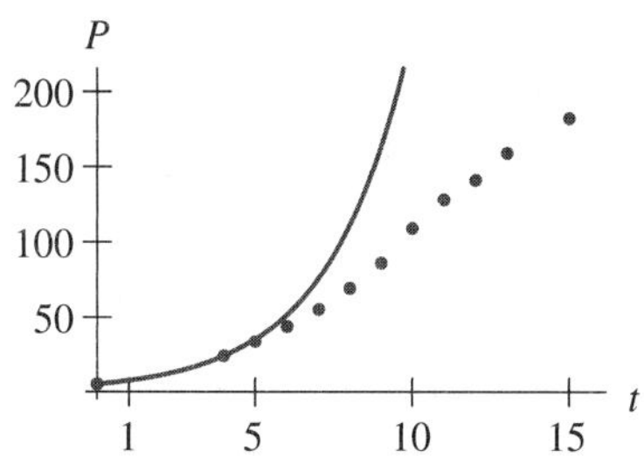

Exponential growth based on $p(0)$ and $p(4)$.

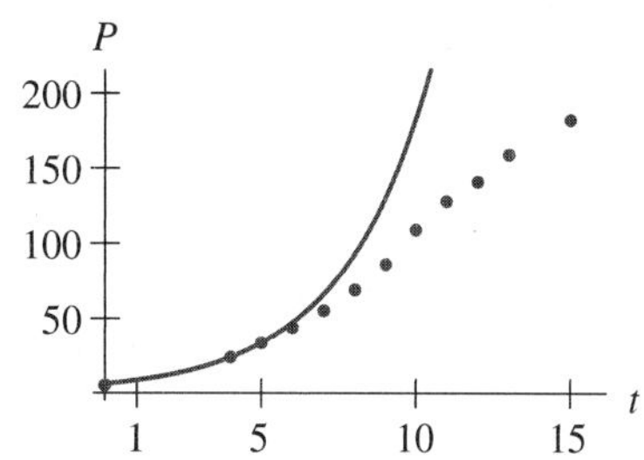

Exponential growth based on $p(4)$ and $p(5)$.

Both of these values of k give exponential growth models that grow much too fast by the year 2000 ($t = 10$).

(b) Using a logistic model

$$\frac{dp}{dt} = kp\left(1 - \frac{p}{N}\right)$$

we get a reasonably good fit to the data with $k = 0.383$

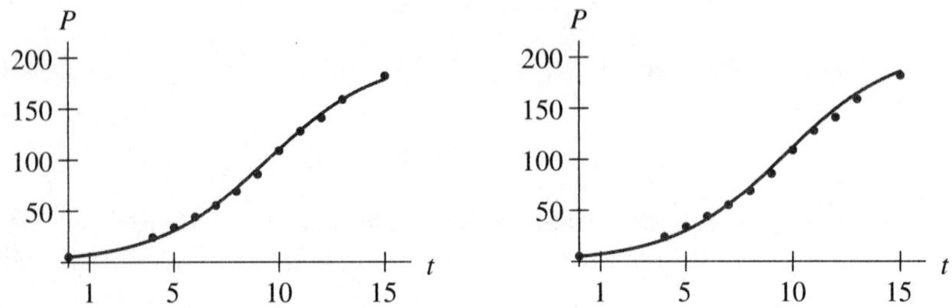

Solution to logistic with $N = 200$. Solution to logistic with $N = 210$.

(c) The exponential model is useless for predicting the value of p in the year 2005.

Using the logistic model with $k = 0.383$ and $N = 200$, we get a prediction of $p = 178.6$ in the year 2005. This prediction is too low by a significant amount. Increasing N to 210 gives a prediction of $p = 186.5$ for the year 2005.

(d) It appears that the number of cell phone subscriptions is approaching its carrying capacity somewhere between 200 and 210. While a gradual increase in the population will certainly have an effect on this capacity, it is likely that there will not be the rapid rate of increase in the number of new subscribers as was experienced between 1990 and 2000. As a result, we can expect the cell phone industry to become even more competitive than it already is. The result will either be lower prices to consumers or industry consolidation (fewer providers).

55. **(a)** The equation is separable. Separating variables and integrating, we obtain

$$\int y^{-2}\,dy = \int -2t\,dt$$
$$-y^{-1} = -t^2 + c,$$

where c is a constant of integration. Multiplying both sides by -1 and inverting yields

$$y(t) = \frac{1}{t^2 + k},$$

where k can be any constant. In addition, the equilibrium solution $y(t) = 0$ for all t is a solution.

(b) If $y(-1) = y_0$, we have

$$y_0 = y(-1) = \frac{1}{1 + k}$$

so

$$k = \frac{1}{y_0} - 1.$$

As long as $k > 0$, the denominator is positive for all t, and the solution is bounded for all t. Hence, for $0 \leq y_0 < 1$, the solution is bounded for all t. (Note that $y_0 = 0$ corresponds to the equilibrium solution.) All other solutions escape to $\pm\infty$ in finite time.

56. **(a)** Let $C(t)$ be the volume of carbon monoxide at time t where t is measured in hours. Initially, the amount of the carbon monoxide is 3% by volume. Since the volume of the room is 1000 cubic feet, there are 30 cubic feet of carbon monoxide in the room at time $t = 0$. Carbon monoxide is being blown into the room at the rate of one cubic foot per hour. The concentration of carbon monoxide is $C/1000$, so carbon monoxide leaves the room at the rate of

$$100\left(\frac{C}{1000}\right).$$

The initial-value problem that models this situation is

$$\frac{dC}{dt} = 1 - \frac{C}{10}, \qquad C(0) = 30.$$

(b) There is one equilibrium point, $C = 10$, and it is a sink. As t increases, $C(t)$ approaches 10, so the concentration approaches 1% carbon monoxide, the concentration of the air being blown into the room.

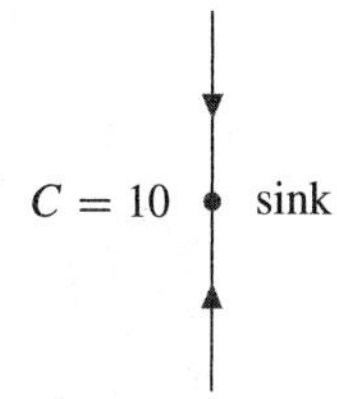

(c) The differential equation is linear. It is also autonomous and, therefore, separable. We can solve the initial-value problem by separating variables, using integrating factors, or by the Extended Linearity Principle. Since we already know one solution to the equation, that is, the equilibrium solution, we use the Extended Linearity Principle.

The associated homogeneous equation is $dC/dt = -C/10$, and its general solution is $ke^{-0.1t}$. Therefore, the general solution of the nonhomogeneous equation is

$$C(t) = 10 + ke^{-0.1t}.$$

Since $C(0) = 30$, $k = 20$.

(d) To find the value of t for which $C(t) = 20$, we solve

$$10 + 20e^{-t/10} = 20$$

We get

$$\begin{aligned} 20e^{-t/10} &= 10 \\ e^{-t/10} &= \tfrac{1}{2} \\ -t/10 &= \ln(\tfrac{1}{2}) \\ t &= 10 \ln 2. \end{aligned}$$

The air in the room is 2% carbon monoxide in approximately 6.93 hours.

57. Let $s(t)$ be the amount (measured in gallons) of cherry syrup in the vat at time t (measured in minutes). Then ds/dt is the difference between the rates at which syrup is added and syrup is withdrawn. Syrup is added at the rate of 2 gallons per minute. Syrup is withdrawn at the rate of

$$5\left(\frac{s}{500+5t}\right)$$

gallons per minute because the well mixed solution is withdrawn at the rate of 5 gallons per minute and the concentration of syrup is the total amount of syrup, s, divided by the total volume, $500+5t$. The differential equation is

$$\frac{ds}{dt} = 2 - \frac{s}{100+t}.$$

We solve this equation using integrating factors. Rewriting the equation as

$$\frac{ds}{dt} + \frac{s}{100+t} = 2,$$

we see that the integrating factor is

$$\mu(t) = e^{\int 1/(100+t)\,dt} = e^{\ln(100+t)} = 100+t.$$

Multiplying both sides of the differential equation by the integrating factor gives

$$(100+t)\frac{ds}{dt} + s = 2(100+t).$$

Using the Product Rule on the left-hand side, we observe that this equation can be rewritten as

$$\frac{d((100+t)s)}{dt} = 2t + 200,$$

and we integrate both sides to obtain

$$(100+t)s = t^2 + 200t + c,$$

where c is a constant that is determined by the initial condition $s(0) = 50$. Since

$$s(t) = \frac{t^2 + 200t + c}{t+100},$$

we see that $c = 5000$. Therefore, the solution of the initial-value problem is

$$s(t) = \frac{t^2 + 200t + 5000}{t+100}.$$

The vat is full when $500 + 5t = 1000$, that is, when $t = 100$ minutes. The amount of cherry syrup in the vat at that time is $s(100) = 175$ gallons, so the concentration is $175/1000 = 17.5\%$.

CHAPTER 2

First-Order Systems

ƆR SECTION 2.1

re it takes many predators to eat one prey, the constant in the negative effect term the prey is small. Therefore, (ii) corresponds the system of large prey and small tors. On the other hand, one predator eats many prey for the system of large predators and small prey, and, therefore, the coefficient of negative effect term on predator-prey interaction on the prey is large. Hence, (i) corresponds to the system of small prey and large predators.

2. For (i), the equilibrium points are $x = y = 0$ and $x = 10$, $y = 0$. For the latter equilibrium point prey alone exist; there are no predators. For (ii), the equilibrium points are $(0, 0)$, $(0, 15)$, and $(3/5, 30)$. For the latter equilibrium point, both species coexist. For $(0, 15)$, the prey are extinct but the predators survive.

3. Substitution of $y = 0$ into the equation for dy/dt yields $dy/dt = 0$ for all t. Therefore, $y(t)$ is constant, and since $y(0) = 0$, $y(t) = 0$ for all t.

Note that to verify this assertion rigorously, we need a uniqueness theorem (see Section 2.4).

4. For (i), the prey obey a logistic model. The population tends to the equilibrium point at $x = 10$. For (ii), the prey obey an exponential growth model, so the population grows unchecked.

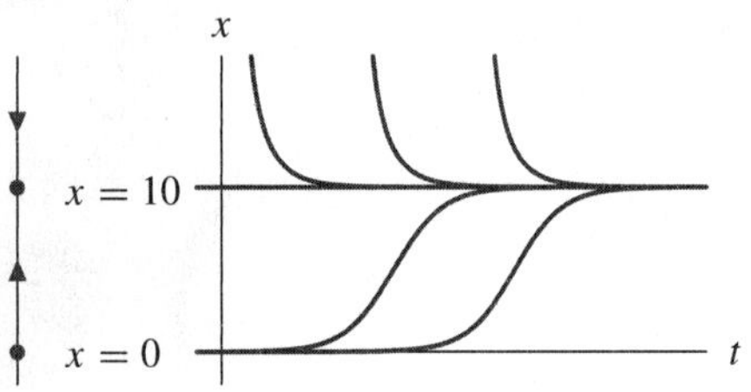

Phase line and graph for (i).

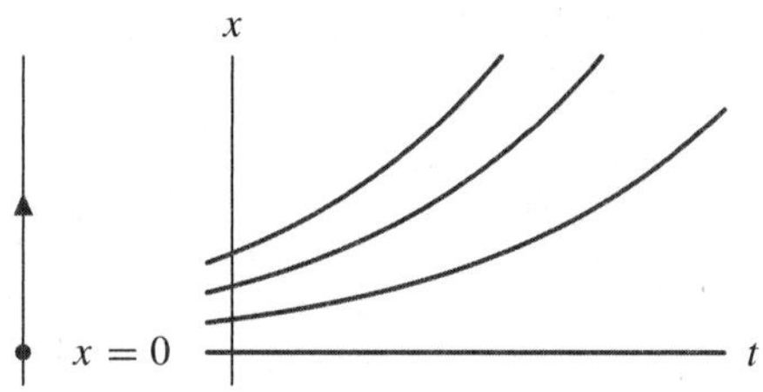

Phase line and graph for (ii).

5. Substitution of $x = 0$ into the equation for dx/dt yields $dx/dt = 0$ for all t. Therefore, $x(t)$ is constant, and since $x(0) = 0$, $x(t) = 0$ for all t.

Note that to verify this assertion rigorously, we need a uniqueness theorem (see Section 2.4).

6. For (i), the predators obey an exponential decay model, so the population tends to 0. For (ii), the predators obey a logistic model. The population tends to the equilibrium point at $y = 15$.

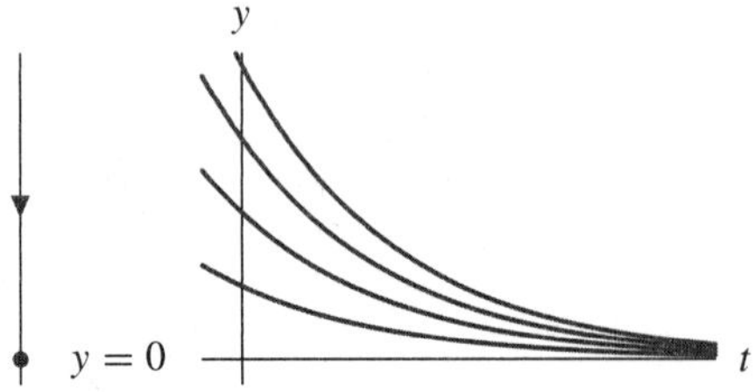

Phase line and graph for (i).

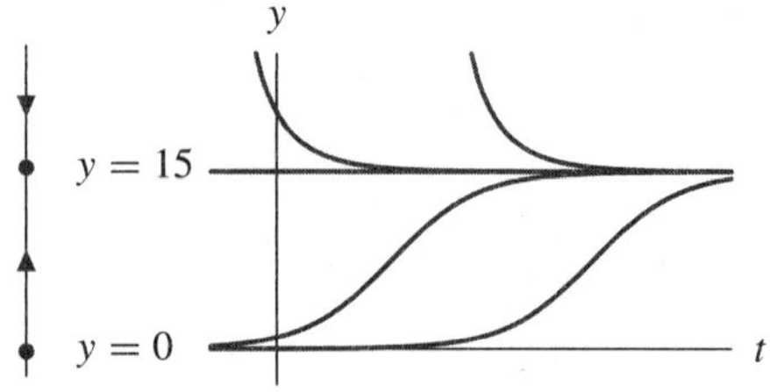

Phase line and graph for (ii).

7. The population starts with a relatively large rabbit (R) and a relatively small fox (F) population. The rabbit population grows, then the fox population grows while the rabbit population decreases. Next the fox population decreases until both populations are close to zero. Then the rabbit population grows again and the cycle starts over. Each repeat of the cycle is less dramatic (smaller total oscillation) and both populations oscillate toward an equilibrium which is approximately $(R, F) = (1/2, 3/2)$.

8. **(a)**

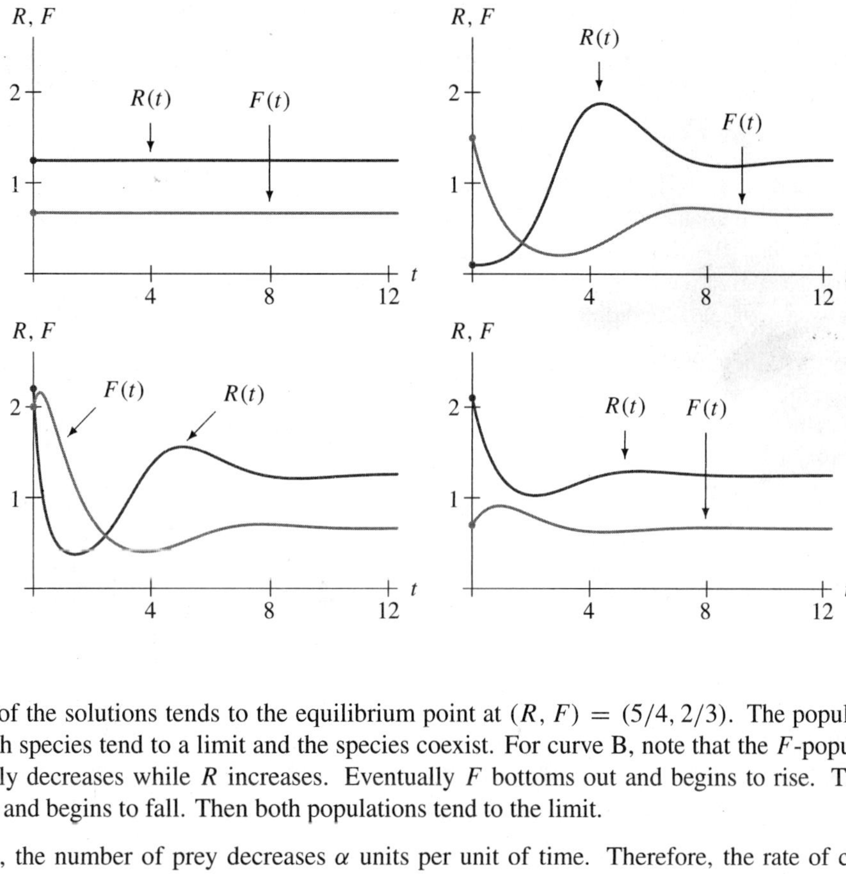

(b) Each of the solutions tends to the equilibrium point at $(R, F) = (5/4, 2/3)$. The populations of both species tend to a limit and the species coexist. For curve B, note that the F-population initially decreases while R increases. Eventually F bottoms out and begins to rise. Then R peaks and begins to fall. Then both populations tend to the limit.

9. By hunting, the number of prey decreases α units per unit of time. Therefore, the rate of change dR/dt of the number of prey has the term $-\alpha$. Only the equation for dR/dt needs modification.

(i) $dR/dt = 2R - 1.2RF - \alpha$

(ii) $dR/dt = R(2 - R) - 1.2RF - \alpha$

10. Hunting decreases the number of predators by an amount proportional to the number of predators alive (that is, by a term of the form $-kF$), so we have $dF/dt = -F + 0.9RF - kF$ in each case.

11. Since the second food source is unlimited, if $R = 0$ and k is the growth parameter for the predator population, F obeys an exponential growth model, $dF/dt = kF$. The only change we have to make is in the rate of F, dF/dt. For both (i) and (ii), $dF/dt = kF + 0.9RF$.

12. In the absence of prey, the predators would obey a logistic growth law. So we could modify both systems by adding a term of the form $-kF/N$, where k is the growth-rate parameter and N is the carrying capacity of predators. That is, we have $dF/dt = kF(1 - F/N) + 0.9RF$.

13. If $R - 5F > 0$, the number of predators increases and, if $R - 5F < 0$, the number of predators decreases. Since the condition on prey is same, we modify only the predator part of the system. the modified rate of change of the predator population is

$$\frac{dF}{dt} = -F + 0.9RF + k(R - 5F)$$

where $k > 0$ is the immigration parameter for the predator population.

14. In both cases the rate of change of population of prey decreases by a factor of kF. Hence we have

(i) $dR/dt = 2R - 1.2RF - kF$
(ii) $dR/dt = 2R - R^2 - 1.2RF - kF$

15. Suppose $y = 1$. If we can find a value of x such that $dy/dt = 0$, then for this x and $y = 1$ the predator population is constant. (This point may not be an equilibrium point because we do not know if $dx/dt = 0$.) The required value of x is $x = 0.05$ in system (i) and $x = 20$ in system (ii). Survival for one unit of predators requires 0.05 units of prey in (i) and 20 units of prey in (ii). Therefore, (i) is a system of inefficient predators and (ii) is a system of efficient predators.

16. At first, the number of rabbits decreases while the number of foxes increases. Then the foxes have too little food, so their numbers begin to decrease. Eventually there are so few foxes that the rabbits begin to multiply. Finally, the foxes become extinct and the rabbit population tends to the constant population $R = 3$.

17. **(a)** For the initial condition close to zero, the pest population increases much more rapidly than the predator. After a sufficient increase in the predator population, the pest population starts to decrease while the predator population keeps increasing. After a sufficient decrease in the pest population, the predator population starts to decrease. Then, the population comes back to the initial point.

(b) After applying the pest control, you may see the increase of the pest population due to the absence of the predator. So in the short run, this sort of pesticide can cause an explosion in the pest population.

18. One way to consider this type of predator-prey interaction is to raise the growth rate of the prey population. If only weak or sick prey are removed, the remaining population may be assumed to be able to reproduce at a higher rate.

19. **(a)** Substituting $y(t) = \sin t$ into the left-hand side of the differential equation gives

$$\frac{d^2y}{dt^2} + y = \frac{d^2(\sin t)}{dt^2} + \sin t$$
$$= -\sin t + \sin t$$
$$= 0,$$

so the left-hand side equals the right-hand side for all t.

(b)

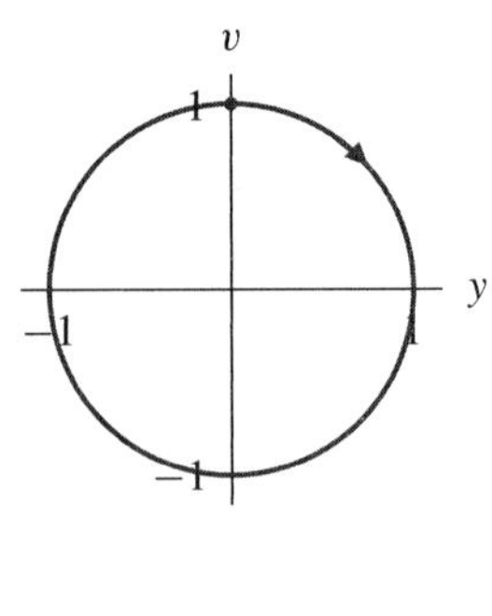

(c) These two solutions trace the same curve in the yv-plane—the unit circle.

(d) The difference in the two solution curves is in how they are parameterized. The solution in this problem is at (0, 1) at time $t = 0$ and hence it lags behind the solution in the section by $\pi/2$. This information cannot be observed solely by looking at the solution curve in the phase plane.

20. **(a)** If we substitute $y(t) = \cos\beta t$ into the left-hand side of the equation, we obtain

$$\begin{aligned}\frac{d^2y}{dt^2} + \frac{k}{m}y &= \frac{d^2(\cos\beta t)}{dt^2} + \frac{k}{m}\cos\beta t \\ &= -\beta^2\cos\beta t + \frac{k}{m}\cos\beta t \\ &= \left(\frac{k}{m} - \beta^2\right)\cos\beta t\end{aligned}$$

Hence, in order for $y(t) = \cos\beta t$ to be a solution we must have $k/m - \beta^2 = 0$. Thus,

$$\beta = \sqrt{\frac{k}{m}}.$$

(b) Substituting $t = 0$ into $y(t) = \cos\beta t$ and $v(t) = y'(t) = -\beta\sin\beta t$ we obtain the initial conditions $y(0) = 1$, $v(0) = 0$.

(c) The solution is $y(t) = \cos((\sqrt{k/m})t)$ and the period of this function is $2\pi/(\sqrt{k/m})$, which simplifies to $2\pi\sqrt{m}/\sqrt{k}$.

(d)

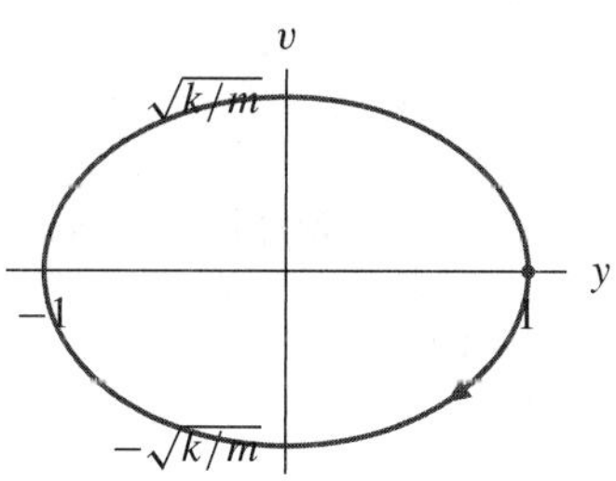

21. Hooke's law tells us that the restoring force exerted by a spring is linearly proportional to the spring's displacement from its rest position. In this case, the displacement is 3 in. while the restoring force is 12 lbs. Therefore, 12 lbs. $= k \cdot 3$ in. or $k = 4$ lbs. per in. $= 48$ lbs. per ft.

22. **(a)** First, we need to determine the spring constant k. Using Hooke's law, we have 4 lbs $= k \cdot 4$ in. Thus, $k = 1$ lbs/in $= 12$ lbs/ft. We will measure distance in feet since the mass is extended 1 foot.

To determine the mass of a 4 lb object, we use the fact that the force due to gravity is mg where $g = 32$ ft/sec^2. Thus, $m = 4/32 = 1/8$.

Using the model

$$\frac{d^2y}{dt^2} + \frac{k}{m}y = 0,$$

for the undamped harmonic oscillator, we obtain

$$\frac{d^2y}{dt^2} + 96y = 0, \quad y(0) = 1, \quad y'(0) = 0$$

as our initial-value problem.

(b) From Exercise 20 we know that $y(t) = \cos \beta t$ is a solution to the differential equation for the simple harmonic oscillator, where $\beta = \sqrt{k/m}$. Since $y(t) = \cos \sqrt{96}\, t$ satisfies both our differential equation and our initial conditions, it is the solution to the initial-value problem.

23. An extra firm mattress does not deform when you lay on it. This means that it takes a great deal of force to compress the springs so the spring constant must be large.

24. **(a)** Let m be the mass of the object, k be the spring constant, and d be the distance the spring is stretched when the mass is attached. Since the force mg stretches the spring a distance d, Hooke's law implies $mg = kd$. Thus, $d = mg/k$. Note that the position $y_1 = 0$ in the first system corresponds to the position $y_2 = -d$ in the second system.

For the first system, the force acting on the mass from the spring is $F_{s_1} = -ky_1$, while in the second system, the force is $F_{s_2} = -k(y_2 + d)$. The reason for the difference is that in the first system the force from the spring is zero when $y_1 = 0$ (the spring has yet to be stretched), while in the second system the force from the spring is zero when $y_2 = -d$. The force due to gravity in either system is mg.

Using Newton's second law of motion, the first system is

$$m\frac{d^2y_1}{dt^2} = -ky_1 + mg,$$

which can be rewritten as

$$\frac{d^2y_1}{dt^2} + \frac{k}{m}y_1 - g = 0.$$

For the second system, we have

$$m\frac{d^2y_2}{dt^2} = -k\left(y_2 + \frac{mg}{k}\right) + mg.$$

This equation can be written as

$$\frac{d^2y_2}{dt^2} + \frac{k}{m}y_2 = 0.$$

(b) Letting $dy_1/dt = v_1$, we have

$$\frac{dv_1}{dt} = \frac{d^2y_1}{dt^2} = -\frac{k}{m}y_1 + g,$$

and the system is

$$\begin{aligned}\frac{dy_1}{dt} &= v_1\\ \frac{dv_1}{dt} &= -\frac{k}{m}y_1 + g.\end{aligned}$$

Letting $dy_2/dt = v_2$, we have

$$\frac{dv_2}{dt} = \frac{d^2y_2}{dt^2} = -\frac{k}{m}y_2.$$

Therefore, the second system is

$$\begin{aligned}\frac{dy_2}{dt} &= v_2\\ \frac{dv_2}{dt} &= -\frac{k}{m}y_2.\end{aligned}$$

The first system has a unique equilibrium point at $(y_1, v_1) = (mg/k, 0)$ while the second has a unique equilibrium point at $(y_2, v_2) = (0, 0)$. The first system is at rest when $y_1 = d = mg/k$ and $v_1 = 0$. The second system is at rest when both $y_2 = 0$ and $v_2 = 0$. The second system is just the standard model of the simple harmonic oscillator while the first system is a translate of this model in the y-coordinate.

(c) Since the first system is just a translation in the y-coordinate of the second system, we can perform a simple change of variables to transform one to the other. (Note that $y_2 = y_1 - d$.) Thus, if $y_1(t)$ is a solution to the first system, then $y_2(t) = y_1(t) - d$ is a solution to the second system.

(d) The second system is easy to work with because it has fewer terms and is the more familiar simple harmonic oscillator.

25. Suppose $\alpha > 0$ is the reaction rate constant for A+B $\rightarrow$ C. The reaction rate is αab at time t, and after the reaction, a and b decrease by αab. We therefore obtain the system

$$\begin{aligned} \frac{da}{dt} &= -\alpha ab \\ \frac{db}{dt} &= -\alpha ab. \end{aligned}$$

26. Measure the amount of C produced during the short time interval from $t = 0$ to $t = \Delta t$. The amount is given by $a(0) - a(\Delta t)$ since one molecule of A yields one molecule of C. Now

$$\frac{a(0) - a(\Delta t)}{\Delta t} \approx -a'(0) = \alpha a(0)b(0).$$

Since we know $a(0)$, $a(\Delta t)$, $b(0)$, and Δt, we can therefore solve for α.

27. Suppose k_1 and k_2 are the rates of increase of A and B respectively. Since A and B are added to the solution at constant rates, k_1 and k_2 are added to da/dt and db/dt respectively. The system becomes

$$\begin{aligned} \frac{da}{dt} &= k_1 - \alpha ab \\ \frac{db}{dt} &= k_2 - \alpha ab. \end{aligned}$$

28. The chance that two A molecules are close is proportional to a^2. Hence, the new system is

$$\begin{aligned} \frac{da}{dt} &= k_1 - \alpha ab - \gamma a^2 \\ \frac{db}{dt} &= k_2 - \alpha ab, \end{aligned}$$

where γ is a parameter that measures the rate at which A combines to make D.

29. Suppose γ is the reaction-rate coefficient for the reaction B + B → A. By the reaction, two B's react with each other to create one A. In other words, B decreases at the rate γb^2 and A increases at the rate $\gamma b^2/2$. The resulting system of the differential equations is

$$\frac{da}{dt} = k_1 - \alpha ab + \frac{\gamma b^2}{2}$$
$$\frac{db}{dt} = k_2 - \alpha ab - \gamma b^2.$$

30. The chance that two B's and an A molecule are close is proportional to ab^2, so

$$\frac{da}{dt} = k_1 - \alpha ab - \gamma ab^2$$
$$\frac{db}{dt} = k_2 - \alpha ab - 2\gamma ab^2,$$

where γ is the reaction-rate parameter for the reaction that produces D from two B's and an A.

EXERCISES FOR SECTION 2.2

1. **(a)** $\mathbf{V}(x, y) = (1, 0)$

(b) See part (c).

(c)

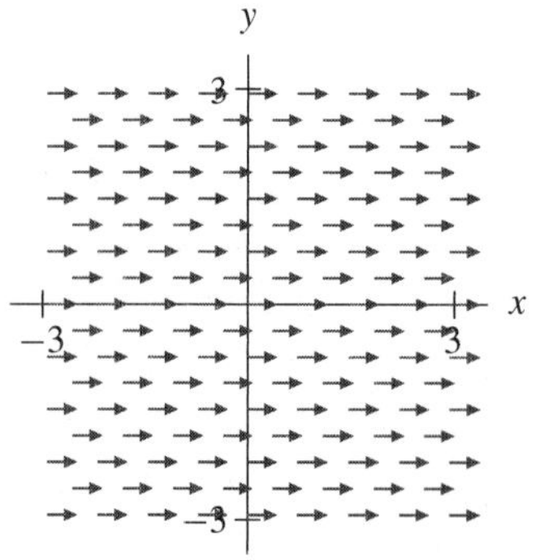

(d)

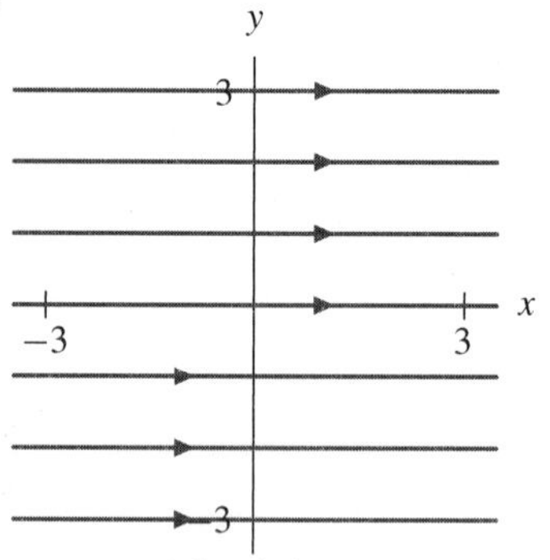

(e) As t increases, solutions move along horizontal lines toward the right.

2. **(a)** $\mathbf{V}(x, y) = (x, 1)$

(b) See part (c).

(c)

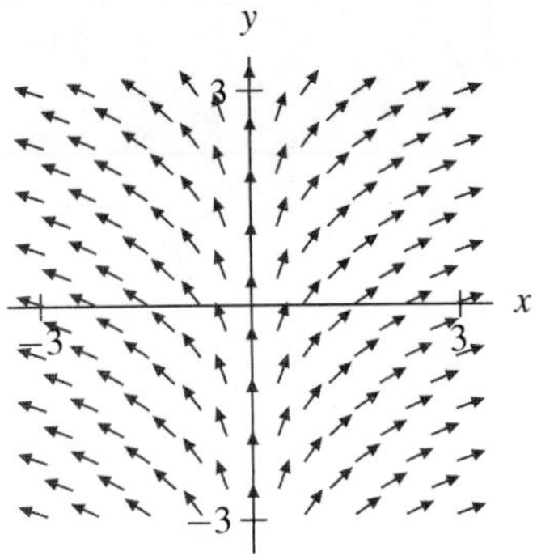

(d)

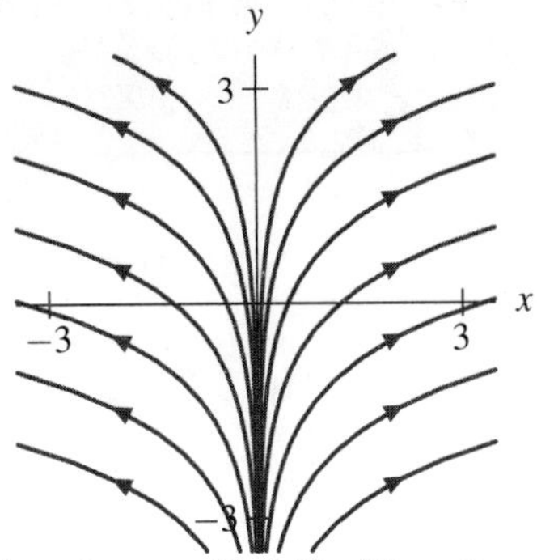

(e) As t increases, solutions move up and right if $x(0) > 0$, up and left if $x(0) < 0$.

3. **(a)** $\mathbf{V}(y, v) = (-v, y)$

(b) See part (c).

(c)

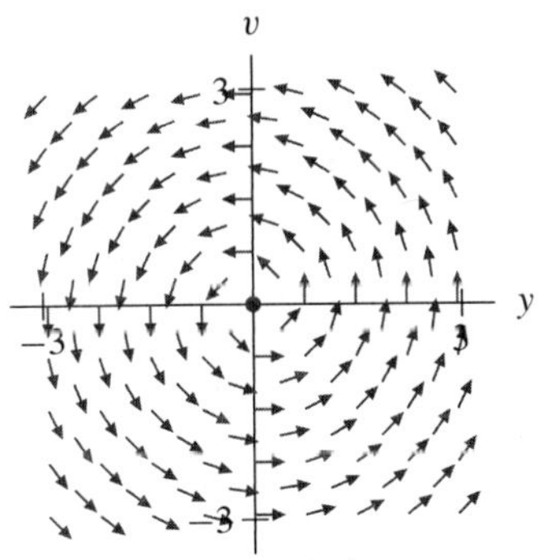

(d)

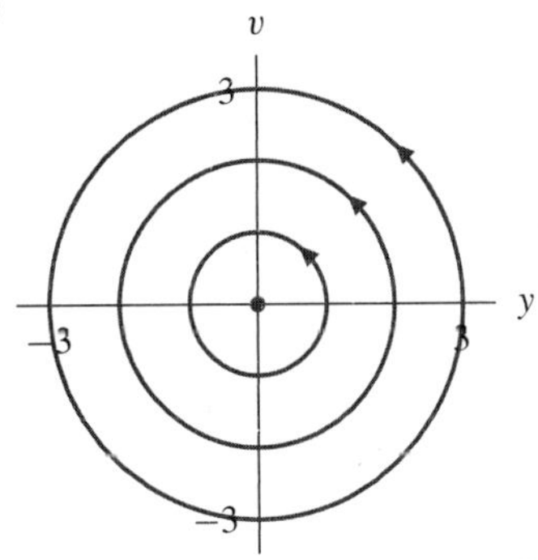

(e) As t increases, solutions move on circles around $(0, 0)$ in the counter-clockwise direction.

4. **(a)** $\mathbf{V}(u, v) = (u - 1, v - 1)$

(b) See part (c).

(c)

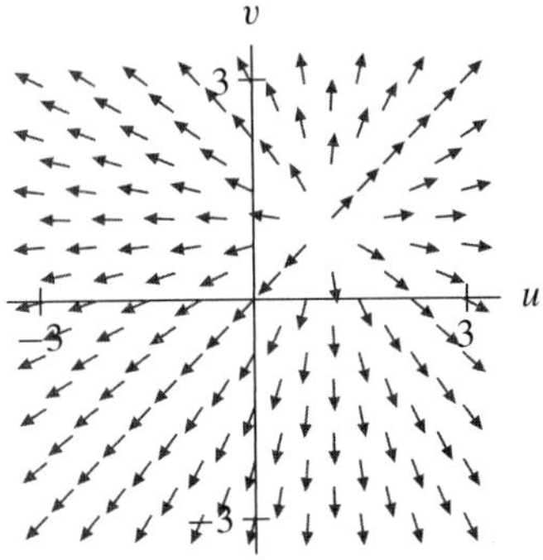

(d)

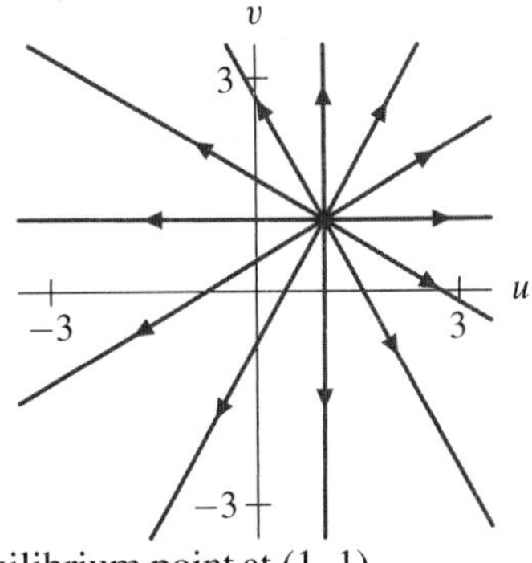

(e) As t increases, solutions move away from the equilibrium point at $(1, 1)$.

5. **(a)** $\mathbf{V}(x, y) = (x, -y)$

(b) See part (c).

(c)

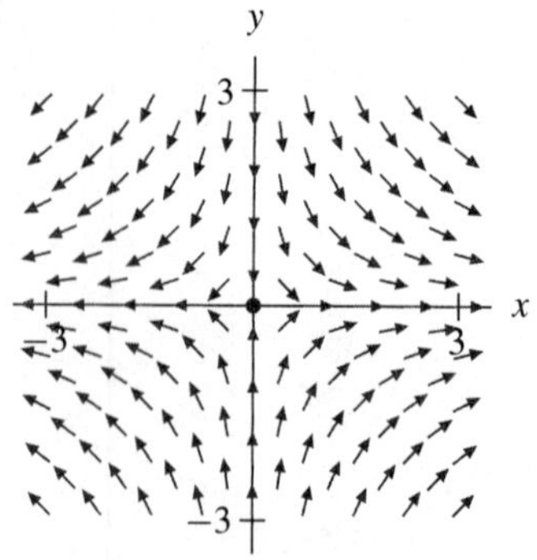

(d)

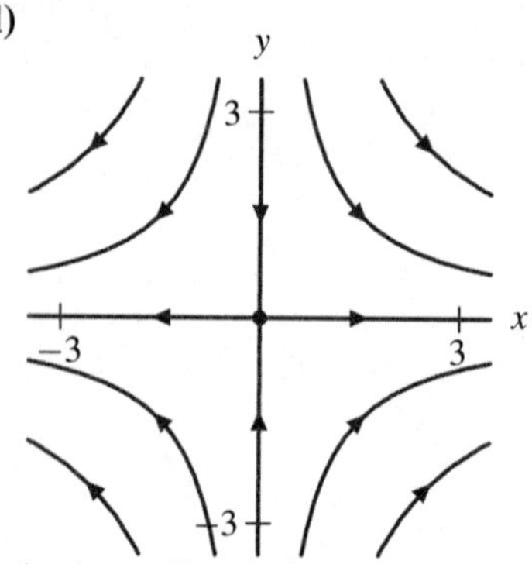

(e) As t increases, solutions move toward the x-axis in the y-direction and away from the y-axis in the x-direction.

6. **(a)** $\mathbf{V}(x, y) = (x, 2y)$

(b) See part (c).

(c)

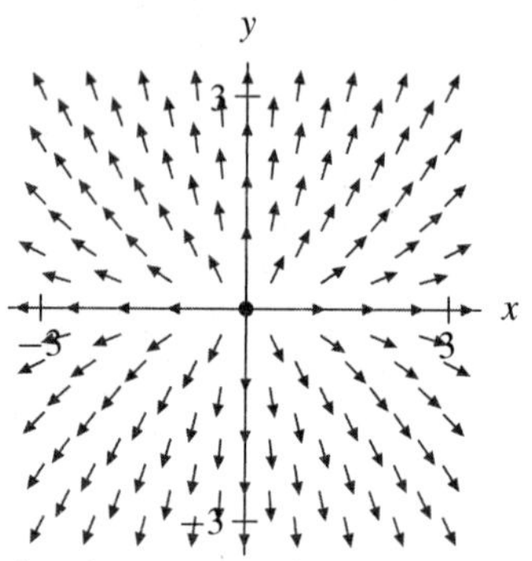

(d)

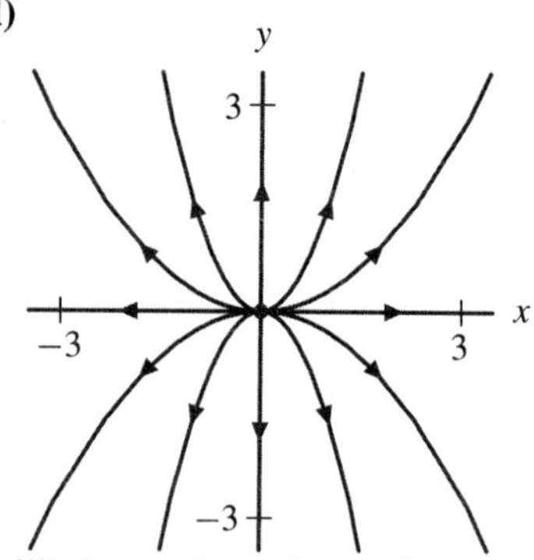

(e) As t increases, solutions move away from the equilibrium point at the origin.

7. **(a)** Let $v = dy/dt$. Then

$$\frac{dv}{dt} = \frac{d^2y}{dt^2} = y.$$

Thus the associated vector field is $\mathbf{V}(y, v) = (v, y)$.

(b) See part (c).

(c)

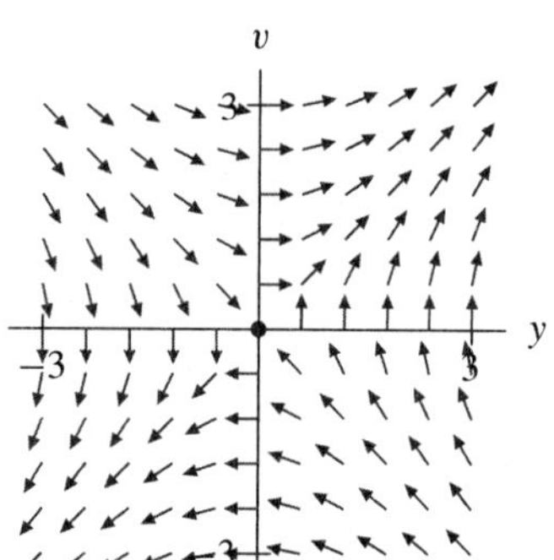

(d)

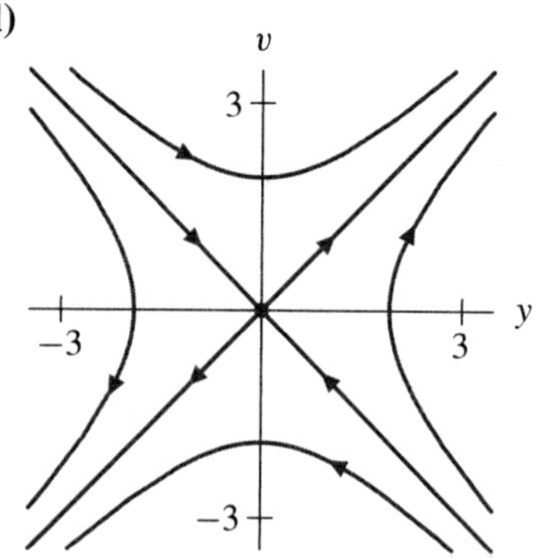

(e) As t increases, solutions in the 2nd and 4th quadrants move toward the origin and away from the line $y = -v$. Solutions in the 1st and 3rd quadrants move away from the origin and toward the line $y = v$.

8. **(a)** Let $v = dy/dt$. Then

$$\frac{dv}{dt} = \frac{d^2y}{dt^2} = -2y.$$

Thus the associated vector field is $\mathbf{V}(y, v) = (v, -2y)$.

(b) See part (c).

(c)

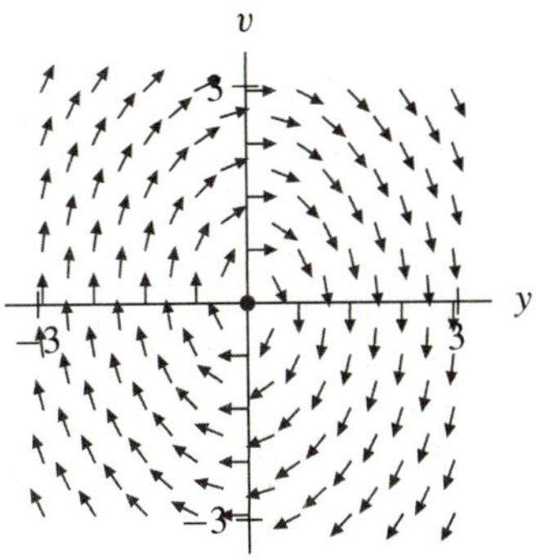

(d)

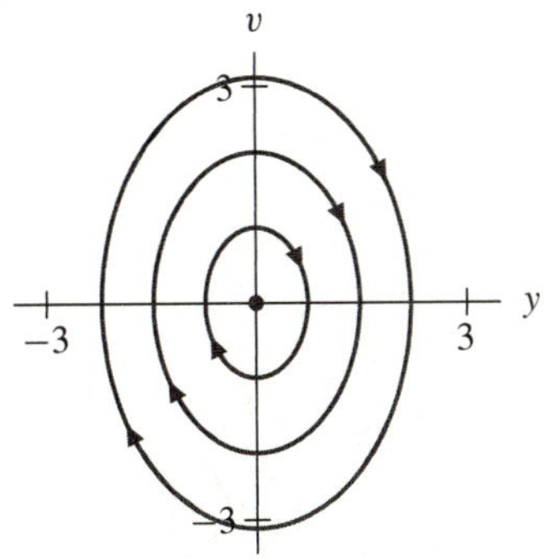

(e) As t increases, solutions move around the origin on ovals in the clockwise direction.

9. **(a)**

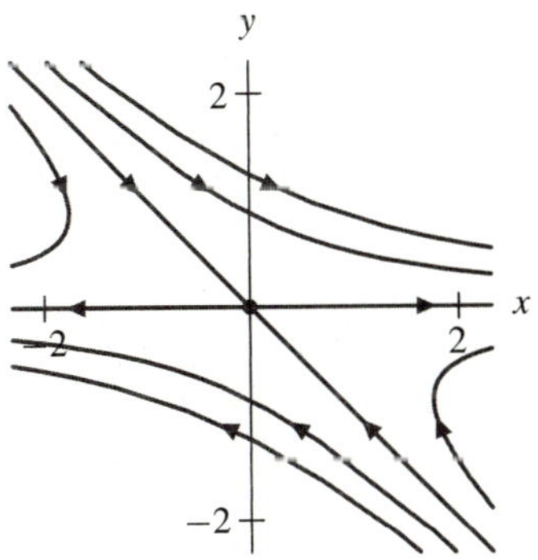

(b) The solution tends to the origin along the line $y = -x$ in the xy-phase plane. Therefore both $x(t)$ and $y(t)$ tend to zero as $t \to \infty$.

10. **(a)**

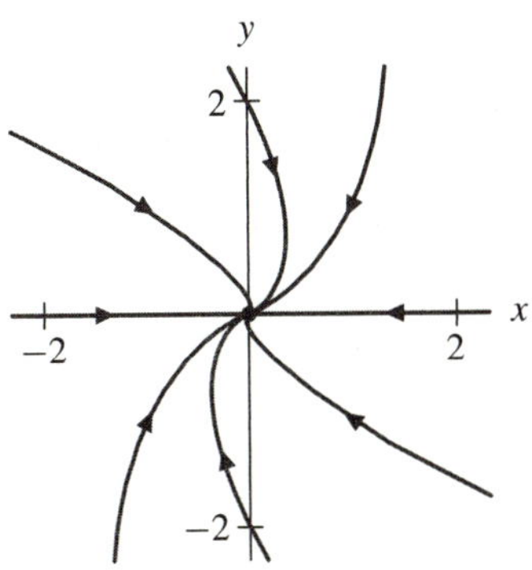

(b) The solution enters the first quadrant and tends to the origin tangent to the positive x-axis. Therefore $x(t)$ initially increases, reaches a maximum value, and then tends to zero as $t \to \infty$. It remains positive for all positive values of t. The function $y(t)$ decreases toward zero as $t \to \infty$.

11. **(a)** There are equilibrium points at $(\pm 1, 0)$, so only systems (ii) and (vii) are possible. Since the direction field points toward the x-axis if $y \neq 0$, the equation $dy/dt = y$ does not match this field. Therefore, system (vii) is the system that generated this direction field.

(b) The origin is the only equilibrium point, so the possible systems are (iii), (iv), (v), and (viii). The direction field is not tangent to the y-axis, so it does not match either system (iv) or (v). Vectors point toward the origin on the line $y = x$, so $dy/dt = dx/dt$ if $y = x$. This condition is not satisfied by system (iii). Consequently, this direction field corresponds to system (viii).

(c) The origin is the only equilibrium point, so the possible systems are (iii), (iv), (v), and (viii). Vectors point directly away from the origin on the y-axis, so this direction field does not correspond to systems (iii) and (viii). Along the line $y = x$, the vectors are more vertical than horizontal. Therefore, this direction field corresponds to system (v) rather than system (iv).

(d) The only equilibrium point is $(-1, 1)$, so the possible systems are (i) and (vi). For system (i), the vector at $(0, 0)$ is $(-1, -1)$, and for system (vi), it is $(1, 1)$. Consequently, this direction field corresponds to system (vi).

12. The equilibrium solutions are those solutions for which $dR/dt = 0$ and $dF/dt = 0$ simultaneously. To find the equilibrium points, we must solve the system of equations

$$\begin{cases} 2R\left(1 - \dfrac{R}{2}\right) - 1.2RF = 0 \\ -F + 0.9RF = 0. \end{cases}$$

The second equation is satisfied if $F = 0$ or if $R = 10/9$, and we consider each case independently. If $F = 0$, then the first equation is satisfied if and only if $R = 0$ or $R = 2$. Thus two equilibrium solutions are $(R, F) = (0, 0)$ and $(R, F) = (2, 0)$.

If $R = 10/9$, we substitute this value into the first equation and obtain $F = 20/27$.

13. **(a)** To find the equilibrium points, we solve the system of equations

$$\begin{cases} 4x - 7y + 2 = 0 \\ 3x + 6y - 1 = 0. \end{cases}$$

These simultaneous equations have one solution, $(x, y) = (-1/9, 2/9)$.

(b)

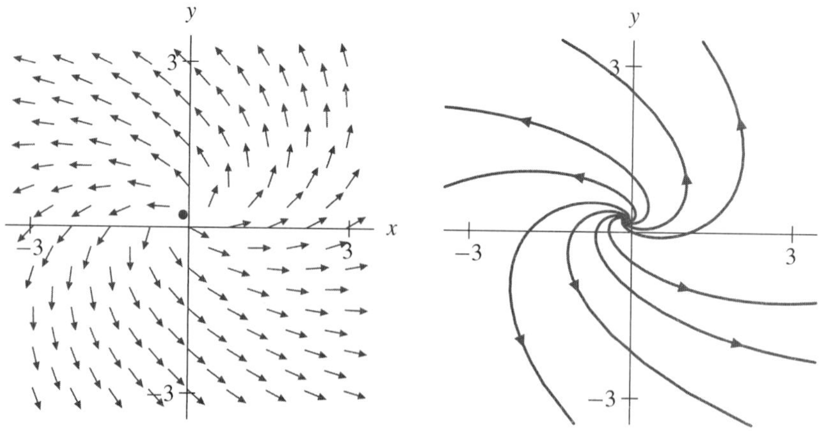

(c) As t increases, typical solutions spiral away from the origin in the counter-clockwise direction.

14. **(a)** To find the equilibrium points, we solve the system of equations

$$\begin{cases} 4R - 7F - 1 = 0 \\ 3R + 6F - 12 = 0. \end{cases}$$

These simultaneous equations have one solution, $(R, F) = (2, 1)$.

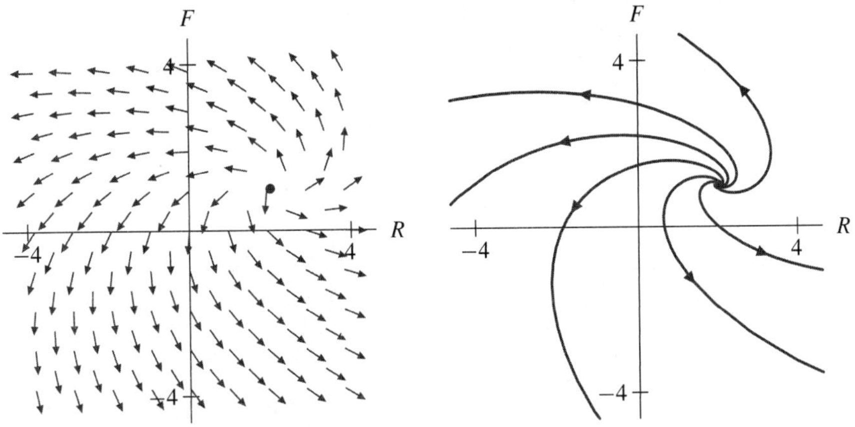

(b) As t increases, typical solutions spiral away from the equilibrium point at (2, 1)

15. **(a)** To find the equilibrium points, we solve the system of equations

$$\begin{cases} \cos w = 0 \\ -z + w = 0. \end{cases}$$

The first equation implies that $w = \pi/2 + k\pi$ where k is any integer, and the second equation implies that $z = w$. The equilibrium points are $(\pi/2 + k\pi, \pi/2 + k\pi)$ for any integer k.

(b)

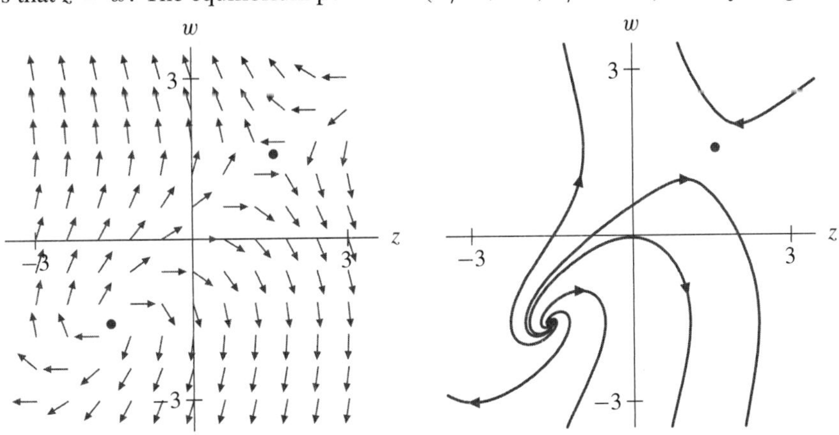

(c) As t increases, typical solutions move away from the line $z = w$, which contains the equilibrium points. The value of w is either increasing or decreasing without bound depending on the initial condition.

16. **(a)** To find the equilibrium points, we solve the system of equations

$$\begin{cases} x - x^3 - y = 0 \\ y = 0. \end{cases}$$

Since $y = 0$, we have $x^3 - x = 0$. If we factor $x - x^3$ into $x(x-1)(x+1)$, we see that there are three equilibrium points, $(0, 0)$, $(1, 0)$, and $(-1, 0)$.

(b)

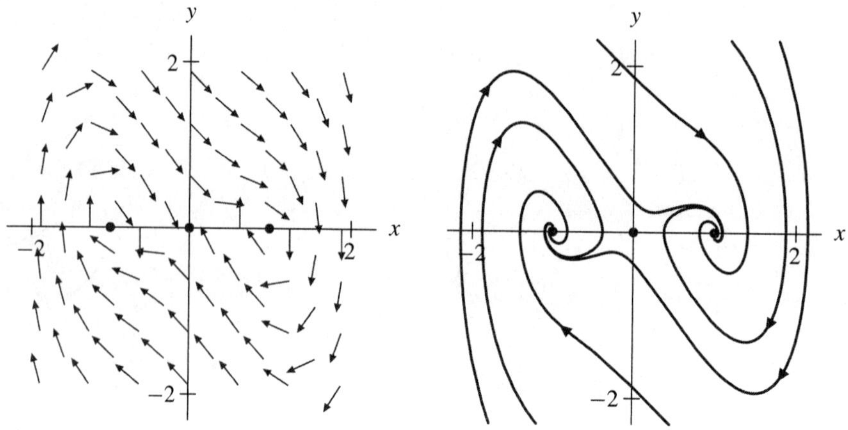

(c) As t increases, typical solutions spiral toward either $(1, 0)$ or $(-1, 0)$ depending on the initial condition.

17. **(a)** To find the equilibrium points, we solve the system of equations

$$\begin{cases} y = 0 \\ -\cos x - y = 0. \end{cases}$$

We see that $y = 0$, and thus $\cos x = 0$. The equilibrium points are $(\pi/2 + k\pi, 0)$ for any integer k.

(b)

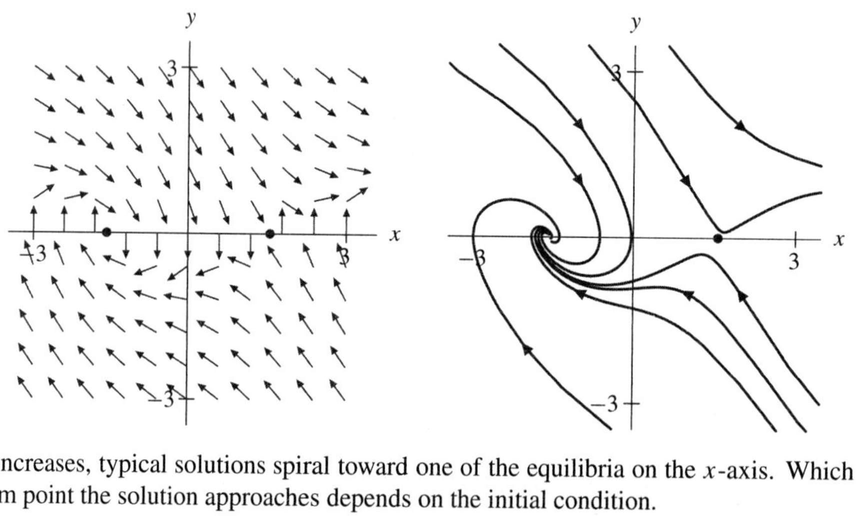

(c) As t increases, typical solutions spiral toward one of the equilibria on the x-axis. Which equilibrium point the solution approaches depends on the initial condition.

18. **(a)** To find the equilibrium points, we solve the system of equations

$$\begin{cases} y(x^2 + y^2 - 1) = 0 \\ -x(x^2 + y^2 - 1) = 0. \end{cases}$$

If $x^2 + y^2 = 1$, then both equations are satisfied. Hence, any point on the unit circle centered at the origin is an equilibrium point. If $x^2 + y^2 \neq 1$, then the first equations implies $y = 0$ and the second equation implies $x = 0$. Hence, the origin is the only other equilibrium point.

(b)

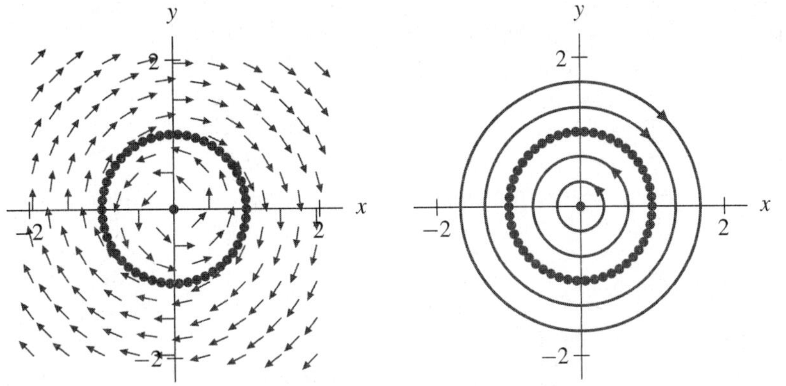

(c) As t increases, typical solutions move on a circle around the origin, either counter-clockwise inside the unit circle, which consists entirely of equilibrium points, or clockwise outside the unit circle.

19. **(a)** Let $v = dx/dt$. Then

$$\frac{dv}{dt} = \frac{d^2x}{dt^2} = 3x - x^3 - 2v.$$

Thus the associated vector field is $\mathbf{V}(x, v) = (v, 3x - x^3 - 2v)$.

(b) Setting $\mathbf{V}(x, v) = (0, 0)$ and solving for (x, v), we get $v = 0$ and $3x - x^3 = 0$. Hence, the equilibria are $(x, v) = (0, 0)$ and $(x, v) = (\pm\sqrt{3}, 0)$.

(c)

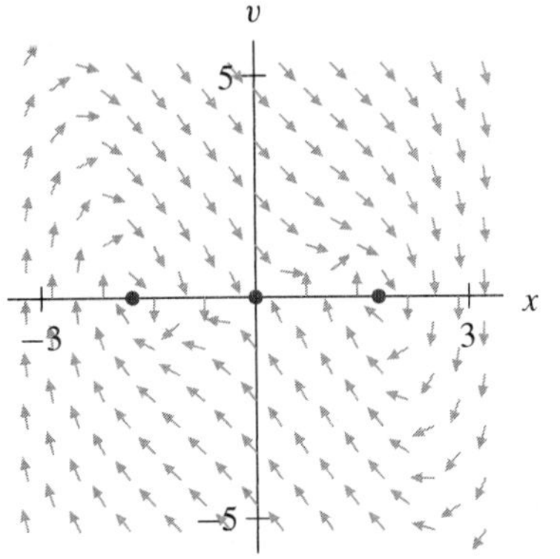

(d)

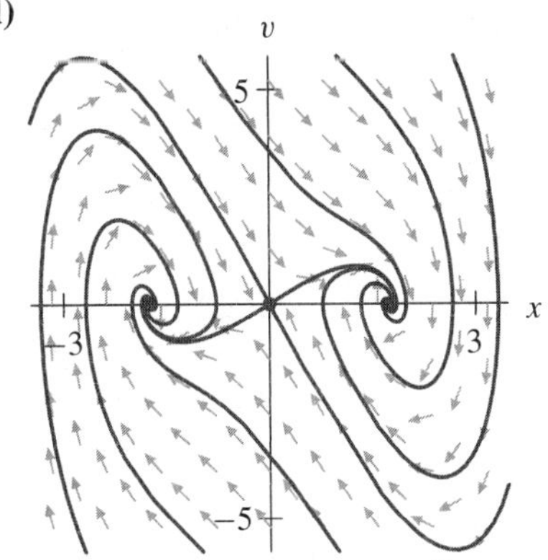

(e) As t increases, almost all solutions spiral to one of the two equilibria $(\pm\sqrt{3}, 0)$. There is a curve of initial conditions that divides these two phenomena. It consists of those initial conditions for which the corresponding solutions tend to the equilibrium point at $(0, 0)$.

20. Consider a point (y, v) on the circle $y^2 + v^2 = r^2$. We can consider this point to be a radius vector—one that starts at the origin and ends at the point (y, v). If we compute the dot product of this vector with the vector field $\mathbf{F}(y, v)$, we obtain

$$(y, v) \cdot \mathbf{F}(y, v) = (y, v) \cdot (v, -y) = yv - vy = 0.$$

Since the dot product of these two vectors is 0, the two vectors are perpendicular. Moreover, we know that any vector that is perpendicular to the radius vector of a circle must be tangent to that circle.

21. **(a)** The $x(t)$- and $y(t)$-graphs are periodic, so they correspond to a solution curve that returns to its initial condition in the phase plane. In other words, its solution curve is a closed curve. Since the amplitude of the oscillation of $x(t)$ is relatively large, these graphs must correspond to the outermost closed solution curve.

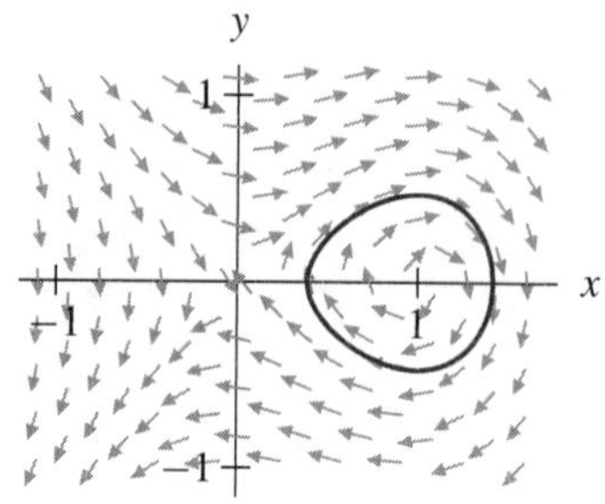

(b) The graphs are not periodic, so they cannot correspond to the two closed solution curves in the phase portrait. Both graphs cross the t-axis. The value of $x(t)$ is initially negative, then becomes positive and reaches a maximum, and finally becomes negative again. Therefore, the corresponding solution curve is the one that starts in the second quadrant, then travels through the first and fourth quadrants, and finally enters the third quadrant.

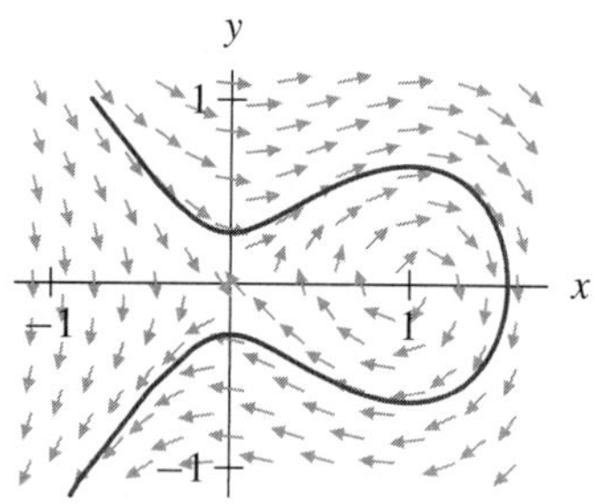

(c) The graphs are not periodic, so they cannot correspond to the two closed solution curves in the phase portrait. Only one graph crosses the t-axis. The other graph remains negative for all time. Note that the two graphs cross.

The corresponding solution curve is the one that starts in the second quadrant and crosses the x-axis and the line $y = x$ as it moves through the third quadrant.

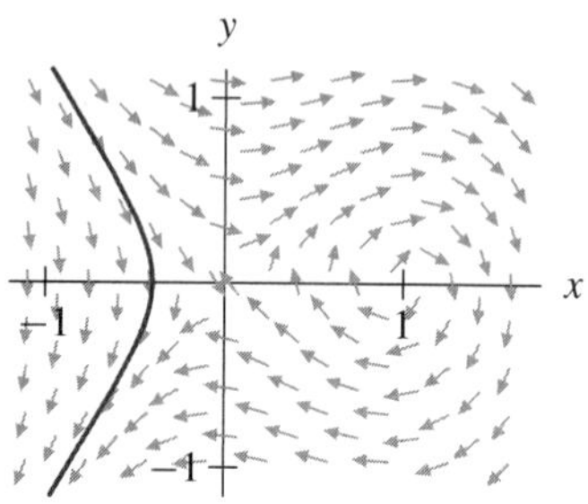

(d) The $x(t)$- and $y(t)$-graphs are periodic, so they correspond to a solution curve that returns to its initial condition in the phase plane. In other words, its solution curve is a closed curve. Since the amplitude of the oscillation of $x(t)$ is relatively small, these graphs must correspond to the intermost closed solution curve.

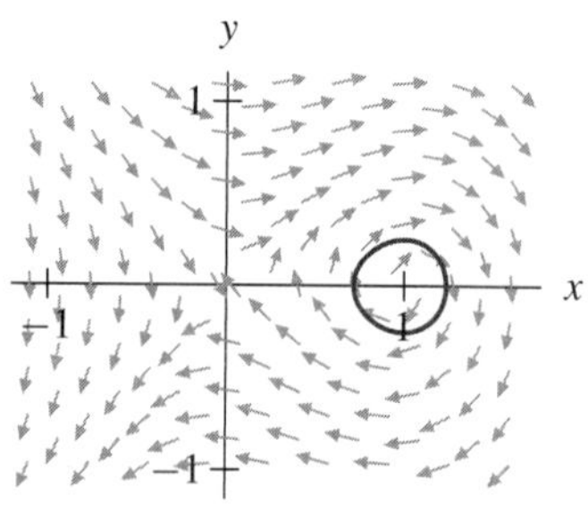

22. Often the solutions in the quiz are over a longer time interval than what is shown in the following graphs.

(a)

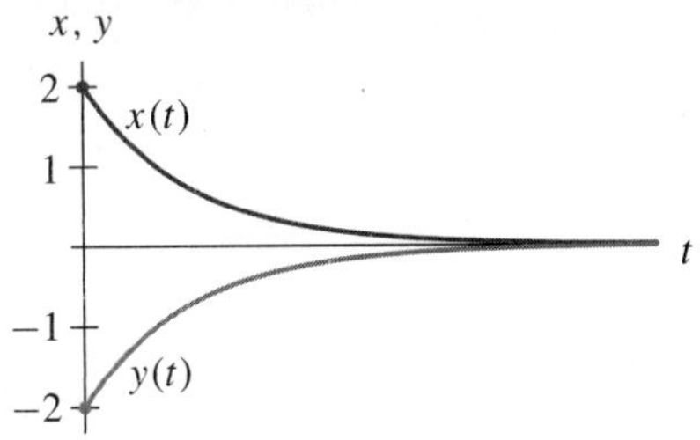

(b)

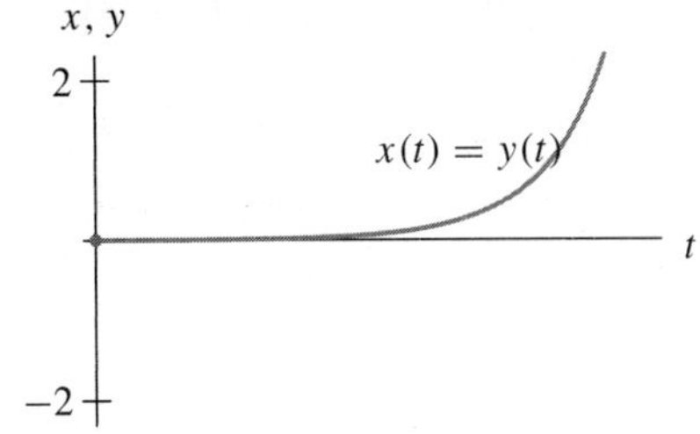

(c)

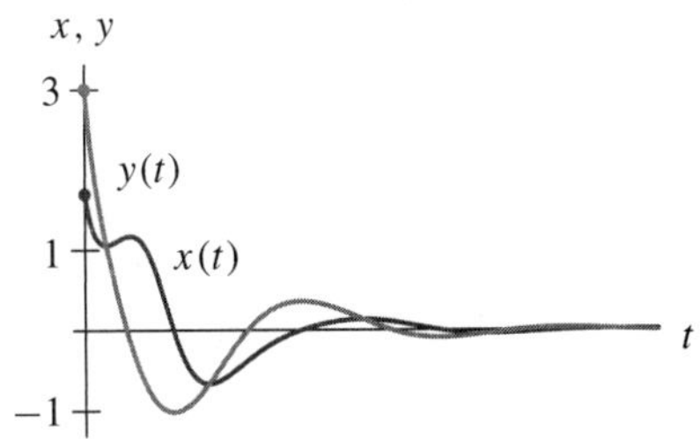

(d)

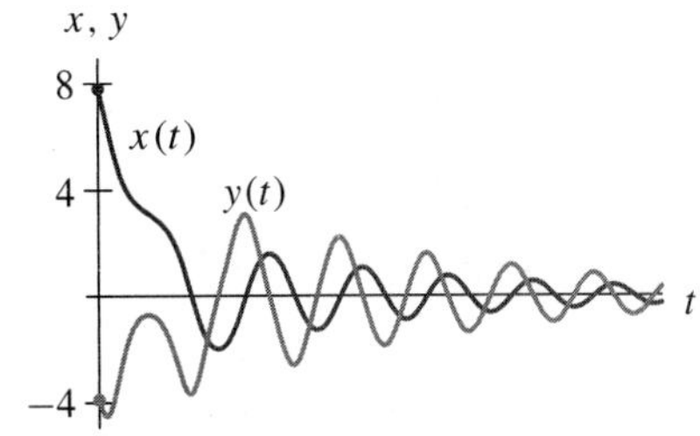

(e)

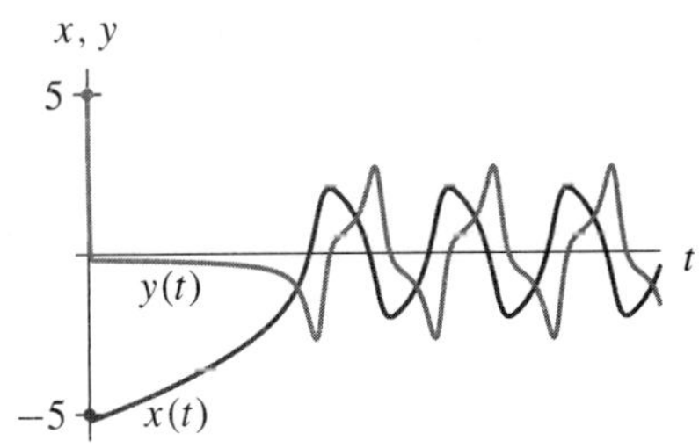

(f)

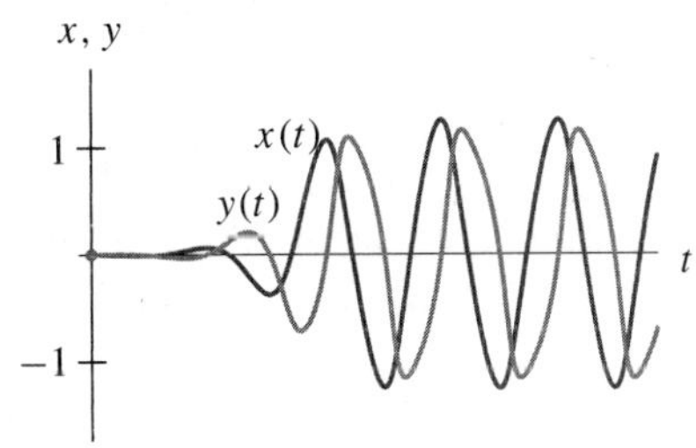

(g)

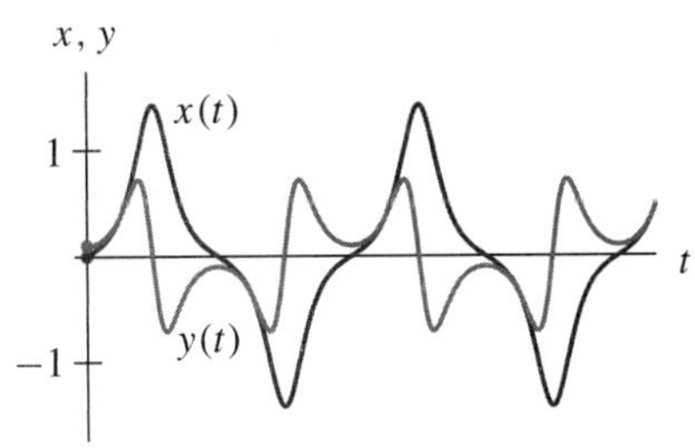

(h)

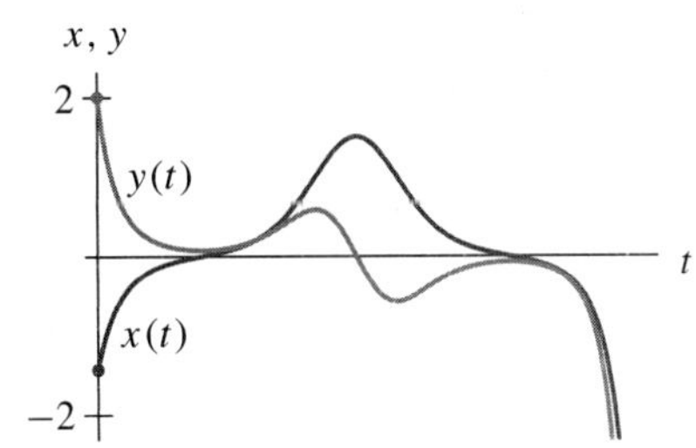

(i)

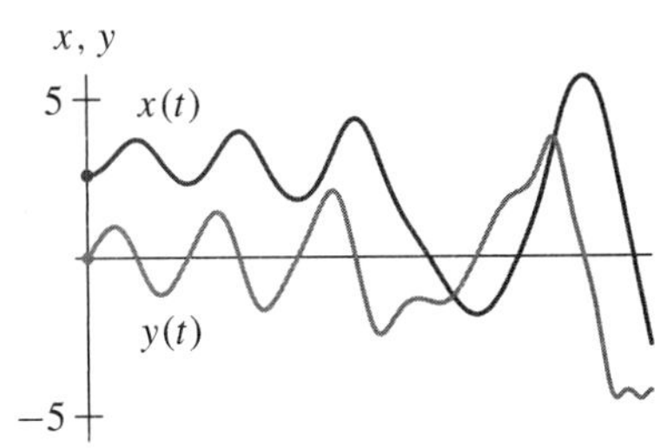

23. Since the solution curve spirals into the origin, the corresponding $x(t)$- and $y(t)$-graphs must oscillate about the t-axis with the decreasing amplitudes.

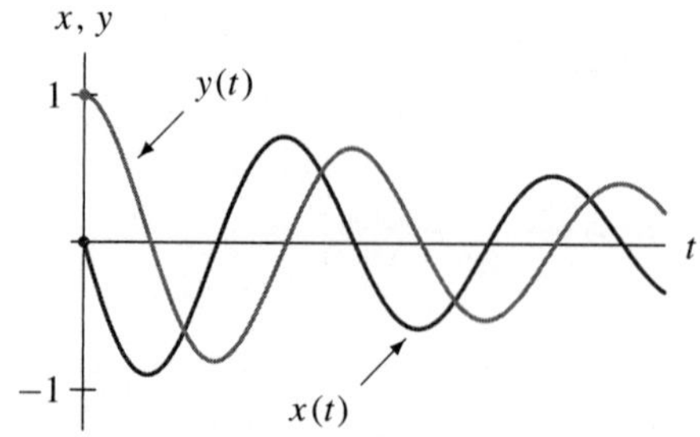

24. Since the solution curve is an ellipse that is centered at (2, 1), the $x(t)$- and $y(t)$-graphs are periodic. They oscillate about the lines $x = 2$ and $y = 1$.

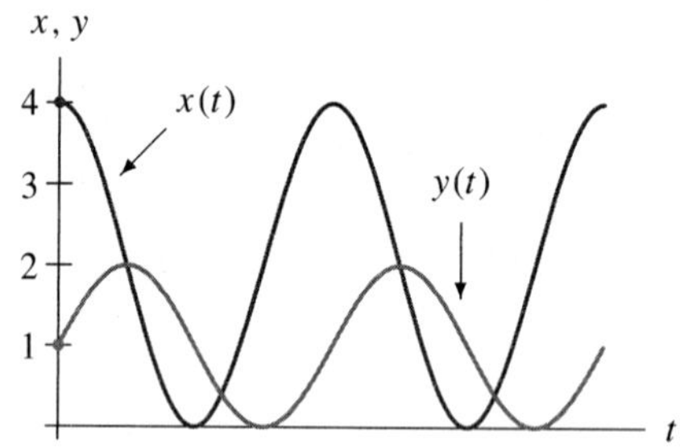

25. The $x(t)$-graph satisfies $-2 < x(0) < -1$ and increases as t increases. The $y(t)$-graph satisfies $1 < y(0) < 2$. Initially it decreases until it reaches its minimum value of $y = 1$ when $x = 0$. Then it increases as t increases.

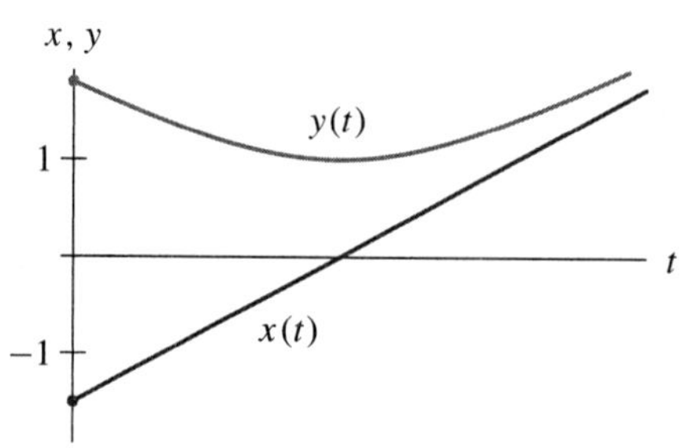

26. The $x(t)$-graph starts with a small positive value and increases as t increases. The $y(t)$-graph starts at approximately 1.6 and decreases as t increases. However, $y(t)$ remains positive for all t.

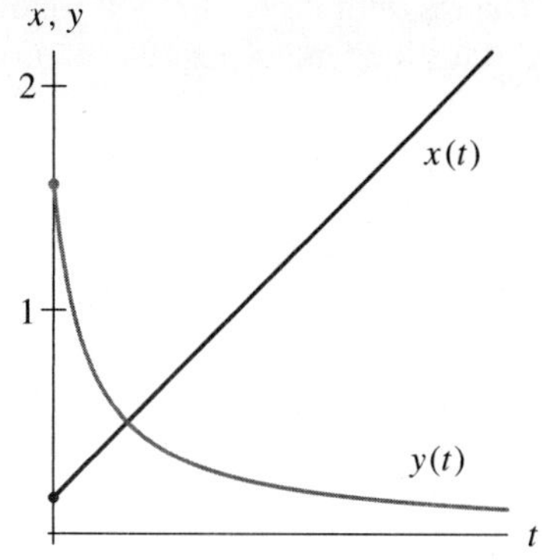

27. From the graphs, we see that $y(0) = 0$ and $x(0)$ is slightly positive. Initially both graphs increase. Then they cross, and slightly later $x(t)$ attains its maximum value. Continuing along we see that $y(t)$ attains its maximum at the same time as $x(t)$ crosses the t-axis.

In the xy-phase plane these graphs correspond to a solution curve that starts on the positive x-axis, enters the first quadrant, crosses the line $y = x$, and eventually crosses the y-axis into the second quadrant exactly when $y(t)$ assumes its maximum value. For this portion of the curve, $y(t)$ is increasing while $x(t)$ assumes a maximum and starts decreasing.

We see that once $y(t)$ attains its maximum, it decreases for a prolonged period of time until it assumes its minimum value. Throughout this interval, $x(t)$ remains negative although it assumes its minimum value twice and a local maximum value once. In the phase plane, the solution curve enters the second quadrant and then crosses into the third quadrant when $y(t) = 0$. The $x(t)$- and $y(t)$ graphs cross precisely when the solution curve crosses the line $y = x$ in the third quadrant.

Finally the $y(t)$-graph is increasing again while the $x(t)$-graph becomes positive and assumes its maximum value once more. The two graphs return to their initial values. In the phase plane this behavior corresponds to the solution curve moving from the third quadrant through the fourth quadrant and back to the original starting point.

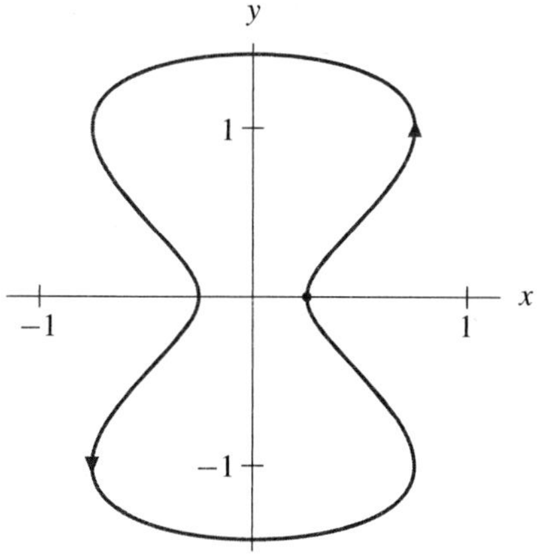

28. Both Gib and Harry will follow the solution curve for their own starting point. A careful proof that distinct solution curves do not touch requires the Uniqueness Theorem (see Section 2.4).

29. Since the vector field does not change with time, Gib will follow the same path as Harry, only one time unit behind. See Section 2.4 for the Uniqueness Theorem that rigorously justifies these statements.

EXERCISES FOR SECTION 2.3

1. To check that $dx/dt = 2x + 2y$, we compute both

$$\frac{dx}{dt} = 2e^t$$

and

$$2x + 2y = 4e^t - 2e^t = 2e^t.$$

To check that $dy/dt = x + 3y$, we compute both

$$\frac{dy}{dt} = -e^t,$$

and

$$x + 3y = 2e^t - 3e^t = -e^t.$$

Both equations are satisfied for all t. Hence $(x(t), y(t))$ is a solution.

2. To check that $dx/dt = 2x + 2y$, we compute both

$$\frac{dx}{dt} = 6e^{2t} + e^t$$

and

$$2x + 2y = 6e^{2t} + 2e^t - 2e^t + 2e^{4t} = 6e^{2t} + 2e^{4t}.$$

Since the results of these two calculations do not agree, the first equation in the system is not satisfied, and $(x(t), y(t))$ is not a solution.

3. To check that $dx/dt = 2x + 2y$, we compute both

$$\frac{dx}{dt} = 2e^t - 4e^{4t}$$

and

$$2x + 2y = 4e^t - 2e^{4t} - 2e^t + 2e^{4t} = 2e^t.$$

Since the results of these two calculations do not agree, the first equation in the system is not satisfied, and $(x(t), y(t))$ is not a solution.

4. To check that $dx/dt = 2x + 2y$, we compute both

$$\frac{dx}{dt} = 4e^t + 4e^{4t}$$

and

$$2x + 2y = 8e^t + 2e^{4t} - 4e^t + 2e^{4t} = 4e^t + 4e^t.$$

To check that $dy/dt = x + 3y$, we compute both

$$\frac{dy}{dt} = -2e^t + 4e^{4t},$$

and

$$x + 3y = 4e^t + e^{4t} - 6e^t + 3e^{4t} = -2e^t + 4e^{4t}.$$

Both equations are satisfied for all t. Hence $(x(t), y(t))$ is a solution.

5. The second equation in the system is $dy/dt = -y$, and from Section 1.1, we know that $y(t)$ must be a function of the form $y_0 e^{-t}$, where y_0 is the initial value.

6. Yes. You can always show that a given function is a solution by verifying the equations directly (as in Exercises 1–4).

To check that $dx/dt = 2x + y$, we compute both

$$\frac{dx}{dt} = 8e^{2t} + e^{-t}$$

and

$$2x + y = 8e^{2t} - 2e^{-t} + 3e^{-t} = 8e^{2t} + e^t.$$

To check that $dy/dt = -y$, we compute both

$$\frac{dy}{dt} = -3e^{-t},$$

and

$$-y = -3e^{-t}.$$

Both equations are satisfied for all t. Hence $(x(t), y(t))$ is a solution.

7. From the second equation, we know that $y(t) = k_1 e^{-t}$ for some constant k_1. Using this observation, the first equation in the system can be rewritten as

$$\frac{dx}{dt} = 2x + k_1 e^{-t}.$$

This equation is a first-order linear equation, and we can derive the general solution using the Extended Linearity Principle from Section 1.8 or integrating factors from Section 1.9.

Using the Extended Linearity Principle, we note that the general solution of the associated homogeneous equation is $x_h(t) = k_2 e^{2t}$.

To find one solution to the nonhomogeneous equation, we guess $x_p(t) = \alpha e^{-t}$. Then

$$\begin{aligned}\frac{dx_p}{dt} - 2x_p &= -\alpha e^{-t} - 2\alpha e^{-t} \\ &= -3\alpha e^{-t}.\end{aligned}$$

Therefore, $x_p(t)$ is a solution if $\alpha = -k_1/3$.

The general solution for $x(t)$ is

$$x(t) = k_2 e^{2t} - \frac{k_1}{3} e^{-t}.$$

8. **(a)** No. Given the general solution

$$\left(k_2 e^{2t} - \frac{k_1}{3} e^{-t}, k_1 e^{-t}\right),$$

the function $y(t) = 3e^{-t}$ implies that $k_1 = 3$. But this choice of k_1 implies that the coefficient of e^{-t} in the formula for $x(t)$ is -1 rather than $+1$.

(b) To determine that $\mathbf{Y}(t)$ is not a solution without reference to the general solution, we check the equation $dx/dt = 2x + y$. We compute both

$$\frac{dx}{dt} = -e^{-t}$$

and

$$2x + y = 2e^{-t} + 3e^{-t}.$$

Since these two functions are not equal, $\mathbf{Y}(t)$ is not a solution.

9. **(a)** Given the general solution

$$\left(k_2 e^{2t} - \frac{k_1}{3} e^{-t}, k_1 e^{-t}\right),$$

we see that $k_1 = 0$, and therefore $k_2 = 1$. We obtain $\mathbf{Y}(t) = (x(t), y(t)) = (e^{2t}, 0)$.

(b)

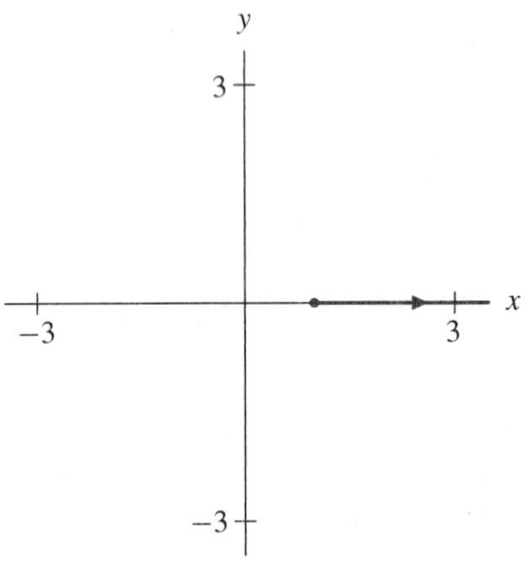

(c)

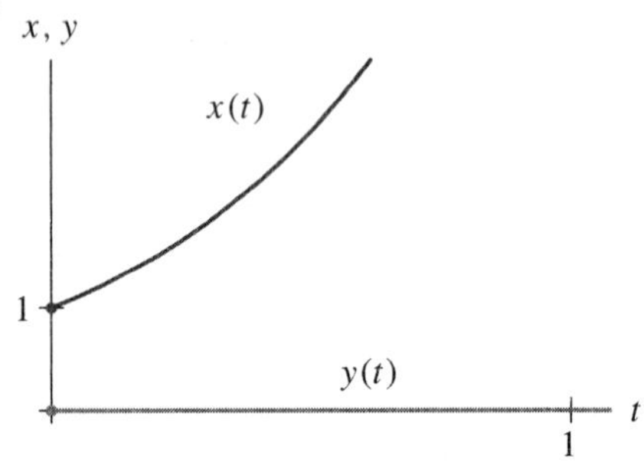

10. **(a)** Given the general solution

$$\left(k_2 e^{2t} - \frac{k_1}{3} e^{-t}, k_1 e^{-t}\right),$$

we see that $k_1 = 3$, and therefore $k_2 = 0$. We obtain $\mathbf{Y}(t) = (x(t), y(t)) = (-e^{-t}, 3e^{-t})$.

(b)

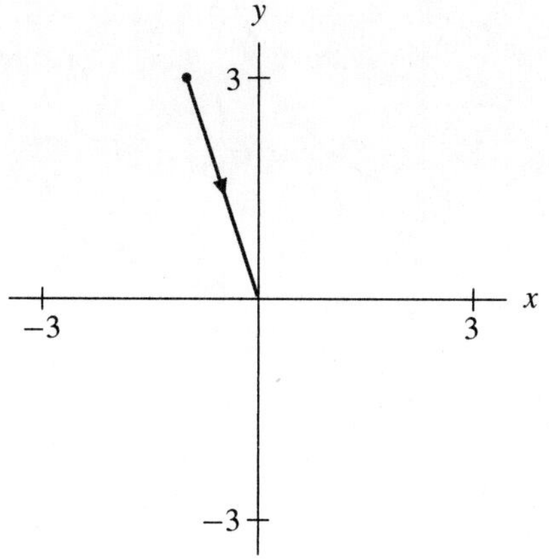

(c)

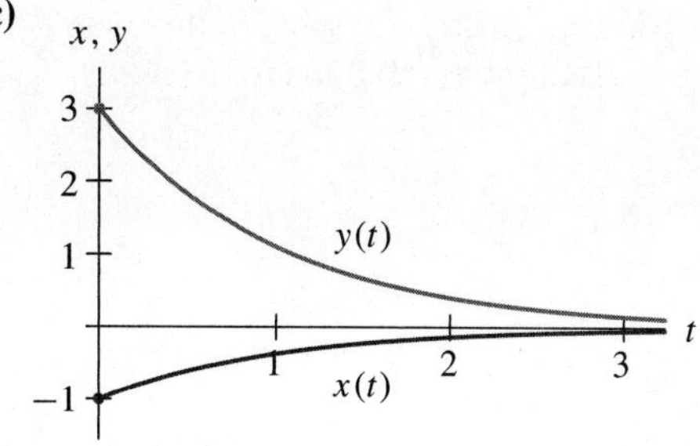

11. **(a)** Given the general solution

$$\left(k_2 e^{2t} - \frac{k_1}{3} e^{-t}, k_1 e^{-t}\right),$$

we see that $k_1 = 1$, and therefore $k_2 = 1/3$. We obtain

$$\mathbf{Y}(t) = (x(t), y(t)) = \left(\tfrac{1}{3}e^{2t} - \tfrac{1}{3}e^{-t}, e^{-t}\right).$$

(b)

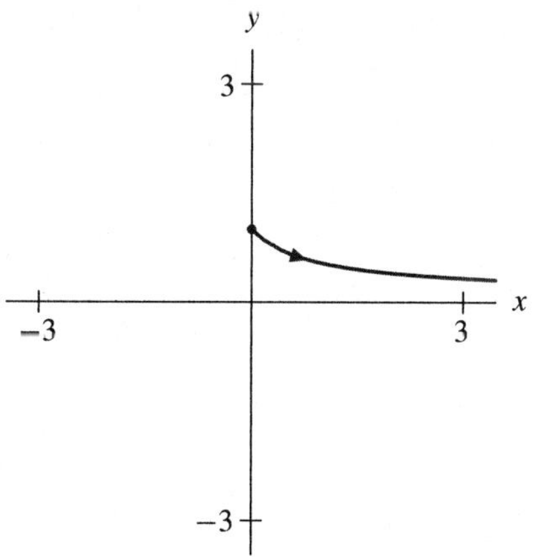

(c)

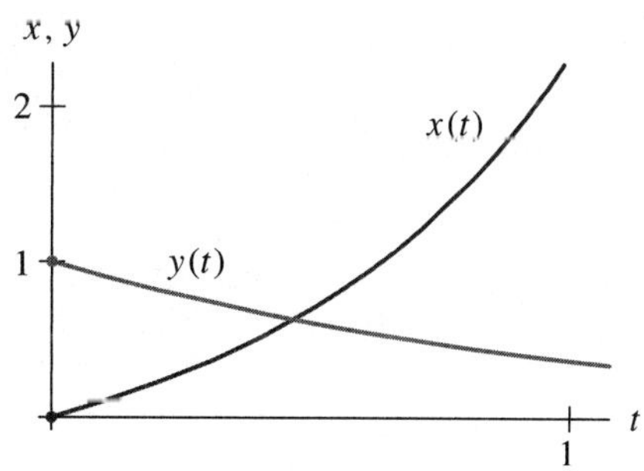

12. **(a)** Given the general solution

$$\left(k_2 e^{2t} - \frac{k_1}{3} e^{-t}, k_1 e^{-t}\right),$$

we see that $k_1 = -1$, and therefore $k_2 = 2/3$. We obtain

$$\mathbf{Y}(t) = (x(t), y(t)) = \left(\tfrac{2}{3}e^{2t} + \tfrac{1}{3}e^{-t}, -e^{-t}\right).$$

(b)

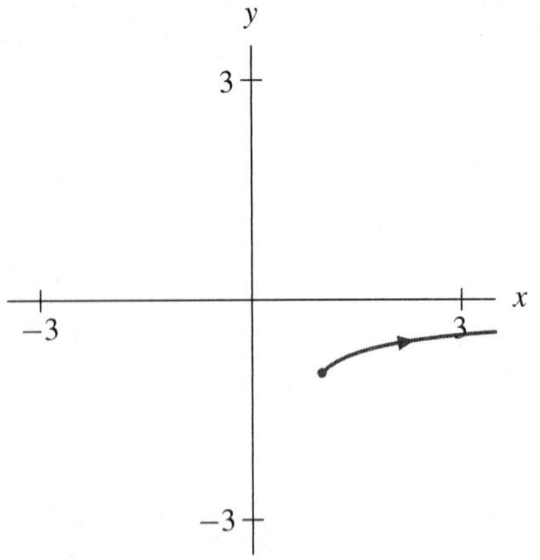

(c)

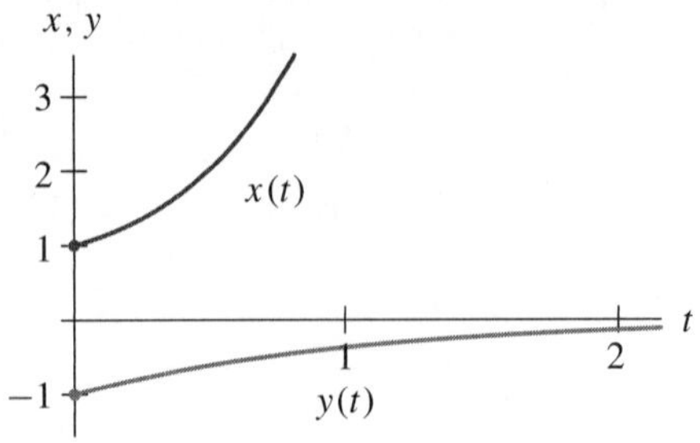

13. We choose the left wall to be the position $x = 0$ with $x > 0$ indicating positions to the right. Each spring exerts a force on the mass. If the position of the mass is x, then the left spring is stretched by the amount $x - L_1$. Therefore, the force F_1 exerted by this spring is

$$F_1 = k_1 (L_1 - x).$$

Similarly, the right spring is stretched by the amount $(1 - x) - L_2$. However, the restoring force F_2 of the right spring acts in the direction of increasing values of x. Therefore, we have

$$F_2 = k_2 ((1 - x) - L_2).$$

Using Newton's second law, we have

$$m\frac{d^2x}{dt^2} = k_1 (L_1 - x) + k_2 ((1 - x) - L_2) - b\frac{dx}{dt},$$

where the term involving dx/dt represents the force due to damping. After a little algebra, we obtain

$$m\frac{d^2x}{dt^2} + b\frac{dx}{dt} + (k_1 + k_2)x = k_1L_1 - k_2L_2 + k_2.$$

14. **(a)** Let $v = dx/dt$ as usual. From Exercise 13, we have

$$\frac{dx}{dt} = v$$
$$\frac{dv}{dt} = -\frac{k_1 + k_2}{m}x - \frac{b}{m}v + \frac{C}{m}$$

where C is the constant $k_1L_1 - k_2L_2 + k_2$.

(b) To find the equilibrium points, we set $dx/dt = 0$ and obtain $v = 0$. Setting $dv/dt = 0$ with $v = 0$, we obtain

$$(k_1 + k_2)x = C.$$

Therefore, this system has one equilibrium point,

$$(x_0, v_0) = \left(\frac{C}{k_1 + k_2}, 0\right).$$

(c) We change coordinates so that the origin corresponds to this equilibrium point. In other words, we reexpress the system in terms of the new variable $y = x - x_0$. Since $dy/dt = dx/dt = v$, we have

$$\begin{aligned}\frac{dv}{dt} &= -\frac{k_1+k_2}{m}x - \frac{b}{m}v + \frac{C}{m}\\ &= -\frac{k_1+k_2}{m}(y+x_0) - \frac{b}{m}v + \frac{C}{m}\\ &= -\frac{k_1+k_2}{m}y - \frac{C}{m} - \frac{b}{m}v + \frac{C}{m},\end{aligned}$$

since $(k_1+k_2)x_0 = C$. In terms of y and v, we have

$$\begin{aligned}\frac{dy}{dt} &= v\\ \frac{dv}{dt} &= -\frac{k_1+k_2}{m}y - \frac{b}{m}v.\end{aligned}$$

(d) In terms of y and v, this system is exactly the same as a damped harmonic oscillator with spring constant $k = k_1 + k_2$ and damping coefficient b.

15. **(a)** See part (c).

(b) We guess that there are solutions of the form $y(t) = e^{st}$ for some choice of the constant s. To determine these values of s, we substitute $y(t) = e^{st}$ into the left-hand side of the differential equation, obtaining

$$\begin{aligned}\frac{d^2y}{dt^2} + 3\frac{dy}{dt} - 10y &= \frac{d^2(e^{st})}{dt^2} + 3\frac{d(e^{st})}{dt} - 10(e^{st})\\ &= s^2e^{st} + 3se^{st} - 10e^{st}\\ &= (s^2+3s-10)e^{st}\end{aligned}$$

In order for $y(t) = e^{st}$ to be a solution, this expression must be 0 for all t. In other words,

$$s^2 + 3s - 10 = 0.$$

This equation is satisfied only if $s = -5$ or $s = 2$. We obtain two solutions, $y_1(t) = e^{-5t}$ and $y_2(t) = e^{2t}$, of this equation.

(c)

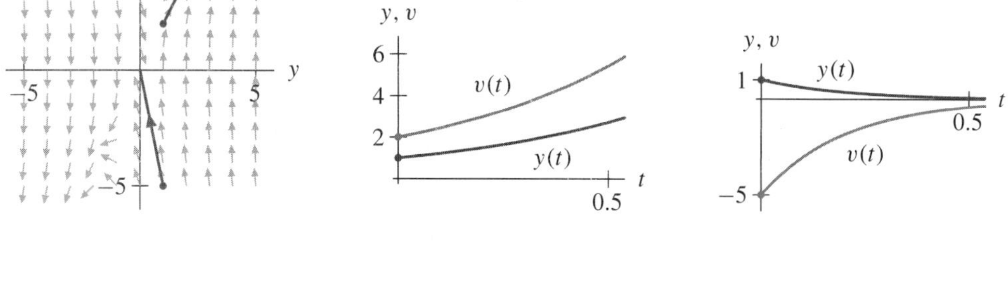

16. **(a)** See part (c).

(b) We guess that there are solutions of the form $y(t) = e^{st}$ for some choice of the constant s. To determine these values of s, we substitute $y(t) = e^{st}$ into the left-hand side of the differential equation, obtaining

$$\begin{aligned}\frac{d^2y}{dt^2} + 3\frac{dy}{dt} + 2y &= \frac{d^2(e^{st})}{dt^2} + 3\frac{d(e^{st})}{dt} + 2(e^{st}) \\ &= s^2e^{st} + 3se^{st} + 2e^{st} \\ &= (s^2 + 3s + 2)e^{st}\end{aligned}$$

In order for $y(t) = e^{st}$ to be a solution, this expression must be 0 for all t. In other words,

$$s^2 + 3s + 2 = 0.$$

This equation is satisfied only if $s = -2$ or $s = -1$. We obtain two solutions, $y_1(t) = e^{-2t}$ and $y_2(t) = e^{-t}$, of this equation.

(c)

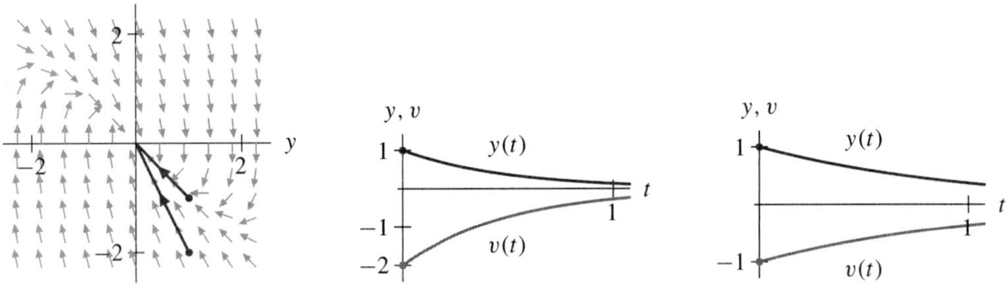

17. **(a)** See part (c).

(b) We guess that there are solutions of the form $y(t) = e^{st}$ for some choice of the constant s. To determine these values of s, we substitute $y(t) = e^{st}$ into the left-hand side of the differential equation, obtaining

$$\begin{aligned}\frac{d^2y}{dt^2} + 3\frac{dy}{dt} + 2y &= \frac{d^2(e^{st})}{dt^2} + 4\frac{d(e^{st})}{dt} + e^{st} \\ &= s^2e^{st} + 4se^{st} + e^{st} \\ &= (s^2 + 4s + 1)e^{st}\end{aligned}$$

In order for $y(t) = e^{st}$ to be a solution, this expression must be 0 for all t. In other words,

$$s^2 + 4s + 1 = 0.$$

Applying the quadratic formula, we obtain the roots $s = -2 \pm \sqrt{3}$ and the two solutions, $y_1(t) = e^{(-2-\sqrt{3})t}$ and $y_2(t) = e^{(-2+\sqrt{3})t}$, of this equation.

(c)

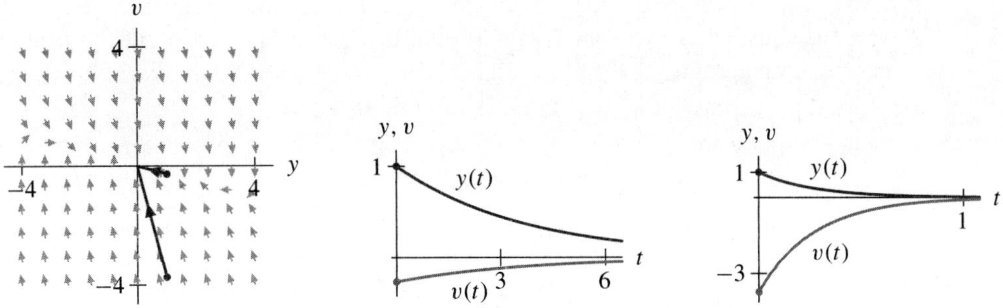

18. **(a)** See part (c).

(b) We guess that there are solutions of the form $y(t) = e^{st}$ for some choice of the constant s. To determine these values of s, we substitute $y(t) = e^{st}$ into the left-hand side of the differential equation, obtaining

$$\begin{aligned}\frac{d^2y}{dt^2} + \frac{dy}{dt} - 2y &= \frac{d^2(e^{st})}{dt^2} + \frac{d(e^{st})}{dt} - 2(e^{st}) \\ &= s^2e^{st} + se^{st} - 2e^{st} \\ &= (s^2 + s - 2)e^{st}\end{aligned}$$

In order for $y(t) = e^{st}$ to be a solution, this expression must be 0 for all t. In other words,

$$s^2 + s - 2 = 0.$$

This equation is satisfied only if $s = -2$ or $s = 1$. We obtain two solutions, $y_1(t) = e^{-2t}$ and $y_2(t) = e^t$, of this equation.

(c)

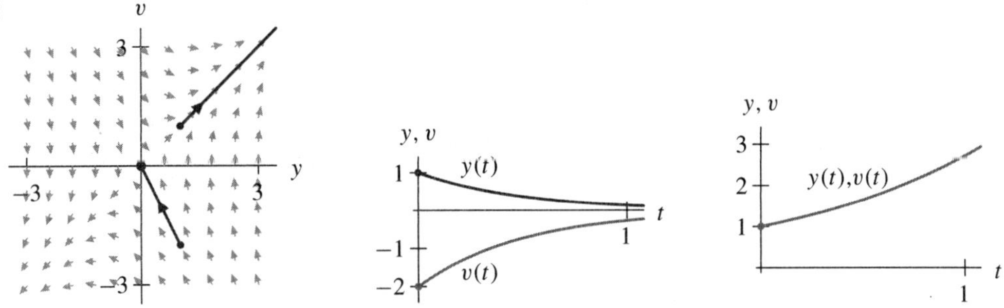

19. **(a)** For this system, we note that the equation for dy/dt is a homogeneous linear equation. Its general solution is

$$y(t) = k_2e^{-3t}.$$

Substituting $y = k_2e^{-3t}$ into the equation for dx/dt, we have

$$\begin{aligned}\frac{dx}{dt} &= 2x - 8(k_2e^{-3t})^2 \\ &= 2x - 8k_2^2e^{-6t}\end{aligned}$$

This equation is a linear and nonhomogeneous. The general solution of the associated homogeneous equation is $x_h(t) = k_1 e^{2t}$. To find one particular solution of the nonhomogeneous equation, we guess

$$x_p(t) = \alpha e^{-6t}.$$

With this guess, we have

$$\begin{aligned} \frac{dx_p}{dt} - 2x_p &= -6\alpha e^{-6t} - 2\alpha e^{-6t} \\ &= -8\alpha e^{-6t}. \end{aligned}$$

Therefore, $x_p(t)$ is a solution if $\alpha = k_2^2$. The general solution for $x(t)$ is $k_1 e^{2t} + k_2^2 e^{-6t}$, and the general solution for the system is

$$(x(t), y(t)) = (k_1 e^{2t} + k_2^2 e^{-6t}, k_2 e^{-3t}).$$

(b) Setting $dy/dt = 0$, we obtain $y = 0$. From $dx/dt = 2x - 8y^2 = 0$, we see that $x = 0$ as well. Therefore, this system has exactly one equilibrium point, $(x, y) = (0, 0)$.

(c) If $(x(0), y(0)) = (0, 1)$, then $k_2 = 1$. We evaluate the expression for $x(t)$ at $t = 0$ and obtain $k_1 + 1 = 0$. Consequently, $k_1 = -1$, and the solution to the initial-value problem is

$$(x(t), y(t)) = (e^{-6t} - e^{2t}, e^{-3t}).$$

(d)

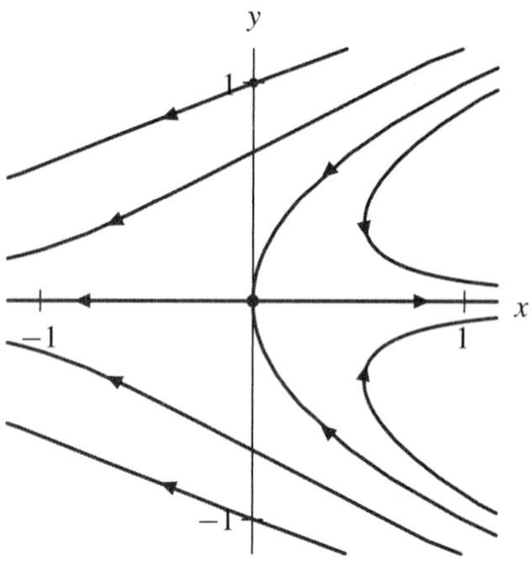

EXERCISES FOR SECTION 2.4

1. **(a)** We compute

$$\frac{dx}{dt} = \frac{d(\cos t)}{dt} = -\sin t = -y \quad \text{and} \quad \frac{dy}{dt} = \frac{d(\sin t)}{dt} = \cos t = x,$$

so $(\cos t, \sin t)$ is a solution.

(b)

Table 2.1

t	Euler's approx.	actual	distance
0	(1, 0)	(1, 0)	
4	(−2.06, −1.31)	(−0.65, −0.76)	1.51
6	(2.87, −2.51)	(0.96, −0.28)	2.94
10	(−9.21, 1.41)	(−0.84, −0.54)	8.59

(c)

Table 2.2

t	Euler's approx.	actual	distance
0	(1, 0)	(1, 0)	
4	(−.81, −.91)	(−0.65, −0.76)	0.22
6	(1.29, −.40)	(0.96, −0.28)	0.35
10	(−1.41, −.85)	(−0.84, −.54)	0.65

(d) The solution curves for this system are all circles centered at the origin. Since Euler's method uses tangent lines to approximate the solution curve and the tangent line to any point on a circle is entirely outside the circle (except at the point of tangency), each step of the Euler approximation takes the approximate solution farther from the origin. So the Euler approximations always spiral away from the origin for this system.

2. **(a)** We compute

$$\frac{dx}{dt} = \frac{d(e^{2t})}{dt} = 2e^{2t} = 2x \quad \text{and} \quad \frac{dy}{dt} = \frac{d(3e^t)}{dt} = 3e^t = y,$$

so $(e^{2t}, 3e^t)$ is a solution.

(b)

Table 2.3

t	Euler's approx.	actual	distance
0	(1, 3)	(1, 3)	
2	(16, 15.1875)	(54.59, 22.17)	39.22
4	(256, 76.88)	(2981, 164)	2726
6	(4096, 389)	(162755, 1210)	158661

(c)

Table 2.4

t	Euler's approx.	actual	distance
0	(1, 3)	(1, 3)	
2	(38.34, 20.18)	(54.59, 22.17)	16.38
4	(1470, 136)	(2981, 164)	1511.4
6	(56347, 913)	(162755, 1210)	106408

(d) The solution curve starts at (1, 3) and tends to infinity in both the x- and y-directions. Because the solution is an exponential, Euler's method has a hard time keeping up with the growth of the solutions.

3. **(a)** Euler approximation yields $(x_5, y_5) \approx (0.65, -0.59)$.

(b)

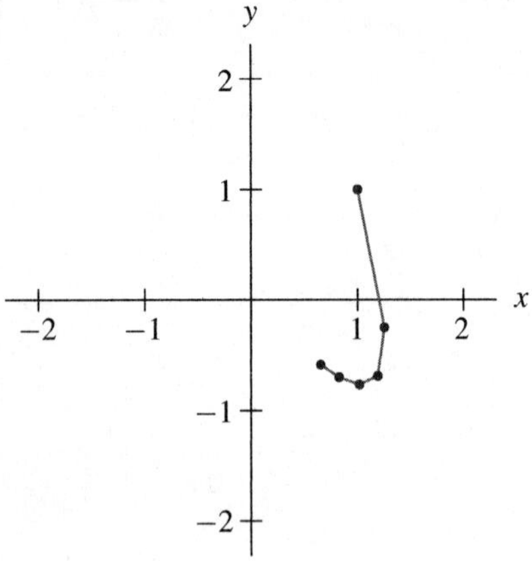

(c)

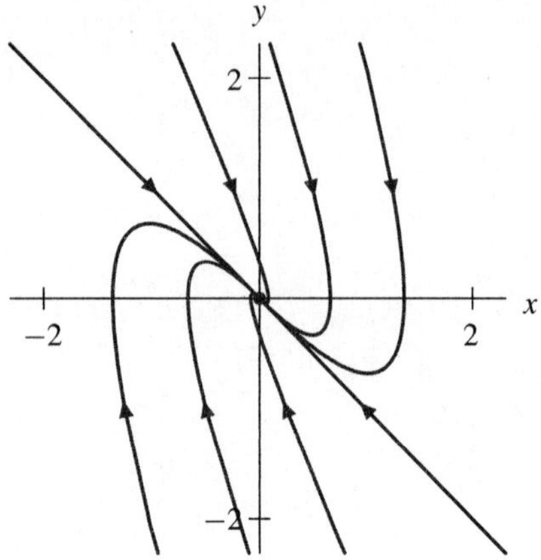

4. **(a)** Euler approximation yields $(x_8, y_8) \approx (3.00, 0.76)$.

(b)

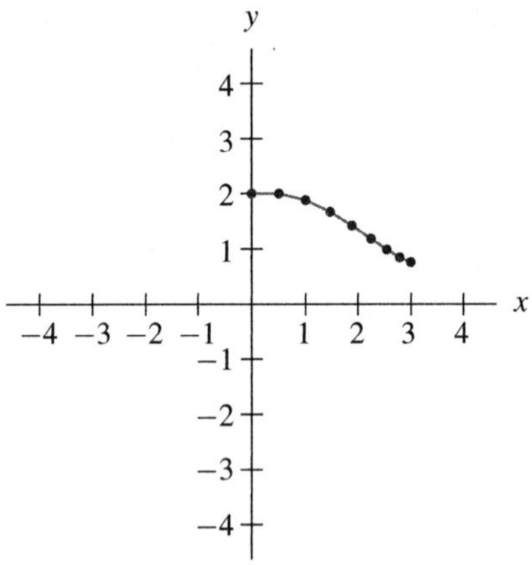

(c)

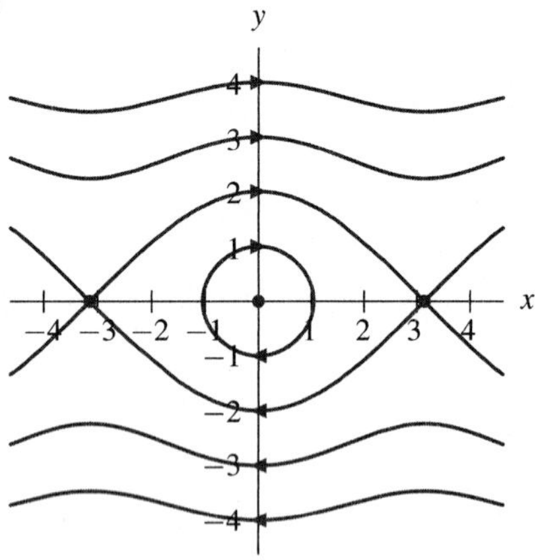

5. **(a)** Euler approximation yields $(x_5, y_5) \approx (1.94, -0.72)$.

(b)

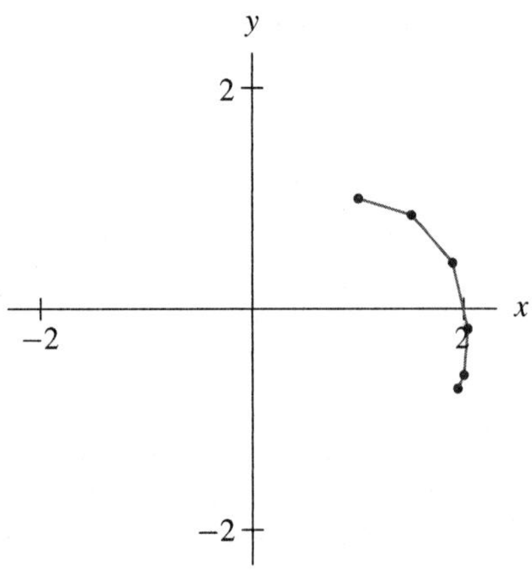

(c)

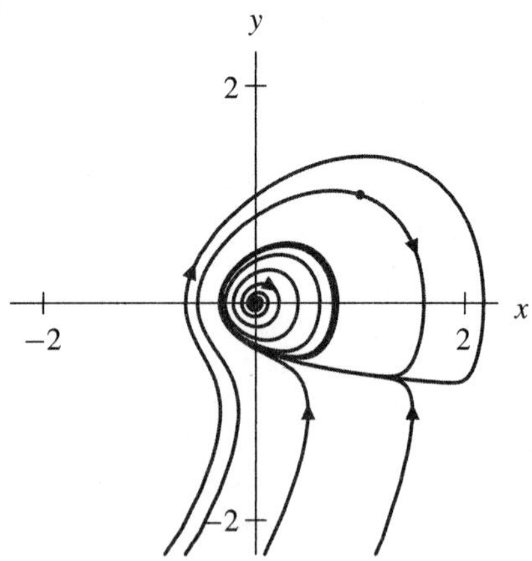

6. **(a)** Euler approximation yields $(x_7, y_7) \approx (0.15, 0.78)$.

(b)

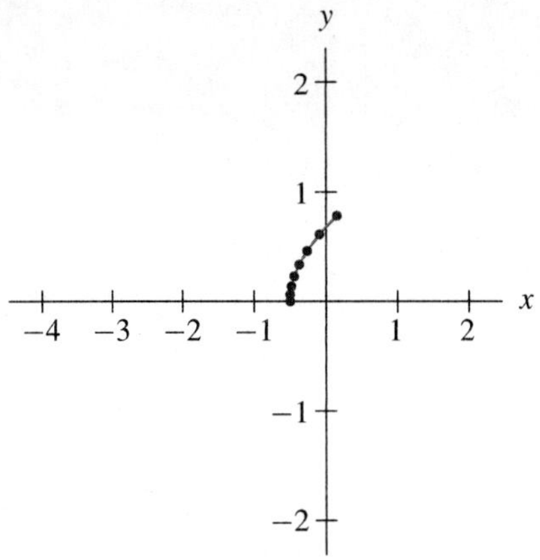

(c)

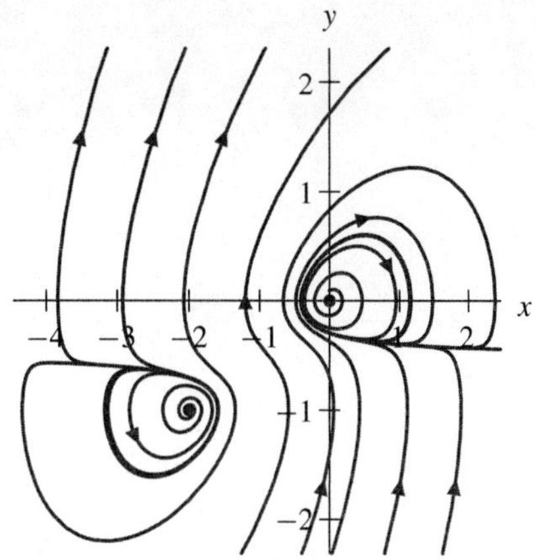

7. With $x(t) = e^{-t}\sin(3t)$ and $y(t) = e^{-t}\cos(3t)$, we have

$$\begin{aligned}\frac{dx}{dt} &= -e^{-t}\sin(3t) + 3e^{-t}\cos(3t)\\ &= -x + 3y\\ \frac{dy}{dt} &= -3e^{-t}\sin(3t) - e^{-t}\cos(3t)\\ &= -3x - y\end{aligned}$$

Therefore, $\mathbf{Y}_1(t)$ is a solution.

8. With $x(t) = e^{-(t-1)}\sin(3(t-1))$ and $y(t) = e^{-(t-1)}\cos(3(t-1))$, we have

$$\begin{aligned}\frac{dx}{dt} &= -e^{-(t-1)}\sin(3(t-1)) + 3e^{-(t-1)}\cos(3(t-1))\\ &= -x + 3y\\ \frac{dy}{dt} &= -3e^{-(t-1)}\sin(3(t-1)) - e^{-(t-1)}\cos(3(t-1))\\ &= -3x - y\end{aligned}$$

Therefore, $\mathbf{Y}_2(t)$ is a solution.

9. If we express the solution curve for $\mathbf{Y}_1(t)$ in terms of polar coordinates (r, θ) where $r^2 = x^2 + y^2$, we obtain $r = e^{-t}$, and

$$\frac{x(t)}{y(t)} = \frac{e^{-t}\sin 3t}{e^{-t}\cos 3t} = \tan 3t.$$

At the same time,

$$\frac{x(t)}{y(t)} = \tan\phi,$$

where $\phi = \pi/2 - \theta$. Therefore, $\tan 3t = \tan\phi$, and $3t = \pi/2 - \theta$. In other words, the angle θ changes according to the relationship $\theta = \pi/2 - 3t$.

These two computations imply that the solution curve for $\mathbf{Y}_1(t)$ spirals into the origin in a clockwise direction.

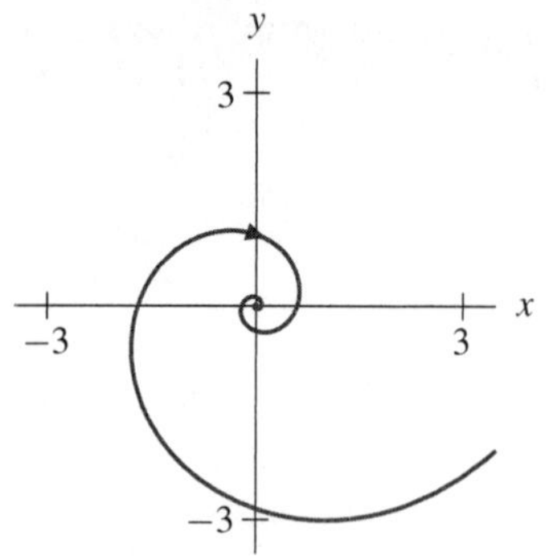

For $\mathbf{Y}_2(t)$, these computations lead to the same conclusions except that, wherever a t appears, it is replaced by $t - 1$. Thus the solution curve for $\mathbf{Y}_2(t)$ is identical to the solution curve for $\mathbf{Y}_1(t)$. In other words, as curves in the plane, they sweep out the same sets of points, but $\mathbf{Y}_1(t-1) = \mathbf{Y}_2(t)$. Whenever $\mathbf{Y}_2(t)$ occupies a certain point in the plane, it turns out that $\mathbf{Y}_1(t)$ occupied that point exactly 1 unit of time earlier. Since these curves never occupy the same point at the same time, they do not violate the Uniqueness Theorem.

10. In order to be able to apply Euler's method to this second-order equation, we reduce the equation to a first-order system using $v = dy/dt$. We obtain

$$\frac{dy}{dt} = v$$
$$\frac{dv}{dt} = -2y - \frac{v}{2}.$$

The choice of Δt has an important effect on the long-term behavior of the approximate solution curve. The approximate solution curve for $\Delta t = 0.25$ seems almost periodic. If $(y_0, v_0) = (2, 0)$, then we obtain $(y_5, v_5) \approx (-0.06, -2.81)$, $(y_{10}, v_{10}) \approx (-1.98, 1.15)$, $(y_{15}, v_{15}) \approx (0.87, 2.34)$, …

However, the approximate solution curve for $\Delta t = 0.1$ spirals toward the origin. If $(y_0, v_0) = (2, 0)$, then we obtain $(y_5, v_5) \approx (1.62, -1.73)$, $(y_{10}, v_{10}) \approx (0.57, -2.44)$, $(y_{15}, v_{15}) \approx (-0.60, -1.94)$, …

The following figure illustrates the results of Euler's method with $\Delta t = 0.1$.

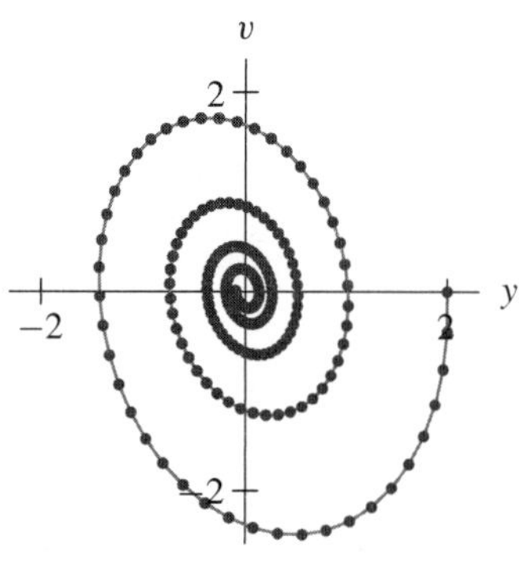

11. In order to be able to apply Euler's method to this second-order equation, we reduce the equation to a first-order system using $v = dy/dt$. We obtain

$$\frac{dy}{dt} = v$$
$$\frac{dv}{dt} = -y - \frac{v}{5}.$$

The choice of Δt has an important effect on the long-term behavior of the approximate solution curve. The curve for $\Delta t = 0.25$ spirals away from the origin. If $(y_0, v_0) = (0, 1)$, then we obtain $(y_5, v_5) \approx (0.98, 0.23)$, $(y_{10}, v_{10}) \approx (0.64, -0.92)$, $(y_{15}, v_{15}) \approx (-0.63, -0.84), \ldots$

The behavior of this approximate solution curve is deceiving. Consider the approximation we obtain if we halve that value of Δt. In other words, let $\Delta t = 0.125$. For $(y_0, v_0) = (2, 0)$, then we obtain $(y_5, v_5) \approx (0.58, 0.73)$, $(y_{10}, v_{10}) \approx (0.91, 0.21)$, $(y_{15}, v_{15}) \approx (0.89, -0.37), \ldots$

The following figure illustrates how this approximate solution curve spirals toward the origin. (As we will see, this second approximation is much better than the first.)

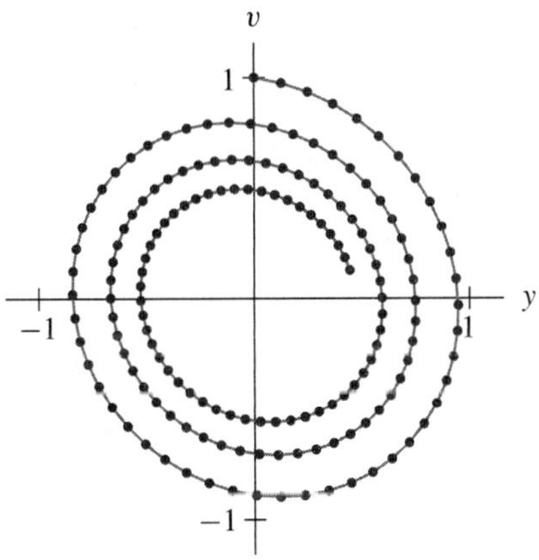

12. We need to assume that the hypotheses of the Uniqueness Theorem apply to the vector field on the parking lot. Then both Gib and Harry will follow the solution curve for their own starting point.

13. Assume the vector field satisfies the hypotheses of the Uniqueness Theorem. Since the vector field does not change with time, Gib will follow the same path as Harry, only one time unit behind.

14. **(a)** Differentiation yields

$$\frac{d\mathbf{Y}_2}{dt} = \frac{d(\mathbf{Y}_1(t + t_0))}{dt} = \mathbf{F}(\mathbf{Y}_1(t + t_0)) = \mathbf{F}(\mathbf{Y}_2(t))$$

where the second equality uses the Chain Rule and the other two equalities involve the definition of $\mathbf{Y}_2(t)$.

(b) They describe the same curve, but differ by a constant shift in parameterization.

15. From Exercise 14 we know that $\mathbf{Y}_1(t-1)$ is a solution of the system and $\mathbf{Y}_1(1-1) = \mathbf{Y}_1(0) = \mathbf{Y}_2(1)$, so both $\mathbf{Y}_2(t)$ and $\mathbf{Y}_1(t - 1)$ occupy the point $\mathbf{Y}_1(0)$ at time $t = 1$. Hence, by the Uniqueness Theorem, they are the same solution. So $\mathbf{Y}_2(t)$ is a reparameterization by a constant time shift of $\mathbf{Y}_1(t)$.

16. As long as $y(t)$ is defined, we have $y(t) \geq 1$ if $t \geq 0$ because dy/dt is nonnegative. Using this observation, we have

$$\frac{dx}{dt} \geq x^2 + 1$$

for all $t \geq 0$ in the domain of $x(t)$. Since $x(t) = \tan t$ satisfies the initial-value problem $dx/dt = x^2 + 1$, $x(0) = 0$, we see that the $x(t)$-function for the solution to our system must satisfy

$$x(t) \geq \tan t.$$

Therefore, since $\tan t \to \infty$ as $t \to \pi/2^-$, $x(t) \to \infty$ as $t \to t_*$, where $0 \leq t_* \leq \pi/2$.

EXERCISES FOR SECTION 2.5

1. **(a)** Substitution of $(0, 0, 0)$ into the given system of differential equations yields $dx/dt = dy/dt = dz/dt = 0$. Similarly, for the case of $(\pm 6\sqrt{2}, \pm 6\sqrt{2}, 27)$, we obtain

$$\begin{aligned} \frac{dx}{dt} &= 10(\pm 6\sqrt{2} - (\pm 6\sqrt{2})) \\ \frac{dy}{dt} &= 28(\pm 6\sqrt{2}) - (\pm 6\sqrt{2}) - 27(\pm 6\sqrt{2}) \\ \frac{dz}{dt} &= -\frac{8}{3}(27) - (\pm 6\sqrt{2})^2. \end{aligned}$$

Therefore, $dx/dt = dy/dt = dz/dt = 0$, and these three points are equilibrium points.

(b) For equilibrium points, we must have $dx/dt = dy/dt = dz/dt = 0$. We therefore obtain the three simultaneous equations

$$\begin{cases} 10(y - x) = 0 \\ 28x - y - xz = 0 \\ -\frac{8}{3}z + xy = 0. \end{cases}$$

From the first equation, $x = y$. Eliminating y, we obtain

$$\begin{cases} x(27 - z) = 0 \\ -\frac{8}{3}z + x^2 = 0 \end{cases}$$

Then, $x = 0$ or $z = 27$. With $x = 0$, $z = 0$. With $z = 27$, $x^2 = 72$, hence $y = x = \pm 6\sqrt{2}$.

2. For equilibrium points, we must have $dx/dt = dy/dt = dz/dt = 0$. We obtain the three simultaneous equations

$$\begin{cases} 10(y - x) = 0 \\ \rho x - y - xz = 0 \\ -\frac{8}{3}z + xy = 0. \end{cases}$$

The first equation implies $x = y$. Eliminating y, we obtain

$$\begin{cases} x(\rho - 1 - z) = 0 \\ -\frac{8}{3}z + x^2 = 0. \end{cases}$$

Thus, $x = 0$, or $z = \rho - 1$. If $x = 0$ and therefore $y = 0$, then $z = 0$ by the last equation. Hence the origin $(0, 0, 0)$ is an equilibrium point for any value of ρ.

If $z = \rho - 1$, the last equation implies that $x^2 = 8(\rho - 1)/3$.

(a) If $\rho < 1$, the equation $x^2 = 8(\rho - 1)/3$ has no solutions. If $\rho = 1$, its only solution is $x = 0$, which corresponds to the equilibrium point at the origin.

(b) If $\rho > 1$, the equation $x^2 = 8(\rho - 1)/3$ has two solutions, $x = \pm\sqrt{8(\rho - 1)/3}$. Hence there are two more equilibrium points, at $x = y = \pm\sqrt{8(\rho - 1)/3}$ and $z = \rho - 1$.

(c) Since the number of equilibrium points jumps from 1 to 3 as ρ passes through the value $\rho = 1$, $\rho = 1$ is a bifurcation value for this system.

3. **(a)** We have

$$\frac{dx}{dt} = 10(y - x) = 0 \quad \text{and} \quad \frac{dy}{dt} = 28x - y = 0,$$

so $x(t) = y(t) = 0$ for all t if $x(0) = y(0) = 0$.

(b) We have

$$\frac{dz}{dt} = -\frac{8}{3}z,$$

so $z(t) = ce^{-8t/3}$. Since $z(0) = 1$, it follows that $c = 1$, and the solution is $x(t) = 0$, $y(t) = 0$, and $z(t) = e^{-8t/3}$.

(c) If $z(0) = z_0$, it follows that $c = z_0$, so the solution is $x(t) = 0$, $y(t) = 0$, and $z(t) = z_0e^{-8t/3}$.

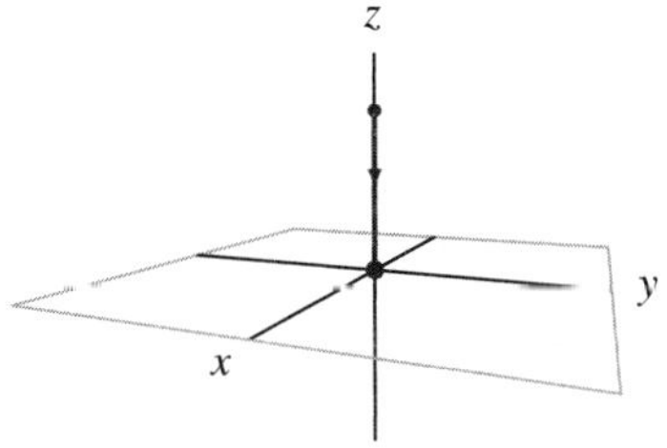

4. Let the parameter $r = 28$. If you select any initial condition that is not an equilibrium point, the solution winds around one of the two nonzero equilibrium points. A second solution whose initial condition differs from the first in the third decimal place is also computed. After a short interval of time, this second solution behaves in a manner that is quite different from the original solution. That is, it winds about the equilibrium points in a completely different pattern. While the two solutions ultimately seem to trace out the same figure, they do so in two very different ways.

No matter which two nearby initial conditions are selected, the result appears to be the same. Within a very short interval of time (usually less than the amount of time it takes the solutions to make twenty revolutions about the equilibrium points), the two solutions have separated and their subsequent trajectories are quite distinct.

5. **(a)** **(b)** **(c)**

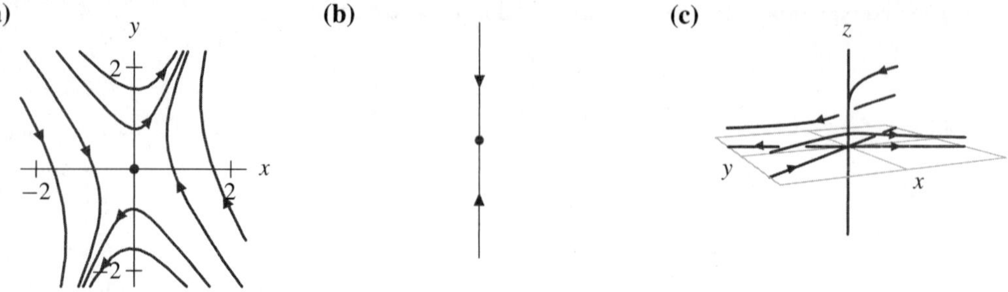

REVIEW EXERCISES FOR CHAPTER 2

1. The simplest solution is an equilibrium solution, and the origin is an equilibrium point for this system. Hence, the equilibrium solution $(x(t), y(t)) = (0, 0)$ for all t is a solution.

2. Note that $dy/dt > 0$ for all (x, y). Hence, there are no equilibrium points for this system.

3. Let $v = dy/dt$. Then $dv/dt = d^2y/dt^2$, and we obtain the system

$$\frac{dy}{dt} = v$$
$$\frac{dv}{dt} = 1.$$

4. First we solve $dv/dt = 1$ and get $v(t) = t + c_1$, where c_1 is an arbitrary constant. Next we solve $dy/dt = v = t + c_1$ and obtain $y(t) = \frac{1}{2}t^2 + c_1 t + c_2$, where c_2 is an arbitrary constant. Therefore, The general solution of the system is

$$y(t) = \tfrac{1}{2}t^2 + c_1 t + c_2$$
$$v(t) = t + c_1.$$

5. The equation for dx/dt gives $y = 0$. If $y = 0$, then $\sin(xy) = 0$, so $dy/dt = 0$. Hence, every point on the x-axis is an equilibrium point.

6. The second-order equation for this harmonic oscillator is

$$\beta\frac{d^2y}{dt^2} + \gamma\frac{dy}{dt} + \alpha y = 0.$$

The corresponding system is

$$\frac{dy}{dt} = v$$
$$\frac{dv}{dt} = -\frac{\alpha}{\beta}y - \frac{\gamma}{\beta}v.$$

7. From the equation for dx/dt, we know that $x(t) = k_1 e^{2t}$, where k_1 is an arbitrary constant, and from the equation for dy/dt, we have $y(t) = k_2 e^{-3t}$, where k_2 is another arbitrary constant. The general solution is $(x(t), y(t)) = (k_1 e^{2t}, k_2 e^{-3t})$.

8. Note that (0, 2) is an equilibrium point for this system. Hence, the solution with this initial condition is an equilibrium solution.

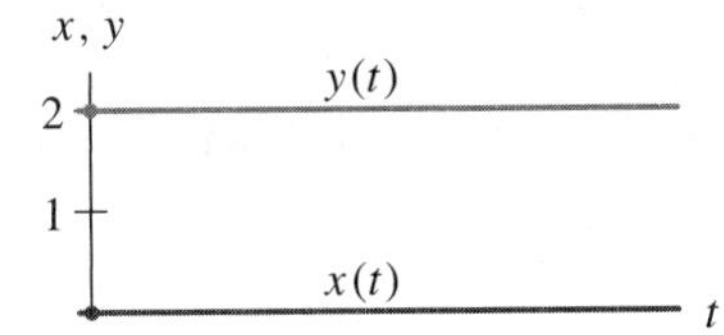

9. There are many examples. One is

$$\begin{aligned}\frac{dx}{dt} &= (x^2-1)(x^2-4)(x^2-9)(x^2-16)(x^2-25)\\ \frac{dy}{dt} &= y.\end{aligned}$$

This system has equilibria at $(\pm 1, 0)$, $(\pm 2, 0)$, $(\pm 3, 0)$, $(\pm 4, 0)$, and $(\pm 5, 0)$.

10. One step of Euler's method is

$$\begin{aligned}(2,1) + \Delta t\,\mathbf{F}(2,1) &= (2,1) + 0.5\,(3,2)\\ &= (3.5, 2).\end{aligned}$$

11. The point (1, 1) is on the line $y = x$. Along this line, the vector field for the system points toward the origin. Therefore, the solution curve consists of the half-line $y = x$ in the first quadrant. Note that the point (0, 0) is not on this curve.

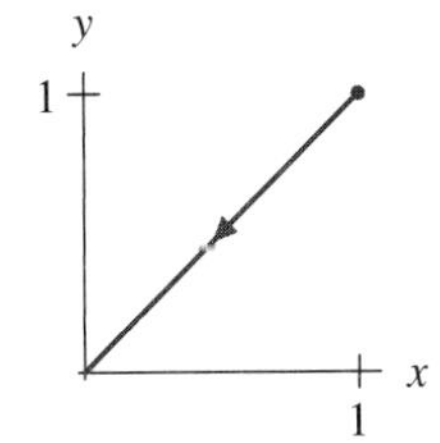

12. Let $\mathbf{F}(x, y) = (f(x, y), g(x, y))$ be the vector field for the original system. The vector field for the new system is

$$\begin{aligned}\mathbf{G}(x,y) &= (-f(x,y), -g(x,y))\\ &= -(f(x,y), g(x,y))\\ &= -\mathbf{F}(x,y).\end{aligned}$$

In other words, the directions of vectors in the new field are the opposite of the directions in the original field. Consequently, the phase portrait of new system has the same solution curves as the original phase portrait except that their directions are reversed. Hence, all solutions tend away from the origin as t increases.

13. True. First, we check the equation for dx/dt. We have

$$\frac{dx}{dt} = \frac{d(e^{-6t})}{dt} = -6e^{-6t},$$

and

$$2x - 2y^2 = 2(e^{-6t}) - 2(2e^{-3t})^2 = 2e^{-6t} - 8e^{-6t} = -6e^{-6t}.$$

Since that equation holds, we check the equation for dy/dt. We have

$$\frac{dy}{dt} = \frac{d(2e^{-3t})}{dt} = -6e^{-3t},$$

and

$$-3y = -3(2e^{-3t}) = -6e^{-3t}.$$

Since the equations for both dx/dt and dy/dt hold, the function $(x(t), y(t)) = (e^{-6t}, 2e^{-3t})$ is a solution of this system.

14. False. A solution to this system must consist of a pair $(x(t), y(t))$ of functions.

15. False. The components of the vector field are the right-hand sides of the equations of the system.

16. True. For example,

$$\begin{aligned} \frac{dx}{dt} &= y \\ \frac{dy}{dt} &= x \end{aligned} \qquad \text{and} \qquad \begin{aligned} \frac{dx}{dt} &= 2y \\ \frac{dy}{dt} &= 2x \end{aligned}$$

have the same direction field. The vectors in their vector fields differ only in length.

17. False. Note that $(x(0), y(0)) = (x(\pi), y(\pi)) = (0, 0)$. However, $(dx/dt, dy/dt) = (1, 1)$ at $t = 0$, and $(dx/dt, dy/dt) = (-1, -1)$ at $t = \pi$. For an autonomous system, the vector in the vector field at any given point does not vary as t varies. This function cannot be a solution of any autonomous system. (This function parameterizes a line segment in the xy-plane from $(1, 1)$ to $(-1, -1)$. In fact, it sweeps out the segment twice for $0 \le t \le 2\pi$.)

18. True. For an autonomous system, the rates of change of solutions depend only on position, not on time. Hence, if a function $(x_1(t), y_1(t))$ satisfies an autonomous system, then the function given by

$$(x_2(t), y_2(t)) = (x_1(t + T), y_1(t + T)),$$

where T is some constant, satisfies the same system.

19. False. The point $(0, 0)$ is an equilibrium point, so the Uniqueness Theorem guarantees that it is not on the solution curve corresponding to $(1, 0)$.

20. False. From the Uniqueness Theorem, we know that the solution curve with initial condition $(1/2, 0)$ is trapped by other solution curves that it cannot cross (or even touch). Hence, $x(t)$ and $y(t)$ must remain bounded for all t.

21. False. These solutions are different because they have different values at $t = 0$. However, they do trace out the same curve in the phase plane.

22. True. The solution curve is in the second quadrant and tends toward the equilibrium point $(0, 0)$ as $t \to \infty$. It never touches $(0, 0)$ by the Uniqueness Theorem.

23. False. The function $y(t)$ decreases monotonically, but $x(t)$ increases until it reaches its maximum at $x = -1$. It decreases monotonically after that.

24. False. The graph of $x(t)$ for this solution has exactly one local maximum and no other critical points. The graph of $y(t)$ has four critical points, two local minimums and two local maximums.

25. **(a)** The equilibrium points satisfy the equations $x = 2y$ and $\cos 2y = 0$. From the second equation, we conclude that

$$2y = \frac{\pi}{2} + k\pi,$$

where $k = 0, \pm 1, \pm 2, \ldots$. Since $2y = x$, we see that the equilibria are

$$(x, y) = \ldots, (-3\pi/2, -3\pi/4), (-\pi/2, -\pi/4), (\pi/2, \pi/4), (3\pi/2, 3\pi/4), (5\pi/2, 5\pi/4), \ldots$$

(b)

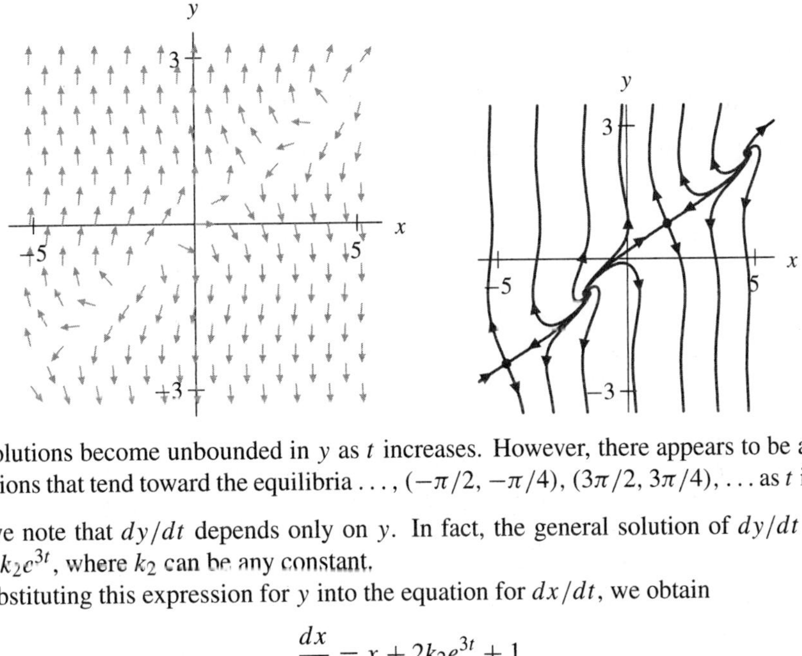

(c) Most solutions become unbounded in y as t increases. However, there appears to be a "curve" of solutions that tend toward the equilibria $\ldots, (-\pi/2, -\pi/4), (3\pi/2, 3\pi/4), \ldots$ as t increases.

26. **(a)** First, we note that dy/dt depends only on y. In fact, the general solution of $dy/dt = 3y$ is $y(t) = k_2e^{3t}$, where k_2 can be any constant.

Substituting this expression for y into the equation for dx/dt, we obtain

$$\frac{dx}{dt} = x + 2k_2e^{3t} + 1.$$

The general solution of the associated homogeneous equation is $x_h(t) = k_1e^t$. To find a particular solution of the nonhomogeneous equation, we guess $x_p(t) = ae^{3t} + b$. Substituting this guess into the equation gives

$$3ae^{3t} = ae^{3t} + b + 2k_2e^{3t} + 1,$$

so if $x_p(t)$ is a solution, we must have $3a = a + 2k_2$ and $b + 1 = 0$. Hence, $a = k_2$ and $b = -1$, and the function $x_p(t) = k_2e^{3t} - 1$ is a solution of the nonhomogeneous equation.

Therefore, the general solution of the system is

$$x(t) = k_1e^t + k_2e^{3t} - 1$$
$$y(t) = k_2e^{3t}.$$

(b) To find the equilibrium points, we solve the system of equations

$$\begin{cases} x + 2y + 1 = 0 \\ 3y = 0, \end{cases}$$

so $(x, y) = (-1, 0)$ is the only equilibrium point.

(c) To find the solution with initial condition $(-1, 3)$, we set

$$\begin{aligned} -1 &= x(0) = k_1 + k_2 - 1 \\ 3 &= y(0) = k_2, \end{aligned}$$

so $k_2 = 3$ and $k_1 = -3$. The solution with the desired initial condition is

$$(x(t), y(t)) = (-3e^t + 3e^{3t} - 1, 3e^{3t}).$$

(d)

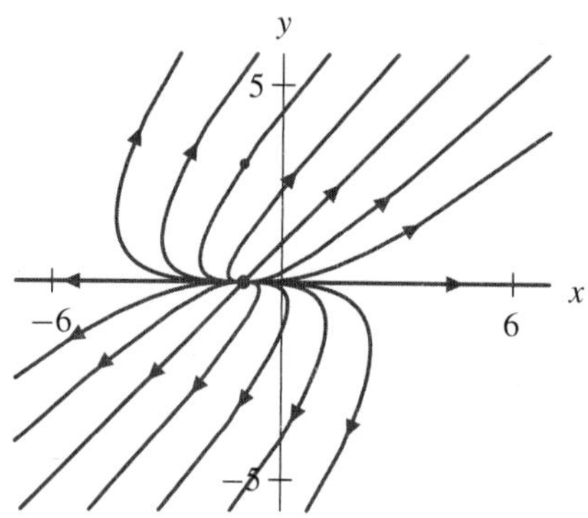

27. **(a)** For this system, we note that the equation for dy/dt depends only on y. In fact, this equation is separable and linear, so we have a choice of techniques for finding the general solution. The general solution for y is $y(t) = -1 + k_1 e^t$, where k_1 can be any constant.

Substituting $y = -1 + k_1 e^t$ into the equation for dx/dt, we have

$$\frac{dx}{dt} = (-1 + k_1 e^t)x.$$

This equation is a homogeneous linear equation, and its general solution is

$$x(t) = k_2 e^{-t + k_1 e^t},$$

where k_2 is any constant. The general solution for the system is therefore

$$(x(t), y(t)) = (k_2 e^{-t + k_1 e^t}, -1 + k_1 e^t),$$

where k_1 and k_2 are constants which we can adjust to satisfy any given initial condition.

(b) Setting $dy/dt = 0$, we obtain $y = -1$. From $dx/dt = xy = 0$, we see that $x = 0$. Therefore, this system has exactly one equilibrium point, $(x, y) = (0, -1)$.

(c) If $(x(0), y(0)) = (1, 0)$, then we must solve the simultaneous equations

$$\begin{cases} k_2 e^{k_1} = 1 \\ -1 + k_1 = 0. \end{cases}$$

Hence, $k_1 = 1$, and $k_2 = 1/e$. The solution to the initial-value problem is

$$(x(t), y(t)) = \left(e^{-1} e^{-t+e^t}, -1 + e^t\right) = \left(e^{e^t - t - 1}, -1 + e^t\right).$$

(d)

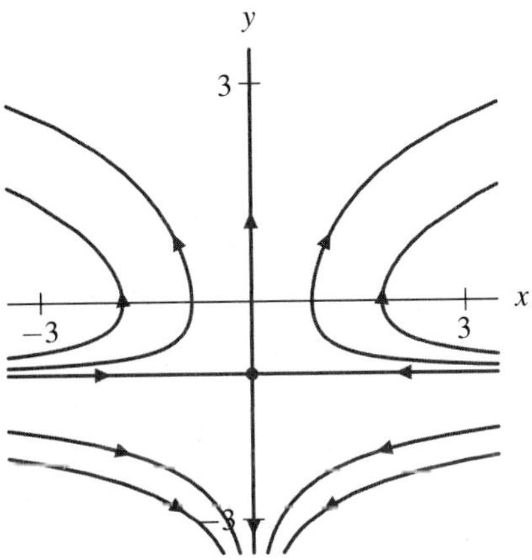

28. If x_1 is a root of $f(x)$ (that is, $f(x_1) = 0$), then the line $x = x_1$ is invariant. In other words, given an initial condition of the form (x_1, y), the corresponding solution curve remains on the line for all t. Along the line $x = x_1$, $y(t)$ obeys $dy/dt = g(y)$, so the line $x = x_1$ looks like the phase line of the equation $dy/dt = g(y)$.

Similarly, if $g(y_1) = 0$, then the line $y = y_1$ looks like the phase line for $dx/dt = f(x)$ except that it is horizontal rather than vertical.

Combining these two observations, we see that there will be vertical phase lines in the phase portrait for each root of $f(x)$ and horizontal phase lines in the phase portrait for each root of $g(y)$.

29.

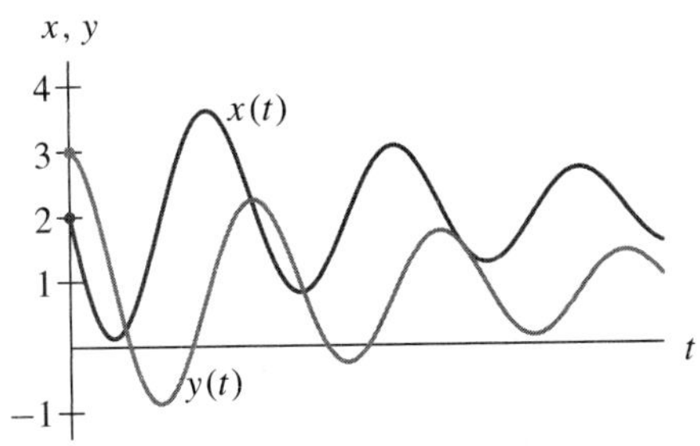

30.

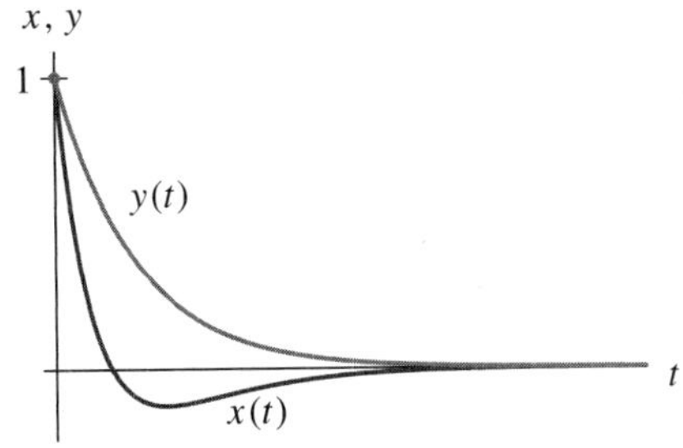

31.

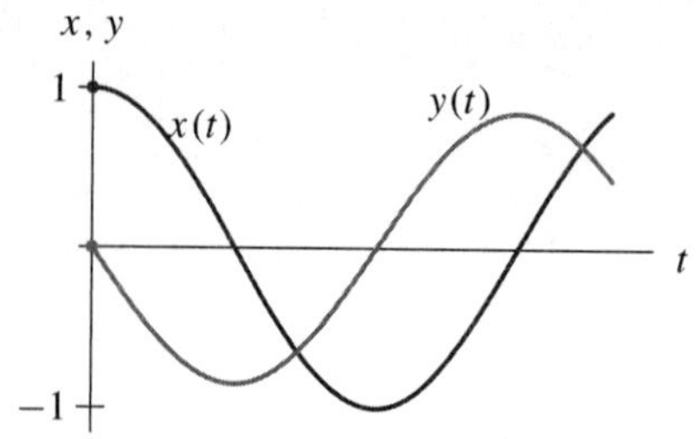

32.

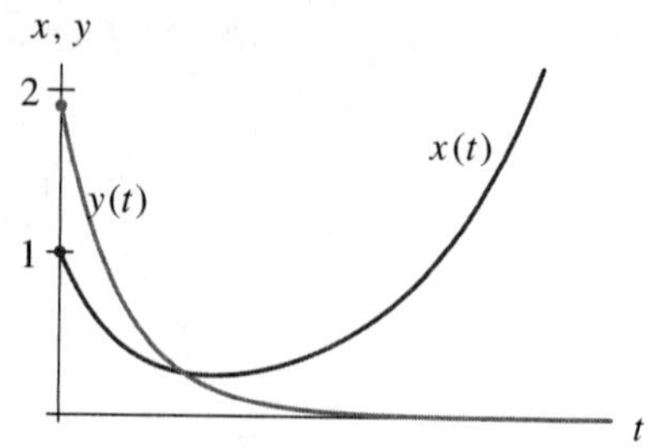

33. **(a)** Since θ represents an angle in this model, we restrict θ to the interval $-\pi < \theta < \pi$. The equilibria must satisfy the equations

$$\begin{cases} \cos\theta = s^2 \\ \sin\theta = -Ds^2. \end{cases}$$

Therefore,

$$\tan\theta = \frac{\sin\theta}{\cos\theta} = \frac{-Ds^2}{s^2} = -D,$$

and consequently, $\theta = -\arctan D$.

To find s, we note that $s^2 = \cos(-\arctan D)$. From trigonometry, we know that

$$\cos(-\arctan D) = \frac{1}{\sqrt{1+D^2}}.$$

If $-\pi < \theta < \pi$, there is a single equilibrium point for each value of the parameter D. It is

$$(\theta, s) = \left(-\arctan D, \frac{1}{\sqrt[4]{1+D^2}}\right).$$

(b) The equilibrium point represents motion along a line at a given angle from the horizon with a constant speed.

34. **(a)**

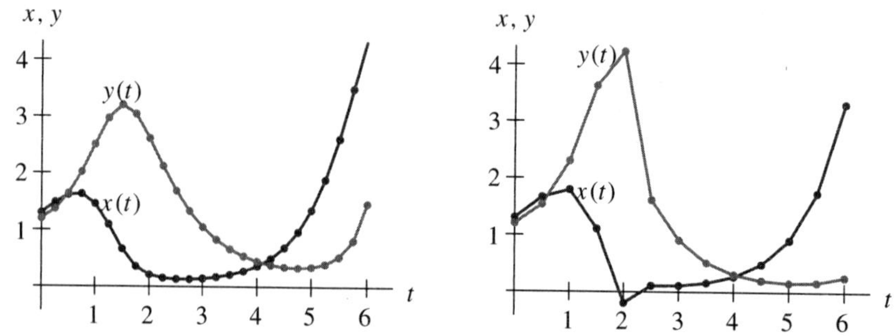

$\Delta t = 0.25$ $\Delta t = 0.50$

(b) As expected, the points that correspond to the $\Delta t = 0.25$ approximate solution are 0.25 apart in the t-direction, and the points that correspond to the $\Delta t = 0.5$ approximation are twice as far apart in the t-direction. Note that the $\Delta t = 0.5$ approximate solution varies much more than the $\Delta t = 0.25$ solution. In fact, its $x(t)$-graph crosses the t-axis. Presumedly, this difference is because the $\Delta t = 0.25$ solution is more accurate.

CHAPTER 3

Linear Systems

OR SECTION 3.1

Paul's making a profit ($x > 0$) has a beneficial effect on Paul's profits in the future x term makes a positive contribution to dx/dt. However, since $b < 0$, Bob's making 0) hinders Paul's ability to make profit because the by term contributes negatively to dx/dt. Roughly speaking, business is good for Paul if his store is profitable and Bob's is not. In fact, since $dx/dt = x - y$, Paul's profits will increase whenever his store is more profitable than Bob's.

Even though $dx/dt = dy/dt = x - y$ for this choice of parameters, the interpretation of the equation is exactly the opposite from Bob's point of view. Since $d < 0$, Bob's future profits are hurt whenever he is profitable because $dy < 0$. But Bob's profits are helped whenever Paul is profitable since $cx > 0$. Once again, since $dy/dt = x - y$, Bob's profits will increase whenever Paul's store is more profitable than his.

Finally, note that both x and y change by identical amounts since dx/dt and dy/dt are always equal.

2. Since $a = 2$, Paul's making a profit ($x > 0$) has a beneficial effect on Paul's future profits because the ax term makes a positive contribution to dx/dt. However, since $b = -1$, Bob's making a profit ($y > 0$) hinders Paul's ability to make profit because the by term contributes negatively to dx/dt. In some sense, Paul's profitability has twice the impact on his profits as does Bob's profitability. For example, Paul's profits will increase whenever his profits are at least one-half of Bob's profits since $dx/dt = 2x - y$.

Since $c = d = 0$, $dy/dt = 0$. Consequently, Bob's profits are not affected by the profitability of either store, and hence his profits are constant in this model.

3. Since $a = 1$ and $b = 0$, we have $dx/dt = x$. Hence, if Paul is making a profit ($x > 0$), then those profits will increase since dx/dt is positive. However, Bob's profits have no effect on Paul's profits. (Note that $dx/dt = x$ is the standard exponential growth model.)

Since $c = 2$ and $d = 1$, profits from both stores have a positive effect on Bob's profits. In some sense, Paul's profits have twice the impact of Bob's profits on dy/dt.

4. Since $a = -1$ and $b = 2$, Paul's making a profit has a negative effect on his future profits. However, if Bob makes a profit, then Paul's profits benefit. Moreover, Bob's profitability has twice the impact as does Paul's. In fact, since $dx/dt = -x + 2y$, Paul's profits will increase if $-x + 2y > 0$ or, in other words, if Bob's profits are at least one-half of Paul's profits.

Since $c = 2$ and $d = -1$, Bob is in the same situation as Paul. His profits contribute negatively to dy/dt since $d = -1$. However, Paul's profitability has twice the positive effect.

Note that this model is symmetric in the sense that both Paul and Bob perceive each others profits in the same way. This symmetry comes from the fact that $a = d$ and $b = c$.

5. $\mathbf{Y} = \begin{pmatrix} x \\ y \end{pmatrix}, \quad \dfrac{d\mathbf{Y}}{dt} = \begin{pmatrix} 2 & 1 \\ 1 & 1 \end{pmatrix}\mathbf{Y}$

6. $\mathbf{Y} = \begin{pmatrix} x \\ y \end{pmatrix}, \quad \dfrac{d\mathbf{Y}}{dt} = \begin{pmatrix} 0 & 3 \\ -0.3 & 3\pi \end{pmatrix}\mathbf{Y}$

7. $\mathbf{Y} = \begin{pmatrix} p \\ q \\ r \end{pmatrix}, \quad \dfrac{d\mathbf{Y}}{dt} = \begin{pmatrix} 3 & -2 & -7 \\ -2 & 0 & 6 \\ 0 & 7.3 & 2 \end{pmatrix}\mathbf{Y}$

8. $$\frac{dx}{dt} = -3x + 2\pi y$$
$$\frac{dy}{dt} = 4x - y$$

9. $$\frac{dx}{dt} = \beta y$$
$$\frac{dy}{dt} = \gamma x - y$$

10. **(a)**

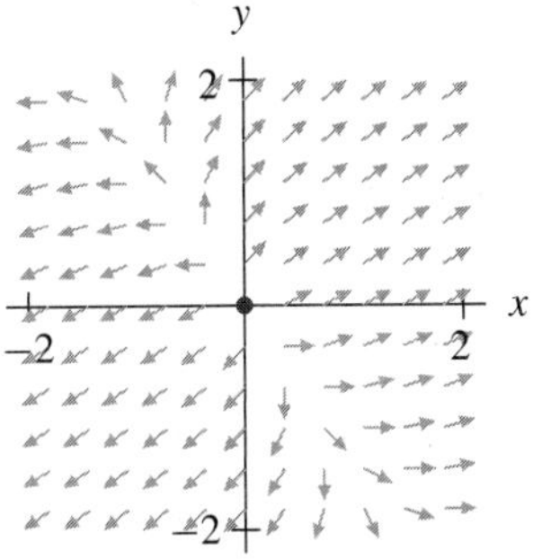

(b)

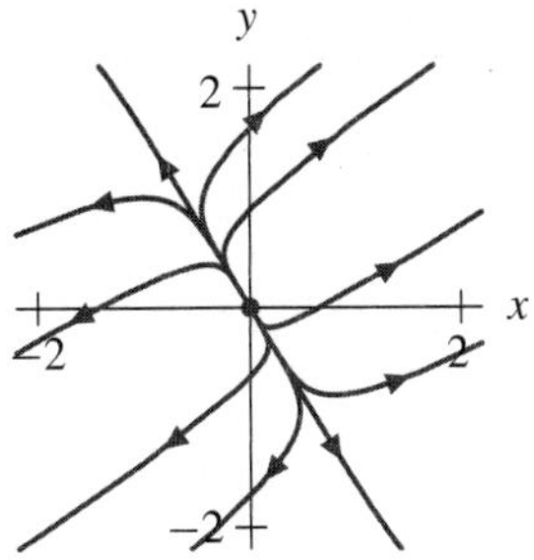

(c)

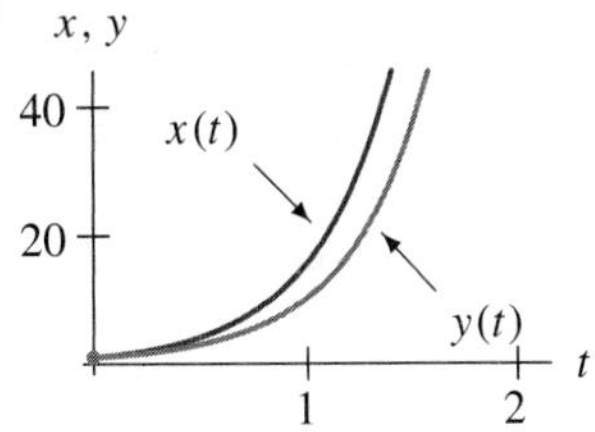

11. **(a)**

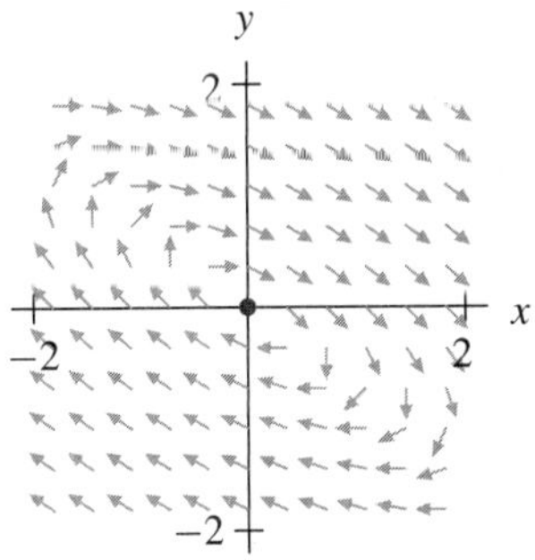

(b)

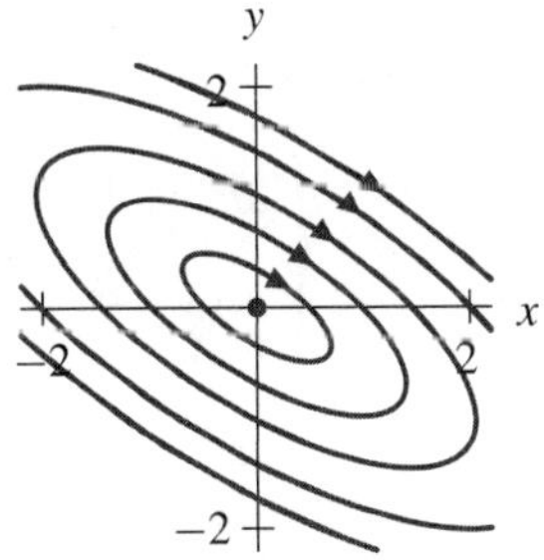

(c)

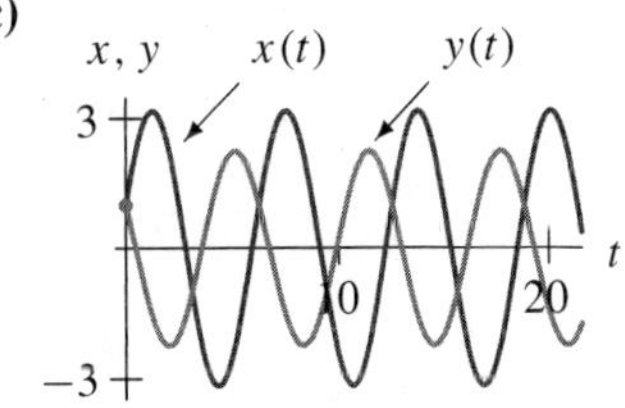

12. **(a)**

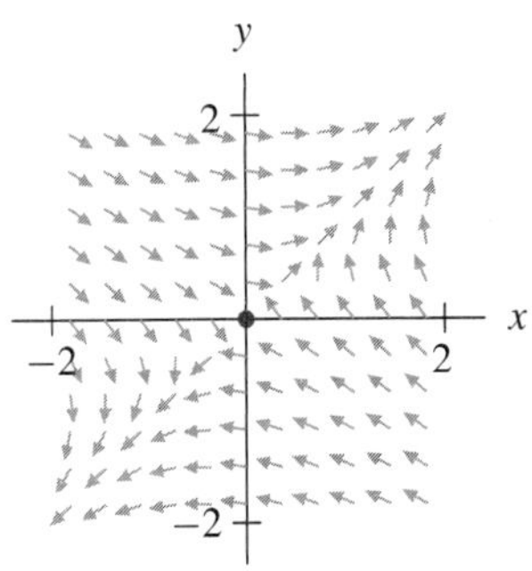

(b)

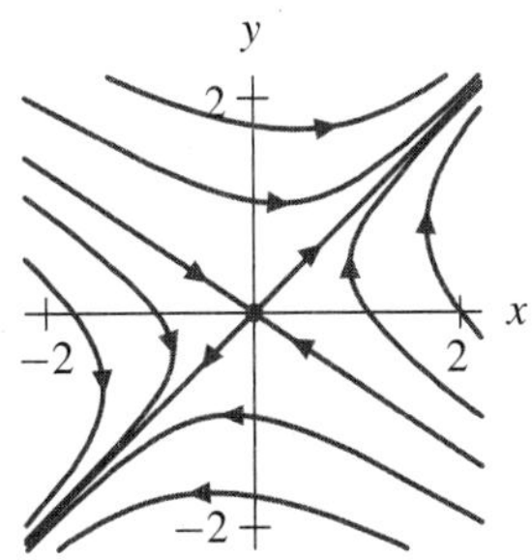

(c)

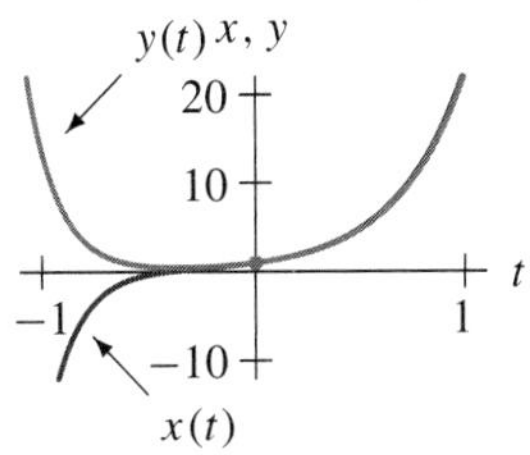

13. **(a)** **(b)** **(c)**

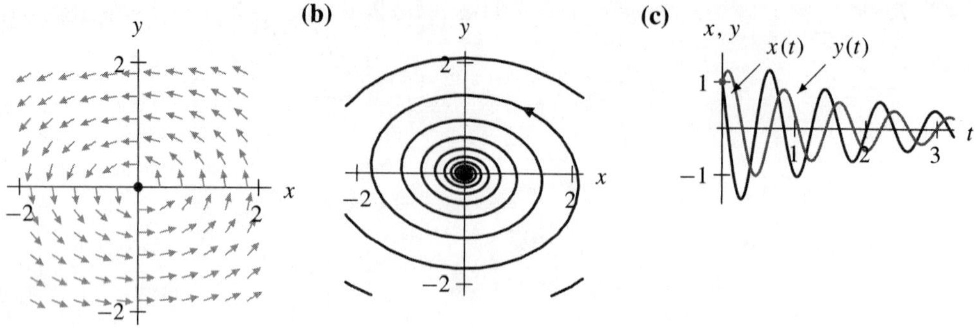

14. **(a)** If $a = 0$, then $\det \mathbf{A} = ad - bc = bc$. Thus both b and c are nonzero if $\det \mathbf{A} \neq 0$.

(b) Equilibrium points (x_0, y_0) are solutions of the simultaneous system of linear equations

$$\begin{cases} ax_0 + by_0 = 0 \\ cx_0 + dy_0 = 0. \end{cases}$$

If $a = 0$, the first equation reduces to $by_0 = 0$, and since $b \neq 0$, $y_0 = 0$. In this case, the second equation reduces to $cx_0 = 0$, so $x_0 = 0$ as well. Therefore, $(x_0, y_0) = (0, 0)$ is the only equilibrium point for the system.

15. The vector field at a point (x_0, y_0) is (ax_0+by_0, cx_0+dy_0), so in order for a point to be an equilibrium point, it must be a solution to the system of simultaneous linear equations

$$\begin{cases} ax_0 + by_0 = 0 \\ cx_0 + dy_0 = 0. \end{cases}$$

If $a \neq 0$, we know that the first equation is satisfied if and only if

$$x_0 = -\frac{b}{a} y_0.$$

Now we see that any point that lies on this line $x_0 = (-b/a)y_0$ also satisfies the second linear equation $cx_0 + dy_0 = 0$. In fact, if we substitute a point of this form into the second component of the vector field, we have

$$\begin{aligned} cx_0 + dy_0 &= c\left(-\frac{b}{a}\right) y_0 + dy_0 \\ &= \left(-\frac{bc}{a} + d\right) y_0 \\ &= \left(\frac{ad - bc}{a}\right) y_0 \\ &= \frac{\det \mathbf{A}}{a} y_0 \\ &= 0, \end{aligned}$$

since we are assuming that $\det \mathbf{A} = 0$. Hence, the line $x_0 = (-b/a)y_0$ consists entirely of equilibrium points.

If $a = 0$ and $b \neq 0$, then the determinant condition $\det \mathbf{A} = ad - bc = 0$ implies that $c = 0$. Consequently, the vector field at the point (x_0, y_0) is (by_0, dy_0). Since $b \neq 0$, we see that we get equilibrium points if and only if $y_0 = 0$. In other words, the set of equilibrium points is exactly the x-axis.

Finally, if $a = b = 0$, then the vector field at the point (x_0, y_0) is $(0, cx_0 + dy_0)$. In this case, we see that a point (x_0, y_0) is an equilibrium point if and only if $cx_0 + dy_0 = 0$. Since at least one of c or d is nonzero, the set of points (x_0, y_0) that satisfy $cx_0 + dy_0 = 0$ is precisely a line through the origin.

16. **(a)** Let $v = dy/dt$. Then $dv/dt = d^2y/dt^2 = -qy - p(dy/dt) = -qy - pv$. Thus we obtain the system

$$\frac{dy}{dt} = v$$
$$\frac{dv}{dt} = -qy - pv.$$

In matrix form, this system is written as

$$\begin{pmatrix} \frac{dy}{dt} \\ \frac{dv}{dt} \end{pmatrix} = \begin{pmatrix} 0 & 1 \\ -q & -p \end{pmatrix} \begin{pmatrix} y \\ v \end{pmatrix}.$$

(b) The determinant of this matrix is q. Hence, if $q \neq 0$, we know that the only equilibrium point is the origin.

(c) If y is constant, then $v = dy/dt$ is identically zero. Hence, $dv/dt = 0$.

Also, the system reduces to

$$\begin{pmatrix} \frac{dy}{dt} \\ \frac{dv}{dt} \end{pmatrix} = \begin{pmatrix} 0 & 1 \\ -q & -p \end{pmatrix} \begin{pmatrix} y \\ 0 \end{pmatrix},$$

which implies that $dv/dt = -qy$.

Combining these two observations, we obtain $dv/dt = -qy = 0$, and if $q \neq 0$, then $y = 0$.

17. The first-order system corresponding to this equation is

$$\frac{dy}{dt} = v$$
$$\frac{dv}{dt} = -qy - pv.$$

(a) If $q = 0$, then the system becomes

$$\frac{dy}{dt} = v$$
$$\frac{dv}{dt} = -pv,$$

and the equilibrium points are the solutions of the system of equations

$$\begin{cases} v = 0 \\ -pv = 0. \end{cases}$$

Thus, the point (y, v) is an equilibrium point if and only if $v = 0$. In other words, the set of all equilibria agrees with the horizontal axis in the yv-plane.

(b) If $p = q = 0$, then the system becomes

$$\frac{dy}{dt} = v$$
$$\frac{dv}{dt} = 0$$

but the equilibrium points are again the points with $v = 0$.

18. In this case, $dv/dt = d^2y/dt^2 = 0$, and the first-order system reduces to

$$\frac{dy}{dt} = v$$
$$\frac{dv}{dt} = 0.$$

(a) Since $dv/dt = 0$, we know that $v(t) = c$ for some constant c.

(b) Since $dy/dt = v = c$, we can integrate to obtain $y(t) = ct + k$ where k is another arbitrary constant. Hence, the general solution of the system consists of all functions of the form $(y(t), v(t)) = (ct + k, c)$ for arbitrary constants c and k.

(c)

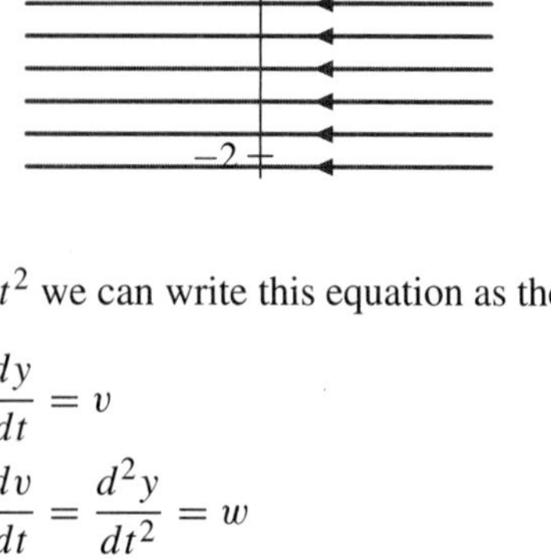

19. Letting $v = dy/dt$ and $w = d^2y/dt^2$ we can write this equation as the system

$$\frac{dy}{dt} = v$$
$$\frac{dv}{dt} = \frac{d^2y}{dt^2} = w$$
$$\frac{dw}{dt} = \frac{d^3y}{dt^3} = -ry - qv - pw.$$

In matrix notation, this system is

$$\frac{d\mathbf{Y}}{dt} = \mathbf{AY}$$

where

$$\mathbf{Y} = \begin{pmatrix} y \\ v \\ w \end{pmatrix} \quad \text{and} \quad \mathbf{A} = \begin{pmatrix} 0 & 1 & 0 \\ 0 & 0 & 1 \\ -r & -q & -p \end{pmatrix}.$$

20. If there are more than the usual number of buyers, then $b > 0$. If this level of buying means that prices will increase and that fewer buyers will enter the market, then the effect on db/dt should be negative. Since $db/dt = \alpha b + \beta s$, we expect that the αb-term will be negative if $b > 0$. Consequently, α should be negative.

21. If there are fewer than the usual number of buyers, then $b < 0$. If this level of b has a negative effect on the number of sellers, we expect the γb-term in ds/dt to be negative. If $\gamma b < 0$ and $b < 0$, then we must have $\gamma > 0$.

22. If $s > 0$, there are more than the usual number of houses for sale and house prices should decline. Declining prices should have a positive effect on the number of buyers and a negative effect on the number of sellers. Since $db/dt = \alpha b + \beta s$, we expect the βs-term to be positive. Since $\beta s > 0$ if $s > 0$, the parameter β should be positive.

23. In the model, $ds/dt = \gamma b + \delta s$. If $s > 0$, then the number of sellers is greater than usual and house prices should decline. Since declining prices should have a negative effect on the number of sellers, we expect the δs-term to be negative. If $\delta s < 0$ when $s > 0$, we should have $\delta < 0$.

24. **(a)** Substituting $\mathbf{Y}_1(t)$ in the left-hand side of the differential equation yields

$$\frac{d\mathbf{Y}_1}{dt} = \begin{pmatrix} 0 \\ e^t \end{pmatrix}.$$

Moreover, the right-hand side becomes

$$\begin{pmatrix} 2 & 0 \\ 1 & 1 \end{pmatrix} \mathbf{Y}_1(t) = \begin{pmatrix} 2 & 0 \\ 1 & 1 \end{pmatrix} \begin{pmatrix} 0 \\ e^t \end{pmatrix} = \begin{pmatrix} 0 \\ e^t \end{pmatrix}.$$

Since the two sides of the differential equation agree, $\mathbf{Y}_1(t)$ is a solution.

Similarly, if we substitute $\mathbf{Y}_2(t)$ in the left-hand side of the differential equation, we get

$$\frac{d\mathbf{Y}_2}{dt} = \begin{pmatrix} 2e^{2t} \\ 2e^{2t} \end{pmatrix}.$$

Moreover, the right-hand side is

$$\begin{pmatrix} 2 & 0 \\ 1 & 1 \end{pmatrix} \mathbf{Y}_2(t) = \begin{pmatrix} 2 & 0 \\ 1 & 1 \end{pmatrix} \begin{pmatrix} e^{2t} \\ e^{2t} \end{pmatrix}$$

$$= \begin{pmatrix} 2e^{2t} \\ e^{2t} + e^{2t} \end{pmatrix}$$

$$= \begin{pmatrix} 2e^{2t} \\ 2e^{2t} \end{pmatrix}.$$

Since the two sides of the differential equation also agree for this function, $\mathbf{Y}_2(t)$ is another solution.

(b) At $t = 0$, $\mathbf{Y}(0) = (-2, -1)$. By the Linearity Principle, any linear combination of two solutions is also a solution. Hence, we solve the given initial-value problem with a function of the form $k_1\mathbf{Y}_1(t) + k_2\mathbf{Y}_2(t)$ where k_1 and k_2 are constants determined by the initial value. That is, we determine k_1 and k_2 via

$$k_1\mathbf{Y}_1(0) + k_2\mathbf{Y}_2(0) = \mathbf{Y}(0) = \begin{pmatrix} -2 \\ -1 \end{pmatrix}.$$

We get

$$k_1 \begin{pmatrix} 0 \\ 1 \end{pmatrix} + k_2 \begin{pmatrix} 1 \\ 1 \end{pmatrix} = \begin{pmatrix} -2 \\ -1 \end{pmatrix}.$$

This vector equation is equivalent to the simultaneous linear equations

$$\begin{cases} k_2 = -2 \\ k_1 + k_2 = -1. \end{cases}$$

From the first equation, we have $k_2 = -2$. Then from the second equation, we obtain $k_1 = 1$. Therefore, the solution to the initial-value problem is

$$\mathbf{Y}(t) = \mathbf{Y}_1(t) - 2\mathbf{Y}_2(t)$$

$$= \begin{pmatrix} 0 \\ e^t \end{pmatrix} - 2 \begin{pmatrix} e^{2t} \\ e^{2t} \end{pmatrix}$$

$$= \begin{pmatrix} -2e^{2t} \\ e^t - 2e^{2t} \end{pmatrix}.$$

Note that (as always) we can check our calculations directly. By direct evaluation, we know that $\mathbf{Y}(0) = (-2, -1)$. Moreover, we can check that $\mathbf{Y}(t)$ satisfies the differential equation. The left-hand side of the differential equation is

$$\frac{d\mathbf{Y}}{dt} = \begin{pmatrix} -4e^{2t} \\ e^t - 4e^{2t} \end{pmatrix},$$

and the right-hand side of the differential equation is

$$\begin{pmatrix} 2 & 0 \\ 1 & 1 \end{pmatrix} \mathbf{Y}(t) = \begin{pmatrix} 2 & 0 \\ 1 & 1 \end{pmatrix} \begin{pmatrix} -2e^{2t} \\ e^t - 4e^{2t} \end{pmatrix}$$

$$= \begin{pmatrix} -4e^{2t} \\ e^t - 4e^{2t} \end{pmatrix}.$$

Since the left-hand side and the right-hand side agree, the function $\mathbf{Y}(t)$ is a solution to the differential equation, and since it assumes the given initial value, this function is the desired solution to the initial-value problem. The Uniqueness Theorem says that this function is the only solution to the initial-value problem.

25. **(a)** Note that substituting $\mathbf{Y}(t)$ into the left-hand side of the differential equation, we get

$$\frac{d\mathbf{Y}}{dt} = \begin{pmatrix} e^{2t} + 2te^{2t} \\ -e^{2t} - 2(t+1)e^{2t} \end{pmatrix}$$

$$= \begin{pmatrix} e^{2t} + 2te^{2t} \\ -3e^{2t} - 2te^{2t} \end{pmatrix}.$$

Substituting $\mathbf{Y}(t)$ into the right-hand side, we get

$$\begin{pmatrix} 1 & -1 \\ 1 & 3 \end{pmatrix} \begin{pmatrix} te^{2t} \\ -(t+1)e^{2t} \end{pmatrix} = \begin{pmatrix} te^{2t} + (t+1)e^{2t} \\ te^{2t} - 3(t+1)e^{2t} \end{pmatrix}$$

$$= \begin{pmatrix} e^{2t} + 2te^{2t} \\ -3e^{2t} - 2te^{2t} \end{pmatrix}.$$

Since the left-hand side of the differential equation equals the right-hand side, the function $\mathbf{Y}(t)$ is a solution.

(b) At $t = 0$, $\mathbf{Y}(0) = (0, -1)$. By the Linearity Principle, any constant multiple of the solution $\mathbf{Y}(t)$ is also a solution. Since the function $-2\mathbf{Y}(t)$ has the desired initial condition, we know that

$$-2\mathbf{Y}(t) = \begin{pmatrix} -2te^{2t} \\ (2t+2)e^{2t} \end{pmatrix}$$

is the desired solution. By the Uniqueness Theorem, this is the only solution with this initial condition. (Given the formula for $-2\mathbf{Y}(t)$ directly above, note that we can directly check our assertion that this function solves the initial-value problem without appealing to the Linearity Principle.)

26. **(a)** Substitute $\mathbf{Y}_1(t)$ into the differential equation and compare the left-hand side to the right-hand side. On the left-hand side, we have

$$\frac{d\mathbf{Y}_1}{dt} = \begin{pmatrix} -3e^{-3t} \\ -3e^{-3t} \end{pmatrix},$$

and on the right-hand side, we have

$$\mathbf{A}\mathbf{Y}_1(t) = \begin{pmatrix} -2 & -1 \\ 2 & -5 \end{pmatrix} \begin{pmatrix} e^{-3t} \\ e^{-3t} \end{pmatrix} = \begin{pmatrix} -2e^{-3t} - e^{-3t} \\ 2e^{-3t} - 5e^{-3t} \end{pmatrix} = \begin{pmatrix} -3e^{-3t} \\ -3e^{-3t} \end{pmatrix}.$$

Since the two sides agree, we know that $\mathbf{Y}_1(t)$ is a solution.

For $\mathbf{Y}_2(t)$,

$$\frac{d\mathbf{Y}_2}{dt} = \begin{pmatrix} -4e^{-4t} \\ -8e^{-4t} \end{pmatrix},$$

and

$$\mathbf{A}\mathbf{Y}_2(t) = \begin{pmatrix} -2 & -1 \\ 2 & -5 \end{pmatrix} \begin{pmatrix} e^{-4t} \\ 2e^{-4t} \end{pmatrix} = \begin{pmatrix} -2e^{-4t} - 2e^{-4t} \\ 2e^{-4t} - 10e^{-4t} \end{pmatrix} = \begin{pmatrix} -4e^{-4t} \\ -8e^{-4t} \end{pmatrix}.$$

Since the two sides agree, the function $\mathbf{Y}_2(t)$ is also a solution.

Both $\mathbf{Y}_1(t)$ and $\mathbf{Y}_2(t)$ are solutions, and we proceed to the next part of the exercise.

(b) Note that $\mathbf{Y}_1(0) = (1, 1)$ and $\mathbf{Y}_2(0) = (1, 2)$. These vectors are not on the same line through the origin, so the initial conditions are linearly independent. If the initial conditions are linearly independent, then the solutions must also be linearly independent. Since the two solutions are linearly independent, we proceed to part (c) of the exercise.

(c) We must find constants k_1 and k_2 such that

$$k_1\mathbf{Y}_1(0) + k_2\mathbf{Y}_2(0) = k_1 \begin{pmatrix} 1 \\ 1 \end{pmatrix} + k_2 \begin{pmatrix} 1 \\ 2 \end{pmatrix} = \begin{pmatrix} 2 \\ 3 \end{pmatrix}.$$

In other words, the constants k_1 and k_2 must satisfy the simultaneous system of linear equations

$$\begin{cases} k_1 + k_2 = 2 \\ k_1 + 2k_2 = 3. \end{cases}$$

It follows that $k_1 = 1$ and $k_2 = 1$. Hence, the required solution is

$$\mathbf{Y}_1(t) + \mathbf{Y}_2(t) = \begin{pmatrix} e^{-3t} + e^{-4t} \\ e^{-3t} + 2e^{-4t} \end{pmatrix}.$$

27. **(a)** Substitute $\mathbf{Y}_1(t)$ into the differential equation and compare the left-hand side to the right-hand side. On the left-hand side, we have

$$\frac{d\mathbf{Y}_1}{dt} = \begin{pmatrix} -3e^{-3t} + 8e^{-4t} \\ -3e^{-3t} + 16e^{-4t} \end{pmatrix},$$

and on the right-hand side, we have

$$\mathbf{A}\mathbf{Y}_1(t) = \begin{pmatrix} -2 & -1 \\ 2 & -5 \end{pmatrix} \begin{pmatrix} e^{-3t} - 2e^{-4t} \\ e^{-3t} - 4e^{-4t} \end{pmatrix} = \begin{pmatrix} -3e^{-3t} + 8e^{-4t} \\ -3e^{-3t} + 16e^{-4t} \end{pmatrix}.$$

Since the two sides agree, we know that $\mathbf{Y}_1(t)$ is a solution.

For $\mathbf{Y}_2(t)$,

$$\frac{d\mathbf{Y}_2}{dt} = \begin{pmatrix} -6e^{-3t} - 4e^{-4t} \\ -6e^{-3t} - 8e^{-4t} \end{pmatrix},$$

and

$$\mathbf{A}\mathbf{Y}_2(t) = \begin{pmatrix} -2 & -1 \\ 2 & -5 \end{pmatrix} \begin{pmatrix} 2e^{-3t} + e{-}4t \\ 2e^{-3t} + 2e^{-4t} \end{pmatrix} = \begin{pmatrix} -6e^{-3t} - 4e^{-4t} \\ -6e^{-3t} - 8e^{-4t} \end{pmatrix}.$$

Since the two sides agree, the function $\mathbf{Y}_2(t)$ is also a solution.

Both $\mathbf{Y}_1(t)$ and $\mathbf{Y}_2(t)$ are solutions, and we proceed to the next part of the exercise.

(b) Note that $\mathbf{Y}_1(0) = (-1, -3)$ and $\mathbf{Y}_2(0) = (3, 4)$. These vectors are not on the same line through the origin, so the initial conditions are linearly independent. If the initial conditions are linearly independent, then the solutions must also be linearly independent. Since the two solutions are linearly independent, we proceed to part (c) of the exercise.

(c) We must find constants k_1 and k_2 such that

$$k_1\mathbf{Y}_1(0) + k_2\mathbf{Y}_2(0) = k_1 \begin{pmatrix} -1 \\ -3 \end{pmatrix} + k_2 \begin{pmatrix} 3 \\ 4 \end{pmatrix} = \begin{pmatrix} 2 \\ 3 \end{pmatrix}.$$

In other words, the constants k_1 and k_2 must satisfy the simultaneous system of linear equations

$$\begin{cases} -k_1 + 3k_2 = 2 \\ -3k_1 + 4k_2 = 3. \end{cases}$$

It follows that $k_1 = -1/5$ and $k_2 = 3/5$. Hence, the required solution is

$$\frac{1}{5}\mathbf{Y}_1(t) + \frac{3}{5}\mathbf{Y}_2(t) = \begin{pmatrix} e^{-3t} + e^{-4t} \\ e^{-3t} + 2e^{-4t} \end{pmatrix}.$$

28. **(a)** First we substitute $\mathbf{Y}_1(t)$ into the differential equation. The left-hand side becomes

$$\begin{aligned} \frac{d\mathbf{Y}_1}{dt} &= -2e^{-2t} \begin{pmatrix} \cos 3t \\ \sin 3t \end{pmatrix} + e^{-2t} \begin{pmatrix} -3\sin 3t \\ 3\cos 3t \end{pmatrix} \\ &= e^{-2t} \begin{pmatrix} -2\cos 3t - 3\sin 3t \\ 3\cos 3t - 2\sin 3t \end{pmatrix}, \end{aligned}$$

and the right-hand side is

$$\mathbf{A}\mathbf{Y}_1(t) = \begin{pmatrix} -2 & -3 \\ 3 & -2 \end{pmatrix} \mathbf{Y}_1(t) = e^{-2t} \begin{pmatrix} -2\cos 3t - 3\sin 3t \\ 3\cos 3t - 2\sin 3t \end{pmatrix}.$$

Since the two sides of the differential equation agree, the function $\mathbf{Y}_1(t)$ is a solution.

Using $\mathbf{Y}_2(t)$, we have

$$\begin{aligned} \frac{d\mathbf{Y}_2}{dt} &= -2e^{-2t} \begin{pmatrix} -\sin 3t \\ \cos 3t \end{pmatrix} + e^{-2t} \begin{pmatrix} -3\cos 3t \\ -3\sin 3t \end{pmatrix} \\ &= e^{-2t} \begin{pmatrix} 2\sin 3t - 3\cos 3t \\ -3\sin 3t - 2\cos 3t \end{pmatrix}, \end{aligned}$$

and

$$\mathbf{A}\mathbf{Y}_2(t) = \begin{pmatrix} -2 & -3 \\ 3 & -2 \end{pmatrix} \mathbf{Y}_2(t) = e^{-2t} \begin{pmatrix} 2\sin 3t - 3\cos 3t \\ -3\sin 3t - 2\cos 3t \end{pmatrix}.$$

The two sides of the differential equation agree. Hence, $\mathbf{Y}_2(t)$ is also a solution.

Since both $\mathbf{Y}_1(t)$ and $\mathbf{Y}_2(t)$ are solutions, we proceed to part (b).

(b) Note that $\mathbf{Y}_1(0) = (1, 0)$ and $\mathbf{Y}_2(0) = (0, 1)$, and these vectors are not on the same line through the origin. Hence, $\mathbf{Y}_1(t)$ and $\mathbf{Y}_2(t)$ are linearly independent, and we proceed to part (c) of the exercise.

(c) To find the solution with the initial condition $\mathbf{Y}(0) = (2, 3)$, we must find constants k_1 and k_2 so that

$$k_1\mathbf{Y}_1(0) + k_2\mathbf{Y}_2(0) = \begin{pmatrix} 2 \\ 3 \end{pmatrix}.$$

We have $k_1 = 2$ and $k_2 = 3$, and the solution with initial condition (2, 3) is

$$\mathbf{Y}(t) = e^{-2t} \begin{pmatrix} 2\cos 3t - 3\sin 3t \\ 2\sin 3t + 3\cos 3t \end{pmatrix}.$$

29. **(a)** First, we check to see if $\mathbf{Y}_1(t)$ is a solution. The left-hand side of the differential equation is

$$\frac{d\mathbf{Y}_1}{dt} = \begin{pmatrix} e^{-t} + 36e^{3t} \\ -e^{-t} + 12e^{3t} \end{pmatrix},$$

and the right-hand side is

$$\mathbf{A}\mathbf{Y}_1(t) = \begin{pmatrix} 2 & 3 \\ 1 & 0 \end{pmatrix} \begin{pmatrix} -e^{-t} + 12e^{3t} \\ e^{-t} + 4e^{3t} \end{pmatrix} = \begin{pmatrix} e^{-t}36e^{3t} \\ -e^{-t} + 12e^{3t} \end{pmatrix}.$$

Consequently, $\mathbf{Y}_1(t)$ is a solution. However,

$$\frac{d\mathbf{Y}_2}{dt} = \begin{pmatrix} e^{-t} \\ -2e^{-t} \end{pmatrix},$$

and

$$\mathbf{A}\mathbf{Y}_2(t) = \begin{pmatrix} 2 & 3 \\ 1 & 0 \end{pmatrix} \begin{pmatrix} -e^{-t} \\ 2e^{-t} \end{pmatrix} = \begin{pmatrix} 4e^{-t} \\ -e^{-t} \end{pmatrix}.$$

Consequently, the function $\mathbf{Y}_2(t)$ is not a solution. In this case, we are not able to solve the given initial-value problem, so we stop here.

30. **(a)** This holds in all dimensions. In two dimensions the computation is

$$\begin{aligned} \mathbf{A}k\mathbf{Y} &= \begin{pmatrix} a & b \\ c & d \end{pmatrix} k \begin{pmatrix} x \\ y \end{pmatrix} \\ &= \begin{pmatrix} a & b \\ c & d \end{pmatrix} \begin{pmatrix} kx \\ ky \end{pmatrix} \end{aligned}$$

$$= \begin{pmatrix} akx + bky \\ ckx + dky \end{pmatrix}$$

$$= k \begin{pmatrix} ax + by \\ cx + dy \end{pmatrix} = k\mathbf{AY}.$$

(b) To verify the first half of the Linearity Principle, we suppose that $\mathbf{Y}_1(t) = (x_1(t), y_1(t))$ is a solution to the system

$$\frac{dx}{dt} = ax + by$$
$$\frac{dy}{dt} = cx + dy$$

and that k is an any constant. In order to verify that the function $\mathbf{Y}_2(t) = k\mathbf{Y}_1(t)$ is also a solution, we need to substitute $\mathbf{Y}_2(t)$ into both sides of the differential equation and check for equality. In other words, after we write $\mathbf{Y}_2(t)$ in scalar notation as $\mathbf{Y}_2(t) = (x_2(t), y_2(t))$, we must show that

$$\frac{dx_2}{dt} = ax_2 + by_2$$
$$\frac{dy_2}{dt} = cx_2 + dy_2$$

given that we know that

$$\frac{dx_1}{dt} = ax_1 + by_1$$
$$\frac{dy_1}{dt} = cx_1 + dy_1.$$

Since $x_2(t) = kx_1(t)$ and $y_2(t) = ky_1(t)$, we can multiply both sides of

$$\frac{dx_1}{dt} = ax_1 + by_1$$
$$\frac{dy_1}{dt} = cx_1 + dy_1.$$

by k to obtain

$$k\frac{dx_1}{dt} = k(ax_1 + by_1)$$
$$k\frac{dy_1}{dt} = k(cx_1 + dy_1).$$

However, using standard algebraic properties and the rules of differentiation, this system is equivalent to

$$\frac{d(k\,x_1)}{dt} = a(kx_1) + b(ky_1)$$
$$\frac{d(k\,y_1)}{dt} = c(kx_1) + d(ky_1),$$

which is the same as the desired equality

$$\begin{aligned}\frac{dx_2}{dt} &= ax_2 + by_2\\ \frac{dy_2}{dt} &= cx_2 + dy_2.\end{aligned}$$

To verify the second half of the Linearity Principle, we suppose that $\mathbf{Y}_1(t) = (x_1(t), y_1(t))$ and $\mathbf{Y}_2(t) = (x_2(t), y_2(t))$ are solutions to the system

$$\begin{aligned}\frac{dx}{dt} &= ax + by\\ \frac{dy}{dt} &= cx + dy.\end{aligned}$$

To verify that the function $\mathbf{Y}_3(t) = \mathbf{Y}_1(t) + \mathbf{Y}_2(t)$ is also a solution, we need to substitute $\mathbf{Y}_3(t)$ into both sides of the differential equation and check for equality. In other words, after we write $\mathbf{Y}_3(t)$ in scalar notation as $\mathbf{Y}_3(t) = (x_3(t), y_3(t))$, we must show that

$$\begin{aligned}\frac{dx_3}{dt} &= ax_3 + by_3\\ \frac{dy_3}{dt} &= cx_3 + dy_3\end{aligned}$$

given that we know that

$$\begin{aligned}\frac{dx_1}{dt} &= ax_1 + by_1\\ \frac{dy_1}{dt} &= cx_1 + dy_1\end{aligned}$$

and

$$\begin{aligned}\frac{dx_2}{dt} &= ax_2 + by_2\\ \frac{dy_2}{dt} &= cx_2 + dy_2.\end{aligned}$$

Adding the two given systems together yields the system

$$\begin{aligned}\frac{dx_1}{dt} + \frac{dx_2}{dt} &= ax_1 + by_1 + ax_2 + by_2\\ \frac{dy_1}{dt} + \frac{dy_2}{dt} &= cx_1 + dy_1 + cx_2 + dy_2,\end{aligned}$$

which can be rewritten as

$$\begin{aligned}\frac{d(x_1 + x_2)}{dt} &= a(x_1 + x_2) + b(y_1 + y_2)\\ \frac{d(y_1 + y_2)}{dt} &= c(x_1 + x_2) + d(y_1 + y_2).\end{aligned}$$

But this last system of equalities is the desired equality that indicates that $\mathbf{Y}_3(t)$ is also a solution.

31. **(a)** If $(x_1, y_1) = (0, 0)$, then (x_1, y_1) and (x_2, y_2) are on the same line through the origin because (x_1, y_1) is the origin. So (x_1, y_1) and (x_2, y_2) are linearly dependent.

(b) If $(x_1, y_1) = \lambda(x_2, y_2)$ for some λ, then (x_1, y_1) and (x_2, y_2) are on the same line through the origin. To see why, suppose that $x_2 \neq 0$ and $\lambda \neq 0$. (The $\lambda = 0$ case was handled in part (a) above.) In this case, $x_1 \neq 0$ as well. Then the slope of the line through the origin and (x_1, y_1) is y_1/x_1, and the slope of the line through the origin and (x_2, y_2) is y_2/x_2. However, because $(x_1, y_1) = \lambda(x_2, y_2)$, we have

$$\frac{y_1}{x_1} = \frac{\lambda y_2}{\lambda x_2} = \frac{y_2}{x_2}.$$

Since these two lines have the same slope and both contain the origin, they are the same line. (The special case where $x_2 = 0$ reduces to considering vertical lines through the origin.)

(c) If $x_1 y_2 - x_2 y_1 = 0$, then $x_1 y_2 = x_2 y_1$. Once again, this condition implies that (x_1, y_1) and (x_2, y_2) are on the same line through the origin. For example, suppose that $x_1 \neq 0$, then

$$y_2 = \frac{x_2 y_1}{x_1} = \frac{x_2}{x_1} y_1.$$

But we already know that

$$x_2 = \frac{x_2}{x_1} x_1,$$

so we have

$$(x_2, y_2) = \frac{x_2}{x_1}(x_1, y_1).$$

By part (b) above (where $\lambda = x_2/x_1$), the two vectors are linearly dependent.

If $x_1 = 0$ but $y_1 \neq 0$, it follows that $x_2 y_1 = 0$, and thus $x_2 = 0$. Thus, both (x_1, y_1) and (x_2, y_2) are on the vertical line through the origin.

Finally, if $x_1 = 0$ and $y_1 = 0$, we can use part (a) to show that the two vectors are linearly dependent.

32. If $x_1 y_2 - x_2 y_1$ is nonzero, then $x_1 y_2 \neq x_2 y_1$. If both $x_1 \neq 0$ and $x_2 \neq 0$, we can divide both sides by $x_1 x_2$, and we obtain

$$\frac{y_2}{x_2} \neq \frac{y_1}{x_1},$$

and therefore, the slope of the line through the origin and (x_2, y_2) is not the same as the slope of the line through the origin and (x_1, y_1).

If $x_1 = 0$, then $x_2 \neq 0$. In this case, the line through the origin and (x_1, y_1) is vertical, and the line through the origin and (x_2, y_2) is not vertical.

33. The initial position of $\mathbf{Y}_1(t)$ is $\mathbf{Y}_1(0) = (-1, 1)$. By the Linearity Principle, we know that $k\mathbf{Y}_1(t)$ is also a solution of the system for any constant k. Hence, for any initial condition of the form $(-k, k)$, the solution is $k\mathbf{Y}_1(t)$.

(a) The curve $2\mathbf{Y}_1(t) = (-2e^{-t}, 2e^{-t})$ is the solution with this initial condition.

(b) We cannot find the solution for this initial condition using only $\mathbf{Y}_1(t)$.

(c) The constant function $0\mathbf{Y}_1(t) = (0, 0)$ (represented by the equilibrium point at the origin) is the solution with this initial condition.

(d) The curve $-3\mathbf{Y}_1(t) = (3e^{-t}, -3e^{-t})$ is the solution with this initial condition.

34. **(a)** If $\mathbf{Y}(t) = (t, t^2/2)$, then $x(t) = t$ and $y(t) = t^2/2$. Then $dx/dt = 1$, and $dy/dt = t = x$. So $\mathbf{Y}(t)$ satisfies the differential equation.

(b) For $2\mathbf{Y}(t)$, we have $x(t) = 2t$, and $y(t) = t^2$. In this case, we need only consider $dx/dt = 2$ to see that the function is not a solution to the system.

35. **(a)** Using the Product Rule we compute

$$\frac{dW}{dt} = \frac{dx_1}{dt}y_2 + x_1\frac{dy_2}{dt} - \frac{dx_2}{dt}y_1 - x_2\frac{dy_1}{dt}.$$

(b) Since $(x_1(t), y_1(t))$ and $(x_2(t), y_2(t))$ are solutions, we know that

$$\frac{dx_1}{dt} = ax_1 + by_1$$
$$\frac{dy_1}{dt} = cx_1 + dy_1$$

and that

$$\frac{dx_2}{dt} = ax_2 + by_2$$
$$\frac{dy_2}{dt} = cx_2 + dy_2.$$

Substituting these equations into the expression for dW/dt, we obtain

$$\frac{dW}{dt} = (ax_1 + by_1)y_2 + x_1(cx_2 + dy_2) - (ax_2 + by_2)y_1 - x_2(cx_1 + dy_1).$$

After we collect terms, we have

$$\frac{dW}{dt} = (a + d)W.$$

(c) This equation is a homogeneous, linear, first-order equation (as such it is also separable—see Sections 1.1, 1.2, and 1.8). Therefore, we know that the general solution is

$$W(t) = Ce^{(a+d)t}$$

where C is any constant (but note that $C = W(0)$).

(d) From Exercises 31 and 32, we know that $\mathbf{Y}_1(t)$ and $\mathbf{Y}_2(t)$ are linearly independent if and only if $W(t) \neq 0$. But, $W(t) = Ce^{(a+d)t}$, so $W(t) = 0$ if and only if $C = W(0) = 0$. Hence, $W(t) = 0$ is zero for some t if and only if $C = W(0) = 0$.

EXERCISES FOR SECTION 3.2

1. **(a)** The characteristic polynomial is

$$(3 - \lambda)(-2 - \lambda) = 0,$$

and therefore the eigenvalues are $\lambda_1 = -2$ and $\lambda_2 = 3$.

(b) To obtain the eigenvectors (x_1, y_1) for the eigenvalue $\lambda_1 = -2$, we solve the tions

$$\begin{cases} 3x_1 + 2y_1 = -2x_1 \\ -2y_1 = -2y_1 \end{cases}$$

and obtain $5x_1 = -2y_1$.

Using the same procedure, we see that the eigenvectors (x_2, y_2) for $\lambda_2 = 3$ must satisfy the equation $y_2 = 0$.

(c)

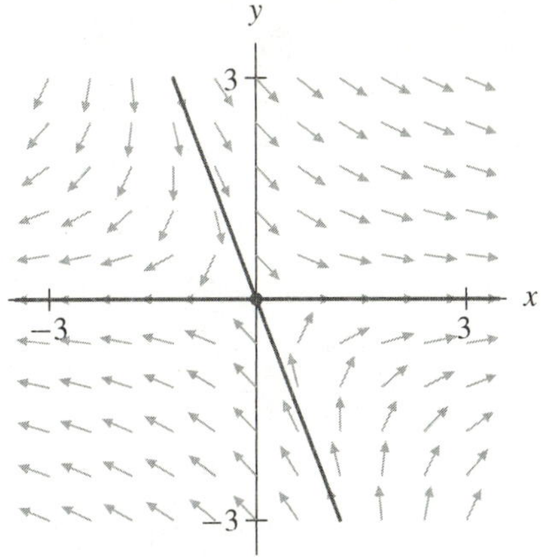

(d) One eigenvector $\mathbf{V}_1$ for λ_1 is $\mathbf{V}_1 = (-2, 5)$, and one eigenvector $\mathbf{V}_2$ for λ_2 is $\mathbf{V}_2 = (1, 0)$.

Given the eigenvalues and these eigenvectors, we have the two linearly independent solutions

$$\mathbf{Y}_1(t) = e^{-2t}\begin{pmatrix} -2 \\ 5 \end{pmatrix} \quad \text{and} \quad \mathbf{Y}_2(t) = e^{3t}\begin{pmatrix} 1 \\ 0 \end{pmatrix}.$$

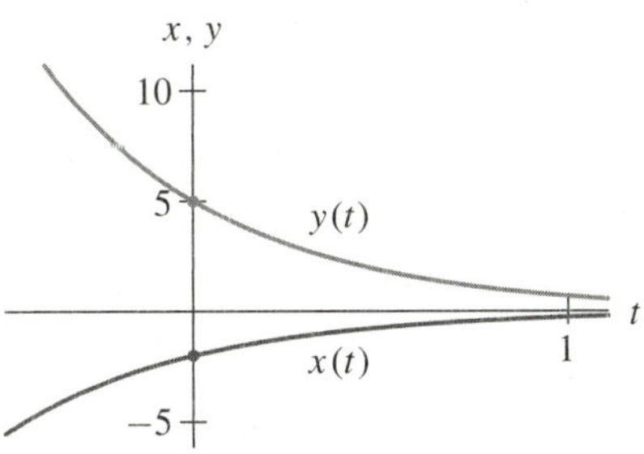

The $x(t)$- and $y(t)$-graphs for $\mathbf{Y}_1(t)$.

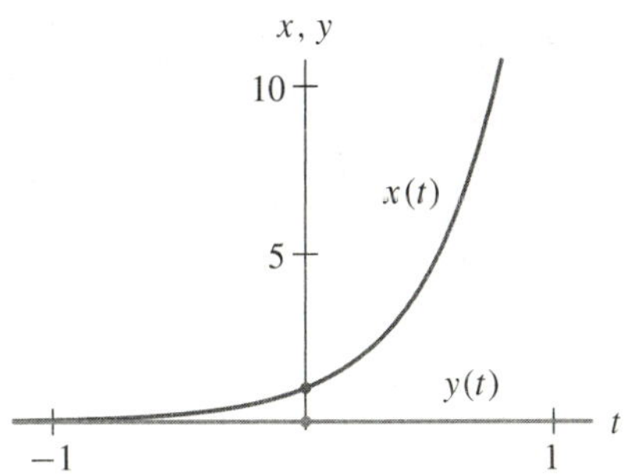

The $x(t)$- and $y(t)$-graphs for $\mathbf{Y}_2(t)$.

(e) The general solution to this linear system is

$$\mathbf{Y}(t) = k_1 e^{-2t}\begin{pmatrix} -2 \\ 5 \end{pmatrix} + k_2 e^{3t}\begin{pmatrix} 1 \\ 0 \end{pmatrix}.$$

2. **(a)** The characteristic polynomial is

$$(-4-\lambda)(-3-\lambda)-2=\lambda^2+7\lambda+10=0,$$

and therefore the eigenvalues are $\lambda_1=-2$ and $\lambda_2=-5$.

(b) To obtain the eigenvectors (x_1, y_1) for the eigenvalue $\lambda_1=-2$, we solve the system of equations

$$\begin{cases} -4x_1-2y_1=-2x_1 \\ -x_1-3y_1=-2y_1 \end{cases}$$

and obtain $y_1=-x_1$.

Using the same procedure, we obtain the eigenvectors (x_2, y_2) where $x_2=2y_2$ for $\lambda_2=-5$.

(c)

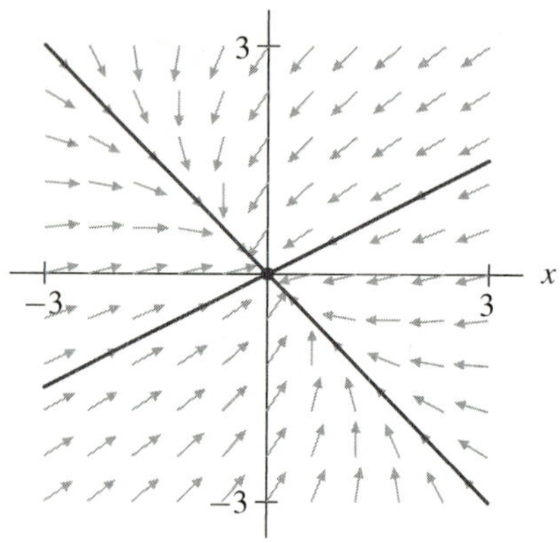

(d) One eigenvector $\mathbf{V}_1$ for λ_1 is $\mathbf{V}_1=(1,-1)$, and one eigenvector $\mathbf{V}_2$ for λ_2 is $\mathbf{V}_2=(2,1)$.

Given the eigenvalues and these eigenvectors, we have two linearly independent solutions

$$\mathbf{Y}_1(t)=e^{-2t}\begin{pmatrix}1\\-1\end{pmatrix} \quad\text{and}\quad \mathbf{Y}_2(t)=e^{-5t}\begin{pmatrix}2\\1\end{pmatrix}.$$

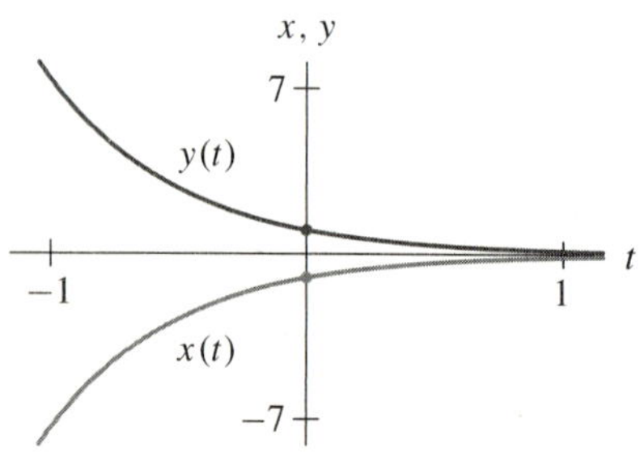

The $x(t)$- and $y(t)$-graphs for $\mathbf{Y}_1(t)$.

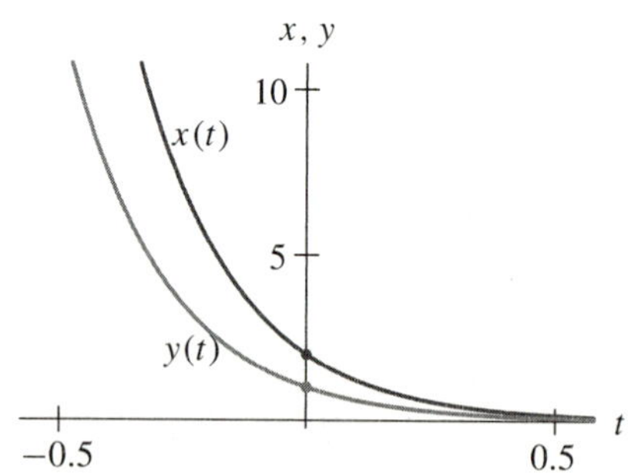

The $x(t)$- and $y(t)$-graphs for $\mathbf{Y}_2(t)$.

(e) The general solution to this linear system is

$$\mathbf{Y}(t) = k_1 e^{-2t} \begin{pmatrix} 1 \\ -1 \end{pmatrix} + k_2 e^{-5t} \begin{pmatrix} 2 \\ 1 \end{pmatrix}.$$

3. **(a)** The eigenvalues are the roots of the characteristic polynomial, so they are the solutions of

$$(-5 - \lambda)(-4 - \lambda) - 2 = \lambda^2 + 9\lambda + 18 = 0.$$

Therefore, the eigenvalues are $\lambda_1 = -3$ and $\lambda_2 = -6$.

(b) To obtain the eigenvectors (x_1, y_1) for the eigenvalue $\lambda_1 = -3$, we solve the system of equations

$$\begin{cases} -5x_1 - 2y_1 = -3x_1 \\ -x_1 - 4y_1 = -3y_1 \end{cases}$$

and obtain $y_1 = -x_1$.

Using the same procedure, we obtain the eigenvalues (x_2, y_2) where $x_2 = 2y_2$ for $\lambda_2 = 6$.

(c)

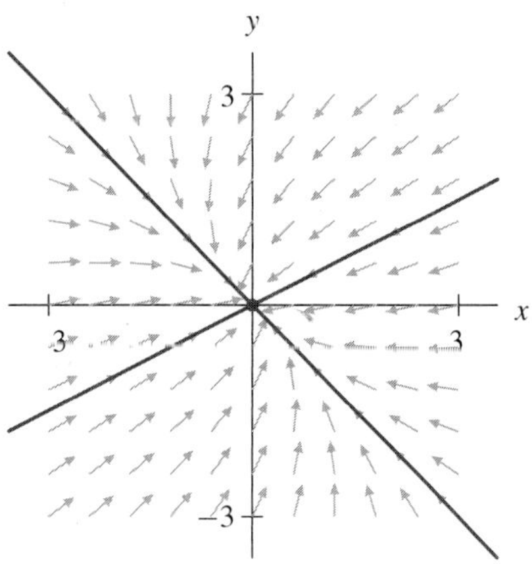

(d) One eigenvector $\mathbf{V}_1$ for $\lambda_1 = -3$ is $\mathbf{V}_1 = (1, -1)$, and one eigenvector $\mathbf{V}_2$ for $\lambda_2 = -6$ is $\mathbf{V}_2 = (2, 1)$.

Given the eigenvalues and these eigenvectors, we have two linearly independent solutions

$$\mathbf{Y}_1(t) = e^{-3t} \begin{pmatrix} 1 \\ -1 \end{pmatrix} \quad \text{and} \quad \mathbf{Y}_2(t) = e^{-6t} \begin{pmatrix} 2 \\ 1 \end{pmatrix}.$$

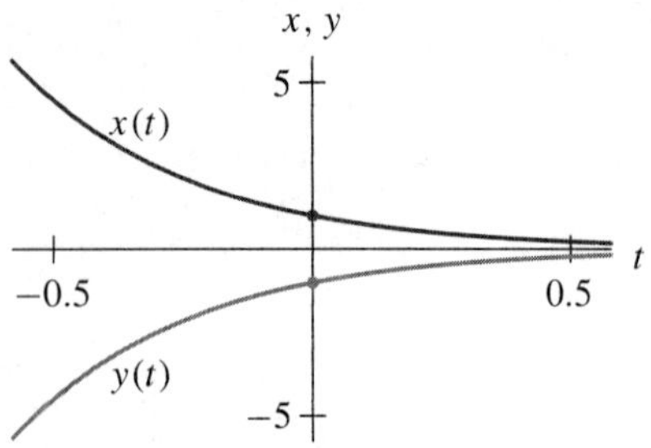

The $x(t)$- and $y(t)$-graphs for $\mathbf{Y}_1(t)$.

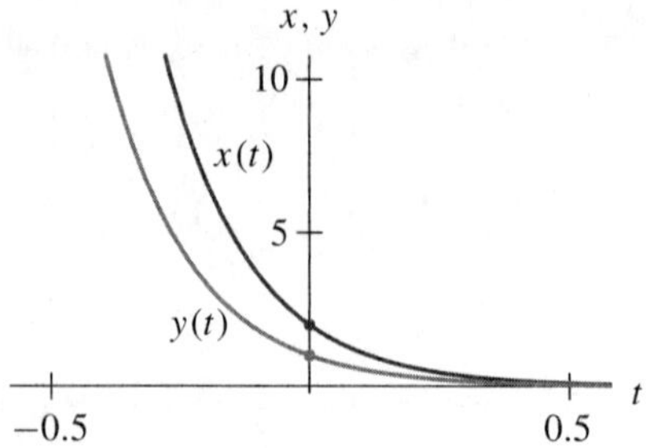

the $x(t)$- and $y(t)$-graphs for $\mathbf{Y}_2(t)$.

(e) The general solution to this linear system is

$$\mathbf{Y}(t) = k_1 e^{-3t} \begin{pmatrix} 1 \\ -1 \end{pmatrix} + k_2 e^{-6t} \begin{pmatrix} 2 \\ 1 \end{pmatrix}.$$

4. **(a)** The characteristic polynomial is

$$(2 - \lambda)(4 - \lambda) + 1 = \lambda^2 - 6\lambda + 9 = 0,$$

and therefore there is only one eigenvalue, $\lambda = 3$.

(b) To obtain the eigenvectors (x_1, y_1) for the eigenvalue $\lambda = 3$, we solve the system of equations

$$\begin{cases} 2x_1 + y_1 = 3x_1 \\ -x_1 + 4y_1 = 3y_1 \end{cases}$$

and obtain $y_1 = x_1$.

(c)

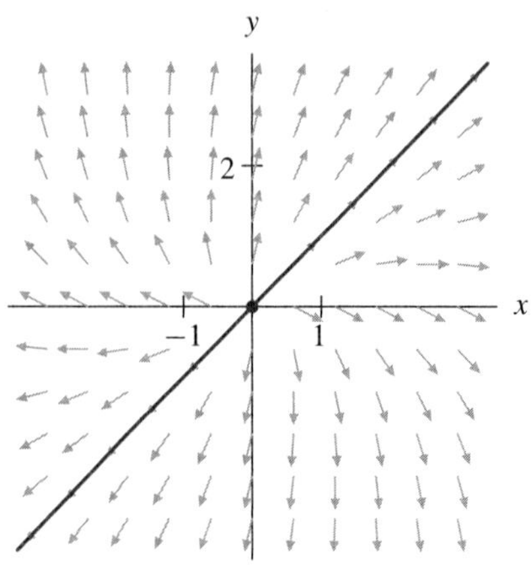

(d) One eigenvector $\mathbf{V}$ for λ is $\mathbf{V} = (1, 1)$. Given this eigenvector, we have the solution

$$\mathbf{Y}(t) = e^{3t} \begin{pmatrix} 1 \\ 1 \end{pmatrix}.$$

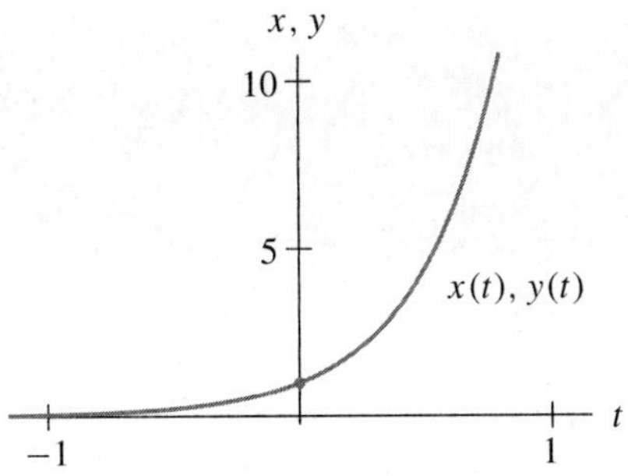

The $x(t)$- and $y(t)$-graphs (which are identical) for $\mathbf{Y}(t)$

(e) Since the method of eigenvalues and eigenvectors does not give us a second solution that is linearly independent from $\mathbf{Y}(t)$, we cannot form the general solution.

5. **(a)** The characteristic polynomial is

$$\left(-\tfrac{1}{2} - \lambda\right)^2 = 0,$$

and therefore there is only one eigenvalue, $\lambda = -1/2$.

(b) To obtain the eigenvectors (x_1, y_1) for the eigenvalue $\lambda = -1/2$, we solve the system of equations

$$\begin{cases} -\frac{1}{2}x_1 = -\frac{1}{2}x_1 \\ x_1 - \frac{1}{2}y_1 = -\frac{1}{2}y_1 \end{cases}$$

and obtain $x_1 = 0$.

(c)

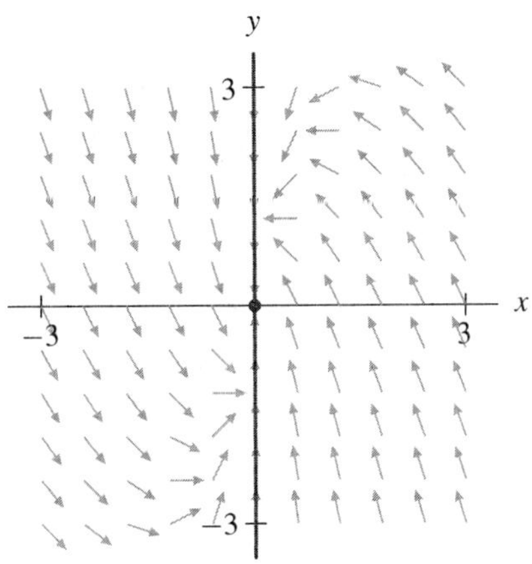

(d) Given the eigenvalue $\lambda = -1/2$ and the eigenvector $\mathbf{V} = (0, 1)$, we have the solution

$$\mathbf{Y}(t) = e^{-t/2}\begin{pmatrix} 0 \\ 1 \end{pmatrix}.$$

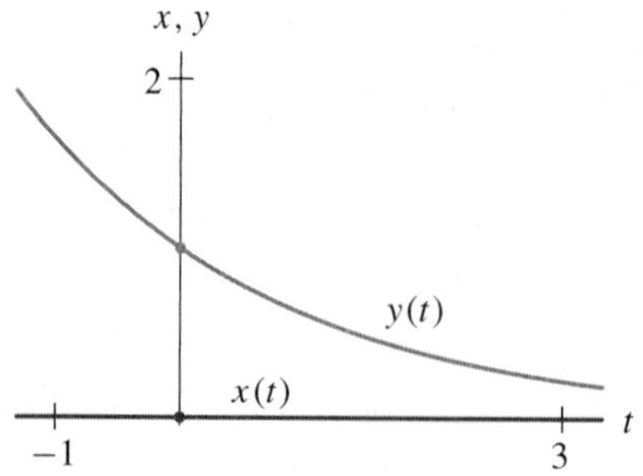

The $x(t)$- and $y(t)$-graphs for $\mathbf{Y}(t)$.

(e) Since the method of eigenvalues and eigenvectors does not give us a second solution that is linearly independent from $\mathbf{Y}(t)$, we cannot form the general solution.

6. **(a)** The characteristic polynomial is

$$(5-\lambda)(-\lambda) - 36 = 0,$$

and therefore the eigenvalues are $\lambda_1 = -4$ and $\lambda_2 = 9$.

(b) To obtain the eigenvectors (x_1, y_1) for the eigenvalue $\lambda_1 = -4$, we solve the system of equations

$$\begin{cases} 5x_1 + 4y_1 = -4x_1 \\ 9x_1 = -4y_1 \end{cases}$$

and obtain $9x_1 = -4y_1$.

Using the same procedure, we see that the eigenvectors (x_2, y_2) for $\lambda_2 = 9$ must satisfy the equation $y_2 = x_2$.

(c)

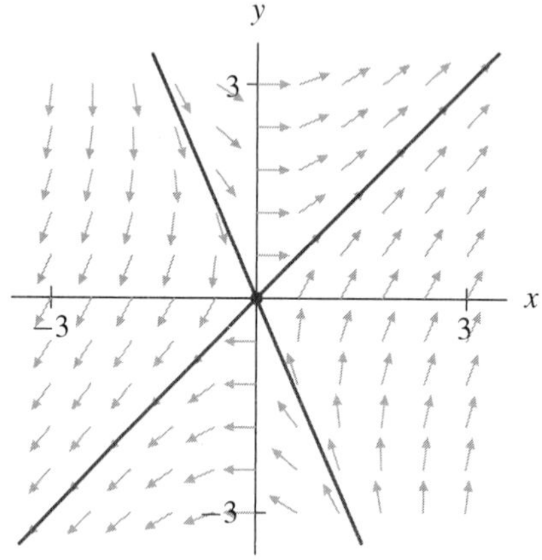

(d) One eigenvector $\mathbf{V}_1$ for λ_1 is $\mathbf{V}_1 = (4, -9)$, and one eigenvector $\mathbf{V}_2$ for λ_2 is $\mathbf{V}_2 = (1, 1)$.

Given the eigenvalues and these eigenvectors, we have the two linearly independent solutions

$$\mathbf{Y}_1(t) = e^{-4t}\begin{pmatrix} 4 \\ -9 \end{pmatrix} \quad \text{and} \quad \mathbf{Y}_2(t) = e^{9t}\begin{pmatrix} 1 \\ 1 \end{pmatrix}.$$

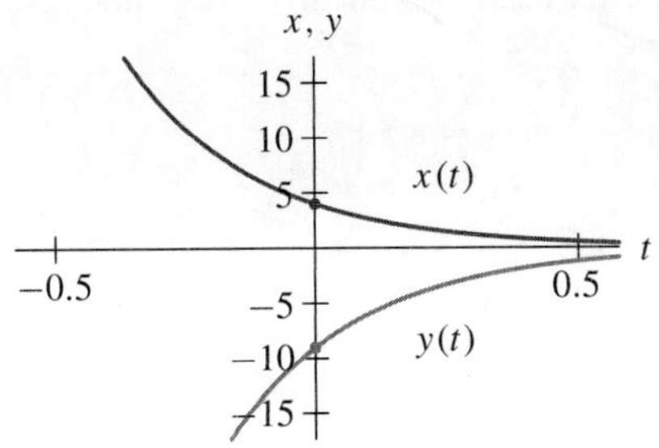

The $x(t)$- and $y(t)$-graphs for $\mathbf{Y}_1(t)$.

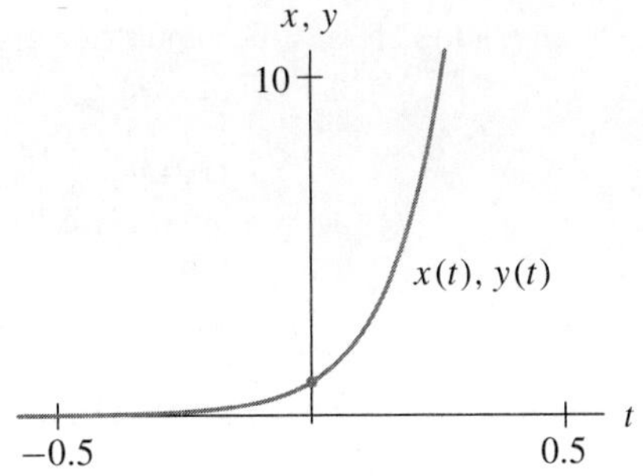

The (identical) $x(t)$- and $y(t)$-graphs for $\mathbf{Y}_2(t)$.

(e) The general solution to this linear system is

$$\mathbf{Y}(t) = k_1 e^{-4t} \begin{pmatrix} 4 \\ -9 \end{pmatrix} + k_2 e^{9t} \begin{pmatrix} 1 \\ 1 \end{pmatrix}.$$

7. **(a)** The characteristic polynomial is

$$(3 - \lambda)(-\lambda) - 4 = \lambda^2 - 3\lambda - 4 = (\lambda - 4)(\lambda + 1) = 0,$$

and therefore the eigenvalues are $\lambda_1 = -1$ and $\lambda_2 = 4$.

(b) To obtain the eigenvectors (x_1, y_1) for the eigenvalue $\lambda_1 = -1$, we solve the system of equations

$$\begin{cases} 3x_1 + 4y_1 = -x_1 \\ x_1 = -y_1 \end{cases}$$

and obtain $y_1 = -x_1$.

Using the same procedure, we obtain the eigenvectors (x_2, y_2) where $x_2 = 4y_2$ for $\lambda_2 = 4$.

(c)

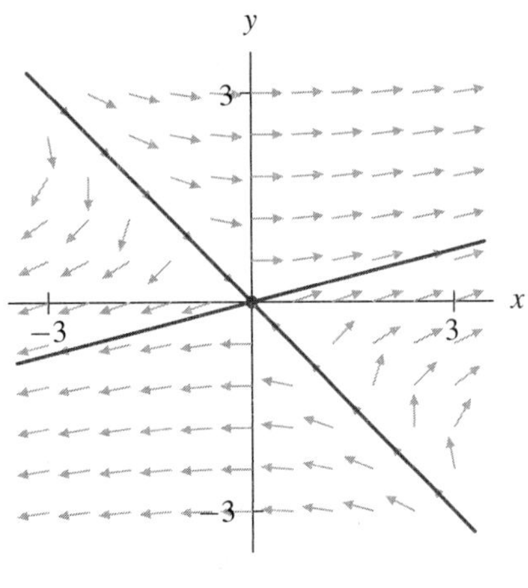

(d) One eigenvector $\mathbf{V}_1$ for λ_1 is $\mathbf{V}_1 = (1, -1)$, and one eigenvector $\mathbf{V}_2$ for λ_2 is $\mathbf{V}_2 = (4, 1)$.

Given the eigenvalues and these eigenvectors, we have two linearly independent solutions

$$\mathbf{Y}_1(t) = e^{-t}\begin{pmatrix} 1 \\ -1 \end{pmatrix} \quad \text{and} \quad \mathbf{Y}_2(t) = e^{4t}\begin{pmatrix} 4 \\ 1 \end{pmatrix}.$$

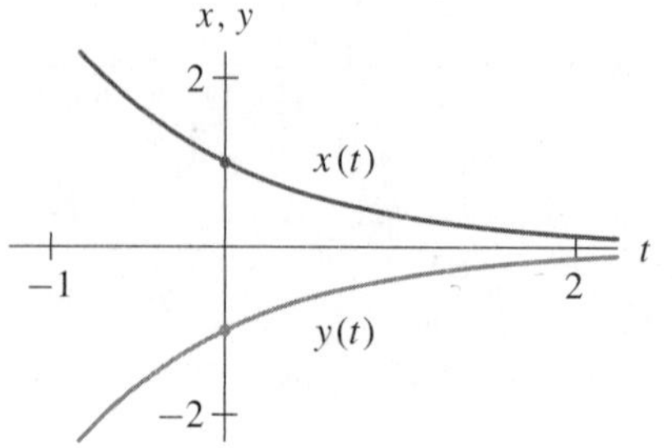

The $x(t)$- and $y(t)$-graphs for $\mathbf{Y}_1(t)$.

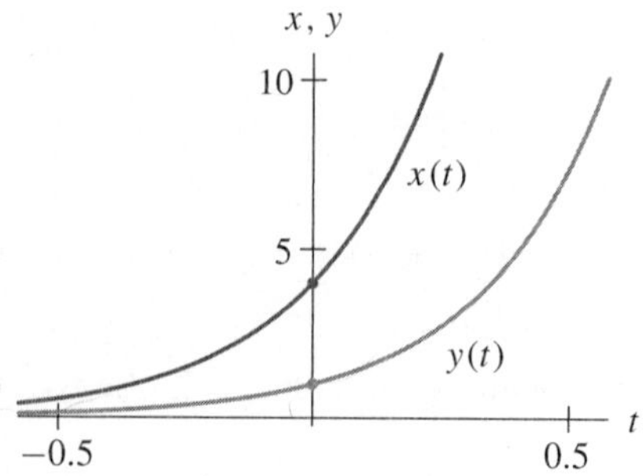

The $x(t)$- and $y(t)$-graphs for $\mathbf{Y}_2(t)$.

(e) The general solution to this linear system is

$$\mathbf{Y}(t) = k_1 e^{-t}\begin{pmatrix} 1 \\ -1 \end{pmatrix} + k_2 e^{4t}\begin{pmatrix} 4 \\ 1 \end{pmatrix}$$

8. **(a)** The characteristic polynomial is

$$(2 - \lambda)(1 - \lambda) - 1 = \lambda^2 - 3\lambda + 1 = 0,$$

and therefore the eigenvalues are

$$\lambda_1 = \frac{3 + \sqrt{5}}{2} \quad \text{and} \quad \lambda_2 = \frac{3 - \sqrt{5}}{2}.$$

(b) To obtain the eigenvectors (x_1, y_1) for the eigenvalue $\lambda_1 = (3 + \sqrt{5})/2$, we solve the system of equations

$$\begin{cases} 2x_1 - y_1 = \dfrac{3 + \sqrt{5}}{2} x_1 \\ -x_1 + y_1 = \dfrac{3 + \sqrt{5}}{2} y_1 \end{cases}$$

and obtain

$$y_1 = \frac{1 - \sqrt{5}}{2} x_1,$$

which is equivalent to the equation $2y_1 = (1 - \sqrt{5})x_1$.

Using the same procedure, we obtain the eigenvectors (x_2, y_2) where $2y_2 = (1 + \sqrt{5})x_2$ for $\lambda_2 = (3 - \sqrt{5})/2$.

(c)

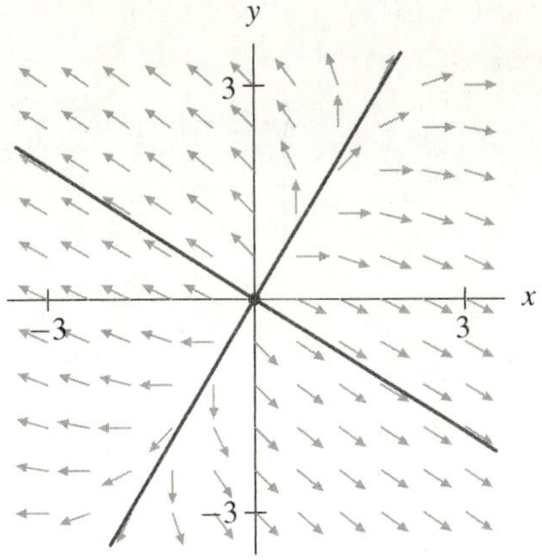

(d) One eigenvector $\mathbf{V}_1$ for the eigenvalue λ_1 is $\mathbf{V}_1 = (2, 1 - \sqrt{5}\,)$, and one eigenvector $\mathbf{V}_2$ for the eigenvalue λ_2 is $\mathbf{V}_2 = (2, 1 + \sqrt{5}\,)$.

Given the eigenvalues and these eigenvectors, we have two linearly independent solutions

$$\mathbf{Y}_1(t) = e^{(3+\sqrt{5})t/2}\begin{pmatrix} 2 \\ 1-\sqrt{5} \end{pmatrix} \quad \text{and} \quad \mathbf{Y}_2(t) = e^{(3-\sqrt{5})t/2}\begin{pmatrix} 2 \\ 1+\sqrt{5} \end{pmatrix}.$$

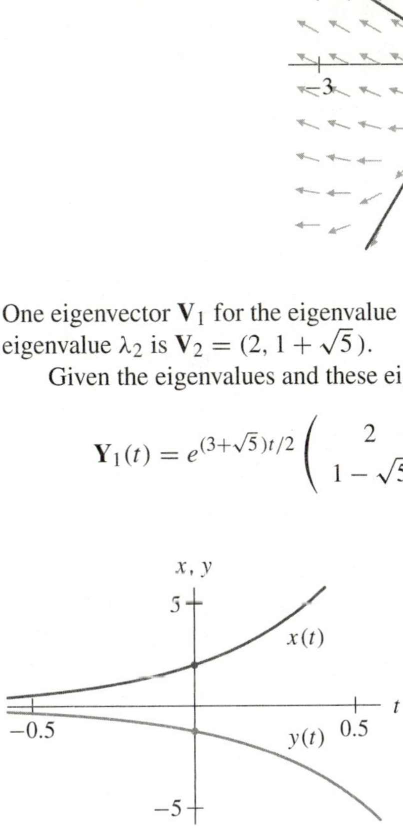

The $x(t)$- and $y(t)$-graphs for $\mathbf{Y}_1(t)$.

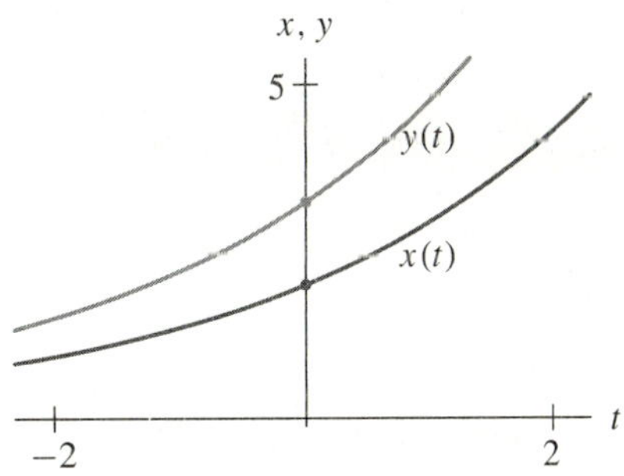

The $x(t)$- and $y(t)$-graphs for $\mathbf{Y}_2(t)$.

(e) The general solution to this linear system is

$$\mathbf{Y}(t) = k_1 e^{(3+\sqrt{5})t/2}\begin{pmatrix} 2 \\ 1-\sqrt{5} \end{pmatrix} + k_2 e^{(3-\sqrt{5})t/2}\begin{pmatrix} 2 \\ 1+\sqrt{5} \end{pmatrix}.$$

9. **(a)** The characteristic polynomial is

$$(2-\lambda)(1-\lambda) - 1 = \lambda^2 - 3\lambda + 1 = 0,$$

and therefore the eigenvalues are

$$\lambda_1 = \frac{3+\sqrt{5}}{2} \quad \text{and} \quad \lambda_2 = \frac{3-\sqrt{5}}{2}.$$

(b) To obtain the eigenvectors (x_1, y_1) for the eigenvalue $\lambda_1 = (3 + \sqrt{5})/2$, we solve the system of equations

$$\begin{cases} 2x_1 + y_1 = \dfrac{3+\sqrt{5}}{2} x_1 \\ x_1 + y_1 = \dfrac{3+\sqrt{5}}{2} y_1 \end{cases}$$

and obtain

$$y_1 = \frac{-1+\sqrt{5}}{2} x_1,$$

which is equivalent to the equation $2y_1 = (-1 + \sqrt{5})x_1$.

Using the same procedure, we obtain the eigenvectors (x_2, y_2) where $2y_2 = (-1 - \sqrt{5})x_2$ for $\lambda_2 = (3 - \sqrt{5})/2$.

(c)

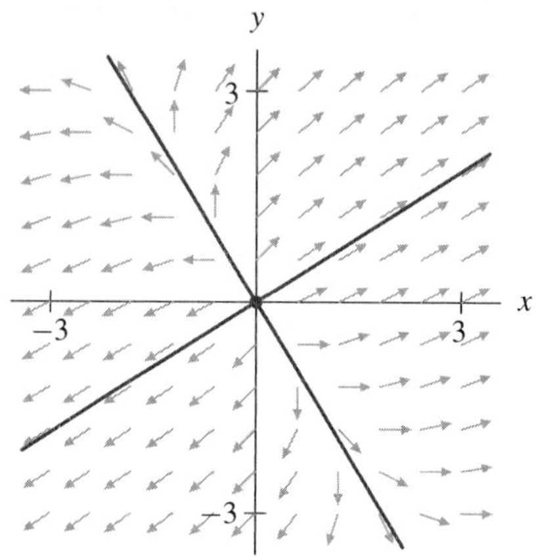

(d) One eigenvector $\mathbf{V}_1$ for the eigenvalue λ_1 is $\mathbf{V}_1 = (2, -1 + \sqrt{5})$, and one eigenvector $\mathbf{V}_2$ for the eigenvalue λ_2 is $\mathbf{V}_2 = (-2, 1 + \sqrt{5})$.

Given the eigenvalues and these eigenvectors, we have two linearly independent solutions

$$\mathbf{Y}_1(t) = e^{(3+\sqrt{5})t/2} \begin{pmatrix} 2 \\ -1+\sqrt{5} \end{pmatrix} \quad \text{and} \quad \mathbf{Y}_2(t) = e^{(3-\sqrt{5})t/2} \begin{pmatrix} -2 \\ 1+\sqrt{5} \end{pmatrix}.$$

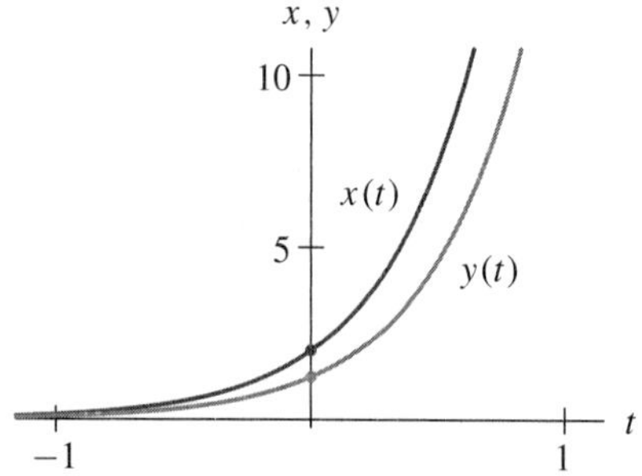

The $x(t)$- and $y(t)$-graphs for $\mathbf{Y}_1(t)$.

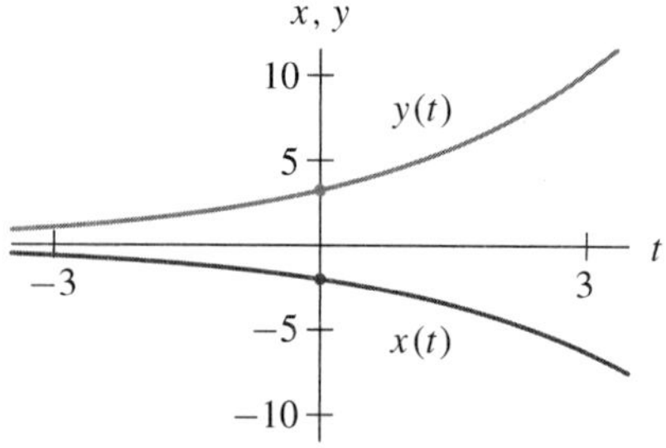

The $x(t)$- and $y(t)$-graphs for $\mathbf{Y}_2(t)$.

(e) The general solution to this linear system is

$$\mathbf{Y}(t) = k_1 e^{(3+\sqrt{5})t/2} \begin{pmatrix} 2 \\ -1+\sqrt{5} \end{pmatrix} + k_2 e^{(3-\sqrt{5})t/2} \begin{pmatrix} -2 \\ 1+\sqrt{5} \end{pmatrix}.$$

10. **(a)** The characteristic polynomial is

$$(-1-\lambda)(-4-\lambda)+2 = \lambda^2+5\lambda+6 = (\lambda+3)(\lambda+2) = 0,$$

and therefore the eigenvalues are $\lambda_1 = -2$ and $\lambda_2 = -3$.

(b) To obtain the eigenvectors (x_1, y_1) for the eigenvalue $\lambda_1 = -2$, we solve the system of equations

$$\begin{cases} -x_1 - 2y_1 = -2x_1 \\ x_1 - 4y_1 = -2y_1 \end{cases}$$

and obtain $x_1 = 2y_1$.

Using the same procedure, we obtain the eigenvectors (x_2, y_2) where $x_2 = y_2$ for $\lambda_2 = -3$.

(c)

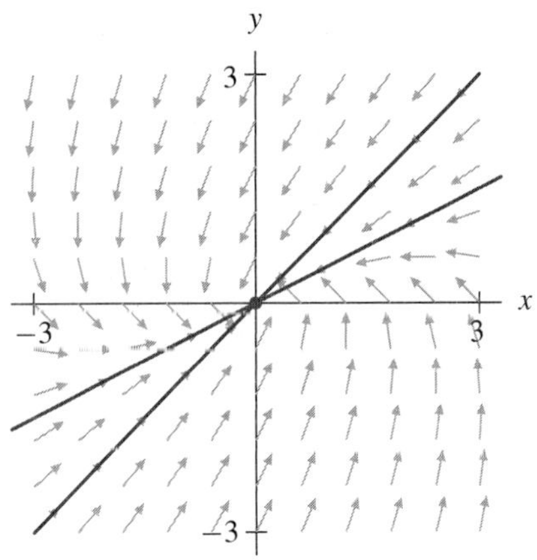

(d) One eigenvector $\mathbf{V}_1$ for λ_1 is $\mathbf{V}_1 = (2, 1)$, and one eigenvector $\mathbf{V}_2$ for λ_2 is $\mathbf{V}_2 = (1, 1)$.

Given the eigenvalues and these eigenvectors, we have two linearly independent solutions

$$\mathbf{Y}_1(t) = e^{-2t} \begin{pmatrix} 2 \\ 1 \end{pmatrix} \quad \text{and} \quad \mathbf{Y}_2(t) = e^{-3t} \begin{pmatrix} 1 \\ 1 \end{pmatrix}.$$

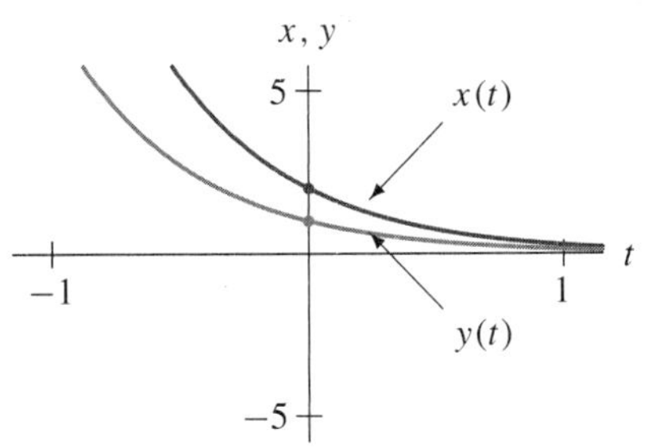

The $x(t)$- and $y(t)$-graphs for $\mathbf{Y}_1(t)$.

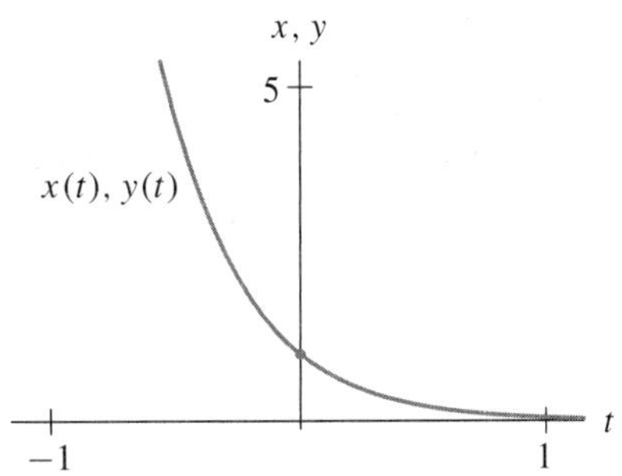

The identical) $x(t)$- and $y(t)$-graphs for $\mathbf{Y}_2(t)$.

(e) The general solution to this linear system is

$$\mathbf{Y}(t) = k_1 e^{-2t} \begin{pmatrix} 2 \\ 1 \end{pmatrix} + k_2 e^{-3t} \begin{pmatrix} 1 \\ 1 \end{pmatrix}$$

11. The eigenvalues are the roots of the characteristic polynomial, so they are solutions of

$$(-2 - \lambda)(1 - \lambda) - 4 = \lambda^2 + \lambda - 6 = 0.$$

Hence, $\lambda_1 = 2$ and $\lambda_2 = -3$ are the eigenvalues.

To find the eigenvectors for the eigenvalue $\lambda_1 = 2$, we solve

$$\begin{cases} -2x_1 - 2y_1 = 2x_1 \\ -2x_1 + y_1 = 2y_1, \end{cases}$$

so $y_1 = -2x_1$ is the line of eigenvectors. In particular, $(1, -2)$ is an eigenvector for $\lambda_1 = 2$.

Similarly, the line of eigenvectors for $\lambda_2 = -3$ is given by $x_1 = 2y_1$. In particular, $(2, 1)$ is an eigenvector for $\lambda_2 = -3$.

Given the eigenvalues and these eigenvectors, we have the two linearly independent solutions

$$\mathbf{Y}_1(t) = e^{2t} \begin{pmatrix} 1 \\ -2 \end{pmatrix} \quad \text{and} \quad \mathbf{Y}_2(t) = e^{-3t} \begin{pmatrix} 2 \\ 1 \end{pmatrix}.$$

The general solution is

$$\mathbf{Y}(t) = k_1 e^{2t} \begin{pmatrix} 1 \\ -2 \end{pmatrix} + k_2 e^{-3t} \begin{pmatrix} 2 \\ 1 \end{pmatrix}.$$

(a) Given the initial condition $\mathbf{Y}(0) = (1, 0)$, we must solve

$$\begin{pmatrix} 1 \\ 0 \end{pmatrix} = \mathbf{Y}(0) = k_1 \begin{pmatrix} 1 \\ -2 \end{pmatrix} + k_2 \begin{pmatrix} 2 \\ 1 \end{pmatrix}$$

for k_1 and k_2. This vector equation is equivalent to the two scalar equations

$$\begin{cases} k_1 + 2k_2 = 1 \\ -2k_1 + k_2 = 0. \end{cases}$$

Solving these equations, we obtain $k_1 = 1/5$ and $k_2 = 2/5$. Thus, the particular solution is

$$\mathbf{Y}(t) = \tfrac{1}{5} e^{2t} \begin{pmatrix} 1 \\ -2 \end{pmatrix} + \tfrac{2}{5} e^{-3t} \begin{pmatrix} 2 \\ 1 \end{pmatrix}.$$

(b) Given the initial condition $\mathbf{Y}(0) = (0, 1)$ we must solve

$$\begin{pmatrix} 0 \\ 1 \end{pmatrix} = \mathbf{Y}(0) = k_1 \begin{pmatrix} 1 \\ -2 \end{pmatrix} + k_2 \begin{pmatrix} 2 \\ 1 \end{pmatrix}$$

for k_1 and k_2. This vector equation is equivalent to the two scalar equations

$$\begin{cases} k_1 + 2k_2 = 0 \\ -2k_1 + k_2 = 1. \end{cases}$$

Solving these equations, we obtain $k_1 = -2/5$ and $k_2 = 1/5$. Thus, the particular solution is

$$\mathbf{Y}(t) = -\tfrac{2}{5}e^{2t}\begin{pmatrix} 1 \\ -2 \end{pmatrix} + \tfrac{1}{5}e^{-3t}\begin{pmatrix} 2 \\ 1 \end{pmatrix}.$$

(c) The initial condition $\mathbf{Y}(0) = (1, -2)$ is an eigenvector for the eigenvalue $\lambda_1 = 2$. Hence, the solution with this initial condition is

$$\mathbf{Y}(t) = e^{2t}\begin{pmatrix} 1 \\ -2 \end{pmatrix}.$$

12. The characteristic polynomial is

$$(3 - \lambda)(-2 - \lambda) = 0,$$

and therefore the eigenvalues are $\lambda_1 = 3$ and $\lambda_2 = -2$.

To obtain the eigenvectors (x_1, y_1) for the eigenvalue $\lambda_1 = 3$, we solve the system of equations

$$\begin{cases} 3x_1 = 3x_1 \\ x_1 - 2y_1 = 3y_1 \end{cases}$$

and obtain

$$5y_1 = x_1.$$

Therefore, an eigenvector for the eigenvalue $\lambda_1 = 3$ is $\mathbf{V}_1 = (5, 1)$.

Using the same procedure, we obtain the eigenvector $\mathbf{V}_2 = (0, 1)$ for $\lambda_2 = -2$.

The general solution to this linear system is therefore

$$\mathbf{Y}(t) = k_1 e^{3t}\begin{pmatrix} 5 \\ 1 \end{pmatrix} + k_2 e^{-2t}\begin{pmatrix} 0 \\ 1 \end{pmatrix}.$$

(a) We have $\mathbf{Y}(0) = (1, 0)$, so we must find k_1 and k_2 so that

$$\begin{pmatrix} 1 \\ 0 \end{pmatrix} = \mathbf{Y}(0) = k_1\begin{pmatrix} 5 \\ 1 \end{pmatrix} + k_2\begin{pmatrix} 0 \\ 1 \end{pmatrix}.$$

This vector equation is equivalent to the simultaneous system of linear equations

$$\begin{cases} 5k_1 = 1 \\ k_1 + k_2 = 0. \end{cases}$$

Solving these equations, we obtain $k_1 = 1/5$ and $k_2 = -1/5$. Thus, the particular solution is

$$\mathbf{Y}(t) = \tfrac{1}{5}e^{3t}\begin{pmatrix} 5 \\ 1 \end{pmatrix} - \tfrac{1}{5}e^{-2t}\begin{pmatrix} 0 \\ 1 \end{pmatrix}.$$

(b) We have $\mathbf{Y}(0) = (0, 1)$. Since this initial condition is an eigenvector associated to the $\lambda = -2$ eigenvalue, we do not need to do any additional calculation. The desired solution to the initial-value problem is

$$\mathbf{Y}(t) = e^{-2t}\begin{pmatrix} 0 \\ 1 \end{pmatrix}.$$

(c) We have $\mathbf{Y}(0) = (2, 2)$, so we must find k_1 and k_2 so that

$$\begin{pmatrix} 2 \\ 2 \end{pmatrix} = \mathbf{Y}(0) = k_1\begin{pmatrix} 5 \\ 1 \end{pmatrix} + k_2\begin{pmatrix} 0 \\ 1 \end{pmatrix}.$$

This vector equation is equivalent to the simultaneous system of linear equations

$$\begin{cases} 5k_1 = 2 \\ k_1 + k_2 = 2. \end{cases}$$

Solving these equations, we obtain $k_1 = 2/5$ and $k_2 = 8/5$. Thus, the particular solution is

$$\mathbf{Y}(t) = \tfrac{2}{5}e^{3t}\begin{pmatrix} 5 \\ 1 \end{pmatrix} + \tfrac{8}{5}e^{-2t}\begin{pmatrix} 0 \\ 1 \end{pmatrix}.$$

13. The characteristic polynomial is

$$(-4 - \lambda)(-3 - \lambda) - 2 = \lambda^2 + 7\lambda + 10 = 0,$$

and therefore the eigenvalues are $\lambda_1 = -5$ and $\lambda_2 = -2$.

To obtain the eigenvectors (x_1, y_1) for the eigenvalue $\lambda_1 = -5$, we solve the system of equations

$$\begin{cases} -4x_1 + y_1 = -5x_1 \\ 2x_1 - 3y_1 = -5y_1 \end{cases}$$

and obtain

$$y_1 = -x_1.$$

Therefore, an eigenvector for the eigenvalue $\lambda_1 = -5$ is $\mathbf{V}_1 = (1, -1)$.

Using the same procedure, we obtain the eigenvector $\mathbf{V}_2 = (1, 2)$ for $\lambda_2 = -2$.

Given the eigenvalues and these eigenvectors, we have two linearly independent solutions

$$\mathbf{Y}_1(t) = e^{-5t}\begin{pmatrix} 1 \\ -1 \end{pmatrix} \quad \text{and} \quad \mathbf{Y}_2(t) = e^{-2t}\begin{pmatrix} 1 \\ 2 \end{pmatrix}.$$

The general solution to this linear system is

$$\mathbf{Y}(t) = k_1e^{-5t}\begin{pmatrix} 1 \\ -1 \end{pmatrix} + k_2e^{-2t}\begin{pmatrix} 1 \\ 2 \end{pmatrix}.$$

(a) We have $\mathbf{Y}(0) = (1, 0)$, so we must find k_1 and k_2 so that

$$\begin{pmatrix} 1 \\ 0 \end{pmatrix} = \mathbf{Y}(0) = k_1 \begin{pmatrix} 1 \\ -1 \end{pmatrix} + k_2 \begin{pmatrix} 1 \\ 2 \end{pmatrix}.$$

This vector equation is equivalent to the simultaneous system of linear equations

$$\begin{cases} k_1 + k_2 = 1 \\ -k_1 + 2k_2 = 0. \end{cases}$$

Solving these equations, we obtain $k_1 = 2/3$ and $k_2 = 1/3$. Thus, the particular solution is

$$\mathbf{Y}(t) = \tfrac{2}{3} e^{-5t} \begin{pmatrix} 1 \\ -1 \end{pmatrix} + \tfrac{1}{3} e^{-2t} \begin{pmatrix} 1 \\ 2 \end{pmatrix}.$$

(b) We have $\mathbf{Y}(0) = (2, 1)$, so we must find k_1 and k_2 so that

$$\begin{pmatrix} 2 \\ 1 \end{pmatrix} = \mathbf{Y}(0) = k_1 \begin{pmatrix} 1 \\ -1 \end{pmatrix} + k_2 \begin{pmatrix} 1 \\ 2 \end{pmatrix}.$$

This vector equation is equivalent to the simultaneous system of linear equations

$$\begin{cases} k_1 + k_2 = 2 \\ -k_1 + 2k_2 = 1. \end{cases}$$

Solving these equations, we obtain $k_1 = 1$ and $k_2 = 1$. Thus, the particular solution is

$$\mathbf{Y}(t) = e^{-5t} \begin{pmatrix} 1 \\ -1 \end{pmatrix} + e^{-2t} \begin{pmatrix} 1 \\ 2 \end{pmatrix}.$$

(c) We have $\mathbf{Y}(0) = (-1, -2)$. Since this initial condition is an eigenvector associated to the $\lambda = -2$ eigenvalue, we do not need to do any additional calculation. The desired solution to the initial-value problem is

$$\mathbf{Y}(t) = -e^{-2t} \begin{pmatrix} 1 \\ 2 \end{pmatrix}.$$

14. The characteristic polynomial is

$$(4 - \lambda)(1 - \lambda) + 2 = \lambda^2 - 5\lambda + 6 = 0,$$

and therefore the eigenvalues are $\lambda_1 = 3$ and $\lambda_2 = 2$.

To obtain the eigenvectors (x_1, y_1) for the eigenvalue $\lambda_1 = 3$, we solve the system of equations

$$\begin{cases} 4x_1 - 2y_1 = 3x_1 \\ x_1 + y_1 = 3y_1 \end{cases}$$

and obtain

$$x_1 = 2y_1.$$

Therefore, an eigenvector for the eigenvalue $\lambda_1 = 3$ is $\mathbf{V}_1 = (2, 1)$.

Using the same procedure, we obtain the eigenvector $\mathbf{V}_2 = (1, 1)$ for $\lambda_2 = 2$.

Given the eigenvalues and these eigenvectors, we have two linearly independent solutions

$$\mathbf{Y}_1(t) = e^{3t}\begin{pmatrix} 2 \\ 1 \end{pmatrix} \quad \text{and} \quad \mathbf{Y}_2(t) = e^{2t}\begin{pmatrix} 1 \\ 1 \end{pmatrix}.$$

The general solution to this linear system is

$$\mathbf{Y}(t) = k_1 e^{3t}\begin{pmatrix} 2 \\ 1 \end{pmatrix} + k_2 e^{2t}\begin{pmatrix} 1 \\ 1 \end{pmatrix}.$$

(a) We have $\mathbf{Y}(0) = (1, 0)$, so we must find k_1 and k_2 so that

$$\begin{pmatrix} 1 \\ 0 \end{pmatrix} = \mathbf{Y}(0) = k_1\begin{pmatrix} 2 \\ 1 \end{pmatrix} + k_2\begin{pmatrix} 1 \\ 1 \end{pmatrix}.$$

This vector equation is equivalent to the simultaneous system of linear equations

$$\begin{cases} 2k_1 + k_2 = 1 \\ k_1 + k_2 = 0. \end{cases}$$

Solving these equations, we obtain $k_1 = 1$ and $k_2 = -1$. Thus, the particular solution is

$$\mathbf{Y}(t) = e^{3t}\begin{pmatrix} 2 \\ 1 \end{pmatrix} - e^{2t}\begin{pmatrix} 1 \\ 1 \end{pmatrix}.$$

(b) We have $\mathbf{Y}(0) = (2, 1)$. Since this initial condition is an eigenvector associated to the $\lambda = 3$ eigenvalue, we do not need to do any additional calculation. The desired solution to the initial-value problem is

$$\mathbf{Y}(t) = e^{3t}\begin{pmatrix} 2 \\ 1 \end{pmatrix}.$$

(c) We have $\mathbf{Y}(0) = (-1, -2)$, so we must find k_1 and k_2 so that

$$\begin{pmatrix} -1 \\ -2 \end{pmatrix} = \mathbf{Y}(0) = k_1\begin{pmatrix} 2 \\ 1 \end{pmatrix} + k_2\begin{pmatrix} 1 \\ 1 \end{pmatrix}.$$

This vector equation is equivalent to the simultaneous system of linear equations

$$\begin{cases} 2k_1 + k_2 = -1 \\ k_1 + k_2 = -2. \end{cases}$$

Solving these equations, we obtain $k_1 = 1$ and $k_2 = -3$. Thus, the particular solution is

$$\mathbf{Y}(t) = e^{3t}\begin{pmatrix} 2 \\ 1 \end{pmatrix} - 3e^{2t}\begin{pmatrix} 1 \\ 1 \end{pmatrix}.$$

15. Given any vector $\mathbf{Y}_0 = (x_0, y_0)$, we have

$$\mathbf{AY}_0 = \begin{pmatrix} a & 0 \\ 0 & a \end{pmatrix} \begin{pmatrix} x_0 \\ y_0 \end{pmatrix} = \begin{pmatrix} ax_0 \\ ay_0 \end{pmatrix} = a \begin{pmatrix} x_0 \\ y_0 \end{pmatrix} = a\mathbf{Y}_0.$$

Therefore, every nonzero vector is an eigenvector associated to the eigenvalue a.

16. The characteristic polynomial of $\mathbf{A}$ is

$$(a - \lambda)(d - \lambda) = 0,$$

and thus the eigenvalues of $\mathbf{A}$ are $\lambda_1 = a$ and $\lambda_2 = d$.

To find the eigenvectors $\mathbf{V}_1 = (x_1, y_1)$ associated to $\lambda_1 = a$, we need to solve the equation

$$\mathbf{AV}_1 = a\mathbf{V}_1$$

for all possible vectors $\mathbf{V}_1$. Rewritten in terms of components, this equation is equivalent to

$$\begin{cases} ax_1 + by_1 = ax_1 \\ dy_1 = ay_1. \end{cases}$$

Since $a \neq d$, the second equation implies that $y_1 = 0$. If so, then the first equation is satisfied for all x_1. In other words, the eigenvectors $\mathbf{V}_1$ associated to the eigenvalue a are the vectors of the form $(x_1, 0)$.

To find the eigenvectors $\mathbf{V}_2 = (x_2, y_2)$ associated to $\lambda_2 = d$, we need to solve the equation

$$\mathbf{AV}_2 = d\mathbf{V}_2$$

for all possible vectors $\mathbf{V}_2$. Rewritten in terms of components, this equation is equivalent to

$$\begin{cases} ax_2 + by_2 - dx_2 \\ dy_2 = dy_2. \end{cases}$$

The second equation always holds, so the eigenvectors $\mathbf{V}_2$ are those vectors that satisfy the equation $ax_2 + by_2 = dx_2$, which can be rewritten as

$$by_2 = (d - a)x_2.$$

These vectors form a line through the origin of slope $(d - a)/b$.

17. The characteristic polynomial of $\mathbf{B}$ is

$$\lambda^2 - (a + d)\lambda + ad - b^2.$$

The roots of this polynomial are

$$\begin{aligned} \lambda &= \frac{a + d \pm \sqrt{(a + d)^2 - 4(ad - b^2)}}{2} \\ &= \frac{a + d \pm \sqrt{a^2 + 2ad + d^2 - 4ad + 4b^2}}{2} \\ &= \frac{a + d \pm \sqrt{(a - d)^2 + 4b^2}}{2}. \end{aligned}$$

Since the discriminant $D = (a-d)^2 + 4b^2$ is always nonnegative, the roots λ are real. Therefore, the matrix $\mathbf{B}$ has real eigenvalues. If $b \neq 0$, then D is positive and hence $\mathbf{B}$ has two distinct eigenvalues. (The only way to have only one eigenvalue is for $D = 0$).

18. The characteristic equation is

$$(a - \lambda)(-\lambda) - bc = \lambda^2 - a\lambda - bc = 0.$$

Finding the roots via the quadratic formula, we obtain the eigenvalues

$$\frac{a \pm \sqrt{a^2 + 4bc}}{2}.$$

Note that these eigenvalues are very different from the case where the matrix is upper triangular (see Exercise 16). For example, they are not necessarily real numbers because $a^2 + 4bc$ can be negative.

19. **(a)** To form the system, we introduce the new dependent variable $v = dy/dt$. Then

$$\frac{dv}{dt} = \frac{d^2y}{dt^2} = -p\frac{dy}{dt} - qy = -pv - qy.$$

Written in matrix form this system where $\mathbf{Y} = (y, v)$, we have

$$\frac{d\mathbf{Y}}{dt} = \begin{pmatrix} 0 & 1 \\ -q & -p \end{pmatrix}\mathbf{Y}.$$

(b) The characteristic polynomial is

$$(0 - \lambda)(-p - \lambda) + q = \lambda^2 + p\lambda + q.$$

(c) The roots of this polynomial (the eigenvalues) are

$$\frac{-p \pm \sqrt{p^2 - 4q}}{2}.$$

(d) The roots are distinct real numbers if the discriminant $D = p^2 - 4q$ is positive. In other words, the roots are distinct real numbers if $p^2 > 4q$.

(e) Since q is positive, $p^2 - 4q < p^2$, so we know that $\sqrt{p^2 - 4q} < \sqrt{p^2} = p$. Since the numerator in the expression for the eigenvalues is $-p \pm \sqrt{p^2 - 4q}$, we see that it must be negative. Since the denominator is positive, the eigenvalues must be negative.

20. **(a)** The parameters $m = 1$, $k = 4$, and $b = 5$ yield the second-order equation

$$\frac{d^2y}{dt^2} + 5\frac{dy}{dt} + 4y = 0.$$

Given $v = dy/dt$, the corresponding system is

$$\frac{dy}{dt} = v$$
$$\frac{dv}{dt} = -4y - 5v.$$

The characteristic polynomial is $\lambda^2 + 5\lambda + 4$, and the eigenvalues are $\lambda_1 = -4$ and $\lambda_2 = -1$.

To find the eigenvectors $\mathbf{V}_1 = (y_1, v_1)$ associated to the eigenvalue $\lambda_1 = -4$, we solve the system of equations.

$$\begin{cases} v_1 = -4y_1 \\ -4y_1 - 5v_1 = -4v_1 \end{cases}$$

and obtain $v_1 = -4y_1$. Thus, one eigenvector for $\lambda_1 = -4$ is $\mathbf{V}_1 = (1, -4)$.

By the same procedure, we can find the eigenvector $\mathbf{V}_2 = (1, -1)$ for the eigenvalue $\lambda_2 = -1$.

(b) Therefore the solution $\mathbf{Y}_1(t)$ that satisfies $\mathbf{Y}_1(0) = \mathbf{V}_1$ is

$$\mathbf{Y}_1(t) = e^{-4t}\begin{pmatrix} 1 \\ -4 \end{pmatrix}.$$

The solution $\mathbf{Y}_2(t)$ that satisfies $\mathbf{Y}_2(0) = \mathbf{V}_2$ is

$$\mathbf{Y}_2(t) = e^{-t}\begin{pmatrix} 1 \\ -1 \end{pmatrix}.$$

(c)

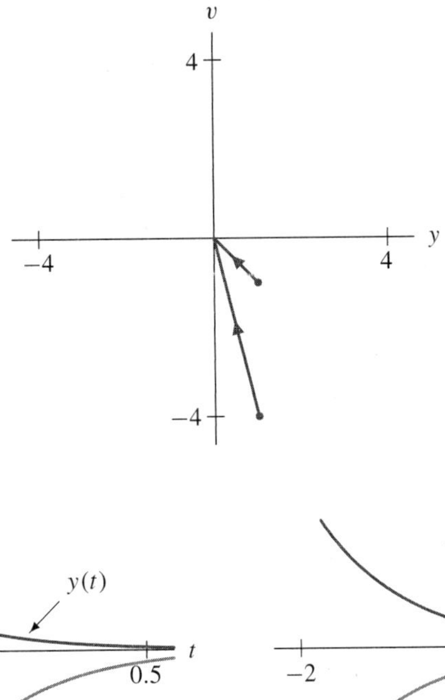

(d)

The $y(t)$- and $v(t)$-graphs for $\mathbf{Y}_1(t)$.

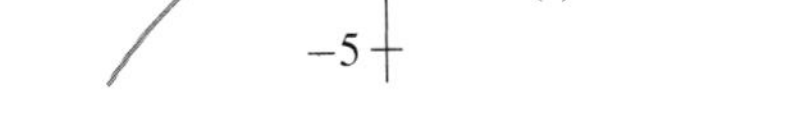

The $y(t)$- and $v(t)$-graphs for $\mathbf{Y}_2(t)$.

(e) The first initial condition $(y_0, v_0) = (1, -4)$ represents a solution whose initial position is 1 unit away from the equilibrium position and whose initial velocity is -4. Note that the solution tends toward the equilibrium point at the origin. Moreover, $y(t)$ is decreasing toward 0, and $v(t)$ is increasing toward 0. Therefore, the mass moves toward the equilibrium position monotonically, and its speed decreases as it approaches the equilibrium position. The mass does not oscillate about the equilibrium position.

The second initial condition $(y_0, v_0) = (1, -1)$ represents a solution whose initial position is 1 unit away from the equilibrium position and whose initial velocity is -1. The behavior of this solution is similar to the first solution.

21. **(a)** Given $v = dy/dt$, the corresponding system is

$$\begin{aligned}\frac{dy}{dt} &= v\\ \frac{dv}{dt} &= 10y - 3v.\end{aligned}$$

(b) The characteristic polynomial is $\lambda^2 + 3\lambda - 10 = (\lambda + 5)(\lambda - 2) = 0$, and the eigenvalues are $\lambda_1 = -5$ and $\lambda_2 = 2$.

To find the eigenvectors $\mathbf{V}_1 = (y_1, v_1)$ associated to the eigenvalue $\lambda_1 = -5$, we solve the system of equations

$$\begin{cases} v_1 = -5y_1 \\ 10y_1 - 3v_1 = -5v_1 \end{cases}$$

and obtain $v_1 = -5y_1$.

By the same procedure, we can find the eigenvectors $\mathbf{V}_2 = (y_2, v_2)$ for the eigenvalue $\lambda_2 = 2$. They consist of all vectors that satisfy the equation $v_2 = 2y_2$.

(c) From part (b) we see that one eigenvector for $\lambda_1 = -5$ is $\mathbf{V}_1 = (1, -5)$. Therefore the solution $\mathbf{Y}_1(t)$ that satisfies $\mathbf{Y}_1(0) = \mathbf{V}_1$ is

$$\mathbf{Y}_1(t) = e^{-5t}\begin{pmatrix} 1 \\ -5 \end{pmatrix}.$$

One eigenvector for $\lambda_2 = 2$ is $\mathbf{V}_2 = (1, 2)$, and the solution $\mathbf{Y}_2(t)$ that satisfies $\mathbf{Y}_2(0) = \mathbf{V}_2$ is

$$\mathbf{Y}_2(t) = e^{2t}\begin{pmatrix} 1 \\ 2 \end{pmatrix}.$$

(d) Note that the solutions obtained here are vector-valued functions of the form $\mathbf{Y}(t) = (y(t), v(t))$. In Section 2.3 we obtained $y_1(t) = e^{2t}$ and $y_2(t) = e^{-5t}$. Using the fact that $v = dy/dt$, we can obtain $\mathbf{Y}_1(t)$ and $\mathbf{Y}_2(t)$ from $y_1(t)$ and $y_2(t)$.

22. **(a)** Given $v = dy/dt$, the corresponding system is

$$\begin{aligned}\frac{dy}{dt} &= v\\ \frac{dv}{dt} &= -2y - 3v.\end{aligned}$$

(b) The characteristic polynomial is $\lambda^2 + 3\lambda + 2 = (\lambda + 2)(\lambda + 1) = 0$, and the eigenvalues are $\lambda_1 = -2$ and $\lambda_2 = -1$.

To find the eigenvectors $\mathbf{V}_1 = (y_1, v_1)$ associated to the eigenvalue $\lambda_1 = -2$, we solve the system of equations

$$\begin{cases} v_1 = -2y_1 \\ -2y_1 - 3v_1 = -2v_1 \end{cases}$$

and obtain $v_1 = -2y_1$.

By the same procedure, we can see that the eigenvectors $\mathbf{V}_2 = (y_2, v_2)$ for the eigenvalue $\lambda_2 = -1$ satisfy $v_2 = -y_2$.

(c) One eigenvector for $\lambda_1 = -2$ is $\mathbf{V}_1 = (1, -2)$. Therefore the solution $\mathbf{Y}_1(t)$ that satisfies $\mathbf{Y}_1(0) = \mathbf{V}_1$ is

$$\mathbf{Y}_1(t) = e^{-2t}\begin{pmatrix} 1 \\ -2 \end{pmatrix}.$$

An eigenvector for $\lambda_2 = -2$ is $(1, -1)$. Therefore the solution $\mathbf{Y}_2(t)$ that satisfies $\mathbf{Y}_2(0) = \mathbf{V}_2$ is

$$\mathbf{Y}_2(t) = e^{-t}\begin{pmatrix} 1 \\ -1 \end{pmatrix}.$$

(d) Note that the solutions obtained here are vector-valued functions of the form $\mathbf{Y}(t) = (y(t), v(t))$. In Section 2.3 we obtained $y_1(t) = e^{-2t}$ and $y_2(t) = e^{-t}$. Using the fact that $v = dy/dt$, we can obtain $\mathbf{Y}_1(t)$ and $\mathbf{Y}_2(t)$ from $y_1(t)$ and $y_2(t)$.

23. **(a)** Given $v = dy/dt$, the corresponding system is

$$\frac{dy}{dt} = v$$
$$\frac{dv}{dt} = -y - 4v.$$

(b) The characteristic polynomial is $\lambda^2 + 4\lambda + 1 = 0$. Using the quadratic formula, we obtain the eigenvalues $\lambda_1 = -2 + \sqrt{3}$ and $\lambda_2 = -2 - \sqrt{3}$.

To find the eigenvectors $\mathbf{V}_1 = (y_1, v_1)$ associated to the eigenvalue $\lambda_1 = -2 + \sqrt{3}$, we solve the system of equations

$$\begin{cases} v_1 = (-2 + \sqrt{3})y_1 \\ -y_1 - 4v_1 = (-2 + \sqrt{3})v_1 \end{cases}$$

and obtain $v_1 = (-2 + \sqrt{3})y_1$.

By the same procedure, we can find the eigenvectors $\mathbf{V}_2 = (y_2, v_2)$ for the eigenvalue $\lambda_2 = -2 - \sqrt{3}$. They consist of all vectors that satisfy the equation $v_2 = (-2 - \sqrt{3})y_2$.

(c) From part (b) we see that one eigenvector for $\lambda_1 = -2 + \sqrt{3}$ is $\mathbf{V}_1 = (1, -2 + \sqrt{3})$. Therefore the solution $\mathbf{Y}_1(t)$ that satisfies $\mathbf{Y}_1(0) = \mathbf{V}_1$ is

$$\mathbf{Y}_1(t) = e^{(-2+\sqrt{3})t}\begin{pmatrix} 1 \\ -2 + \sqrt{3} \end{pmatrix}.$$

One eigenvector for $\lambda_2 = -2 - \sqrt{3}$ is $\mathbf{V}_2 = (1, -2 - \sqrt{3})$, and the solution $\mathbf{Y}_2(t)$ that satisfies $\mathbf{Y}_2(0) = \mathbf{V}_2$ is

$$\mathbf{Y}_2(t) = e^{(-2-\sqrt{3})t} \begin{pmatrix} 1 \\ -2 - \sqrt{3} \end{pmatrix}.$$

(d) Note that the solutions obtained here are vector-valued functions of the form $\mathbf{Y}(t) = (y(t), v(t))$. In Section 2.3 we obtained $y_1(t) = e^{(-2+\sqrt{3})t}$ and $y_2(t) = e^{(-2-\sqrt{3})t}$. Using the fact that $v = dy/dt$, we can obtain $\mathbf{Y}_1(t)$ and $\mathbf{Y}_2(t)$ from $y_1(t)$ and $y_2(t)$.

24. **(a)** Given $v = dy/dt$, the corresponding system is

$$\frac{dy}{dt} = v$$
$$\frac{dv}{dt} = 2y - v.$$

(b) The characteristic polynomial is $\lambda^2 + \lambda - 2 = (\lambda + 2)(\lambda - 1) = 0$, and the eigenvalues are $\lambda_1 = -2$ and $\lambda_2 = 1$.

To find the eigenvectors $\mathbf{V}_1 = (y_1, v_1)$ associated to the eigenvalue $\lambda_1 = -2$, we solve the system of equations

$$\begin{cases} v_1 = -2y_1 \\ 2y_1 - v_1 = -2v_1 \end{cases}$$

and obtain $v_1 = -2y_1$.

By the same procedure, we can find the eigenvectors $\mathbf{V}_2 = (y_2, v_2)$ for the eigenvalue $\lambda_2 = 1$. They consist of all vectors that satisfy the equation $v_2 = y_2$.

(c) From part (b) we see that one eigenvector for $\lambda_1 = -2$ is $\mathbf{V}_1 = (1, -2)$. Therefore the solution $\mathbf{Y}_1(t)$ that satisfies $\mathbf{Y}_1(0) = \mathbf{V}_1$ is

$$\mathbf{Y}_1(t) = e^{-2t} \begin{pmatrix} 1 \\ -2 \end{pmatrix}.$$

One eigenvector for $\lambda_2 = 1$ is $\mathbf{V}_2 = (1, 1)$, and the solution $\mathbf{Y}_2(t)$ that satisfies $\mathbf{Y}_2(0) = \mathbf{V}_2$ is

$$\mathbf{Y}_2(t) = e^{t} \begin{pmatrix} 1 \\ 1 \end{pmatrix}.$$

(d) Note that the solutions obtained here are vector-valued functions of the form $\mathbf{Y}(t) = (y(t), v(t))$. In Section 2.3 we obtained $y_1(t) = e^{-2t}$ and $y_2(t) = e^{t}$. Using the fact that $v = dy/dt$, we can obtain $\mathbf{Y}_1(t)$ and $\mathbf{Y}_2(t)$ from $y_1(t)$ and $y_2(t)$.

25. With $m = 1$, $k = 4$, and $b = 1$, the system is

$$\frac{dy}{dt} = v$$
$$\frac{dv}{dt} = -4y - v.$$

The characteristic polynomial is $\lambda^2 + \lambda + 4$, and its roots are the complex numbers $(-1 \pm \sqrt{15}\,i)/2$. Therefore there are no straight-line solutions. According to the direction field, the solution curves seem to spiral around the origin.

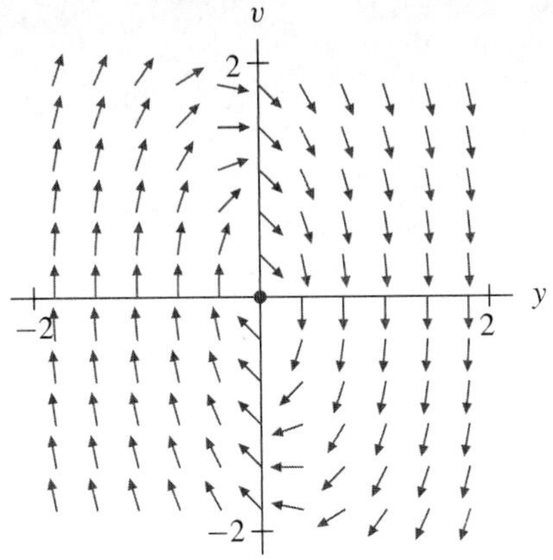

EXERCISES FOR SECTION 3.3

1. As we computed in Exercise 1 of Section 3.2, the eigenvalues are $\lambda_1 = -2$ and $\lambda_2 = 3$. The eigenvectors (x_1, y_1) for the eigenvalue $\lambda_1 = -2$ satisfy $5x_1 = -2y_1$, and the eigenvectors (x_2, y_2) for $\lambda_2 = 3$ satisfy the equation $y_2 = 0$. The equilibrium point at the origin is a saddle.

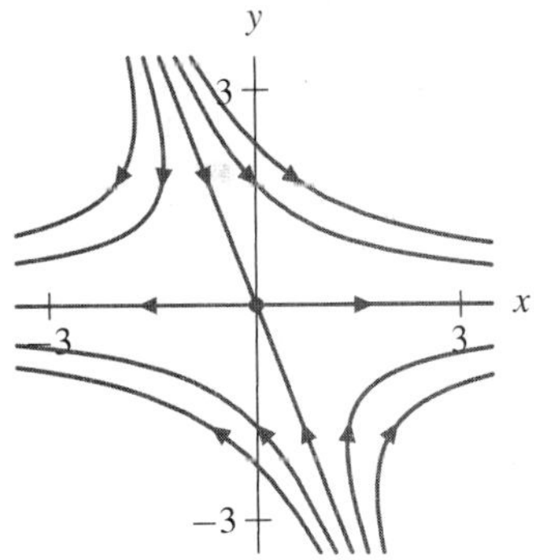

2. As we computed in Exercise 2 of Section 3.2, the eigenvalues are $\lambda_1 = -2$ and $\lambda_2 = -5$. The eigenvectors (x_1, y_1) for the eigenvalue $\lambda_1 = -2$ satisfy $y_1 = -x_1$, and the eigenvectors (x_2, y_2) for $\lambda_2 = -5$ satisfy $x_2 = 2y_2$. The equilibrium point at the origin is a sink.

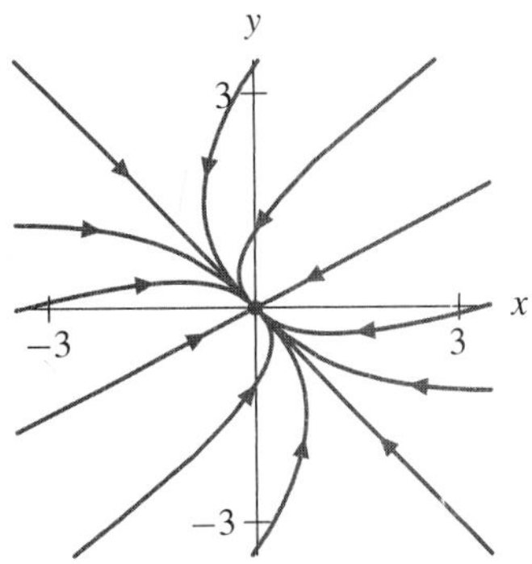

3. As we computed in Exercise 3 of Section 3.2, the eigenvalues are $\lambda_1 = -3$ and $\lambda_2 = -6$. The eigenvectors (x_1, y_1) for the eigenvalue $\lambda_1 = -3$ satisfy $y_1 = -x_1$, and the eigenvectors for $\lambda_2 = -6$ satisfy $x_2 = 2y_2$. The equilibrium point at the origin is a sink.

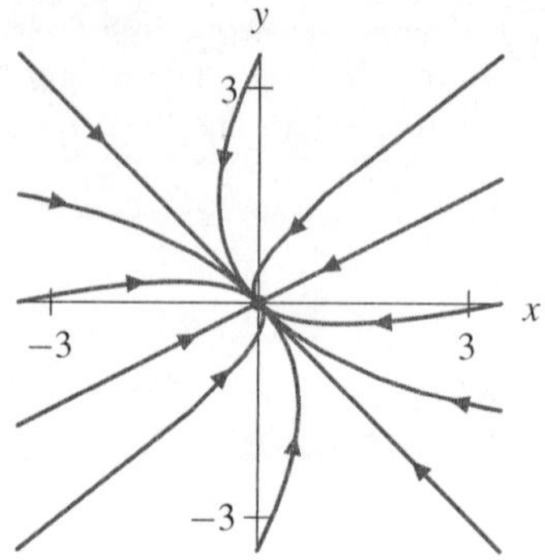

4. As we computed in Exercise 6 of Section 3.2, the eigenvalues are $\lambda_1 = -4$ and $\lambda_2 = 9$. The eigenvectors (x_1, y_1) for the eigenvalue $\lambda_1 = -4$ satisfy $9x_1 = -4y_1$, and the eigenvectors (x_2, y_2) for $\lambda_2 = 9$ satisfy the equation $y_2 = x_2$. The equilibrium point at the origin is a saddle.

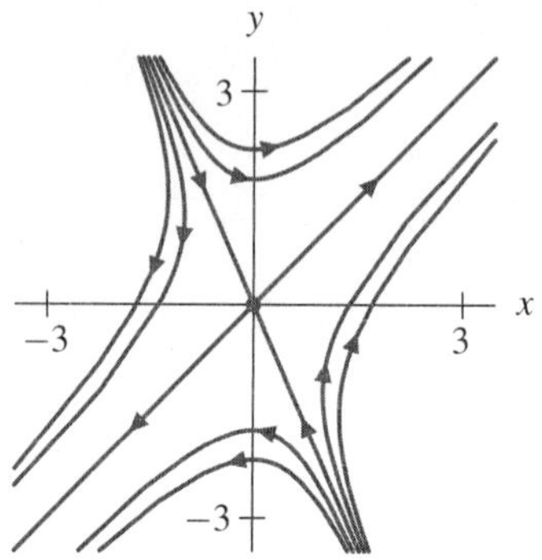

5. As we computed in Exercise 7 of Section 3.2, the eigenvalues are $\lambda_1 = -1$ and $\lambda_2 = 4$. The eigenvectors (x_1, y_1) for the eigenvalue $\lambda_1 = -1$ satisfy $y_1 = -x_1$, and the eigenvectors (x_2, y_2) for $\lambda_2 = 4$ satisfy $x_2 = 4y_2$. The equilibrium point at the origin is a saddle.

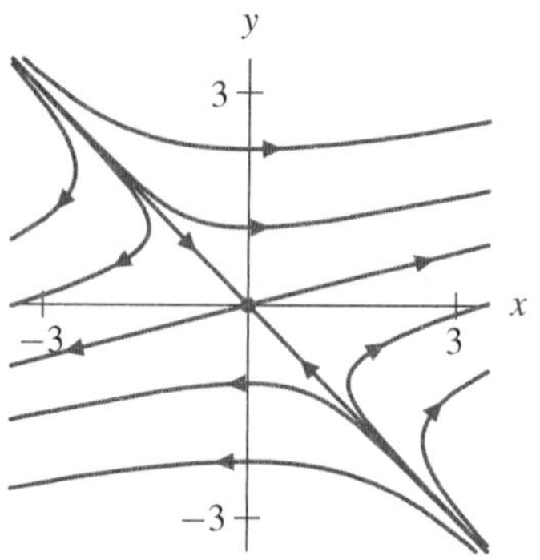

6. As we computed in Exercise 8 of Section 3.2, the eigenvalues are

$$\lambda_1 = \frac{3 + \sqrt{5}}{2} \quad \text{and} \quad \lambda_2 = \frac{3 - \sqrt{5}}{2}.$$

The eigenvectors (x_1, y_1) for the eigenvalue λ_1 satisfy $y_1 = (1 - \sqrt{5})x_1/2$, and the eigenvectors (x_2, y_2) for the eigenvalue λ_2 satisfy $y_2 = (1 + \sqrt{5})x_2/2$. The equilibrium point at the origin is a source.

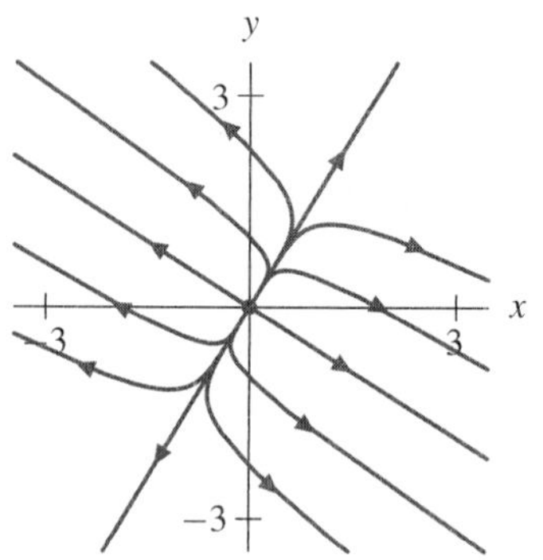

7. As we computed in Exercise 9 of Section 3.2, the eigenvalues are

$$\lambda_1 = \frac{3 + \sqrt{5}}{2} \quad \text{and} \quad \lambda_2 = \frac{3 - \sqrt{5}}{2}.$$

The eigenvectors (x_1, y_1) for the eigenvalue λ_1 satisfy $y_1 = (-1 + \sqrt{5})x_1/2$, and the eigenvectors (x_2, y_2) for λ_2 satisfy $y_2 = (-1 - \sqrt{5})x_2/2$. The equilibrium point at the origin is a source.

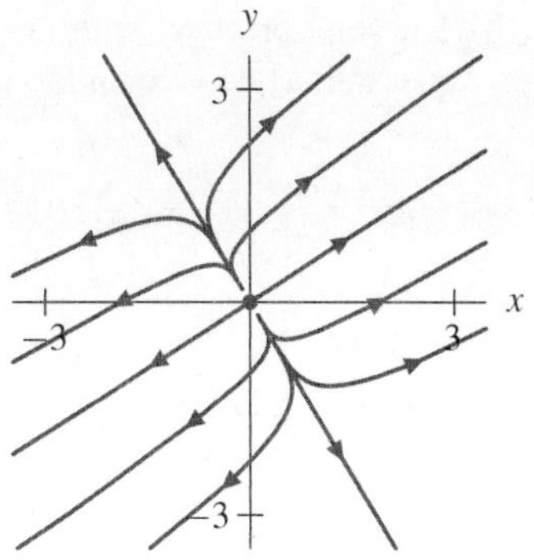

8. As we computed in Exercise 10 of Section 3.2, the eigenvalues are $\lambda_1 = -2$ and $\lambda_2 = -3$. The eigenvectors (x_1, y_1) for the eigenvalue $\lambda_1 = -2$ satisfy $x_1 = 2y_1$, and the eigenvectors (x_2, y_2) for $\lambda_2 = -3$ satisfy $x_2 = y_2$. The equilibrium point at the origin is a sink.

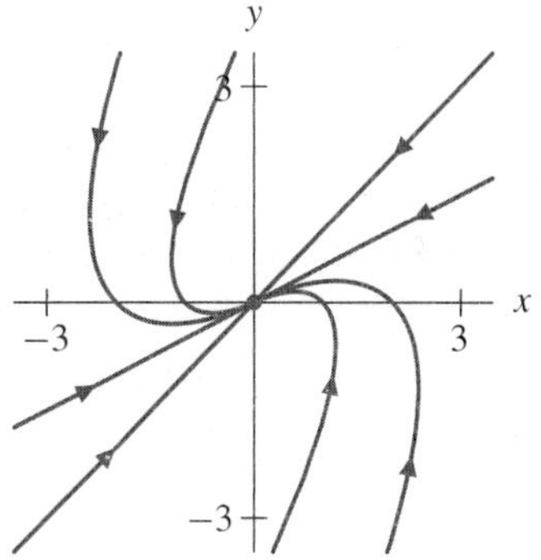

9. As we computed in Exercise 11 of Section 3.2, the eigenvalues are $\lambda_1 = 2$ and $\lambda_2 = -3$. The eigenvectors (x_1, y_1) for the eigenvalue $\lambda_1 = 2$ satisfy $y_1 = -2x_1$, and the eigenvectors for the eigenvalue $\lambda_2 = -3$ satisfy $x_1 = 2y_1$. The equilibrium point at the origin is a saddle. The solution curves in the phase plane for the initial conditions (1, 0), (0, 1), and (1, −2) are shown in the figure on the right.

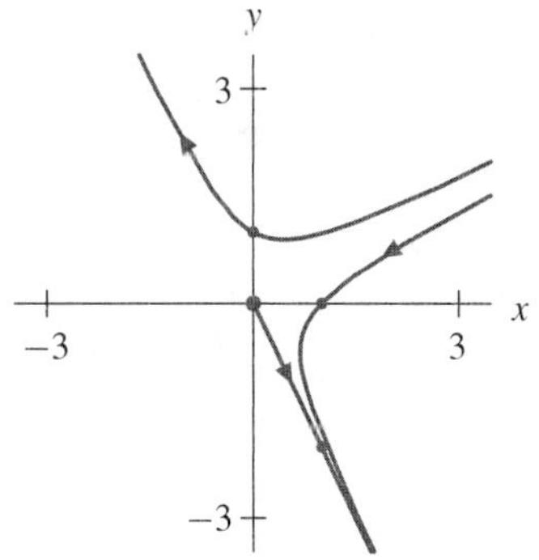

(a) The solution with initial condition (1, 0) is asymptotic to the line $y = -2x$ in the fourth quadrant as $t \to \infty$ and to the line $x = 2y$ in the first quadrant as $t \to -\infty$.

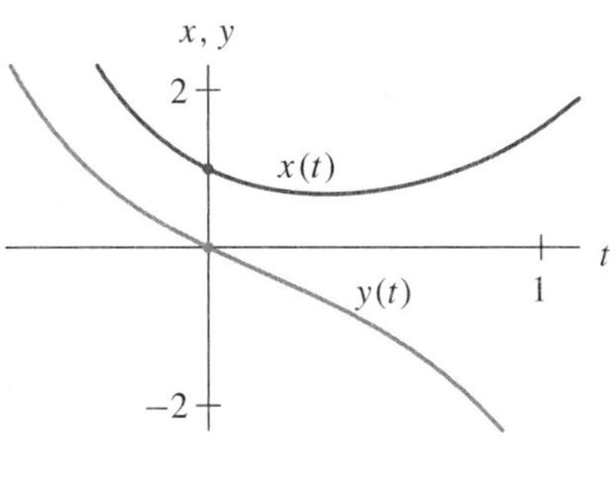

(b) The solution curve with initial condition $(1, 0)$ is asymptotic to the line $y = -2x$ in the second quadrant as $t \to \infty$ and to the line $x = 2y$ in the first quadrant as $t \to -\infty$.

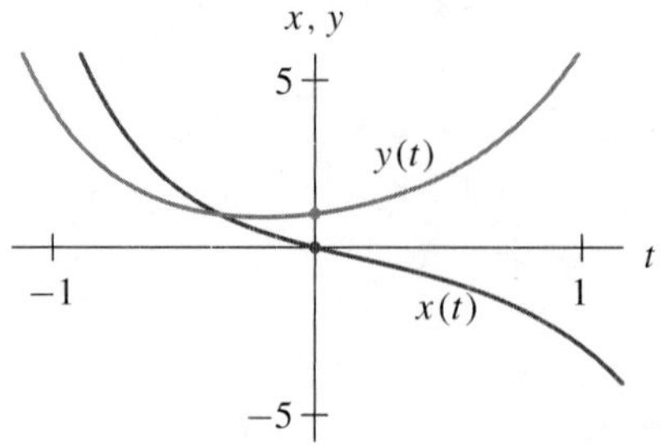

(c) The solution curve with initial condition $(1, -2)$ is on the line of eigenvectors for the eigenvalue $\lambda_1 = 2$. Hence, this solution curve stays on the line $y = -2x$. It approaches the origin as $t \to -\infty$, and it tends to ∞ in the fourth quadrant as $t \to \infty$.

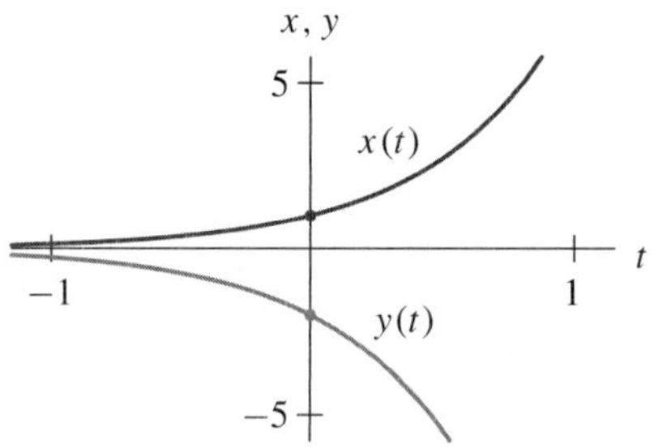

10. As we computed in Exercise 12 of Section 3.2, the eigenvalues are $\lambda_1 = 3$ and $\lambda_2 = -2$. The eigenvectors (x_1, y_1) for the eigenvalue $\lambda_1 = 3$ satisfy $y_1 = x_1/5$, and the eigenvectors (x_2, y_2) for the eigenvalue $\lambda_2 = -2$ satisfy $x_2 = 0$. The equilibrium point at the origin is a saddle. Therefore, the solution curves in the phase plane for the initial conditions $(1, 0)$, $(0, 1)$, and $(2, 2)$ are shown in the figure on the right.

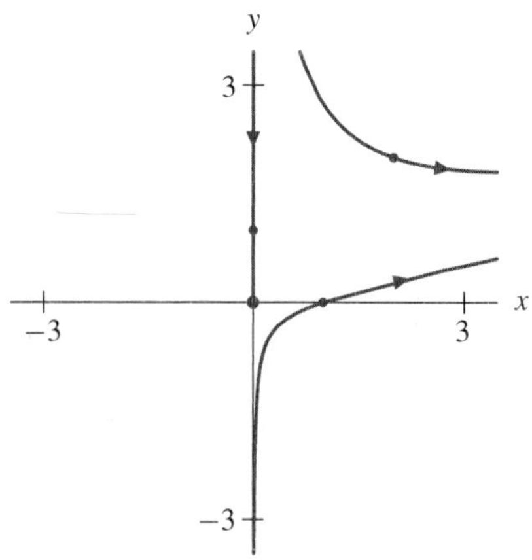

(a) The solution curve with initial condition $(1, 0)$ is asymptotic to the negative y-axis as $t \to -\infty$ and is asymptotic to the line $y = x/5$ in the first quadrant as $t \to \infty$.

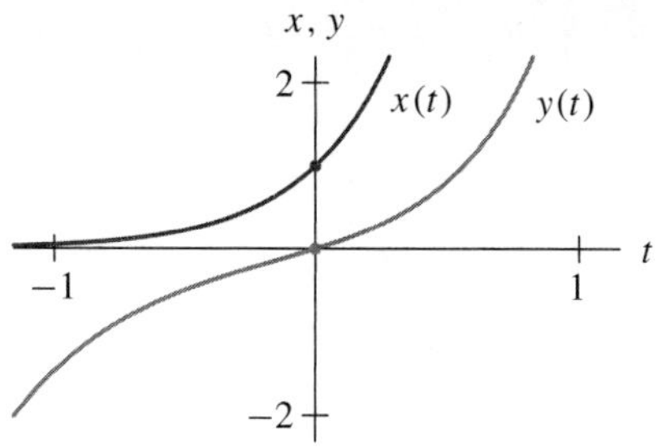

(b) The solution curve with initial condition $(0, 1)$ lies entirely on the positive y-axis, and $y(t) \to 0$ in an exponential fashion as $t \to \infty$.

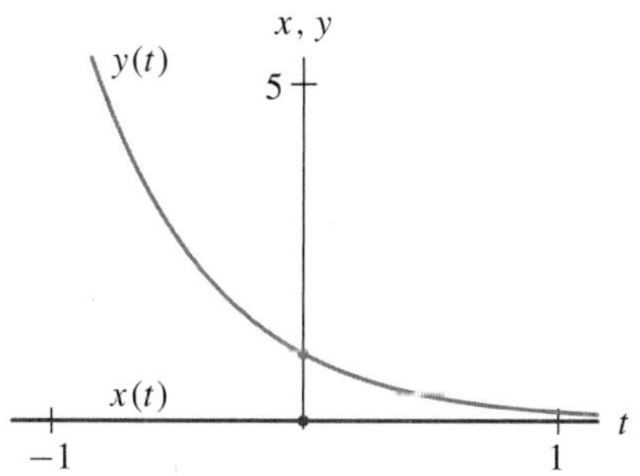

(c) The solution curve with initial condition $(2, 2)$ lies entirely in the first quadrant. It is asymptotic to the positive y-axis as $t \to -\infty$ and asymptotic to the line $y = x/5$ as $t \to \infty$.

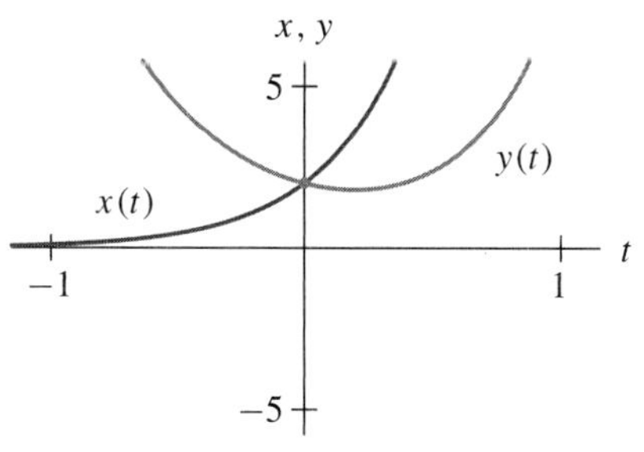

11. As we computed in Exercise 13 of Section 3.2, the eigenvalues are $\lambda_1 = -5$ and $\lambda_2 = -2$. The eigenvectors (x_1, y_1) for the eigenvalue $\lambda_1 = -5$ satisfy $y_1 = -x_1$, and the eigenvectors (x_2, y_2) for the eigenvalue $\lambda_2 = -2$ satisfy $y_2 = 2x_2$. The equilibrium point at the origin is a sink. The solution curves in the phase plane for the initial conditions $(1, 0)$, $(2, 1)$, and $(-1, -2)$ are shown in the following figure.

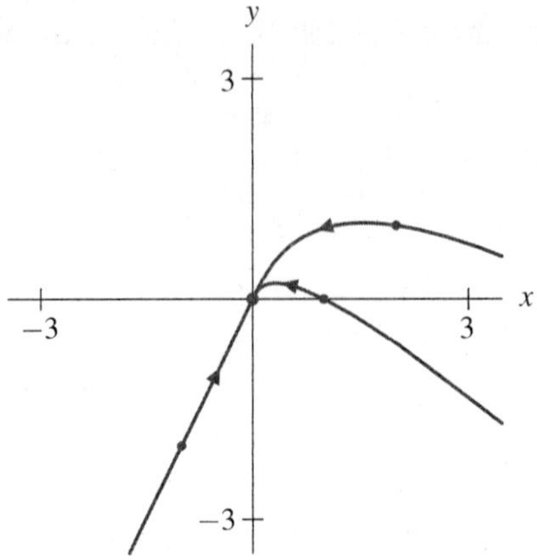

(a) The solution curve with initial condition $(1, 0)$ approaches the origin tangent to the line $y = 2x$.

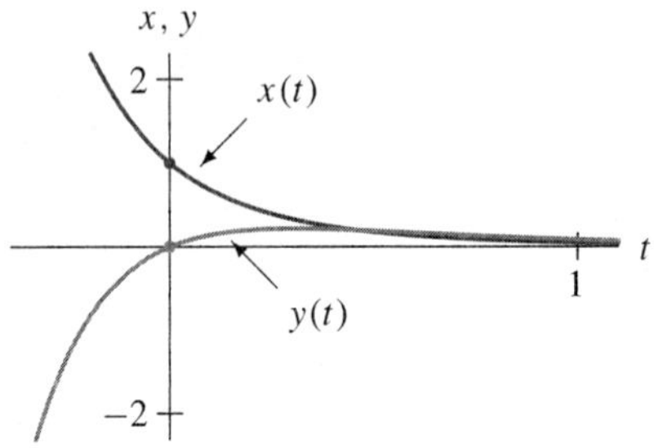

(b) The solution curve with initial condition $(2, 1)$ approaches the origin tangent to the line $y = 2x$.

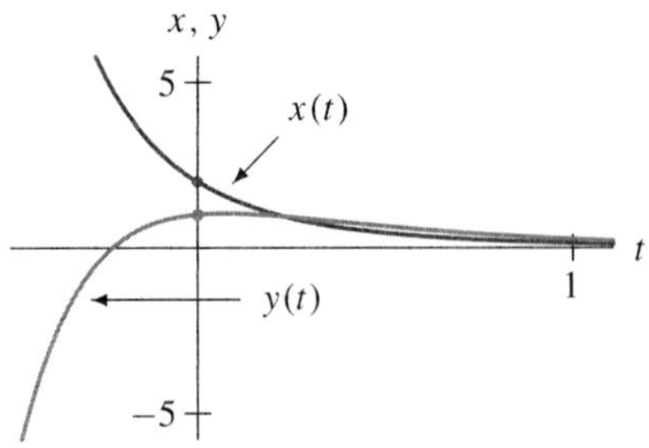

(c) The initial condition $(-1, -2)$ is an eigenvector associated to the eigenvalue $\lambda_2 = -2$. The corresponding solution curve approaches the origin along the line $y = 2x$ as $t \to \infty$.

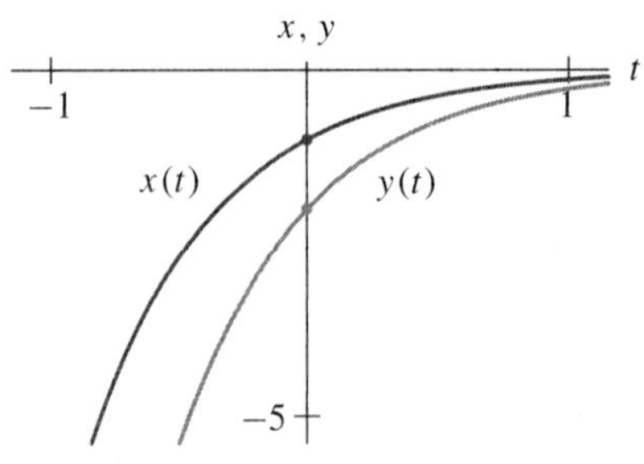

12. As we computed in Exercise 14 of Section 3.2, the eigenvalues are $\lambda_1 = 3$ and $\lambda_2 = 2$. The eigenvectors (x_1, y_1) for the eigenvalue $\lambda_1 = 3$ satisfy $x_1 = 2y_1$, and the eigenvectors (x_2, y_2) for the eigenvalue $\lambda_2 = 3$ satisfy $x_2 = y_2$. The equilibrium point at the origin is a source. The solution curves in the phase plane for the initial conditions $(1, 0)$, $(2, 1)$, and $(-1, -2)$ are shown in the figure on the right.

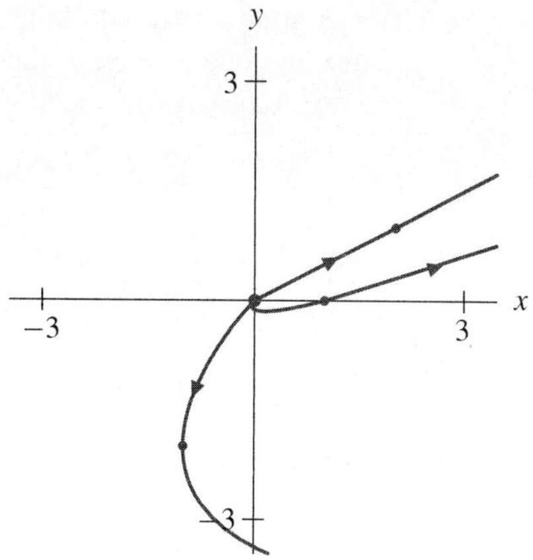

(a) The solution curve with initial condition $(1, 0)$ leaves the origin tangent to the line $y = x$. It grows without bound as $t \to \infty$, "almost parallel" to the line $y = x/2$.

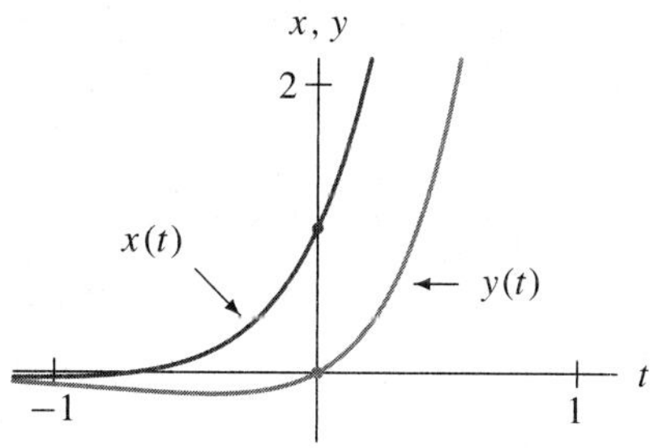

(b) The initial condition $(2, 1)$ is an eigenvector associated to the eigenvalue $\lambda_1 = 3$. The corresponding solution curve increases without bound along the line $y = x/2$ as $t \to \infty$.

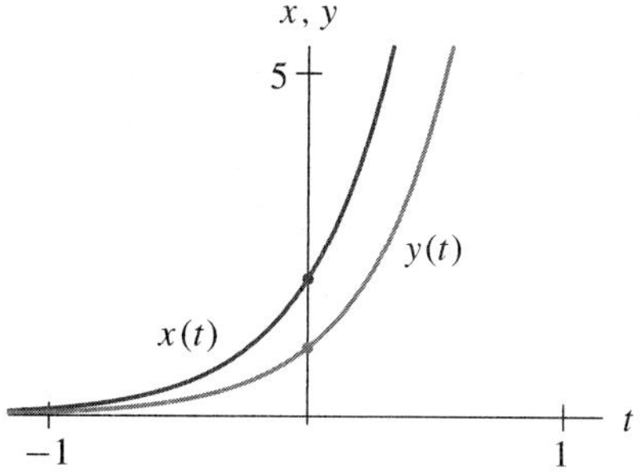

(c) The solution curve with initial condition $(-1, -2)$ leaves the origin in the third quadrant tangent to the line $y = x$. It then turns and crosses the fourth quadrant and eventually enters the first quadrant. It grows without bound as $t \to \infty$, "almost parallel" to the line $y = x/2$.

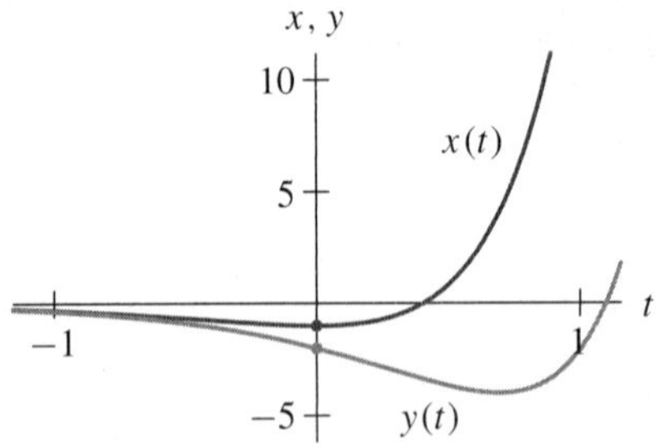

13. As we computed in Exercise 21 of Section 3.2, the eigenvalues are $\lambda_1 = -5$ and $\lambda_2 = 2$. The eigenvectors (y_1, v_1) associated to the eigenvalue $\lambda_1 = -5$ satisfy $v_1 = -5y_1$, and the eigenvectors (y_2, v_2) for the eigenvalue $\lambda_2 = 2$ satisfy the equation $v_2 = 2y_2$. The equilibrium point at the origin is a saddle.

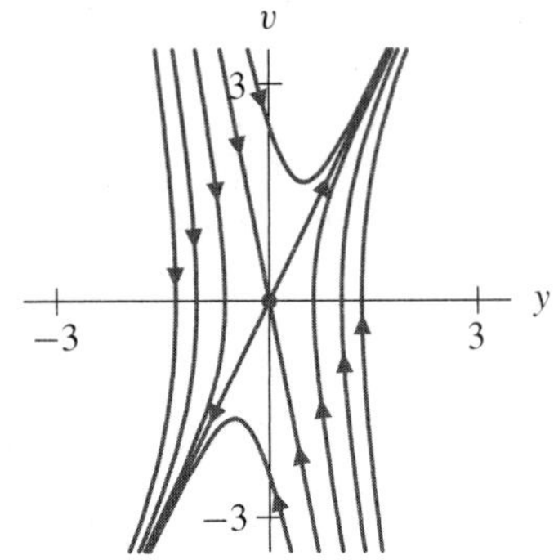

14. As we computed in Exercise 22 of Section 3.2, the eigenvalues are $\lambda_1 = -2$ and $\lambda_2 = -1$. The eigenvectors (y_1, v_1) associated to the eigenvalue $\lambda_1 = -2$ satisfy $v_1 = -2y_1$, and the eigenvectors (y_2, v_2) for the eigenvalue $\lambda_2 = -1$ satisfy $v_2 = -y_2$. The equilibrium point at the origin is a sink.

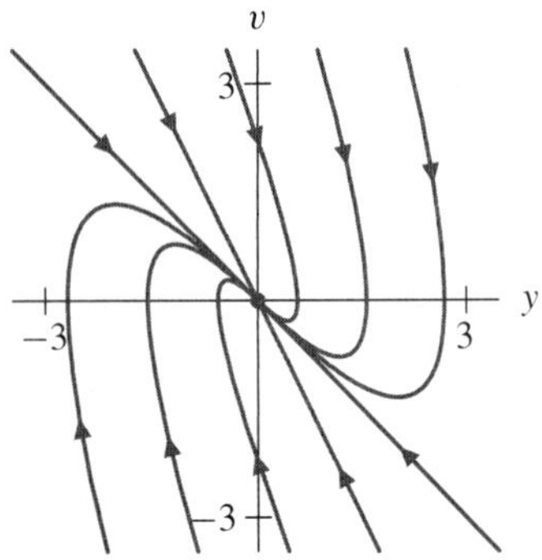

15. As we computed in Exercise 23 of Section 3.2, the eigenvalues are $\lambda_1 = -2 + \sqrt{3}$ and $\lambda_2 = -2 - \sqrt{3}$. The eigenvectors (y_1, v_1) associated to the eigenvalue $\lambda_1 = -2+\sqrt{3}$ satisfy the equation $v_1 = (-2+\sqrt{3})y_1$, and the eigenvectors (y_2, v_2) for the eigenvalue $\lambda_2 = -2-\sqrt{3}$ satisfy the equation $v_2 = (-2-\sqrt{3})y_2$. The equilibrium point at the origin is a sink.

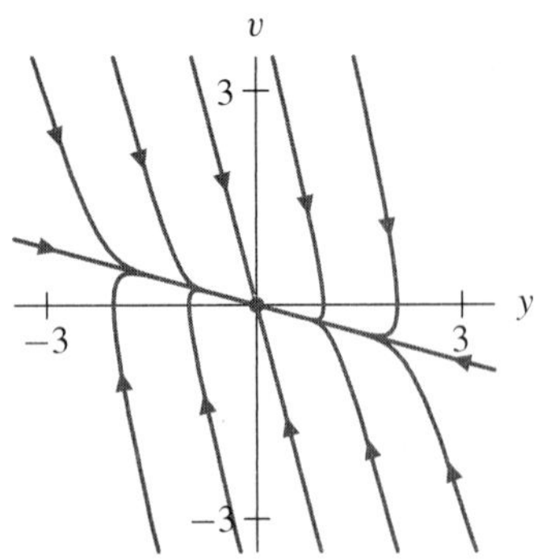

16. As we computed in Exercise 24 of Section 3.2, the eigenvalues are $\lambda_1 = -2$ and $\lambda_2 = 1$. The eigenvectors (y_1, v_1) associated to the eigenvalue $\lambda_1 = -2$ satisfy $v_1 = -2y_1$, and the eigenvectors (y_2, v_2) for the eigenvalue $\lambda_2 = 1$ satisfy the equation $v_2 = y_2$. The equilibrium point at the origin is a saddle.

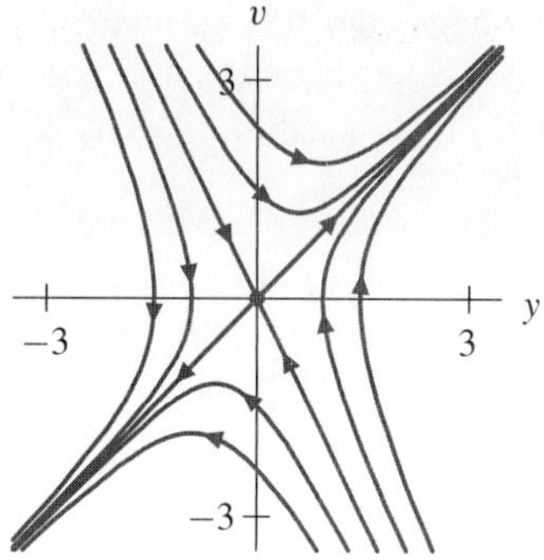

17. The characteristic equation is

$$(2 - \lambda)(-1 - \lambda) = 0,$$

and therefore, the eigenvalues are $\lambda_1 = 2$ and $\lambda_2 = -1$. The equilibrium point at the origin is a saddle.

To compute the eigenvectors associated to $\lambda_1 = 2$, we must solve the simultaneous equations

$$\begin{cases} 2x + y = 2x \\ -y = 2y. \end{cases}$$

Therefore, any vector of the form $(x, 0)$ is an eigenvector associated to the eigenvalue $\lambda_1 = 2$.

Similarly, for $\lambda_1 = -1$, the eigenvectors (x, y) associated to the eigenvalue $\lambda_2 = -1$ must satisfy thc cquation $y = \ 3x$. Therefore, we know that the phase portrait is

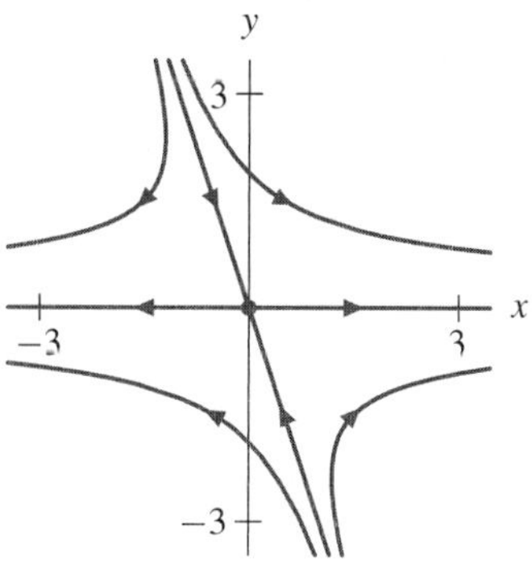

Given an initial condition on the line $y = -3x$, the corresponding solution is a straight-line solution that is asymptotic to the origin. For any other initial condition, the corresponding solution is asymptotic to the x-axis. Therefore, for any initial condition, Bob's profits, $y(t)$, eventually tend to 0 as $t \to \infty$.

To see what happens to Paul's profits, we must locate the initial condition relative to the line $y = -3x$. As stated above, if the initial condition (x_0, y_0) lies on the line $y = -3x$, then Paul's profits will also tend to 0 eventually. However, if (x_0, y_0) lies to the left of the line $y = -3x$, Paul goes broke. On the other hand, if (x_0, y_0) lies to the right of the line $y = -3x$, then Paul makes a fortune.

18. The characteristic equation is

$$(-2-\lambda)(-1-\lambda)-1=\lambda^2+3\lambda+1=0,$$

and therefore, the eigenvalues are $\lambda_1=(-3-\sqrt{5})/2$ and $\lambda_2=(-3+\sqrt{5})/2$. Since both of these eigenvalues are negative, the equilibrium point at the origin is a sink. Therefore, we know that the profits of both stores will eventually approach 0 for any given initial condition.

To compute the eigenvectors associated to $\lambda_1=(-3-\sqrt{5})/2$, we must solve the simultaneous equations

$$\begin{cases} -2x-y=\dfrac{-3-\sqrt{5}}{2}x \\ -x-y=\dfrac{-3-\sqrt{5}}{2}y. \end{cases}$$

Therefore, any vector on the line

$$y=\frac{-1+\sqrt{5}}{2}x$$

is an eigenvector associated to the eigenvalue λ_1.

To compute the eigenvectors associated to $\lambda_2=(-3+\sqrt{5})/2$, we must solve the simultaneous equations

$$\begin{cases} -2x-y=\dfrac{-3+\sqrt{5}}{2}x \\ -x-y=\dfrac{-3+\sqrt{5}}{2}y. \end{cases}$$

Therefore, any vector on the line

$$y=\frac{-1-\sqrt{5}}{2}x$$

is an eigenvector associated to the eigenvalue λ_2.

Therefore, we know that the phase portrait is

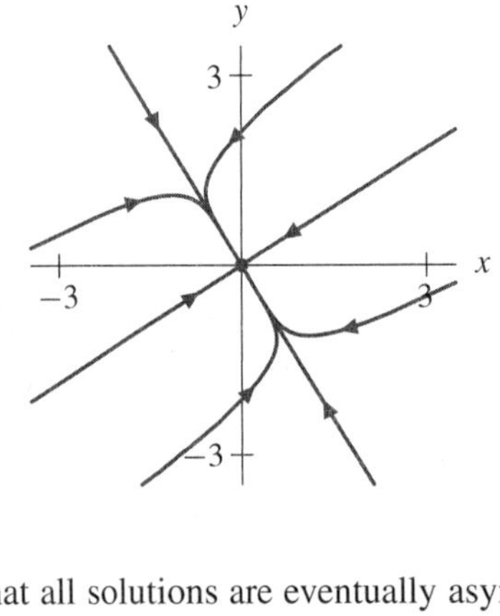

Even though we know that all solutions are eventually asymptotic to the origin, the location of the initial condition relative to the lines of eigenvectors has important qualitative significance for the behavior of the solution. For example, suppose that the initial condition is located in the first quadrant and above the line

$$y = \frac{-1 + \sqrt{5}}{2} x.$$

Then Bob's profits will monotonically decrease toward 0. Similarly, Paul's profits will decrease and become negative at some time. Eventually they will reach a minimum, and then they will increase monotonically and approach 0. Paul's profits never become positive once they pass from positive to negative.

19. **(a)** The characteristic equation is

$$(-2 - \lambda)(-1 - \lambda) = 0,$$

so the eigenvalues are $\lambda_1 = -2$ and $\lambda_2 = -1$. Therefore, the equilibrium point at the origin is a sink.

(b) To find all the straight-line solutions, we must calculate the eigenvectors. For the eigenvalue $\lambda_1 = -2$, we have the simultaneous equations

$$\begin{cases} -2x_1 + \frac{1}{2}y_1 = -2x_1 \\ -y_1 = -2y_1. \end{cases}$$

The second equation implies that $y_1 = 0$. In other words, all vectors on the x-axis are eigenvectors for λ_1. Therefore, any solution of the form $e^{-2t}(x_1, 0)$ for any x_1 is a straight-line solution corresponding to the eigenvalue $\lambda_1 = -2$.

To calculate the eigenvectors associated to the eigenvalue $\lambda_2 = -1$, we must solve the equations

$$\begin{cases} -2x_2 + \frac{1}{2}y_2 = -x_2 \\ -y_2 = -y_2. \end{cases}$$

From the first equation, we see that $y_2 = 2x_2$. Therefore, any solution of the form $e^{-t}(x_2, 2x_2)$ for any x_2 is a straight-line solution corresponding to the eigenvalue $\lambda_2 = -1$.

(c) In the phase plane, all solution curves approach the origin as $t \to \infty$. If the initial condition is on the x-axis, it yields a straight-line solution that remains on the x-axis as $t \to \infty$. For any other initial condition, the solution approaches the origin tangent to the line $y = 2x$.

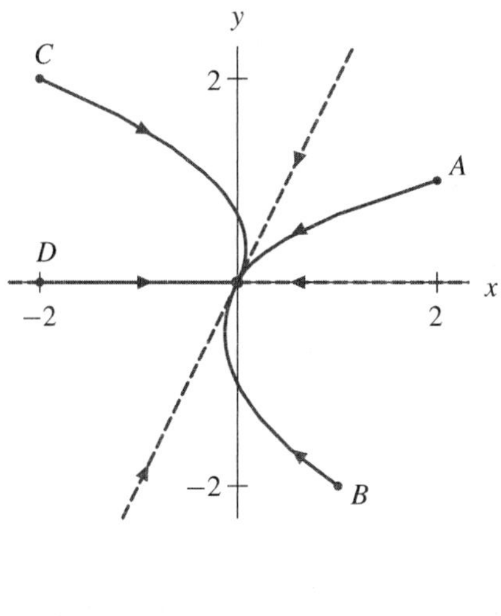

For the initial condition $A = (2, 1)$, the solution curve remains in the first quadrant. Since it approaches the origin tangent to the line $y = 2x$, it must cross the line $y = x$ at some time. Therefore, the $x(t)$- and $y(t)$-graphs are positive for all t, but they cross at some time $t > 0$.

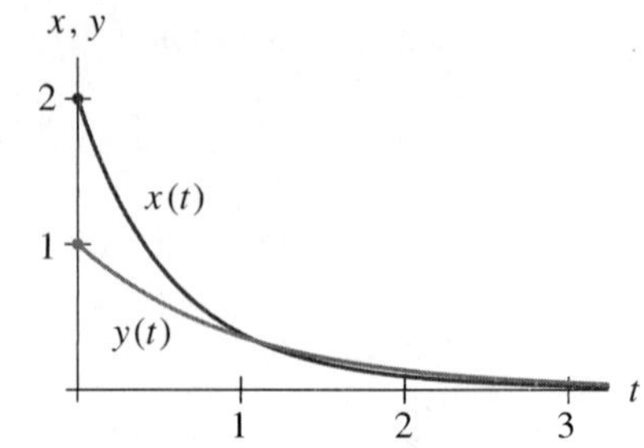

For the initial condition $B = (1, -2)$, we see that $y(t)$ is increasing but negative for all t. We also see that $x(t)$ is decreasing initially. It becomes negative, reaches a minimum, and then increases as it approaches 0. Note that $x(t) \neq y(t)$ for all t.

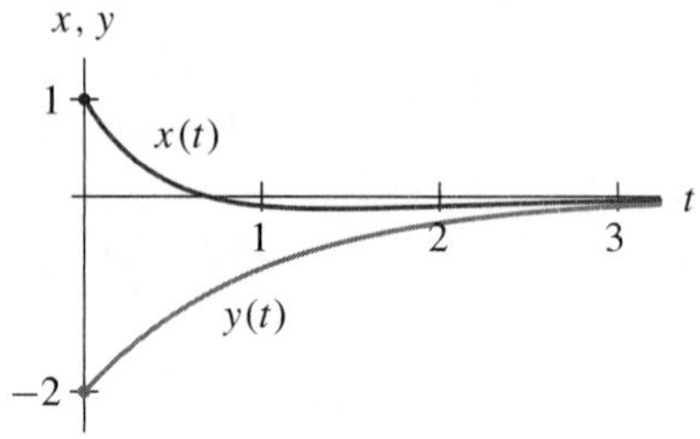

For the initial condition $C = (-2, 2)$, we see that $y(t)$ is decreasing but positive for all t. We also see that $x(t)$ is increasing initially. It becomes positive, reaches a maximum, and then decreases as it approaches 0. Again, these two graphs do not cross at any time.

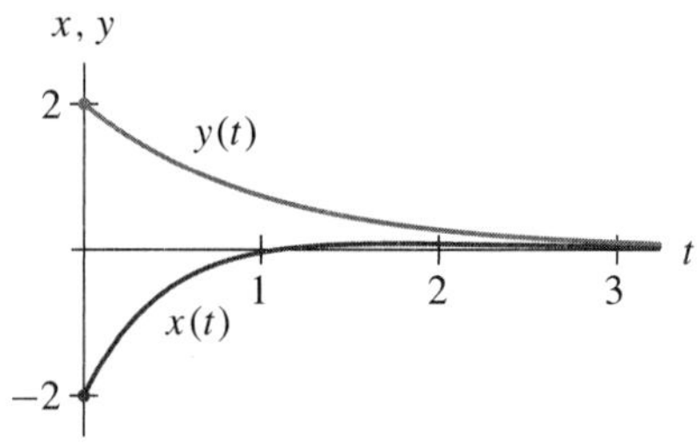

The initial condition $D = (-2, 0)$ lies on the line of eigenvectors associated to the eigenvalue $\lambda_1 = -1$. Therefore, the solution curve remains on the x-axis for all t. Hence, $y(t) = 0$ for all t, and $x(t)$ is the exponential function $-2e^{-2t}$.

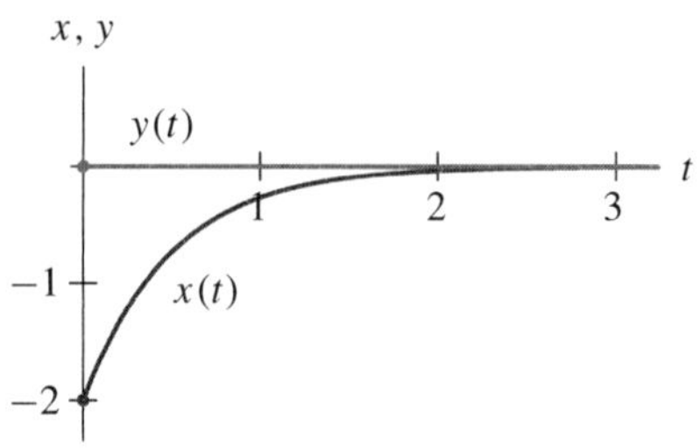

20. **(a)** The characteristic equation is

$$(2 - \lambda)(-2 - \lambda) - 12 = \lambda^2 - 16 = 0,$$

so the eigenvalues are $\lambda_1 = -4$ and $\lambda_2 = 4$. Therefore, the equilibrium point at the origin is a saddle.

(b) To find all the straight-line solutions, we must calculate the eigenvectors. For the eigenvalue $\lambda_1 = -4$, we have the simultaneous equations

$$\begin{cases} 2x_1 + 6y_1 = -4x_1 \\ 2x_1 - 2y_1 = -4y_1, \end{cases}$$

and we obtain $y_1 = -x_1$. In other words, all vectors on the line $y_1 = -x_1$ are eigenvectors for λ_1. Therefore, any solution of the form $e^{-4t}(x_1, -x_1)$ for any x_1 is a straight-line solution corresponding to the eigenvalue $\lambda_1 = -4$.

To calculate the eigenvectors associated to the eigenvalue $\lambda_2 = 4$, we must solve the equations

$$\begin{cases} 2x_2 + 6y_2 = 4x_2 \\ 2x_2 - 2y_2 = 4y_2, \end{cases}$$

and we obtain $x_2 = 3y_2$. Therefore, any solution of the form $e^{4t}(3y_2, y_2)$ for any y_2 is a straight-line solution corresponding to the eigenvalue $\lambda_2 = 4$.

(c) In the phase plane, the only solution curves that approach the origin are those whose initial conditions lie on the line $y = -x$. All other solution curves eventually approach those that correspond to the line $x = 3y$.

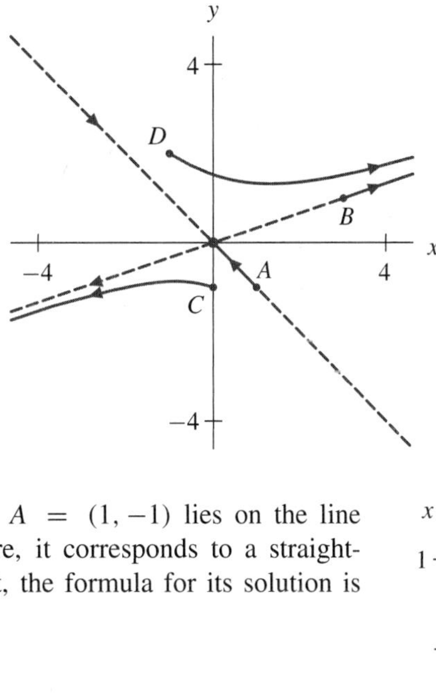

The initial condition $A = (1, -1)$ lies on the line $y = -x$. Therefore, it corresponds to a straight-line solution. In fact, the formula for its solution is $e^{-4t}(1, -1)$.

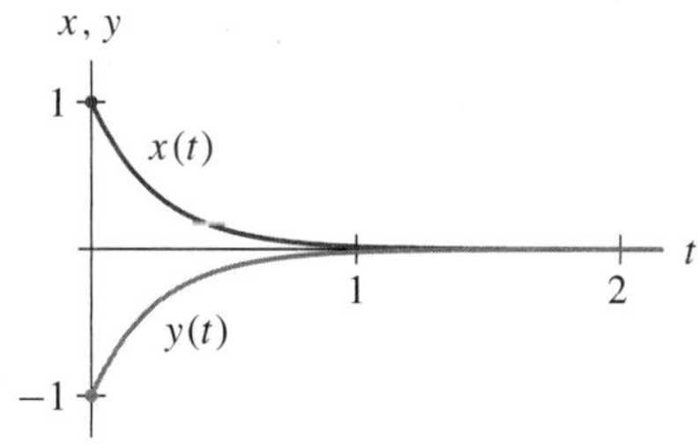

The initial condition $B = (3, 1)$ lies on the line $x = 3y$. Therefore, it corresponds to a straight-line solution, and the formula is $e^{4t}(3, 1)$.

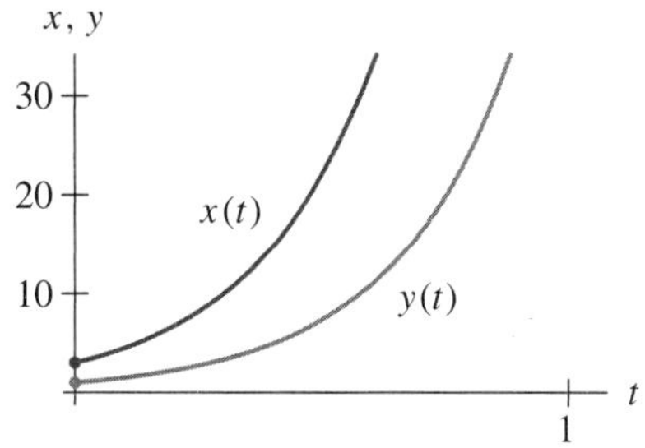

The solution curve that corresponds to the initial condition $C = (0, -1)$ enters the third quadrant and eventually approaches line $x = 3y$. From the phase plane, we see that $x(t)$ is decreasing for all $t > 0$. We also see that $y(t)$ increases initially, reaches a negative maximum value, and then decreases in an exponential fashion. Since the solution curve crosses the line $y = x$, we know that these two graphs cross. By examining the line where $dy/dt = 0$, we see that these two graphs cross at precisely the same time as $y(t)$ attains its maximum value.

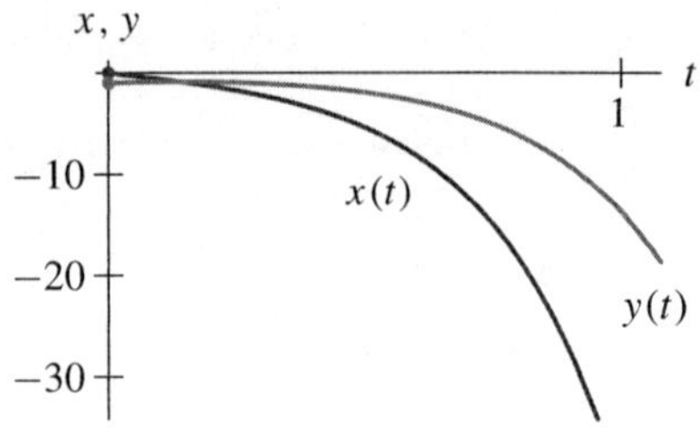

The solution curve that corresponds to the initial condition $D = (-1, 2)$ moves from the second quadrant into the first quadrant and eventually approaches the line $x = 3y$. From the phase plane, we see that $x(t)$ is increasing for all $t > 0$. We also see that $y(t)$ decreases initially, reaches a positive minimum value, and then increases in an exponential fashion. Since this solution curve crosses the line $y = x$, we know that these two graphs cross. By examining the line for which $dy/dt = 0$, we see that these two graphs cross at precisely the same time as $y(t)$ attains its minimum value.

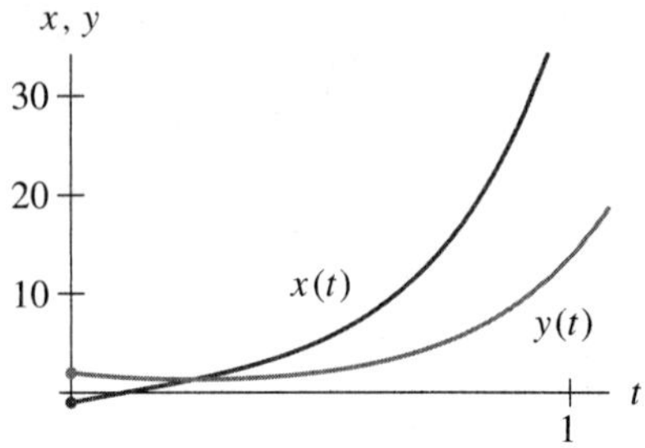

21. **(a)** The second-order equation is

$$\frac{d^2y}{dt^2} + 7\frac{dy}{dt} + 6y = 0.$$

Introducing $v = dy/dt$, we obtain the system

$$\begin{aligned}\frac{dy}{dt} &= v \\ \frac{dv}{dt} &= -6y - 7v.\end{aligned}$$

(b) The characteristic polynomial is

$$\lambda^2 + 7\lambda + 6,$$

which factors into $(\lambda + 6)(\lambda + 1)$.

(c) From the characteristic polynomial, we obtain the eigenvalues $\lambda_1 = -6$ and $\lambda_2 = -1$.

(d) To compute the eigenvectors associated to $\lambda_1 = -6$, we solve the simultaneous equations

$$\begin{cases} v = -6y \\ -6y - 7v = -6v. \end{cases}$$

Therefore, any vector on the line $v = -6y$ is an eigenvector associated to the eigenvalue λ_1.

To compute the eigenvectors associated to $\lambda_2 = -1$, we must solve the simultaneous equations

$$\left\{\begin{aligned} v &= -y \\ -6y - 7v &= -v. \end{aligned}\right.$$

Therefore, any vector on the line $y = -v$ is an eigenvector associated to the eigenvalue λ_2.

Since both eigenvalues are real and negative, we know that origin is a sink, and the solution curve corresponding to the initial condition $(y(0), v(0)) = (2, 0)$ tends toward the origin tangent to the line $y = -v$ in the yv-plane.

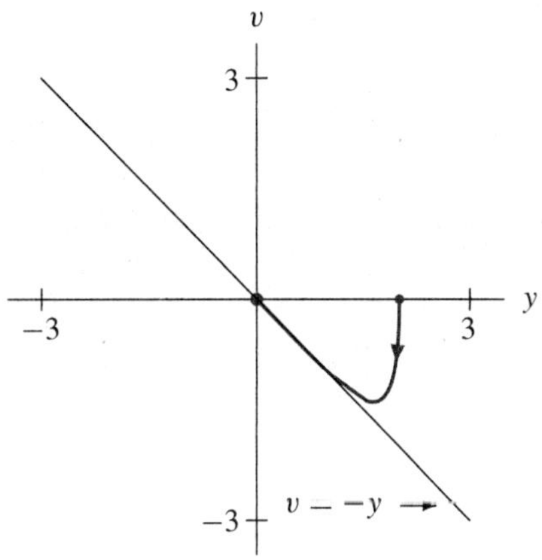

From the phase portrait, we see that the solution curve remains in the fourth quadrant for all $t > 0$. Consequently, it does not cross the line $y = 0$, and the mass cannot cross the equilibrium position. The solution approaches the origin at the rate that is determined by the eigenvalue $\lambda_2 = -1$. In other words, it approaches the origin at the rate of e^{-t}.

22. The differential equation is

$$\frac{d^2y}{dt^2} + 4\frac{dy}{dt} + y = 0,$$

which corresponds to the system

$$\begin{aligned} \frac{dy}{dt} &= v \\ \frac{dv}{dt} &= -y - 4v. \end{aligned}$$

The characteristic polynomial is $\lambda^2 + 4\lambda + 1$, and consequently the eigenvalues are $\lambda = -2 \pm \sqrt{3}$.

The eigenvectors for $\lambda = -2 + \sqrt{3}$ satisfy $v = (-2 + \sqrt{3})y$, and the eigenvectors for $\lambda = -2 - \sqrt{3}$ satisfy $v = (-2 - \sqrt{3})y$. Looking at the phase plane, the line $y = 2$ crosses each line of eigenvectors once. The line of eigenvectors corresponding to $\lambda = -2 - \sqrt{3}$ is crossed at $v = -4 - 2\sqrt{3}$ while the line of eigenvectors corresponding to $\lambda = -2 + \sqrt{3}$ is crossed at $v = -4 + 2\sqrt{3}$.

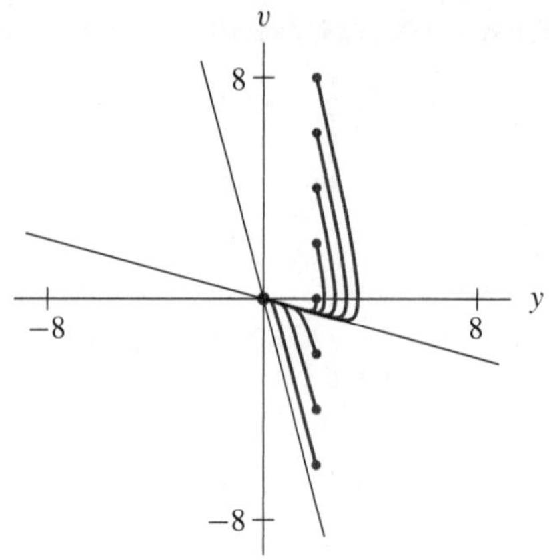

The solutions with $y = 2$, $v < -4 - 2\sqrt{3}$ all cross into the left-half ($y < 0$ half) of the phase plane. In other words, if the initial velocity is sufficiently negative, then y overshoots $y = 0$. For $v \geq -4 - 2\sqrt{3}$, $y(t)$ remains positive for all t. Solutions tending toward the origin most quickly are those on the line of eigenvectors corresponding to the more negative eigenvalue, so the solution that reaches 0.1 quickest is the one whose initial velocity is $v = -4 - 2\sqrt{3}$.

23. **(a)** Written in terms of its components, the system is

$$\frac{dx}{dt} = -0.2x - 0.1y$$
$$\frac{dy}{dt} = -0.1y.$$

Since the coefficient of y in dx/dt is negative, the introduction of new fish ($y > 0$) contributes negatively to dx/dt. Hence, the new fish have a negative effect on the native fish population.

Since the equation for dy/dt involves only y, the native fish have no effect on the population of the new fish.

(b) If the new species of fish is not introduced (that is, if $y(t) = 0$ for all t), then the system reduces to $dx/dt = -0.2x$. In this case, we have an exponential decay model as in Section 1.1, and the native fish population tends to its equilibrium level. (Remember: the quantity $x(t)$ is the difference between the native fish population and its equilibrium level, not the actual of fish population.) Thus, the model agrees with the system as described.

(c) There are two lines consisting of straight-line solutions, and the solutions with initial conditions on these lines are asymptotic to the origin as $t \to 0$. To find these lines, we must compute the eigenvalues and eigenvectors.

The characteristic polynomial of the given matrix is

$$(-0.2 - \lambda)(-0.1 - \lambda),$$

and hence the eigenvalues are $\lambda_1 = -0.2$ and $\lambda_2 = -0.1$.

To find an eigenvector for $\lambda_1 = -0.2$, we must solve

$$\begin{pmatrix} -0.2 & -0.1 \\ 0.0 & -0.1 \end{pmatrix} \begin{pmatrix} x_0 \\ y_0 \end{pmatrix} = -0.2 \begin{pmatrix} x_0 \\ y_0 \end{pmatrix}.$$

Rewritten in terms of components, this equation becomes

$$\begin{cases} -0.2x_0 - 0.1y_0 = -0.2x_0 \\ -0.1y_0 = -0.2y_0, \end{cases}$$

which is equivalent to

$$\begin{cases} -0.1y_0 = 0 \\ 0.1y_0 = 0. \end{cases}$$

If we multiply the second equation by -1, we obtain the first equation. Therefore, the equations are redundant and any vector (x_0, y_0) that satisfies the first equation is an eigenvector. Setting $x_0 = 1$ yields the eigenvector $\mathbf{V}_1 = (1, 0)$.

To find an eigenvector for $\lambda_2 = -0.1$, we repeat the process with λ_2 in replace of λ_1, and we obtain the eigenvector $\mathbf{V}_2 = (1, -1)$

Since both eigenvalues are negative, solutions with initial conditions that lie on the lines through the origin determined by the eigenvectors (the x-axis and the line $y = -x$) tend toward the origin.

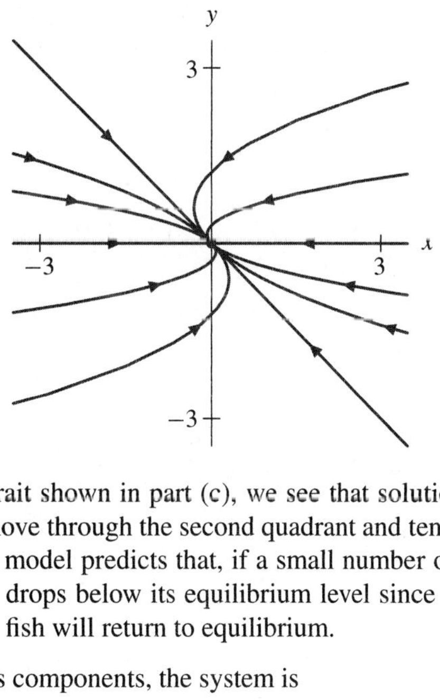

(d) Using the phase portrait shown in part (c), we see that solutions with initial conditions of the form $(0, y)$, $y > 0$, move through the second quadrant and tend toward the equilibrium point at the origin. Thus, our model predicts that, if a small number of new fish are added to the lake, the native population drops below its equilibrium level since x is negative. The new fish will die out and the native fish will return to equilibrium.

24. **(a)** Written in terms of its components, the system is

$$\frac{dx}{dt} = -0.1x + 0.2y$$
$$\frac{dy}{dt} = 1.0y.$$

Since the coefficient of y in dx/dt is positive, the introduction of new fish ($y > 0$) contributes positively to dx/dt. Hence, the new fish have a positive effect on the native fish population.

Since the equation for dy/dt involves only y, the native fish have no effect on the population of the new fish.

(b) If the new species of fish is not introduced (that is, if $y(t) = 0$ for all t), then the system reduces to $dx/dt = -0.1x$. In this case, we have an exponential decay model as in Section 1.1, and the native fish population tends to its equilibrium level. (Remember: the quantity $x(t)$ is the difference between the native fish population and its equilibrium level, not the actual of fish population.) Thus, the model agrees with the system as described.

(c) There are two lines consisting of straight-line solutions. To find these lines, we must compute the eigenvalues and eigenvectors.

The characteristic polynomial of the given matrix is

$$(-0.1 - \lambda)(1.0 - \lambda),$$

and hence the eigenvalues are $\lambda_1 = -0.1$ and $\lambda_2 = 1.0$.

To find an eigenvector for $\lambda_1 = -0.1$, we must solve

$$\begin{pmatrix} -0.1 & 0.2 \\ 0.0 & 1.0 \end{pmatrix} \begin{pmatrix} x_0 \\ y_0 \end{pmatrix} = -0.1 \begin{pmatrix} x_0 \\ y_0 \end{pmatrix}.$$

Rewritten in terms of components, this equation becomes

$$\begin{cases} -0.1x_0 + 0.2y_0 = -0.1x_0 \\ 1.0y_0 = -0.1y_0, \end{cases}$$

which is equivalent to $y_0 = 0$. Thus, the x-axis consists of straight-line solutions that correspond to the eigenvalue $\lambda_1 = -0.1$.

To find an eigenvector for $\lambda_2 = 1.0$, we repeat the process with λ_2 in replace of λ_1, and we obtain the line $y = 5.5x$. Since λ_2 is positive, solutions with initial conditions on this line tend away from the origin as t increases.

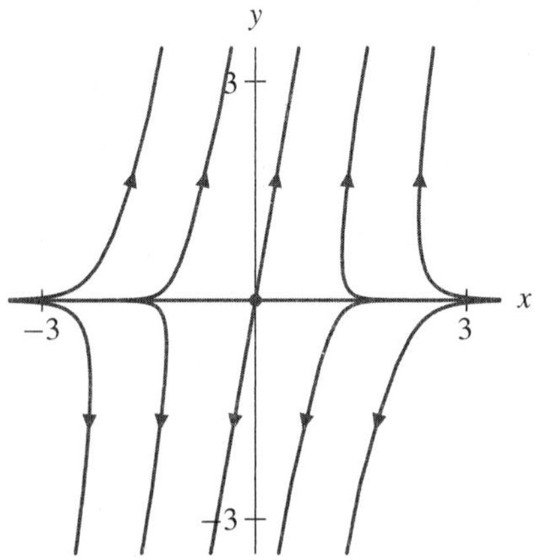

(d) Using the phase portrait shown in part (c), we see that solutions with initial conditions of the form $(0, y)$, $y > 0$, move into the first quadrant and are asymptotic to the straight-line solutions on the line $y = 5.5x$. Since the associated eigenvalue is positive, we conclude that both $x(t)$ and $y(t)$ increase without bound as $t \to \infty$. In other words, we have population explosions for both populations.

25. **(a)** Written in terms of its components, the system is

$$\frac{dx}{dt} = -0.2x + 0.1y$$
$$\frac{dy}{dt} = -0.1y.$$

Since the dy/dt equation does not depend on x, the native fish have no effect on the population of the new fish. Since the coefficient of y in the equation for dx/dt is positive, the introduction of new fish increases the population of the native fish.

(b) If the new species of fish is not introduced (that is, if $y(t) = 0$ for all t), then the above system simply becomes $dx/dt = -0.2x$. In this case, we have an exponential decay model as in Section 1.1, and the native fish population tends to its equilibrium level. (Remember: the quantity $x(t)$ is the difference between the native fish population and its equilibrium level, not the actual of fish population.) Thus, the model agrees with the system as described.

(c) The characteristic equation is

$$(-0.2 - \lambda)(-0.1 - \lambda) = 0,$$

and the eigenvalues are $\lambda_1 = -0.2$ and $\lambda_2 = -0.1$.

To find an eigenvector for $\lambda_1 = -0.2$, we must solve the simultaneous equations

$$\begin{cases} -0.2x + 0.1y = -0.2x \\ -0.1y = -0.2y. \end{cases}$$

Therefore, $y = 0$. One such eigenvector $\mathbf{V}_1$ is $(1, 0)$.

For $\lambda = -0.1$, the simultaneous equations are

$$\begin{cases} -0.2x + 0.1y = -0.1x \\ -0.1y = -0.1y. \end{cases}$$

Any vector that satisfies $y = x$ satisfies these equations. One such eigenvector is $\mathbf{V}_2 = (1, 1)$.

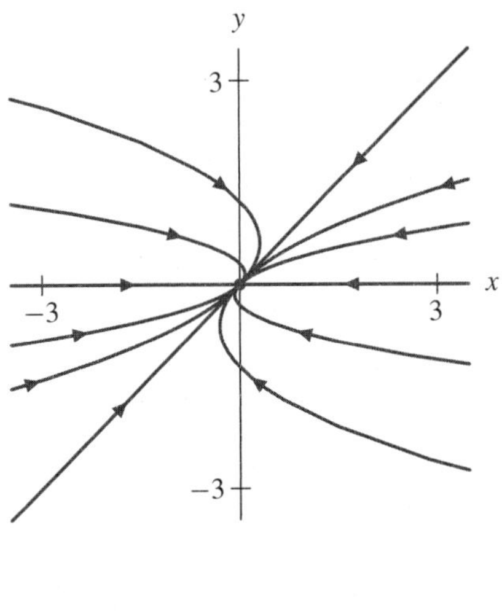

(d) Using the phase portrait shown in part (c), we see that solutions with initial conditions of the form $(0, y)$, $y > 0$, move through the first quadrant and tend toward the equilibrium point at the origin. Thus, our model predicts that, if a small number of new fish are added to the lake, the population of native fish increases above its equilibrium value and the population of new fish decreases. Eventually, the new fish head toward extinction, and the native fish return to their equilibrium population.

26. **(a)** Written in terms of its components, the system is

$$\frac{dx}{dt} = 0.1x$$
$$\frac{dy}{dt} = -0.2x + 0.2y.$$

Since dx/dt does not depend on y, the new fish have no effect on the population of the native fish.

Since the coefficient of x in dy/dt is negative, an increase in the population of the native fish above their equilibrium level ($x > 0$) has a negative effect on the population of the new fish.

(b) Since $dx/dt = 0.1x$, we have an exponential growth model for x. In other words, if we consider the x-axis as a phase line corresponding to the absence of new fish, we see that the equilibrium point at the origin is a source. In this model, the native fish population does not tend toward an equilibrium level, and therefore the model does not agree with the stated assumptions.

(c) The characteristic equation is

$$(0.1 - \lambda)(0.2 - \lambda) = 0,$$

and the eigenvalues are $\lambda_1 = 0.1$ and $\lambda_2 = 0.2$. Therefore, the origin is a source, and all solutions tend away from the origin as t increases.

To find the eigenvectors for the eigenvalue $\lambda_1 = 0.1$, we must solve the simultaneous equations

$$\begin{cases} 0.1x = 0.1x \\ -0.2x + 0.2y = 0.1y. \end{cases}$$

Then, $y = 2x$. One such eigenvector is $\mathbf{V}_1 = (1, 2)$. Similarly, for $\lambda_2 = 0.2$, $x = 0$, and an eigenvector is $\mathbf{V}_2 = (0, 1)$. Solutions with initial conditions on these lines are straight-line solutions that tend away from the origin as t increases.

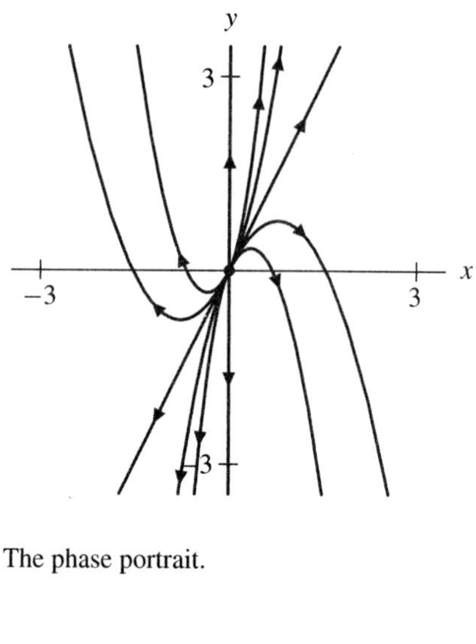

The phase portrait.

(d) Since the positive y-axis consists of straight-line solutions with initial conditions for the form $(0, y_0)$, the population of native species does not change. If a small number of new fish are added to the lake, the native species population does not change, and the new fish population grows exponentially.

27. **(a)** The characteristic equation is

$$(-2-\lambda)(2-\lambda)=0.$$

Therefore, the eigenvalues are $\lambda_1 = -2$ and $\lambda_2 = 2$, and the equilibrium point at the origin is a saddle.

(b) To find the eigenvectors (x_1, y_1) corresponding to $\lambda_1 = -2$, we solve the simultaneous linear equations

$$\begin{cases} -2x + y = -2x \\ 2y = -2y. \end{cases}$$

Therefore, the eigenvectors lie on the line $y = 0$, the x-axis. Similarly, the eigenvectors associated to the eigenvalue $\lambda_2 = 2$ lie on the line $y = 4x$.

Using these eigenvalues and eigenvectors, we can give a rough sketch of the phase portrait.

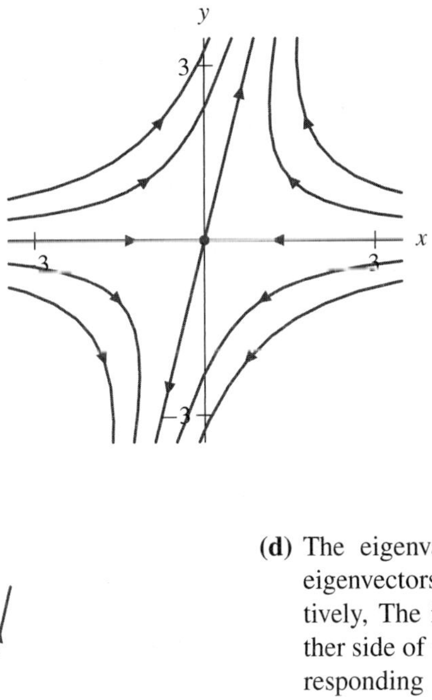

(c)

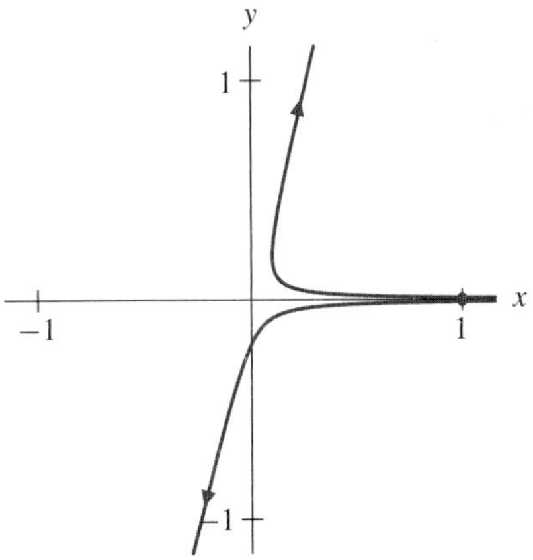

(d) The eigenvalues are -2 and 2 with eigenvectors $(1, 0)$ and $(1, 4)$ respectively, The initial conditions are on either side of the line of eigenvectors corresponding to eigenvalue -2. Hence, these two solutions will approach the origin along the x-axis, but the y coordinates will grow, at approximately the rate e^{2t}. Since the initial separation is 0.02 and we seek the approximate time t when the separation is 1, we must solve $0.02e^{2t} = 1$, which yields $t = \ln(50)/2 \approx 1.96$.

EXERCISES FOR SECTION 3.4

1. Using Euler's formula, we can write the complex-valued solution $\mathbf{Y}_c(t)$ as

$$\begin{aligned}\mathbf{Y}_c(t) &= e^{(1+3i)t}\begin{pmatrix} 2+i \\ 1 \end{pmatrix} \\ &= e^t e^{3it}\begin{pmatrix} 2+i \\ 1 \end{pmatrix} \\ &= e^t(\cos 3t + i\sin 3t)\begin{pmatrix} 2+i \\ 1 \end{pmatrix} \\ &= e^t\begin{pmatrix} 2\cos 3t - \sin 3t \\ \cos 3t \end{pmatrix} + ie^t\begin{pmatrix} 2\sin 3t + \cos 3t \\ \sin 3t \end{pmatrix}.\end{aligned}$$

Hence, we have

$$\mathbf{Y}_{\text{re}}(t) = e^t\begin{pmatrix} 2\cos 3t - \sin 3t \\ \cos 3t \end{pmatrix} \quad \text{and} \quad \mathbf{Y}_{\text{im}}(t) = e^t\begin{pmatrix} \cos 3t + 2\sin 3t \\ \sin 3t \end{pmatrix}.$$

The general solution is

$$\mathbf{Y}(t) = k_1 e^t\begin{pmatrix} 2\cos 3t - \sin 3t \\ \cos 3t \end{pmatrix} + k_2 e^t\begin{pmatrix} \cos 3t + 2\sin 3t \\ \sin 3t \end{pmatrix}.$$

2. The complex solution is

$$\mathbf{Y}_c(t) = e^{(-2+5i)t}\begin{pmatrix} 1 \\ 4-3i \end{pmatrix},$$

so we can use Euler's formula to write

$$\begin{aligned}\mathbf{Y}_c(t) &= e^{(-2+5i)t}\begin{pmatrix} 1 \\ 4-3i \end{pmatrix} \\ &= e^{-2t}e^{5it}\begin{pmatrix} 1 \\ 4-3i \end{pmatrix} \\ &= e^{-2t}(\cos 5t + i\sin 5t)\begin{pmatrix} 1 \\ 4-3i \end{pmatrix} \\ &= e^{-2t}\begin{pmatrix} \cos 5t \\ 4\cos 5t + 3\sin 5t \end{pmatrix} + ie^{-2t}\begin{pmatrix} \sin 5t \\ 4\sin 5t - 3\cos 5t \end{pmatrix}.\end{aligned}$$

Hence, we have

$$\mathbf{Y}_{\text{re}}(t) = e^{-2t}\begin{pmatrix} \cos 5t \\ 4\cos 5t + 3\sin 5t \end{pmatrix} \quad \text{and} \quad \mathbf{Y}_{\text{im}}(t) = e^{-2t}\begin{pmatrix} \sin 5t \\ 4\sin 5t - 3\cos 5t \end{pmatrix}.$$

The general solution is

$$\mathbf{Y}(t) = k_1 e^{-2t} \begin{pmatrix} \cos 5t \\ 4\cos 5t + 3\sin 5t \end{pmatrix} + k_2 e^{-2t} \begin{pmatrix} \sin 5t \\ 4\sin 5t - 3\cos 5t \end{pmatrix}.$$

3. **(a)** The characteristic equation is

$$(-\lambda)^2 + 4 = \lambda^2 + 4 = 0,$$

and the eigenvalues are $\lambda = \pm 2i$.

(b) Since the real part of the eigenvalues are 0, the origin is a center.

(c) Since $\lambda = \pm 2i$, the natural period is $2\pi/2 = \pi$, and the natural frequency is $1/\pi$.

(d) At $(1, 0)$, the tangent vector is $(-2, 0)$. Therefore, the direction of oscillation is clockwise.

(e) According to the phase plane, $x(t)$ and $y(t)$ are periodic with period π. At the initial condition $(1, 0)$, both $x(t)$ and $y(t)$ are initially decreasing.

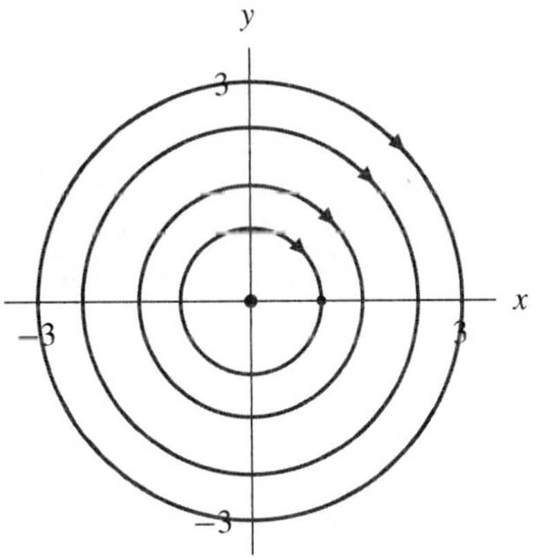

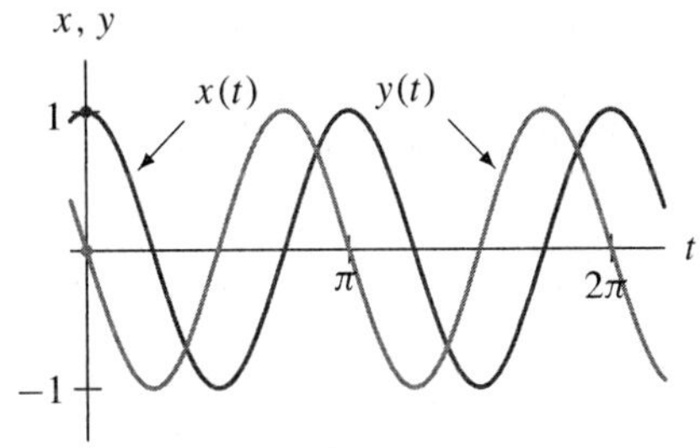

4. **(a)** The characteristic equation is

$$(2 - \lambda)(6 - \lambda) + 8 = \lambda^2 - 8\lambda + 20,$$

and the eigenvalues are $\lambda = 4 \pm 2i$.

(b) Since the real part of the eigenvalues is positive, the origin is a spiral source.

(c) Since $\lambda = 4 \pm 2i$, the natural period is $2\pi/2 = \pi$, and the natural frequency is $1/\pi$.

(d) At the point $(1, 0)$, the tangent vector is $(2, -4)$. Therefore, the solution curves spiral around the origin in a clockwise fashion.

(e) Since $d\mathbf{Y}/dt = (4, 2)$ at $\mathbf{Y}_0 = (1, 1)$, both $x(t)$ and $y(t)$ increase initially. The distance between successive zeros is π, and the amplitudes of both $x(t)$ and $y(t)$ are increasing.

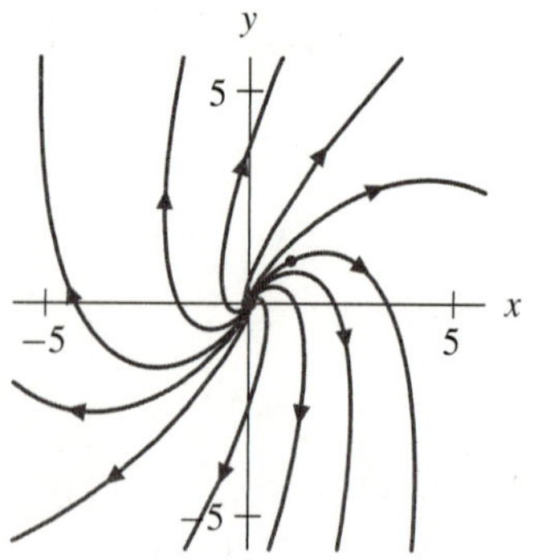

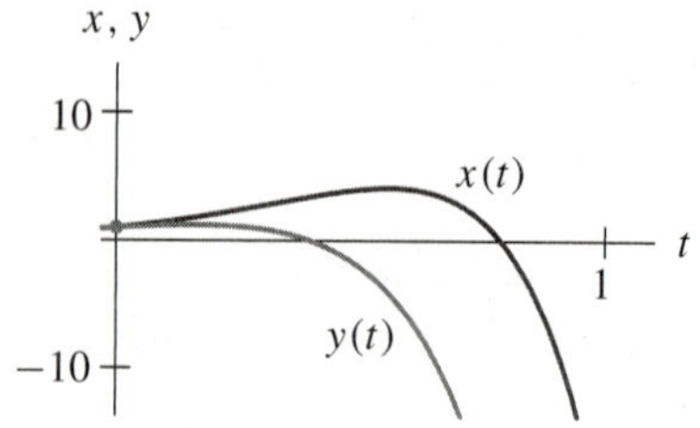

5. **(a)** The characteristic polynomial is

$$(-3-\lambda)(1-\lambda)+15 = \lambda^2 + 2\lambda + 12,$$

so the eigenvalues are $\lambda = -1 \pm i\sqrt{11}$.

(b) The eigenvalues are complex and the real part is negative, so the origin is a spiral sink.

(c) The natural period is $2\pi/\sqrt{11}$. The natural frequency is $\sqrt{11}/(2\pi)$.

(d) At the point $(1, 0)$, the vector field is $(-3, 3)$. Hence, the solution curves must spiral in a counterclockwise fashion.

(e)

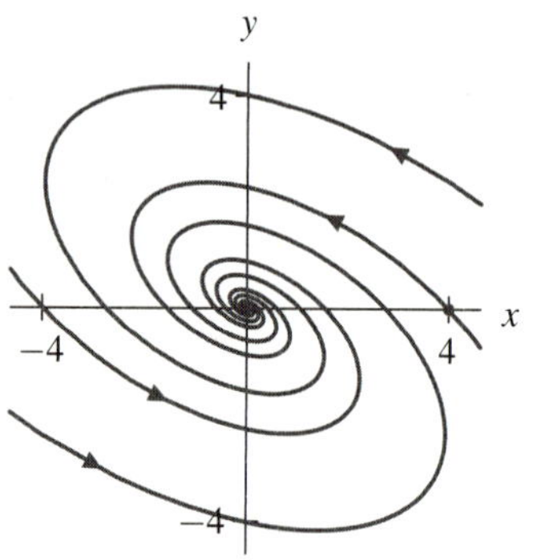

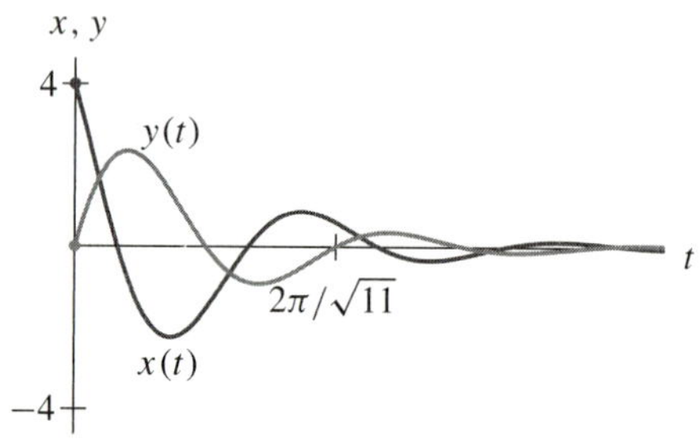

6. **(a)** The characteristic polynomial is

$$(-\lambda)(-1-\lambda)+4 = \lambda^2 + \lambda + 4,$$

so the eigenvalues are $\lambda = (-1 \pm i\sqrt{15})/2$.

(b) The eigenvalues are complex and the real part is negative, so the origin is a spiral sink.

(c) The natural period is $2\pi/(\sqrt{15}/2) = 4\pi/\sqrt{15}$. The natural frequency is $\sqrt{15}/(4\pi)$.

(d) The vector field at $(1, 0)$ is $(0, -2)$. Hence, solution curves spiral about the origin in a clockwise fashion.

(e) From the phase plane, we see that both $x(t)$ and $y(t)$ are initially increasing. However, $y(t)$ quickly reaches a local maximum. Although both functions oscillate, each successive oscillation has a smaller amplitude.

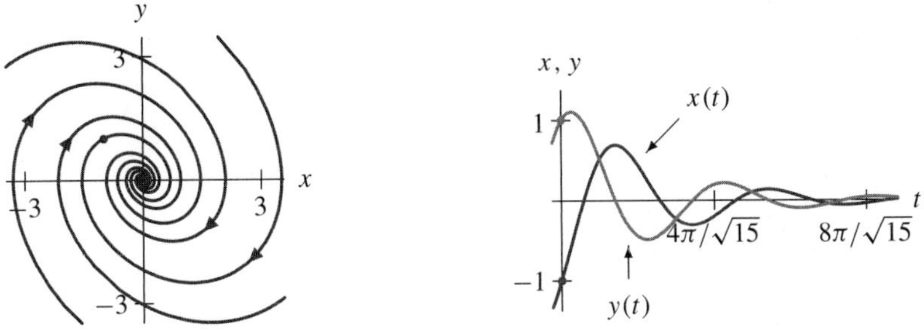

7. **(a)** The characteristic equation is

$$(2 - \lambda)(1 - \lambda) + 12 = \lambda^2 + 3\lambda + 14 = 0,$$

and the eigenvalues are $\lambda = (3 \pm \sqrt{47}\,i)/2$.

(b) Since the real part of the eigenvalues is positive, the origin is a spiral source.

(c) Since $\lambda = (3 \pm \sqrt{47}i)/2$, natural period is $4\pi/\sqrt{47}$, and natural frequency is $\sqrt{47}/(4\pi)$

(d) At the point $(1, 0)$, the tangent vector is $(2, 2)$. Therefore, the solution curves spiral about the origin in a counterclockwise fashion.

(e) From the phase plane, we see that both $x(t)$ and $y(t)$ oscillate and that the amplitude of these oscillations increases rapidly.

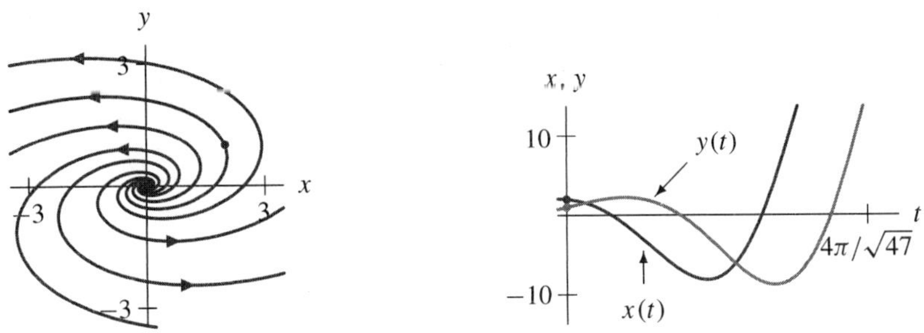

8. **(a)** The characteristic polynomial is

$$(1 - \lambda)(2 - \lambda) + 12 = \lambda^2 - 3\lambda + 14,$$

so the eigenvalues are $\lambda = (3 \pm i\sqrt{47})/2$.

(b) The eigenvalues are complex and the real part is positive, so the origin is a spiral source.

(c) The natural period is $2\pi/(\sqrt{47}/2) = 4\pi/\sqrt{47}$. The natural frequency is $\sqrt{47}/(4\pi)$.

(d) The vector field at $(1, 0)$ is $(1, -3)$. Hence, the solution curves spiral about the origin in a clockwise fashion.

(e) From the phase plane, we see that both $x(t)$ and $y(t)$ oscillate about 0 and that the amplitude of these oscillations grows quickly.

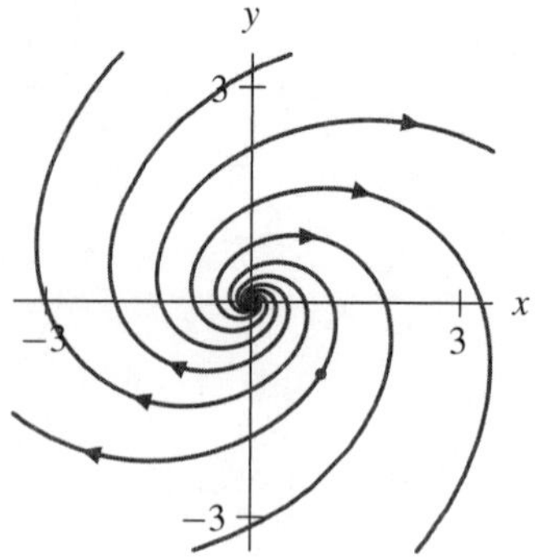

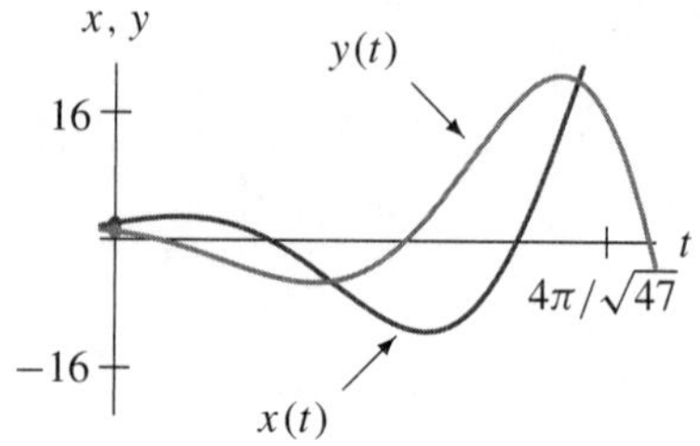

9. **(a)** According to Exercise 3, $\lambda = \pm 2i$. The eigenvectors (x, y) associated to eigenvalue $\lambda = 2i$ must satisfy the equation $2y = 2ix$, which is equivalent to $y = ix$. One such eigenvector is $(1, i)$, and thus we have the complex solution

$$\mathbf{Y}(t) = e^{2it}\begin{pmatrix} 1 \\ i \end{pmatrix} = \begin{pmatrix} \cos 2t \\ -\sin 2t \end{pmatrix} + i\begin{pmatrix} \sin 2t \\ \cos 2t \end{pmatrix}.$$

Taking real and imaginary parts, we obtain the general solution

$$\mathbf{Y}(t) = k_1\begin{pmatrix} \cos 2t \\ -\sin 2t \end{pmatrix} + k_2\begin{pmatrix} \sin 2t \\ \cos 2t \end{pmatrix}.$$

(b) From the initial condition, we obtain

$$k_1\begin{pmatrix} 1 \\ 0 \end{pmatrix} + k_2\begin{pmatrix} 0 \\ 1 \end{pmatrix} = \begin{pmatrix} 1 \\ 0 \end{pmatrix},$$

and therefore, $k_1 = 1$ and $k_2 = 0$. Consequently,the solution with the initial condition $(1, 0)$ is

$$\mathbf{Y}(t) = \begin{pmatrix} \cos 2t \\ -\sin 2t \end{pmatrix}.$$

(c)

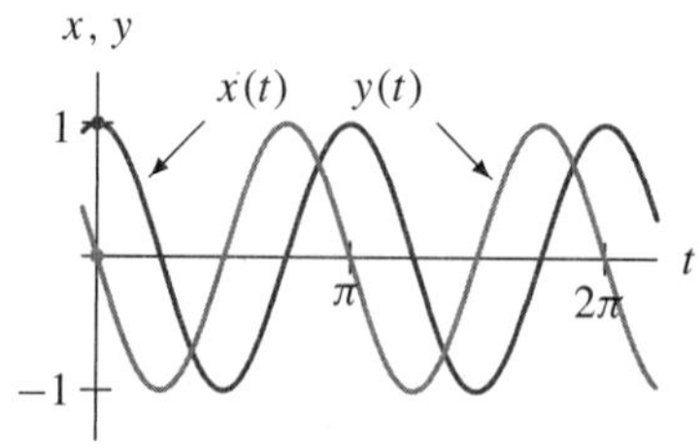

10. **(a)** According to Exercise 4, the eigenvalues are $\lambda = 4 \pm 2i$. The eigenvectors (x, y) associated to the eigenvalue $4 + 2i$ must satisfy the equation $y = (1 + i)x$. Hence, using the eigenvector $(1, 1 + i)$, we obtain the complex-valued solution

$$\mathbf{Y}(t) = e^{(4+2i)t} \begin{pmatrix} 1 \\ 1+i \end{pmatrix} = e^{4t} \begin{pmatrix} \cos 2t \\ \cos 2t - \sin 2t \end{pmatrix} + i e^{4t} \begin{pmatrix} \sin 2t \\ \cos 2t + \sin 2t \end{pmatrix}.$$

From the real and imaginary parts of $\mathbf{Y}(t)$, we obtain the general solution

$$\mathbf{Y}(t) = k_1 e^{4t} \begin{pmatrix} \cos 2t \\ \cos 2t - \sin 2t \end{pmatrix} + k_2 e^{4t} \begin{pmatrix} \sin 2t \\ \cos 2t + \sin 2t \end{pmatrix}.$$

(b) Using the initial condition, we have

$$k_1 \begin{pmatrix} 1 \\ 1 \end{pmatrix} + k_2 \begin{pmatrix} 0 \\ 1 \end{pmatrix} = \begin{pmatrix} 1 \\ 1 \end{pmatrix},$$

and thus $k_1 = 1$ and $k_2 = 0$. The desired solution is

$$\mathbf{Y}(t) = e^{4t} \begin{pmatrix} \cos 2t \\ \cos 2t - \sin 2t \end{pmatrix}.$$

(c)

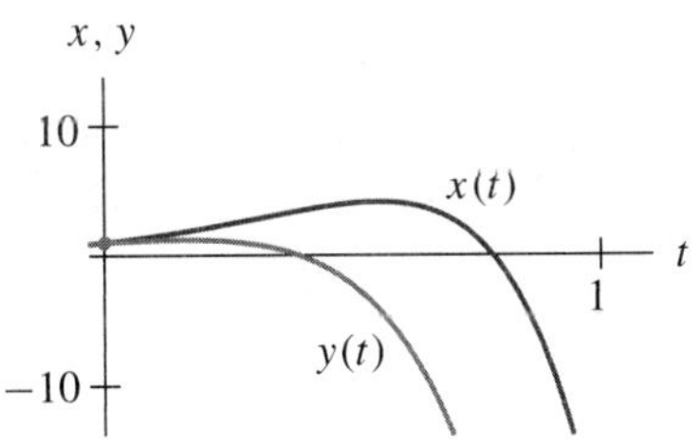

11. **(a)** To find the general solution, we find the eigenvectors from the characteristic polynomial

$$(-3 - \lambda)(1 - \lambda) + 15 = \lambda^2 + 2\lambda + 12.$$

The eigenvalues are $\lambda = -1 \pm i\sqrt{11}$. To find an eigenvector associated to the eigenvector $-1 + i\sqrt{11}$, we must solve the equations

$$\begin{cases} -3x - 5y = \left(-1 + i\sqrt{11}\right) x \\ 3x + y = \left(-1 + i\sqrt{11}\right) y. \end{cases}$$

We see that the eigenvectors must satisfy the equation $3x = (-2 + i\sqrt{11})y$. Using the eigenvector $(-2 + i\sqrt{11}, 3)$, we obtain the complex-valued solution

$$\mathbf{Y}(t) = e^{\left(-1+i\sqrt{11}\right)t} \begin{pmatrix} -2 + i\sqrt{11} \\ 3 \end{pmatrix}.$$

Using Euler's formula, we write $\mathbf{Y}(t)$ as

$$\mathbf{Y}(t) = e^{-t}\left(\cos\sqrt{11}\,t + i\sin\sqrt{11}\,t\right)\begin{pmatrix} -2+i\sqrt{11} \\ 3 \end{pmatrix},$$

which can be expressed as

$$\mathbf{Y}(t) = e^{-t}\begin{pmatrix} -2\cos\sqrt{11}\,t - \sqrt{11}\sin\sqrt{11}\,t \\ 3\cos\sqrt{11}\,t \end{pmatrix} + ie^{-t}\begin{pmatrix} \sqrt{11}\cos\sqrt{11}\,t - 2\sin\sqrt{11}\,t \\ 3\sin\sqrt{11}\,t \end{pmatrix}.$$

Taking real and imaginary parts, we can form the general solution

$$k_1e^{-t}\begin{pmatrix} -2\cos\sqrt{11}\,t - \sqrt{11}\sin\sqrt{11}\,t \\ 3\cos\sqrt{11}\,t \end{pmatrix} + k_2e^{-t}\begin{pmatrix} \sqrt{11}\cos\sqrt{11}\,t - 2\sin\sqrt{11}\,t \\ 3\sin\sqrt{11}\,t \end{pmatrix}.$$

(b) To find the particular solution with initial condition (4, 0), we solve for k_1 and k_2 and obtain

$$\begin{cases} -2k_1 + \sqrt{11}\,k_2 = 4 \\ 3k_1 = 0. \end{cases}$$

We have $k_1 = 0$ and $k_2 = 4/\sqrt{11}$.

The desired solution is

$$\mathbf{Y}(t) = e^{-t}\begin{pmatrix} 4\cos\sqrt{11}\,t - \frac{8}{\sqrt{11}}\sin\sqrt{11}\,t \\ \frac{12}{\sqrt{11}}\sin\sqrt{11}\,t \end{pmatrix}.$$

(c)

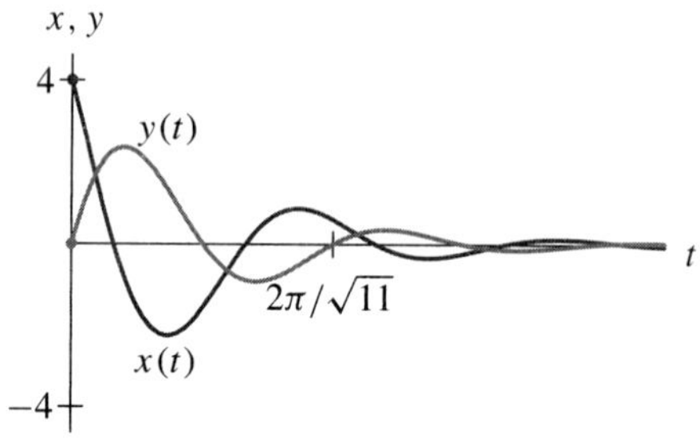

12. **(a)** The eigenvalues are the roots of the characteristic polynomial

$$(-\lambda)(-1-\lambda) + 4 = \lambda^2 + \lambda + 4.$$

So $\lambda = (-1 \pm i\sqrt{15})/2$. The eigenvectors (x, y) associated to the eigenvalue $\lambda = (-1 + i\sqrt{15})/2$ must satisfy the equation $4y = (-1 + i\sqrt{15})x$. Hence, $(4, -1 + i\sqrt{15})$ is an eigenvector for this eigenvalue, and we have the complex-valued solution

$$\mathbf{Y}(t) = e^{(-1+i\sqrt{15})t/2}\begin{pmatrix} 4 \\ -1+i\sqrt{15} \end{pmatrix}$$

$$= e^{-t/2}\left(\cos\left(\tfrac{\sqrt{15}}{2}t\right) + i\sin\left(\tfrac{\sqrt{15}}{2}t\right)\right)\begin{pmatrix} 4 \\ -1+i\sqrt{15} \end{pmatrix}$$

$$= e^{-t/2}\begin{pmatrix} 4\cos\left(\tfrac{\sqrt{15}}{2}t\right) \\ -\cos\left(\tfrac{\sqrt{15}}{2}t\right) - \sqrt{15}\sin\left(\tfrac{\sqrt{15}}{2}t\right) \end{pmatrix} +$$

$$ie^{-t/2}\begin{pmatrix} 4\sin\left(\tfrac{\sqrt{15}}{2}t\right) \\ -\sin\left(\tfrac{\sqrt{15}}{2}t\right) + \sqrt{15}\cos\left(\tfrac{\sqrt{15}}{2}t\right) \end{pmatrix}.$$

By taking real and imaginary parts

$$\mathbf{Y}_{\text{re}}(t) = e^{-t/2}\begin{pmatrix} 4\cos\left(\tfrac{\sqrt{15}}{2}t\right) \\ -\cos\left(\tfrac{\sqrt{15}}{2}t\right) - \sqrt{15}\sin\left(\tfrac{\sqrt{15}}{2}t\right) \end{pmatrix}$$

and

$$\mathbf{Y}_{\text{im}}(t) = e^{-t/2}\begin{pmatrix} 4\sin\left(\tfrac{\sqrt{15}}{2}t\right) \\ -\sin\left(\tfrac{\sqrt{15}}{2}t\right) + \sqrt{15}\cos\left(\tfrac{\sqrt{15}}{2}t\right) \end{pmatrix},$$

we form the general solution $k_1\mathbf{Y}_{\text{re}}(t) + k_2\mathbf{Y}_{\text{im}}(t)$.

(b) To find the particular solution with the initial condition $(-1, 1)$ we set $t = 0$ in the general solution and solve for k_1 and k_2. We get

$$\begin{cases} 4k_1 = -1 \\ -k_1 + \sqrt{15}k_2 = 1, \end{cases}$$

which yields $k_1 = -1/4$ and $k_2 = 3/(\sqrt{15}4) = \sqrt{15}/20$. The desired solution is

$$\mathbf{Y}(t) = e^{-t/2}\begin{pmatrix} -\cos\left(\tfrac{\sqrt{15}}{2}t\right) + \tfrac{\sqrt{15}}{5}\sin\left(\tfrac{\sqrt{15}}{2}t\right) \\ \cos\left(\tfrac{\sqrt{15}}{2}t\right) + \tfrac{\sqrt{15}}{5}\sin\left(\tfrac{\sqrt{15}}{2}t\right) \end{pmatrix}.$$

(c)

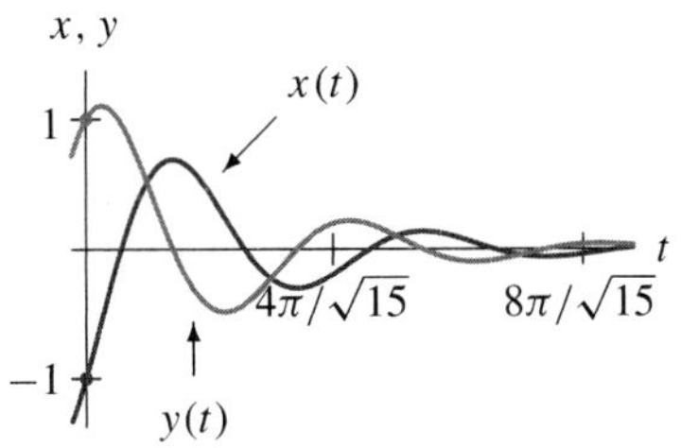

13. **(a)** According to Exercise 7, the eigenvalues are $\lambda = (3 \pm i\sqrt{47})/2$. The eigenvalues (x, y) associated to the eigenvector $(3 + i\sqrt{47})/2$ must satisfy the equation $12y = (1 - i\sqrt{47})x$. Hence, one eigenvector is $(12, 1 - i\sqrt{47})$, and we have the complex-valued solution

$$\mathbf{Y}(t) = e^{(3+i\sqrt{47})t/2}\begin{pmatrix} 12 \\ 1 - i\sqrt{47} \end{pmatrix}$$

$$= e^{3t/2}\begin{pmatrix} 12\cos\left(\frac{\sqrt{47}}{2}t\right) \\ \cos\left(\frac{\sqrt{47}}{2}t\right) + \sqrt{47}\sin\left(\frac{\sqrt{47}}{2}t\right) \end{pmatrix} +$$

$$ie^{3t/2}\begin{pmatrix} 12\sin\left(\frac{\sqrt{47}}{2}t\right) \\ -\sqrt{47}\cos\left(\frac{\sqrt{47}}{2}t\right) + \sin\left(\frac{\sqrt{47}}{2}t\right) \end{pmatrix}.$$

Taking real and imaginary parts

$$\mathbf{Y}_{\text{re}}(t) = e^{3t/2}\begin{pmatrix} 12\cos\left(\frac{\sqrt{47}}{2}t\right) \\ \cos\left(\frac{\sqrt{47}}{2}t\right) + \sqrt{47}\sin\left(\frac{\sqrt{47}}{2}t\right) \end{pmatrix}$$

and

$$\mathbf{Y}_{\text{im}}(t) = e^{3t/2}\begin{pmatrix} 12\sin\left(\frac{\sqrt{47}}{2}t\right) \\ -\sqrt{47}\cos\left(\frac{\sqrt{47}}{2}t\right) + \sin\left(\frac{\sqrt{47}}{2}t\right) \end{pmatrix},$$

we obtain the general solution $k_1\mathbf{Y}_{\text{re}}(t) + k_2\mathbf{Y}_{\text{im}}(t)$.

(b) From the initial condition, we have

$$k_1\begin{pmatrix} 12 \\ 1 \end{pmatrix} + k_2\begin{pmatrix} 0 \\ -\sqrt{47} \end{pmatrix} = \begin{pmatrix} 2 \\ 1 \end{pmatrix}.$$

Thus, $k_1 = 1/6$ and $k_2 = -5/(6\sqrt{47})$, and the desired solution is

$$\mathbf{Y}(t) = e^{3t/2}\begin{pmatrix} 2\cos\left(\frac{\sqrt{47}}{2}t\right) - \frac{10}{\sqrt{47}}\sin\left(\frac{\sqrt{47}}{2}t\right) \\ \cos\left(\frac{\sqrt{47}}{2}t\right) + \frac{7}{\sqrt{47}}\sin\left(\frac{\sqrt{47}}{2}t\right) \end{pmatrix}.$$

(c)

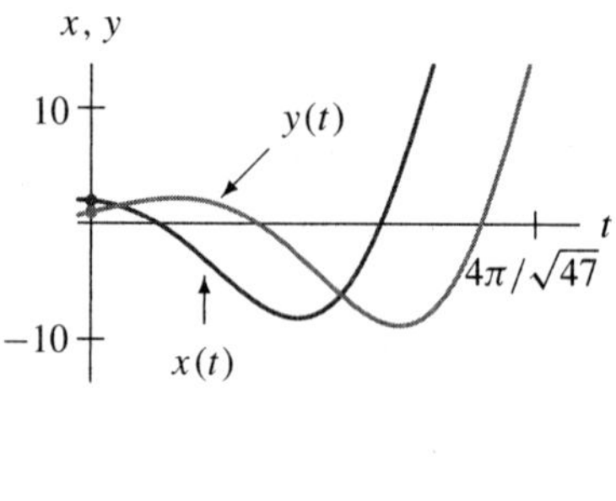

14. **(a)** The characteristic polynomial is

$$(1-\lambda)(2-\lambda)+12=\lambda^2-3\lambda+14,$$

so the eigenvalues are $\lambda = 3/2 \pm i\sqrt{47}/2$. The eigenvectors (x, y) associated to the eigenvalue $(3+i\sqrt{47})/2$ must satisfy the equation $8y = (1+i\sqrt{47})x$. Hence, $(8, 1+i\sqrt{47})$ is an eigenvector, and we obtain the complex-valued solution

$$\begin{aligned}
\mathbf{Y}(t) &= e^{(3+i\sqrt{47})t/2}\begin{pmatrix} 8 \\ 1+i\sqrt{47} \end{pmatrix} \\
&= e^{3t/2}\left(\cos\left(\tfrac{\sqrt{47}}{2}t\right)+i\sin\left(\tfrac{\sqrt{47}}{2}t\right)\right)\begin{pmatrix} 8 \\ 1+i\sqrt{47} \end{pmatrix} \\
&= e^{3t/2}\begin{pmatrix} 8\cos\left(\frac{\sqrt{47}}{2}t\right) \\ \cos\left(\frac{\sqrt{47}}{2}t\right)-\sqrt{47}\sin\left(\frac{\sqrt{47}}{2}t\right) \end{pmatrix}+ \\
&\quad ie^{3t/2}\begin{pmatrix} 8\sin\left(\frac{\sqrt{47}}{2}t\right) \\ \sin\left(\frac{\sqrt{47}}{2}t\right)+\sqrt{47}\cos\left(\frac{\sqrt{47}}{2}t\right) \end{pmatrix}.
\end{aligned}$$

Taking real and imaginary parts

$$\mathbf{Y}_{\text{re}}(t) = e^{3t/2}\begin{pmatrix} 8\cos\left(\frac{\sqrt{47}}{2}t\right) \\ \cos\left(\frac{\sqrt{47}}{2}t\right)-\sqrt{47}\sin\left(\frac{\sqrt{47}}{2}t\right) \end{pmatrix}$$

and

$$\mathbf{Y}_{\text{im}}(t) = e^{3t/2}\begin{pmatrix} 8\sin\left(\frac{\sqrt{47}}{2}t\right) \\ \sin\left(\frac{\sqrt{47}}{2}t\right)+\sqrt{47}\cos\left(\frac{\sqrt{47}}{2}t\right) \end{pmatrix},$$

we obtain the general solution $k_1\mathbf{Y}_{\text{re}}(t) + k_2\mathbf{Y}_{\text{im}}(t)$.

(b) To find the particular solution, we solve

$$\begin{cases} 8k_1 = 1 \\ k_1 + \sqrt{47}k_2 = -1 \end{cases}$$

and obtain $k_1 = 1/8$ and $k_1 = -9/(8\sqrt{47})$. The desired solution is

$$\mathbf{Y}(t) = e^{3t/2}\begin{pmatrix} \cos\left(\frac{\sqrt{47}}{2}t\right)-\frac{9}{\sqrt{47}}\sin\left(\frac{\sqrt{47}}{2}t\right) \\ -\cos\left(\frac{\sqrt{47}}{2}t\right)-\frac{7}{\sqrt{47}}\sin\left(\frac{\sqrt{47}}{2}t\right) \end{pmatrix}.$$

(c)

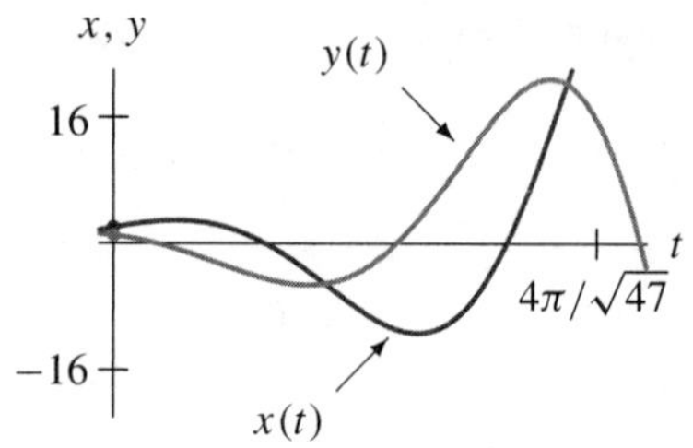

15. **(a)** In the case of complex eigenvalues, the function $x(t)$ oscillates about $x = 0$ with constant period, and the amplitude of successive oscillations is either increasing, decreasing, or constant depending on the sign of the real part of the eigenvalue. The graphs that satisfy these properties are (iv) and (v).

(b) For (iv), the natural period is approximately 1.5, and since the amplitude tends toward zero as t increases, the origin is a sink. For (v), the natural period is approximately 1.25, and since the amplitude increases as t increases, the origin is a source.

(c) (i) The time between successive zeros is not constant.
(ii) Oscillation stops at some t.
(iii) The amplitude is not monotonically decreasing or increasing.
(vi) Oscillation starts at some t. There was no prior oscillation.

16. The characteristic polynomial is

$$(a - \lambda)(a - \lambda) + b^2 = \lambda^2 - 2a\lambda + (a^2 + b^2),$$

so the eigenvalues are

$$\lambda = \frac{2a \pm \sqrt{4a^2 - 4(a^2 + b^2)}}{2} = a \pm \frac{\sqrt{-4b^2}}{2} = a \pm \sqrt{-b^2}.$$

Since $b^2 > 0$, the eigenvalues are complex. In fact, they are $a \pm bi$.

17. We know that $\lambda_1 = \alpha + i\beta$ satisfies the equation $\lambda_1^2 + a\lambda_1 + b = 0$. Therefore, if we take the complex conjugate all of the terms in this equation, we obtain

$$(\alpha - i\beta)^2 + a(\alpha - i\beta) + b = 0,$$

since a and b are real. The complex conjugate of λ_1 is $\lambda_2 = \alpha - i\beta$, and we have

$$\lambda_2^2 + a\lambda_2 + b = 0.$$

Therefore, λ_2 is also a root.

18. Let

$$\mathbf{A} = \begin{pmatrix} a & b \\ c & d \end{pmatrix}.$$

If (x_0, y_0) is an eigenvector associated to the eigenvalue $\alpha + i\beta$, we have

$$\begin{pmatrix} a & b \\ c & d \end{pmatrix} \begin{pmatrix} x_0 \\ y_0 \end{pmatrix} = \begin{pmatrix} (\alpha + i\beta)x_0 \\ (\alpha + i\beta)y_0 \end{pmatrix}.$$

Then $ax_0 + by_0 = (\alpha + i\beta)x_0$, which is equivalent to

$$y_0 = \frac{a - \alpha + i\beta}{b} x_0.$$

Suppose x_0 is real and nonzero, then the imaginary part of y_0 is $\beta x_0/b$. Since $\beta \neq 0$, the imaginary part of y_0 must be nonzero. (Note: If $b = 0$, then the eigenvalues are a and d. In other words, they are not complex, so the hypothesis of the exercise is not satisfied).

19. Suppose $\mathbf{Y}_2 = k\mathbf{Y}_1$ for some constant k. Then, $\mathbf{Y}_0 = (1 + ik)\mathbf{Y}_1$. Since $\mathbf{A}\mathbf{Y}_0 = \lambda\mathbf{Y}_0$, we have

$$(1 + ik)\mathbf{A}\mathbf{Y}_1 = \lambda(1 + ik)\mathbf{Y}_1.$$

Thus, $\mathbf{A}\mathbf{Y}_1 = \lambda\mathbf{Y}_1$. Now note that the left-hand side, $\mathbf{A}\mathbf{Y}_1$, is a real vector. However, since λ is complex and $\mathbf{Y}_1$ is real, the right-hand side is complex (that is, it has a nonzero imaginary part). Thus, we have a contradiction, and $\mathbf{Y}_1$ and $\mathbf{Y}_2$ must be linearly independent.

20. If $\mathbf{A}\mathbf{Y}_0 = \lambda\mathbf{Y}_0$, then we can take complex conjugates of both sides to obtain $\overline{\mathbf{A}\mathbf{Y}_0} = \overline{\lambda\mathbf{Y}_0}$ (where the complex conjugate of a vector or matrix is the complex conjugate of its entries). But $\overline{\mathbf{A}\mathbf{Y}_0} = \overline{\mathbf{A}}\,\overline{\mathbf{Y}_0} = \mathbf{A}\overline{\mathbf{Y}_0}$ because $\mathbf{A}$ is real. Also, $\overline{\lambda\mathbf{Y}_0} = \overline{\lambda}\,\overline{\mathbf{Y}_0}$. Hence, $\mathbf{A}\overline{\mathbf{Y}_0} = \overline{\lambda}\overline{\mathbf{Y}_0}$. In other words, $\overline{\lambda}$ is an eigenvalue of $\mathbf{A}$ with eigenvector $\overline{\mathbf{Y}_0}$.

21. **(a)** The factor $e^{-\alpha t}$ is positive for all t. Hence, the zeros of $x(t)$ are exactly the zeros of $\sin\beta t$. Suppose t_1 and t_2 are successive zeros (that is, $t_1 < t_2$, $x(t_1) = x(t_2) = 0$, and $x(t) \neq 0$ for $t_1 < t < t_2$), then $\beta t_2 - \beta t_1 = \pi$. In other words, $t_2 - t_1 = \pi/\beta$.

(b) By the nature of sine function, local maxima and local minima appear alternately. Therefore, we look for t_1 and t_2 such that $x'(t_1) = x'(t_2) = 0$ and $x'(t) \neq 0$ for $t_1 < t < t_2$. From

$$x'(t) - e^{-\alpha t}(-\alpha\sin\beta t + \beta\cos\beta t) = 0,$$

we know that $\tan\beta t = \beta/\alpha$ if t is corresponds to a local extremum. Since the tangent function is periodic with period π, $\beta(t_2 - t_1) = \pi$. Hence, $t_2 - t_1 = \pi/\beta$. Note that the distance between a local minimum and the following local maximum of $x(t)$ is constant over t.

(c) From part (b), we know that the distance between the first local maximum and the first local minimum is π/β and the distance between the first local minimum and the second local maximum is π/β. Therefore, the distance between the first two local maxima of $x(t)$ is $2\pi/\beta$.

(d) From part (b), we know that the first local maximum of $x(t)$ occurs at $t = (\arctan(\beta/\alpha))/\beta$.

22. Consider the point in the plane determined by the coordinates (k_1, k_2), and let ϕ be an angle such that $K\cos\phi = k_1$ and $K\sin\phi = k_2$. (Such an angle exists since $(K\cos\phi, K\sin\phi)$ parameterizes the circle through (k_1, k_2) centered at the origin. In fact, there are infinitely many such ϕ, all differing by integer multiples of 2π.) Then

$$\begin{aligned} x(t) &= k_1\cos\beta t + k_2\sin\beta t \\ &= K\cos\phi\cos\beta t + K\sin\phi\sin\beta t \\ &= K\cos(\beta t - \phi). \end{aligned}$$

The last equality comes from the trigonometric identity for the cosine of the difference of two angles.

23. **(a)** The corresponding first-order system is

$$\frac{dy}{dt} = v$$
$$\frac{dv}{dt} = -qy - pv.$$

(b) The characteristic polynomial is

$$(-\lambda)(-p - \lambda) + q = \lambda^2 + p\lambda + q,$$

so the eigenvalues are $\lambda = (-p \pm \sqrt{p^2 - 4q})/2$. Hence, the eigenvalues are complex if and only if $p^2 < 4q$. Note that q must be positive for this condition to be satisfied.

(c) In order to have a spiral sink, we must have $p^2 < 4q$ (to make the eigenvalues complex) and $p > 0$ (to make the real part of the eigenvalues negative). In other words, the origin is a spiral sink if and only if $q > 0$ and $0 < p < 2\sqrt{q}$. The origin is a center if and only if $q > 0$ and $p = 0$. Finally, the origin is a spiral source if and only if $q > 0$ and $-2\sqrt{q} < p < 0$.

(d) The vector field at $(1, 0)$ is $(0, -q)$. Hence, if $q > 0$, then the vector field points down along the entire y-axis, and the solution curves spiral about the origin in a clockwise fashion. Note that q must be positive for the eigenvalues to be complex, so the solution curves always spiral about the origin in a clockwise fashion as long as the eigenvalues are complex.

24. Note that the graphs have the same period and exponential rate of growth.

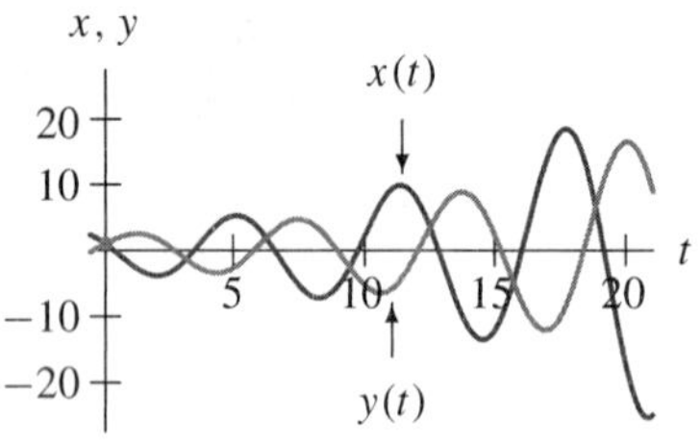

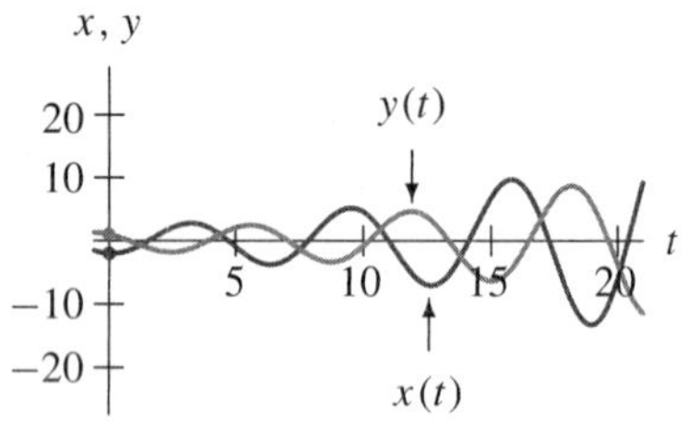

25. There is no spiral saddle because a linear saddle is a linear system where some solutions approach the origin and some move away. If one solution spirals toward (or away from) the origin, then we can multiply that solution by any constant, scaling it so that it goes through any point in the plane. This scaled solution is still a solution of the system (recall the Linearity Principle), so every solution spirals in the same way, either toward or away from the origin.

26. The eigenvalues are $\pm i$. Using the usual method to find eigenvectors, we see that the eigenvectors corresponding to the eigenvalue i satisfy the equation $10y = (3 + i)x$. We use the eigenvector $\mathbf{V}_0 = (10, 3 + i)$ to determine the general solution. It is

$$\mathbf{Y}(t) = k_1 \begin{pmatrix} 10\cos t \\ 3\cos t - \sin t \end{pmatrix} + k_2 \begin{pmatrix} 10\sin t \\ \cos t + 3\sin t \end{pmatrix}.$$

In terms of components, we have

$$x(t) = 10k_1 \cos t + 10k_2 \sin t$$
$$y(t) = (3k_1 + k_2)\cos t + (3k_2 - k_1)\sin t.$$

To show that the solution curves are ellipses, we need to find an "elliptical" relationship that $x(t)$ and $y(t)$ satisfy. In this case, it turns out that

$$[x(t)]^2 - 6x(t)\,y(t) + 10[y(t)]^2 = 10(k_1^2 + k_2^2).$$

In particular, the value of $x^2 - 6xy + 10y^2$ does not depend on t. It only depends on k_1 and k_2, which are, in turn, determined by the initial condition. It is an exercise in analytic geometry to show that the curves that satisfy

$$x^2 - 6xy + 10y^2 = K$$

are ellipses for any positive constant K.

You may wonder where $x^2 - 6xy + 10y^2$ comes from. See the technique for constructing Hamiltonian functions described in Section 5.3.

EXERCISES FOR SECTION 3.5

1. **(a)** The characteristic equation is

$$(-3 - \lambda)^2 = 0,$$

and the eigenvalue is $\lambda = -3$.

(b) To find an eigenvector, we solve the simultaneous equations

$$\begin{cases} -3x = -3x \\ x - 3y = -3y. \end{cases}$$

Then, $x = 0$, and one eigenvector is $(0, 1)$.

(c) Note the straight-line solutions along the y-axis.

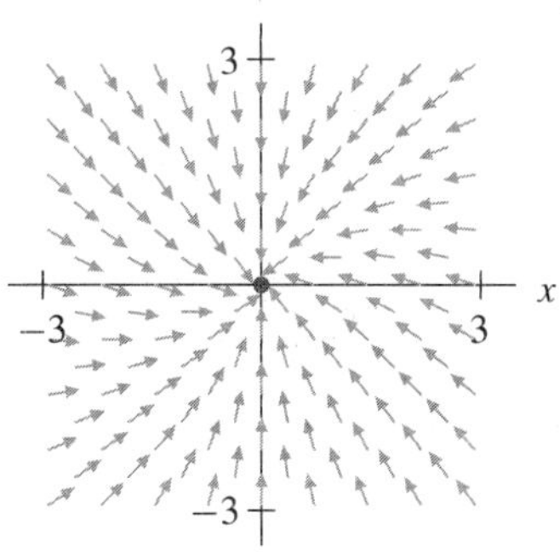

(d) Since the eigenvalue is negative, any solution with an initial condition on the y-axis tends toward the origin as t increases. According to the direction field, every solution tends to the origin as t increases. The solutions with initial conditions in the half-plane $x > 0$ eventually approach the origin along the positive y-axis. Similarly, the solutions with initial conditions in the half-plane $x < 0$ eventually approach the origin along the negative y-axis.

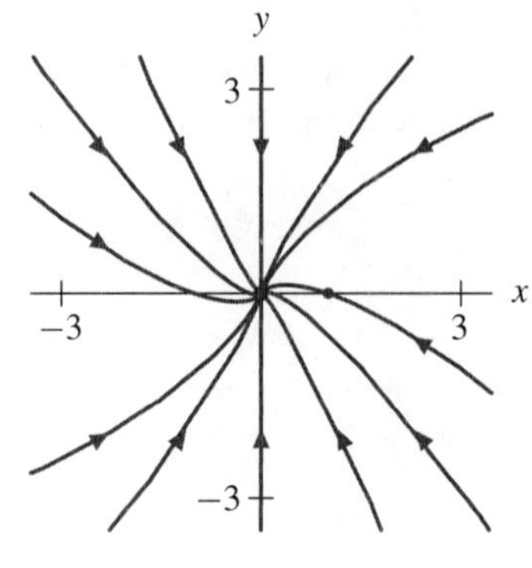

(e) At the point $\mathbf{Y}_0 = (1, 0)$, $d\mathbf{Y}/dt = (-3, 1)$. Therefore, $x(t)$ decreases initially and $y(t)$ increases initially. The solution eventually approaches the origin tangent to the positive y-axis. Therefore, $x(t)$ monotonically decreases to zero and $y(t)$ eventually decreases toward zero. Since the solution with the initial condition $\mathbf{Y}_0$ never crosses y-axis in the phase plane, the function $x(t) > 0$ for all t.

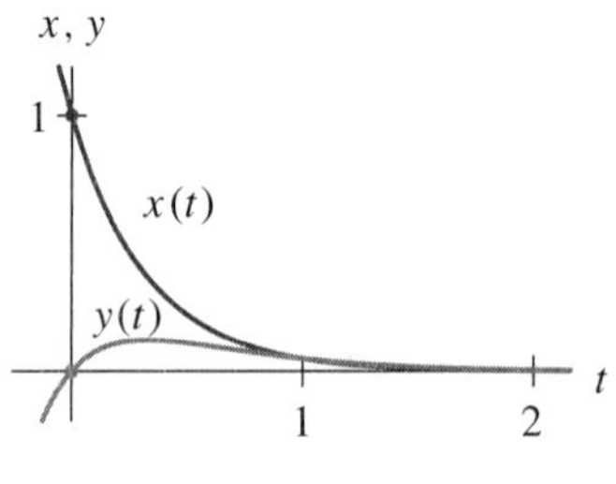

2. **(a)** The characteristic polynomial is

$$(2-\lambda)(4-\lambda)+1 = \lambda^2 - 6\lambda + 9 = (\lambda - 3)^2,$$

so there is only one eigenvalue, $\lambda = 3$.

(b) To find an eigenvector, we solve the equations

$$\begin{cases} 2x + y = 3x \\ -x + 4y = 3y. \end{cases}$$

Both equations simplify to $y = x$, so $(1, 1)$ is one eigenvector.

(c) Note the straight-line solutions along the line $y = x$.

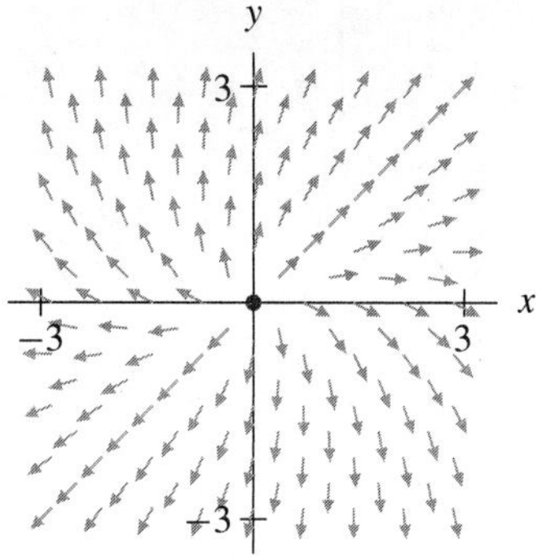

(d) Since the sole eigenvalue is positive, all solutions except the equilibrium solution are unbounded as t increases. As $t \to -\infty$, the solutions with initial conditions in the half-plane $y > x$ tend to the origin tangent to the half-line $y = x$ with $y < 0$. Similarly, solutions with initial conditions in the half-plane $y < x$ tend to the origin tangent to the half-line $y = x$ with $y > 0$. Note the solution curve that goes through the initial condition $(1, 0)$.

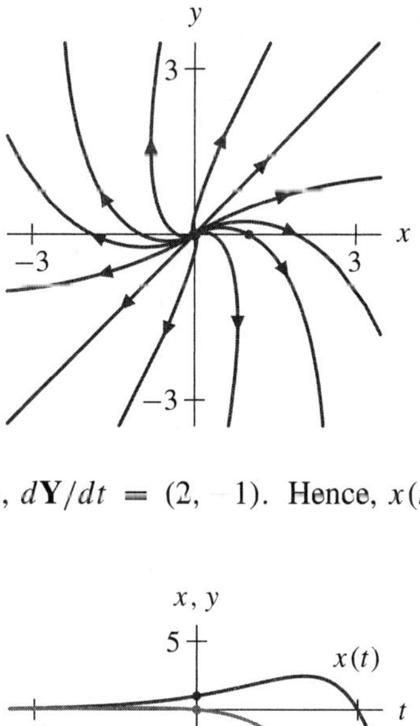

(e) At the point $\mathbf{Y}_0 = (1, 0)$, $d\mathbf{Y}/dt = (2, -1)$. Hence, $x(t)$ is initially increasing, and $y(t)$ is initially decreasing.

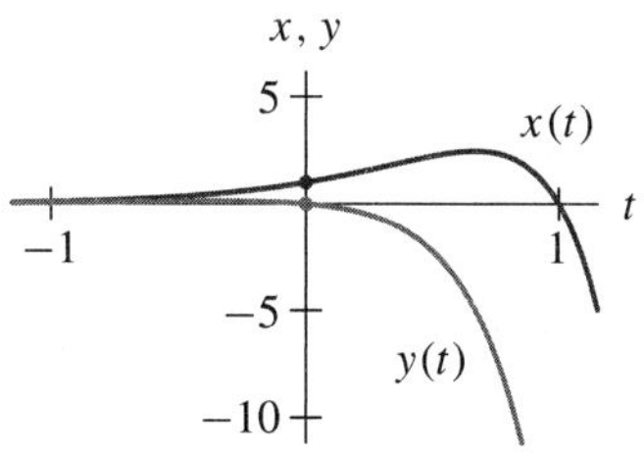

3. **(a)** The characteristic equation is

$$(-2 - \lambda)(-4 - \lambda) + 1 = (\lambda + 3)^2 = 0,$$

and the eigenvalue is $\lambda = -3$.

(b) To find an eigenvector, we solve the simultaneous equations

$$\begin{cases} -2x - y = -3x \\ x - 4y = -3y. \end{cases}$$

Then, $y = x$, and one eigenvector is $(1, 1)$.

(c) Note the straight-line solutions along the line $y = x$.

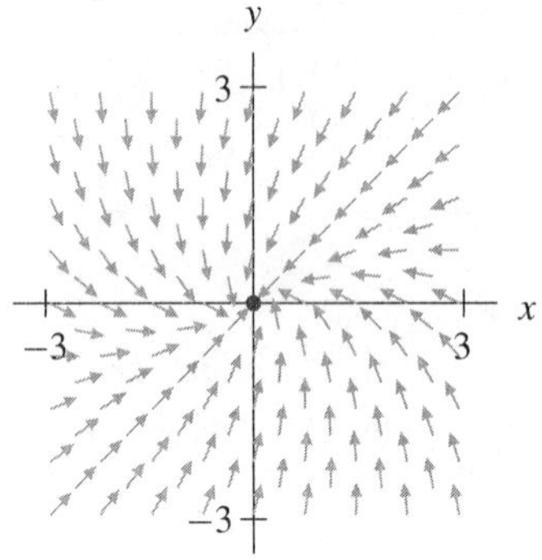

(d) Since the eigenvalue is negative, any solution on the line $y = x$ tends toward the origin along $y = x$ as t increases. According to the direction field, every solution tends to the origin as t increases. The solutions with initial conditions that lie in the half-plane $y > x$ eventually approach the origin tangent to the half-line $y = x$ with $y < 0$. Similarly, the solutions with initial conditions that lie in the half-plane $y < x$ eventually approach the origin tangent to the line $y = x$ with $y > 0$.

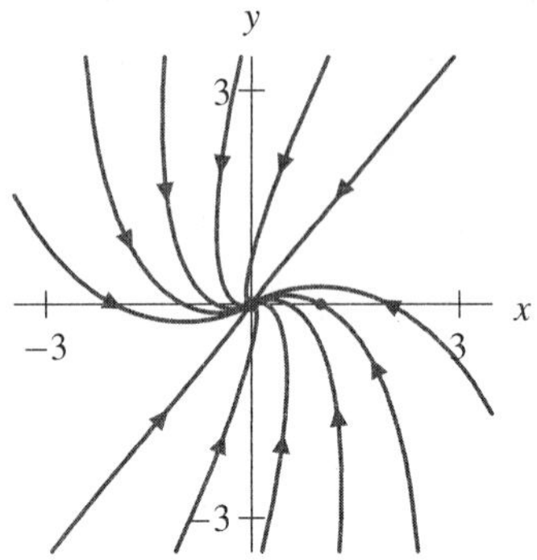

(e) At the point $\mathbf{Y}_0 = (1, 0)$, $d\mathbf{Y}/dt = (-2, 1)$. Therefore, $x(t)$ initially decreases and $y(t)$ initially increases. The solution eventually approaches the origin tangent to the line $y = x$. Since the solution curve never crosses the line $y = x$, the graphs of $x(t)$ and $y(t)$ do not cross.

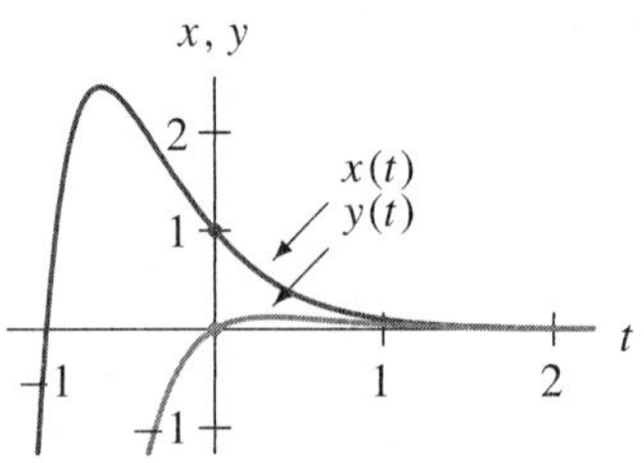

4. **(a)** The characteristic polynomial is

$$(-\lambda)(-2-\lambda)+1=\lambda^2+2\lambda+1=(\lambda+1)^2,$$

so there is only one eigenvalue, $\lambda=-1$.

(b) To find an eigenvalue we solve

$$\begin{cases} y=-x \\ -x-2y=-y. \end{cases}$$

These equations both simplify to $y=-x$, so $(1,-1)$ is one eigenvector.

(c) Note the straight-line solutions along the line $y=-x$.

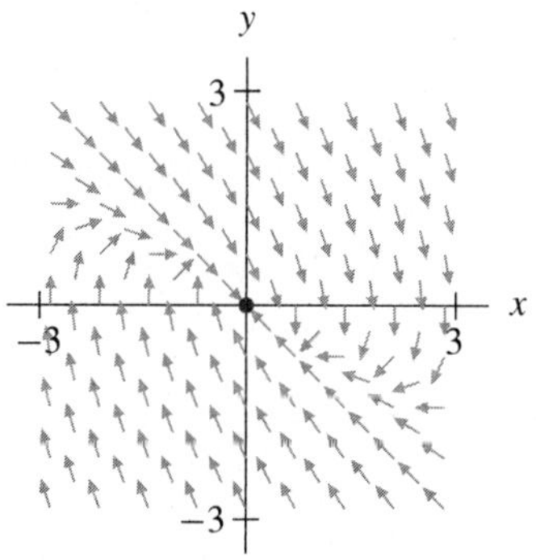

(d) Since the eigenvalue is negative, all solutions approach the origin as t increases. Solutions with initial conditions on the line $y=-x$ approach the origin along $y=-x$. Solutions with initial conditions that lie in the half-plane $y>-x$ approach the origin tangent to the half-line $y=-x$ with $y<0$. Solutions with initial conditions that lie in the half-plane $y<-x$ approach the origin tangent to the half-line $y=-x$ with $y>0$.

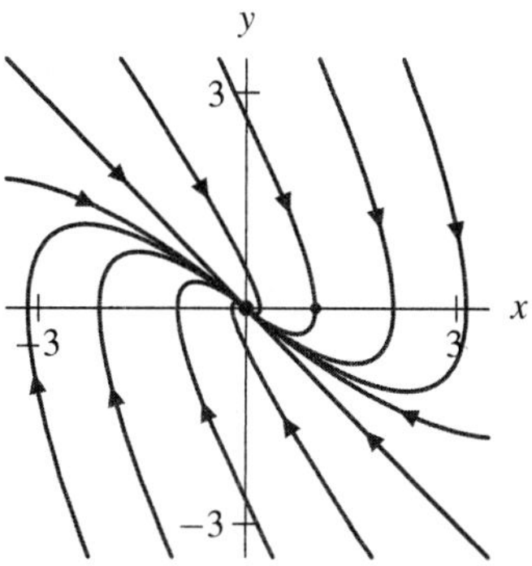

(e) At the point $\mathbf{Y}_0 = (1, 0)$, $d\mathbf{Y}/dt = (0, -1)$. Therefore, $x(t)$ assumes a maximum at $t = 0$ and then decreases toward 0. Also, $y(t)$ becomes negative. Then, it assumes a (negative) minimum, and finally it is asymptotic to 0 without crossing $y = 0$.

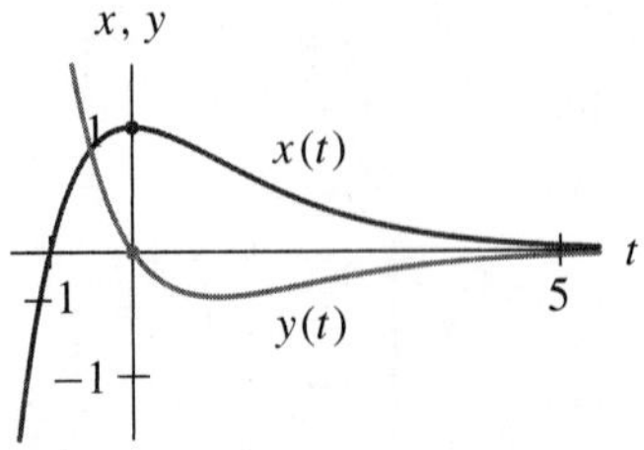

5. **(a)** According to Exercise 1, there is one eigenvalue, -3, with eigenvectors of the form $(0, y_0)$, where $y_0 \neq 0$.

To find the general solution, we start with an arbitrary initial condition $\mathbf{V}_0 = (x_0, y_0)$. Then

$$\mathbf{V}_1 = \left[\begin{pmatrix} -3 & 0 \\ 1 & -3 \end{pmatrix} + 3\begin{pmatrix} 1 & 0 \\ 0 & 1 \end{pmatrix}\right]\mathbf{V}_0$$

$$= \begin{pmatrix} 0 & 0 \\ 1 & 0 \end{pmatrix}\begin{pmatrix} x_0 \\ y_0 \end{pmatrix}$$

$$= \begin{pmatrix} 0 \\ x_0 \end{pmatrix}.$$

We obtain the general solution

$$\mathbf{Y}(t) = e^{-3t}\begin{pmatrix} x_0 \\ y_0 \end{pmatrix} + te^{-3t}\begin{pmatrix} 0 \\ x_0 \end{pmatrix}.$$

(b) The solution that satisfies the initial condition $(x_0, y_0) = (1, 0)$ is

$$\mathbf{Y}(t) = e^{-3t}\begin{pmatrix} 1 \\ 0 \end{pmatrix} + te^{-3t}\begin{pmatrix} 0 \\ 1 \end{pmatrix}.$$

Hence, $x(t) = e^{-3t}$ and $y(t) = te^{-3t}$.

(c) Compare the graphs of $x(t) = e^{-3t}$ and $y(t) = te^{-3t}$ with the sketches obtained in part (e) of Exercise 1.

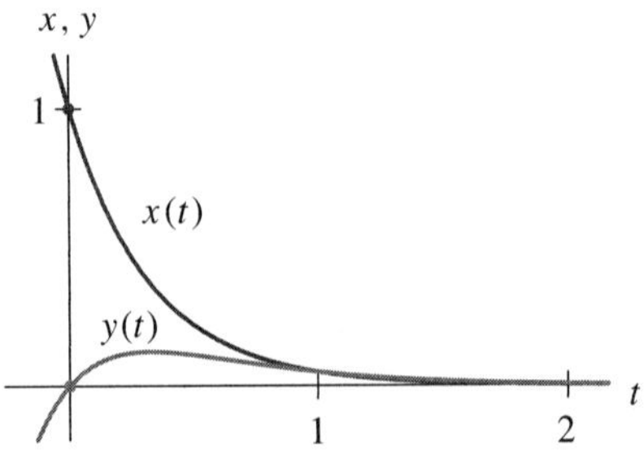

6. **(a)** From Exercise 2, we know that there is only one eigenvalue, $\lambda = 3$, and the eigenvectors (x_0, y_0) satisfy the equation $y_0 = x_0$.

To find the general solution, we start with an arbitrary initial condition $\mathbf{V}_0 = (x_0, y_0)$. Then

$$\begin{aligned}\mathbf{V}_1 &= \left[\begin{pmatrix} 2 & 1 \\ -1 & 4 \end{pmatrix} - 3\begin{pmatrix} 1 & 0 \\ 0 & 1 \end{pmatrix}\right]\mathbf{V}_0 \\ &= \begin{pmatrix} -1 & 1 \\ -1 & 1 \end{pmatrix}\begin{pmatrix} x_0 \\ y_0 \end{pmatrix} \\ &= \begin{pmatrix} y_0 - x_0 \\ y_0 - x_0 \end{pmatrix}.\end{aligned}$$

We obtain the general solution

$$\mathbf{Y}(t) = e^{3t}\begin{pmatrix} x_0 \\ y_0 \end{pmatrix} + te^{3t}\begin{pmatrix} y_0 - x_0 \\ y_0 - x_0 \end{pmatrix}.$$

(b) The solution that satisfies the initial condition $(x_0, y_0) = (1, 0)$ is

$$\mathbf{Y}(t) = e^{3t}\begin{pmatrix} 1 \\ 0 \end{pmatrix} + te^{3t}\begin{pmatrix} -1 \\ -1 \end{pmatrix}.$$

Hence, $x(t) = e^{3t}(1 - t)$ and $y(t) = -te^{3t}$.

(c) Compare the graphs of $x(t) = e^{3t}(1-t)$ and $y(t) = -te^{3t}$ with the sketches obtained in part (e) of Exercise 2.

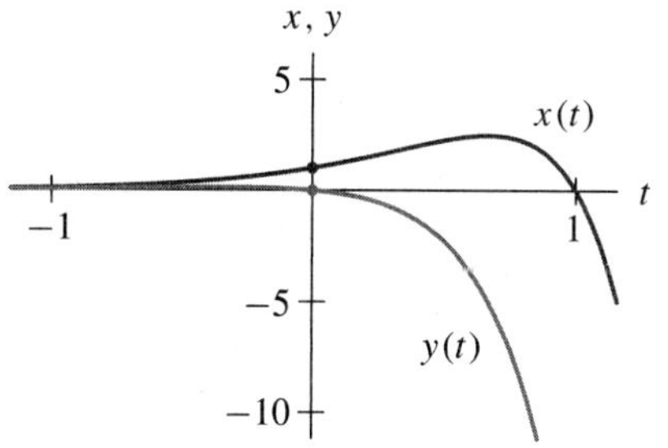

7. **(a)** From Exercise 3, we know that there is only one eigenvalue, $\lambda = -3$, and the eigenvectors (x_0, y_0) satisfy the equation $y_0 = x_0$.

To find the general solution, we start with an arbitrary initial condition $\mathbf{V}_0 = (x_0, y_0)$. Then

$$\begin{aligned}\mathbf{V}_1 &= \left[\begin{pmatrix} -2 & -1 \\ 1 & -4 \end{pmatrix} + 3\begin{pmatrix} 1 & 0 \\ 0 & 1 \end{pmatrix}\right]\mathbf{V}_0 \\ &= \begin{pmatrix} 1 & -1 \\ 1 & -1 \end{pmatrix}\begin{pmatrix} x_0 \\ y_0 \end{pmatrix}\end{aligned}$$

$$= \begin{pmatrix} x_0 - y_0 \\ x_0 - y_0 \end{pmatrix}.$$

We obtain the general solution

$$\mathbf{Y}(t) = e^{-3t} \begin{pmatrix} x_0 \\ y_0 \end{pmatrix} + te^{-3t} \begin{pmatrix} x_0 - y_0 \\ x_0 - y_0 \end{pmatrix}.$$

(b) The solution that satisfies the initial condition $(x_0, y_0) = (1, 0)$ is

$$\mathbf{Y}(t) = e^{-3t} \begin{pmatrix} 1 \\ 0 \end{pmatrix} + te^{-3t} \begin{pmatrix} 1 \\ 1 \end{pmatrix}.$$

Hence, $x(t) = e^{-3t}(t + 1)$ and $y(t) = te^{-3t}$.

(c) Compare the graphs of $x(t) = e^{-3t}(t + 1)$ and $y(t) = te^{-3t}$ with the sketches obtained in part (e) of Exercise 3.

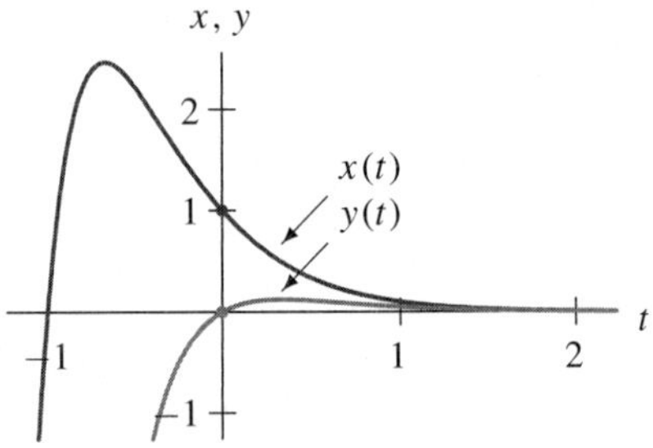

8. **(a)** From Exercise 4, we know that there is only one eigenvalue, $\lambda = -1$, and the eigenvectors (x_0, y_0) satisfy the equation $y_0 = -x_0$.

To find the general solution, we start with an arbitrary initial condition $\mathbf{V}_0 = (x_0, y_0)$. Then

$$\mathbf{V}_1 = \left[\begin{pmatrix} 0 & 1 \\ -1 & -2 \end{pmatrix} + 1 \begin{pmatrix} 1 & 0 \\ 0 & 1 \end{pmatrix} \right] \mathbf{V}_0$$

$$= \begin{pmatrix} 1 & 1 \\ -1 & -1 \end{pmatrix} \begin{pmatrix} x_0 \\ y_0 \end{pmatrix}$$

$$= \begin{pmatrix} x_0 + y_0 \\ -x_0 - y_0 \end{pmatrix}.$$

We obtain the general solution

$$\mathbf{Y}(t) = e^{-t} \begin{pmatrix} x_0 \\ y_0 \end{pmatrix} + te^{-t} \begin{pmatrix} x_0 + y_0 \\ -x_0 - y_0 \end{pmatrix}.$$

(b) The solution that satisfies the initial condition $(x_0, y_0) = (1, 0)$ is

$$\mathbf{Y}(t) = e^{-t}\begin{pmatrix} 1 \\ 0 \end{pmatrix} + te^{-t}\begin{pmatrix} 1 \\ -1 \end{pmatrix}.$$

Hence, $x(t) = e^{-t}(t+1)$ and $y(t) = -te^{-t}$.

(c) Compare the graphs of $x(t) = e^{-t}(t+1)$ and $y(t) = -te^{-t}$ with those obtained in part (e) of Exercise 4.

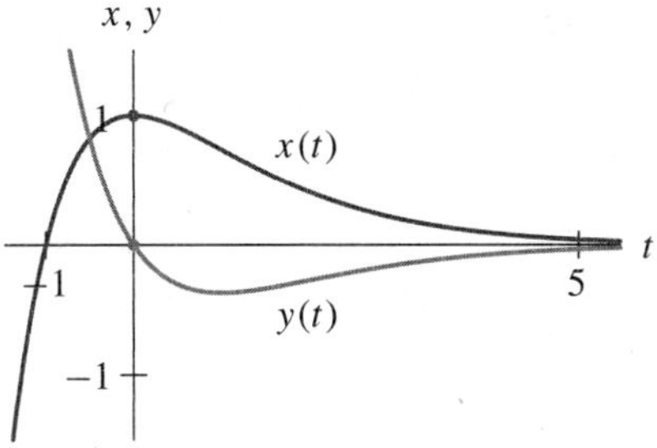

9. **(a)** By solving the quadratic equation, we obtain

$$\lambda = \frac{-\alpha \pm \sqrt{\alpha^2 - 4\beta}}{2}.$$

Therefore, for the quadratic to have a double root, we must have

$$\alpha^2 - 4\beta = 0.$$

(b) If zero is a root, we set $\lambda = 0$ in $\lambda^2 + \alpha\lambda + \beta = 0$, and we obtain $\beta = 0$.

10. **(a)** To compute the limit of $te^{\lambda t}$ as $t \to \infty$ if $\lambda > 0$, we note that both t and $e^{\lambda t}$ go to infinity as t goes to infinity. So $te^{\lambda t}$ blows up as t tends to infinity, and the limit does not exist.

(b) To compute the limit of $te^{\lambda t}$ as $t \to \infty$ if $\lambda < 0$, we write

$$\lim_{t\to\infty} te^{\lambda t} = \lim_{t\to\infty} \frac{t}{e^{-\lambda t}} = \lim_{t\to\infty} \frac{1}{-\lambda e^{-\lambda t}}$$

where the last equality follows from L'Hôpital's Rule. Because $e^{-\lambda t}$ tends to infinity as $t \to \infty$ $(-\lambda > 0)$, the fraction tends to 0.

11. The characteristic equation is

$$-\lambda(-p - \lambda) + q = \lambda^2 + p\lambda + q = 0.$$

Solving the quadratic equation, one obtains

$$\lambda = \frac{-p \pm \sqrt{p^2 - 4q}}{2}.$$

(a) Therefore, in order for **A** to have two real eigenvalues, p and q must satisfy $p^2 - 4q > 0$.

(b) In order for **A** to have complex eigenvalues, p and q must satisfy $p^2 - 4q < 0$.

(c) In order for **A** to have only one eigenvalue, p and q must satisfy $p^2 - 4q = 0$.

12. The characteristic polynomial of $\mathbf{A}$ is

$$\det(\mathbf{A} - \lambda\mathbf{I}) = \lambda^2 - (a+d)\lambda + (ad - bc) = \lambda^2 - \mathrm{tr}(\mathbf{A})\lambda + \det(\mathbf{A})$$

(see Section 3.2). A quadratic polynomial has only one root if and only if its discriminant is 0. In this case, the discriminant of $\det(\mathbf{A} - \lambda\mathbf{I})$ is $\mathrm{tr}(\mathbf{A})^2 - 4\det(\mathbf{A})$.

13. Since every vector is an eigenvector with eigenvalue λ, we substitute $\mathbf{Y} = (1, 0)$ into the equation $\mathbf{AY} = \lambda\mathbf{Y}$ and get

$$\mathbf{A}\begin{pmatrix} 1 \\ 0 \end{pmatrix} = \begin{pmatrix} a \\ c \end{pmatrix} = \lambda\begin{pmatrix} 1 \\ 0 \end{pmatrix}.$$

Hence, $a = \lambda$ and $c = 0$. Similarly, letting $\mathbf{Y} = (0, 1)$, we have

$$\begin{pmatrix} b \\ d \end{pmatrix} = \lambda\begin{pmatrix} 0 \\ 1 \end{pmatrix}.$$

Therefore, $b = 0$ and $d = \lambda$.

14. First note that, because $\mathbf{Y}_1$ and $\mathbf{Y}_2$ are independent, any vector $\mathbf{Y}_3$ can be written as a linear combination of $\mathbf{Y}_1$ and $\mathbf{Y}_2$. In other words, there exists k_1 and k_2 such that

$$\mathbf{Y}_3 = k_1\mathbf{Y}_1 + k_1\mathbf{Y}_2.$$

But then

$$\begin{aligned}
\mathbf{AY}_3 &= \mathbf{A}(k_1\mathbf{Y}_1 + k_2\mathbf{Y}_2) \\
&= k_1\mathbf{AY}_1 + k_2\mathbf{AY}_2 \\
&= k_1\lambda\mathbf{Y}_1 + k_2\lambda\mathbf{Y}_2 \\
&= \lambda(k_1\mathbf{Y}_1 + k_2\mathbf{Y}_2) \\
&= \lambda\mathbf{Y}_3.
\end{aligned}$$

That is, any $\mathbf{Y}_3$ is an eigenvector with eigenvalue λ.

Now use the result of Exercise 13 to conclude that $a = d = \lambda$ and $b = c = 0$.

15. Since $\mathbf{Y}_1(0) = \mathbf{V}_0$ and $\mathbf{Y}_2(0) = \mathbf{W}_0$, we see that $\mathbf{V}_0 = \mathbf{W}_0$.

Evaluating at $t = 1$ yields

$$\mathbf{Y}_1(1) = e^{\lambda}(\mathbf{V}_0 + \mathbf{V}_1) \quad \text{and} \quad \mathbf{Y}_2(1) = e^{\lambda}(\mathbf{W}_0 + \mathbf{W}_1).$$

Since $\mathbf{Y}_1(1) = \mathbf{Y}_2(1)$ and $\mathbf{V}_0 = \mathbf{W}_0$, we see that $\mathbf{V}_1 = \mathbf{W}_1$.

16. **(a)** Suppose that

$$\mathbf{A} = \begin{pmatrix} a & b \\ c & d \end{pmatrix}.$$

By assumption, we know that the characteristic polynomial of $\mathbf{A}$ has λ_0 as a root of multiplicity two. That is,

$$\begin{aligned}
\lambda^2 - (a+d)\lambda + (ad - bc) &= (\lambda - \lambda_0)^2 \\
&= \lambda^2 - (2\lambda_0)\lambda + \lambda_0^2
\end{aligned}$$

Therefore, $a + d = 2\lambda_0$, and $ad - bc = \lambda_0^2$.

Now we compute $(\mathbf{A} - \lambda_0\mathbf{I})^2$ using the definition of matrix multiplication. We have

$$(\mathbf{A} - \lambda_0\mathbf{I})^2 = \begin{pmatrix} a - \lambda_0 & b \\ c & d - \lambda_0 \end{pmatrix}\begin{pmatrix} a - \lambda_0 & b \\ c & d - \lambda_0 \end{pmatrix}$$

$$= \begin{pmatrix} (a - \lambda_0)^2 + bc & b(a + d - 2\lambda_0) \\ c(a + d - 2\lambda_0) & bc + (d - \lambda_0)^2 \end{pmatrix}.$$

Since $a + d = 2\lambda_0$, we see that the bottom-left and top-right entries are zero.

Now consider the top-left entry $(a - \lambda_0)^2 + bc$. We have

$$\begin{aligned}(a - \lambda_0)^2 + bc &= a^2 - 2a\lambda_0 + \lambda_0^2 + bc \\ &= a^2 - 2a\lambda_0 + ad - bc + bc,\end{aligned}$$

because $ad - bc = \lambda_0^2$. The right-hand side simplifies to

$$a^2 - 2a\lambda_0 + ad = a(a - 2\lambda_0 + d) = 0$$

because $a + d = 2\lambda_0$.

A similar argument is used to show that the bottom-right entry is zero.

(b) If $\mathbf{V}_0$ is an eigenvector, then $\mathbf{V}_1 = (\mathbf{A} - \lambda_0\mathbf{I})\mathbf{V}_0$ is the zero vector. If not, we use the result of part (a) to compute

$$(\mathbf{A} - \lambda_0\mathbf{I})\mathbf{V}_1 = (\mathbf{A} - \lambda_0\mathbf{I})^2\mathbf{V}_0 = 0 \text{ (the zero vector).}$$

Consequently, $\mathbf{V}_1$ is an eigenvector.

17. **(a)** The characteristic polynomial is

$$(-\lambda)(-1 - \lambda) + 0 = \lambda^2 + \lambda,$$

so the eigenvalues are $\lambda = 0$ and $\lambda = -1$.

(b) To find the eigenvectors $\mathbf{V}_1$ associated to the eigenvalue $\lambda = 0$, we must solve $\mathbf{AV}_1 = 0\mathbf{V}_1 = 0$ where $\mathbf{A}$ is the matrix that defines this linear system. (Note that this is the same calculation we do if we want to locate the equilibrium points.) We get

$$\begin{cases} 2y_1 = 0 \\ -y_1 = 0, \end{cases}$$

where $\mathbf{V}_1 = (x_1, y_1)$. Hence, the eigenvectors associated to $\lambda = 0$ (as well as the equilibrium points) must satisfy the equation $y_1 = 0$.

To find the eigenvectors $\mathbf{V}_2$ associated to the eigenvalue $\lambda = -1$, we must solve $\mathbf{AV}_2 = -\mathbf{V}_2$. We get

$$\begin{cases} 2y_2 = -x_2 \\ -y_2 = -y_2. \end{cases}$$

where $\mathbf{V}_2 = (x_2, y_2)$. Hence, the eigenvectors associated to $\lambda = -1$ must satisfy $2y_2 = -x_2$.

(c) The equation $y_1 = 0$ specifies a line of equilibrium points. Since the other eigenvalue is negative, solution curves not corresponding to equilibria move toward this line as t increases.

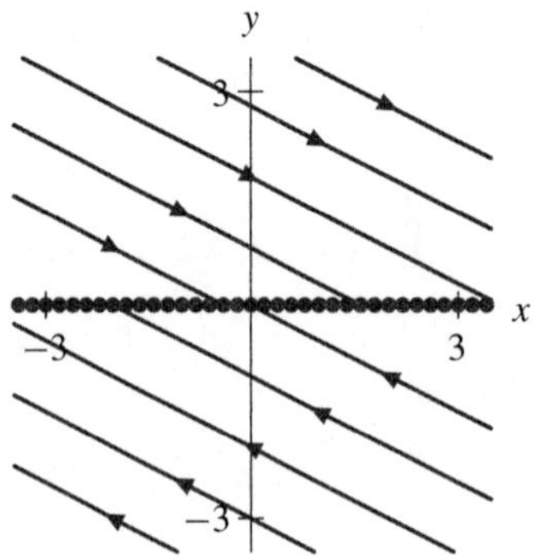

(d) Since $(1, 0)$ is an equilibrium point, it is easy to sketch the corresponding $x(t)$- and $y(t)$-graphs.

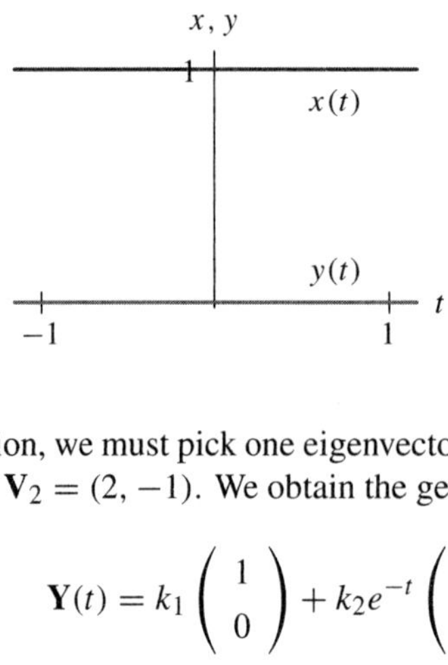

(e) To form the general solution, we must pick one eigenvector for each eigenvalue. Using part (b), we pick $\mathbf{V}_1 = (1, 0)$, and $\mathbf{V}_2 = (2, -1)$. We obtain the general solution

$$\mathbf{Y}(t) = k_1 \begin{pmatrix} 1 \\ 0 \end{pmatrix} + k_2 e^{-t} \begin{pmatrix} 2 \\ -1 \end{pmatrix}.$$

(f) To determine the solution whose initial condition is $(1, 0)$, we can substitute $t = 0$ in the general solution and solve for k_1 and k_2. However, since this initial condition is an equilibrium point, we need not make the effort. We simply observe that

$$\mathbf{Y}(t) = \begin{pmatrix} 1 \\ 0 \end{pmatrix}$$

is the desired solution.

18. **(a)** The characteristic equation is

$$(2 - \lambda)(6 - \lambda) - 12 = \lambda^2 - 8\lambda = 0.$$

Therefore, the eigenvalues are $\lambda = 0$ and $\lambda = 8$.

(b) To find the eigenvectors $\mathbf{V}_1$ associated to the eigenvalue $\lambda = 0$, we must solve $\mathbf{AV}_1 = 0\mathbf{V}_1 = \mathbf{0}$ where $\mathbf{A}$ is the matrix that defines this linear system. (Note that this is the same calculation we do if we want to locate the equilibrium points.) We get

$$\begin{cases} 2x_1 + 4y_1 = 0 \\ 3x_1 + 6y_1 = 0, \end{cases}$$

where $\mathbf{V}_1 = (x_1, y_1)$. Hence, the eigenvectors associated to $\lambda = 0$ (as well as the equilibrium points) must satisfy the equation $x_1 + 2y_1 = 0$.

To find the eigenvectors $\mathbf{V}_2$ associated to the eigenvalue $\lambda = 8$, we must solve $\mathbf{AV}_2 = 8\mathbf{V}_2$. We get

$$\begin{cases} 2x_2 + 4y_2 = 8x_2 \\ 3x_2 + 6y_2 = 8y_2, \end{cases}$$

where $\mathbf{V}_2 = (x_2, y_2)$. Hence, the eigenvectors associated to $\lambda = 8$ must satisfy $2y_2 = 3x_2$.

(c) The equation $x_1 + 2y_1 = 0$ specifies a line of equilibrium points. Since the other eigenvalue is positive, solution curves not corresponding to equilibria move away from this line as t increases.

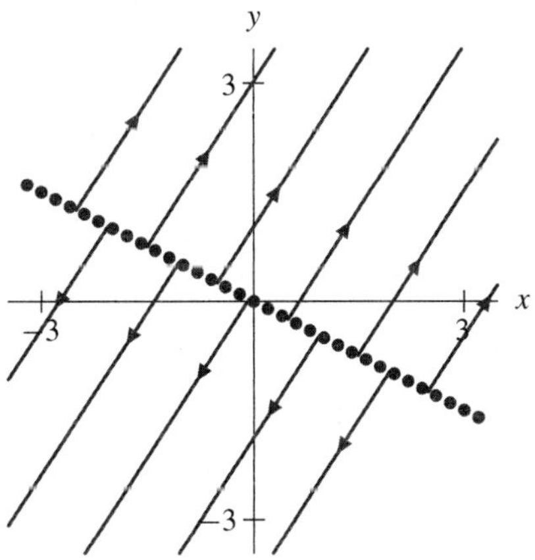

(d) As t increases, both $x(t)$ and $y(t)$ increase exponentially. As t decreases, both x and y approach constants that are determined by the line of equilibrium points.

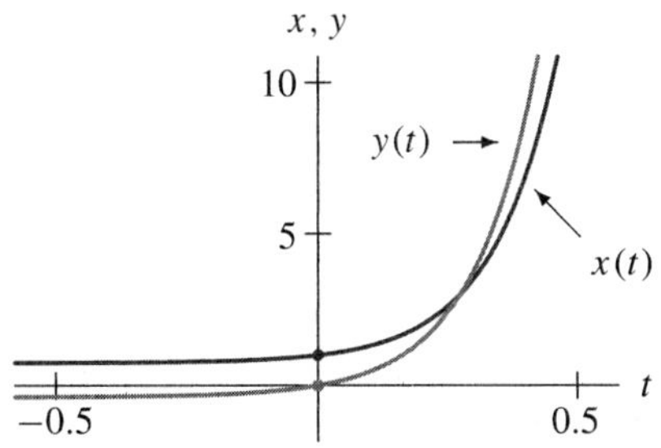

(e) To form the general solution, we must pick one eigenvector for each eigenvalue. Using part (b), we pick $\mathbf{V}_1 = (-2, 1)$, and $\mathbf{V}_2 = (2, 3)$. We obtain the general solution

$$\mathbf{Y}(t) = k_1 \begin{pmatrix} -2 \\ 1 \end{pmatrix} + k_2 e^{8t} \begin{pmatrix} 2 \\ 3 \end{pmatrix}.$$

(f) To determine the solution whose initial condition is $(1, 0)$, we let $t = 0$ in the general solution and obtain the equations

$$k_1 \begin{pmatrix} -2 \\ 1 \end{pmatrix} + k_2 \begin{pmatrix} 2 \\ 3 \end{pmatrix} = \begin{pmatrix} 1 \\ 0 \end{pmatrix}.$$

Therefore, $k_1 = -3/8$ and $k_2 = 1/8$. The particular solution is

$$\mathbf{Y}(t) = \begin{pmatrix} \frac{3}{4} + \frac{1}{4}e^{8t} \\ -\frac{3}{8} + \frac{3}{8}e^{8t} \end{pmatrix}.$$

19. **(a)** The characteristic polynomial is

$$(4 - \lambda)(1 - \lambda) - 4 = \lambda^2 - 5\lambda,$$

so the eigenvalues are $\lambda = 0$ and $\lambda = 5$.

(b) To find the eigenvectors $\mathbf{V}_1$ associated to the eigenvalue $\lambda = 0$, we must solve $\mathbf{AV}_1 = 0\mathbf{V}_1 = \mathbf{0}$ where $\mathbf{A}$ is the matrix that defines this linear system. (Note that this is the same calculation we do if we want to locate the equilibrium points.) We get

$$\begin{cases} 4x_1 + 2y_1 = 0 \\ 2x_1 + y_1 = 0, \end{cases}$$

where $\mathbf{V}_1 = (x_1, y_1)$. Hence, the eigenvectors associated to $\lambda = 0$ (as well as the equilibrium points) must satisfy the equation $y_1 = -2x_1$.

To find the eigenvectors $\mathbf{V}_2$ associated to the eigenvalue $\lambda = 5$, we must solve $\mathbf{AV}_2 = 5\mathbf{V}_2$. We get

$$\begin{cases} 4x_2 + 2y_2 = 5x_2 \\ 2x_2 + y_2 = 5y_2. \end{cases}$$

where $\mathbf{V}_2 = (x_2, y_2)$. Hence, the eigenvectors associated to $\lambda = 5$ must satisfy $x_2 = 2y_2$.

(c) The equation $y_1 = -2x_1$ specifies a line of equilibrium points. Since the other eigenvalue is positive, solution curves not corresponding to equilibria move away from this line as t increases.

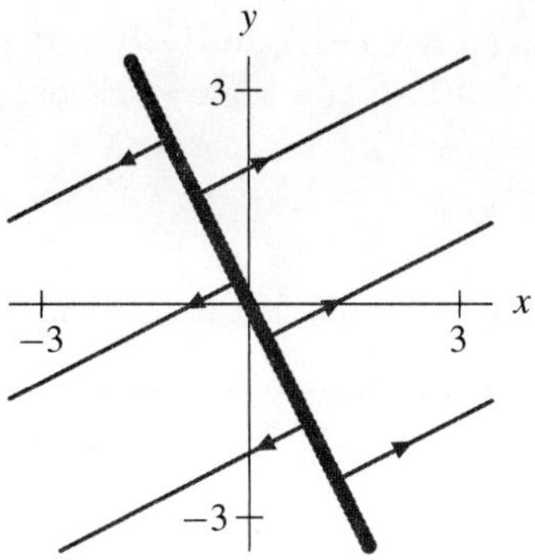

(d) As t increases, both $x(t)$ and $y(t)$ increase exponentially. As t decreases, both x and y approach constants that are determined by the line of equilibrium points.

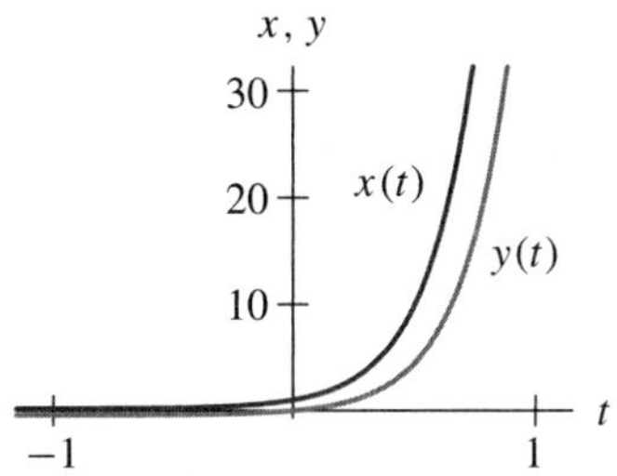

(e) To form the general solution, we must pick one eigenvector for each eigenvalue. Using part (b), we pick $\mathbf{V}_1 = (1, -2)$, and $\mathbf{V}_2 = (2, 1)$. We obtain the general solution

$$\mathbf{Y}(t) = k_1 \begin{pmatrix} 1 \\ -2 \end{pmatrix} + k_2 e^{5t} \begin{pmatrix} 2 \\ 1 \end{pmatrix}.$$

(f) To determine the solution whose initial condition is $(1, 0)$, we let $t = 0$ in the general solution and obtain the equations

$$k_1 \begin{pmatrix} 1 \\ -2 \end{pmatrix} + k_2 \begin{pmatrix} 2 \\ 1 \end{pmatrix} = \begin{pmatrix} 1 \\ 0 \end{pmatrix}.$$

Therefore, $k_1 = 1/5$ and $k_2 = 2/5$, and the particular solution is

$$\mathbf{Y}(t) = \begin{pmatrix} \frac{1}{5} + \frac{4}{5}e^{5t} \\ -\frac{2}{5} + \frac{2}{5}e^{5t} \end{pmatrix}.$$

20. **(a)** The characteristic equation is $\lambda^2 - (a + d)\lambda + ad - bc = 0$. If 0 is an eigenvalue of $\mathbf{A}$, then 0 is a root of the characteristic polynomial. Thus, the constant term in the above equation must be 0—that is, $ad - bc = \det \mathbf{A} = 0$.

(b) If $\det \mathbf{A} = 0$, then the characteristic equation becomes $\lambda^2 - (a + d)\lambda = 0$, and this equation has 0 as a root. Therefore 0 is an eigenvalue of $\mathbf{A}$.

21. **(a)** The characteristic polynomial is $\lambda^2 = 0$, so $\lambda = 0$ is the sole eigenvalue. To sketch the phase portrait we note that $dy/dt = 0$, so $y(t)$ is always a constant function. Moreover, $dx/dt = 2y$, so $x(t)$ is increasing if $y > 0$, and it is decreasing if $y < 0$.

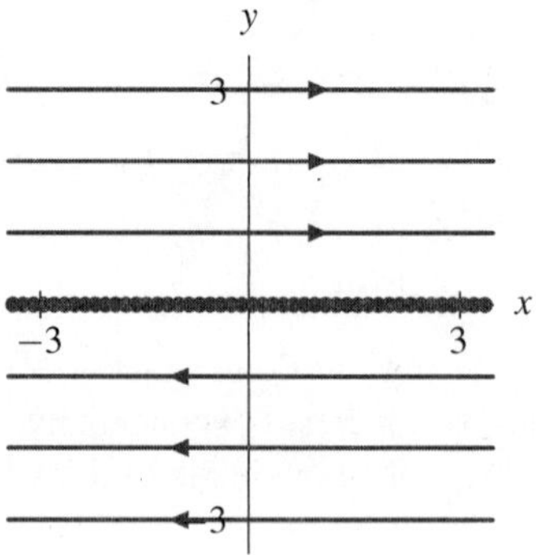

(b) This system is exactly the same as the one in part (a) except that the sign of dx/dt has changed. Hence, the phase portrait is the identical except for the fact that the arrows point the other way.

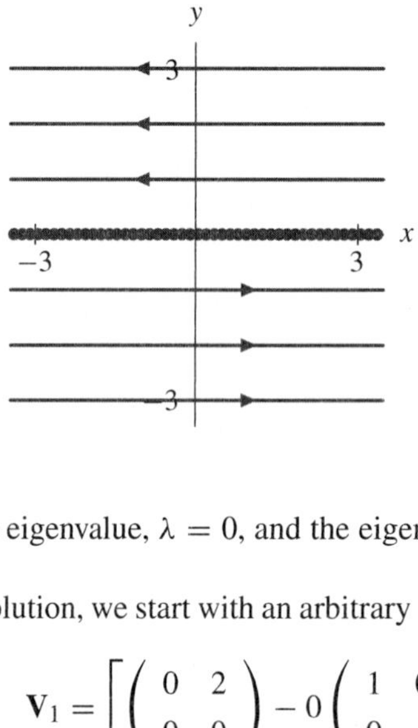

22. **(a)** This system has only one eigenvalue, $\lambda = 0$, and the eigenvectors lie along the x-axis (the line $y = 0$).

To find the general solution, we start with an arbitrary initial condition $\mathbf{V}_0 = (x_0, y_0)$. Then

$$\mathbf{V}_1 = \left[\begin{pmatrix} 0 & 2 \\ 0 & 0 \end{pmatrix} - 0\begin{pmatrix} 1 & 0 \\ 0 & 1 \end{pmatrix}\right]\mathbf{V}_0$$

$$= \begin{pmatrix} 0 & 2 \\ 0 & 0 \end{pmatrix}\begin{pmatrix} x_0 \\ y_0 \end{pmatrix} = \begin{pmatrix} 2y_0 \\ 0 \end{pmatrix}.$$

We obtain the general solution

$$\mathbf{Y}(t) = \begin{pmatrix} x_0 \\ y_0 \end{pmatrix} + t\begin{pmatrix} 2y_0 \\ 0 \end{pmatrix}.$$

(b) Following the procedure in part (a) we obtain

$$\mathbf{V}_1 = \begin{pmatrix} -2y_0 \\ 0 \end{pmatrix},$$

and consequently, the general solution is

$$\mathbf{Y}(t) = \begin{pmatrix} x_0 \\ y_0 \end{pmatrix} + t \begin{pmatrix} -2y_0 \\ 0 \end{pmatrix}.$$

23. **(a)** The characteristic polynomial is $(a - \lambda)(d - \lambda)$, so the eigenvalues are a and d.

(b) If $a \neq d$, the lines of eigenvectors for a and d are the x- and y-axes respectively.

(c) If $a = d < 0$, every nonzero vector is an eigenvector (see Exercise 14), and all the vectors point toward the origin. Hence, every solution curve is asymptotic to the origin along a straight line.

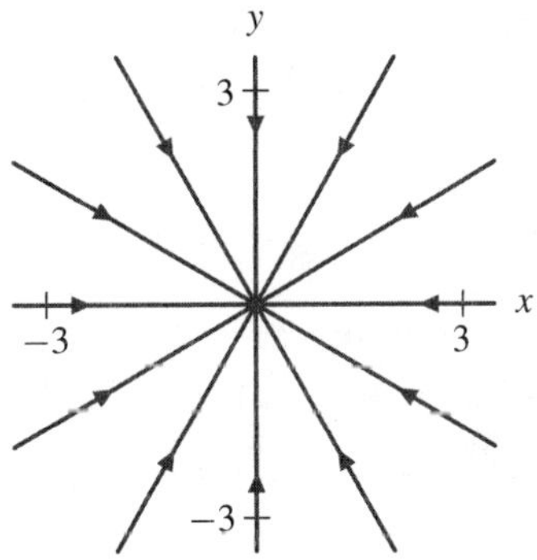

The general solution is $\mathbf{Y}(t) = e^{at}\mathbf{Y}_0$, where $\mathbf{Y}_0$ is the initial condition.

(d) The only difference between this case and part (c) is that the arrows in the vector field are reversed. Every solution tends away from the origin along a straight line.

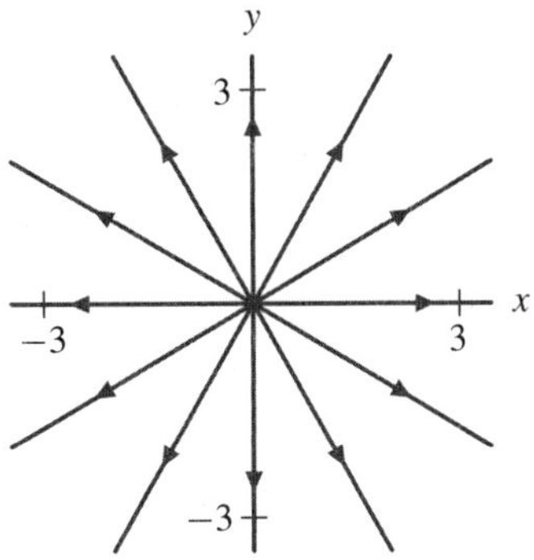

Again the general solution is $\mathbf{Y}(t) = e^{at}\mathbf{Y}_0$, where $\mathbf{Y}_0$ is the initial condition.

24. **(a)** The characteristic equation is $\lambda^2+2\lambda+1 = (\lambda+1)^2 = 0$, so the eigenvalue $\lambda = -1$ is repeated. The equilibrium point at the origin is a sink.

(b) To find the associated eigenvectors $\mathbf{V}$, we must solve $\mathbf{AV} = -\mathbf{V}$ where $\mathbf{A}$ is the matrix that defines this linear system. This vector equation is equivalent to the system of scalar equations

$$\begin{cases} -2x - y = 0 \\ 4x + 2y = 0, \end{cases}$$

so the eigenvectors must satisfy $y = -2x$. One such eigenvector is therefore $(1, -2)$, and all straight-line solutions are of the form

$$\mathbf{Y}(t) = ke^{-t}\begin{pmatrix} 1 \\ -2 \end{pmatrix},$$

where k is an arbitrary constant.

(c) Since this system has only one eigenvalue $\lambda = -1$, we know that the origin is a sink and that all solution curves in the phase plane approach the origin tangent to the line $y = -2x$ of eigenvectors. The direction of approach is determined by the direction field for the system. Solutions with initial conditions that satisfy $y > -2x$ move in a "counter-clockwise" direction and approach the origin in the second quadrant, and solutions with initial conditions that satisfy $y < -2x$ also move in a "counter-clockwise" direction and approach the origin in the fourth quadrant.

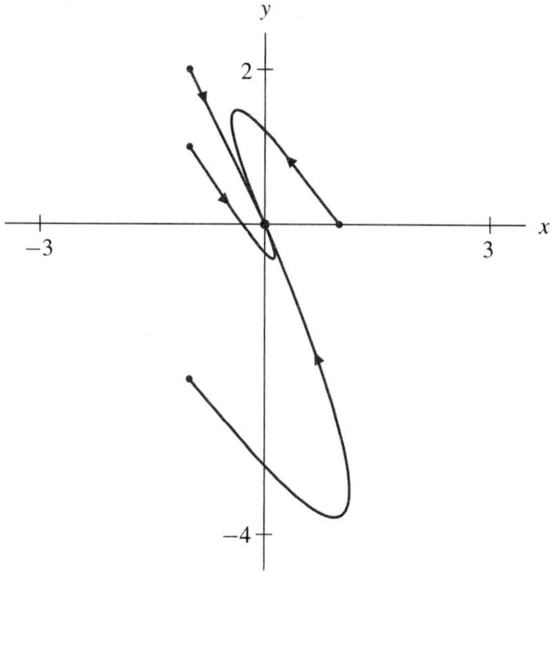

The initial condition $A = (-1, 2)$ is an eigenvector, so the corresponding solution is a straight-line solution. Its $x(t)$- and $y(t)$-graphs are therefore simple exponentials that approach 0 at the rate e^{-t}. We have $y(t) = -2x(t)$ for all t.

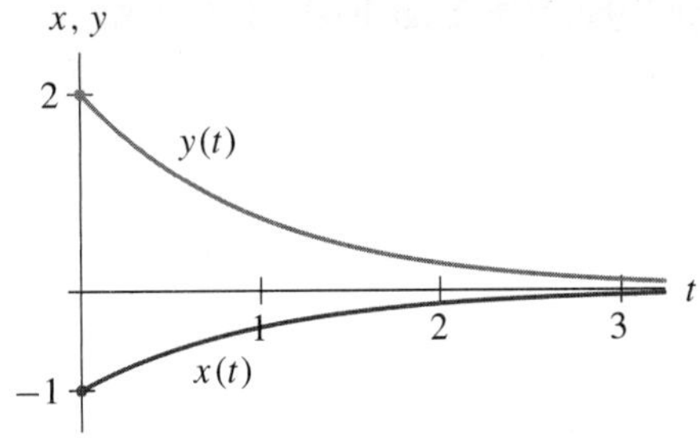

The initial condition $B = (-1, 1)$ lies to the left of the line of eigenvectors. Therefore, its solution curve heads down through the third quadrant and enters the fourth quadrant before it tends to the origin tangent to the line $y = -2x$. The $y(t)$-graph decreases as the $x(t)$-graph increases. We note that $y(t) = 0$ when the solution curve crosses the x-axis, and the two graphs cross when the solution curve crosses the line $y = x$. The function $x(t)$ continues to increase as it becomes positive and attains its maximum value before it tends to 0. The function $y(t)$ assumes a minimum value before it tends to 0.

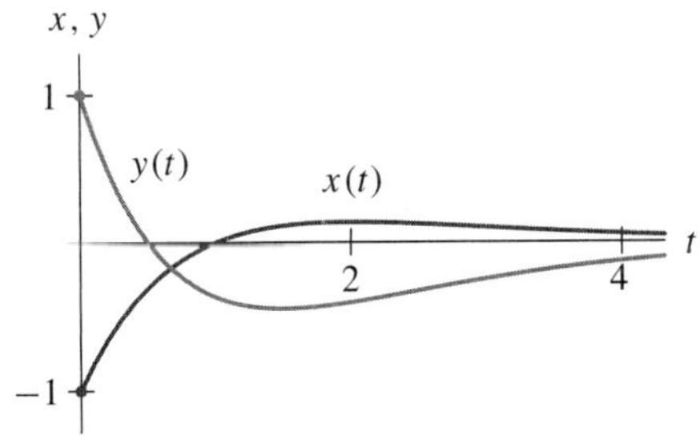

The solution corresponding to the initial condition $C = (-1, -2)$ behaves in a similar fashion to the solution with initial condition B. The only significant difference is that C is below the line $y = x$ in the third quadrant. Therefore the $x(t)$- and $y(t)$-graphs do not cross as they tend toward 0. However, they do exhibit the remaining aspects of the graphs that correspond to the initial condition B.

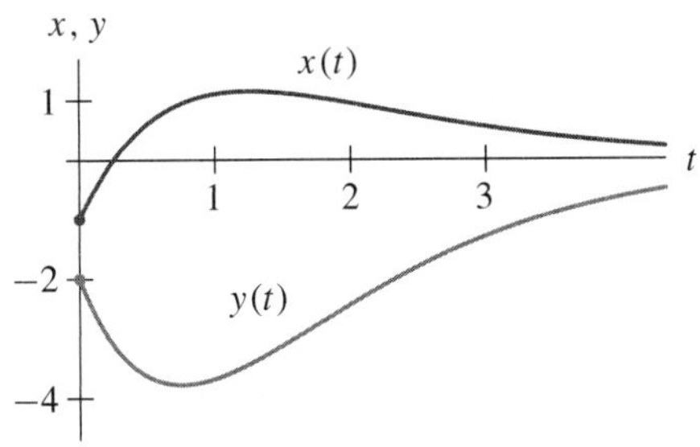

The solution corresponding to the initial condition $D = (1, 0)$ moves to the left and up through the first quadrant in the phase plane before it enters the second quadrant and heads toward the origin tangent to the line $y = -2x$. Thus the $y(t)$-graph is always positive for $t > 0$, and it attains a unique maximum value before it tends to 0. Initially the $x(t)$-graph decreases. It crosses the $y(t)$-graph, becomes negative, and attains a minimum value before it tends to 0 as $t \to \infty$.

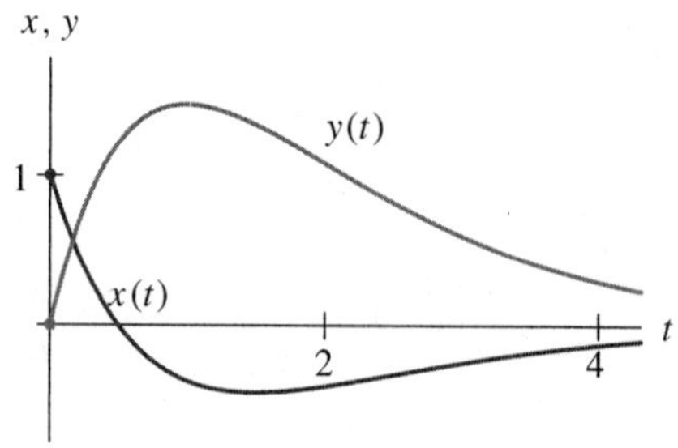

EXERCISES FOR SECTION 3.6

1. The characteristic polynomial is

$$s^2 - 6s - 7,$$

so the eigenvalues are $s = -1$ and $s = 7$. Hence, the general solution is

$$y(t) = k_1 e^{-t} + k_2 e^{7t}.$$

2. The characteristic polynomial is

$$s^2 - s - 12,$$

so the eigenvalues are $s = -3$ and $s = 4$. Hence, the general solution is

$$y(t) = k_1 e^{-3t} + k_2 e^{4t}.$$

3. The characteristic polynomial is

$$s^2 + 6s + 9,$$

so $s = -3$ is a repeated eigenvalue. Hence, the general solution is

$$y(t) = k_1 e^{-3t} + k_2 t e^{-3t}.$$

4. The characteristic polynomial is

$$s^2 - 4s + 4,$$

so $s = 2$ is a repeated eigenvalue. Hence, the general solution is

$$y(t) = k_1 e^{2t} + k_2 t e^{2t}.$$

5. The characteristic polynomial is

$$s^2 + 8s + 25,$$

so the complex eigenvalues are $s = -4 \pm 3i$. Hence, the general solution is

$$y(t) = k_1 e^{-4t} \cos 3t + k_2 e^{-4t} \sin 3t.$$

6. The characteristic polynomial is

$$s^2 - 4s + 29,$$

so the complex eigenvalues are $s = 2 \pm 5i$. Hence, the general solution is

$$y(t) = k_1 e^{2t} \cos 5t + k_2 e^{2t} \sin 5t.$$

7. The characteristic polynomial is

$$s^2 + 2s - 3,$$

so the eigenvalues are $s = 1$ and $s = -3$. Hence, the general solution is

$$y(t) = k_1 e^t + k_2 e^{-3t},$$

and we have

$$y'(t) = k_1 e^t - 3k_2 e^{-3t}.$$

From the initial conditions, we obtain the simultaneous equations

$$\begin{cases} k_1 + k_2 = 6 \\ k_1 - 3k_2 = -2. \end{cases}$$

Solving for k_1 and k_2 yields $k_1 = 4$ and $k_2 = 2$. Hence, the solution to our initial-value problem is $y(t) = 4e^t + 2e^{-3t}$.

8. The characteristic polynomial is

$$s^2 + 4s - 5,$$

so the eigenvalues are $s = 1$ and $s = -5$. Hence, the general solution is

$$y(t) = k_1 e^t + k_2 e^{-5t},$$

and we have

$$y'(t) = k_1 e^t - 5k_2 e^{-5t}.$$

From the initial conditions, we obtain the simultaneous equations

$$\begin{cases} k_1 + k_2 = 11 \\ k_1 - 5k_2 = -7. \end{cases}$$

Solving for k_1 and k_2 yields $k_1 = 8$ and $k_2 = 3$. Hence, the solution to our initial-value problem is $y(t) = 8e^t + 3e^{-5t}$.

9. The characteristic polynomial is

$$s^2 - 4s + 13,$$

so the eigenvalues are $s = 2 \pm 3i$. Hence, the general solution is

$$y(t) = k_1 e^{2t} \cos 3t + k_2 e^{2t} \sin 3t.$$

From the initial condition $y(0) = 1$, we see that $k_1 = 1$. Differentiating

$$y(t) = e^{2t} \cos 3t + k_2 e^{2t} \sin 3t$$

and evaluating $y'(t)$ at $t = 0$ yields $y'(0) = 2 + 3k_2$. Since $y'(0) = -4$, we have $k_2 = -2$. Hence, the solution to our initial-value problem is

$$y(t) = e^{2t} \cos 3t - 2e^{2t} \sin 3t.$$

10. The characteristic polynomial is

$$s^2 + 4s + 20,$$

so the eigenvalues are $s = -2 \pm 4i$. Hence, the general solution is

$$y(t) = k_1 e^{-2t} \cos 4t + k_2 e^{-2t} \sin 4t.$$

From the initial condition $y(0) = 2$, we see that $k_1 = 2$. Differentiating

$$y(t) = 2e^{-2t} \cos 4t + k_2 e^{-2t} \sin 4t$$

and evaluating $y'(t)$ at $t = 0$ yields $y'(0) = -4 + 4k_2$. Since $y'(0) = -8$, we have $k_2 = -1$. Hence, the solution to our initial-value problem is

$$y(t) = 2e^{-2t} \cos 4t - e^{-2t} \sin 4t.$$

11. The characteristic polynomial is

$$s^2 - 8s + 16,$$

so $s = 4$ is a repeated eigenvalue. Hence, the general solution is

$$y(t) = k_1 e^{4t} + k_2 t e^{4t}.$$

From the initial condition $y(0) = 3$, we see that $k_1 = 3$. Differentiating

$$y(t) = 3e^{4t} + k_2 t e^{4t}$$

and evaluating $y'(t)$ at $t = 0$ yields $y'(0) = 12 + k_2$. Since $y'(0) = 11$, we have $k_2 = -1$. Hence, the solution to our initial-value problem is

$$y(t) = 3e^{4t} - te^{4t}.$$

12. The characteristic polynomial is

$$s^2 - 4s + 4,$$

so $s = 2$ is a repeated eigenvalue. Hence, the general solution is

$$y(t) = k_1 e^{2t} + k_2 t e^{2t}.$$

From the initial condition $y(0) = 1$, we see that $k_1 = 1$. Differentiating $y(t) = e^{2t} + k_2 t e^{2t}$ and evaluating $y'(t)$ at $t = 0$ yields $y'(0) = 2 + k_2$. Since $y'(0) = 1$, we have $k_2 = -1$. Hence, the solution to our initial-value problem is

$$y(t) = e^{2t} - te^{2t}.$$

13. **(a)** The resulting second-order equation is

$$\frac{d^2y}{dt^2} + 8\frac{dy}{dt} + 7y = 0,$$

and the corresponding system is

$$\begin{aligned}\frac{dy}{dt} &= v\\ \frac{dv}{dt} &= -7y - 8v.\end{aligned}$$

(b) Recall that we can read off the characteristic equation of the second-order equation straight from the equation without having to revert to the corresponding system. We obtain

$$\lambda^2 + 8\lambda + 7 = 0.$$

Therefore, the eigenvalues are $\lambda_1 = -1$ and $\lambda_2 = -7$.

To find the eigenvectors associated to the eigenvalue λ_1, we solve the simultaneous system of equations

$$\begin{cases} v = -y \\ -7y - 8v = -v. \end{cases}$$

From the first equation, we immediately see that the eigenvectors associated to this eigenvalue must satisfy $v = -y$. Similarly, the eigenvectors associated to the eigenvalue $\lambda_2 = -7$ must satisfy the equation $v = -7y$.

(c) Since the eigenvalues are real and negative, the equilibrium point at the origin is a sink, and the system is overdamped.

(d) We know that all solution curves approach the origin as $t \to \infty$ and, with the exception of those whose initial conditions lie on the line $v = -7y$, these solution curves approach the origin tangent to the line $v = -y$.

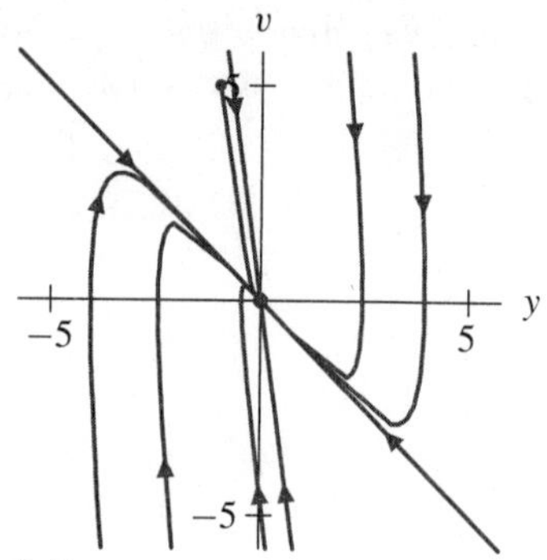

(e) From the phase portrait, we see that $y(t)$ increases monotonically toward 0 as $t \to \infty$. Also, $v(t)$ decreases monotonically toward 0. It is useful to remember that $v = dy/dt$.

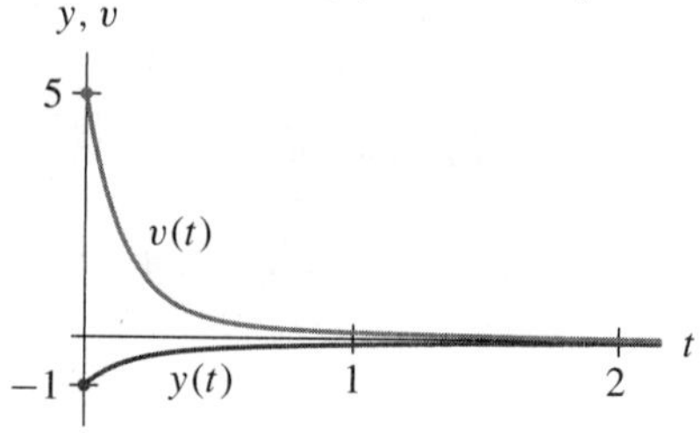

14. **(a)** The resulting second-order equation is

$$\frac{d^2y}{dt^2} + 6\frac{dy}{dt} + 8y = 0,$$

and the corresponding system is

$$\frac{dy}{dt} = v$$
$$\frac{dv}{dt} = -8y - 6v.$$

(b) Recall that we can read off the characteristic equation of the second-order equation straight from the equation without having to revert to the corresponding system. We obtain

$$\lambda^2 + 6\lambda + 8 = 0.$$

Therefore, the eigenvalues are $\lambda_1 = -4$ and $\lambda_2 = -2$.

To find the eigenvectors associated to the eigenvalue λ_1, we solve the simultaneous system of equations

$$\begin{cases} v = -4y \\ -8y - 6v = -4v. \end{cases}$$

From the first equation, we immediately see that the eigenvectors associated to this eigenvalue must satisfy $v = -4y$. Similarly, the eigenvectors associated to the eigenvalue $\lambda_2 = -2$ must satisfy the equation $v = -2y$.

(c) Since the eigenvalues are real and negative, the equilibrium point at the origin is a sink, and the system is overdamped.

(d) We know that all solution curves approach the origin as $t \to \infty$ and, with the exception of those whose initial conditions lie on the line $v = -4y$, these solution curves approach the origin tangent to the line $v = -2y$.

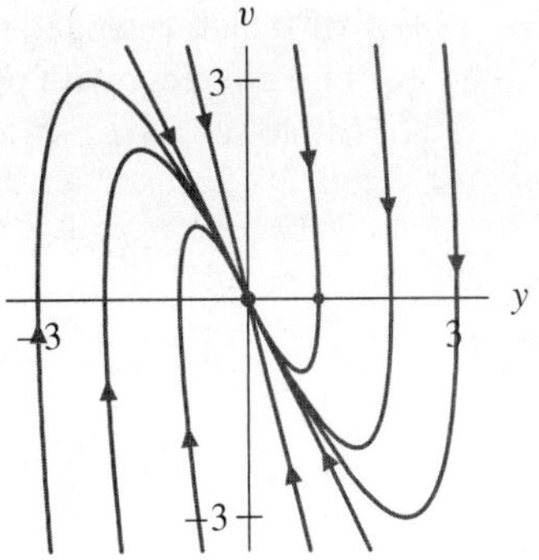

(e) From the phase portrait, we see that $v(t)$ initially decreases from 0 and then increases and tends toward 0 as $t \to \infty$. Also, $y(t)$ decreases monotonically toward 0. It is useful to remember that $v = dy/dt$.

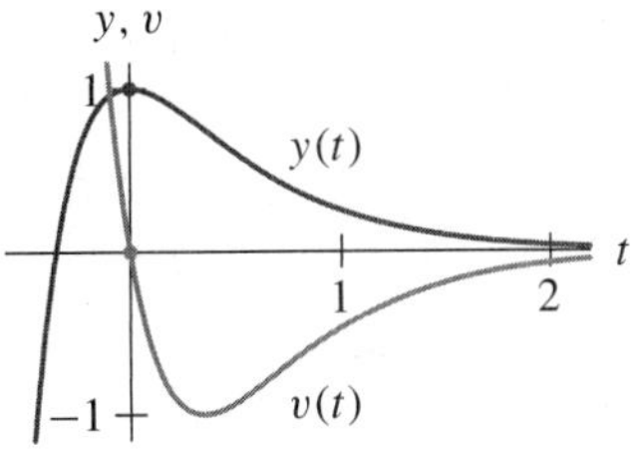

15. **(a)** The resulting second-order equation is

$$\frac{d^2y}{dt^2} + 4\frac{dy}{dt} + 5y = 0,$$

and the corresponding system is

$$\begin{aligned}\frac{dy}{dt} &= v \\ \frac{dv}{dt} &= -5y - 4v.\end{aligned}$$

(b) Recall that we can read off the characteristic equation of the second-order equation straight from the equation without having to revert to the corresponding system. We obtain

$$\lambda^2 + 4\lambda + 5 = 0.$$

Therefore, the eigenvalues are $\lambda_1 = -2 + i$ and $\lambda_2 = -2 - i$.

To find the eigenvectors associated to the eigenvalue λ_1, we solve the simultaneous system of equations

$$\begin{cases} v = (-2+i)y \\ -8y - 6v = (-2+i)v. \end{cases}$$

From the first equation, we immediately see that the eigenvectors associated to this eigenvalue must satisfy $v = (-2+i)y$. Similarly, the eigenvectors associated to the eigenvalue $\lambda_2 = -2-i$ must satisfy the equation $v = (-2-i)y$.

(c) Since the eigenvalues are complex with negative real part, the equilibrium point at the origin is a spiral sink, and the system is underdamped.

(d) All solutions tend to the origin spiralling in the clockwise direction with period 2π. Admittedly, it is difficult to see these oscillations in the picture.

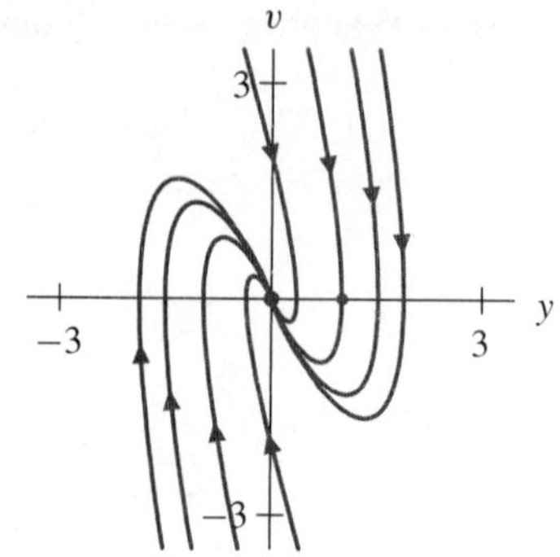

(e) The graph of $y(t)$ initially decreases then oscillates with decreasing amplitude as it tends to 0. Similarly, $v(t)$ initially decreases and becomes negative, then oscillates with decreasing amplitude as it tends to 0.

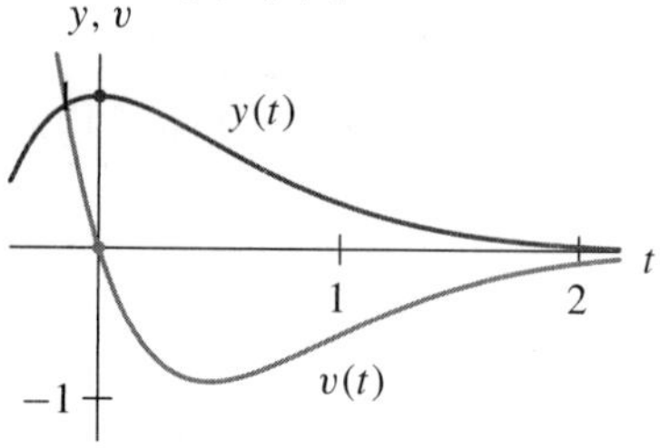

16. **(a)** The resulting second-order equation is

$$\frac{d^2y}{dt^2} + 8y = 0,$$

and the corresponding system is

$$\frac{dy}{dt} = v$$
$$\frac{dv}{dt} = -8y.$$

(b) Recall that we can read off the characteristic equation of the second-order equation straight from the equation without having to revert to the corresponding system. We obtain

$$\lambda^2 + 8 = 0.$$

Therefore, the eigenvalues are $\lambda_1 = 2\sqrt{2}\,i$ and $\lambda_2 = -2\sqrt{2}\,i$.

To find the eigenvectors associated to the eigenvalue λ_1, we solve the simultaneous system of equations

$$\begin{cases} v = 2\sqrt{2}\,iy \\ -8y = 2\sqrt{2}\,iv. \end{cases}$$

From the first equation, we immediately see that the eigenvectors associated to this eigenvalue must satisfy $v = 2\sqrt{2}\,iy$. Similarly, the eigenvectors associated to the eigenvalue $\lambda_2 = -2\sqrt{2}\,i$ must satisfy the equation $v = -2\sqrt{2}\,iy$.

(c) Since the eigenvalues are pure imaginary the equilibrium point at the origin is a center with natural period $\pi/\sqrt{2}$, and the system is undamped.

(d) All solutions move clockwise.

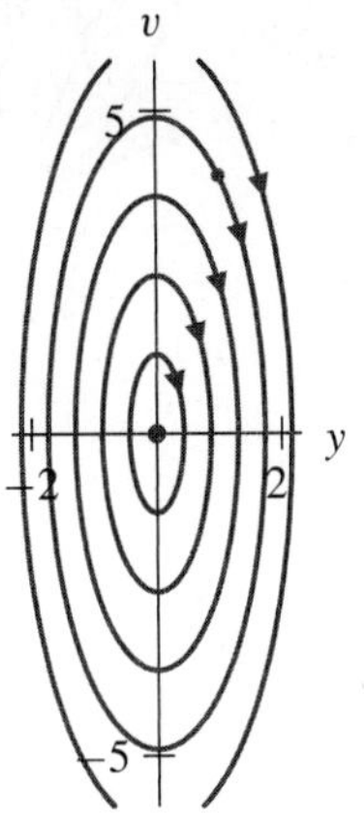

(e) Each graph is periodic with period $\pi/\sqrt{2}$.

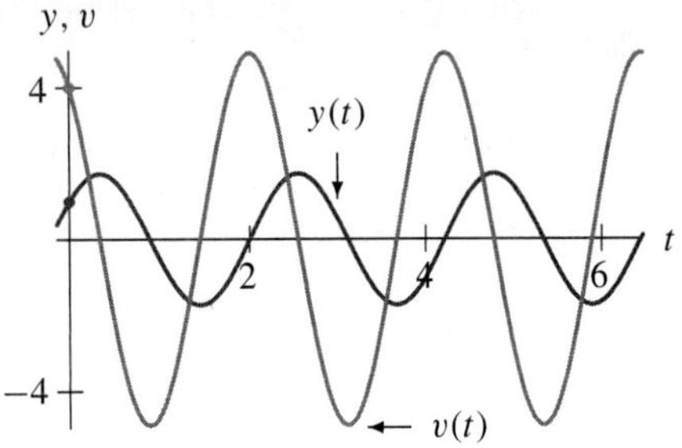

17. **(a)** The resulting second-order equation is

$$2\frac{d^2y}{dt^2} + 3\frac{dy}{dt} + y = 0,$$

and the corresponding system is

$$\begin{aligned}\frac{dy}{dt} &= v \\ \frac{dv}{dt} &= -\frac{1}{2}y - \frac{3}{2}v.\end{aligned}$$

(b) Recall that we can read off the characteristic equation of the second-order equation straight from the equation without having to revert to the corresponding system. We obtain

$$2\lambda^2 + 3\lambda + 1 = 0.$$

Therefore, the eigenvalues are $\lambda_1 = -1$ and $\lambda_2 = -1/2$.

To find the eigenvectors associated to the eigenvalue λ_1, we solve the simultaneous system of equations

$$\begin{cases} v = -y \\ -\frac{1}{2}y - \frac{3}{2}v = -v. \end{cases}$$

From the first equation, we immediately see that the eigenvectors associated to this eigenvalue must satisfy $v = -y$. Similarly, the eigenvectors associated to the eigenvalue $\lambda_2 = -1/2$ must satisfy the equation $v = -y/2$.

(c) Since the eigenvalues are real and negative, the equilibrium point at the origin is a sink, and the system is overdamped.

(d) We know that all solution curves approach the origin as $t \to \infty$ and, with the exception of those whose initial conditions lie on the line $v = -y$, these solution curves approach the origin tangent to the line $v = -y/2$.

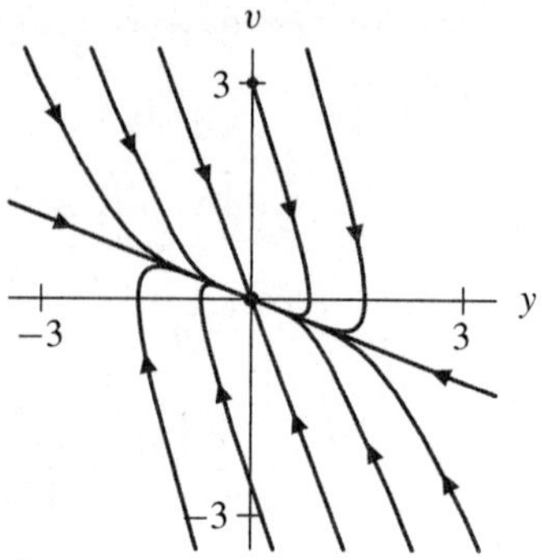

(e) According to the phase plane, $y(t)$ increases initially. Eventually it reaches a maximum value. Then it approaches 0 as $t \to \infty$. Also, $v(t)$ decreases, becomes negative, and then approaches 0 from below. While sketching these graphs, it is useful to remember that $v = dy/dt$.

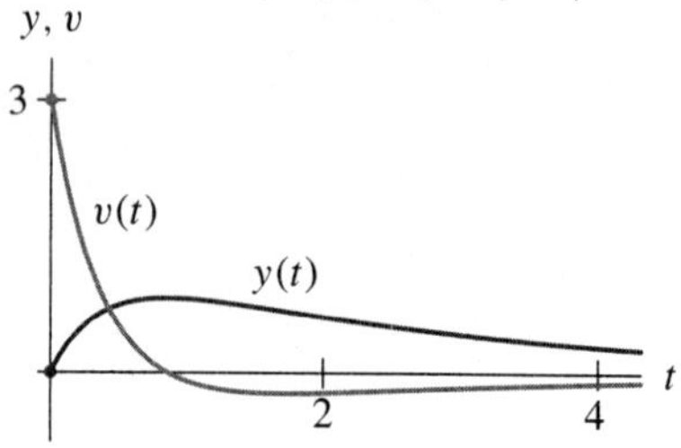

18. **(a)** The resulting second-order equation is

$$9\frac{d^2y}{dt^2} + 6\frac{dy}{dt} + y = 0,$$

and the corresponding system is

$$\begin{aligned} \frac{dy}{dt} &= v \\ \frac{dv}{dt} &= -\tfrac{1}{9}y - \tfrac{2}{3}v. \end{aligned}$$

(b) Recall that we can read off the characteristic equation of the second-order equation straight from the equation without having to revert to the corresponding system. We obtain

$$9\lambda^2 + 6\lambda + 1 = (3\lambda + 1)^2 = 0.$$

Therefore, we have repeated eigenvalues. In other words, there is only one eigenvalue, $\lambda = -1/3$.

To find the eigenvectors associated to this eigenvalue, we solve the simultaneous system of equations

$$\begin{cases} v = -\tfrac{1}{3}y \\ -\tfrac{1}{9}y - \tfrac{2}{3}v = -\tfrac{1}{3}v. \end{cases}$$

From the first equation, we immediately see that the eigenvectors associated to this eigenvalue must satisfy $v = -y/3$.

(c) Since there is only one eigenvalue and it is negative, the equilibrium point at the origin is a sink, and the system is critically damped.

(d) We know that all solution curves approach the origin as $t \to \infty$ and they all do so tangent to the line $v = -y/3$. To determine the direction in which they approach the origin, we must evaluate $d\mathbf{Y}/dt$ at one point that is not an eigenvector. In this case, it makes sense to pick the initial condition $(1, 1)$. At $(1, 1)$, we have $d\mathbf{Y}/dt = (1, -7/9)$. Thus, the solution curve that starts at $(1, 1)$ initially moves to the right and down. Combining this simple observation with what we already know about the phase portraits of critically damped oscillators, we see that any solution curve that starts in the first quadrant must enter the fourth quadrant before it approaches the origin.

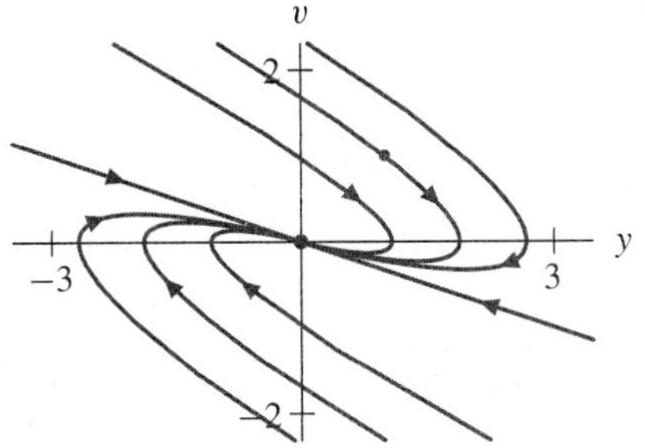

(e) According to the phase plane, $y(t)$ initially increases. Eventually it reaches a maximum value and then it approaches 0 as $t \to \infty$. Also, $v(t)$ decreases, becomes negative, and then approaches 0 from below. While sketching these graphs, it is useful to remember that $v = dy/dt$.

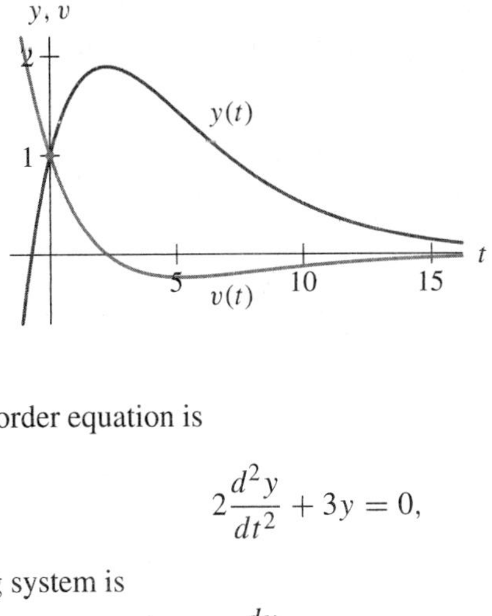

19. **(a)** The resulting second-order equation is

$$2\frac{d^2y}{dt^2} + 3y = 0,$$

and the corresponding system is

$$\frac{dy}{dt} = v$$
$$\frac{dv}{dt} = -\frac{3}{2}y.$$

(b) Recall that we can read off the characteristic equation of the second-order equation straight from the equation without having to revert to the corresponding system. We obtain

$$2\lambda^2 + 3 = 0.$$

Therefore, we have pure imaginary eigenvalues, $\lambda = \pm i\sqrt{3/2}$.

To find the eigenvectors associated to the eigenvalue $\lambda = i\sqrt{3/2}$, we solve the simultaneous system of equations

$$\begin{cases} v = i\sqrt{\frac{3}{2}}\, y \\ -\frac{3}{2}y = i\sqrt{\frac{3}{2}}\, v. \end{cases}$$

From the first equation, we immediately see that the eigenvectors associated to this eigenvalue must satisfy $v = i\sqrt{3/2}\, y$. Similarly, the eigenvectors associated to the eigenvalue $\lambda = -i\sqrt{3/2}$ must satisfy the equation $v = -i\sqrt{3/2}\, y$.

(c) Since the eigenvalues are pure imaginary, the system is undamped. (Of course, we already knew this because $b = 0$.) The natural period is $2\pi/\sqrt{3/2} = 4\pi/\sqrt{6}$.

(d) Since the eigenvalues are pure imaginary, we know that the solution curves are ellipses. At the point $(1, 0)$, $d\mathbf{Y}/dt = (0, -3/2)$. Therefore, we know that the oscillation is clockwise.

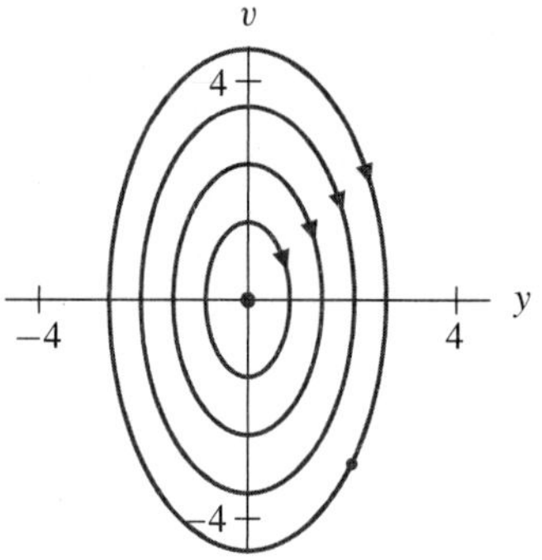

(e)

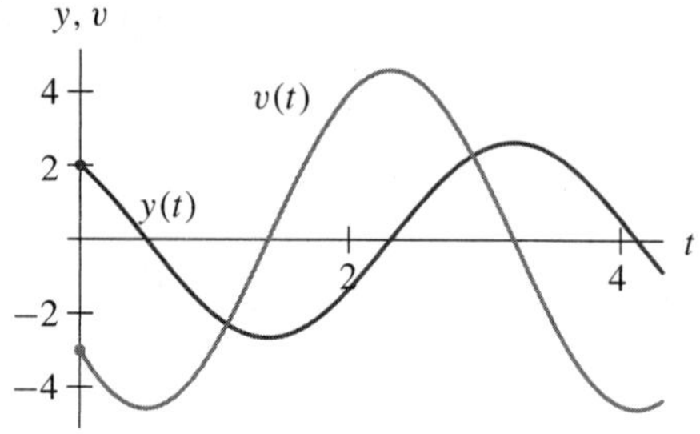

20. **(a)** The resulting second-order equation is

$$2\frac{d^2y}{dt^2} + \frac{dy}{dt} + 3y = 0,$$

and the corresponding system is

$$\begin{aligned} \frac{dy}{dt} &= v \\ \frac{dv}{dt} &= -\frac{3}{2}y - \frac{1}{2}v. \end{aligned}$$

(b) Recall that we can read off the characteristic equation of the second-order equation straight from the equation without having to revert to the corresponding system. We obtain

$$2\lambda^2 + \lambda + 3 = 0.$$

Therefore, the eigenvalues are $\lambda_1 = (-1 + \sqrt{23}\,i)/4$ and $\lambda_2 = (-1 - \sqrt{23}\,i)/4$.

To find the eigenvectors associated to the eigenvalue λ_1, we solve the simultaneous system of equations

$$\begin{cases} v = \dfrac{1+\sqrt{23}\,i}{4}\,y \\ -\dfrac{3}{2}y - \dfrac{1}{2}v = \dfrac{1+\sqrt{23}\,i}{4}\,v. \end{cases}$$

From the first equation, we immediately see that the eigenvectors associated to this eigenvalue must satisfy $v = (1+\sqrt{23}\,i)y/4$. Similarly, the eigenvectors associated to the eigenvalue $\lambda_2 = -2 - i$ must satisfy the equation $v = (1-\sqrt{23}\,i)y/4$.

(c) Since the eigenvalues are complex with negative real part, the equilibrium point at the origin is a spiral sink, and the system is underdamped.

(d) All solutions tend to the origin spiralling in the clockwise direction with period $8\pi/\sqrt{23}$.

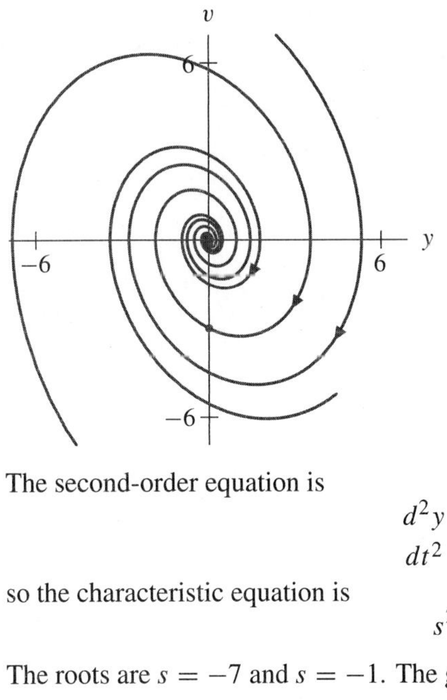

(e) The graph of $y(t)$ becomes negative then oscillates with decreasing amplitude as it tends to 0. Similarly, $v(t)$ initially increases, then oscillates with decreasing amplitude as it tends to 0.

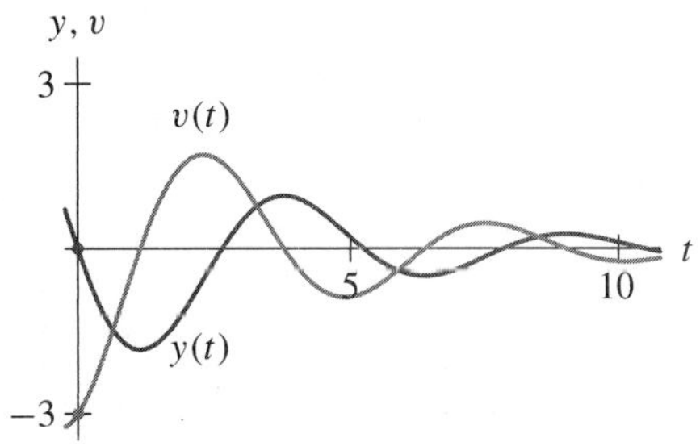

21. **(a)** The second-order equation is

$$\frac{d^2y}{dt^2} + 8\frac{dy}{dt} + 7y = 0,$$

so the characteristic equation is

$$s^2 + 8s + 7 = 0.$$

The roots are $s = -7$ and $s = -1$. The general solution is

$$y(t) = k_1 e^{-7t} + k_2 e^{-t}.$$

(b) To find the particular solution we compute

$$v(t) = -7k_1 e^{-7t} - k_2 e^{-t}.$$

The particular solution satisfies

$$\begin{cases} -1 = y(0) = k_1 + k_2 \\ 5 = v(0) = -7k_1 - k_2. \end{cases}$$

The first equation yields $k_1 = -k_2 - 1$. Substituting into the second we obtain $5 = 6k_2 + 7$, which implies $k_2 = -1/3$. The first equation then yields $k_1 = -2/3$. The particular solution is

$$y(t) = -\tfrac{2}{3}e^{-7t} - \tfrac{1}{3}e^{-t}.$$

(c) The $y(t)$- and $v(t)$-graphs are displayed in the solution of Exercise 13.

22. **(a)** The second-order equation is

$$\frac{d^2y}{dt^2} + 6\frac{dy}{dt} + 8y = 0,$$

so the characteristic equation is

$$s^2 + 6s + 8 = 0.$$

The roots are $s = -4$ and $s = -2$. The general solution is

$$y(t) = k_1 e^{-4t} + k_2 e^{-2t}.$$

(b) To find the particular solution we compute

$$v(t) = -4k_1 e^{-4t} - 2k_2 e^{-2t}.$$

The particular solution satisfies

$$\begin{cases} 1 = y(0) = k_1 + k_2 \\ 0 = v(0) = -4k_1 - 2k_2. \end{cases}$$

The first equation yields $k_1 = -k_2 + 1$. Substituting into the second we obtain $0 = 2k_2 - 4$, which implies $k_2 = 2$. The first equation then yields $k_1 = -1$. The particular solution is

$$y(t) = -e^{-4t} + 2e^{-2t}.$$

(c) The $y(t)$- and $v(t)$-graphs are displayed in the solution of Exercise 14.

23. **(a)** The second-order equation is

$$\frac{d^2y}{dt^2} + 4\frac{dy}{dt} + 5y = 0,$$

so the characteristic equation is

$$s^2 + 4s + 5 = 0.$$

The roots are $s = -2 + i$ and $s = -2 - i$. A complex-valued solution is

$$y_c(t) = e^{(-2+i)t} = e^{-2t}\cos t + ie^{-2t}\sin t.$$

Therefore the general solution is

$$y(t) = k_1 e^{-2t}\cos t + k_2 e^{-2t}\sin t.$$

(b) To find the particular solution we compute

$$v(t) = (-2k_1 + k_2)e^{-2t}\cos t + (-k_1 - 2k_2)e^{-2t}\sin t.$$

The particular solution satisfies

$$\begin{cases} 1 = y(0) = k_1 \\ 0 = v(0) = -2k_1 + k_2. \end{cases}$$

The first equation yields $k_1 = 1$. Substituting into the second we obtain $k_2 = 2$. The particular solution is

$$y(t) = e^{-2t} \cos t + 2e^{-2t} \sin t.$$

(c) The $y(t)$- and $v(t)$-graphs are displayed in the solution of Exercise 15.

24. **(a)** The second-order equation is

$$\frac{d^2y}{dt^2} + 8y = 0,$$

so the characteristic equation is

$$s^2 + 8 = 0.$$

The roots are $s = 2\sqrt{2}\,i$ and $s = -2\sqrt{2}\,i$. A complex-valued solution is

$$y_c(t) = e^{(2\sqrt{2}\,i)t} = \cos(2\sqrt{2}\,t) + i \sin(2\sqrt{2}\,t).$$

Therefore the general solution is

$$y(t) = k_1 \cos(2\sqrt{2}\,t) + k_2 \sin(2\sqrt{2}\,t).$$

(b) To find the particular solution we compute

$$v(t) = -2\sqrt{2}\,k_1 \sin(2\sqrt{2}\,t) + 2\sqrt{2}\,k_2 \cos(2\sqrt{2}\,t).$$

The particular solution satisfies

$$\begin{cases} 1 = y(0) = k_1 \\ 4 = v(0) = 2\sqrt{2}\,k_2 \end{cases}$$

The first equation yields $k_1 = 1$. Substituting into the second we find $k_2 = \sqrt{2}$. So the particular solution is

$$y(t) = \cos(2\sqrt{2}\,t) + \sqrt{2} \sin(2\sqrt{2}\,t).$$

(c) The $y(t)$- and $v(t)$-graphs are displayed in the solution of Exercise 16.

25. **(a)** The second-order equation is

$$2\frac{d^2y}{dt^2} + 3\frac{dy}{dt} + y = 0,$$

so the characteristic equation is

$$2s^2 + 3s + 1 = 0.$$

The roots are $s = -1$ and $s = -1/2$. So the general solution is

$$y(t) = k_1 e^{-t} + k_2 e^{-t/2}.$$

(b) To find the particular solution we compute

$$v(t) = -k_1 e^{-t} - \frac{k_2}{2} e^{-t/2}.$$

The particular solution satisfies

$$\begin{cases} 0 = y(0) = k_1 + k_2 \\ 3 = v(0) = -k_1 - \frac{k_2}{2}. \end{cases}$$

The first equation yields $k_1 = -k_2$. Substituting into the second we obtain $3 = k_2 - k_2/2$, which implies that $k_2 = 6$. The first equation then yields $k_1 = -6$. The particular solution is

$$y(t) = -6e^{-t} + 6e^{-t/2}.$$

(c) The $y(t)$- and $v(t)$-graphs are displayed in the solution of Exercise 17.

26. **(a)** The second-order equation is

$$9\frac{d^2y}{dt^2} + 6\frac{dy}{dt} + y = 0,$$

so the characteristic equation is

$$9s^2 + 6s + 1 = 0.$$

This quadratic has the repeated root $s = -1/3$. The general solution is

$$y(t) = k_1 e^{-t/3} + k_2 t e^{-t/3}.$$

(b) To find the particular solution we compute

$$v(t) = -\frac{k_1}{3} e^{-t/3} + k_2 e^{-t/3} - \frac{k_2}{3} t e^{-t/3}.$$

The particular solution satisfies

$$\begin{cases} 1 = y(0) = k_1 \\ 1 = v(0) = -\frac{k_1}{3} + k_2. \end{cases}$$

The first equation yields $k_1 = 1$. Substituting into the second we find $k_2 = 4/3$. The particular solution is

$$y(t) = e^{-t/3} + \tfrac{4}{3} t e^{-t/3}.$$

(c) The $y(t)$- and $v(t)$- graphs are displayed in the solution of Exercise 18.

27. **(a)** The second-order equation is

$$2\frac{d^2y}{dt^2} + 3y = 0,$$

so the characteristic equation is

$$2s^2 + 3 = 0.$$

The roots are $s = \sqrt{3/2}\, i$ and $s = -\sqrt{3/2}\, i$. A complex-valued solution is

$$y_c(t) = e^{(\sqrt{3/2}\, i)t} = \cos(\sqrt{3/2}\, t) + i \sin(\sqrt{3/2}\, t).$$

Therefore the general solution is

$$y(t) = k_1 \cos(\sqrt{3/2}\, t) + k_2 \sin(\sqrt{3/2}\, t).$$

(b) To find the particular solution we compute

$$v(t) = -\sqrt{3/2}\,k_1 \sin(\sqrt{3/2}\,t) + \sqrt{3/2}\,k_2 \cos(\sqrt{3/2}\,t).$$

The particular solution satisfies

$$\begin{cases} 2 = y(0) = k_1 \\ -3 = v(0) = \sqrt{3/2}\,k_2. \end{cases}$$

The first equation yields $k_1 = 2$. Substituting into the second we find $k_2 = -\sqrt{6}$. So the particular solution is

$$y(t) = 2\cos(\sqrt{3/2}\,t) - \sqrt{6}\sin(\sqrt{3/2}\,t).$$

(c) The $y(t)$- and $v(t)$-graphs are displayed in the solution of Exercise 19.

28. **(a)** The second-order equation is

$$2\frac{d^2y}{dt^2} + \frac{dy}{dt} + 3y = 0,$$

so the characteristic equation is

$$2s^2 + s + 3 = 0.$$

The roots are

$$s = -\frac{1}{4} \pm i\frac{\sqrt{23}}{4}.$$

A complex-valued solution is

$$y_c(t) = e^{(\ 1\ |\ \sqrt{23}\,i)t/4} = e^{\ t/4}\cos\left(\frac{\sqrt{23}}{4}t\right) + ie^{\ t/4}\sin\left(\frac{\sqrt{23}}{4}t\right).$$

Therefore the general solution is

$$y(t) = k_1 e^{-t/4}\cos\left(\frac{\sqrt{23}}{4}t\right) + k_2 e^{-t/4}\sin\left(\frac{\sqrt{23}}{4}t\right).$$

(b) To find the particular solution we compute

$$v(t) = \frac{-k_1 + \sqrt{23}\,k_2}{4}e^{-t/4}\cos\left(\frac{\sqrt{23}}{4}t\right) - \frac{\sqrt{23}\,k_1 + k_2}{4}e^{-t/4}\sin\left(\frac{\sqrt{23}}{4}t\right).$$

The particular solution satisfies

$$\begin{cases} 0 = y(0) = k_1 \\ -3 = v(0) = \dfrac{-k_1 + \sqrt{23}\,k_2}{4}. \end{cases}$$

The first equation yields $k_1 = 0$. Substituting into the second we find $k_2 = -12/\sqrt{23}$. So the particular solution is

$$y(t) = -\frac{12}{\sqrt{23}}e^{-t/4}\sin\left(\frac{\sqrt{23}}{4}t\right).$$

(c) The $y(t)$- and $v(t)$-graphs are displayed in the solution of Exercise 20.

29. Note: We assume that m, k and b are nonnegative—the physically relevant case. All references to graphs and phase portraits are from Sections 3.5 and 3.6.

Table 3.1
Possible harmonic oscillators.

name	eigenvalues	parameters	decay rate	phase portrait and graphs
undamped	pure imaginary	$b = 0$	no decay	Figure 3.41
underdamped	complex with negative real part	$b^2 - 4mk < 0$	$e^{-bt/(2m)}$	Figure 3.42
critically damped	only one eigenvalue	$b^2 - 4mk = 0$	$e^{-bt/(2m)}$	Figure 3.34
overdamped	two negative real	$b^2 - 4mk > 0$	$e^{\lambda t}$ where $\lambda = \dfrac{-b + \sqrt{b^2 - 4mk}}{2m}$	Figures 3.43–3.45 and Exercise 13

30. Note that

$$\frac{dy}{dt} = \frac{d}{dt}(k_1 y_1 + k_2 y_2) = k_1 \frac{dy_1}{dt} + k_2 \frac{dy_2}{dt}$$

and

$$\frac{d^2y}{dt^2} = \frac{d^2}{dt^2}(k_1 y_1 + k_2 y_2) = k_1 \frac{d^2y_1}{dt^2} + k_2 \frac{d^2y_2}{dt^2}.$$

Therefore,

$$\begin{aligned}\frac{d^2y}{dt^2} + p\frac{dy}{dt} + qy &= k_1 \frac{d^2y_1}{dt^2} + k_2 \frac{d^2y_2}{dt^2} + p\left(k_1 \frac{dy_1}{dt} + k_2 \frac{dy_2}{dt}\right) + q\,(k_1 y_1 + k_2 y_2)\\ &= k_1\left(\frac{d^2y_1}{dt^2} + p\frac{dy_1}{dt} + qy_1\right) + k_2\left(\frac{d^2y_2}{dt^2} + p\frac{dy_2}{dt} + qy_2\right)\\ &= 0.\end{aligned}$$

31. Note that

$$\frac{dy}{dt} = \frac{d}{dt}(y_{\text{re}} + i y_{\text{im}}) = \frac{dy_{\text{re}}}{dt} + i\frac{dy_{\text{im}}}{dt}$$

and

$$\frac{d^2y}{dt^2} = \frac{d^2}{dt^2}(y_{\text{re}} + i y_{\text{im}}) = \frac{d^2y_{\text{re}}}{dt^2} + i\frac{d^2y_{\text{im}}}{dt^2}.$$

Then note that

$$\frac{d^2y}{dt^2} + p\frac{dy}{dt} + qy = \left(\frac{d^2y_{\text{re}}}{dt^2} + p\frac{dy_{\text{re}}}{dt} + qy_{\text{re}}\right) + i\left(\frac{d^2y_{\text{im}}}{dt^2} + p\frac{dy_{\text{im}}}{dt} + qy_{\text{im}}\right).$$

Both

$$\frac{d^2y_{\text{re}}}{dt^2} + p\frac{dy_{\text{re}}}{dt} + qy_{\text{re}} = 0 \quad \text{and} \quad \frac{d^2y_{\text{im}}}{dt^2} + p\frac{dy_{\text{im}}}{dt} + qy_{\text{im}} = 0$$

because a complex number is zero only if both its real and imaginary parts vanish. In other words, $y_{\text{re}}(t)$ and $y_{\text{im}}(t)$ are solutions of the original equation.

32. If we let $v = dy/dt$, then the corresponding first-order system is

$$\frac{dy}{dt} = v$$
$$\frac{dv}{dt} = -qy - pv,$$

and the corresponding matrix is

$$\mathbf{A} = \begin{pmatrix} 0 & 1 \\ -q & -p \end{pmatrix}.$$

If λ is an eigenvalue, then it is a root of the characteristic polynomial. In other words,

$$\lambda^2 + p\lambda + q = 0.$$

Now consider

$$\mathbf{A}\begin{pmatrix} 1 \\ \lambda \end{pmatrix} = \begin{pmatrix} \lambda \\ -q - p\lambda \end{pmatrix}$$
$$= \begin{pmatrix} \lambda \\ \lambda^2 \end{pmatrix}$$
$$= \lambda \begin{pmatrix} 1 \\ \lambda \end{pmatrix}.$$

33. **(a)** If we let $v = dy/dt$, then the corresponding first-order system is

$$\frac{dy}{dt} = v$$
$$\frac{dv}{dt} = -qy - pv,$$

and the corresponding matrix is

$$\mathbf{A} = \begin{pmatrix} 0 & 1 \\ -q & -p \end{pmatrix}.$$

If λ_0 is a repeated eigenvalue, then the characteristic polynomial is

$$\lambda^2 + p\lambda + q = (\lambda - \lambda_0^2) = \lambda^2 - 2\lambda_0\lambda + \lambda_0^2.$$

Consequently, $p = -2\lambda_0$, $q = \lambda_0^2$, and

$$\mathbf{A} = \begin{pmatrix} 0 & 1 \\ -\lambda_0^2 & 2\lambda_0 \end{pmatrix}.$$

(b) To compute the general solution of the corresponding first-order system, we consider an arbitrary initial condition $\mathbf{V}_0 = (y_0, v_0)$ and calculate

$$(\mathbf{A} - \lambda_0 \mathbf{I})\mathbf{V}_0 = \begin{pmatrix} -\lambda_0 & 1 \\ -\lambda_0^2 & \lambda_0 \end{pmatrix} \begin{pmatrix} y_0 \\ v_0 \end{pmatrix}$$

$$= \begin{pmatrix} -\lambda_0 y_0 + v_0 \\ -\lambda_0^2 y_0 + \lambda_0 v_0 \end{pmatrix}.$$

The general solution of the first-order system is

$$\mathbf{Y}(t) = e^{\lambda_0 t} \begin{pmatrix} y_0 \\ v_0 \end{pmatrix} + t e^{\lambda_0 t} \begin{pmatrix} -\lambda_0 y_0 + v_0 \\ -\lambda_0^2 y_0 + \lambda_0 v_0 \end{pmatrix}.$$

(c) From the first component of the result in part (b), we obtain the general solution of the original second-order equation in the form

$$y(t) = y_0 e^{\lambda_0 t} + (-\lambda_0 y_0 + v_0) t e^{\lambda_0 t}.$$

(d) Let $k_1 = y_0$ and $k_2 = -\lambda_0 y_0 + v_0$. Clearly, all k_1 are possible. Moreover, once the value of k_1 is determined, k_2 can be determined from v_0 using $k_2 = -\lambda_0 k_1 + v_0$, and v_0 can be determined by k_2 using $v_0 = k_2 + \lambda_0 k_1$. Hence, k_1 and k_2 are arbitrary constants because y_0 and v_0 are arbitrary.

34. We must first find out how fast the "typical" solution of this equation approaches the origin.

The characteristic equation for this harmonic oscillator is

$$s^2 + bs + 3 = 0,$$

and the roots are

$$\frac{-b \pm \sqrt{b^2 - 12}}{2}.$$

These roots are complex if $b^2 < 12$, and all solutions tend to the equilibrium at the rate of $e^{(-b/2)t}$.

If $b^2 > 12$, the roots are real, and the general solution is

$$y(t) = k_1 e^{((-b+\sqrt{b^2-12})/2)t} + k_2 e^{((-b-\sqrt{b^2-12})/2)t}.$$

For the typical solution, both k_1 and k_2 are nonzero, so the typical solution tends to the origin at a rate determined by the slower of these two exponentials. The second of these exponential terms tends to 0 most quickly since $-b - \sqrt{b^2 - 12} < -b + \sqrt{b^2 - 12} < 0$. So the typical solution tends to 0 at the rate determined by the exponential of the form $e^{((-b+\sqrt{b^2-12})/2)t}$.

We must determine which of the two exponentials

$$e^{(-b/2)t}$$

(for $b < 2\sqrt{3}$) and

$$e^{((-b+\sqrt{b^2-12})/2)t}$$

(if $b > 2\sqrt{3}$) tends to 0 most quickly. This determination depends upon the value of b for which $-b/2$ (if $b < 2\sqrt{3}$) or $(-b + \sqrt{b^2 - 12})/2$ (if $b > 2\sqrt{3}$) is most negative.

For $0 < b < 2\sqrt{3}$, $-b/2$ is decreasing. Using calculus, we can show that $(-b + \sqrt{b^2 - 12})/2$ is increasing for $b > 2\sqrt{3}$. Therefore, we must examine the rate if $b = 2\sqrt{3}$. In this case, we have repeated eigenvalues, and the typical solution is a linear combination of terms of the form $e^{-\sqrt{3}t}$ and $te^{-\sqrt{3}t}$. Again using calculus, we can check that both of these solutions tend to 0 faster than $e^{-\alpha t}$ where $\alpha \neq 2\sqrt{3}$.

35. The characteristic equation for this harmonic oscillator is

$$s^2 + bs + 3 = 0,$$

and the roots are

$$s_1 = \frac{-b - \sqrt{b^2 - 12}}{2} \quad \text{and} \quad s_2 = \frac{-b + \sqrt{b^2 - 12}}{2}.$$

If $b^2 < 12$, these roots are complex. In this case, all solutions include a factor of the form $e^{(-b/2)t}$, and they tend to the equilibrium at this rate.

If $b^2 > 12$, the roots are real, and the general solution is

$$y(t) = k_1 e^{s_1 t} + k_2 e^{s_2 t}.$$

The first exponential in this expression tends to 0 most quickly, so if $k_2 = 0$, we have solutions that tend to 0 at the rate of $e^{s_1 t}$. This rate is the quickest approach to 0.

The roots are repeated if $b^2 - 12 = 0$, that is, if $b = 2\sqrt{3}$. The fastest approach is then given by a term of the form $e^{-\sqrt{3}t}$.

36. **(a)**

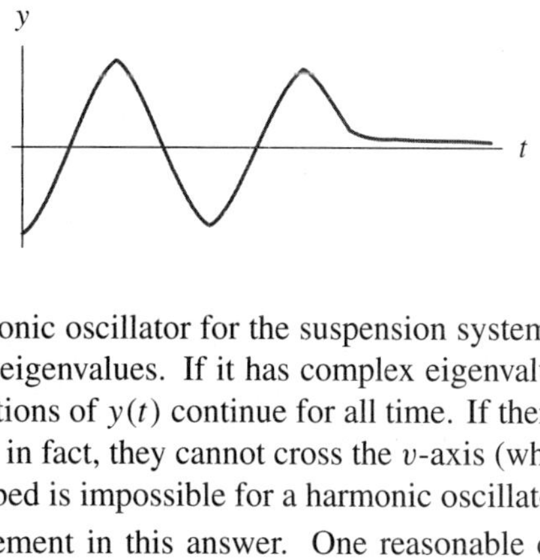

(b) Using the model of a harmonic oscillator for the suspension system, the corresponding system has either real or complex eigenvalues. If it has complex eigenvalues, then solutions spiral in the phase plane and oscillations of $y(t)$ continue for all time. If there are real eigenvalues, then solutions do not spiral, and in fact, they cannot cross the v-axis (where $y = 0$) more than once. Hence, the behavior described is impossible for a harmonic oscillator.

(c) There is room for disagreement in this answer. One reasonable choice is an oscillator with complex eigenvalues and some damping so that the system does oscillate, but the amplitude of the oscillations decays sufficiently rapidly so that only the first two "bounces" are of significant size.

37. **(a)** Since the fluid causes the object to accelerate as it moves and the force causing this acceleration is proportional to the velocity, the force equation for this "mass-spring" system is

$$m\frac{d^2y}{dt^2} = -ky + b_{mf}\frac{dy}{dt},$$

which can be written as

$$m\frac{d^2y}{dt^2} - b_{mf}\frac{dy}{dt} + ky = 0.$$

(b) The equivalent first-order system is

$$\frac{dy}{dt} = v$$
$$\frac{dv}{dt} = -\frac{k}{m}y + \frac{b_{mf}}{m}v.$$

(c) The characteristic equation is

$$m\lambda^2 - b_{mf}\lambda + k = 0,$$

and the eigenvalues are

$$\frac{b_{mf} \pm \sqrt{b_{mf}^2 - 4mk}}{2m}.$$

Since m, b_{mf}, and k are all positive parameters, the eigenvalues are either positive real numbers or complex numbers with a positive real part. If both eigenvalues are real, then the origin is called an "overstimulated" source. The magnitudes of $y(t)$ and $v(t)$ tend to infinity without oscillation. If the eigenvalues are complex, then the origin is a spiral source and the oscillator is called understimulated. The solutions spiral away from the origin with natural period $4m\pi/\sqrt{b_{mf}^2 - 4mk}$.

38. We have the second-order differential equation

$$m\frac{d^2y}{dt^2} + b\frac{dy}{dt} + ky = 0.$$

The characteristic polynomial is

$$m\lambda^2 + b\lambda + k,$$

and the eigenvalues are

$$\frac{-b \pm \sqrt{b^2 - 4mk}}{2m}.$$

In our case, $b^2 - 4mk < 0$, so the eigenvalues can be written as

$$\frac{-b \pm i\sqrt{4mk - b^2}}{2m}.$$

Using this expression for the eigenvalues, we obtain the natural period P as

$$P = \frac{2\pi}{\frac{\sqrt{4mk - b^2}}{2m}} = \frac{4m\pi}{\sqrt{4mk - b^2}}.$$

(a) If $m = 1$, $k = 2$, and $b = 1$, we have a natural period of $4\pi/\sqrt{7}$.

(b) To see how the period changes as m changes, we compute

$$\frac{\partial P}{\partial m} = 4\pi(4mk - b^2)^{-3/2}(2mk - b^2).$$

In our case, $m = 1$, $k = 2$, and $b = 1$. Hence, we have $\partial P/\partial m = 12\pi/(7\sqrt{7})$, and the period increases as the mass increases. The speed that it increases is given by the value of $\partial P/\partial m$, which is $12\pi/(7\sqrt{7})$.

(c) To see how the period changes as k changes, we compute

$$\frac{\partial P}{\partial k} = -8\pi m^2(4mk - b^2)^{-3/2}.$$

In our case, $m = 1$, $k = 2$, and $b = 1$. Hence, we have $\partial P/\partial k = -8\pi/(7\sqrt{7})$, and the period decreases as the spring constant increases. The speed that it increases is given by the value of $\partial P/\partial k$, which is $-8\pi/(7\sqrt{7})$.

(d) To see how the period changes as b changes, we compute

$$\frac{\partial P}{\partial b} = 4\pi mb(4mk - b^2)^{-3/2}.$$

In our case, $m = 1$, $k = 2$, and $b = 1$. Hence, we have $\partial P/\partial b = 4\pi/(7\sqrt{7})$, and the period increases as the damping increases. The speed that it increases is given by the value of $\partial P/\partial b$, which is $4\pi/(7\sqrt{7})$.

39. The differential equation is

$$m\frac{d^2y}{dt^2} + 2y = 0,$$

and the characteristic equation is $m\lambda^2 + 2 = 0$. Hence, the eigenvalues are $\lambda = \pm i\sqrt{2/m}$. The natural period is $2\pi/\sqrt{2/m} = \pi\sqrt{2m}$. For natural period to be 1, we must have $m = 1/(2\pi^2)$.

40. We have the second-order differential equation

$$m\frac{d^2y}{dt^2} + b\frac{dy}{dt} + ky = 0.$$

The characteristic polynomial is $m\lambda^2 + b\lambda + k$, and the eigenvalues are $(-b \pm \sqrt{b^2 - 4mk})/(2m)$. In our case, $b^2 - 4mk < 0$, so the eigenvalues can be written as

$$\frac{-b \pm i\sqrt{4mk - b^2}}{2m}.$$

Using this expression for the eigenvalues, we obtain the natural period P as

$$P = \frac{2\pi}{\dfrac{\sqrt{4mk - b^2}}{2m}} = \frac{4\pi m}{\sqrt{4mk - b^2}}.$$

Each tick of the clock takes one-half of the period. Consequently, if the period gets longer, the time between ticks gets longer and the clock runs slow. Note that the period is inversely proportional to the quantity $\gamma = \sqrt{4mk - b^2}$.

(a) If b increases slightly, then γ decreases slightly. Hence, the period P increases slightly, and the clock runs slow.

(b) If the spring provides slightly less force for a given compression or extension, k decreases slightly. Then, γ decreases slightly, the period increases slightly, and the clock runs slow.

(c) The behavior of the period with respect to the mass is more complicated. We compute the derivative of P with respect to m and obtain

$$\frac{dP}{dm} = 4\pi(4mk - b^2)^{-3/2}(2mk - b^2).$$

Hence, the sign of dP/dm is determined by the sign of $(2mk - b^2)$. If $b < \sqrt{2mk}$, then P increases if m increases slightly. If $b > \sqrt{2mk}$, the P decreases if m increases slightly.

(d) It is possible to have the effects cancel out, so any result (fast, slow or unchanged) is possible.

EXERCISES FOR SECTION 3.7

1.

Table 3.2
Possibilities for linear systems

type	condition on λ	examples
sink	$\lambda_1 < \lambda_2 < 0$	Sec. 3.7, Fig. 3.52
saddle	$\lambda_1 < 0 < \lambda_2$	Sec. 3.3, Fig. 3.12–3.14
source	$0 < \lambda_1 < \lambda_2$	Sec. 3.3, Fig. 3.19
spiral sink	$\lambda = \alpha \pm i\beta$, $\alpha < 0$, $\beta \neq 0$	Sec. 3.1, Fig. 3.2 and 3.4
spiral source	$\lambda = \alpha \pm i\beta$, $\alpha > 0$, $\beta \neq 0$	Sec. 3.4, Fig. 3.29–3.30
center	$\lambda_1 = \pm i\beta$, $\beta \neq 0$	Sec. 3.1, Fig. 3.1 and 3.3
		Sec. 3.4, Fig. 3.28
sink	$\lambda_1 = \lambda_2 < 0$	Sec. 3.5, Fig. 3.35–3.36
(special case)	One line of eigenvectors	
source	$0 < \lambda_1 = \lambda_2$	Sec. 3.5, Ex. 2
(special case)	One line of eigenvectors	
sink	$\lambda_1 = \lambda_2 < 0$	Sec. 3.5, Ex. 23
(special case)	Every vector is eigenvector	
source	$0 < \lambda_1 = \lambda_2$	Sec. 3.5, Ex. 23
(special case)	Every vector is eigenvector	
no name	$\lambda_1 < \lambda_2 = 0$	Sec. 3.5, Fig. 3.39–3.40
no name	$0 = \lambda_1 < \lambda_2$	Sec. 3.5, Ex. 19
no name	$\lambda_1 = \lambda_2 = 0$	Sec. 3.5, Ex. 21
	One line of eigenvectors	
no name	$\lambda_1 = \lambda_2 = 0$	entire plane of equilibrium points
	Every vector is an eigenvector	

2. **(a)**

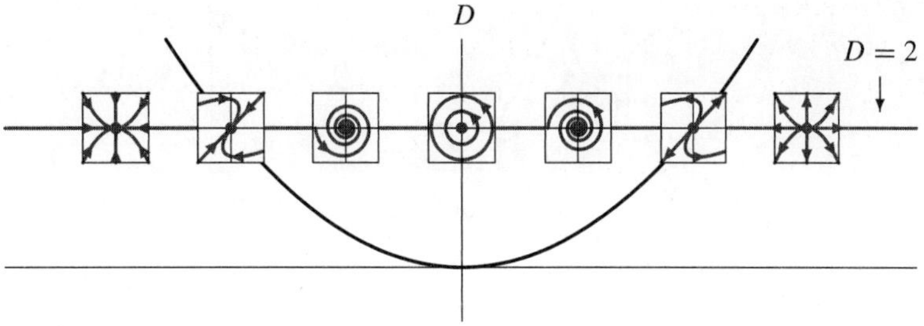

(b) The curve in the trace-determinant plane is the horizontal line given by $D = 2$. The eigenvalues are the roots of $\lambda^2 - a\lambda + 2$, which are

$$\frac{a}{2} \pm \frac{\sqrt{a^2 - 8}}{2}.$$

So we have complex eigenvalues if $|a| < 2\sqrt{2}$, real eigenvalues if $|a| > 2\sqrt{2}$, and repeated eigenvalues if $a = \pm 2\sqrt{2}$. Glancing at the trace-determinant plane, we see that we have a sink with real eigenvalues if $a < -2\sqrt{2}$, a spiral sink if $-2\sqrt{2} < a < 0$, and spiral source if $0 < a < 2\sqrt{2}$, and a source with real eigenvalues if $a > 2\sqrt{2}$.

(c) Bifurcations occur at $a = -2\sqrt{2}$, where we have a sink with repeated eigenvalues, at $a = 2\sqrt{2}$, where we have a source with repeated eigenvalues, and at $a = 0$, where we have a center.

3. **(a)**

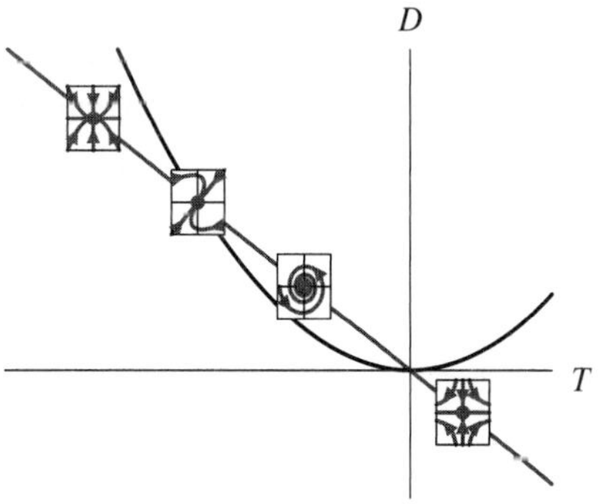

(b) The trace T is $2a$, and the determinant D is $-a$. Therefore, the curve in the trace-determinant plane is $D = -T/2$. This line crosses the parabola $T^2 - 4D = 0$ at two points—at $(T, D) = (0, 0)$ if $a = 0$ and at $(T, D) = (-2, 1)$ if $a = -1$.

The portion of the line for which $a < -1$ corresponds to a positive determinant and a negative trace such that $T^2 - 4D > 0$. The corresponding phase portraits are real sinks. If $a = -1$, we have a sink with repeated eigenvalues. If $-1 < a < 0$, we have complex eigenvalues with negative real parts. Therefore, the phase portraits are spiral sinks. If $a = 0$, we have a degenerate case with an entire line of equilibrium points. Finally, if $a > 0$, the corresponding portion of the line is below the T-axis, and the phase portraits are saddles.

(c) Bifurcations occur at $a = -1$, where we have a sink with repeated eigenvalues, and at $a = 0$, where we have zero as a repeated eigenvalue. For $a = 0$, the y-axis is entirely composed of equilibrium points.

4. **(a)**

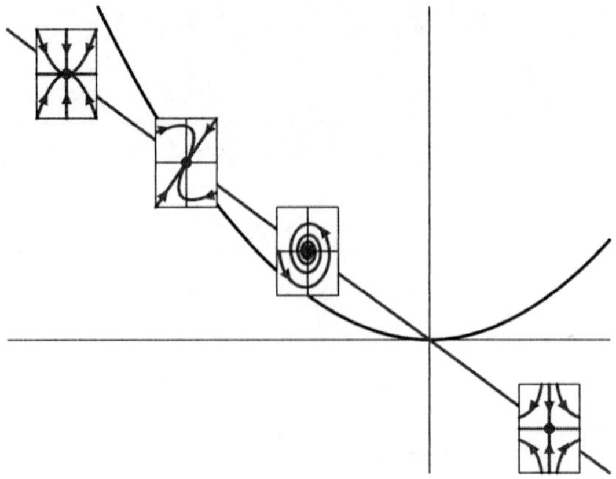

(b) The trace T is a, and the determinant D is $-a$. Therefore, the curve in the trace-determinant plane is $D = -T$. This line crosses the parabola $T^2 - 4D = 0$ at two points—at $(T, D) = (0, 0)$ if $a = 0$ and at $(T, D) = (-4, 4)$ if $a = -4$.

The portion of the line for which $a < -4$ corresponds to a positive determinant and a negative trace such that $T^2 - 4D < 0$. The corresponding phase portraits are real sinks. If $a = -4$, we have a sink with repeated eigenvalues. If $-4 < a < 0$, we have complex eigenvalues with negative real parts. Therefore, the phase portraits are spiral sinks. If $a = 0$, we have a degenerate case with an entire line of equilibrium points. Finally, if $a > 0$, the corresponding portion of the line is below the T-axis, and the phase portraits are saddles.

(c) Bifurcations occur at $a = -4$, where we have a sink with repeated eigenvalues, and at $a = 0$, where we have zero as a repeated eigenvalue. For $a = 0$, the y-axis is entirely composed of equilibrium points.

5. **(a)**

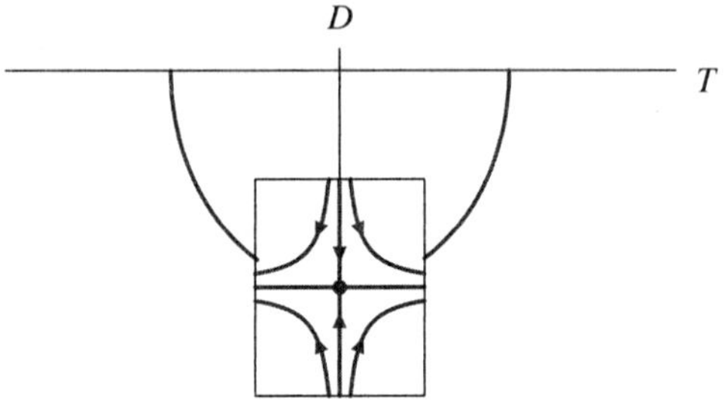

(b) The curve in the trace-determinant plane is the portion of the unit circle centered at 0 that lies in the half-plane $y \leq 0$.

A glance at the trace-determinant plane shows that for $-1 < a < 1$, we have a saddle. If $a = 1$, the eigenvalues are 0 and 1. If $a = -1$, the eigenvalues are 0 and -1.

(c) Bifurcations occur only at $a = \pm 1$. For these two special values of a, we have a line of equilibrium points. The nonzero equilibrium points disappear if $-1 < a < 1$.

6. **(a)**

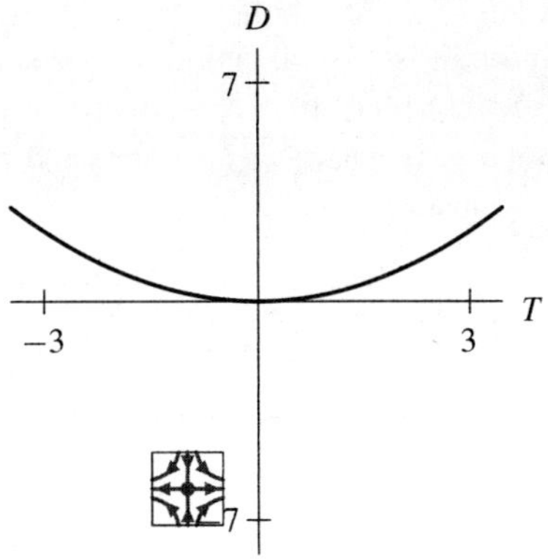

(b) The curve in the trace-determinant plane is not a curve at all. For all values of a, $T = -1$ and $D = -6$. So the curve is simply a point in the trace-determinant plane. For all a, we have a saddle.

(c) There are no bifurcations, since the origin is always a saddle. (There is nothing special about $a = 0$, by the way.)

7. **(a)**

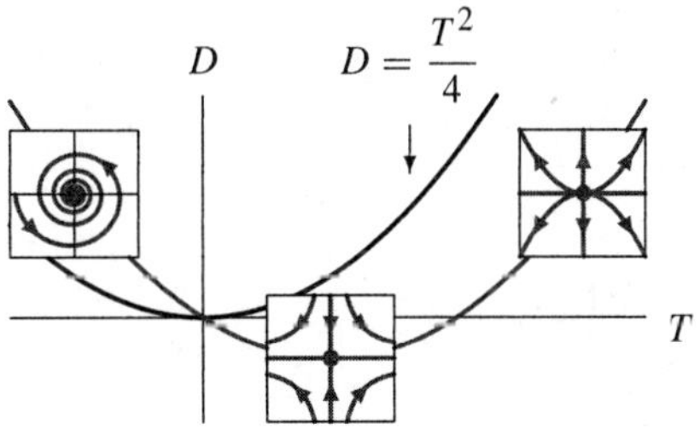

(b) The trace T is $2a$, and the determinant D is $a^2 - a$. Therefore, the curve in the trace-determinant plane is

$$\begin{aligned} D &= a^2 - a \\ &= \left(\frac{T}{2}\right)^2 - \frac{T}{2} \\ &= \frac{T^2}{4} - \frac{T}{2}. \end{aligned}$$

This curve is a parabola. It meets the repeated-eigenvalue parabola (the parabola $D = T^2/4$) if

$$\frac{T^2}{4} - \frac{T}{2} = \frac{T^2}{4}.$$

Solving this equation yields $T = 0$, which corresponds to $a = 0$.

This curve also meets the T-axis (the line $D = 0$) if

$$\frac{T^2}{4} - \frac{T}{2} = 0,$$

so if $T = 0$ or $T = 2$, then $D = 0$.

From the location of the parabola $D = T^2/4 - T/2$ in the trace-determinant plane, we see that the phase portrait is a spiral sink if $a < 0$ since $T < 0$, a saddle if $0 < a < 1$ since $0 < T < 2$, and a source with distinct real eigenvalues if $a > 1$ since $T > 2$.

(c) Bifurcations occur at $a = 0$, where we have repeated zero eigenvalues, and at $a = 1$, where we have a single zero eigenvalue.

8. The eigenvalues are roots of $\lambda^2 - (a+1)\lambda + a - b = 0$, which are

$$\frac{a+1}{2} \pm \frac{1}{2}\sqrt{(a-1)^2 + 4b}.$$

So the eigenvalues are complex if $(a-1)^2 + 4b < 0$, repeated if $(a-1)^2 + 4b = 0$, and real if $(a-1)^2 + 4b > 0$.

The curve $(a-1)^2 + 4b = 0$ is the curve of repeated eigenvalues, analogous to the parabola $T^2 - 4D = 0$ in the usual trace-determinant plane. Note that this curve is a parabola opening downward in the ab-plane with vertex at $(a, b) = (1, 0)$. On this parabola, if $a < -1$, then both eigenvalues are negative; if $a > -1$, then both eigenvalues are positive; if $a = -1$, both eigenvalues are 0.

If $(a-1)^2 + 4b < 0$, we have complex eigenvalues with real parts $(a+1)/2$. So if $a < -1$, we have a spiral sink; if $a > -1$, we have a spiral source; and if $a = -1$, we have a center.

The systems with zero determinant for this family satisfies $a = b$ since $D = a - b$. If $a > b$, $D > 0$, and if $a < b$, we have $D < 0$. So in the case of real eigenvalues ($(a-1)^2 + 4b > 0$), we have a saddle if $a < b$ because $D < 0$. If we graph the line $b = a$ together with the parabola $(a-1)^2 + 4b = 0$, we see that they are tangent at the point $(-1, -1)$. The regions between the line $a = b$ and the parabola $(a-1)^2 + 4b = 0$ give the places where we have sinks or sources with real eigenvalues. If $a > -1$ in this region, then both eigenvalues are positive, so we have a source. If $a < -1$ in this region, then both eigenvalues are negative (a sink). If $(a, b) = (-1, -1)$, we have repeated zero eigenvalues. Whew! That was a toughie! It is worthwhile to draw a picture of the ab-plane.

9. The eigenvalues are roots of the equation $\lambda^2 - 2a\lambda + a^2 - b^2 = 0$. These roots are

$$a \pm \sqrt{b^2} = a \pm |b|.$$

So we have a repeated zero eigenvalue if $a = b = 0$.

If $a = \pm b$, then one of the eigenvalues is 0, and as long as $a \neq 0$ (so $b \neq 0$), the other eigenvalue is nonzero.

The eigenvalues are repeated (both equal to a) if $b = 0$. The eigenvalues are never complex since $\sqrt{b^2} \geq 0$.

If $a > |b|$, then $a \pm |b| > 0$, so we have a source with real eigenvalues. If $a < 0$ and $-a > |b|$, then $a \pm |b| < 0$, so we have a sink with real eigenvalues. In all other cases we have a saddle.

10. The eigenvalues are roots of the equation $\lambda^2 - 2a\lambda + a^2 + b^2 = 0$. They are $a \pm ib$. Hence we have complex roots if $b \neq 0$. If $b \neq 0$ and $a < 0$, the phase portrait is a spiral sink; if $a = 0$, the phase portrait is a center; and if $a > 0$, the phase portrait is a spiral source. If $b = 0$, a is a repeated eigenvalue (repeated 0 eigenvalue if $a = 0$, source if $a > 0$, and sink if $a < 0$).

11. **(a)** This second-order equation is equivalent to the system

$$\frac{dy}{dt} = v$$
$$\frac{dv}{dt} = -3y - bv.$$

Therefore, $T = -b$ and $D = 3$. So the corresponding curve in the trace-determinant plane is $D = 3$.

(b)

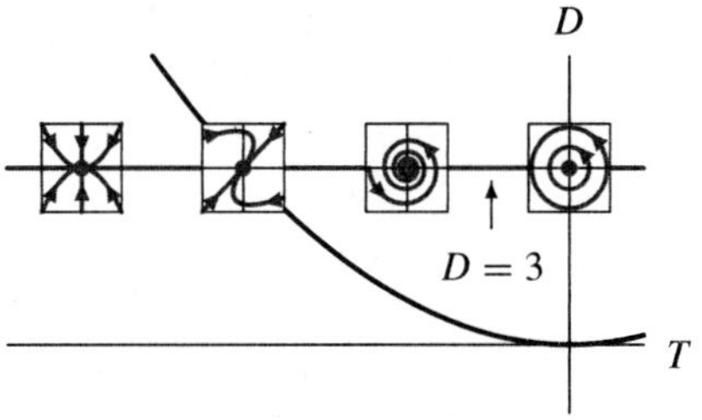

(c) The line $D = 3$ in the trace-determinant plane crosses the repeated-eigenvalue parabola $D = T^2/4$ if $b^2 = 12$, which implies that $b = 2\sqrt{3}$ since b is a nonnegative parameter. If $b = 0$, we have pure imaginary eigenvalues—the undamped case. If $0 < b < 2\sqrt{3}$, the eigenvalues are complex with a negative real part—the underdamped case. If $b = 2\sqrt{3}$, the eigenvalues are repeated and negative—the critically damped case. Finally, if $b > 2\sqrt{3}$, the eigenvalues are real and negative—the overdamped case.

12. **(a)** This second-order equation is equivalent to the system

$$\frac{dy}{dt} = v$$
$$\frac{dv}{dt} = -ky - 2v.$$

Therefore, $T = -2$ and $D = k$. So the curve in the trace-determinant plane is the vertical line $T = -2$.

(b)

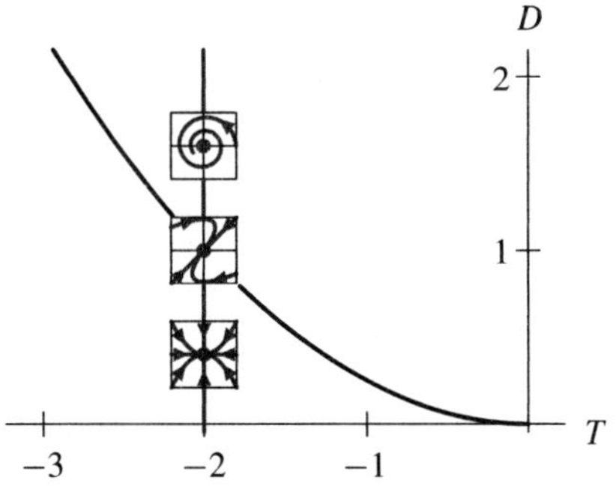

(c) The line $T = -2$ meets the parabola $D = T^2/4$ at $(T, D) = (-2, 1)$, which corresponds to $k = 1$. From the trace-determinant plane, we see that we have a sink with real eigenvalues if $0 < k \leq 1$, repeated eigenvalues if $k = 1$, and complex eigenvalues if $k > 1$. Therefore, the oscillator is overdamped if $0 < k \leq 1$, critically damped if $k = 1$, and underdamped if $k > 1$.

13. **(a)** The second-order equation reduces to the first-order system

$$\frac{dy}{dt} = v$$
$$\frac{dv}{dt} = -\frac{2}{m}y + \frac{1}{m}v.$$

Hence $T = -1/m$, $D = 2/m$, and as the parameter m varies, the systems move along the line $D = -2T$ in the second quadrant of the trace-determinant plane.

(b)

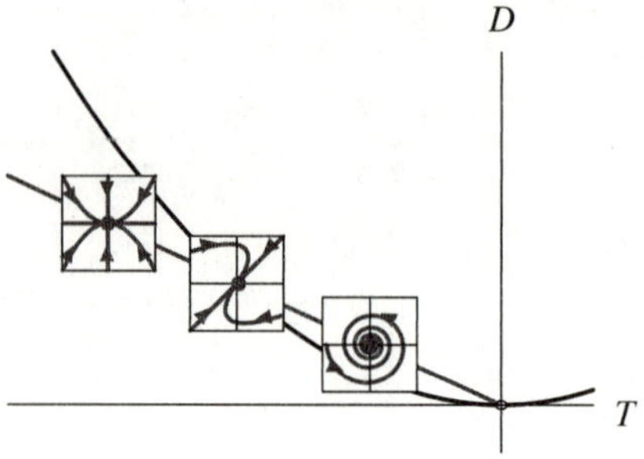

(c) The line $D = -2T$ intersects the repeated-eigenvalue parabola $D = T^2/4$ at the point (T, D) that satisfies $-2T = T^2/4$. We have

$$\frac{T^2}{4} + 2T = T\left(\frac{T}{4} + 2\right) = 0,$$

which yields $T = 0$ or $T = -8$.

For $-8 < T < 0$, the system is underdamped; for $T = -8$, the system is critically damped; and for $T < -8$, the system is overdamped. Since $T = -1/m$, the system is overdamped if $0 < m < 1/8$; it is critically damped if $m = 1/8$; and it is underdamped if $m > 1/8$.

14. **(a)** In Animation A, slides 0–11 are saddles, and slides 12–20 include a line of equilibrium points. Slides 21–23 are sources with distinct real eigenvalues, slide 24 is a source with repeated eigenvalues, and slides 25–32 are spiral sources.

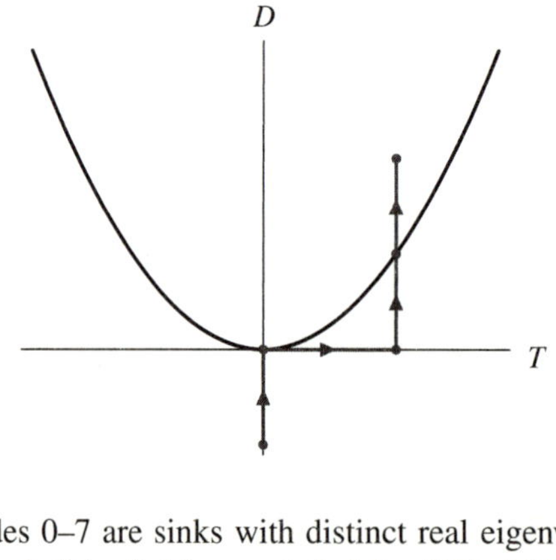

(b) In Animation B, slides 0–7 are sinks with distinct real eigenvalues, slide 8 is a sink with repeated eigenvalues, and slides 9–15 are spiral sinks. Slide 16 is a center. Then slides 17–23 are spiral sources, slide 24 is a source with repeated eigenvalues, and slides 25–32 are sources with distinct real eigenvalues.

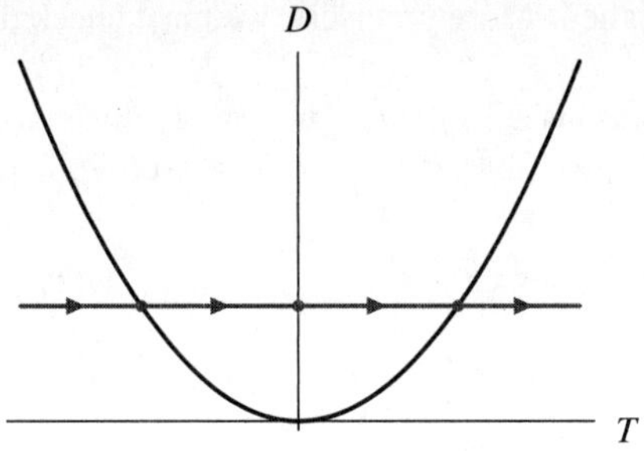

(c) In Animation C, slides 0–15 are saddles, and slide 16 includes a line of equilibrium points. Slides 17–19 are sinks with distinct real eigenvalues, slide 20 is a sink with repeated eigenvalues, and slides 21–32 are spiral sinks.

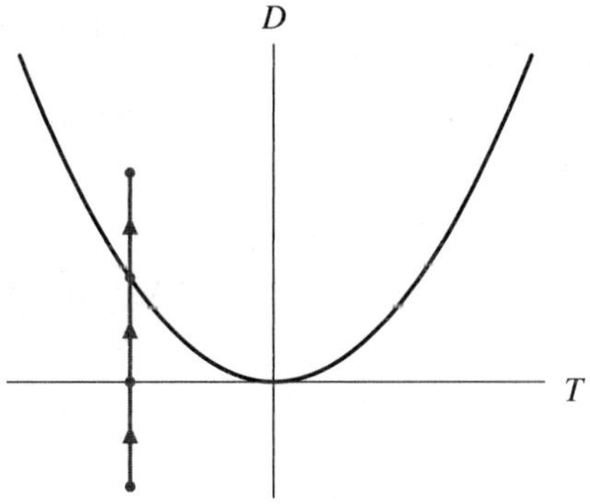

(d) In Animation D, slides 0–7 are sinks with distinct real eigenvalues, slide 8 is a sink with repeated eigenvalues, and slides 9–15 are spiral sinks. Slides 16–19 are centers, slide 20 includes a line of equilibrium points, and slides 21–32 are saddles.

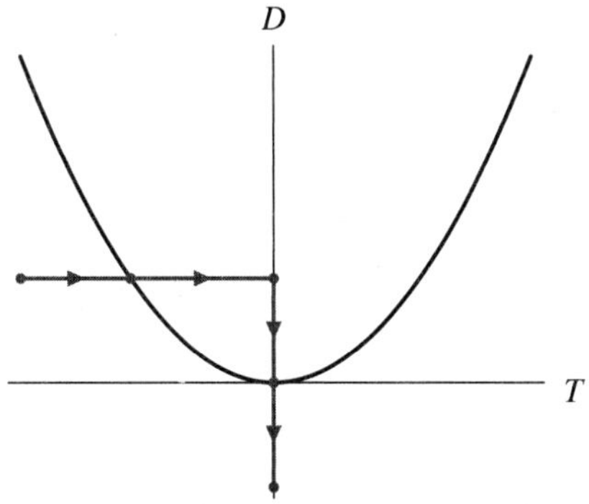

EXERCISES FOR SECTION 3.8

1. To check that the given vector-valued functions are solutions, we differentiate each coordinate and check that the system is satisfied. More precisely, we must check that $dx/dt = 0.1y$, $dy/dt = 0.2z$, and $dz/dt = 0.4x$.

To check that these equations are satisfied, we simply differentiate and simplify. For example, to check that $\mathbf{Y}_2(t)$ is a solution, we first differentiate the x-coordinate, and we obtain

$$\begin{aligned}\frac{dx}{dt} &= -0.1\,e^{-0.1t}\left(-\cos\left(\sqrt{0.03}\,t\right) - \sqrt{3}\,\sin\left(\sqrt{0.03}\,t\right)\right) + \\ &\qquad e^{-0.1t}\left(\sqrt{0.03}\,\sin\left(\sqrt{0.03}\,t\right) - \sqrt{3}\,\sqrt{0.03}\,\cos\left(\sqrt{0.03}\,t\right)\right) \\ &= e^{-0.1t}\left(\left(0.1 - \sqrt{0.09}\right)\cos\left(\sqrt{0.03}\,t\right) + \left(0.1\sqrt{3} + \sqrt{0.03}\right)\sin\left(\sqrt{0.03}\,t\right)\right) \\ &= e^{-0.1t}\left(-0.2\cos\left(\sqrt{0.03}\,t\right) + 0.2\sqrt{3}\,\sin\left(\sqrt{0.03}\,t\right)\right) \\ &= 0.1\,e^{-0.1t}\left(-2\cos\left(\sqrt{0.03}\,t\right) + 2\sqrt{3}\,\sin\left(\sqrt{0.03}\,t\right)\right).\end{aligned}$$

Note that this last expression is just $0.1y(t)$. Hence, the first component of the differential equation is satisfied.

In order to complete the verification that $\mathbf{Y}_2(t)$ is a solution, we must also verify the equations for dy/dt and dz/dt. The calculations are similar.

2. Suppose that the equation

$$k_1\mathbf{Y}_1 + k_2\mathbf{Y}_2 + k_3\mathbf{Y}_3 = 0$$

holds and $k_3 \neq 0$. Then

$$\mathbf{Y}_3 = -\frac{k_1}{k_3}\mathbf{Y}_1 - \frac{k_2}{k_3}\mathbf{Y}_2,$$

and $\mathbf{Y}_3$ is a linear combination of $\mathbf{Y}_1$ and $\mathbf{Y}_2$. Thus, $\mathbf{Y}_3$ is in the plane determined by $\mathbf{Y}_1$ and $\mathbf{Y}_2$.

The other two cases are analogous. For example, suppose that the equation

$$k_1\mathbf{Y}_1 + k_2\mathbf{Y}_2 + k_3\mathbf{Y}_3 = 0$$

holds and $k_1 \neq 0$. Then we can write

$$\mathbf{Y}_1 = -\frac{k_2}{k_1}\mathbf{Y}_2 - \frac{k_3}{k_1}\mathbf{Y}_3.$$

Hence, $\mathbf{Y}_1$ is in the plane formed by $\mathbf{Y}_2$ and $\mathbf{Y}_3$.

In all cases, we can solve for one of the $\mathbf{Y}$'s as a linear combination of the other two vectors, so the three $\mathbf{Y}$'s do not all point in all "possible" directions. They are linearly dependent.

3. **(a)** Suppose

$$k_1\begin{pmatrix}1\\2\\1\end{pmatrix} + k_2\begin{pmatrix}1\\3\\1\end{pmatrix} + k_3\begin{pmatrix}1\\4\\1\end{pmatrix} = \begin{pmatrix}0\\0\\0\end{pmatrix}.$$

We obtain the simultaneous equations

$$\begin{cases} k_1 + k_2 + k_3 = 0 \\ 2k_1 + 3k_2 + 4k_3 = 0. \end{cases}$$

We have two equations with three unknowns. Therefore, we cannot uniquely determine the values of k_1, k_2, and k_3. In other words, we can find infinitely many triples (k_1, k_2, k_3) such that $k_1\mathbf{Y}_1 + k_2\mathbf{Y}_2 + k_3\mathbf{Y}_3 = 0$. For example, $k_1 = -1$, $k_2 = 2$, and $k_3 = -1$ is one such triple.

(b) Suppose

$$k_1 \begin{pmatrix} 2 \\ 0 \\ 1 \end{pmatrix} + k_2 \begin{pmatrix} 3 \\ 2 \\ 2 \end{pmatrix} + k_3 \begin{pmatrix} 1 \\ -2 \\ -3 \end{pmatrix} = \begin{pmatrix} 0 \\ 0 \\ 0 \end{pmatrix}.$$

In matrix notation, we can write this as

$$\mathbf{A} \begin{pmatrix} k_1 \\ k_2 \\ k_3 \end{pmatrix} = \begin{pmatrix} -2 & 3 & 0 \\ 3 & -2 & 0 \\ 0 & 0 & -1 \end{pmatrix} \begin{pmatrix} k_1 \\ k_2 \\ k_3 \end{pmatrix} = \begin{pmatrix} 0 \\ 0 \\ 0 \end{pmatrix}.$$

Since $\det \mathbf{A} = -12 \neq 0$, $\mathbf{A}$ has an inverse matrix, and by multiplying the inverse matrix, we obtain $(k_1, k_2, k_3) = (0, 0, 0)$. Hence, the vectors are linearly independent.

(c) Suppose

$$k_1 \begin{pmatrix} 1 \\ 2 \\ 0 \end{pmatrix} + k_2 \begin{pmatrix} 0 \\ 1 \\ 2 \end{pmatrix} + k_3 \begin{pmatrix} 2 \\ 0 \\ 1 \end{pmatrix} = \begin{pmatrix} 0 \\ 0 \\ 0 \end{pmatrix}.$$

In scalar form, this vector equation is equivalent to the scalar equations

$$\begin{aligned} k_1 + k_3 &= 0 \\ 2k_1 + k_2 &= 0 \\ 2k_2 + k_3 &= 0. \end{aligned}$$

From the third equation, $-2k_3 = k_1$, and substitution of this equation into the first equation yields $k_1 = 0$. Then, $k_3 = 0$, and using the second equation, we conclude that $k_2 = 0$. Therefore, the three vectors are independent.

(d) Suppose

$$k_1 \begin{pmatrix} -3 \\ \pi \\ 1 \end{pmatrix} + k_2 \begin{pmatrix} 0 \\ 1 \\ 0 \end{pmatrix} + k_3 \begin{pmatrix} -2 \\ -2 \\ -2 \end{pmatrix} = \begin{pmatrix} 0 \\ 0 \\ 0 \end{pmatrix}.$$

This vector equation can be written as the simultaneous equations

$$\begin{aligned} -3k_1 - 2k_3 &= 0 \\ \pi k_1 + k_2 - 2k_3 &= 0 \\ k_1 - 2k_3 &= 0. \end{aligned}$$

From the third equation, $2k_3 = k_1$ and substitution of this equation into the first equation yields $k_1 = 0$. Then, $k_3 = 0$, and using the second equation, we obtain $k_2 = 0$. Therefore, the three vectors are independent.

4. **(a)** The characteristic equation is

$$\det(\mathbf{A} - \lambda\mathbf{I}) = (\lambda^2 + 1)(2 - \lambda) = 0.$$

Therefore, the eigenvalues are $\lambda = \pm i$ and $\lambda = 2$.

(b) Writing the differential equation in coordinates, one obtains

$$\frac{dx}{dt} = y$$
$$\frac{dy}{dt} = -x$$
$$\frac{dz}{dt} = 2z.$$

Since dx/dt and dy/dt do not depend on z and dz/dt does not depend on x or y, the system decouples into a two-dimensional system in the xy-plane and a one-dimensional system on the z-axis.

(c) In the xy-plane, the characteristic equation is $\lambda^2 + 1 = 0$, and the eigenvalues are $\lambda = \pm i$. Therefore, the system is a center in the xy-plane.

The z-axis is the phase line for the equation $dz/dt = 2z$. Therefore, there is a single equilibrium point at the origin, and every solution curve lying on the z-axis is asymptotic to 0 as $t \to -\infty$.

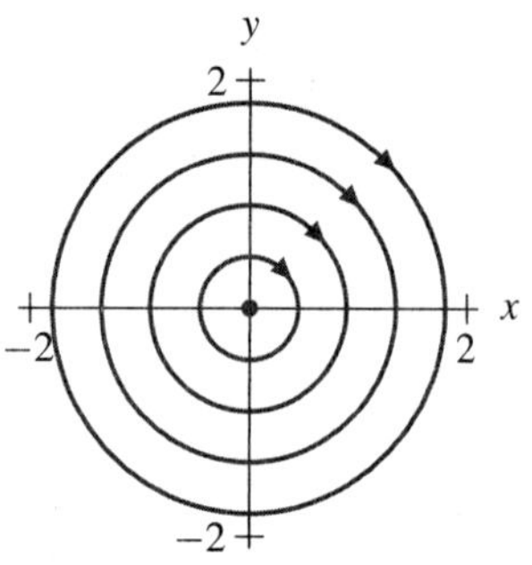

xy-phase plane

z-phase line

(d) The xy-phase plane and the z-phase line can be combined to obtain the xyz-phase space.

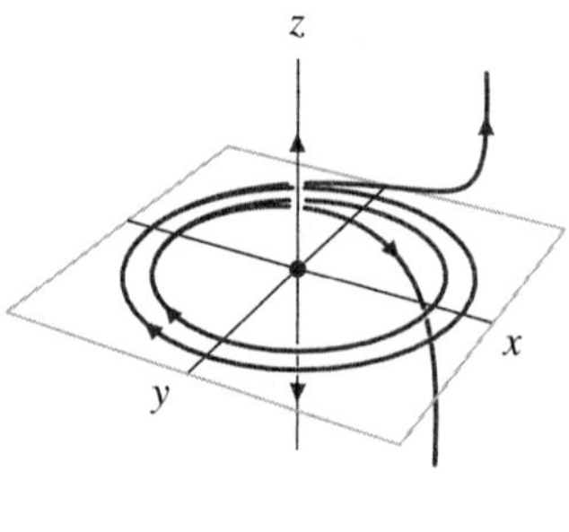

5. **(a)** The characteristic equation is

$$\det(\mathbf{A} - \lambda\mathbf{I}) = (-2-\lambda)(-2-\lambda)(-1-\lambda) - (3)(3)(-1-\lambda) = 0,$$

which reduces to

$$-(\lambda+1)(\lambda+5)(\lambda-1) = 0.$$

Therefore, the eigenvalues are $\lambda = \pm 1$ and $\lambda = -5$.

(b) Writing the differential equation in coordinates, one obtains

$$\begin{aligned}\frac{dx}{dt} &= -2x + 3y\\ \frac{dy}{dt} &= 3x - 2y\\ \frac{dz}{dt} &= -z.\end{aligned}$$

Since dx/dt and dy/dt do not depend on z and dz/dt does not depend on x or y, the system decouples into a two-dimensional system in the xy-plane and a one-dimensional system on the z-axis.

(c) In the xy-plane, the characteristic equation is $(-2-\lambda)^2 - 9 = \lambda^2 + 4\lambda - 5 = 0$, and the eigenvalues are $\lambda = -5$ and $\lambda = 1$. Therefore, the system is a saddle in the xy-plane. The eigenvectors (x, y, z) for $\lambda = -5$ satisfy the equations $y = -x$ and $z = 0$, and the eigenvectors for $\lambda = 1$ satisfy the equations $x = y$ and $z = 0$.

The z-axis is the phase line for the equation $dz/dt = -z$. Therefore, there is a single equilibrium point at the origin, and every solution curve lying on the z-axis is asymptotic to 0 as $t \to \infty$.

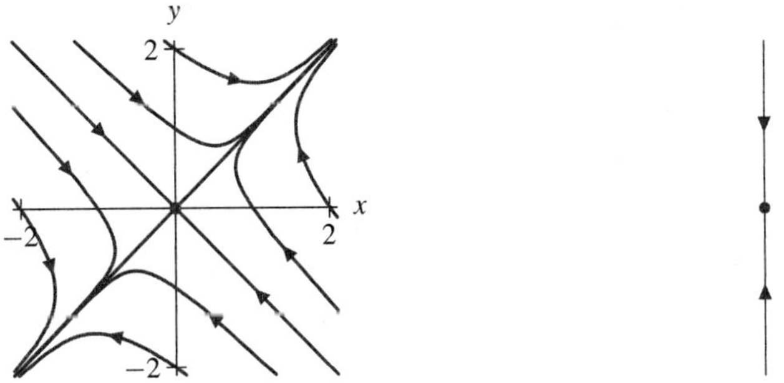

xy-phase plane $\qquad$ z-phase line

(d) The xy-phase plane and the z-phase line can be combined to obtain the xyz-phase space.

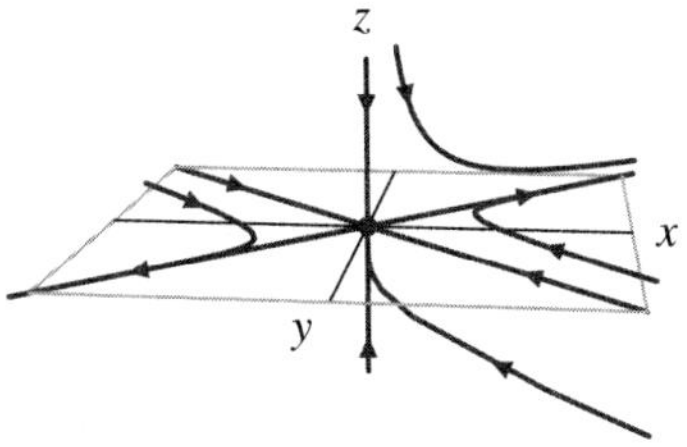

6. **(a)** The characteristic equation is

$$\det(\mathbf{A} - \lambda\mathbf{I}) = (1 - \lambda)(\lambda^2 - 2\lambda + 10) = 0.$$

Therefore, the eigenvalues are $\lambda = 1 \pm 3i$ and $\lambda = 1$.

(b) Writing the differential equation in coordinates, one obtains

$$\frac{dx}{dt} = x + 3z$$
$$\frac{dy}{dt} = -y$$
$$\frac{dz}{dt} = -3x + z.$$

Since dx/dt and dz/dt do not depend on y and dy/dt does not depend on either x or z, the system decouples into a two-dimensional system in the xz-plane and a one-dimensional system on the y-axis.

(c) In the xz-plane, the characteristic equation is $\lambda^2 - 2\lambda + 10 = 0$, and the eigenvalues are $\lambda = 1 \pm 3i$. Therefore, the system is a spiral source in the xz-plane.

The y-axis is the phase line for the equation $dy/dt = -y$. Therefore, there is a single equilibrium point at the origin, and every solution curve lying on the y-axis is asymptotic to 0 as $t \to \infty$.

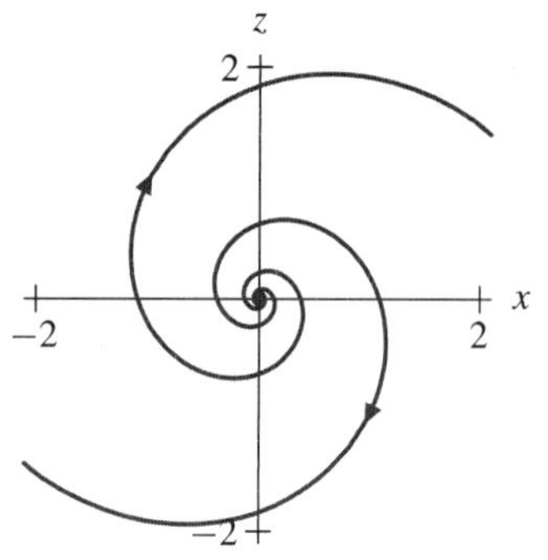

xz-phase plane

z-phase line

(d) The xz-phase plane and the y-phase line can be combined to obtain the xyz-phase space.

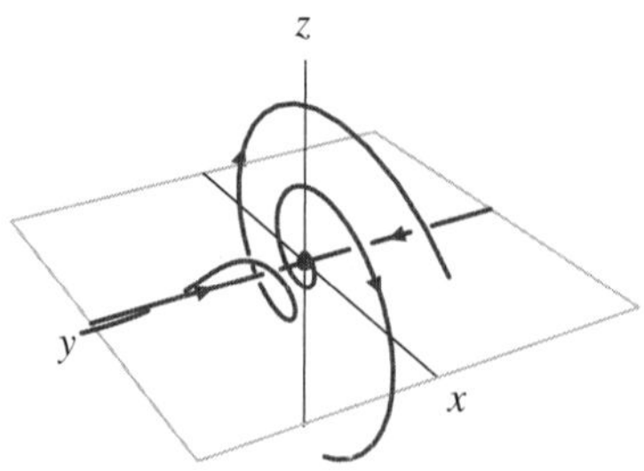

7. (a) The characteristic equation is

$$\det(\mathbf{A} - \lambda\mathbf{I}) = (1 - \lambda)((2 - \lambda)(2 - \lambda) - 1) = 0,$$

which simplifies to

$$-(\lambda - 3)(\lambda - 1)(\lambda - 1) = 0.$$

Therefore, the eigenvalues are $\lambda = 1$ and $\lambda = 3$.

(b) Writing the differential equation in coordinates, one obtains

$$\frac{dx}{dt} = x$$
$$\frac{dy}{dt} = 2y - z$$
$$\frac{dz}{dt} = -y + 2z.$$

Since dy/dt and dz/dt do not depend on x, and dx/dt does not depend on y or z, the system decouples into a two-dimensional system in the yz-plane and a one-dimensional system on the x-axis.

(c) In yz-plane, the characteristic equation is $(2-\lambda)^2-1 = (\lambda-3)(\lambda-1) = 0$, and the eigenvalues are $\lambda = 1$ and $\lambda = 3$. Therefore, in the yz-phase plane, the system is a source. The eigenvectors (x, y, z) for $\lambda = 1$ satisfy the equations $x = 0$ and $y = z$, and the eigenvectors for $\lambda = 3$ satisfy the equations $x = 0$ and $y = -z$.

The x-axis is the phase line for the equation $dx/dt = x$. Therefore, there is a single equilibrium point at the origin, and every solution curve lying on the x-axis tends to infinity as $t \to \infty$.

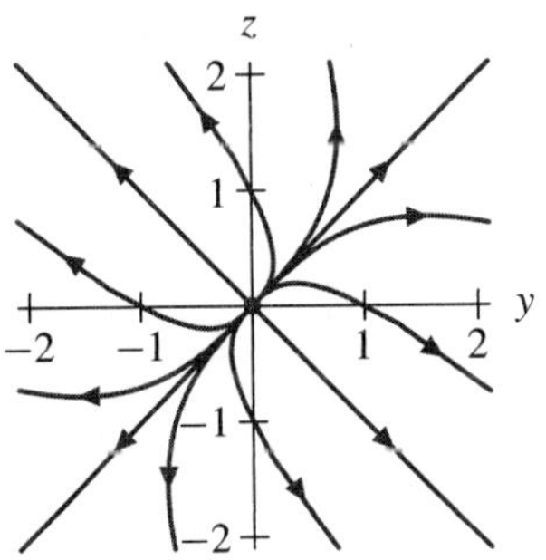

yz-phase plane

x-phase line

(d) The x-phase line and the yz-phase plane can be combined to obtain the xyz-phase space.

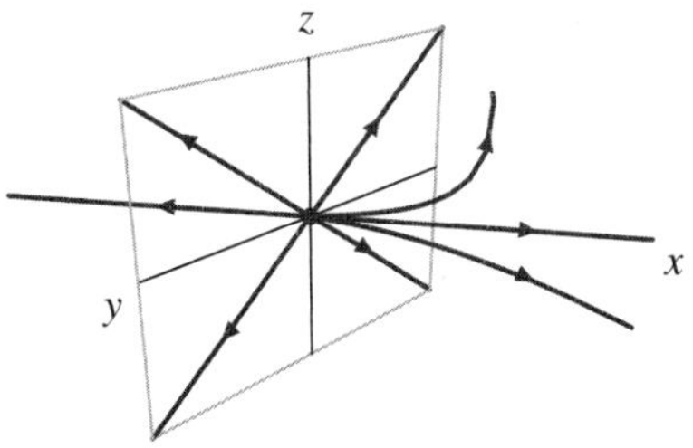

8. **(a)** The λ^3 term dominates the others when λ is large. We can write

$$p(\lambda) = \lambda^3 \left(\alpha + \frac{\beta}{\lambda} + \frac{\gamma}{\lambda^2} + \frac{\delta}{\lambda^3} \right).$$

As λ gets large, the term in the parentheses tends to α. Because $\alpha > 0$, $p(\lambda) \to \infty$ as $t \to \infty$. Similarly, $p(\lambda) \to -\infty$ as $t \to -\infty$.

(b) Same as part (a) except that the sign of α is negative.

(c) Because the $p(\lambda) \to \infty$ in one direction and $p(\lambda) \to -\infty$ in the other direction and because $p(\lambda)$ is continuous, we can use the Intermediate Value Theorem to show that the graph must cross the axis at at least one number λ_0.

9. (Note: This exercise can also be done by taking the complex conjugate of both sides of the equation $p(a + ib) = 0$.)

Using the definition of $p(x)$, we compute

$$p(a+ib) = (\alpha a^3 - 3\alpha ab^2 + \beta a^2 - \beta b^2 + \gamma a + \delta) + i(3\alpha a^2 b - \alpha b^3 + 2\beta ab + \gamma b)$$

and

$$p(a-bi) = (\alpha a^3 - 3\alpha ab^2 + \beta a^2 - \beta b^2 + \gamma a + \delta) - i(3\alpha a^2 b - \alpha b^3 + 2\beta ab + \gamma b).$$

Since $p(a+bi) = 0$,

$$\alpha a^3 - 3\alpha ab^2 + \beta a^2 - \beta b^2 + \gamma a + \delta = 3\alpha a^2 b - \alpha b^3 + 2\beta ab + \gamma b = 0.$$

Therefore, $p(a - ib) = 0$, and $a - ib$ is also a root.

10. **(a)** To compute the eigenvalues, we first compute the characteristic polynomial

$$(-2-\lambda)(-2-\lambda)(-1-\lambda),$$

so the eigenvalues are -2 and -1.

(b) In the xy-plane, the system has a repeated eigenvalue associated to eigenvectors along the x-axis. The z-phase line has a sink at the origin.

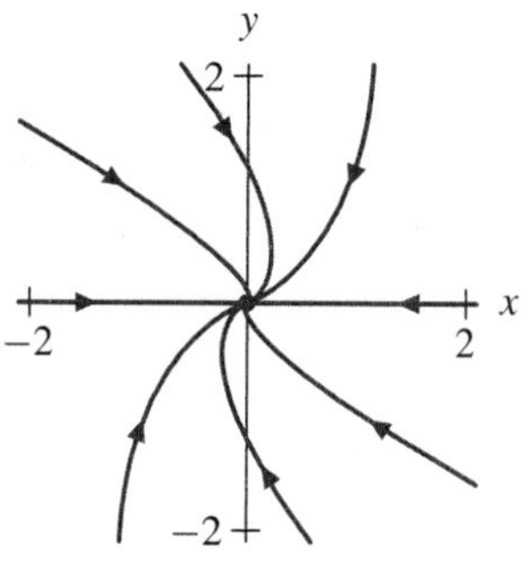

xy-phase plane

z-phase line

(c) Combining the xy-phase plane and z-phase line, we obtain a picture of the xyz-phase space.

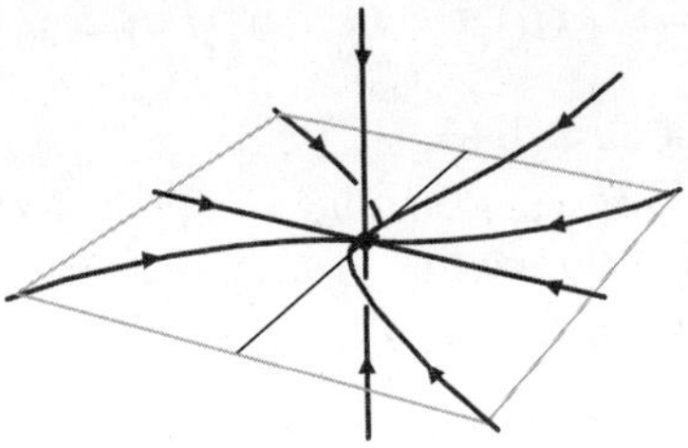

11. **(a)** The characteristic equation is

$$(-2-\lambda)(-2-\lambda)(1-\lambda)=0,$$

and the eigenvalues are $\lambda=-2$ and $\lambda=1$.

(b) In the xy-plane, the system has a repeated eigenvalue, $\lambda=-2$, and the eigenvectors (x, y, z) associated to that eigenvalue satisfy the equations $y=z=0$. The origin is a sink in the xy-phase plane.

The z-axis is the phase line for the equation $dz/dt=z$. Therefore, there is a single equilibrium point at the origin, and every solution curve lying on the z-axis tends to infinity as $t\to\infty$.

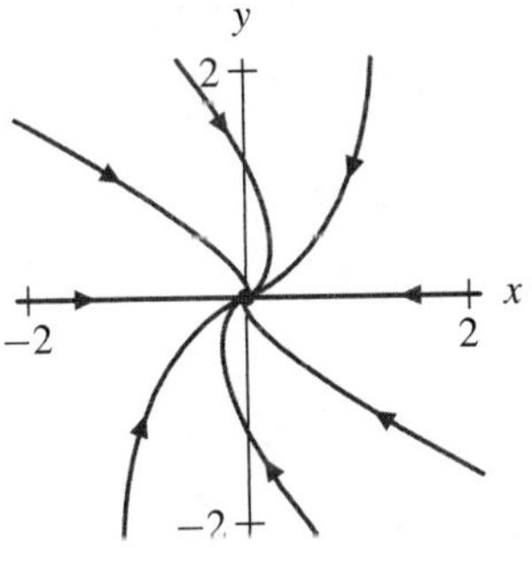

xy-phase plane

z-phase line

(c) Combining the xy-phase plane and z-phase line, we obtain a picture of the xyz-phase space.

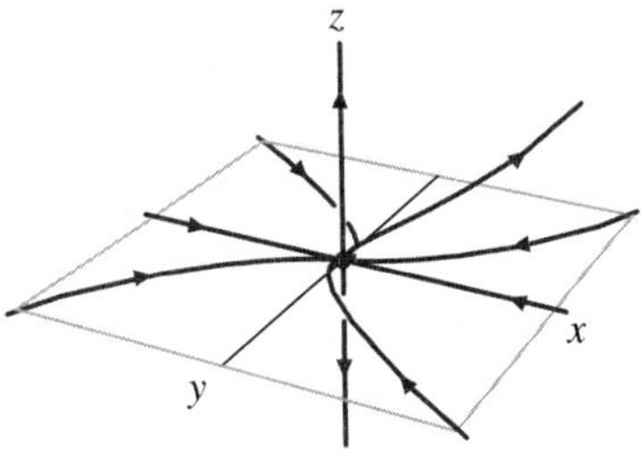

12. **(a)** The characteristic polynomial is

$$(-1-\lambda)(-4-\lambda)(-1-\lambda) - 2(2)(-1-\lambda) = (-1-\lambda)(\lambda^2+5\lambda),$$

so the eigenvalues are -1, 0, and -5.

(b) The xy-phase plane has 0 as an eigenvalue with the line of equilibrium points $y = x/2$. The eigenvalue -5 corresponds to the line of eigenvectors $y = -2x$. The phase line in the z-direction has a sink at the origin.

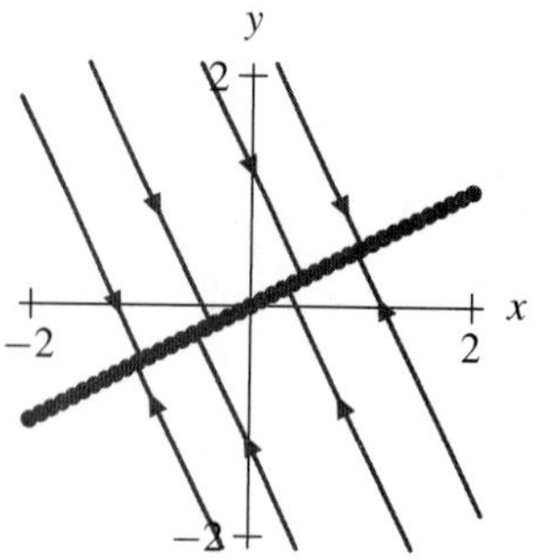

xy-phase plane

z-phase line

(c) Combining the xy-phase plane and z-phase line, we obtain a picture of the xyz-phase space.

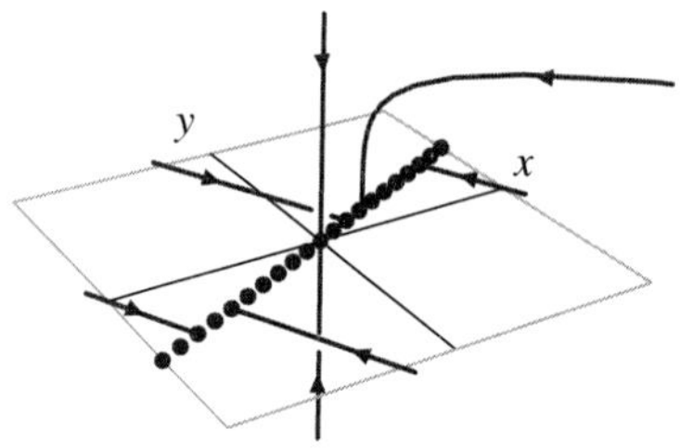

13. **(a)** The characteristic equation is

$$(-1-\lambda)(-4-\lambda)(-\lambda) - 4(-\lambda) = -\lambda^2(\lambda+5) = 0.$$

One eigenvalue is $\lambda = -5$, and the other eigenvalue, $\lambda = 0$, is a repeated eigenvalue.

(b) In xy-plane, the eigenvalues are $\lambda = 0$ and $\lambda = -5$. The eigenvectors (x, y, z) associated to the eigenvalue $\lambda = 0$ satisfy the equations $z = 0$ and $-x + 2y = 0$. The eigenvectors associated to the eigenvalue $\lambda = -5$ satisfy the equations $z = 0$ and $4x + 2y = 0$. Hence, the line $y = x/2$ is a line of equilibrium points, and every solution curve lying in the xy-plane tends toward one of the equilibrium points on $y = x/2$ as $t \to \infty$.

The z-axis is the phase line for $dz/dt = 0$. Note that this line consists entirely of equilibrium points.

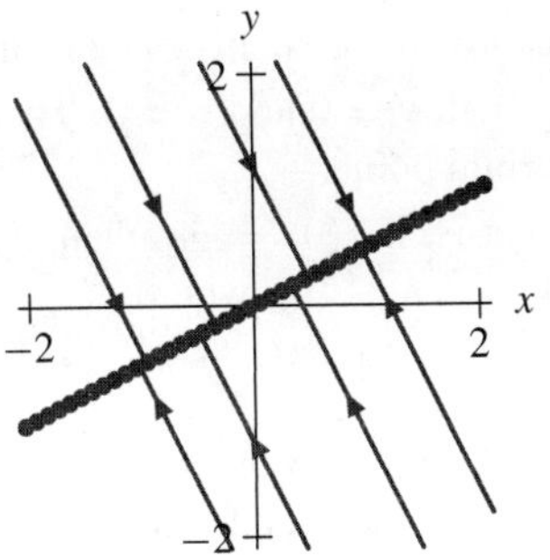

xy-phase plane

z-phase line

(c) Combining the xy-phase plane and z-phase line, we obtain a picture of the xyz-phase space.

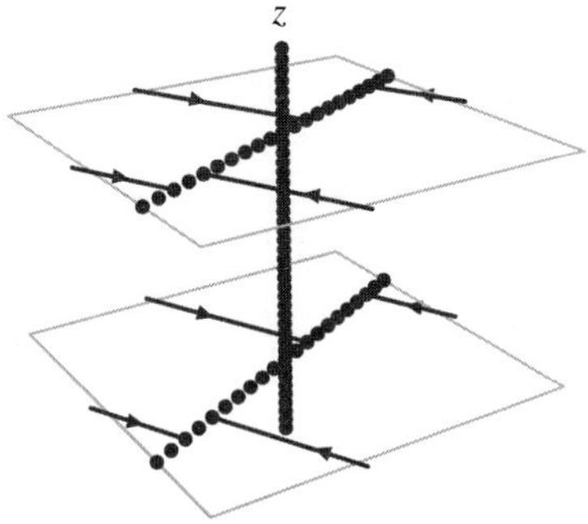

14. **(a)** The characteristic polynomial is

$$(-2-\lambda)(-2-\lambda)(-2-\lambda),$$

so there is only one eigenvalue, $\lambda = -2$.

(b) To compute the eigenvectors, we must solve

$$\begin{cases} -2x + y = -2x \\ -2y + z = -2y \\ -2z = -2z. \end{cases}$$

We conclude that $y = z = 0$. Hence, the only eigenvectors lie on the x-axis.

(c) There is one line — the line of eigenvectors — that consists of straight-line solutions.

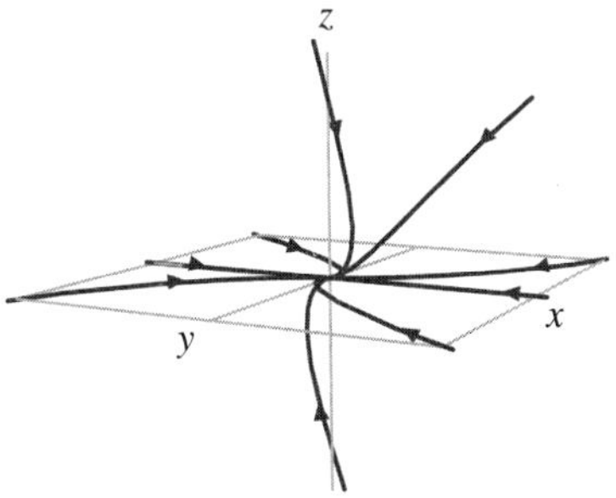

15. **(a)** The characteristic equation is $-\lambda^3 = 0$. Consequently, there is only one eigenvalue, $\lambda = 0$.

(b) For $\lambda = 0$, the eigenvectors (x, y, z) must satisfy both $y = 0$ and $z = 0$. Therefore, the x-axis is both a line of eigenvectors and a line of equilibrium points.

(c) Since $dz/dt = 0$, $z(t)$ is a constant function. That is, if $z(0) = z_0$, then $z(t) = z_0$ for all t. Since $dy/dt = z$ and z is constant, we have $y(t) = z_0 t + y_0$, where $y(0) = y_0$ is the initial condition for $y(t)$. Finally, since $dx/dt = y = z_0 t + y_0$, $x(t) = z_0 t^2/2 + y_0 t + x_0$, where $x(0) = x_0$ is the initial condition for $x(t)$.

For $z_0 = 0$, $y(t)$ is constant, and the solution curves lie on straight lines parallel to the x-axis. For $y_0 > 0$, $x(t)$ is increasing, and for $y_0 < 0$, $x(t)$ is decreasing. For $z_0 \neq 0$, $x(t)$ is quadratic in y. Therefore, solution curves that satisfy the initial condition $z(0) = z_0$ stay on the plane $z = z_0$ and lie on a parabola.

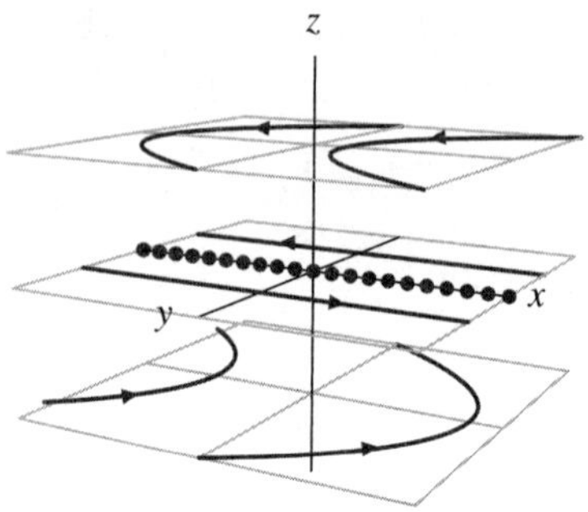

16. **(a)** To show that $\mathbf{V}_1 = (1, 1, 1)$ is an eigenvector, we compute $\mathbf{A}\mathbf{V}_1$. We have

$$\mathbf{A}\mathbf{V}_1 = \begin{pmatrix} 2 & -1 & 0 \\ 0 & -2 & 3 \\ -1 & 3 & -1 \end{pmatrix} \begin{pmatrix} 1 \\ 1 \\ 1 \end{pmatrix} = \begin{pmatrix} 1 \\ 1 \\ 1 \end{pmatrix}.$$

Consequently, the vector $\mathbf{V}_1 = (1, 1, 1)$ is an eigenvector associated to the eigenvalue 1.

(b) To find the other eigenvalues, we first compute the characteristic polynomial, which is

$$-\lambda^3 - \lambda^2 + 13\lambda - 11.$$

Since we know that $\lambda = 1$ is an eigenvalue, we can divide this polynomial by $\lambda - 1$. We obtain $11 - 2\lambda - \lambda^2$. The roots of this quadratic are $\lambda = -1 \pm 2\sqrt{3}$.

(c) The system has three real eigenvalues — two are positive and one is negative. Hence, the equilibrium point at the origin is a saddle.

(d) One eigenvector for $\lambda = -1 - 2\sqrt{3}$ is the vector

$$\left(2\sqrt{3} - \frac{9}{-1 + 2\sqrt{3}}, \frac{-3}{-1 + 2\sqrt{3}}, 1\right),$$

and one eigenvector for $\lambda = -1 + 2\sqrt{3}$ is

$$\left(-2\sqrt{3} - \frac{9}{-1 - 2\sqrt{3}}, \frac{3}{1 + 2\sqrt{3}}, 1\right).$$

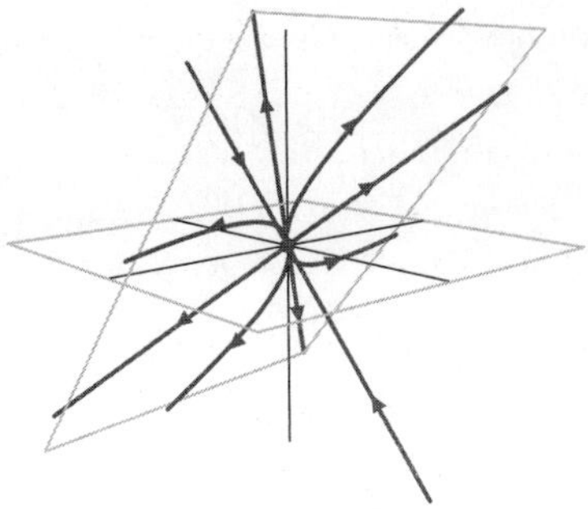

17. **(a)** We compute

$$\mathbf{AV}_1 = \mathbf{A}\begin{pmatrix} 1 \\ 1 \\ 0 \end{pmatrix} = \begin{pmatrix} -1 \\ -1 \\ 0 \end{pmatrix} = -1\begin{pmatrix} 1 \\ 1 \\ 0 \end{pmatrix}.$$

Hence, $\mathbf{V}_1$ is an eigenvector associated to the eigenvalue -1.

(b) The characteristic equation is

$$(-4-\lambda)((-1-\lambda)(-\lambda)+5)+15 = (\lambda+1)(\lambda^2+4\lambda+5) = 0.$$

The eigenvalues are $\lambda = -1$ and $\lambda = -2 \pm i$.

(c) Since one eigenvalue is a negative real number and the other two are complex with a negative real part, the system is a spiral sink.

(d) To determine all eigenvectors $\mathbf{V}_2$ associated to the eigenvalue $\lambda = -2+i$, we must solve the vector equation $\mathbf{AV}_2 = (-2+i)\mathbf{V}_2$. This vector equation is equivalent to the three simultaneous equations

$$\begin{cases} -4x+3y = (-2+i)x \\ -y+z = (-2+i)y \\ -x+3y-z = (-2+i)z. \end{cases}$$

Then, all complex eigenvectors $\mathbf{V}_2 = (x, y, z)$ must satisfy the equations $y = (2+i)x/3$ and $z = (-3+i)x/3$. One such eigenvector is $(3, 2+i, -3+i) = (3, 2, -3) + i(0, 1, 1)$. Taking real and imaginary parts, we obtain two vectors, $(3, 2, -3)$ and $(0, 1, 1)$, that determine a plane on which the solutions spiral toward the origin with natural period 2π.

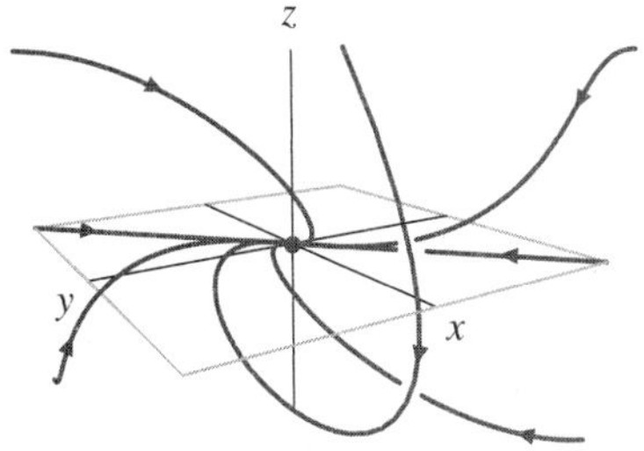

18. **(a)** The characteristic polynomial is

$$(-10-\lambda)(-1-\lambda)(-8/3-\lambda) - 10(28)(-8/3-\lambda) = (-8/3-\lambda)(\lambda^2+11\lambda-270).$$

The eigenvalues are $-8/3$ and $(-11 \pm \sqrt{1201})/2$.

(b) One eigenvector for $-8/3$ is $(0, 0, 1)$. One eigenvector for $(-11-\sqrt{1201})/2$ is

$$\left(\frac{-9-\sqrt{1201}}{56}, 1, 0\right),$$

and one eigenvector for $(-11+\sqrt{1201})/2$ is

$$\left(\frac{-9+\sqrt{1201}}{56}, 1, 0\right).$$

(c) Along the z-axis, the equilibrium point at the origin is a one-dimensional sink. In the xy-phase plane, the equilibrium point at the origin is a two-dimensional saddle.

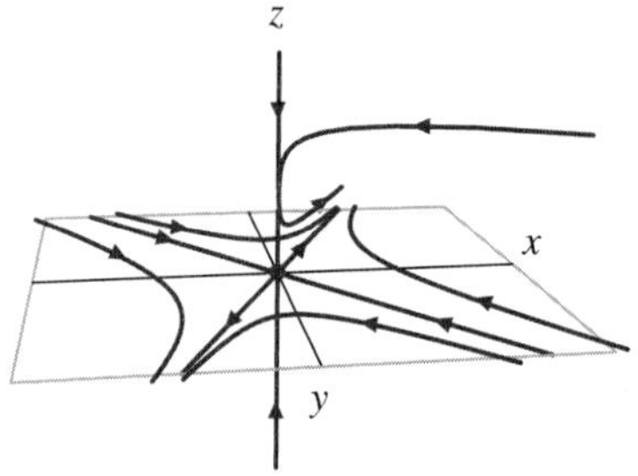

(d) Since the system decouples, we can treat the z-axis and the xy-plane separately, and then combine the pictures. Of course, visualizing this combination is easier said than done. Nevertheless, splitting the picture into vertical and horizontal ones does help.

19. **(a)** If Glen makes a profit, z is positive. Since the coefficients of z in the equations for dx/dt and dy/dt are positive, these terms contribute positively to dx/dt and dy/dt. In other words, Glen's profitability helps Paul and Bob be profitable (and they need all the help they can get).

(b) Since dz/dt does not have either an x or a y term, the values of x and y do not contribute to dz/dt. Hence, the profitability of either Paul or Bob makes no difference to Glen's future profits.

20. In matrix form, the system is

$$\frac{d\mathbf{Y}}{dt} = \begin{pmatrix} 0 & -1 & 1 \\ -1 & 0 & 1 \\ 0 & 0 & 1 \end{pmatrix} \mathbf{Y}.$$

The characteristic polynomial is

$$-\lambda(-\lambda)(1-\lambda) + 1(-1)(1-\lambda) + 1(0) = (1-\lambda)(\lambda^2-1).$$

Hence, there are only two eigenvalues, 1 and -1. Since some of the eigenvalues are positive and some are negative, this system is a saddle.

21. **(a)** For $z = 0$, $dz/dt = 0$ and, therefore, $z(t) = 0$ for all t. Consequently, we can analyze this system as if it has only two dependent variables, x and y. We have

$$\frac{dx}{dt} = -y$$
$$\frac{dy}{dt} = -x.$$

This system is a saddle in the xy-plane.

The eigenvalues for the xy-system are ± 1, and the eigenvectors for the eigenvalue -1 satisfy the equation $x = y$. Since we are assuming $x(0) = y(0)$, the given solution tends to the origin along the line $x = y$ in the xy-plane.

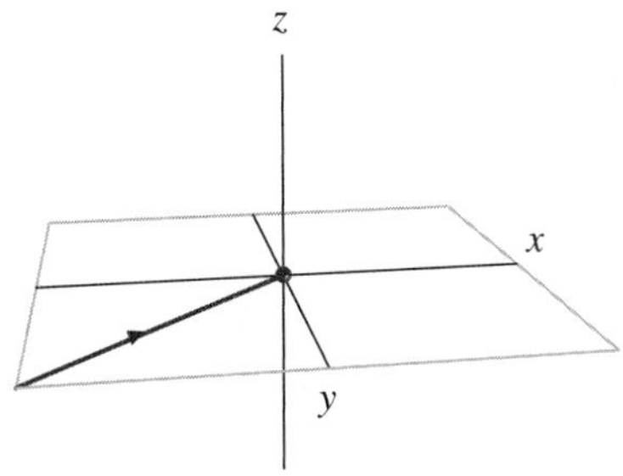

(b) Since $x(0) = y(0)$ and $z(0) = 0$ is an initial condition that is an eigenvector associated to the eigenvalue -1, we know that $x(t) = y(t)$ and $z(t) = 0$ for all t. We also know that both $x(t)$ and $y(t)$ decay to 0 like the function e^{-t}.

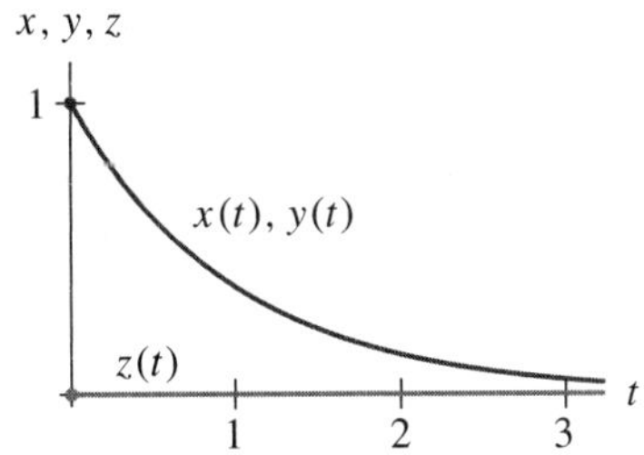

The graphs of $x(t)$ and $y(t)$ are identical, and $z(t) = 0$ for all t.

(c) Glen continues to break even. Both Paul's and Bob's profits tend to the break-even point as $t \to \infty$.

22. **(a)** The equation for dz/dt does not depend on x or y. Hence, we can solve for $z(t)$ using our knowledge of first-order equations. The general solution is $z(t) = z_0 e^t$. Thus, $z(t)$ grows exponentially.

Unfortunately, it is not so easy to see what happens to $x(t)$ and $y(t)$. One way to study these solutions is to use a numerical method (such as Euler's method) to approximate $x(t)$ and $y(t)$ using the equations

$$\frac{dx}{dt} = -y + z_0 e^t$$
$$\frac{dy}{dt} = -x + z_0 e^t.$$

Using a numerical solver, we obtain the solution curve

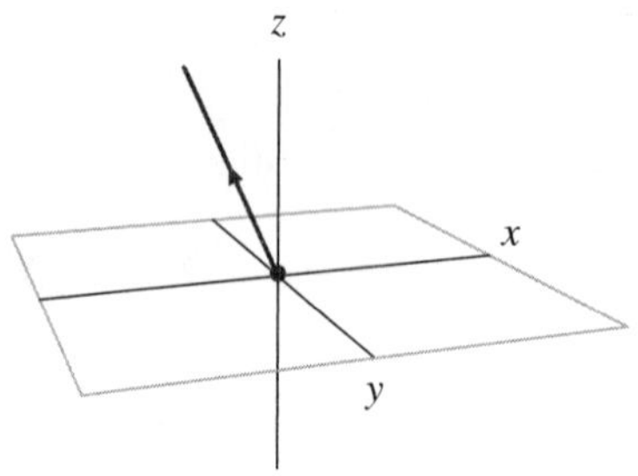

(b) These numerical results can also be used to create the $x(t)$-, $y(t)$- and $z(t)$-graphs for this solution.

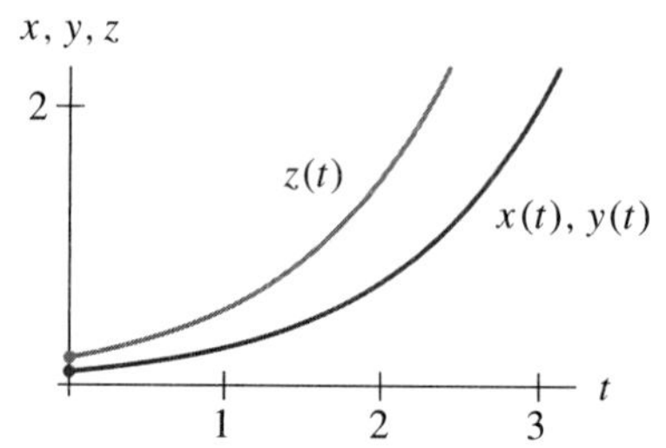

By examining the equations and the initial conditions, we see that $x(t) = y(t)$, and both functions are roughly equal to $z(t)/2$. This observation leads us to consider a function of the form $\mathbf{Y}(t) = (x(t), y(t), z(t)) = z_0 e^t (1, 1, 2)$. By substituting this function into the equations, we see that it is a solution. The given initial conditions $(x(0), y(0), z(0)) = (0, 0, \epsilon)$ where ϵ is small are not satisfied by $\mathbf{Y}(t)$ no matter what z_0 is, but both our desired solution and $\mathbf{Y}(t)$ behave similarly as $t \to \infty$.

(c) All three stores have exponentially growing profits. Also, Glen's profits grow roughly twice as fast as those of the other two bozos.

REVIEW EXERCISES FOR CHAPTER 3

1. The characteristic polynomial is $(1-\lambda)(2-\lambda)$, so the eigenvalues are $\lambda = 1$ and $\lambda = 2$.

2. The characteristic polynomial is

$$(-\lambda)(-\lambda) - (1)(2) = \lambda^2 - 2,$$

so the eigenvalues are $\lambda = \pm\sqrt{2}$.

3. The system has eigenvalues -2 and 3. One eigenvector associated with $\lambda = 3$ is $(1, 0)$, and one eigenvector associated with $\lambda = -2$ is $(0, 1)$. The general solution is

$$\mathbf{Y}(t) = k_1 e^{3t} \begin{pmatrix} 1 \\ 0 \end{pmatrix} + k_2 e^{-2t} \begin{pmatrix} 0 \\ 1 \end{pmatrix}.$$

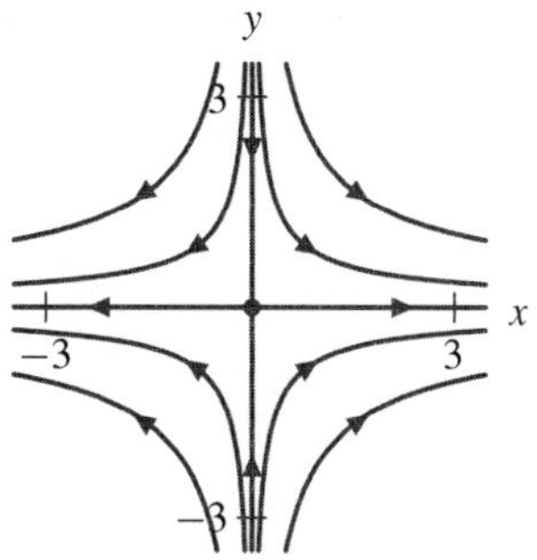

4. By definition, the zero vector, $\mathbf{Y}_1$, is never an eigenvector. We can check the others by computing $\mathbf{AY}$. For example,

$$\mathbf{AY}_2 = \mathbf{A}\begin{pmatrix} 2 \\ -2 \end{pmatrix} = \begin{pmatrix} 2 \\ -2 \end{pmatrix} = \mathbf{Y}_2,$$

so $\mathbf{Y}_2$ is an eigenvector (with eigenvalue $\lambda = 1$). On the other hand,

$$\mathbf{AY}_3 = \begin{pmatrix} 1 \\ 5 \end{pmatrix},$$

which is not a scalar multiple of $\mathbf{Y}_3$, so $\mathbf{Y}_3$ is not an eigenvector. Also, $\mathbf{AY}_4 = 3\mathbf{Y}_4$, so $\mathbf{Y}_4$ is an eigenvector (with eigenvalue $\lambda = 3$). Since we know that $\mathbf{Y}_2$ is an eigenvector and $\mathbf{Y}_5 = -2\mathbf{Y}_2$, $\mathbf{Y}_5$ is also an eigenvector. The vectors $\mathbf{Y}_2$ and $\mathbf{Y}_4$ are two linearly independent eigenvectors corresponding to different eigenvalues. Therefore, $\mathbf{Y}_6$ cannot be an eigenvector because it is neither a scalar multiple of $\mathbf{Y}_2$ nor $\mathbf{Y}_4$.

5. Note that $b \geq 0$ by assumption. The characteristic polynomial is

$$s^2 + bs + 5,$$

so the eigenvalues are

$$s = \frac{-b \pm \sqrt{b^2 - 20}}{2}.$$

If $b > \sqrt{20}$, the harmonic oscillator is overdamped. If $b = \sqrt{20}$, the harmonic oscillator is critically damped. If $0 < b < \sqrt{20}$, the harmonic oscillator is underdamped, and if $b = 0$, the harmonic oscillator is undamped.

6. Written in coordinates, the system is $dx/dt = 0$ and $dy/dt = x - y$. Hence, the equilibrium points are all points on the line $y = x$.

7. Every linear system has the origin as an equilibrium point, so the solution to the initial-value problem is the equilibrium solution $\mathbf{Y}(t) = (0, 0)$ for all t.

8. If $k > 0$, the general solution is

$$y(t) = c_1 \sin \sqrt{k}\, t + c_2 \cos \sqrt{k}\, t.$$

If $k < 0$, the general solution is

$$y(t) = c_1 e^{\sqrt{-k}\, t} + c_2 e^{-\sqrt{-k}\, t}.$$

The general solution for $k = 0$ is $y(t) = c_1 t + c_2$. Hence, only (b) and (d) are solutions under the given assumptions on k.

9. Letting $x(t) = 2\cos t$ and $y(t) = \sin t$, we have

$$\frac{dx}{dt} = \frac{d(2\cos t)}{dt} = -2\sin t = -2y$$

and

$$\frac{dy}{dt} = \frac{d(\sin t)}{dt} = \cos t = \frac{x}{2}.$$

Hence, $\mathbf{Y}(t)$ satisfies the linear system

$$\frac{d\mathbf{Y}}{dt} = \begin{pmatrix} 0 & -2 \\ 1/2 & 0 \end{pmatrix} \mathbf{Y}.$$

10. Written in terms of coordinates, the system is $dx/dt = y$ and $dy/dt = 0$. From the second equation, we see that $y(t) = k_2$, where k_2 is an arbitrary constant. Then $x(t) = k_2 t + k_1$, where k_1 is another arbitrary constant. In vector notation, the general solution is

$$\mathbf{Y}(t) = \begin{pmatrix} k_2 t + k_1 \\ k_2 \end{pmatrix}.$$

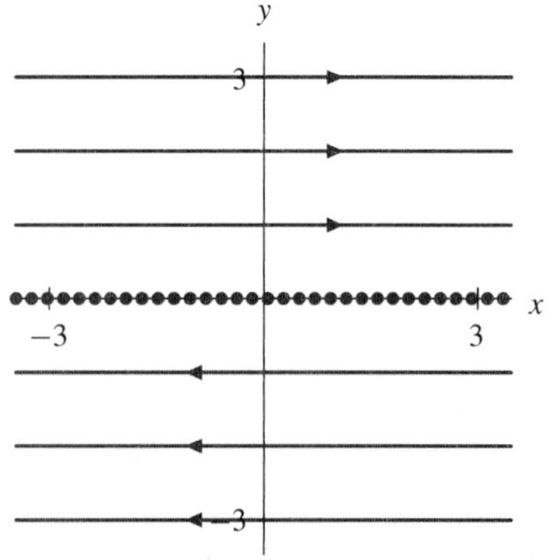

11. False. For example, the linear system

$$\frac{d\mathbf{Y}}{dt} = \begin{pmatrix} 3 & 0 \\ 0 & 0 \end{pmatrix} \mathbf{Y}$$

has a line of equilibria (the y-axis). Another example is the linear system

$$\frac{d\mathbf{Y}}{dt} = \begin{pmatrix} 0 & 0 \\ 0 & 0 \end{pmatrix} \mathbf{Y}.$$

Every point is an equilibrium point for this system.

12. True. If $\mathbf{A}$ is the matrix and λ is the eigenvalue associated to $\mathbf{Y}_0$, then

$$\mathbf{A}(k\mathbf{Y}_0) = k\mathbf{A}\mathbf{Y}_0 = k\lambda\mathbf{Y}_0 = \lambda(k\mathbf{Y}_0).$$

Consequently, $k\mathbf{Y}_0$ is an eigenvector as long as $k \neq 0$. (Note that $k = 0$ is excluded because the zero vector is never an eigenvector by definition.)

13. True. Linear systems have solutions that consist of just sine and cosine functions only when the eigenvalues are purely imaginary (that is, of the form $\pm i\omega$). In this case, the sine and cosine terms are of the form $\sin \omega t$ and $\cos \omega t$. For the first coordinate of $\mathbf{Y}(t)$ to be part of a solution, we would have to have $\omega = 2$, but the second coordinate would force $\omega = 1$. So this function cannot be the solution of a linear system.

14. False. The graph has $y(0) = 0$ and $y'(0) = 0$. However, these values are the initial conditions for the equilibrium solution $y(t) = 0$ for all t.

15. False. In the graph, the amount of time between consecutive crossings of the t-axis decreases as t increases. Even though solutions of underdamped harmonic oscillators oscillate, the amount of time between consecutive crossings of the t-axis is constant.

16. True. The period of solutions is $2\pi/\sqrt{k}$, so if k increases, then the period decreases. Consequently, the time between successive maxima decreases.

17. True. If the matrix has a real eigenvalue, then there the corresponding system has at least a line of eigenvectors. Any initial condition that is an eigenvector corresponds to a solution that stays on the line of eigenvectors for all time. Hence, a system with a real eigenvalue has infinitely many straight-line solutions.

18. True. The functions that arise in solutions of a linear system are linear combinations of $e^{\lambda t}$, $\cos \beta t$, $\sin \beta t$, and $te^{\lambda t}$. All of these functions are defined for all t. Consequently, all solutions of $d\mathbf{Y}/dt = \mathbf{AY}$ are defined for all t.

19. First, we compute the characteristic polynomials and eigenvalues for each matrix.

(i) The characteristic polynomial is $\lambda^2 + 1$, and the eigenvalues are $\lambda = \pm i$. Center.

(ii) The characteristic polynomial is $\lambda^2 + 2\lambda - 2$, and the eigenvalues are $\lambda = 1 \pm \sqrt{3}$. Saddle.

(iii) The characteristic polynomial is $\lambda^2 + 3\lambda + 1$, and the eigenvalues are $\lambda = (-3 \pm \sqrt{5})/2$. Sink.

(iv) The characteristic polynomial is $\lambda^2 + 1$, and the eigenvalues are $\lambda = \pm i$. Center.

(v) The characteristic polynomial is $\lambda^2 - \lambda - 2$, and the eigenvalues are $\lambda = -1$ and $\lambda = 2$. Saddle.

(vi) The characteristic polynomial is $\lambda^2 - 3\lambda + 1$, and the eigenvalues are $\lambda = (3 \pm \sqrt{5})/2$. Source.

(vii) The characteristic polynomial is $\lambda^2 + 4\lambda + 4$. The eigenvalue $\lambda = -2$ is a repeated eigenvalue. Sink.

(viii) The characteristic polynomial is $\lambda^2 + 2\lambda + 3$, and the eigenvalues are $\lambda = -1 \pm i\sqrt{2}$. Spiral sink.

Given this information, we can match the matrices with the phase portraits.

(a) This portrait is a center. There are two possibilities, (i) and (iv). At $(1, 0)$, the vector for (i) is $(1, -2)$, and the vector for (iv) is $(-1, -2)$. This phase portrait corresponds to matrix (iv).

(b) This portrait is a sink with two lines of eigenvectors. The only possibility is matrix (iii).

(c) This portrait is a saddle. The only possibilities are (ii) and (v). However, in (v), all vectors on the y-axis are eigenvectors corresponding to the eigenvalue $\lambda = -1$. Therefore, the phase portrait cannot correspond to (v).

(d) This portrait is a sink with a single line of eigenvectors. The only possibility is matrix (vii).

20. **(a)** The trace T is a, and the determinant D is $-3a$. Therefore, the curve in the trace-determinant plane is $D = -3T$.

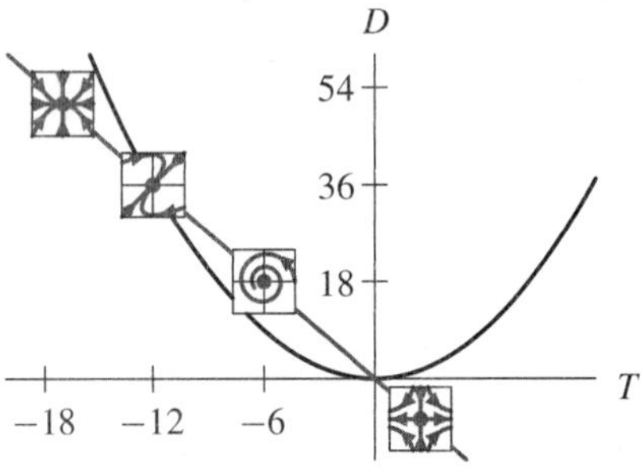

(b) The line $D = -3T$ crosses the parabola $T^2 - 4D = 0$ at two points—at $(T, D) = (-12, 36)$ if $a = -12$ and at $(T, D) = (0, 0)$ if $a = 0$. Therefore, bifucations occur at $a = -12$ and at $a = 0$. The portion of the line for which $a < -12$ corresponds to a positive determinant and a negative trace such that $T^2 - 4D < 0$. The corresponding phase portraits are real sinks. If $a = -12$, we have a sink with repeated eigenvalues. If $-12 < a < 0$, we have complex eigenvalues with negative real parts. Therefore, the phase portraits are spiral sinks. If $a = 0$, we have a degenerate case where the y-axis is an entire line of equilibrium points. Finally, if $a > 0$, the corresponding portion of the line is below the T-axis, and the phase portraits are saddles.

21. First, we compute the characteristic polynomials and eigenvalues for each matrix.

(i) The characteristic polynomial is $\lambda^2 - 3\lambda - 4$, and the eigenvalues are $\lambda = -1$ and $\lambda = 4$. Saddle.

(ii) The characteristic polynomial is $\lambda^2 - 7\lambda + 10$, and the eigenvalues are $\lambda = 2$ and $\lambda = 5$. Source.

(iii) The characteristic polynomial is $\lambda^2 + 4\lambda + 3$, and the eigenvalues are $\lambda = -3$ and $\lambda = -1$. Sink.

(iv) The characteristic polynomial is $\lambda^2 + 4$, and the eigenvalues are $\lambda = \pm 2i$. Center.

(v) The characteristic polynomial is $\lambda^2 + 9$, and the eigenvalues are $\lambda = \pm 3i$. Center.

(vi) The characteristic polynomial is $\lambda^2 - 2\lambda + \frac{15}{16}$, and the eigenvalues are $\lambda = 3/4$ and $\lambda = 5/4$. Source.

(vii) The characteristic polynomial is $\lambda^2 + 2.2\lambda + 5.21$. The eigenvalues are $\lambda = -1.1 \pm 2i$. Spiral sink.

(viii) The characteristic polynomial is $\lambda^2 + 0.2\lambda + 4.01$, and the eigenvalues are $\lambda = -0.1 \pm 2i$. Spiral sink.

Given this information, we can match the matrices with the $x(t)$- and $y(t)$-graphs.

(a) This solution approaches equilibrium without oscillating. Therefore, the system has at least one negative real eigenvalue. Matrices (i) and (iii) are the only matrices with a negative eigenvalue. Since matrix (i) corresponds to a saddle, its only solutions that approach equilibrium are straight line-solutions. However, this solution is not a straight-line solution because $y(t)/x(t)$ is not constant. This solution must correspond to matrix (iii).

(b) Note that $y(t) = -x(t)$ for all t. Therefore, this solution corresponds to a straight-line solution for a source or a saddle with eigenvector $(1, -1)$. Direct computation shows that $(1, -1)$ is not an eigenvector for matrices (i) and (ii), and it is an eigenvector corresponding to eigenvalue $\lambda = 3/4$ for matrix (vi).

(c) This solution is periodic. Therefore, the corresponding matrix has purely imaginary eigenvalues. Matrices (iv) and (v) are the only matrices with purely imaginary eigenvalues. The solution oscillates three times over any interval of length 2π. Hence, the period of the solution is $2\pi/3$. Therefore, the eigenvalues must be $\pm 3i$, and this solution corresponds to matrix (v).

(d) This solution oscillates as it approaches equilibrium. Therefore, the corresponding matrix has complex eigenvalues with a negative real part. Matrices (vii) and (viii) are the only possibilities. Since the real part of the eigenvalues for matrix (vii) is -1.1, its solutions decay at a rate of $e^{-1.1t}$. Similarly, the real part of the eigenvalues for matrix (viii) is -0.1, so its solutions decay at a rate of $e^{-0.1t}$. The rate of decay of the solution graphed is $e^{-0.1t}$. Consequently, these graphs correspond to matrix (viii).

22. **(a)** The system must have the line $y = 2x$ as a line of equilibria. There are infinitely many such systems. One is

$$\frac{dx}{dt} = 2x - y$$
$$\frac{dy}{dt} = 0.$$

(b) The matrix for the system must have the line $y = 2x$ as a line of eigenvectors. There are infinitely many such matrices. One is

$$\begin{pmatrix} 1 & 1 \\ 2 & 2 \end{pmatrix}.$$

Note that the eigenvalue corresponding to this line is 3, so the corresponding straight-line solution is

$$\mathbf{Y}(t) = e^{3t} \begin{pmatrix} -1 \\ -2 \end{pmatrix}.$$

23. The characteristic polynomial is

$$s^2 + 5s + 6,$$

so the eigenvalues are $s = -2$ and $s = -3$. Hence, the general solution is

$$y(t) = k_1 e^{-2t} + k_2 e^{-3t},$$

and we have

$$y'(t) = -2k_1e^{-2t} - 3k_2e^{-3t}.$$

From the initial conditions, we obtain the simultaneous equations

$$\begin{cases} k_1 + k_2 = 0 \\ -2k_1 - 3k_2 = 2. \end{cases}$$

Solving for k_1 and k_2 yields $k_1 = 2$ and $k_2 = -2$. Hence, the solution to our initial-value problem is $y(t) = 2e^{-2t} - 2e^{-3t}$.

24. The characteristic polynomial is

$$s^2 + 2s + 5,$$

so the eigenvalues are $s = -1 \pm 2i$. Hence, the general solution is

$$y(t) = k_1e^{-t}\cos 2t + k_2e^{-t}\sin 2t.$$

From the initial condition $y(0) = 3$, we see that $k_1 = 3$. Differentiating

$$y(t) = 3e^{-t}\cos 2t + k_2e^{-t}\sin 2t$$

and evaluating $y'(t)$ at $t = 0$ yields $y'(0) = -3 + 2k_2$. Since $y'(0) = -1$, we have $k_2 = 1$. Hence, the solution to our initial-value problem is

$$y(t) = 3e^{-t}\cos 2t + e^{-t}\sin 2t.$$

25. The characteristic polynomial is

$$s^2 + 2s + 1,$$

so $s = -1$ is a repeated eigenvalue. Hence, the general solution is

$$y(t) = k_1e^{-t} + k_2te^{-t}.$$

From the initial condition $y(0) = 1$, we see that $k_1 = 1$. Differentiating

$$y(t) = e^{-t} + k_2te^{-t}$$

and evaluating $y'(t)$ at $t = 0$ yields $y'(0) = -1 + k_2$. Since $y'(0) = 1$, we have $k_2 = 2$. Hence, the solution to our initial-value problem is

$$y(t) = e^{-t} + 2te^{-t}.$$

26. The characteristic polynomial is s^2+2, so the eigenvalues are $s = \pm i\sqrt{2}$. Hence, the general solution is

$$y(t) = k_1\cos\sqrt{2}\,t + k_2\sin\sqrt{2}\,t.$$

From the initial condition $y(0) = 3$, we see that $k_1 = 3$. Differentiating $y(t)$ and evaluating at $t = 0$, we get $y'(0) = \sqrt{2}\,k_2$. Since $y'(0) = -\sqrt{2}$, we have $k_2 = -1$. The solution to our initial-value problem is

$$y(t) = 3\cos\sqrt{2}\,t - \sin\sqrt{2}\,t.$$

27. **(a)** The characteristic polynomial is

$$(1-\lambda)(-1-\lambda)-3=\lambda^2-4$$

so the eigenvalues are $\lambda_1 = 2$ and $\lambda_2 = -2$. The equilibrium point at the origin is a saddle.

The eigenvectors for $\lambda_1 = 2$ satisfy the equations

$$\begin{cases} x+3y=2x \\ x-y=2y. \end{cases}$$

Consequently, the eigenvectors (x, y) for this eigenvalue satisfy $x = 3y$. The eigenvector $(3, 1)$ is one such point.

The eigenvectors for $\lambda_2 = -2$ satisfy the equations

$$\begin{cases} x+3y=-2x \\ x-y=-2y. \end{cases}$$

Consequently, the eigenvectors (x, y) for this eigenvalue satisfy $y = -x$. The eigenvector $(1, -1)$ is one such point.

Hence, the general solution of the system is

$$\mathbf{Y}(t) = k_1 e^{2t}\begin{pmatrix} 3 \\ 1 \end{pmatrix} + k_2 e^{-2t}\begin{pmatrix} 1 \\ -1 \end{pmatrix}.$$

(b)

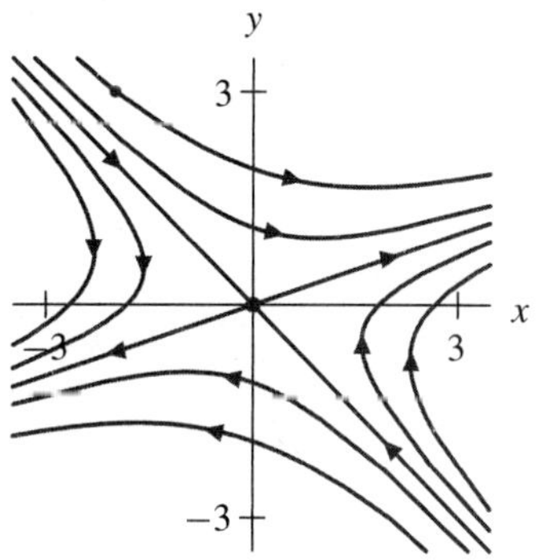

(c) To solve the initial-value problem, we solve for k_1 and k_2 in the equation

$$\begin{pmatrix} -2 \\ 3 \end{pmatrix} = \mathbf{Y}(0) = k_1\begin{pmatrix} 3 \\ 1 \end{pmatrix} + k_2\begin{pmatrix} 1 \\ -1 \end{pmatrix}.$$

This vector equation is equivalent to the two scalar equations

$$\begin{cases} 3k_1 + k_2 = -2 \\ k_1 - k_2 = 3, \end{cases}$$

so $k_1 = 1/4$ and $k_2 = -11/4$. The solution to the initial-value problem is

$$\mathbf{Y}(t) = \tfrac{1}{4}e^{2t}\begin{pmatrix} 3 \\ 1 \end{pmatrix} - \tfrac{11}{4}e^{-2t}\begin{pmatrix} 1 \\ -1 \end{pmatrix}$$

(d)

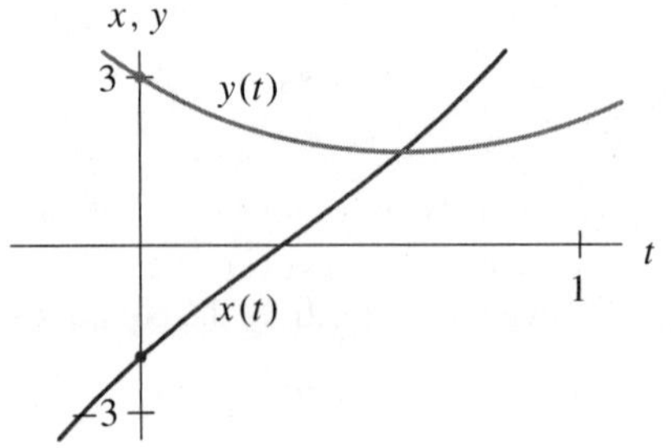

28. **(a)** The characteristic polynomial is

$$(4-\lambda)(3-\lambda) - 2 = \lambda^2 - 7\lambda + 10 = (\lambda - 5)(\lambda - 2),$$

so the eigenvalues are $\lambda_1 = 5$ and $\lambda_2 = 2$. The equilibrium point at the origin is a source.

The eigenvectors for $\lambda_1 = 5$ satisfy the equations

$$\begin{cases} 4x + 2y = 5x \\ x + 3y = 5y. \end{cases}$$

Consequently, the eigenvectors (x, y) for this eigenvalue satisfy $x = 2y$. The eigenvector $(2, 1)$ is one such point.

The eigenvectors for $\lambda_2 = 2$ satisfy the equations

$$\begin{cases} 4x + 2y = 2x \\ x + 3y = 2y. \end{cases}$$

Consequently, the eigenvectors (x, y) for this eigenvalue satisfy $y = -x$. The eigenvector $(1, -1)$ is one such point.

Hence, the general solution is

$$\mathbf{Y}(t) = k_1 e^{5t}\begin{pmatrix} 2 \\ 1 \end{pmatrix} + k_2 e^{2t}\begin{pmatrix} 1 \\ -1 \end{pmatrix}.$$

(b)

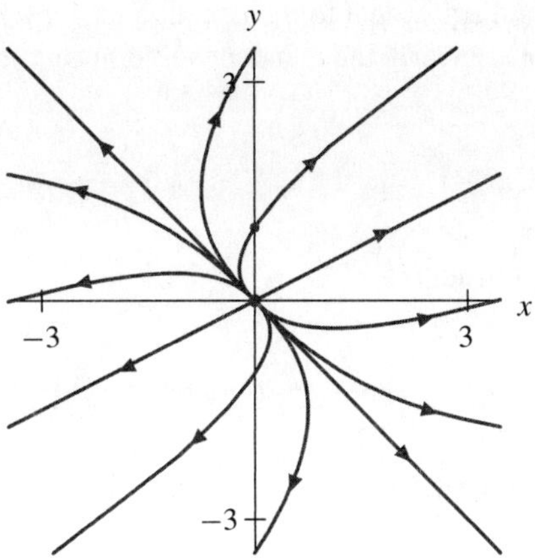

(c) To solve the initial-value problem, we solve for k_1 and k_2 in the equation

$$\begin{pmatrix} 0 \\ 1 \end{pmatrix} = \mathbf{Y}(0) = k_1 \begin{pmatrix} 2 \\ 1 \end{pmatrix} + k_2 \begin{pmatrix} 1 \\ -1 \end{pmatrix}.$$

This vector equation is equivalent to the two scalar equations

$$\begin{cases} 2k_1 + k_2 = 0 \\ k_1 - k_2 = 1, \end{cases}$$

so $k_1 = 1/3$ and $k_2 = -2/3$. The solution of the initial-value problem is

$$\mathbf{Y}(t) = \tfrac{1}{3}e^{5t} \begin{pmatrix} 2 \\ 1 \end{pmatrix} - \tfrac{2}{3}e^{2t} \begin{pmatrix} 1 \\ -1 \end{pmatrix}.$$

(d)

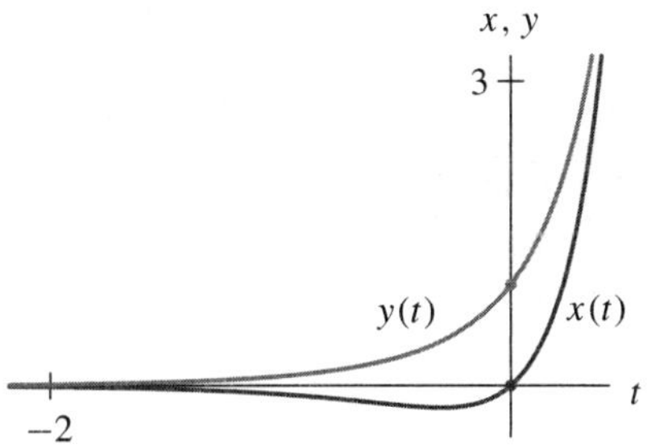

29. **(a)** The characteristic polynomial is

$$(-2 - \lambda)(2 - \lambda) + 6 = \lambda^2 + 2,$$

so the eigenvalues are $\lambda = \pm i\sqrt{2}$. The equilibrium point at the origin is a center.

The eigenvectors (x, y) corresponding to $\lambda = i\sqrt{2}$ are solutions of the equations

$$\begin{cases} -2x + 3y = i\sqrt{2}\,x \\ -2x + 2y = i\sqrt{2}\,y. \end{cases}$$

These equations are equivalent to the equation $(2 + i\sqrt{2})x = 3y$. Consequently, $(3, 2 + i\sqrt{2})$ is one eigenvector. Linearly independent solutions are given by the real and imaginary parts of

$$e^{i\sqrt{2}t}\begin{pmatrix} 3 \\ 2 + i\sqrt{2} \end{pmatrix},$$

which by Euler's formula is

$$(\cos\sqrt{2}t + i\sin\sqrt{2}t)\begin{pmatrix} 3 \\ 2 + i\sqrt{2} \end{pmatrix}.$$

Hence, the general solution is

$$\mathbf{Y}(t) = k_1\begin{pmatrix} 3\cos\sqrt{2}t \\ 2\cos\sqrt{2}t - \sqrt{2}\sin\sqrt{2}t \end{pmatrix} + k_2\begin{pmatrix} 3\sin\sqrt{2}t \\ \sqrt{2}\cos\sqrt{2}t + 2\sin\sqrt{2}t \end{pmatrix}.$$

(b)

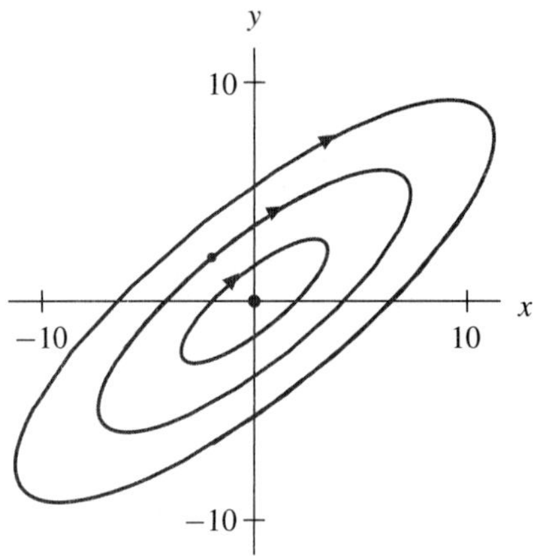

(c) To satisfy the initial condition, we solve

$$\begin{pmatrix} -2 \\ 2 \end{pmatrix} = \mathbf{Y}(0) = k_1\begin{pmatrix} 3 \\ 2 \end{pmatrix} + k_2\begin{pmatrix} 0 \\ \sqrt{2} \end{pmatrix},$$

which is equivalent to the two scalar equations

$$\begin{cases} 3k_1 = -2 \\ 2k_1 + \sqrt{2}k_2 = 2. \end{cases}$$

We get $k_1 = -2/3$ and $k_2 = 5\sqrt{2}/3$, and the solution to the initial-value problem is

$$\mathbf{Y}(t) = \begin{pmatrix} -2\cos\sqrt{2}t + 5\sqrt{2}\sin\sqrt{2}t \\ 2\cos\sqrt{2}t + 4\sqrt{2}\sin\sqrt{2}t \end{pmatrix}$$

(d)

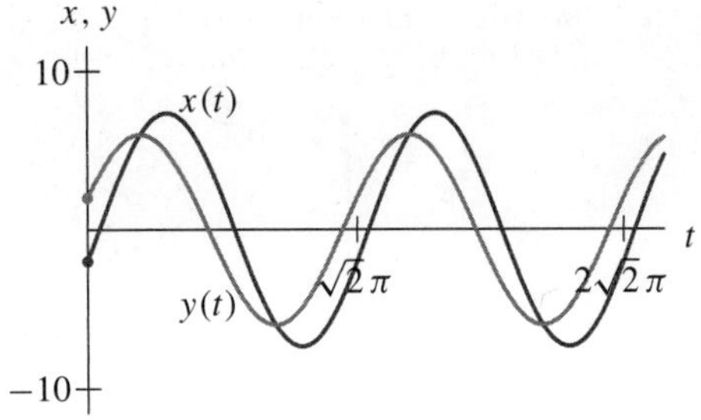

30. **(a)** The characteristic polynomial is

$$(-3-\lambda)(1-\lambda)+12=\lambda^2+2\lambda+9,$$

so the eigenvalues are $\lambda = -1 \pm 2\sqrt{2}\,i$. The equilibrium point at the origin is a spiral sink.

The eigenvectors (x, y) corresponding to $\lambda = -1 + 2\sqrt{2}\,i$ are solutions of the equations

$$\begin{cases} -3x + 6y = (-1 + 2\sqrt{2}\,i)x \\ -2x + y = (-1 + 2\sqrt{2}\,i)y. \end{cases}$$

These equations are equivalent to the equation $3y = (1 + \sqrt{2}\,i)x$. Consequently, $(3, 1 + \sqrt{2}\,i)$ is one eigenvector. Linearly independent solutions are given by the real and imaginary parts of

$$\mathbf{Y}_c(t) = e^{(-1+2\sqrt{2}\,i)t}\begin{pmatrix} 3 \\ 1 + \sqrt{2}\,i \end{pmatrix}.$$

Hence, the general solution is

$$\mathbf{Y}(t) = k_1 e^{-t}\begin{pmatrix} 3\cos 2\sqrt{2}\,t \\ \cos 2\sqrt{2}\,t - \sqrt{2}\sin 2\sqrt{2}\,t \end{pmatrix} + k_2 e^{-t}\begin{pmatrix} 3\sin 2\sqrt{2}\,t \\ \sqrt{2}\cos 2\sqrt{2}\,t + \sin 2\sqrt{2}\,t \end{pmatrix}.$$

(b)

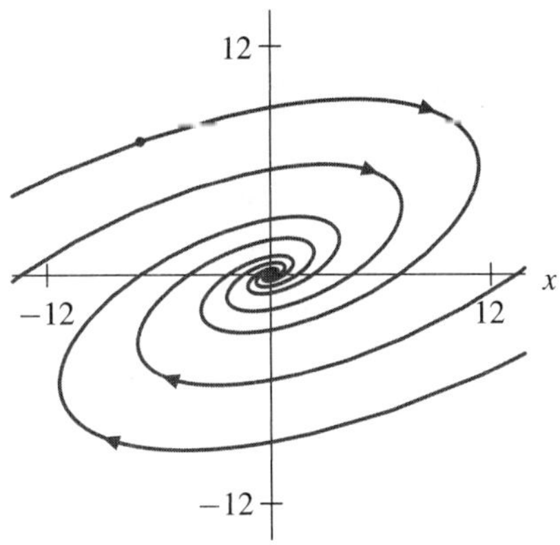

(c) To satisfy the initial condition, we solve

$$\begin{pmatrix} -7 \\ 7 \end{pmatrix} = \mathbf{Y}(0) = k_1\begin{pmatrix} 3 \\ 1 \end{pmatrix} + k_2\begin{pmatrix} 0 \\ \sqrt{2} \end{pmatrix},$$

which is equivalent to the two scalar equations

$$\begin{cases} 3k_1 = -7 \\ k_1 + \sqrt{2}k_2 = 7. \end{cases}$$

We get $k_1 = -7/3$ and $k_2 = 14\sqrt{2}/3$. The solution of the initial-value problem is

$$\mathbf{Y}(t) = -\frac{7}{3}e^{-t}\begin{pmatrix} 3\cos 2\sqrt{2}\,t \\ \cos 2\sqrt{2}\,t - \sqrt{2}\sin 2\sqrt{2}\,t \end{pmatrix} + \frac{14\sqrt{2}}{3}e^{-t}\begin{pmatrix} 3\sin 2\sqrt{2}\,t \\ \sqrt{2}\cos 2\sqrt{2}\,t + \sin 2\sqrt{2}\,t \end{pmatrix}$$

$$= e^{-t}\begin{pmatrix} -7\cos 2\sqrt{2}\,t + 14\sqrt{2}\,\sin 2\sqrt{2}\,t \\ 7\cos 2\sqrt{2}\,t + 7\sqrt{2}\,\sin 2\sqrt{2}\,t \end{pmatrix}.$$

(d)

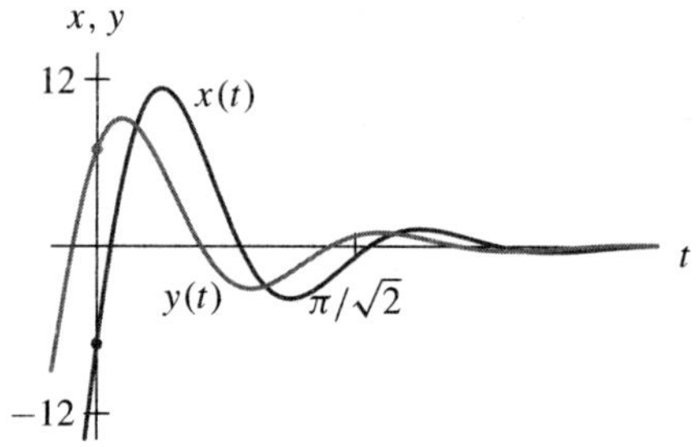

31. **(a)** The characteristic polynomial is

$$(-3 - \lambda)(-1 - \lambda) + 1 = (\lambda + 2)^2,$$

so $\lambda = -2$ is a repeated eigenvalue. The equilibrium point at the origin is a sink.

To find the general solution, we start with an arbitrary initial condition (x_0, y_0), and we calculate

$$\begin{pmatrix} -1 & 1 \\ -1 & 1 \end{pmatrix}\begin{pmatrix} x_0 \\ y_0 \end{pmatrix} = \begin{pmatrix} y_0 - x_0 \\ y_0 - x_0 \end{pmatrix}.$$

We obtain the general solution

$$\mathbf{Y}(t) = e^{-2t}\begin{pmatrix} x_0 \\ y_0 \end{pmatrix} + te^{-2t}\begin{pmatrix} y_0 - x_0 \\ y_0 - x_0 \end{pmatrix}.$$

(b)

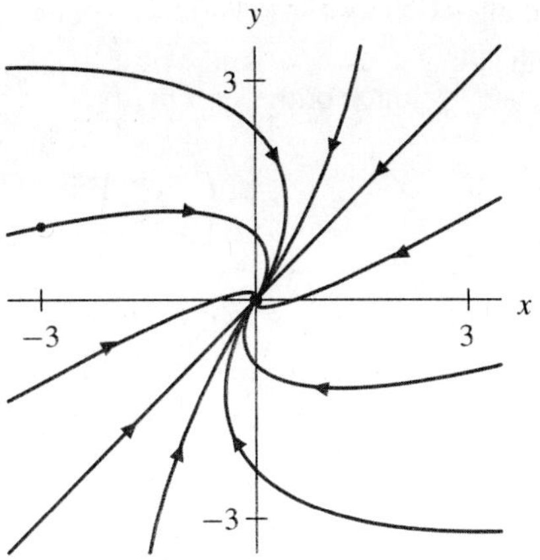

(c) The solution that satisfies the initial condition $(x_0, y_0) = (-3, 1)$ is

$$\mathbf{Y}(t) = e^{-2t}\begin{pmatrix} -3 \\ 1 \end{pmatrix} + te^{-2t}\begin{pmatrix} 4 \\ 4 \end{pmatrix} = e^{-2t}\begin{pmatrix} 4t - 3 \\ 4t + 1 \end{pmatrix}.$$

(d)

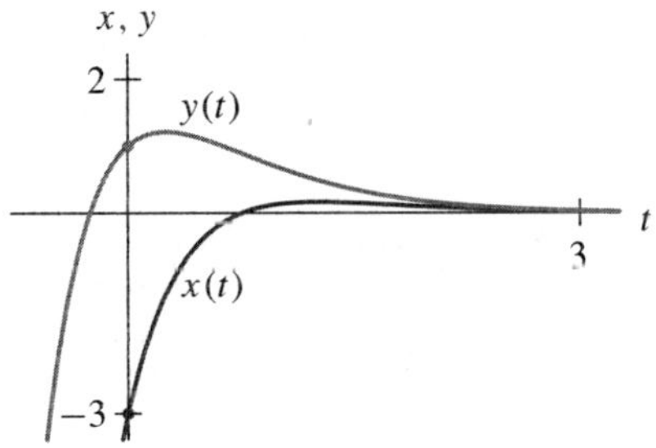

32. **(a)** The characteristic polynomial is

$$(0 - \lambda)(-3 - \lambda) + 2 = \lambda^2 + 3\lambda + 2,$$

so the eigenvalues are $\lambda_1 = -2$ and $\lambda_2 = -1$. The equilibrium point at the origin is a sink.

The eigenvectors for $\lambda_1 = -2$ satisfy the equations

$$\begin{cases} y = -2x \\ -2x - 3y = -2y. \end{cases}$$

Consequently, the eigenvectors (x, y) for this eigenvalue satisfy $y = -2x$. The eigenvector $(1, -2)$ is one such point.

The eigenvectors for $\lambda_2 = -1$ satisfy the equations

$$\begin{cases} y = -x \\ -2x - 3y = -y. \end{cases}$$

Consequently, the eigenvectors (x, y) for this eigenvalue satisfy $y = -x$. The eigenvector $(1, -1)$ is one such point.

Hence, the general solution of the system is

$$\mathbf{Y}(t) = k_1 e^{-2t} \begin{pmatrix} 1 \\ -2 \end{pmatrix} + k_2 e^{-t} \begin{pmatrix} 1 \\ -1 \end{pmatrix}.$$

(b)

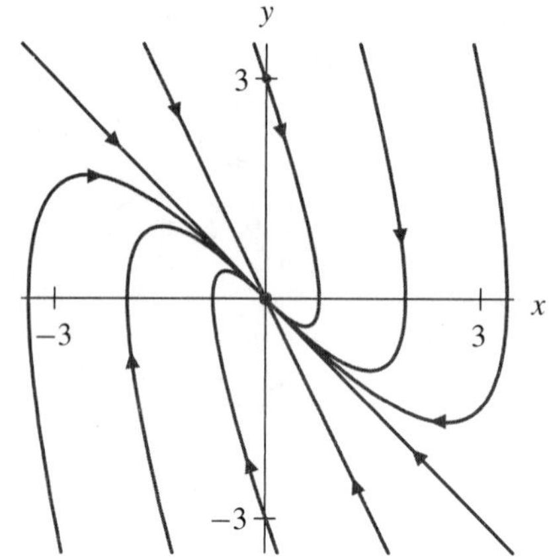

(c) To solve the initial-value problem, we solve for k_1 and k_2 in the equation

$$\begin{pmatrix} 0 \\ 3 \end{pmatrix} = \mathbf{Y}(0) = k_1 \begin{pmatrix} 1 \\ -2 \end{pmatrix} + k_2 \begin{pmatrix} 1 \\ -1 \end{pmatrix}.$$

This vector equation is equivalent to the two scalar equations

$$\begin{cases} k_1 + k_2 = 0 \\ -2k_1 - k_2 = 3, \end{cases}$$

so $k_1 = -3$ and $k_2 = 3$. The solution of the initial-value problem is

$$\mathbf{Y}(t) = -3e^{-2t} \begin{pmatrix} 1 \\ -2 \end{pmatrix} + 3e^{-t} \begin{pmatrix} 1 \\ -1 \end{pmatrix}.$$

(d)

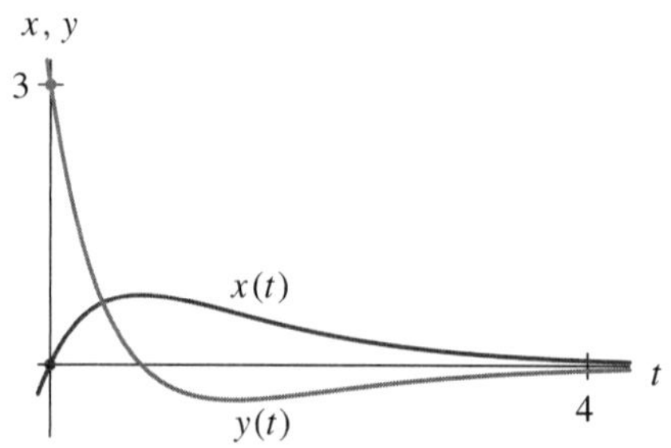

CHAPTER 4

Forcing and Resonance

EXERCISES FOR SECTION 4.1

1. To compute the general solution of the unforced equation, we use the method of Section 3.6. The characteristic polynomial is

$$s^2 - s - 6,$$

so the eigenvalues are $s = -2$ and $s = 3$. Hence, the general solution of the homogeneous equation is

$$k_1 e^{-2t} + k_2 e^{3t}.$$

To find a particular solution of the forced equation, we guess $y_p(t) = ke^{4t}$. Substituting into the left-hand side of the differential equation gives

$$\begin{aligned} \frac{d^2 y_p}{dt^2} - \frac{dy_p}{dt} - 6y_p &= 16ke^{4t} - 4ke^{4t} - 6ke^{4t} \\ &= 6ke^{4t}. \end{aligned}$$

In order for $y_p(t)$ to be a solution of the forced equation, we must take $k = 1/6$. The general solution of the forced equation is

$$y(t) = k_1 e^{-2t} + k_2 e^{3t} + \tfrac{1}{6} e^{4t}.$$

2. To compute the general solution of the unforced equation, we use the method of Section 3.6. The characteristic polynomial is

$$s^2 + 6s + 8,$$

so the eigenvalues are $s = -2$ and $s = -4$. Hence, the general solution of the homogeneous equation is

$$k_1 e^{-2t} + k_2 e^{-4t}.$$

To find a particular solution of the forced equation, we guess $y_p(t) = ke^{-3t}$. Substituting into the left-hand side of the differential equation gives

$$\begin{aligned} \frac{d^2 y_p}{dt^2} + 6\frac{dy_p}{dt} + 8y_p &= 9ke^{-3t} - 18ke^{-3t} + 8ke^{-3t} \\ &= -ke^{-3t}. \end{aligned}$$

In order for $y_p(t)$ to be a solution of the forced equation, we must take $k = -2$. The general solution of the forced equation is

$$y(t) = k_1 e^{-2t} + k_2 e^{-4t} - 2e^{-3t}.$$

3. To compute the general solution of the unforced equation, we use the method of Section 3.6. The characteristic polynomial is

$$s^2 - s - 2,$$

so the eigenvalues are $s = -1$ and $s = 2$. Hence, the general solution of the homogeneous equation is

$$k_1 e^{-t} + k_2 e^{2t}.$$

To find a particular solution of the forced equation, we guess $y_p(t) = ke^{3t}$. Substituting into the left-hand side of the differential equation gives

$$\begin{aligned} \frac{d^2 y_p}{dt^2} - \frac{dy_p}{dt} - 2y_p &= 9ke^{3t} - 3ke^{3t} - 2ke^{3t} \\ &= 4ke^{3t}. \end{aligned}$$

In order for $y_p(t)$ to be a solution of the forced equation, we must take $k = 5/4$. The general solution of the forced equation is

$$y(t) = k_1 e^{-t} + k_2 e^{2t} + \tfrac{5}{4} e^{3t}.$$

4. To compute the general solution of the unforced equation, we use the method of Section 3.6. The characteristic polynomial is

$$s^2 + 4s + 13,$$

so the eigenvalues are $s = -2 \pm 3i$. Hence, the general solution of the homogeneous equation is

$$k_1 e^{-2t} \cos 3t + k_2 e^{-2t} \sin 3t.$$

To find a particular solution of the forced equation, we guess $y_p(t) = ke^{-t}$. Substituting into the left-hand side of the differential equation gives

$$\begin{aligned} \frac{d^2 y_p}{dt^2} + 4\frac{dy_p}{dt} + 13y_p &= ke^{-t} - 4ke^{-t} + 13ke^{-t} \\ &= 10ke^{-t}. \end{aligned}$$

In order for $y_p(t)$ to be a solution of the forced equation, we must take $k = 1/10$. The general solution of the forced equation is

$$y(t) = k_1 e^{-2t} \cos 3t + k_2 e^{-2t} \sin 3t + \tfrac{1}{10} e^{-t}.$$

5. To compute the general solution of the unforced equation, we use the method of Section 3.6. The characteristic polynomial is

$$s^2 + 4s + 13,$$

so the eigenvalues are $s = -2 \pm 3i$. Hence, the general solution of the homogeneous equation is

$$k_1 e^{-2t} \cos 3t + k_2 e^{-2t} \sin 3t.$$

To find a particular solution of the forced equation, we guess $y_p(t) = ke^{-2t}$. Substituting into the left-hand side of the differential equation gives

$$\begin{aligned} \frac{d^2 y_p}{dt^2} + 4\frac{dy_p}{dt} + 13y_p &= 4ke^{-2t} - 8ke^{-2t} + 13ke^{-2t} \\ &= 9ke^{-2t}. \end{aligned}$$

In order for $y_p(t)$ to be a solution of the forced equation, we must take $k = -1/3$. The general solution of the forced equation is

$$y(t) = k_1 e^{-2t} \cos 3t + k_2 e^{-2t} \sin 3t - \tfrac{1}{3} e^{-2t}.$$

6. To compute the general solution of the unforced equation, we use the method of Section 3.6. The characteristic polynomial is

$$s^2 + 7s + 10,$$

so the eigenvalues are $s = -2$ and $s = -5$. Hence, the general solution of the homogeneous equation is

$$k_1 e^{-2t} + k_2 e^{-5t}.$$

To find a particular solution of the forced equation, a reasonable looking guess is $y_p(t) = ke^{-2t}$. However, this guess is a solution of the homogeneous equation, so it is doomed to fail. We make the standard second guess of $y_p(t) = kte^{-2t}$. Substituting into the left-hand side of the differential equation gives

$$\begin{aligned}\frac{d^2y_p}{dt^2} + 7\frac{dy_p}{dt} + 10y_p &= (-4ke^{-2t} + 4kte^{-2t}) + 7(ke^{-2t} - 2kte^{-2t}) + 10kte^{-2t} \\ &= 3ke^{-2t}.\end{aligned}$$

In order for $y_p(t)$ to be a solution of the forced equation, we must take $k = 1/3$. The general solution of the forced equation is

$$y(t) = k_1 e^{-2t} + k_2 e^{-5t} + \tfrac{1}{3}te^{-2t}.$$

7. To compute the general solution of the unforced equation, we use the method of Section 3.6. The characteristic polynomial is

$$s^2 - 5s + 4,$$

so the eigenvalues are $s = 1$ and $s = 4$. Hence, the general solution of the homogeneous equation is

$$k_1 e^t + k_2 e^{4t}.$$

To find a particular solution of the forced equation, a reasonable looking guess is $y_p(t) = ke^{4t}$. However, this guess is a solution of the homogeneous equation, so it is doomed to fail. We make the standard second guess of $y_p(t) = kte^{4t}$. Substituting into the left-hand side of the differential equation gives

$$\begin{aligned}\frac{d^2y_p}{dt^2} - 5\frac{dy_p}{dt} + 4y_p &= (8ke^{4t} + 16kte^{4t}) - 5(ke^{4t} + 4kte^{4t}) + 4kte^{4t} \\ &= 3ke^{4t}.\end{aligned}$$

In order for $y_p(t)$ to be a solution of the forced equation, we must take $k = 1/3$. The general solution of the forced equation is

$$y(t) = k_1 e^t + k_2 e^{4t} + \tfrac{1}{3}te^{4t}.$$

8. To compute the general solution of the unforced equation, we use the method of Section 3.6. The characteristic polynomial is

$$s^2 + s - 6,$$

so the eigenvalues are $s = -3$ and $s = 2$. Hence, the general solution of the homogeneous equation is

$$k_1 e^{-3t} + k_2 e^{2t}.$$

To find a particular solution of the forced equation, a reasonable looking guess is $y_p(t) = ke^{-3t}$. However, this guess is a solution of the homogeneous equation, so it is doomed to fail. We make the standard second guess of $y_p(t) = kte^{-3t}$. Substituting into the left-hand side of the differential equation gives

$$\begin{aligned}\frac{d^2y_p}{dt^2} + \frac{dy_p}{dt} - 6y_p &= (-6ke^{-3t} + 9kte^{-3t}) + (ke^{-3t} - 3kte^{-3t}) - 6kte^{-3t} \\ &= -5ke^{-3t}.\end{aligned}$$

In order for $y_p(t)$ to be a solution of the forced equation, we must take $k = -4/5$. The general solution of the forced equation is

$$y(t) = k_1e^{-3t} + k_2e^{2t} - \tfrac{4}{5}te^{-3t}.$$

9. First we derive the general solution. The characteristic polynomial is

$$s^2 + 6s + 8,$$

so the eigenvalues are $s = -2$ and $s = -4$. To find the general solution of the forced equation, we also need a particular solution. We guess $y_p(t) = ke^{-t}$ and find that $y_p(t)$ is a solution only if $k = 1/3$. Therefore, the general solution is

$$y(t) = k_1e^{-2t} + k_2e^{-4t} + \tfrac{1}{3}e^{-t}.$$

To find the solution with the initial conditions $y(0) = y'(0) = 0$, we compute

$$y'(t) = -2k_1e^{-2t} - 4k_2e^{-4t} - \tfrac{1}{3}e^{-t}.$$

Then we evaluate at $t = 0$ and obtain the simultaneous equations

$$\begin{cases} k_1 + k_2 + \frac{1}{3} = 0 \\ -2k_1 - 4k_2 - \frac{1}{3} = 0. \end{cases}$$

Solving, we have $k_1 = -1/2$ and $k_2 = 1/6$, so the solution of the initial-value problem is

$$y(t) = -\tfrac{1}{2}e^{-2t} + \tfrac{1}{6}e^{-4t} + \tfrac{1}{3}e^{-t}.$$

10. First we derive the general solution. The characteristic polynomial is

$$s^2 + 7s + 12,$$

so the eigenvalues are $s = -3$ and $s = -4$. To find the general solution of the forced equation, we also need a particular solution. We guess $y_p(t) = ke^{-t}$ and find that $y_p(t)$ is a solution only if $k = 1/2$. Therefore, the general solution is

$$y(t) = k_1e^{-3t} + k_2e^{-4t} + \tfrac{1}{2}e^{-t}.$$

To find the solution with the initial conditions $y(0) = 2$ and $y'(0) = 1$, we compute

$$y'(t) = -3k_1 e^{-3t} - 4k_2 e^{-4t} - \tfrac{1}{2}e^{-t}.$$

Then we evaluate at $t = 0$ and obtain the simultaneous equations

$$\begin{cases} k_1 + k_2 + \frac{1}{2} = 2 \\ -3k_1 - 4k_2 - \frac{1}{2} = 1. \end{cases}$$

Solving, we have $k_1 = 15/2$ and $k_2 = -6$, so the solution of the initial-value problem is

$$y(t) = \tfrac{15}{2}e^{-3t} - 6e^{-4t} + \tfrac{1}{2}e^{-t}.$$

11. This is the same equation as Exercise 5. The general solution is

$$y(t) = k_1 e^{-2t}\cos 3t + k_2 e^{-2t}\sin 3t - \tfrac{1}{3}e^{-2t}.$$

To find the solution with the initial conditions $y(0) = y'(0) = 0$, we compute

$$y'(t) = -2k_1 e^{-2t}\cos 3t - 3k_1 e^{-2t}\sin 3t - 2k_2 e^{-2t}\sin 3t + 3k_2 e^{-2t}\cos 3t + \tfrac{2}{3}e^{-2t}.$$

Then we evaluate at $t = 0$ and obtain the simultaneous equations

$$\begin{cases} k_1 - \frac{1}{3} = 0 \\ -2k_1 + 3k_2 + \frac{2}{3} = 0. \end{cases}$$

Solving, we have $k_1 = 1/3$ and $k_2 = 0$, so the solution of the initial-value problem is

$$y(t) = \tfrac{1}{3}e^{-2t}\cos 3t - \tfrac{1}{3}e^{-2t}.$$

12. This is the same equation as Exercise 6. The general solution is

$$y(t) = k_1 e^{-2t} + k_2 e^{-5t} + \tfrac{1}{3}te^{-2t}.$$

To find the solution with the initial conditions $y(0) = y'(0) = 0$, we compute

$$y'(t) = -2k_1 e^{-2t} - 5k_2 e^{-5t} + \tfrac{1}{3}e^{-2t} - \tfrac{2}{3}te^{-2t}.$$

Then we evaluate at $t = 0$ and obtain the simultaneous equations

$$\begin{cases} k_1 + k_2 = 0 \\ -2k_1 - 5k_2 + \frac{1}{3} = 0. \end{cases}$$

Solving, we have $k_1 = -1/9$ and $k_2 = 1/9$, so the solution of the initial-value problem is

$$y(t) = -\tfrac{1}{9}e^{-2t} + \tfrac{1}{9}e^{-5t} + \tfrac{1}{3}te^{-2t}.$$

13. **(a)** The characteristic polynomial of the unforced equation is

$$s^2 + 4s + 3.$$

So the eigenvalues are $s = -1$ and $s = -3$, and the general solution of the unforced equation is

$$k_1 e^{-t} + k_2 e^{-3t}.$$

To find a particular solution of the forced equation, we guess $y_p(t) = ke^{-t/2}$. Substituting $y_p(t)$ into the left-hand side of the differential equation gives

$$\begin{aligned}\frac{d^2 y_p}{dt^2} + 4\frac{dy_p}{dt} + 3y_p &= \tfrac{1}{4}ke^{-t/2} - 2ke^{-t/2} + 3ke^{-t/2} \\ &= \tfrac{5}{4}ke^{-t/2}.\end{aligned}$$

So $k = 4/5$ yields a solution of the forced equation.

The general solution of the forced equation is therefore

$$y(t) = k_1 e^{-t} + k_2 e^{-3t} + \tfrac{4}{5}e^{-t/2}.$$

(b) The derivative of the general solution is

$$y'(t) = -k_1 e^{-t} - 3k_2 e^{-3t} - \tfrac{2}{5}e^{-t/2}.$$

To find the solution with $y(0) = y'(0) = 0$, we evaluate at $t = 0$ and obtain the simultaneous equations

$$\begin{cases} k_1 + k_2 + \frac{4}{5} = 0 \\ -k_1 - 3k_2 - \frac{2}{5} = 0. \end{cases}$$

Solving, we find that $k_1 = -1$ and $k_2 = 1/5$, so the solution of the initial-value problem is

$$y(t) = -e^{-t} + \tfrac{1}{5}e^{-3t} + \tfrac{4}{5}e^{-t/2}.$$

(c) Every solution tends to zero as t increases. Of the three terms that sum to the general solution, $\frac{4}{5}e^{-t/2}$ dominates when t is large, so all solutions are approximately $\frac{4}{5}e^{-t/2}$ for t large.

14. **(a)** The characteristic polynomial of the unforced equation is

$$s^2 + 4s + 3.$$

So the eigenvalues are $s = -1$ and $s = -3$, and the general solution of the unforced equation is

$$k_1 e^{-t} + k_2 e^{-3t}.$$

To find a particular solution of the forced equation, we guess $y_p(t) = ke^{-2t}$. Substituting $y_p(t)$ into the left-hand side of the differential equation gives

$$\begin{aligned}\frac{d^2 y_p}{dt^2} + 4\frac{dy_p}{dt} + 3y_p &= 4ke^{-2t} - 8ke^{-2t} + 3ke^{-2t} \\ &= -ke^{-2t}.\end{aligned}$$

So $k = -1$ yields a solution of the forced equation.

The general solution of the forced equation is therefore

$$y(t) = k_1 e^{-t} + k_2 e^{-3t} - e^{-2t}.$$

(b) The derivative of the general solution is

$$y'(t) = -k_1 e^{-t} - 3k_2 e^{-3t} + 2e^{-2t}.$$

To find the solution with $y(0) = y'(0) = 0$, we evaluate at $t = 0$ and obtain the simultaneous equations

$$\begin{cases} k_1 + k_2 - 1 = 0 \\ -k_1 - 3k_2 + 2 = 0. \end{cases}$$

Solving, we find that $k_1 = 1/2$ and $k_2 = 1/2$, so the solution of the initial-value problem is

$$y(t) = \tfrac{1}{2}e^{-t} + \tfrac{1}{2}e^{-3t} - e^{-2t}.$$

(c) In the general solution, all three terms tend to zero, so the solution tends to zero. We can say a little more by noting that the term $k_1 e^{-t}$ is much larger (provided $k_1 \neq 0$). Hence, most solutions tend to zero at the rate of e^{-t}. If $k_1 = 0$, then solutions tend to zero at the rate of e^{-3t} provided $k_2 \neq 0$.

15. **(a)** The characteristic polynomial of the unforced equation is

$$s^2 + 4s + 3.$$

So the eigenvalues are $s = -1$ and $s = -3$, and the general solution of the unforced equation is

$$k_1 e^{-t} + k_2 e^{-3t}.$$

To find a particular solution of the forced equation, we guess $y_p(t) = ke^{-4t}$. Substituting $y_p(t)$ into the left-hand side of the differential equation gives

$$\begin{aligned} \frac{d^2y_p}{dt^2} + 4\frac{dy_p}{dt} + 3y_p &= 16ke^{-4t} - 16ke^{-4t} + 3ke^{-4t} \\ &= 3ke^{-4t}. \end{aligned}$$

So $k = 1/3$ yields a solution of the forced equation.

The general solution of the forced equation is therefore

$$y(t) = k_1 e^{-t} + k_2 e^{-3t} + \tfrac{1}{3}e^{-4t}.$$

(b) The derivative of the general solution is

$$y'(t) = -k_1 e^{-t} - 3k_2 e^{-3t} - \tfrac{4}{3}e^{-4t}.$$

To find the solution with $y(0) = y'(0) = 0$, we evaluate at $t = 0$ and obtain the simultaneous equations

$$\begin{cases} k_1 + k_2 + \frac{1}{3} = 0 \\ -k_1 - 3k_2 - \frac{4}{3} = 0. \end{cases}$$

Solving, we find that $k_1 = 1/6$ and $k_2 = -1/2$, so the solution of the initial-value problem is

$$y(t) = \tfrac{1}{6}e^{-t} - \tfrac{1}{2}e^{-3t} + \tfrac{1}{3}e^{-4t}.$$

(c) In the general solution, all three terms tend to zero, so the solution tends to zero. We can say a little more by noting that the term $k_1 e^{-t}$ is much larger (provided $k_1 \neq 0$). Hence, most solutions tend to zero at the rate of e^{-t}. If $k_1 = 0$, then solutions tend to zero at the rate of e^{-3t} provided $k_2 \neq 0$.

16. **(a)** The characteristic polynomial of the unforced equation is

$$s^2 + 4s + 20.$$

So the eigenvalues are $s = -2 \pm 4i$, and the general solution of the unforced equation is

$$k_1 e^{-2t} \cos 4t + k_2 e^{-2t} \sin 4t.$$

To find a particular solution of the forced equation, we guess $y_p(t) = ke^{-t/2}$. Substituting $y_p(t)$ into the left-hand side of the differential equation gives

$$\begin{aligned} \frac{d^2 y_p}{dt^2} + 4\frac{dy_p}{dt} + 20y_p &= \tfrac{1}{4}ke^{-t/2} - 2ke^{-t/2} + 20ke^{-t/2} \\ &= \tfrac{73}{4}ke^{-t/2}. \end{aligned}$$

So $k = 4/73$ yields a solution of the forced equation.

The general solution of the forced equation is therefore

$$y(t) = k_1 e^{-2t} \cos 4t + k_2 e^{-2t} \sin 4t + \tfrac{4}{73}e^{-t/2}.$$

(b) The derivative of the general solution is

$$\begin{aligned} y'(t) = -k_1 e^{-2t} \cos 4t - 4k_1 e^{-2t} \sin 4t \\ -2k_2 e^{-2t} \sin 4t + 4k_2 e^{-2t} \cos 4t - \tfrac{2}{73}e^{-t/2}. \end{aligned}$$

To find the solution with $y(0) = y'(0) = 0$, we evaluate at $t = 0$ and obtain the simultaneous equations

$$\begin{cases} k_1 + \frac{4}{73} = 0 \\ -2k + 4k_2 - \frac{2}{73} - 0. \end{cases}$$

Solving, we find that $k_1 = -4/73$ and $k_2 = -3/146$, so the solution of the initial-value problem is

$$y(t) = -\tfrac{4}{73}e^{-2t} \cos 4t - \tfrac{3}{146}e^{-2t} \sin 4t + \tfrac{4}{73}e^{-t/2}.$$

(c) Every solution tends to zero at the rate $e^{-t/2}$. The terms involving sine and cosine have e^{-4t} as a coefficient, so they tend to zero much more quickly than the exponential $\frac{4}{73}e^{-t/2}$.

17. **(a)** The characteristic polynomial of the unforced equation is

$$s^2 + 4s + 20.$$

So the eigenvalues are $s = -2 \pm 4i$, and the general solution of the unforced equation is

$$k_1 e^{-2t} \cos 4t + k_2 e^{-2t} \sin 4t.$$

To find a particular solution of the forced equation, we guess $y_p(t) = ke^{-2t}$. Substituting $y_p(t)$ into the left-hand side of the differential equation gives

$$\begin{aligned} \frac{d^2 y_p}{dt^2} + 4\frac{dy_p}{dt} + 20y_p &= 4ke^{-2t} - 8ke^{-2t} + 20ke^{-2t} \\ &= 16ke^{-2t}. \end{aligned}$$

So $k = 1/16$ yields a solution of the forced equation.

The general solution of the forced equation is therefore

$$y(t) = k_1 e^{-2t} \cos 4t + k_2 e^{-2t} \sin 4t + \tfrac{1}{16} e^{-2t}.$$

(b) The derivative of the general solution is

$$y'(t) = -2k_1 e^{-2t} \cos 4t - 4k_1 e^{-2t} \sin 4t - 2k_2 e^{-2t} \sin 4t + 4k_2 e^{-2t} \cos 4t - \tfrac{1}{8} e^{-2t}.$$

To find the solution with $y(0) = y'(0) = 0$, we evaluate at $t = 0$ and obtain the simultaneous equations

$$\begin{cases} k_1 + \frac{1}{16} = 0 \\ -2k_1 + 4k_2 - \frac{1}{8} = 0. \end{cases}$$

Solving, we find that $k_1 = -1/16$ and $k_2 = 0$, so the solution of the initial-value problem is

$$y(t) = -\tfrac{1}{16} e^{-2t} \cos 4t + \tfrac{1}{16} e^{-2t}.$$

(c) Every solution tends to zero like e^{-2t} and all but one exponential solution oscillates with frequency $2/\pi$.

18. **(a)** The characteristic polynomial of the unforced equation is

$$s^2 + 4s + 20.$$

So the eigenvalues are $s = -2 \pm 4i$, and the general solution of the unforced equation is

$$k_1 e^{-2t} \cos 4t + k_2 e^{-2t} \sin 4t.$$

To find a particular solution of the forced equation, we guess $y_p(t) = ke^{-4t}$. Substituting $y_p(t)$ into the left-hand side of the differential equation gives

$$\begin{aligned} \frac{d^2 y_p}{dt^2} + 4\frac{dy_p}{dt} + 20y_p &= 16ke^{-4t} - 16ke^{-4t} + 20ke^{-4t} \\ &= 20ke^{-4t}. \end{aligned}$$

So $k = 1/20$ yields a solution of the forced equation.

The general solution of the forced equation is therefore

$$y(t) = k_1 e^{-2t} \cos 4t + k_2 e^{-2t} \sin 4t + \tfrac{1}{20} e^{-4t}.$$

(b) The derivative of the general solution is

$$\begin{aligned} y'(t) = -k_1 e^{-2t} \cos 4t - 4k_1 e^{-2t} \sin 4t \\ -2k_2 e^{-2t} \sin 4t + 4k_2 e^{-2t} \cos 4t - \tfrac{1}{5} e^{-4t}. \end{aligned}$$

To find the solution with $y(0) = y'(0) = 0$, we evaluate at $t = 0$ and obtain the simultaneous equations

$$\begin{cases} k_1 + \frac{1}{20} = 0 \\ -2k_1 + 4k_2 - \frac{1}{5} = 0. \end{cases}$$

Solving, we find that $k_1 = -1/20$ and $k_2 = 1/40$, so the solution of the initial-value problem is

$$y(t) = -\tfrac{1}{20} e^{-2t} \cos 4t + \tfrac{1}{40} e^{-2t} \sin 4t + \tfrac{1}{20} e^{-4t}.$$

(c) From the formula for the general solution, we see that every solution tends to zero. The e^{-4t} term in the general solution tends to zero quickest, so for large t, the solution is very close to the unforced solution. All solutions tend to zero and all but the purely exponential one oscillates with frequency $2/\pi$ and an amplitude that decreases at the rate of e^{-2t}.

19. The natural guesses of $y_p(t) = ke^{-t}$ and $y_p(t) = kte^{-t}$ fail to be solutions of the forced equation because they are both solutions of the unforced equation. (The characteristic polynomial of the unforced equation is

$$s^2 + 2s + 1,$$

which has -1 as a double root.)

So we guess $y_p(t) = kt^2e^{-t}$. Substituting this guess into the left-hand side of the differential equation gives

$$\begin{aligned}\frac{d^2y_p}{dt^2} + 2\frac{dy_p}{dt} + y_p &= (2ke^{-t} - 4kte^{-t} + kt^2e^{-t}) + 2(2kte^{-t} - kt^2e^{-t}) + kt^2e^{-t} \\ &= 2ke^{-t}.\end{aligned}$$

So $k = 1/2$ yields the solution

$$y_p(t) = \tfrac{1}{2}t^2e^{-t}.$$

From the characteristic polynomial, we know that the general solution of the unforced equation is

$$k_1e^{-t} + k_2te^{-t}.$$

Consequently, the general solution of the forced equation is

$$y(t) = k_1e^{-t} + k_2te^{-t} + \tfrac{1}{2}t^2e^{-t}.$$

20. If we guess a constant function of the form $y_p(t) = k$, then substituting $y_p(t)$ into the left-hand side of the differential equation yields

$$\begin{aligned}\frac{d^2(k)}{dt^2} + p\frac{d(k)}{dt} + qk &= 0 + 0 + qk \\ &= qk.\end{aligned}$$

Since the right-hand side of the differential equation is simply the constant c, $k = c/q$ yields a constant solution.

21. **(a)** The characteristic polynomial of the unforced equation is

$$s^2 - 5s + 4.$$

So the eigenvalues are $s = 1$ and $s = 4$, and the general solution of the unforced equation is

$$k_1e^t + k_2e^{4t}.$$

To find one solution of the forced equation, we guess the constant function $y_p(t) = k$. Substituting $y_p(t)$ into the left-hand side of the differential equation, we obtain

$$\frac{d^2y_p}{dt^2} - 5\frac{dy_p}{dt} + 4y_p = 0 - 5 \cdot 0 + 4k = 4k.$$

Hence, $k = 5/4$ yields a solution of the forced equation. The general solution of the forced equation is

$$y(t) = k_1 e^t + k_2 e^{4t} + \tfrac{5}{4}.$$

(b) To find the solution satisfying the initial conditions $y(0) = y'(0) = 0$, we compute the derivative of the general solution

$$y'(t) = k_1 e^t + 4k_2 e^{4t}.$$

Using the initial conditions and evaluating $y(t)$ and $y'(t)$ at $t = 0$, we obtain the simultaneous equations

$$\begin{cases} k_1 + k_2 + \frac{5}{4} = 0 \\ k_1 + 4k_2 = 0. \end{cases}$$

Solving for k_1 and k_2 gives $k_1 = -5/3$ and $k_2 = 5/12$. The solution of the initial-value problem is

$$y(t) = \tfrac{5}{4} - \tfrac{5}{3}e^t + \tfrac{5}{12}e^{4t}.$$

22. **(a)** The characteristic polynomial of the unforced equation is

$$s^2 + 5s + 6.$$

So the eigenvalues are $s = -2$ and $s = -3$, and the general solution of the unforced equation is

$$k_1 e^{-2t} + k_2 e^{-3t}.$$

To find one solution of the forced equation, we guess the constant function $y_p(t) = k$. Substituting $y_p(t)$ into the left-hand side of the differential equation, we obtain

$$\frac{d^2 y_p}{dt^2} + 5\frac{dy_p}{dt} + 6y_p = 0 + 5 \cdot 0 + 6k = 6k.$$

Hence, $k = 1/3$ yields a solution of the forced equation. The general solution of the forced equation is

$$y(t) = k_1 e^{-2t} + k_2 e^{-3t} + \tfrac{1}{3}.$$

(b) To find the solution satisfying the initial conditions $y(0) = y'(0) = 0$, we compute the derivative of the general solution

$$y'(t) = -2k_1 e^{-2t} - 3k_2 e^{-3t}.$$

Using the initial conditions and evaluating $y(t)$ and $y'(t)$ at $t = 0$, we obtain the simultaneous equations

$$\begin{cases} k_1 + k_2 + \frac{1}{3} = 0 \\ -2k_1 - 3k_2 = 0. \end{cases}$$

Solving for k_1 and k_2 gives $k_1 =\ 1$ and $k_2 - 2/3$. The solution of the initial-value problem is

$$y(t) = -e^{-2t} + \tfrac{2}{3}e^{-3t} + \tfrac{1}{3}.$$

23. **(a)** The characteristic polynomial of the unforced equation is

$$s^2 + 2s + 10.$$

So the eigenvalues are $s = -1 \pm 3i$, and the general solution of the unforced equation is

$$k_1 e^{-t} \cos 3t + k_2 e^{-t} \sin 3t.$$

To find one solution of the forced equation, we guess the constant function $y_p(t) = k$. Substituting $y_p(t)$ into the left-hand side of the differential equation, we obtain

$$\frac{d^2 y_p}{dt^2} + 2\frac{dy_p}{dt} + 10y_p = 0 + 2 \cdot 0 + 10k = 10k.$$

Hence, $k = 1$ yields a solution of the forced equation. The general solution of the forced equation is

$$y(t) = k_1 e^{-t} \cos 3t + k_2 e^{-t} \sin 3t + 1.$$

(b) To find the solution satisfying the initial conditions $y(0) = y'(0) = 0$, we compute the derivative of the general solution

$$y'(t) = -k_1 e^{-t} \cos 3t - 3k_1 e^{-t} \sin 3t - k_2 e^{-t} \sin 3t + 3k_2 e^{-t} \cos 3t.$$

Using the initial conditions and evaluating $y(t)$ and $y'(t)$ at $t = 0$, we obtain the simultaneous equations

$$\begin{cases} k_1 + 1 = 0 \\ -k_1 + 3k_2 = 0. \end{cases}$$

Solving for k_1 and k_2 gives $k_1 = -1$ and $k_2 = -1/3$. The solution of the initial-value problem is

$$y(t) = -e^{-t} \cos 3t - \tfrac{1}{3} e^{-t} \sin 3t + 1.$$

24. **(a)** The characteristic polynomial of the unforced equation is

$$s^2 + 4s + 6.$$

So the eigenvalues are $s = -2 \pm i\sqrt{2}$, and the general solution of the unforced equation is

$$k_1 e^{-2t} \cos \sqrt{2}\, t + k_2 e^{-2t} \sin \sqrt{2}\, t.$$

To find one solution of the forced equation, we guess the constant function $y_p(t) = k$. Substituting $y_p(t)$ into the left-hand side of the differential equation, we obtain

$$\frac{d^2 y_p}{dt^2} + 4\frac{dy_p}{dt} + 6y_p = 0 + 4 \cdot 0 + 6k = 6k.$$

Hence, $k = -4/3$ yields a solution of the forced equation. The general solution of the forced equation is

$$y(t) = k_1 e^{-2t} \cos \sqrt{2}\, t + k_2 e^{-2t} \sin \sqrt{2}\, t - \tfrac{4}{3}.$$

(b) To find the solution satisfying the initial conditions $y(0) = y'(0) = 0$, we compute the derivative of the general solution

$$y'(t) = -2k_1e^{-2t}\cos\sqrt{2}\,t - \sqrt{2}\,k_1e^{-2t}\sin\sqrt{2}\,t$$
$$-2k_2e^{-2t}\sin\sqrt{2}\,t + \sqrt{2}\,k_2e^{-2t}\cos\sqrt{2}\,t.$$

Using the initial conditions and evaluating $y(t)$ and $y'(t)$ at $t = 0$, we obtain the simultaneous equations

$$\begin{cases} k_1 - \frac{4}{3} = 0 \\ -2k_1 + \sqrt{2}\,k_2 = 0. \end{cases}$$

Solving for k_1 and k_2 gives $k_1 = 4/3$ and $k_2 = 4\sqrt{2}/3$. The solution of the initial-value problem is

$$y(t) = \tfrac{4}{3}e^{-2t}\cos\sqrt{2}\,t - \tfrac{4\sqrt{2}}{3}e^{-2t}\sin\sqrt{2}\,t - \tfrac{4}{3}.$$

25. **(a)** The characteristic polynomial of the unforced equation is

$$s^2 + 9.$$

So the eigenvalues are $s = \pm 3i$, and the general solution of the unforced equation is

$$k_1\cos 3t + k_2\sin 3t.$$

To find one solution of the forced equation, we guess $y_p(t) = ke^{-t}$. Substituting $y_p(t)$ into the left-hand side of the differential equation, we obtain

$$\frac{d^2y_p}{dt^2} + 9y_p = ke^{-t} + 9ke^{-t}$$
$$= 10ke^{-t}.$$

Hence, $k = 1/10$ yields a solution of the forced equation. The general solution of the forced equation is

$$y(t) = k_1\cos 3t + k_2\sin 3t + \tfrac{1}{10}e^{-t}.$$

(b) To find the solution satisfying the initial conditions $y(0) = y'(0) = 0$, we compute the derivative of the general solution

$$y'(t) = -3k_1\sin 3t + 3k_2\cos 3t - \tfrac{1}{10}e^{-t}.$$

Using the initial conditions and evaluating $y(t)$ and $y'(t)$ at $t = 0$, we obtain the simultaneous equations

$$\begin{cases} k_1 + \frac{1}{10} = 0 \\ 3k_2 - \frac{1}{10} = 0. \end{cases}$$

Solving for k_1 and k_2 gives $k_1 = -1/10$ and $k_2 = 1/30$. The solution of the initial-value problem is

$$y(t) = -\tfrac{1}{10}\cos 3t + \tfrac{1}{30}\sin 3t + \tfrac{1}{10}e^{-t}.$$

(c) Since the function $e^{-t}/10 \to 0$ quickly, the solution quickly approaches a solution of the unforced oscillator.

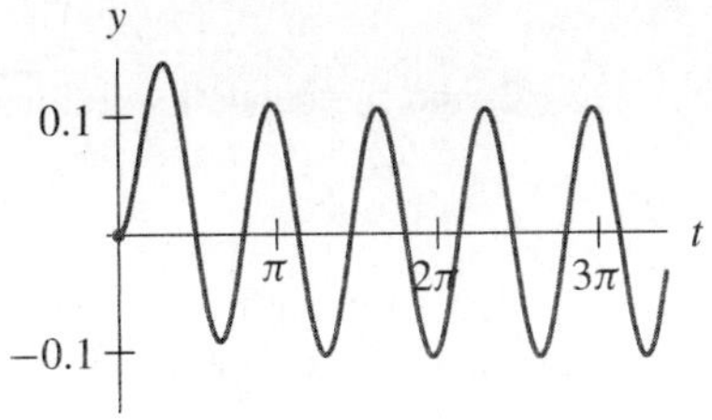

26. **(a)** The characteristic polynomial of the unforced equation is

$$s^2 + 4.$$

So the eigenvalues are $s = \pm 2i$, and the general solution of the unforced equation is

$$k_1 \cos 2t + k_2 \sin 2t.$$

To find one solution of the forced equation, we guess $y_p(t) = ke^{-2t}$. Substituting into the left-hand side of the differential equation, we obtain

$$\begin{aligned} \frac{d^2y_p}{dt^2} + 4y_p &= 4ke^{-2t} + 4ke^{-2t} \\ &= 8ke^{-2t}. \end{aligned}$$

Hence, $k = 1/4$ yields a solution of the forced equation. The general solution of the forced equation is

$$y(t) = k_1 \cos 2t + k_2 \sin 2t + \tfrac{1}{4}e^{-2t}.$$

(b) To find the solution satisfying the initial conditions $y(0) = y'(0) = 0$, we compute the derivative of the general solution

$$y'(t) = -2k_1 \sin 2t + 2k_2 \cos 2t - \tfrac{1}{2}e^{-2t}.$$

Using the initial conditions and evaluating $y(t)$ and $y'(t)$ at $t = 0$, we obtain the simultaneous equations

$$\begin{cases} k_1 + \frac{1}{4} = 0 \\ 2k_2 - \frac{1}{2} = 0. \end{cases}$$

Solving for k_1 and k_2 gives $k_1 = -1/4$ and $k_2 = 1/4$. The solution of the initial-value problem is

$$y(t) = -\tfrac{1}{4}\cos 2t + \tfrac{1}{4}\sin 2t + \tfrac{1}{4}e^{-2t}.$$

(c) Since $e^{-2t}/4 \to 0$ quickly, the solution quickly approaches a solution of the unforced oscillator.

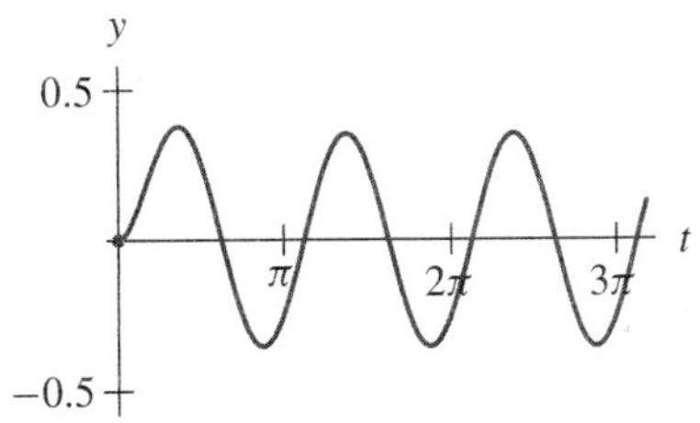

27. **(a)** The characteristic polynomial of the unforced equation is

$$s^2 + 2.$$

So the eigenvalues are $s = \pm i\sqrt{2}$, and the general solution of the unforced equation is

$$k_1 \cos \sqrt{2}\,t + k_2 \sin \sqrt{2}\,t.$$

To find one solution of the forced equation, we guess $y_p(t) = k$. Substituting into the left-hand side of the differential equation, we obtain

$$\begin{aligned} \frac{d^2y_p}{dt^2} + 2y_p &= 0 + 2k \\ &= 2k. \end{aligned}$$

Hence, $k = -3/2$ yields a solution of the forced equation. The general solution of the forced equation is

$$y(t) = k_1 \cos \sqrt{2}\,t + k_2 \sin \sqrt{2}\,t - \tfrac{3}{2}.$$

(b) To find the solution satisfying the initial conditions $y(0) = y'(0) = 0$, we compute the derivative of the general solution

$$y'(t) = -\sqrt{2}\,k_1 \sin \sqrt{2}\,t + \sqrt{2}\,k_2 \cos \sqrt{2}\,t.$$

Using the initial conditions and evaluating $y(t)$ and $y'(t)$ at $t = 0$, we obtain the simultaneous equations

$$\begin{cases} k_1 - \tfrac{3}{2} = 0 \\ \sqrt{2}\,k_2 = 0. \end{cases}$$

Solving for k_1 and k_2 gives $k_1 = 3/2$ and $k_2 = 0$. The solution of the initial-value problem is

$$y(t) = \tfrac{3}{2} \cos \sqrt{2}\,t - \tfrac{3}{2}.$$

(c) The solution oscillates about the constant $y = -3/2$ with oscillations of amplitude $3/2$.

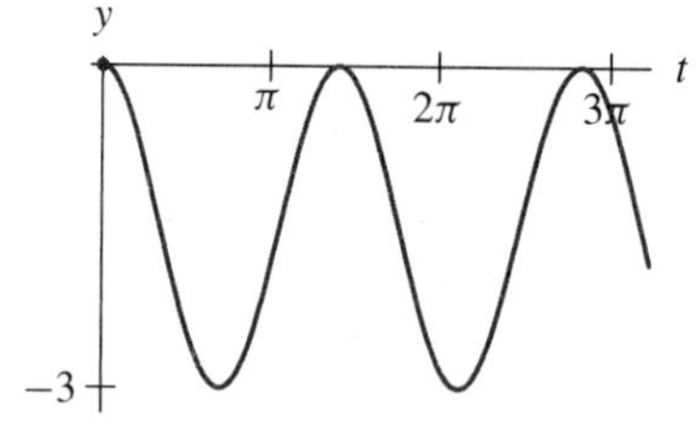

28. **(a)** The characteristic polynomial of the unforced equation is

$$\lambda^2 + 4 = 0.$$

So the eigenvalues are $\lambda = \pm 2i$, and the general solution of the unforced equation is

$$k_1 \cos 2t + k_1 \sin 2t.$$

To find a particular solution of the forced equation, we guess $y_p(t) = ke^t$. Substituting into the differential equation, we obtain

$$ke^t + 4ke^t = e^t,$$

which is satisfied if $5k = 1$. Hence, $k = 1/5$ yields a solution of the forced equation.

The general solution of the forced equation is

$$y(t) = k_1 \cos 2t + k_2 \sin 2t + \tfrac{1}{5}e^t.$$

(b) To find the solution with $y(0) = y'(0) = 0$, we note that

$$y'(t) = -2k_1 \sin 2t + 2k_2 \cos 2t + \tfrac{1}{5}e^t.$$

Using the initial conditions and evaluating $y(t)$ and $y'(t)$ at $t = 0$, we obtain the simultaneous equations

$$\begin{cases} k_1 + \frac{1}{5} = 0 \\ 2k_2 + \frac{1}{5} = 0. \end{cases}$$

Solving for k_1 and k_2 gives $k_1 = -1/5$ and $k_2 = -1/10$. The solution of the initial-value problem is

$$y(t) = -\tfrac{1}{5}\cos 2t - \tfrac{1}{10}\sin 2t + \tfrac{1}{5}e^t.$$

(c) Since $e^t \to \infty$, the solution tends to infinity, but it oscillates about the values of $\frac{1}{5}e^t$ as it does so.

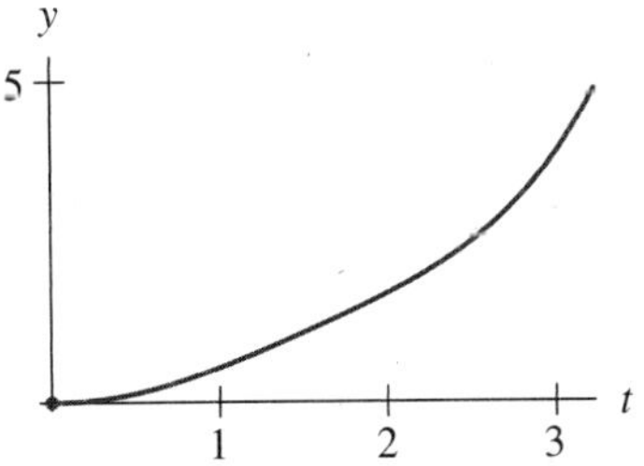

29. **(a)** The characteristic polynomial of the unforced equation is

$$s^2 + 9.$$

So the eigenvalues are $s = \pm 3i$, and the general solution of the unforced equation is

$$k_1 \cos 3t + k_2 \sin 3t.$$

To find one solution of the forced equation, we guess $y_p(t) = k$, where k is a constant. Substituting this guess into the left-hand side of the differential equation, we obtain

$$\frac{d^2 y_p}{dt^2} + 9y_p = 9k.$$

Hence, $k = 2/3$ yields a solution of the forced equation. The general solution of the forced equation is

$$y(t) = k_1 \cos 3t + k_2 \sin 3t + \tfrac{2}{3}.$$

(b) To find the solution satisfying the initial conditions $y(0) = y'(0) = 0$, we compute the derivative of the general solution

$$y'(t) = -3k_1 \sin 3t + 3k_2 \cos 3t.$$

Using the initial conditions and evaluating $y(t)$ and $y'(t)$ at $t = 0$, we obtain the simultaneous equations

$$\begin{cases} k_1 + \frac{2}{3} = 0 \\ 3k_2 = 0. \end{cases}$$

Solving for k_1 and k_2 gives $k_1 = -2/3$ and $k_2 = 0$. The solution of the initial-value problem is

$$y(t) = -\tfrac{2}{3} \cos 3t + \tfrac{2}{3}.$$

(c) The solution oscillates about the constant function $y = 2/3$ with amplitude $2/3$.

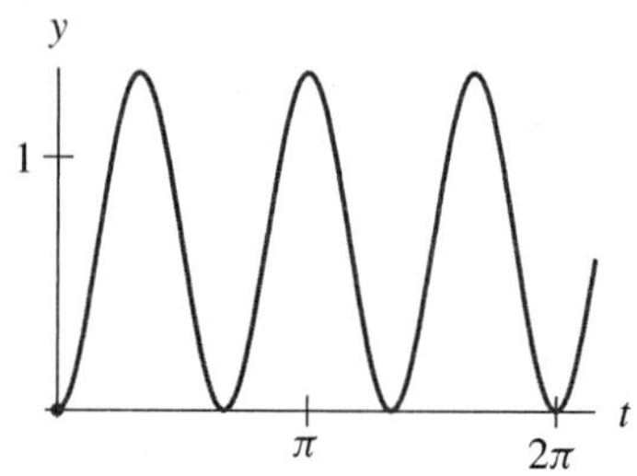

30. **(a)** The characteristic polynomial of the unforced equation is

$$s^2 + 2 = 0.$$

So $s = \pm i\sqrt{2}$ are the eigenvalues, and the general solution of the unforced equation is

$$k_1 \cos \sqrt{2}\,t + k_2 \sin \sqrt{2}\,t.$$

To find a particular solution of the forced equation, we guess $y_p(t) = ke^t$. Substituting this guess into the differential equation yields

$$ke^t + 2ke^t = -e^t,$$

which is satisfied if $3k = -1$. Hence, $k = -1/3$ yields a solution of the forced equation. The general solution of the forced equation is

$$y(t) = k_1 \cos \sqrt{2}\,t + k_2 \sin \sqrt{2}\,t - \tfrac{1}{3}e^t.$$

(b) To satisfy the initial conditions $y(0) = y'(0) = 0$, we note that

$$y'(t) = -\sqrt{2}\,k_1 \sin \sqrt{2}\,t + \sqrt{2}\,k_2 \cos \sqrt{2}\,t - \tfrac{1}{3}e^t.$$

Using the initial conditions and evaluating $y(t)$ and $y'(t)$ at $t = 0$, we obtain the simultaneous equations

$$\begin{cases} k_1 - \frac{1}{3} = 0 \\ \sqrt{2}\,k_2 - \frac{1}{3} = 0. \end{cases}$$

Solving for k_1 and k_2 gives $k_1 = 1/3$ and $k_2 = \sqrt{2}/6$. The solution of the initial-value problem is

$$y(t) = \tfrac{1}{3}\cos\sqrt{2}\,t + \tfrac{\sqrt{2}}{6}\sin\sqrt{2}\,t - \tfrac{1}{3}e^t.$$

(c) Since $e^t \to \infty$ quickly, the solution tends to $-\infty$ at an exponential rate.

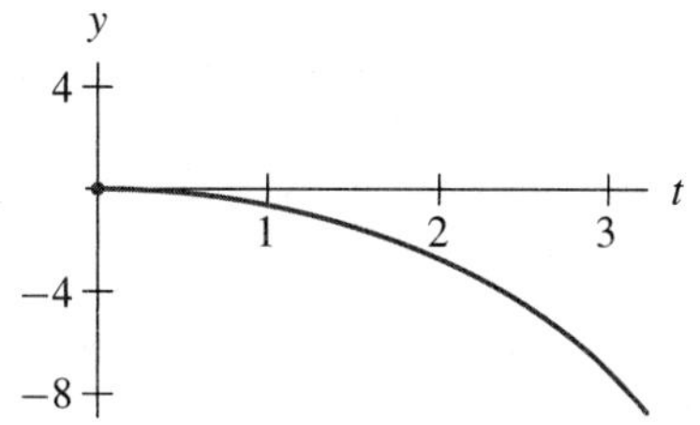

31. **(a)** The general solution for the homogeneous equation is

$$k_1 \cos 2t + k_2 \sin 2t.$$

Suppose $y_p(t) = at^2 + bt + c$. Substituting $y_p(t)$ into the differential equation, we get

$$\frac{d^2 y_p}{dt^2} + 4y_p = -3t^2 + 2t + 3$$

$$2a + 4(at^2 + bt + c) = -3t^2 + 2t + 3$$

$$4at^2 + 4bt + (2a + 4c) = -3t^2 + 2t + 3.$$

Therefore, $y_p(t)$ is a solution if and only if

$$\begin{cases} 4a = -3 \\ 4b = 2 \\ 2a + 4c = 3. \end{cases}$$

Therefore, $a = -3/4$, $b = 1/2$, and $c = 9/8$. The general solution is

$$y(t) = k_1 \cos 2t + k_2 \sin 2t - \tfrac{3}{4}t^2 + \tfrac{1}{2}t + \tfrac{9}{8}.$$

(b) To solve the initial-value problem, we use the initial conditions $y(0) = 2$ and $y'(0) = 0$ along with the general solution to form the simultaneous equations

$$\begin{cases} k_1 + \frac{9}{8} = 2 \\ 2k_2 + \frac{1}{2} = 0. \end{cases}$$

Therefore, $k_1 = 7/8$ and $k_2 = -1/4$. The solution is

$$y(t) = \tfrac{7}{8}\cos 2t - \tfrac{1}{4}\sin 2t - \tfrac{3}{4}t^2 + \tfrac{1}{2}t + \tfrac{9}{8}.$$

32. **(a)** The general solution for the homogeneous equation is

$$k_1 + k_2 e^{-2t}.$$

Suppose $y_p(t) = at^2 + bt$. Substituting $y_p(t)$ into the differential equation, we get

$$\begin{aligned} \frac{d^2 y_p}{dt^2} + 2\frac{dy_p}{dt} &= 3t + 2 \\ 2a + 2(2at + b) &= 3t + 2 \\ 4at + (2a + 2b) &= 3t + 2. \end{aligned}$$

Therefore, $y_p(t)$ is a solution only if $4a = 3$ and $2a + 2b = 2$. These two equations imply that $a = 3/4$ and $b = 1/4$. The general solution is

$$y(t) = k_1 + k_2 e^{-2t} + \tfrac{3}{4}t^2 + \tfrac{1}{4}t.$$

(b) To solve the initial-value problem, we compute

$$y'(t) = -2k_2 e^{-2t} + \tfrac{3}{2}t + \tfrac{1}{4}.$$

Evaluating $y(t)$ and $y'(t)$ at $t = 0$ and using the initial conditions, we obtain the simultaneous equations

$$\begin{cases} k_1 + k_2 = 0 \\ -2k_2 + \tfrac{1}{4} = 0. \end{cases}$$

Hence, $k_1 = -1/8$ and $k_2 = 1/8$ provide the desired solution

$$y(t) = -\tfrac{1}{8} + \tfrac{1}{8}e^{-2t} + \tfrac{3}{4}t^2 + \tfrac{1}{4}t.$$

(c) Since $e^{-2t} \to 0$ quickly, the solution tends to infinity at a rate that is determined by $\frac{3}{4}t^2$.

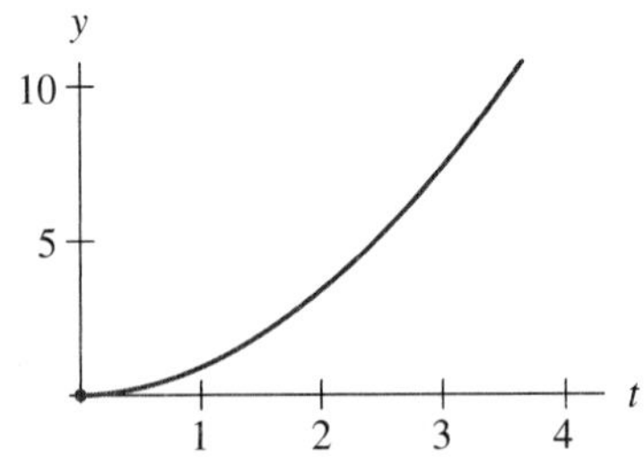

33. **(a)** For the unforced equation, the general solution is

$$k_1 \cos 2t + k_2 \sin 2t.$$

To find a particular solution of the forced equation, we guess $y_p(t) = at + b$. Substituting this guess into the differential equation, we get

$$\begin{aligned}\frac{d^2y_p}{dt} + 4y_p &= 3t + 2\\ 0 + 4(at + b) &= 3t + 2\\ 4at + 4b &= 3t + 2.\end{aligned}$$

Therefore, $a = 3/4$ and $b = 1/2$ yield a solution. The general solution for the forced equation is

$$y(t) = k_1 \cos 2t + k_2 \sin 2t + \tfrac{3}{4}t + \tfrac{1}{2}.$$

(b) To solve the initial-value problem, we compute

$$y'(t) = -2k_1 \sin 2t + 2k_2 \cos 2t + \tfrac{3}{4}.$$

Evaluating $y(t)$ and $y'(t)$ at $t = 0$ and using the initial conditions, we obtain the simultaneous equations

$$\begin{cases} k_1 + \frac{1}{2} = 0 \\ 2k_2 + \frac{3}{4} = 0. \end{cases}$$

Hence, $k_1 = -1/2$ and $k_2 = -3/8$ provide the desired solution

$$y(t) = -\tfrac{1}{2} \cos 2t - \tfrac{3}{8} \cos 2t + \tfrac{3}{4}t + \tfrac{1}{2}.$$

(c) The solution tends to ∞ as it oscillates about the line $y = \frac{3}{4}t + \frac{1}{2}$.

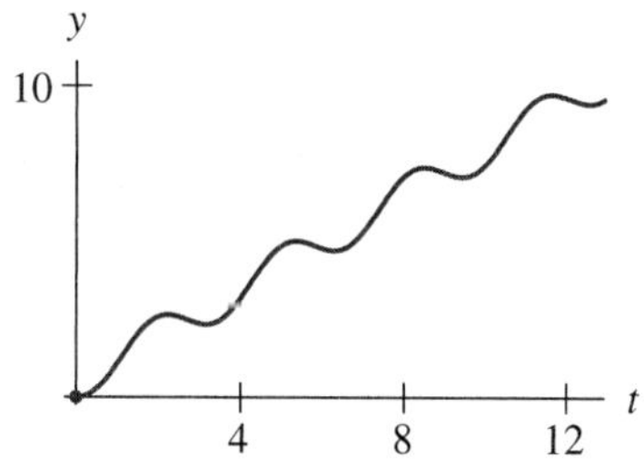

34. **(a)** To find a particular solution of the forced equation, we guess

$$y_p(t) = at^2 + bt + c.$$

Substituting this guess into the equation yields

$$(2a) + 3(2at + b) + 2(at^2 + bt + c) = t^2,$$

which can be rewritten as

$$(2a)t^2 + (6a + 2b)t + (2a + 3b + 2c) = t^2.$$

Equating coefficients, we have

$$\begin{cases} 2a = 1 \\ 6a + 2b = 0 \\ 2a + 3b + 2c = 0, \end{cases}$$

which gives $a = 1/2$, $b = -3/2$ and $c = 7/4$. So

$$y_p(t) = \tfrac{1}{2}t^2 - \tfrac{3}{2}t + \tfrac{7}{4}.$$

To find the general solution of the unforced equation, we note that the characteristic polynomial

$$s^2 + 3s + 2$$

has roots $s = -2$ and $s = -1$, so the general solution for the forced equation is

$$y(t) = k_1 e^{-2t} + k_2 e^{-t} + \tfrac{1}{2}t^2 - \tfrac{3}{2}t + \tfrac{7}{4}.$$

(b) Note that

$$y'(t) = -2k_1 e^{-2t} - k_2 e^{-t} + t - \tfrac{3}{2}.$$

To satisfy the desired initial conditions, we compute

$$y(0) = k_1 + k_2 + \tfrac{7}{4}$$

and

$$y'(0) = -2k_1 - k_2 - \tfrac{3}{2}.$$

Using the initial conditions $y(0) = 0$ and $y'(0) = 0$, we have $k_1 = 1/4$ and $k_2 = -2$. So the desired solution is

$$y(t) = \tfrac{1}{4}e^{-2t} - 2e^{-t} + \tfrac{1}{2}t^2 - \tfrac{3}{2}t + \tfrac{7}{4}.$$

(c) The solution $\frac{1}{4}e^{-2t} - 2e^{-t}$ of the unforced equation tends to zero quickly, so the solution of the original equation tends to infinity at a rate that is determined by the quadratic $t^2/2 - 3t/2 + 7/4$. This rate is essentially the same as that of t^2.

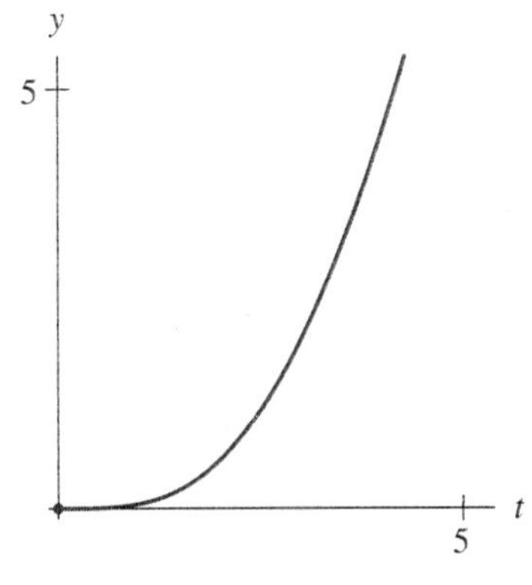

35. **(a)** The general solution of the homogeneous equation is

$$k_1 \cos 2t + k_2 \sin 2t.$$

To find a particular solution to the nonhomogeneous equation, we guess

$$y_p(t) = at^2 + bt + c.$$

Substituting $y_p(t)$ into the differential equation yields

$$\begin{aligned} \frac{d^2 y_p}{dt^2} + 4y_p &= t - \frac{t}{20} \\ 2a + 4(at^2 + bt + c) &= t - \frac{t}{20} \\ (4a)t^2 + (4b)t + (2a + 4c) &= t - \frac{t}{20}. \end{aligned}$$

Equating coefficients, we obtain the simultaneous equations

$$\begin{cases} 4a = -\frac{1}{20} \\ 4b = 1 \\ 2a + 4c = 0. \end{cases}$$

Therefore, $a = -1/80$, $b = 1/4$, and $c = 1/160$ yield a solution to the nonhomogeneous equation, and the general solution of the nonhomogeneous equation is

$$y(t) = k_1 \cos 2t + k_2 \sin 2t - \tfrac{1}{80}t^2 + \tfrac{1}{4}t + \tfrac{1}{160}.$$

(b) To solve the initial-value problem with $y(0) = 0$ and $y'(0) - 0$, we have

$$\begin{cases} k_1 + \frac{1}{160} = 0 \\ 2k_2 + \frac{1}{4} = 0. \end{cases}$$

Therefore, $k_1 = -1/160$ and $k_2 = -1/8$, and the solution is

$$y(t) = -\tfrac{1}{160}\cos 2t - \tfrac{1}{8}\sin 2t - \tfrac{1}{80}t^2 + \tfrac{1}{4}t + \tfrac{1}{160}.$$

(c) Since the solution to the homogeneous equation is periodic with a small amplitude and since the solution to the nonhomogeneous equation goes to $-\infty$ at a rate determined by $-t^2/80$, the solution tends to $-\infty$.

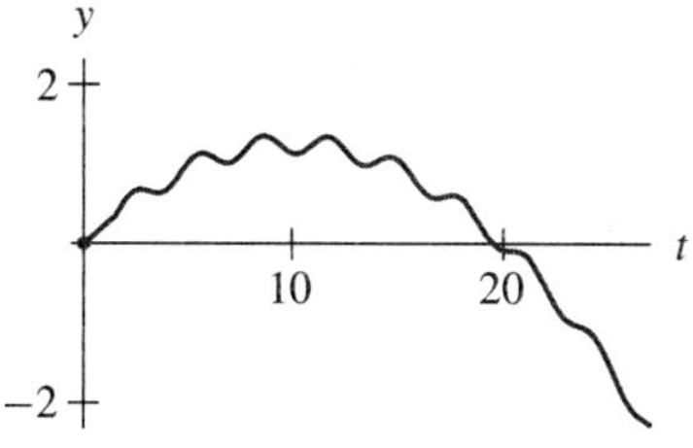

36. Substituting $y_1 + y_2$ into the differential equation, we obtain

$$\frac{d^2(y_1+y_2)}{dt^2} + p\frac{d(y_1+y_2)}{dt} + q(y_1+y_2)$$

$$= \left(\frac{d^2y_1}{dt^2} + p\frac{dy_1}{dt} + qy_1\right) + \left(\frac{d^2y_2}{dt^2} + p\frac{dy_2}{dt} + qy_2\right)$$

$$= g(t) + h(t)$$

since y_1 and y_2 are solutions of $y'' + py' + qy = g(t)$ and $y'' + py' + qy = h(t)$ respectively. Therefore, $y_1 + y_2$ is a solution of

$$\frac{d^2y}{dt^2} + p\frac{dy}{dt} + qy = g(t) + h(t).$$

37. **(a)** We must find a particular solution. Using the result of Exercise 36, we guess $y_p(t) = ae^{-t} + b$, where a and b are constants to be determined. (We could solve two separate problems and add the answers, but this approach is more efficient.) Hence we have $dy_p/dt = -ae^{-t}$ and $d^2y_p/dt^2 = ae^{-t}$. Substituting these derivatives into the differential equation, we obtain

$$(a - 5a + 6a)e^{-t} + 6b = e^{-t} + 4,$$

which is satisfied if $2a = 1$ and $6b = 4$. Hence, $a = 1/2$ and $b = 2/3$ yield the particular solution $y_p(t) = e^{-t}/2 + 2/3$.

The general solution of the homogeneous equation is obtained from the characteristic polynomial

$$s^2 + 5s + 6,$$

whose roots are $s = -2$ and $s = -3$.

Hence the general solution is

$$y(t) = k_1e^{-2t} + k_2e^{-3t} + \tfrac{1}{2}e^{-t} + \tfrac{2}{3}.$$

(b) To obtain the solution to the initial-value problem specified, we note that

$$y(0) = k_1 + k_2 + 1/2 + 2/3 \quad \text{and} \quad y'(0) = -2k_1 - 3k_2 - 1/2.$$

Using the initial conditions $y(0) = 0$ and $y'(0) = 0$, we have $k_1 = -3$ and $k_2 = 11/6$. The solution is

$$y(t) = -3e^{-2t} + \tfrac{11}{6}e^{-3t} + \tfrac{1}{2}e^{-t} + \tfrac{2}{3}.$$

(c) All of the exponential terms in the solution to the initial-value problem tend to 0. Hence, the solution tends to the constant $y = 2/3$. The rate that this solution tends to the constant is determined by $e^{-t}/2$, which is the largest of the terms that tend to zero when t is large.

38. **(a)** By Exercise 34, the general solution of the unforced equation is

$$k_1e^{-2t} + k_2e^{-t}.$$

To find a particular solution of the forced equation, we guess $y_p(t) = ae^{-t} + b$. Substituting this guess into the equation yields

$$ae^{-t} + 3(-ae^{-t}) + 2(ae^{-t} + b) = e^{-t} - 4,$$

which unfortunately reduces to

$$0 \cdot e^{-t} + 2b = e^{-t} - 4.$$

This guess does not produce a solution to the forced equation. (The difficulty is caused by the fact that ae^{-t} is a solution of the unforced equation.)

We must make a second guess of $y_p(t) = ate^{-t} + b$. Substituting this second guess into the forced equation yields

$$(-2ae^{-t} + ate^{-t}) + 3(ae^{-t} - ate^{-t}) + 2(ate^{-t} + b) = e^{-t} - 4,$$

which can be simplified to

$$ae^{-t} + 2b = e^{-t} - 4.$$

Hence, $a = 1$ and $b = -2$ yield the solution

$$y_p(t) = te^{-t} - 2,$$

and

$$y(t) = k_1e^{-2t} + k_2e^{-t} + te^{-t} - 2$$

is the general solution of the forced equation.

(b) Note that

$$y'(t) = -2k_1e^{-2t} - k_2e^{-t} + e^{-t} - te^{-t}.$$

To satisfy the initial conditions $y(0) = 0$ and $y'(0) = 0$, we must have

$$\begin{cases} k_1 + k_2 - 2 = 0 \\ -2k_1 - k_2 + 1 = 0, \end{cases}$$

which hold if $k_1 = -1$ and $k_2 = 3$. Hence, the solution of the initial-value problem is

$$y(t) = -e^{-2t} + 3e^{-t} + te^{-t} - 2.$$

(c) Since all three terms that include an exponential tend to 0 relatively quickly, the solution tends to $y = -2$.

39. **(a)** First, to find a particular solution of the forced equation, we guess

$$y_p(t) = at + b + ce^{-t}.$$

For y_p, $dy_p/dt = a - ce^{-t}$ and $d^2y_p/dt^2 = ce^{-t}$. Substituting these derivatives into the differential equation and collecting terms gives

$$(c - 6c + 8c)e^{-t} + (8a)t + (6a + 8b) = 2t + e^{-t},$$

which holds if $3c = 1$, $8a = 2$, and $6a + 8b = 0$. Hence, $c = 1/3$, $a = 1/4$, and $b = -3/16$ yield the solution

$$y_p(t) = -\tfrac{3}{16} + \tfrac{1}{4}t + \tfrac{1}{3}e^{-t}.$$

The characteristic polynomial of the homogeneous equation is

$$s^2 + 6s + 8,$$

which has roots $s = -4$ and $s = -2$, so the general solution of the forced equation is

$$y(t) = k_1 e^{-4t} + k_2 e^{-2t} - \tfrac{3}{16} + \tfrac{1}{4}t + \tfrac{1}{3}e^{-t}.$$

(b) To find the solution for the initial conditions $y(0) = 0$ and $y'(0) = 0$, we solve

$$\begin{cases} k_1 + k_2 - \frac{3}{16} + \frac{1}{3} = 0 \\ -4k_1 - 2k_2 + \frac{1}{4} - \frac{1}{3} = 0. \end{cases}$$

Thus, $k_1 = 5/48$ and $k_2 = -1/4$ yield the solution

$$y(t) = \tfrac{5}{48}e^{-4t} - \tfrac{1}{4}e^{-2t} - \tfrac{3}{16} + \tfrac{1}{4}t + \tfrac{1}{3}e^{-t}$$

of the initial-value problem.

(c) All exponential terms in the solution tend to zero. Hence, the solution tends to infinity linearly in t and is close to $t/4$ for large t.

40. **(a)** From Exercise 39 we know that the general solution of the unforced equation is

$$y(t) = k_1 e^{-4t} + k_2 e^{-2t}.$$

To find a particular solution of the forced equation, we guess

$$y_p(t) = at + b + ce^t.$$

Substituting this guess into the differential equation, we obtain

$$ce^t + 6(a + ce^t) + 8(at + b + ce^t) = 2t + e^t,$$

which simplifies to

$$(15c)e^t + (8a)t + (6a + 8b) = 2t + e^t.$$

Hence, $c = 1/15$, $a = 1/4$, and $b = -3/16$ yield the solution

$$y_p(t) = \tfrac{1}{4}t - \tfrac{3}{16} + \tfrac{1}{15}e^t.$$

The general solution of the forced equation is

$$y(t) = k_1 e^{-4t} + k_2 e^{-2t} + \tfrac{1}{4}t - \tfrac{3}{16} + \tfrac{1}{15}e^t.$$

(b) Note that

$$y'(t) = -4k_1 e^{-4t} - 2k_2 e^{-2t} + \tfrac{1}{4} + \tfrac{1}{15}e^t.$$

Hence, to obtain the desired initial conditions we must solve

$$\begin{cases} k_1 + k_2 - \frac{3}{16} + \frac{1}{15} = 0 \\ -4k_1 - 2k_2 + \frac{1}{4} + \frac{1}{15} = 0. \end{cases}$$

We obtain $k_1 = 3/80$ and $k_2 = 1/12$. Hence, the desired solution is

$$y(t) = \tfrac{3}{80}e^{-4t} + \tfrac{1}{12}e^{-2t} + \tfrac{1}{4}t - \tfrac{3}{16} + \tfrac{1}{15}e^{t}.$$

(c) For large t, the term $e^t/15$ dominates, so the solution tends to infinity at a rate determined by $e^t/15$.

41. **(a)** To find the general solution, we first guess

$$y_p(t) = ae^{-t} + bt + c,$$

where a, b and c are constants to be determined. For y_p,

$$\frac{dy_p}{dt} = -ae^{-t} + b \quad \text{and} \quad \frac{d^2y_p}{dt^2} = ae^{-t}.$$

Substituting these derivatives into the differential equation and collecting terms gives

$$(a + 4a)e^{-t} + (4b)t + (4c) = t + e^{-t},$$

which is satisfied if $5a = 1$, $4b = 1$, and $4c = 0$. Hence, a solution is

$$y_p(t) = \tfrac{1}{5}e^{-t} + \tfrac{1}{4}t.$$

To find the general solution of the homogeneous equation, we note that the characteristic polynomial $s^2 + 4$ has roots $s = \pm 2i$. Hence, the general solution of the forced equation is

$$y(t) = k_1 \cos 2t + k_2 \sin 2t + \tfrac{1}{5}e^{-t} + \tfrac{1}{4}t.$$

(b) To find the solution with the desired initial conditions, we note that $y(0) = k_1 + 0 + 1/5$ and $y'(0) = 2k_2 - 1/5 + 1/4$. We must solve the simultaneous equations

$$\begin{cases} k_1 + \frac{1}{5} = 0 \\ 2k_2 + \frac{1}{20} = 0. \end{cases}$$

Thus, $k_1 = -1/5$ and $k_2 = -1/40$ yield the solution

$$y(t) = -\tfrac{1}{5}\cos 2t - \tfrac{1}{40}\sin 2t + \tfrac{1}{5}e^{-t} + \tfrac{1}{4}t.$$

(c) Since all of the terms in the solution except $t/4$ are bounded for $t > 0$, the solution tends to infinity at a rate that is determined by $t/4$.

42. **(a)** To find the general solution of the unforced equation, we note that the characteristic polynomial is $s^2 + 4$, which has roots $s = \pm 2i$. So the general solution of the unforced equation is

$$k_1 \cos 2t + k_2 \sin 2t.$$

To find a particular solution of the forced equation we guess

$$y_p(t) = a + bt + ct^2 + de^t.$$

Substituting this guess into the differential equation yields

$$(2c + de^t) + 4(a + bt + ct^2 + de^t) = 6 + t^2 + e^t,$$

which simplifies to

$$(2c + 4a) + (4b)t + (4c)t^2 + (5d)e^t = 6 + t^2 + e^t.$$

So $d = 1/5$, $c = 1/4$, $b = 0$, and $a = 1/8$ yield a solution, and the general solution of the forced equation is

$$y(t) = k_1 \cos 2t + k_2 \sin 2t + \tfrac{11}{8} + \tfrac{1}{4}t^2 + \tfrac{1}{5}e^t.$$

(b) Note that

$$y'(t) = -2k_1 \sin 2t + 2k_2 \cos 2t + \tfrac{1}{2}t + \tfrac{1}{5}e^t.$$

To obtain the desired initial conditions we must solve

$$\begin{cases} k_1 + \frac{11}{8} + \frac{1}{5} = 0 \\ 2k_2 + \frac{1}{5} = 0, \end{cases}$$

which yields $k_1 = -63/40$ and $k_2 = -1/10$. The solution of the initial-value problem is

$$y(t) = -\tfrac{63}{40} \cos 2t - \tfrac{1}{10} \sin 2t + \tfrac{11}{8} + \tfrac{1}{4}t^2 + \tfrac{1}{5}e^t.$$

(c) This solution tends to infinity at a rate that is determined by $e^t/5$ because this term dominates when t is large.

EXERCISES FOR SECTION 4.2

1. Recalling that the real part of e^{it} is $\cos t$, we see that the complex version of this equation is

$$\frac{d^2y}{dt^2} + 3\frac{dy}{dt} + 2y = e^{it}.$$

To find a particular solution, we guess $y_c(t) = ae^{it}$. Then $dy_c/dt = iae^{it}$ and $d^2y_c/dt^2 = -ae^{it}$. Substituting these derivatives into the equation and collecting terms yields

$$(-a + 3ia + 2a)e^{-it} = e^{it},$$

which is satisfied if

$$(1 + 3i)a = 1.$$

Hence, we must have

$$a = \frac{1}{1 + 3i} = \frac{1}{10} - \frac{3}{10}i.$$

So

$$y_c(t) = \frac{1 - 3i}{10}e^{it} = \frac{1 - 3i}{10}(\cos t + i \sin t)$$

is a particular solution of the complex version of the equation. Taking the real part, we obtain the solution

$$y(t) = \tfrac{1}{10} \cos t + \tfrac{3}{10} \sin t.$$

To produce the general solution of the homogeneous equation, we note that the characteristic polynomial $s^2 + 3s + 2$ has roots $s = -2$ and $s = -1$. So the general solution is

$$y(t) = k_1 e^{-2t} + k_2 e^{-t} + \tfrac{1}{10} \cos t + \tfrac{3}{10} \sin t.$$

2. The only difference between this exercise and Exercise 1 is the coefficient of 5 on the right-hand side. Hence, the complex version is

$$\frac{d^2y}{dt^2} + 3\frac{dy}{dt} + 2y = 5e^{it}.$$

The guess for the particular solution is the same $y_c(t) = ae^{it}$, and the same calculation yields that $y_c(t)$ is a solution if

$$a = \frac{5}{1 + 3i} = \frac{1}{2} - \frac{3}{2}i.$$

Hence,

$$y_c(t) = \frac{1 - 3i}{2}e^{it} = \frac{1 - 3i}{2}(\cos t + i \sin t).$$

Taking the real part and adding the general solution of the homogeneous equation (see Exercise 1), we obtain the general solution

$$y(t) = k_1 e^{-2t} + k_2 e^{-t} + \tfrac{1}{2} \cos t + \tfrac{3}{2} \sin t.$$

3. Recalling that the imaginary part of e^{it} is $\sin t$, the complex version of the equation is

$$\frac{d^2y}{dt^2} + 3\frac{dy}{dt} + 2y = e^{it}.$$

This equation is precisely the same complex equation as in Exercise 1. Hence, we have already computed the solution

$$y_c(t) = \frac{1 - 3i}{10}(\cos t + i \sin t).$$

In this case we take the imaginary part

$$y(t) = -\tfrac{3}{10} \cos t + \tfrac{1}{10} \sin t$$

to obtain a solution of the original differential equation.

The general solution of the homogeneous equation is the same as in Exercise 1, so the general solution is

$$y(t) = k_1 e^{-2t} + k_2 e^{-t} - \tfrac{3}{10}\cos t + \tfrac{1}{10}\sin t.$$

4. This equation is the same as the equation in Exercise 3 except for the coefficient of 2 on the right-hand side. The complex version of the equation is

$$\frac{d^2y}{dt^2} + 3\frac{dy}{dt} + 2y = 2e^{it},$$

and the guess of the particular solution $y_c(t) = ae^{it}$ yields

$$a = \frac{1-3i}{5}$$

via the same steps as in Exercise 3. Taking the imaginary part of

$$y_c(t) = \frac{1-3i}{5}e^{it} = \frac{1-3i}{5}(\cos t + i\sin t)$$

and adding the general solution of the homogeneous equation yields

$$y(t) = k_1 e^{-2t} + k_2 e^{-t} + \tfrac{1}{5}\sin t - \tfrac{3}{5}\cos t.$$

5. The complex version of this equation is

$$\frac{d^2y}{dt^2} + 6\frac{dy}{dt} + 8y = e^{it}.$$

We guess a particular solution of the form $y_c(t) = ae^{it}$. Then $dy_c/dt = iae^{it}$ and $d^2y/dt^2 = -ae^{it}$. Substituting these derivatives into the complex differential equation yields

$$(-a + 6ia + 8a)e^{it} = e^{it},$$

which is satisfied if $(7+6i)a = 1$. Then $a = 1/(7+6i)$, and

$$y_c(t) = \frac{7-6i}{85}e^{it} = \frac{7-6i}{85}(\cos t + i\sin t).$$

The real part

$$y(t) = \tfrac{7}{85}\cos t + \tfrac{6}{85}\sin t$$

is a solution of the original differential equation.

To find the general solution of the homogeneous equation, we note that the characteristic polynomial $s^2 + 6s + 8$ has roots $s = -4$ and $s = -2$. Consequently, the general solution of the original nonhomogeneous equation is

$$y(t) = k_1 e^{-4t} + k_2 e^{-2t} + \tfrac{7}{85}\cos t + \tfrac{6}{85}\sin t.$$

6. The complex version of the equation is

$$\frac{d^2y}{dt^2} + 6\frac{dy}{dt} + 8y = -4e^{3it},$$

and to find a particular solution, we guess $y_c(t) = ae^{3it}$. Substituting this guess into the equation, we obtain

$$-9ae^{3it} + 18aie^{3it} + 8ae^{3it} = -4e^{3it},$$

which can be simplified to

$$(-1 + 18i)ae^{3it} = -4e^{3it}.$$

Thus, $y_c(t)$ is a solution if $a = -4/(-1 + 18i)$. We have

$$y_c(t) = \frac{-4}{-1 + 18i}e^{3it} = \frac{4 + 72i}{325}(\cos 3t + i \sin 3t),$$

and we take the real part to obtain a solution

$$y(t) = \tfrac{4}{325} \cos 3t - \tfrac{72}{325} \sin 3t$$

of the original equation.

To find the general solution of the unforced equation, we note that the characteristic polynomial is $s^2 + 6s + 8$, which has roots $s = -4$ and $s = -2$. Hence, the general solution of the original forced equation is

$$y(t) = k_1e^{-4t} + k_2e^{-2t} + \tfrac{4}{325} \cos 3t - \tfrac{72}{325} \sin 3t.$$

7. The complex version of this equation is

$$\frac{d^2y}{dt^2} + 4\frac{dy}{dt} + 13y = 3e^{2it},$$

so we guess $y_c(t) = ae^{2it}$ to find a particular solution. Substituting $y_c(t)$ into the differential equation gives

$$(-4a + 8ai + 13a)e^{2it} = 3e^{2it},$$

which is satisfied if $(9 + 8i)a = 3$. Thus, $y_c(t)$ is a solution if

$$y_c(t) = \frac{3}{9 + 8i}e^{2it} = \frac{27 - 24i}{145}(\cos 2t + i \sin 2t).$$

A particular solution of the original equation is the real part

$$y(t) = \tfrac{27}{145} \cos 2t + \tfrac{24}{145} \sin 2t.$$

To find the general solution of the homogeneous equation, we note that the characteristic polynomial $s^2+4s+13$ has roots $s = -2 \pm 3i$. Hence, the general solution of the original forced equation is

$$y(t) = k_1e^{-2t} \cos 3t + k_2e^{-2t} \sin 3t + \tfrac{27}{145} \cos 2t + \tfrac{24}{145} \sin 2t.$$

8. The complex version of the equation is

$$\frac{d^2y}{dt^2} + 4\frac{dy}{dt} + 20y = -e^{5it},$$

and we guess that there is a solution of the form $y_c(t) = ae^{5it}$. Substituting this guess into the differential equation yields

$$-25ae^{5it} + 20aie^{5it} + 20ae^{5it} = -e^{5it},$$

which can be simplified to

$$(-5 + 20i)ae^{5it} = -e^{5it}.$$

Thus, $y_c(t)$ is a solution if $a = -1/(-5 + 20i)$. We have

$$y_c(t) = \frac{-1}{-5 + 20i}\, e^{5it} = \frac{1 + 4i}{85}(\cos 5t + i \sin 5t).$$

We take the real part to obtain a solution

$$y(t) = \tfrac{1}{85} \cos 5t - \tfrac{4}{85} \sin 5t$$

of the original equation.

To find the general solution of the homogeneous equation, we note that the characteristic polynomial is $s^2 + 4s + 20$, which has roots $s = -2 \pm 4i$. Hence, the general solution of the original equation is

$$y(t) = k_1e^{-2t} \cos 4t + k_2e^{-2t} \sin 4t + \tfrac{1}{85} \cos 5t - \tfrac{4}{85} \sin 5t.$$

9. The complex version of this equation is

$$\frac{d^2y}{dt^2} + 4\frac{dy}{dt} + 20y = -3e^{2it},$$

and we guess that there is a solution of the form $y_c(t) = ae^{2it}$. Substituting this guess into the differential equation yields

$$(-4a + 8ia + 20a)e^{2it} = -3e^{2it},$$

which can be simplified to

$$(16 + 8i)ae^{2it} = -3e^{2it}.$$

Thus, $y_c(t)$ is a solution if $a = -3/(16 + 8i)$. We have

$$y_c(t) = \frac{-3}{16 + 8i}\, e^{2it} = \left(-\frac{3}{20} + \frac{3}{40}i\right)\left(\cos 2t + i \sin 2t\right).$$

We take the imaginary part to obtain a solution

$$y(t) = \tfrac{3}{40} \cos 2t - \tfrac{3}{20} \sin 2t$$

of the original equation.

To find the general solution of the homogeneous equation, we note that the characteristic polynomial is $s^2 + 4s + 20$, whose roots are $s = -2 \pm 4i$. Hence, the general solution of the original equation is

$$y(t) = k_1e^{-2t} \cos 4t + k_2e^{-2t} \sin 4t - \tfrac{3}{20} \sin 2t + \tfrac{3}{40} \cos 2t.$$

10. The complex version of the equation is

$$\frac{d^2y}{dt^2} + 2\frac{dy}{dt} + y = e^{3it},$$

and we guess there is a particular solution of the form $y_c(t) = ae^{3it}$. Substituting this guess into the differential equation yields

$$-9ae^{3it} + 6iae^{3it} + ae^{3it} = e^{3it},$$

which can be simplified to

$$(-8 + 6i)ae^{3it} = e^{3it}.$$

Thus, $y_c(t)$ is a solution if $a = 1/(-8 + 16i)$. We have

$$y_c(t) = \frac{1}{-8 + 16i}e^{3it} = -\left(\frac{2}{25} + \frac{3}{50}i\right)\left(\cos 3t + i \sin 3t\right).$$

We take the real part to obtain a solution

$$y(t) = -\tfrac{2}{25}\cos 3t + \tfrac{3}{50}\sin 3t$$

of the original equation.

To find the general solution of the unforced equation we note that the characteristic polynomial is s^2+2s+1, which has $s = -1$ as a double root. Hence, the general solution of the original equation is

$$y(t) = k_1e^{-t} + k_2te^{-t} - \tfrac{2}{25}\cos 3t + \tfrac{3}{50}\sin 3t.$$

11. From Exercise 5, we know that the general solution of this equation is

$$y(t) = k_1e^{-4t} + k_2e^{-2t} + \tfrac{7}{85}\cos t + \tfrac{6}{85}\sin t.$$

To find the desired solution, we must solve for k_1 and k_2 using the initial conditions. We have

$$\begin{cases} k_1 + k_2 + \tfrac{7}{85} = 0 \\ -4k_1 - 2k_2 + \tfrac{6}{85} = 0. \end{cases}$$

We obtain $k_1 = 2/17$ and $k_2 = -1/5$. The desired solution is

$$y(t) = \tfrac{2}{17}e^{-4t} - \tfrac{1}{5}e^{-2t} + \tfrac{7}{85}\cos t + \tfrac{6}{85}\sin t.$$

12. We can solve this initial-value problem by finding the general solution in many ways. One method involves producing a particular solution to the differential equation using the guess-and-test technique described in the text. Another way to find a particular solution is to note that the left-hand side of this equation is the same as the left-hand side of the equation in Exercise 6 and the right-hand side of this equation differs from the right-hand side of that equation by a factor of $-1/2$. Since these equations are linear, we can use the Linearity Principle to derive a particular solution to this equation by multiplying the particular solution we found in Exercise 6 by $-1/2$. Hence the general solution of the differential equation in this exercise is

$$y(t) = k_1 e^{-4t} + k_2 e^{-2t} - \tfrac{2}{325}\cos 3t + \tfrac{36}{325}\sin 3t.$$

To obtain the desired initial conditions, we must solve for k_1 and k_2. We have

$$\begin{cases} k_1 + k_2 - \tfrac{2}{325} = 0 \\ -4k_1 - 2k_2 + \tfrac{108}{324} = 0. \end{cases}$$

We obtain $k_1 = 4/25$ and $k_2 = -2/13$. The desired solution is

$$y(t) = \tfrac{4}{25}e^{-4t} - \tfrac{2}{13}e^{-2t} - \tfrac{2}{325}\cos 3t + \tfrac{36}{325}\sin 3t.$$

13. From Exercise 9, we know that the general solution of this equation is

$$y(t) = k_1 e^{-2t}\cos 4t + k_2 e^{-2t}\sin 4t - \tfrac{3}{20}\sin 2t + \tfrac{3}{40}\cos 2t.$$

To find the desired solution, we must solve for k_1 and k_2 using the initial conditions. We have

$$\begin{cases} k_1 + \tfrac{3}{40} = 0 \\ -2k_1 + 4k_2 - \tfrac{6}{20} = 0. \end{cases}$$

We obtain $k_1 = -3/40$ and $k_2 = 3/80$. The desired solution is

$$y(t) = -\tfrac{3}{40}e^{-2t}\cos 4t + \tfrac{3}{80}e^{-2t}\sin 4t - \tfrac{3}{20}\sin 2t + \tfrac{3}{40}\cos 2t.$$

14. First we find the general solution of the differential equation using the Extended Linearity Principle and the standard guess-and-test technique. The complex version of the equation is

$$\frac{d^2y}{dt^2} + 2\frac{dy}{dt} + y = 2e^{2it},$$

and we guess $y_c(t) = ae^{2it}$ as a particular solution. Substituting this guess into the equation yields

$$a = \frac{2}{-3+4i} = \frac{-6-8i}{25}.$$

Hence, a particular solution is the real part of

$$y_c(t) = \frac{-6-8i}{25}(\cos 2t + i\sin 2t).$$

We have

$$y(t) = -\tfrac{6}{25}\cos 2t + \tfrac{8}{25}\sin 2t.$$

To find the general solution of the homogeneous equation, we note that the characteristic polynomial is $s^2 + 2s + 1$, which has $s = -1$ as a double root. Hence, the general solution of the original equation is

$$y(t) = k_1 e^{-t} + k_2 t e^{-t} - \tfrac{6}{25}\cos 2t + \tfrac{8}{25}\sin 2t.$$

To obtain the desired initial conditions, we solve for k_1 and k_2 using

$$\begin{cases} k_1 - \tfrac{6}{25} = 0 \\ -k_1 + k_2 + \tfrac{16}{25} = 0. \end{cases}$$

We see that $k_1 = 6/25$ and $k_2 = -2/5$, so the desired solution is

$$y(t) = \tfrac{6}{25}e^{-t} - \tfrac{2}{5}te^{-t} - \tfrac{6}{25}\cos 2t + \tfrac{8}{25}\sin 2t.$$

15. **(a)** If we guess

$$y_p(t) = a\cos 3t + b\sin 3t,$$

then

$$y_p'(t) = -3a\sin 3t + 3b\cos 3t$$

and

$$y_p''(t) = -9a\cos 3t - 9b\sin 3t.$$

Substituting this guess and its derivatives into the differential equation gives

$$(-8a + 9b)\cos 3t + (-9a - 8b)\sin 3t = \cos 3t.$$

Thus $y_p(t)$ is a solution if a and b satisfy the simultaneous equations

$$\begin{cases} -8a + 9b = 1 \\ -9a - 8b = 0. \end{cases}$$

Solving these equations for a and b, we obtain $a = -8/145$ and $b = 9/145$, so

$$y_p(t) = -\tfrac{8}{145}\cos 3t + \tfrac{9}{145}\sin 3t$$

is a solution.

(b) If we guess

$$y_p(t) = A\cos(3t + \phi),$$

then

$$y_p'(t) = -3A\sin(3t + \phi)$$

and

$$y_p''(t) = -9A\cos(3t + \phi).$$

Substituting this guess and its derivatives into the differential equation yields

$$-8A\cos(3t + \phi) - 9A\sin(3t + \phi) = \cos 3t.$$

Using the trigonometric identities for the sine and cosine of the sum of two angles, we have

$$-8A\,(\cos 3t\cos\phi - \sin 3t\sin\phi) - 9A\,(\sin 3t\cos\phi + \cos 3t\sin\phi) = \cos 3t.$$

This equation can be rewritten as

$$(-8A\cos\phi - 9A\sin\phi)\cos 3t + (8A\sin\phi - 9A\cos\phi)\sin 3t = \cos 3t.$$

It holds if

$$\begin{cases} -8A\cos\phi - 9A\sin\phi = 1 \\ 9A\cos\phi - 8A\sin\phi = 0. \end{cases}$$

Multiplying the first equation by 9 and the second by 8 and adding yields

$$145A\sin\phi = -9.$$

Similarly, multiplying the first equation by -8 and the second by 9 and adding yields

$$145A\cos\phi = -8.$$

Taking the ratio gives

$$\frac{\sin\phi}{\cos\phi} = \tan\phi = \frac{9}{8}.$$

Also, squaring both $145A\sin\phi = -9$ and $145A\cos\phi = -8$ yields

$$145^2A^2\cos^2\phi + 145^2A^2\sin^2\phi = 145,$$

so $A^2 = 1/145$.

We can use either $A = 1/\sqrt{145}$ or $A = -1/\sqrt{145}$, but this choice of sign for A effects the value of ϕ. If we pick $A = -1/\sqrt{145}$, then $\sqrt{145}\sin\phi = 9$, $\sqrt{145}\cos\phi = 8$, and $\tan\phi = 9/8$. In this case, $\phi = \arctan(9/8)$. Hence, a particular solution of the original equation is

$$y_p(t) = \frac{1}{\sqrt{145}}\cos\left(3t + \arctan\frac{9}{8}\right).$$

16. **(a)** Substituting $ky_p(t)$ into the left-hand side of the differential equation and simplifying yields

$$\frac{d^2(ky_p)}{dt^2} + p\frac{d(ky_p)}{dt} + q(ky_p) = k\frac{d^2y_p}{dt^2} + pk\frac{dy_p}{dt} + qky_p$$

since the derivative of ky is $k(dy/dt)$ if k is a constant. Consequently,

$$\begin{aligned}\frac{d^2(ky_p)}{dt^2} + p\frac{d(ky_p)}{dt} + q(ky_p) &= k\left(\frac{d^2y_p}{dt^2} + p\frac{dy_p}{dt} + qy_p\right)\\ &= k\,g(t)\end{aligned}$$

because $y_p(t)$ is a solution of

$$\frac{d^2y}{dt^2} + p\frac{dy}{dt} + qy = g(t).$$

(b) By Exercise 5 we know that one solution of

$$\frac{d^2y}{dt^2} + 6\frac{dy}{dt} + 8y = \cos t$$

is

$$y_1(t) = \tfrac{7}{85}\cos t + \tfrac{6}{85}\sin t.$$

Using the result of part (a), a particular solution of the given equation is $y_2(t) = 5y_1(t)$. In other words,

$$y_2(t) = \tfrac{7}{17}\cos t + \tfrac{6}{17}\sin t$$

is a particular solution to the equation in this exercise.

The general solution of the homogeneous equation is the same as in Exercise 5, so the general solution for this exercise is

$$y(t) = k_1e^{-4t} + k_2e^{-2t} + \tfrac{7}{17}\cos t + \tfrac{6}{17}\sin t.$$

17. Since p and q are both positive, the solution of the homogeneous equation (the unforced response) tends to zero. Hence, we can match solutions to equations by considering the period (or frequency) and the amplitude of the steady-state solution (forced response). We also need to consider the rate at which solutions tend to the steady-state solution.

(a) The steady-state solution has period $2\pi/3$, and since the period of the steady-state solution is the same as the period of the forcing function, these solutions correspond to equations (v) or (vi). Moreover, this observation applies to the solutions in part (d) as well. Therefore, we need to match equations (v) and (vi) with the solutions in parts (a) and (d).

Solutions approach the steady-state faster in (d) than in (a). To distinguish (v) from (vi), we consider their characteristic polynomials. The characteristic polynomial for (v) is

$$s^2 + 5s + 1,$$

which has eigenvalues $(-5 \pm \sqrt{21})/2$. The characteristic polynomial for (vi) is

$$s^2 + s + 1,$$

which has eigenvalues $(-1 \pm i\sqrt{3})/2$. The rate of approach to the steady-state for (v) is determined by the slow eigenvalue $(-5+\sqrt{21})/2 \approx -0.21$. The rate of approach to the steady-state for (vi) is determined by the real part of the eigenvalue, -0.5. Therefore, the graphs in part (a) come from equation (v), and the graphs in part (d) come from equation (vi).

(b) The steady-state solution has period 2π, and since the period of the steady-state solution is the same as the period of the forcing function, these solutions correspond to equations (i) or (ii). Moreover, this observation applies to the solutions in part (c) as well. Therefore, we need to match equations (i) and (ii) with the solutions in parts (b) and (c).

The amplitude of the steady-state solution is larger in (b) than in (c). To distinguish (i) from (ii), we calculate the amplitudes of the steady-state solutions for (i) and (ii). If we complexify these equations, we get

$$\frac{d^2y}{dt^2} + p\frac{dy}{dt} + qy = e^{it}.$$

Guessing a solution of the form $y_c(t) = ae^{it}$, we see that

$$a = \frac{1}{(q-1) + pi}.$$

The amplitude of the steady-state solution is $|a|$. For equation (i), $|a| = 1/\sqrt{29} \approx 0.19$, and for equation (ii), it is $1/\sqrt{5} \approx 0.44$. Therefore, the graphs in part (b) correspond to equation (ii), and the graphs in part (c) correspond to equation (i).

(c) See the answer to part (b).

(d) See the answer to part (a).

18. **(a)** Due to the result in Exercise 36 in Section 4.1, we consider each forcing term separately. That is, we consider the two differential equations

$$\frac{d^2y}{dt^2} + 4\frac{dy}{dt} + 20y = 3 \quad \text{and} \quad \frac{d^2y}{dt^2} + 4\frac{dy}{dt} + 20y = 2\cos 2t.$$

To find a particular solution of the first equation, we guess a constant function $y_1(t) = a$. Substituting this guess into the equation yields $20a = 3$, so $y_1(t) = 3/20$ is a solution.

To find a particular solution of the second equation, we consider the complex version

$$\frac{d^2y}{dt^2} + 4\frac{dy}{dt} + 20y = 2e^{2it}$$

and guess a solution of the form $y_c(t) = ae^{2it}$. Substituting $y_c(t)$ into the equation yields

$$(16 + 8i)ae^{2it} = 2e^{2it},$$

which is satisfied if $a = 2/(16 + 8i)$. A solution $y_2(t)$ of the second equation is obtained by taking the real part of $y_c(t)$. Since

$$y_c(t) = \left(\tfrac{1}{10} - \tfrac{1}{20}i\right)\left(\cos 2t + i \sin 2t\right),$$

we have

$$y_2(t) = \tfrac{1}{10}\cos 2t + \tfrac{1}{20}\sin 2t.$$

To find the solution of the unforced equation, we note that the characteristic polynomial is $s^2 + 4s + 20$, which has roots $s = -2 \pm 4i$.

Hence, the general solution is

$$y(t) = k_1 e^{-2t}\cos 4t + k_2 e^{-2t}\sin 4t + \tfrac{3}{20} + \tfrac{1}{10}\cos 2t + \tfrac{1}{20}\sin 2t.$$

(b) The first two terms of the general solution tend quickly to zero, so all solutions eventually oscillate about $y = 3/20$. The period and amplitude of the oscillations is determined by the period and amplitude of the oscillations of $y_2(t)$. The period of $y_2(t)$ is π.

19. **(a)** Using the fact that the real part of $e^{(-1+i)t}$ is $e^{-t}\cos t$, the complex version of this equation is

$$\frac{d^2y}{dt^2} + 4\frac{dy}{dt} + 20y = e^{(-1+i)t}.$$

Guessing $y_c(t) = ae^{(-1+i)t}$ yields

$$a(-1+i)^2 e^{(-1+i)t} + 4a(-1+i)e^{(-1+i)t} + 20ae^{(-1+i)t} = e^{(-1+i)t}.$$

Simplifying we have

$$a(16+2i)e^{(-1+i)t} = e^{(-1+i)t}.$$

Thus, $y_c(t)$ is a solution of the complex differential equation if $a = 1/(16+2i)$, and we have

$$y_c(t) = \left(\tfrac{4}{65} - \tfrac{1}{130}i\right) e^{-t} (\cos t + i \sin t).$$

So one solution of the original equation is

$$y_p(t) = \tfrac{4}{65}e^{-t} \cos t + \tfrac{1}{130}e^{-t} \sin t.$$

To find the general solution of the homogeneous equation, we note that the characteristic polynomial $s^2 + 4s + 20$ has roots $s = -2 \pm 4i$.

Hence, the general solution of the original equation is

$$y(t) = k_1 e^{-2t} \cos 4t + k_2 e^{-2t} \sin 4t + \tfrac{4}{65}e^{-t} \cos t + \tfrac{1}{130}e^{-t} \sin t.$$

(b) All four terms in the general solution tend to zero as $t \to \infty$. Hence, all solutions tend to zero as $t \to \infty$. The terms with factors of e^{-2t} tend to zero very quickly, which leaves the terms of the particular solution $y_p(t)$ as the largest terms, so all solutions are asymptotic to $y_p(t)$. Since the solution $y_p(t)$ oscillates with period 2π and the amplitude of its oscillations decreases at the rate of e^{-t}, all solutions oscillate with this period and decaying amplitude.

20. **(a)** To find the general solution of the unforced equation, we note that the characteristic polynomial is $s^2 + 4s + 20$, which has roots $s = -2 \pm 4i$.

To find a particular solution of the forced equation, we note that the complex version of the equation is

$$\frac{d^2y}{dt^2} + 4\frac{dy}{dt} + 20y = e^{(-2+4i)t}.$$

We could guess $y_c(t) = ae^{(-2+4i)t}$ as a particular solution, but with perfect hindsight, we recall that the roots of the characteristic polynomial of the unforced equation are $-2 \pm 4i$. Hence, $e^{(-2+4i)t}$ is already a solution of the homogeneous equation. In other words, no value of a will make this $y_c(t)$ a solution. (Why?)

So we second guess $y_c(t) = ate^{(-2+4i)t}$ and substitute this guess into the equation to obtain

$$a\left([-4+8i+(-12-16i)t] + 4[1+(-2+4i)t] + 20t\right) e^{(-2+4i)t} = e^{(-2+4i)t},$$

which simplifies to

$$a(8i)e^{(-2+4i)t} = e^{(-2+4i)t}.$$

Hence, $y_c(t) = ate^{(-2+4i)t}$ is a solution if $a = 1/(8i) = -i/8$. To find a particular solution of the original equation, we compute the imaginary part of

$$y_c(t) = -\tfrac{1}{8}ite^{-2t} (\cos 4t + i \sin 4t).$$

We have

$$y(t) = -\tfrac{1}{8}te^{-2t}\cos 4t.$$

The general solution of the original equation is

$$y(t) = k_1e^{-2t}\cos 4t + k_2e^{-2t}\sin 4t - \tfrac{1}{8}te^{-2t}\cos 4t.$$

(b) All terms of the general solution tend to zero. The term that tends to zero most slowly is $-(te^{-2t}\cos 4t)/8$, so for large t, all solutions are approximately equal to this term.

21. Note that the real part of

$$(a - bi)(\cos\omega t + i\sin\omega t)$$

is $g(t)$. Hence, we must find k and ϕ such that

$$ke^{i\phi} = a - bi.$$

Using the polar form of the complex number $z = a - bi$, we see that

$$ke^{i\phi} = a - bi = z = |z|e^{i\theta},$$

where θ is the polar angle for z (see Appendix C). Therefore, we can choose

$$k = |z| = \sqrt{a^2 + b^2} \quad \text{and} \quad \phi = \theta.$$

22. Note that $g_1(t)$ is the real part of $k_1e^{i\phi_1}e^{i\omega t}$ and $g_2(t)$ is the real part of $k_2e^{i\phi_2}e^{i\omega t}$, so $g_1(t) + g_2(t)$ is the real part of

$$k_1e^{i\phi_1}e^{i\omega t} + k_2e^{i\phi_2}e^{i\omega t},$$

which can be rewritten as

$$\left(k_1e^{i\phi_1} + k_2e^{i\phi_2}\right)e^{i\omega t}.$$

Thus, the phasor for $g_1(t) + g_2(t)$ is $k_1e^{i\phi_1} + k_2e^{i\phi_2}$.

23. Note that the real part of

$$(k_1 - ik_2)e^{i\beta t} = (k_1 - ik_2)(\cos\beta t + i\sin\beta t)$$

is

$$y(t) = k_1\cos\beta t + k_2\sin\beta t.$$

Let $Ke^{i\phi}$ be the polar form of the complex number $k_1 + ik_2$. Then the polar form of $k_1 - ik_2$ is $Ke^{-i\phi}$. Using the Laws of Exponents and Euler's formula, we have

$$\begin{aligned}(k_1 - ik_2)e^{i\beta t} &= Ke^{-i\phi}e^{i\beta t}\\ &= Ke^{i(\beta t - \phi)}\\ &= K(\cos(\beta t - \phi) + i\sin(\beta t - \phi)),\end{aligned}$$

and the real part is $K\cos(\beta t - \phi)$. Hence, we see that

$$y(t) = k_1\cos\beta t + k_2\sin\beta t$$

can be rewritten as

$$y(t) = K\cos(\beta t - \phi).$$

EXERCISES FOR SECTION 4.3

1. The complex version of this equation is

$$\frac{d^2y}{dt^2} + 9y = e^{it}.$$

Guessing $y_c(t) = ae^{it}$ as a particular solution and substituting this guess into the left-hand side of the differential equation yields

$$8ae^{it} = e^{it}.$$

Thus, $y_c(t)$ is a solution if $8a = 1$. The real part of

$$y_c(t) = \tfrac{1}{8}e^{it} = \tfrac{1}{8}(\cos t + i \sin t)$$

is $y(t) = \frac{1}{8}\cos t$. This $y(t)$ is a solution to the original differential equation. [Because there is no dy/dt-term (no damping), we could have guessed a solution of the form $y(t) = a\cos t$ instead of using the complex version of the equation.]

To find the general solution of the homogeneous equation, we note that the characteristic polynomial is $s^2 + 9$, which has roots $s = \pm 3i$. So the general solution of the original equation is

$$y(t) = k_1 \cos 3t + k_2 \sin 3t + \tfrac{1}{8}\cos t.$$

2. The complex version of this equation is

$$\frac{d^2y}{dt^2} + 9y = 5e^{2it}.$$

Guessing $y_c(t) = ae^{2it}$ as a particular solution and substituting this guess into the left-hand side of the differential equation yields

$$5ae^{2it} = 5e^{2it}.$$

Thus, $y_c(t)$ is a solution if $5a = 5$. The imaginary part of

$$y_c(t) = \cos 2t + i \sin 2t$$

is $y(t) = \sin 2t$. This $y(t)$ is a solution to the original differential equation. [Because there is no dy/dt-term (no damping), we could have guessed a solution of the form $y(t) = a\sin 2t$ instead of using the complex version of the equation.]

To find the general solution of the homogeneous equation, we note that the characteristic polynomial is $s^2 + 9$, which has roots $s = \pm 3i$. So the general solution of the original equation is

$$y(t) = k_1 \cos 3t + k_2 \sin 3t + \sin 2t.$$

3. The complex version of this equation is

$$\frac{d^2y}{dt^2} + 4y = -e^{it/2}.$$

Guessing $y_c(t) = ae^{it/2}$ as a particular solution and substituting into the equation yields

$$\tfrac{15}{4}ae^{it/2} = -e^{it/2}.$$

Thus, $y_c(t)$ is a solution if $\frac{15}{4}a = -1$. The real part of

$$y_c(t) = -\frac{4}{15}e^{it/2} = -\frac{4}{15}\left(\cos\frac{t}{2} + i\sin\frac{t}{2}\right)$$

is

$$y(t) = -\frac{4}{15}\cos\frac{t}{2}.$$

This $y(t)$ is a solution to the original differential equation. [Because there is no dy/dt-term (no damping), we could have guessed a solution of the form $y(t) = a\cos t/2$ instead of using the complex version of the equation.]

To find the general solution of the homogeneous equation, we note that the characteristic polynomial is $s^2 + 4$, which has roots $s = \pm 2i$. So the general solution of the original equation is

$$y(t) = k_1\cos 2t + k_2\sin 2t - \frac{4}{15}\cos\frac{t}{2}.$$

4. The complex version of the equation is

$$\frac{d^2y}{dt^2} + 4y = 3e^{2it}.$$

Guessing $y_c(t) = ae^{2it}$ as a particular solution and substituting this guess into the left-hand side of the differential equation, we see that a must satisfy

$$-4a + 4a = 3,$$

which is impossible. Hence, the forcing is in resonance with the associated homogeneous equation.

We must make a second guess of $y_c(t) = ate^{2it}$. This guess gives

$$y_c'(t) = a(1 + 2it)e^{2it}$$

and

$$y_c''(t) = 4a(i - t)e^{2it}.$$

Substituting $y_c(t)$ and its second derivative into the differential equation, we obtain

$$4a(i - t)e^{2it} + 4ate^{2it} = 3e^{2it},$$

which simplifies to

$$4aie^{2it} = 3e^{2it}.$$

Thus, $y_c(t)$ is a solution if $a = 3/(4i) = -3i/4$. Taking the real part of

$$y_c(t) = -\tfrac{3}{4}it(\cos 2t + i\sin 2t),$$

we obtain the solution

$$y(t) = \tfrac{3}{4}t \sin 2t$$

of the original equation.

To find the general solution of the homogeneous equation, we note that the characteristic polynomial is $s^2 + 4$, which has roots $s = \pm 2i$.

Hence, the general solution of the original equation is

$$y(t) = k_1 \cos 2t + k_2 \sin 2t + \tfrac{3}{4}t \sin 2t.$$

5. The complex version of the equation is

$$\frac{d^2y}{dt^2} + 9y = 2e^{3it}.$$

Guessing $y_c(t) = ae^{3it}$ as a particular solution and substituting this guess into the left-hand side of the differential equation, we see that a must satisfy

$$-9a + 9a = 2,$$

which is impossible. Hence, the forcing is in resonance with the associated homogeneous equation.

We must make a second guess of $y_c(t) = ate^{3it}$. This guess gives

$$y_c'(t) = a(1 + 3it)e^{3it}$$

and

$$y_c''(t) = a(6i - 9t)e^{3it}.$$

Substituting $y_c(t)$ and its second derivative into the differential equation, we obtain

$$a(6i - 9t)e^{3it} + 9ate^{3it} = 2e^{3it},$$

which simplifies to

$$6aie^{3it} = 2e^{3it}.$$

Thus, $y_c(t)$ is a solution if $a = 2/(6i) = -i/3$. Taking the real part of

$$y_c(t) = -\tfrac{1}{3}it(\cos 3t + i \sin 3t),$$

we obtain the solution

$$y(t) = \tfrac{1}{3}t \sin 3t$$

of the original equation.

To find the general solution of the homogeneous equation, we note that the characteristic polynomial is $s^2 + 9$, which has roots $s = \pm 3i$.

Hence, the general solution of the original equation is

$$y(t) = k_1 \cos 3t + k_2 \sin 3t + \tfrac{1}{3}t \sin 3t.$$

6. The complex version of this equation is

$$\frac{d^2y}{dt^2} + 3y = 2e^{9it}.$$

Guessing $y_c(t) = ae^{9it}$ as a particular solution and substituting this guess into the left-hand side of the differential equation yields

$$-78ae^{9it} = 2e^{9it}.$$

Thus, $y_c(t)$ is a solution if $-78a = 2$, which yields $a = -1/39$. The real part of

$$y_c(t) = -\tfrac{1}{39}e^{9it} = -\tfrac{1}{39}(\cos 9t + i \sin 9t)$$

is $y(t) = -\frac{1}{39}\cos 9t$. This $y(t)$ is a solution to the original differential equation. [Because there is no dy/dt-term (no damping), we could have guessed a solution of the form $y(t) = a\cos 9t$ instead of using complex version of the equation.]

To find the general solution of the homogeneous equation, we note that the characteristic polynomial is $s^2 + 3$, which has roots $s = \pm\sqrt{3}\,i$. So the general solution of the original equation is

$$y(t) = k_1 \cos\sqrt{3}\,t + k_2 \sin\sqrt{3}\,t - \tfrac{1}{39}\cos 9t.$$

7. The complex version of this equation is

$$\frac{d^2y}{dt^2} + 3y = e^{3it}.$$

Guessing $y_c(t) = ae^{3it}$ as a particular solution and substituting this guess into the left-hand side of the differential equation yields

$$-6ae^{3it} = e^{3it}.$$

Thus, $y_c(t)$ is a solution if $-6a = 1$. The real part of

$$y_c(t) = -\tfrac{1}{6}e^{3it} = -\tfrac{1}{6}(\cos 3t + i \sin 3t)$$

is $y(t) = -\frac{1}{6}\cos 3t$. This $y(t)$ is a solution to the original differential equation. [Because there is no dy/dt-term (no damping), we could have guessed a solution of the form $y(t) = a\cos 3t$ instead of using the complex version of the equation.]

To find the general solution of the homogeneous equation, we note that the characteristic polynomial is $s^2 + 3$, which has roots $s = \pm\sqrt{3}\,i$.

Hence, the general solution of the original equation is

$$y(t) = k_1 \cos\sqrt{3}\,t + k_2 \sin\sqrt{3}\,t - \tfrac{1}{6}\cos 3t.$$

8. The complex version of this equation is

$$\frac{d^2y}{dt^2} + 5y = 5e^{5it}.$$

Guessing $y_c(t) = ae^{5it}$ as a particular solution and substituting this guess into the left-hand side of the differential equation yields

$$-20ae^{5it} = 5e^{5it}.$$

Thus, $y_c(t)$ is a solution if $-20a = 5$. The imaginary part of

$$y_c(t) = -\tfrac{1}{4}(\cos 5t + i \sin 5t)$$

is $y(t) = -\frac{1}{4} \sin 5t$. This $y(t)$ is a solution to the original differential equation. [Because there is no dy/dt-term (no damping), we could have guessed a solution of the form $y(t) = a \sin 5t$ instead of using the complex version of the equation.]

To find the general solution of the homogeneous equation, we note that the characteristic polynomial is $s^2 + 5$, which has roots $s = \pm\sqrt{5}\, i$.

Hence, the general solution of the original problem is

$$y(t) = k_1 \cos \sqrt{5}\, t + k_2 \sin \sqrt{5}\, t - \tfrac{1}{4} \sin 5t.$$

9. From Exercise 1, we know that the general solution is

$$y(t) = k_1 \cos 3t + k_2 \sin 3t + \tfrac{1}{8} \cos t.$$

So

$$y'(t) = -3k_1 \sin 3t + 3k_2 \cos 3t - \tfrac{1}{8} \sin t.$$

Using the initial conditions $y(0) = 0$ and $y'(0) = 0$, we obtain the simultaneous equations

$$\begin{cases} k_1 + \frac{1}{8} = 0 \\ 3k_2 = 0, \end{cases}$$

which imply that $k_1 = -1/8$ and $k_2 = 0$. The solution to the initial-value problem is

$$y(t) = -\tfrac{1}{8} \cos 3t + \tfrac{1}{8} \cos t.$$

10. From Exercise 4, we know that the general solution of this equation is

$$y(t) = k_1 \cos 2t + k_2 \sin 2t + \tfrac{3}{4} t \sin 2t.$$

Hence,

$$y'(t) = -2k_1 \sin 2t + 2k_2 \cos 2t + \tfrac{3}{4} \sin 2t + \tfrac{3}{2} t \cos 2t.$$

Using the initial conditions $y(0) = 0$ and $y'(0) = 0$, we obtain the simultaneous equations

$$\begin{cases} k_1 = 0 \\ 2k_2 = 0, \end{cases}$$

which imply that $k_1 = 0$ and $k_2 = 0$. The solution to the initial-value problem is

$$y(t) = \tfrac{3}{4} t \sin 2t.$$

11. First we find the general solution by considering the complex version of the equation

$$\frac{d^2y}{dt^2} + 5y = 3e^{2it}.$$

Guessing a particular solution of the form $y_c(t) = ae^{2it}$ and substituting this guess into the left-hand side of the equation yields

$$ae^{2it} = 3e^{2it}.$$

Thus, $y_c(t)$ is a solution if $a = 3$. The real part of

$$y_c(t) = 3(\cos 2t + i \sin 2t)$$

is $y(t) = 3\cos 2t$. This $y(t)$ is a solution to the original differential equation. [Because there is no dy/dt-term (no damping), we could have guessed a solution of the form $y(t) = a\cos 2t$ instead of using the complex version of the equation.]

To find the general solution of the homogeneous equation, we note that the characteristic polynomial is $s^2 + 5$, which has roots $s = \pm\sqrt{5}\,i$. Hence, the general solution is

$$y(t) = k_1 \cos \sqrt{5}\,t + k_2 \sin \sqrt{5}\,t + 3\cos 2t.$$

Note that

$$y'(t) = -\sqrt{5}\,k_1 \sin \sqrt{5}\,t + \sqrt{5}\,k_2 \cos \sqrt{5}\,t - 6\sin 2t.$$

Using the initial conditions $y(0) = 0$ and $y'(0) = 0$, we obtain the simultaneous equations

$$\begin{cases} k_1 + 3 = 0 \\ \sqrt{5}\,k_2 = 0, \end{cases}$$

which imply that $k_1 = -3$ and $k_2 = 0$. The solution to the initial-value problem is

$$y(t) = -3\cos \sqrt{5}\,t + 3\cos 2t.$$

12. First to find the general solution, we consider the complex version of the equation

$$\frac{d^2y}{dt^2} + 9y = e^{3it}.$$

Guessing a particular solution of the form $y_c(t) = ae^{3it}$ and substituting this guess into the left-hand side of the equation yields

$$0 = 3e^{2it}.$$

This initial guess for $y_c(t)$ does not provide a solution. (The forcing function is in resonance with the homogeneous part of the differential equation.)

We make a second guess of the form $y_c(t) = ate^{3it}$, so

$$y'(t) = a(1 + 3it)e^{3it}$$

and

$$y''(t) = a(6i - 9t)e^{3it}.$$

Substituting $y_c(t)$ and its second derivative into the differential equation yields

$$a(6i - 9t)e^{3it} + 9ate^{3it} = e^{3it},$$

which simplifies to

$$a(6i)e^{3it} = e^{3it}.$$

Thus, $y_c(t)$ is a solution if $a = 1/(6i) = -i/6$. The imaginary part of

$$y_c(t) = -\tfrac{1}{6}it(\cos 3t + i \sin 3t)$$

is $y(t) = -\frac{1}{6}t \cos 3t$. This $y(t)$ is a solution to the original differential equation.

To find the general solution of the homogeneous equation we note that the characteristic polynomial is $s^2 + 9$, which has roots $s = \pm 3i$. Hence, the general solution of the original equation is

$$y(t) = k_1 \cos 3t + k_2 \sin 3t - \tfrac{1}{6}t \cos 3t.$$

Note that

$$y'(t) = -3k_1 \sin 3t + 3k_2 \cos 3t - \tfrac{1}{6} \cos 3t + \tfrac{1}{2}t \sin 3t.$$

Using the initial conditions $y(0) = 1$ and $y'(0) = -1$, we obtain the simultaneous equations

$$\begin{cases} k_1 = 1 \\ 3k_2 - \frac{1}{6} = -1, \end{cases}$$

which imply $k_1 = 1$ and $k_2 = -5/18$. The solution to the initial-value problem is

$$y(t) = \cos 3t - \tfrac{5}{18} \sin 3t - \tfrac{1}{6}t \cos 3t.$$

13. From Exercise 5, we know that the general solution is

$$y(t) = k_1 \cos 3t + k_2 \sin 3t + \tfrac{1}{3}t \sin 3t.$$

So

$$y'(t) = -3k_1 \sin 3t + 3k_2 \cos 3t + \tfrac{1}{3} \sin 3t + t \cos 3t.$$

From the initial condition $y(0) = 2$, we see that $k_1 = 2$. Using the initial condition $y'(0) = -9$, we have $3k_2 = -9$. Hence, $k_2 = -3$. The solution to the initial-value problem is

$$y(t) = 2\cos 3t - 3 \sin 3t + \tfrac{1}{3}t \sin 3t.$$

14. First we find the general solution by considering the complex version of the equation

$$\frac{d^2y}{dt^2} + 4y = e^{3it}.$$

Guessing a particular solution of the form $y_c(t) = ae^{3it}$ and substituting this guess into the left-hand side of the equation yields

$$-5ae^{3it} = e^{3it}.$$

Thus, $y_c(t)$ is a solution if $a = -1/5$. The imaginary part of

$$y_c(t) = -\tfrac{1}{5}(\cos 3t + i \sin 3t)$$

is $y(t) = -\frac{1}{5}\sin 3t$. This $y(t)$ is a solution to the original differential equation. [Because there is no dy/dt-term (no damping), we could have guessed a solution of the form $y(t) = a \sin 3t$ instead of using the complex version of the equation.]

To find the general solution of the homogeneous equation, we note that the characteristic polynomial is $s^2 + 4$, which has roots $s = \pm 2i$. Hence, the general solution is

$$y(t) = k_1 \cos 2t + k_2 \sin 2t - \tfrac{1}{5}\sin 3t.$$

Note that

$$y'(t) = -2k_1 \sin 2t + 2k_2 \cos 2t - \tfrac{3}{5}\cos 3t.$$

From the initial condition $y(0) = 2$, we have $k_1 = 2$. Using the initial condition $y'(0) = 0$, we obtain the equation $2k_2 - \frac{3}{5} = 0$, which implies that $k_2 = \frac{3}{10}$. The solution to the initial-value problem is

$$y(t) = 2\cos 2t + \tfrac{3}{10}\sin 2t - \tfrac{1}{5}\sin 3t.$$

15. The characteristic polynomial of the unforced equation is $s^2 + 4$, which has roots $s = \pm 2i$. So the natural frequency is $2/(2\pi)$, and the forcing frequency is $9/(8\pi)$.

(a) The frequency of the beats is

$$\frac{\frac{9}{4} - 2}{4\pi} = \frac{1}{16\pi},$$

and therefore, the period of one beat is $16\pi \approx 50$.

(b) The frequency of the rapid oscillations is

$$\frac{\frac{9}{4} + 2}{4\pi} = \frac{17}{16\pi}.$$

Therefore, there are 17 rapid oscillations in each beat.

(c)

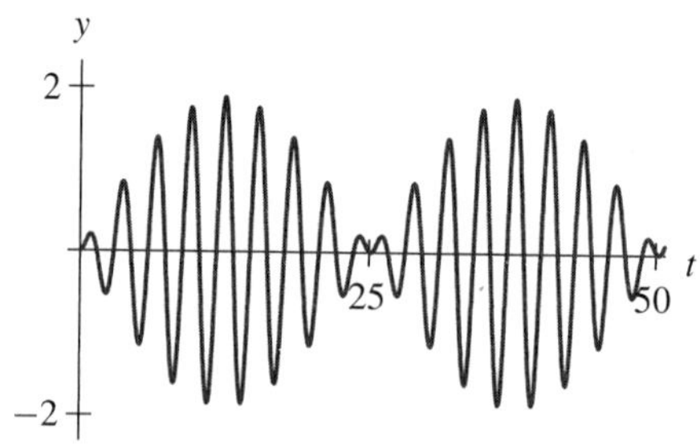

16. The characteristic polynomial of the unforced equation is $s^2 + 11$, which has roots $s = \pm i\sqrt{11}$. So the natural frequency is $\sqrt{11}/(2\pi)$, and the forcing frequency is $3/(2\pi)$.

(a) The frequency of the beats is

$$\frac{\sqrt{11} - 3}{4\pi},$$

and therefore, the period of one beat is $4\pi/(\sqrt{11} - 3) \approx 40$.

(b) The frequency of the rapid oscillations is

$$\frac{\sqrt{11}+3}{4\pi}.$$

Therefore, there are approximately 20 rapid oscillations in each beat.

(c)

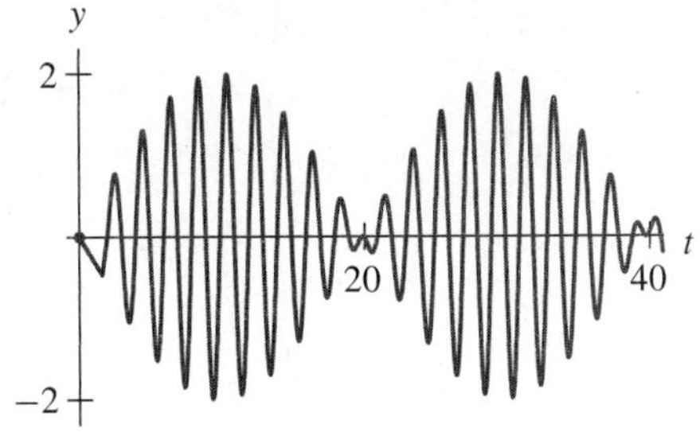

17. The characteristic polynomial of the unforced equation is $s^2 + 5$, which has roots $s = \pm i\sqrt{5}$. So the natural frequency is $\sqrt{5}/(2\pi)$, and the forcing frequency is $2/(2\pi)$.

(a) The frequency of the beats is

$$\frac{\sqrt{5}-2}{4\pi},$$

and therefore, the period of one beat is approximately 53.

(b) The frequency of the rapid oscillations is

$$\frac{\sqrt{5}+2}{4\pi}.$$

Therefore, there are approximately 18 rapid oscillations in each beat.

(c)

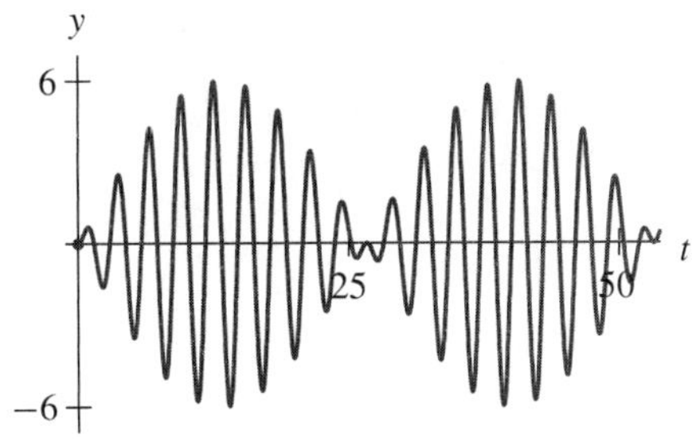

18. Note that the characteristic polynomial of the homogeneous equation is $s^2 + 6$, which has roots $s = \pm i\sqrt{6}$. So the natural frequency is $\sqrt{6}/(2\pi)$, and the forcing frequency is $2/(2\pi)$.

(a) The frequency of the beats is

$$\frac{\sqrt{6}-2}{4\pi},$$

and therefore, the period of one beat is approximately 28.

(b) The frequency of the rapid oscillations is

$$\frac{\sqrt{6}+2}{4\pi}.$$

Therefore, there are approximately 10 rapid oscillations in each beat.

(c)

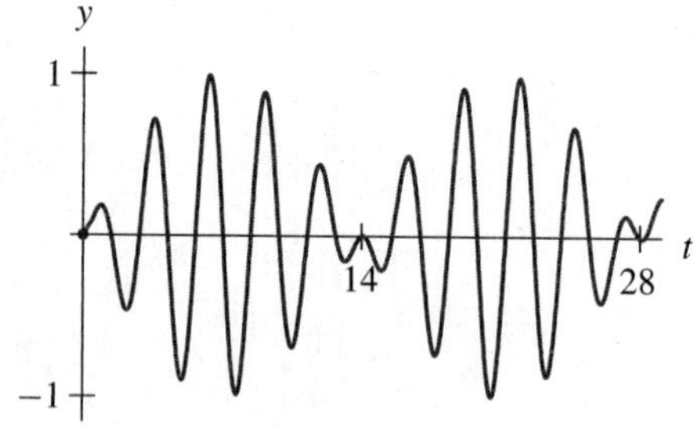

19. **(a)** To find the general solution, we deal with each of the forcing terms separately. In other words, we find solutions to

$$\frac{d^2y}{dt^2} + 12y = 3\cos 4t \quad \text{and} \quad \frac{d^2y}{dt^2} + 12y = 2\sin t$$

separately and add them to get a solution to the original equation (see Exercise 36 in Section 4.1).

First consider the equation whose forcing term is $3\cos 4t$. The complex version of the equation is

$$\frac{d^2y}{dt^2} + 12y = 3e^{4it}.$$

We guess a solution of the form $y_c(t) = ae^{4it}$ and substitute it into the differential equation to obtain

$$a(-16+12)e^{4it} = 3e^{4it},$$

which is satisfied if $a = -3/4$. Hence, by taking the real part of

$$y_c(t) = -\tfrac{3}{4}(\cos 4t + i\sin 4t),$$

we obtain the solution $y_1(t) = -\frac{3}{4}\cos 4t$.

Similarly, to find a solution of the equation whose forcing term is $2\sin t$, we consider

$$\frac{d^2y}{dt^2} + 12y = 2e^{it}$$

and guess $y_c(t) = be^{it}$. Substituting this guess into the equation yields

$$b(-1+12)e^{it} = 2e^{it},$$

which is satisfied if $b = 2/11$. So, by taking the imaginary part of

$$y_c(t) = \tfrac{2}{11}(\cos t + i\sin t),$$

we obtain the solution $y_2(t) = \frac{2}{11} \sin t$.

To obtain the general solution of the unforced equation, we note that the characteristic polynomial is $s^2 + 12$, which has roots $s = \pm i\sqrt{12}$. So the general solution of the original differential equation is

$$y(t) = k_1 \cos \sqrt{12}\,t + k_2 \sin \sqrt{12}\,t - \tfrac{3}{4} \cos 4t + \tfrac{2}{11} \sin t.$$

(b) To obtain the initial conditions $y(0) = 0$ and $y'(0) = 0$, we note that

$$y'(t) = -\sqrt{12}\,k_1 \sin \sqrt{12}\,t + \sqrt{12}\,k_2 \cos \sqrt{12}\,t + 3 \sin 4t + \tfrac{2}{11} \cos t.$$

Hence, we must solve the simultaneous equations

$$\begin{cases} k_1 - \frac{3}{4} = 0 \\ \sqrt{12}\,k_2 + \frac{2}{11} = 0 \end{cases}$$

for k_1 and k_2. We obtain $k_1 = 3/4$ and $k_2 = -2/(11\sqrt{12})$. Hence, the solution to the initial-value problem is

$$y(t) = \tfrac{3}{4} \cos \sqrt{12}\,t - \tfrac{2}{11\sqrt{12}} \sin \sqrt{12}\,t - \tfrac{3}{4} \cos 4t + \tfrac{2}{11} \sin t.$$

(c)

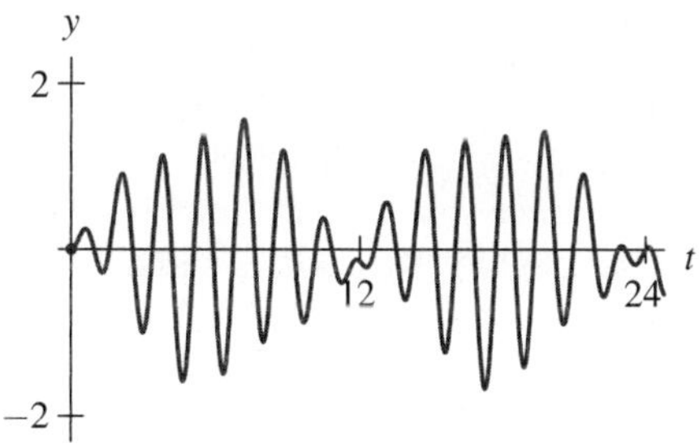

(d) The solution is the sum of the general solution of the unforced equation and particular solutions of the forced equations with each forcing term considered separately. Since the forcing function $\cos 4t$ is close to resonance, we expect to see large amplitude beats with frequency $(4 - \sqrt{12})/(4\pi)$ and rapid oscillations with frequency $(4 + \sqrt{12})/(4\pi)$. The particular solution for the $2 \sin t$ forcing term has a relatively small amplitude, and therefore it does not have a significant effect on the final oscillations most of the time.

20. The crystal glass and the opera singer's voice have similar frequencies. The singer's voice becomes a driving force, and the glass is shattered due to resonance. If the recorded voice has the same effect on the glass, the recorded voice also has a frequency similar to the glass's frequency. Thus, the recorded sound must have a frequency that is very close to the frequency of the original sound.

21. **(a)** The graph shows either the solution of a resonant equation or one with beats whose period is very large. The period of the beats in equation (iii) is 4π, and the period of the beats in equation (iv) is $4\pi/(4 - \sqrt{14}) \approx 48.6$. Hence this graph must correspond to a solution of the resonant equation—equation (v).

(b) The graph has beats with period 4π. Therefore, this graph corresponds to equation (iii).

(c) This solution has no beats and no change in amplitude. Therefore, it corresponds to either (i), (ii), or (vi). Note that the general solution of equation (i) is

$$k_1 \cos 4t + k_2 \sin 4t + \tfrac{5}{8},$$

and the general solution of equation (ii) is

$$k_1 \cos 4t + k_2 \sin 4t - \tfrac{5}{8}.$$

Equation (iv) has a steady-state solution whose oscillations are centered about $y = 0$. Since the oscillations shown are centered around a positive constant, this function is a solution of equation (i).

(d) The graph has beats with a period that is approximately 50. Therefore, this graph corresponds to equation (iv) (see part (a)).

22. The frequency of the stomping is almost the same as the natural frequency of the swaying motion of the stadium. Therefore, the stadium structure reacted violently due to the resonant effects of the stomping.

23. The equation of motion for the unforced mass-spring system is

$$\frac{d^2y}{dt^2} + 16y = 0,$$

so the natural period is $2\pi/4 = \pi/2 \approx 1.57$.

Tapping with the hammer as shown increases the velocity if the mass is moving to the right at the time of the tap and decreases the velocity if the mass is moving to the left at the time of the tap. Faster motion results in higher amplitude oscillations. Since none of the tapping periods is exactly $\pi/2$, the taps sometimes increase the amplitude and sometimes decrease the amplitude of the oscillations (that is, resonance does not occur).

The period $T = 3/2$ is closest to the natural period and hence for taps with this period we expect the largest amplitude oscillations.

24. To produce the most dramatic effect, the forcing frequency due to the speed bumps must agree with the natural frequency of the suspension system of the average car. Therefore, the speed bumps should be spaced so that the amount of time between bumps is exactly the same as the natural period of the oscillator. Since the natural period of the oscillator is 2 seconds, we compute the distance that the car travels in 2 seconds. At 10 miles per hour, the car travels $1/180$ miles in 2 seconds, and $1/180$ miles is 29 feet, 4 inches.

EXERCISES FOR SECTION 4.4

1. Rubbing a finger around the edge of the glass starts the glass vibrating. The finger then skips along the glass, giving a forcing term which has the same frequency as the natural frequency of vibration of the glass. Pressing harder changes the size of the forcing. Since the frequency is determined by the motion of the glass itself, pushing harder does not alter the frequency.

2. Suppose that the vibrations of the glass are modeled by an underdamped system of the form

$$\frac{d^2y}{dt^2} + b\frac{dy}{dt} + ky = 0.$$

Then the characteristic equation is $s^2 + bs + k = 0$, and the natural frequency for this system is

$$\frac{\sqrt{4k - b^2}}{4\pi}.$$

(a) When water is added to the glass, the damping coefficient b increases, and therefore, the natural frequency decreases. The singer has to sing a lower note to break the glass half-filled with water.

(b) If the damping is increased but the amplitude of the forcing term remains the same, then the amplitude of the forced response decreases. Hence, to break the glass the singer will have to sing louder.

3. Given that

$$y_p''(t) + py_p'(t) + qy_p(t) = g(t)$$

for all t, we have

$$y_p''(t+\theta) + py_p'(t+\theta) + qy_p(t+\theta) = g(t+\theta).$$

Let $z(t) = y_p(t+\theta)$. Then

$$z'(t) = y_p'(t+\theta) \quad \text{and} \quad z''(t) = y_p''(t+\theta)$$

by the Chain Rule. Thus,

$$z''(t) + pz'(t) + qz(t) = g(t+\theta),$$

and $z(t) = y_p(t+\theta)$ is a solution of

$$\frac{d^2y}{dt^2} + p\frac{dy}{dt} + qy = g(t+\theta).$$

4. By the Extended Linearity Principle, the forced response for this equation is the sum of the forced responses for

$$\frac{d^2y}{dt^2} + p\frac{dy}{dt} + qy = \cos\omega_1 t$$

and

$$\frac{d^2y}{dt^2} + p\frac{dy}{dt} + qy = \cos\omega_2 t,$$

that is, it is

$$y(t) = A_1\cos\omega_1 t + A_2\cos\omega_2 t$$

where

$$A_i = \frac{1}{\sqrt{(q-\omega_i^2)^2 + p^2\omega_i^2}}$$

for $i = 1$ and 2.

(a) The maximum value of A_1 occurs where

$$\frac{\partial A_1}{\partial q} = \frac{\omega_1^2 - q}{\sqrt{(q - \omega_1^2)^2 + p^2\omega_1^2}} = 0,$$

which implies $q = \omega_1^2$. This value is a local maximum by the First Derivative Test.

(b) For $q = \omega_1^2$, the amplitude of the other term is given by

$$A_2 = \frac{1}{\sqrt{(\omega_1^2 - \omega_2^2)^2 + p^2\omega_2^2}}.$$

5. As in Exercise 4, the forced response is

$$y(t) = A_1 \cos \omega_1 t + A_2 \cos \omega_2 t.$$

where

$$A_i = \frac{1}{\sqrt{(q - \omega_i^2)^2 + p^2\omega_i^2}}$$

for $i = 1$ and 2.

(a) Setting $q = \omega_1^2$, we have

$$\frac{A_1}{A_2} = \frac{\sqrt{(\omega_1^2 - \omega_2^2)^2 + p^2\omega_2^2}}{p\omega_1}.$$

(b) Let $R(p) = A_1/A_2$. We see that $R(p) \to \infty$ as $p \to 0$ and $R(p) \to \omega_2/\omega_1$ as $p \to \infty$. Moreover,

$$\frac{dR}{dp} = \frac{-(\omega_1^2 - \omega_2^2)^2}{p^2\omega_1^2\sqrt{(\omega_1^2 - \omega_2^2)^2 + p^2\omega_2^2}},$$

which is negative for all $p > 0$. Hence, $R(p)$ is monotonically decreasing.

6. **(a)** The partial is

$$\begin{aligned}\frac{\partial A}{\partial \omega} &= -\frac{1}{2}\frac{2(q - \omega^2)(-2\omega) + 2p^2\omega}{\left((q - \omega^2)^2 + p^2\omega^2\right)^{3/2}} \\ &= \frac{\omega(2q - p^2 - 2\omega^2)}{\left((q - \omega^2)^2 + p^2\omega^2\right)^{3/2}}\end{aligned}$$

(b) To find the maximum value, we let $\partial A/\partial\omega = 0$. This partial vanishes if $\omega = 0$ or if $\omega^2 = q - p^2/2$ (assuming that $q - p^2/2 > 0$). If $q - p^2/2 \le 0$, then

$$M(p, q) = A(0, p, q) = \frac{1}{\sqrt{q^2}} = \frac{1}{|q|}.$$

If $q - p^2/2 > 0$, then we must compare the value of $A(0, p, q)$ with the value of $A(\omega, p, q)$ where $\omega^2 = q - p^2/2$. From the formula for $A(\omega, p, q)$, we have

$$\begin{aligned} A(\omega, p, q) &= \frac{1}{\sqrt{(q-\omega^2)^2 + p^2\omega^2}} \\ &= \frac{1}{\sqrt{q^2 - 2q\omega^2 + \omega^4 + p^2\omega^2}} \\ &= \frac{1}{\sqrt{q^2 - (2q - p^2)\omega^2 + \omega^4}} \\ &= \frac{1}{\sqrt{q^2 - (2\omega^2)\omega^2 + \omega^4}} \\ &= \frac{1}{\sqrt{q^2 - \omega^4}} \\ &> \frac{1}{\sqrt{q^2}}. \end{aligned}$$

We have

$$M(p, q) = \begin{cases} \dfrac{1}{|q|} & \text{if } q \le \dfrac{p^2}{2}; \\[2ex] \dfrac{1}{\sqrt{p^2 q - p^4/4}} & \text{if } q \ge \dfrac{p^2}{2}. \end{cases}$$

7. **(a)**

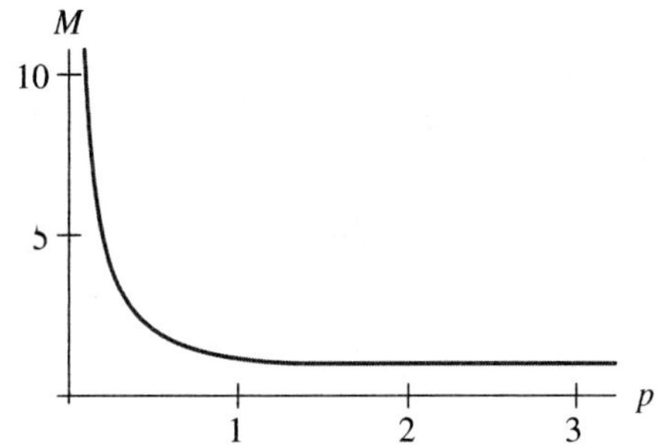

(b) If q is fixed and $p \to 0$, we need only consider the formula

$$M(p, q) = \frac{1}{\sqrt{p^2 q - p^4/4}}.$$

For small $p > 0$,

$$M(p, q) \approx \frac{1}{\sqrt{p^2 q}} = \frac{1}{\sqrt{q}}\,\frac{1}{p}.$$

8. **(a)** Note that $d^2w/dt^2 = d^2y/dt^2$ and $dw/dt = dy/dt$. We obtain

$$\frac{d^2w}{dt^2} + b\frac{dw}{dt} + k\left(w + \frac{y_0}{k}\right) = g(t) + y_0,$$

which simplifies to

$$\frac{d^2w}{dt^2} + b\frac{dw}{dt} + kw = g(t).$$

(b) The solutions differ only by a translation. The amplitudes, periods, and other qualitative features stay the same.

9. We guess a solution of the form $y(t) = \alpha \sin \omega t + \beta \cos \omega t$. Substituting $y(t)$ into the left-hand side of the equation, we obtain

$$-\omega^2\alpha \sin \omega t - \omega^2\beta \cos \omega t + b(\omega\alpha \cos \omega t - \omega\beta \sin \omega t) + k(\alpha \sin \omega t + \beta \cos \omega t).$$

Our guess is a solution if this expression agrees with the right-hand side of the equation, namely

$$kA \cos \omega t - b\omega A \sin \omega t.$$

Equating the coefficients of sine and cosine from both sides of the equation, we obtain

$$\begin{aligned} -\omega^2\beta + b\omega\alpha + k\beta &= kA \\ -\omega^2\alpha - b\omega\beta + k\alpha &= -b\omega A, \end{aligned}$$

which can be rewritten as

$$\begin{aligned} b\omega\alpha + (k - \omega^2)\beta &= kA \\ (k - \omega^2)\alpha - b\omega\beta &= -b\omega A. \end{aligned}$$

Solving this system of two equations for α and β yields

$$\alpha = A\frac{b\omega^3}{(k - \omega^2)^2 + b^2\omega^2} \quad \text{and} \quad \beta = A\frac{b\omega^3}{(k - \omega^2)^2 + b^2\omega^2}.$$

So the particular solution is

$$y(t) = A\frac{b\omega^3}{(k - \omega^2)^2 + b^2\omega^2} \sin \omega t + A\frac{b\omega^3}{(k - \omega^2)^2 + b^2\omega^2} \cos \omega t.$$

10. **(a)** If k is large and the others small, the coefficient of the sine term in the particular solution is small while the coefficient of the cosine term is close to one. So the oscillations of the mass match the forcing. A very strong spring acts like a rigid metal bar in this case and the oscillations are in phase with the forcing.

(b) If b is large and the others small, the coefficient of the sine term in the particular solution is small while the coefficient of the cosine term is near one. A very high damping coefficient dash pot acts like a rigid metal bar in this case.

11. If ω is large, the coefficients of both terms in the particular solution are small because they both have ω^4 in the denominator and at most ω^3 in the numerator. Hence, the oscillations are small. The weak spring and dash pot do not transmit the forcing to the mass fast enough and the pushes and pulls "average" out before they can effect the motion of the mass.

12. **(a)** Differentiating

$$\tan\phi = \frac{-p\omega}{q-\omega^2}$$

implicitly, we obtain

$$\left(\sec^2\phi\right)\frac{\partial\phi}{\partial\omega} = \frac{-pq - p\omega^2}{(q-\omega^2)^2}.$$

Since $\sec^2\phi = 1 + \tan^2\phi$, we have

$$\frac{\partial\phi}{\partial\omega} = -\frac{p(q+\omega^2)}{(q-\omega^2)^2 + p^2\omega^2}.$$

(b) Using the Quotient Rule on

$$-\frac{p(q+\omega^2)}{(q-\omega^2)^2 + p^2\omega^2}$$

and simplifying (perhaps with the help of a computer algebra program), we get

$$\frac{\partial^2\phi}{\partial\omega^2} = \frac{2p\omega(p^q - 3q^2 + 2q\omega^2 + \omega^4)}{\left((q-\omega^2)^2 + p^2\omega^2\right)^2}.$$

(c) To find the maximum value of $\partial\phi/\partial\omega$, we find the critical points of $\partial\phi/\partial\omega$ by solving $\partial^2\phi/\partial\omega^2 = 0$. One critical point is $w = 0$. The other relevant critical point comes from the other factor, $p^q - 3q^2 + 2q\omega^2 + \omega^4$, in the numerator of $\partial^2\phi/\partial\omega^2$. Note that this expression is quadratic in ω^2, so we can use the quadratic formula to solve for ω^2. We obtain

$$\omega^2 = -q \pm \sqrt{q(4q-p)}.$$

The relevant critical point is therefore

$$\omega = \sqrt{-q + \sqrt{q(4q-p)}}.$$

If $q = 2$ and p is near zero, this critical point is near $\omega = \sqrt{2}$ (see Figure 4.25 in the section).

EXERCISES FOR SECTION 4.5

1. **(a)** The stiffness of the roadbed is measured by the coefficient of y. Increasing the stiffness corresponds to increasing β.

(b) As β increases, the term corresponding to the stretch in cable becomes less important. Therefore, the system behaves more like a linear system.

2. **(a)** The damping coefficient of damping is α, so α is increased.

(b) Since the damping is increased, the system tends to the equilibrium faster. The equilibrium point is $y = -g/(\beta + \gamma)$, and the damping causes y to become negative more quickly. If y remains negative, the system behaves like the linear system.

3. **(a)** As the strength of the cable is increased, the spring constant γ for the cables (which appears in the definition of $c(y)$) increases.

(b) Increasing γ in the definition of $c(y)$ has several effects. First, it makes the jump between the behavior of the system for $y < 0$ and $y > 0$ (that is, between taut and loose cables) more pronounced. This effect increases the system's "nonlinearity." Also, the equilibrium point moves closer to $y = 0$. On the other hand, larger γ makes it harder to displace the bridge from its rest position, so in most situations, we only see small amplitude oscillations.

4. The supports beneath the bridge provide a restoring force if the bridge moves above the rest position. Therefore, the sudden change in direction for the original suspension bridge is reduced. If the material of the support is same as the suspension, then the spring constant is same. For $y < 0$, restoring force is only due to the suspension because the cables beneath the bridge are slack. Similarly, for $y \leq 0$, the restoring force is only due to the cable beneath the bridge. Therefore, the differential equation is

$$\frac{d^2y}{dt^2} + \alpha\frac{dy}{dt} + \beta y + \gamma y = -g.$$

5. The buoyancy force is proportional to the volume of the water displaced. Suppose p is that proportionality constant and a is the length of one side of a cube. Let y be the distance between the bottom of the cube and the surface of the water. Since the cube always stays in contact with the water and is never completely submerged, the buoyancy force is $-pa^2y$, and the force equation $F = ma$ is

$$mg - pa^2y - \epsilon\frac{dy}{dt} = m\frac{d^2y}{dt^2},$$

where the term $\epsilon(dy/dt)$ measures the damping. This equation can be rewritten as

$$m\frac{d^2y}{dt^2} + \epsilon\frac{dy}{dt} + pa^2y = mg.$$

6. Once the cube is completely submerged ($y > a$), the buoyancy is always $-pa^3$. If we represent the buoyancy force by the function $b(y)$, we have

$$b(y) = \begin{cases} -pa^2y & \text{if } y < a; \\ -pa^3 & \text{if } y \geq a. \end{cases}$$

Therefore, the differential equation is

$$m\frac{d^2y}{dt^2} + \epsilon\frac{dy}{dt} = mg + b(y).$$

7. If the cube rises completely out of the water, the buoyancy force vanishes. If we represent the buoyancy force by the function $b(y)$, we have

$$b(y) = \begin{cases} -pa^2 y & \text{if } y > 0; \\ 0 & \text{if } y < 0. \end{cases}$$

Therefore, the differential equation is

$$m\frac{d^2y}{dt^2} + \epsilon\frac{dy}{dt} = mg + b(y).$$

8. **(a)** The wave provides an extra force to the cube, and that force is expressed as $q \sin \omega t$ where q is a constant and ω is the frequency of the wave. Then, in Exercise 5, the buoyancy force becomes

$$b(y) = -pa^2 y + pa^2 q \sin \omega t.$$

In fact, in each of the models in Exercises 5–7, the buoyancy force is adjusted in this manner.

(b) Suppose the cube stays in contact with the water and is never completely submerged before the waves play any role. If the natural frequency $\sqrt{pa^2/m}$ is close to the frequency of the wave, the oscillations of the cube become large and the cube is submerged completely under water or rises completely out of the water. If the natural frequency is different from the frequency of the wave, the oscillation with the natural frequency dies out due to the damping, and the cube oscillates with the frequency of the forcing term.

In the case that the cube can rise completely out of the water, the differential equation becomes

$$m\frac{d^2y}{dt^2} + \epsilon\frac{dy}{dt} = mg + b(y),$$

where

$$b(y) = \begin{cases} pa^2 q \sin \omega t - pa^2 y & \text{if } y > 0; \\ pa^2 q \sin \omega t & \text{if } y < 0. \end{cases}$$

In the case that the cube can be completely submerged in the water, the differential equation becomes

$$m\frac{d^2y}{dt^2} + \epsilon\frac{dy}{dt} = mg + b(y),$$

where

$$b(y) = \begin{cases} pa^2 q \sin \omega t - pa^2 y & \text{if } y < a; \\ pa^2 q \sin \omega t - pa^3 & \text{if } y > a. \end{cases}$$

In both cases, we obtain the similar type of differential equations as in the text, and in either case, the similar type of analysis should be required to describe the motion of the cube.

REVIEW EXERCISES FOR CHAPTER 4

1. The natural guess for a particular solution is a constant function $y_p(t) = a$. For this function, $dy_p/dt = d^2y_p/dt^2 = 0$, so it is a solution if and only if $ka = 1$. Hence, the constant function $y_p(t) = 1/k$ for all t is a solution.

2. The angular frequency ω of the forcing function is the same as the natural frequency of the oscillator if $\omega^2 = 4$. Hence the oscillator is in resonance if $\omega = \pm 2$. (Note that $\cos(-2t) = \cos(2t)$, so including the $\pm$ sign is optional.)

3. The frequency of the steady-state solution is the same as the frequency of the forcing function. The frequency of $4\cos 2t$ is $2/(2\pi) = 1/\pi$.

4. The system corresponding to this equation is

$$\begin{aligned}\frac{dy}{dt} &= v\\ \frac{dv}{dt} &= -4y + \sin t,\end{aligned}$$

and there are no points (y, v) for which the right-hand sides of both equations are zero for all t. Hence there are no equilibrium solutions of this equation.

5. Yes, there is a steady-state response for this equation. Note that the general solution to the associated homogeneous equation is $ke^{-\lambda t}$, which tends to zero as $t \to \infty$.

We can compute one particular solution to forced equation using the techniques of Section 1.8, or we can complexify the equation and guess a solution to

$$\frac{dy}{dt} + \lambda y = e^{i\omega t}.$$

To find one solution to the complexified equation, we guess $y_c(t) = ae^{i\omega t}$. Then

$$\begin{aligned}\frac{dy_c}{dt} + \lambda y_c &= a(i\omega)e^{i\omega t} + \lambda(ae^{i\omega t})\\ &= a(\lambda + i\omega)e^{i\omega t}.\end{aligned}$$

We have a solution if $a = 1/(\lambda + i\omega)$.

To obtain the steady-state solution of the original equation, we take the real part of

$$y_c(t) = \frac{1}{\lambda + i\omega}e^{i\omega t} = \frac{\lambda - i\omega}{\lambda^2 + \omega^2}e^{i\omega t}.$$

We obtain

$$\frac{\lambda}{\lambda^2 + \omega^2}\cos\omega t + \frac{\omega}{\lambda^2 + \omega^2}\sin\omega t.$$

See the discussion and example on page 120 of Section 1.8.

6. No. Resonance occurs if the frequency of the forcing matches the natural frequency of the equation. Solutions of

$$\frac{dy}{dt} + \lambda y = 0$$

do not oscillate, so this equation has no natural frequency and no matching of frequencies can occur. The general solution of the forced equation has the form

$$y(t) = ke^{-\lambda t} + \alpha \sin \omega t + \beta \cos \omega t,$$

where α and β are constants that are determined by the values of λ and ω. Consequently, all solutions remain bounded regardless of the relative values of λ and ω.

7. The eigenvalues of the associated homogeneous equation are $-1 \pm i\sqrt{3}$. Hence, the difference $|y_1(t) - y_2(t)|$ between solutions decays at a rate of e^{-t}. To get a rough estimate of T, we solve $e^{-T} = 1/100$ and obtain $T \approx \ln 100 \approx 4.6$

8. The displacement of the coffee from level in the cup can be modeled by a harmonic oscillator. (The restoring force when it sloshes to one side of the cup is proportional to the displacement from level.) Walking down the stairs provides periodic forcing. If the frequency of the forcing is close to the resonant frequency (the natural sloshing frequency of the coffee), then the amplitude of the sloshing is high and the coffee spills. The forcing frequencies that produce large amplitude responces are only those that are close to resonance.

9. By observing the amplitude of the oscillations for different forcing frequencies, you can determine the resonant frequency. Since the damping is very small, the model is approximately

$$m\frac{d^2y}{dt^2} + ky = \cos \omega t,$$

and resonance occurs if $\omega = \sqrt{k/m}$. Hence, you can determine the ratio of the spring constant and the mass.

10. The unforced system is overdamped with eigenvalues $s = -1$ and $s = -4$. Consequently, the general solution of the homogeneous equation dies out relatively quickly, and the solutions approach the steady-state solution. The steady-state solution has a relatively small amplitude because the damping is large.

11. True. The eigenvalues of the associated homogeneous equation are $s = (-1 \pm i\sqrt{23})/2$. Hence, the general solution has the form

$$y(t) = k_1 e^{-t/2} \cos \sqrt{23}\, t + k_2 e^{-t/2} \sin \sqrt{23}\, t + \alpha \cos t + \beta \sin t,$$

where α and β are determined by the equation and k_1 and k_2 are determined by the initial condition. Unless both k_1 and k_2 are zero, $y(t)$ is unbounded as $t \to -\infty$. If $k_1 = k_2 = 0$, then $y(t)$ is the steady-state solution. Using the standard way to compute the amplitude of the steady-state solution, we see that it is $1/\sqrt{26}$, which is definitely less than 1.

12. True. See Exercise 16 in Section 4.2.

13. False. The frequency of the forcing function and the natural frequency of the oscillator agree. Consequently, the system is in resonance, and the amplitude of the forced response increases without bound as $t \to \infty$. However, this amplitude grows linearly. Repeatedly doubling over equal intervals of time results in exponential growth rather than linear growth.

14. False. Resonance occurs only when the forcing frequency is exactly the natural frequency. In order to get large amplitude responce, the natural and forcing frequencies must be very close. It is very unlikely that the bumps in a bumpy road are evenly spaced, so it is unlikely that resonance plays a role in the vibrations you experience.

15. To compute the general solution of the unforced equation, we use the method of Section 3.6. The characteristic polynomial is $s^2 + 6s + 8$, so the eigenvalues are $s = -2$ and $s = -4$. Hence, the general solution of the homogeneous equation is

$$k_1 e^{-2t} + k_2 e^{-4t}.$$

To find a particular solution of the forced equation, we guess $y_p(t) = ke^{-t}$. Substituting into the left-hand side of the differential equation gives

$$\begin{aligned}\frac{d^2 y_p}{dt^2} + 6\frac{dy_p}{dt} + 8y_p &= ke^{-t} - 6ke^{-t} + 8ke^{-t} \\ &= 3ke^{-t}.\end{aligned}$$

In order for $y_p(t)$ to be a solution of the forced equation, we must take $k = 1/3$. The general solution of the forced equation is

$$y(t) = k_1 e^{-2t} + k_2 e^{-4t} + \tfrac{1}{3} e^{-t}.$$

16. To compute the general solution of the unforced equation, we use the method of Section 3.6. The characteristic polynomial is $s^2 + 7s + 12$, so the eigenvalues are $s = -3$ and $s = -4$. Hence, the general solution of the homogeneous equation is

$$k_1 e^{-3t} + k_2 e^{-4t}.$$

To find a particular solution of the forced equation, we guess $y_p(t) = ke^{-2t}$. Substituting into the left-hand side of the differential equation gives

$$\begin{aligned}\frac{d^2 y_p}{dt^2} + 7\frac{dy_p}{dt} + 12y_p &= 4ke^{-2t} - 14ke^{-2t} + 12ke^{-2t} \\ &= 2ke^{-2t}.\end{aligned}$$

In order for $y_p(t)$ to be a solution of the forced equation, we must take $k = 3/2$. The general solution of the forced equation is

$$y(t) = k_1 e^{-3t} + k_2 e^{-4t} + \tfrac{3}{2} e^{-2t}.$$

17. To compute the general solution of the unforced equation, we use the method of Section 3.6. The characteristic polynomial is $s^2 - 2s - 3$, so the eigenvalues are $s = -1$ and $s = 3$. Hence, the general solution of the homogeneous equation is

$$k_1 e^{-t} + k_2 e^{3t}.$$

To find a particular solution of the forced equation, a reasonable looking guess is $y_p(t) = ke^{3t}$. However, this guess is a solution of the homogeneous equation, so it is doomed to fail. We make the standard second guess of $y_p(t) = kte^{3t}$. Substituting into the left-hand side of the differential equation gives

$$\begin{aligned}\frac{d^2 y_p}{dt^2} - 2\frac{dy_p}{dt} - 3y_p &= (6ke^{3t} + 9kte^{3t}) - 2(ke^{3t} + 3kte^{3t}) - 3kte^{3t} \\ &= 4ke^{3t}.\end{aligned}$$

In order for $y_p(t)$ to be a solution of the forced equation, we must take $k = 1/4$. The general solution of the forced equation is

$$y(t) = k_1 e^{-t} + k_2 e^{3t} + \tfrac{1}{4} t e^{3t}.$$

18. To compute the general solution of the unforced equation, we use the method of Section 3.6. The characteristic polynomial is $s^2 + s - 2$, so the eigenvalues are $s = 1$ and $s = -2$. Hence, the general solution of the homogeneous equation is

$$k_1 e^{t} + k_2 e^{-2t}.$$

To find a particular solution of the forced equation, a reasonable looking guess is $y_p(t) = ke^{-2t}$. However, this guess is a solution of the homogeneous equation, so it is doomed to fail. We make the standard second guess of $y_p(t) = kte^{-2t}$. Substituting into the left-hand side of the differential equation gives

$$\begin{aligned} \frac{d^2 y_p}{dt^2} + \frac{dy_p}{dt} - 2y_p &= (-4ke^{-2t} + 4kte^{-2t}) + (ke^{-2t} - 2kte^{-2t}) - 2kte^{-2t} \\ &= -3ke^{-2t}. \end{aligned}$$

In order for $y_p(t)$ to be a solution of the forced equation, we must take $k = -5/3$. The general solution of the forced equation is

$$y(t) = k_1 e^{t} + k_2 e^{-2t} - \tfrac{5}{3} t e^{-2t}.$$

19. To compute the general solution of the unforced equation, we use the method of Section 3.6. The characteristic polynomial is $s^2 + 6s + 8$. So the eigenvalues are $s = -2$ and $s = -4$, and the general solution of the homogeneous equation is

$$k_1 e^{-2t} + k_2 e^{-4t}.$$

To find one solution of the forced equation, we guess the constant function $y_p(t) = k$. Substituting $y_p(t)$ into the left-hand side of the differential equation, we obtain

$$\frac{d^2 y_p}{dt^2} + 6\frac{dy_p}{dt} + 8y_p = 0 + 6 \cdot 0 + 8k = 8k.$$

Hence, $k = 5/8$ yields a solution of the forced equation. The general solution of the forced equation is

$$y(t) = k_1 e^{-2t} + k_2 e^{-4t} + \tfrac{5}{8}.$$

20. To compute the general solution of the unforced equation, we use the method of Section 3.6. The characteristic polynomial is $s^2 - s - 6$, so the eigenvalues are $s = -2$ and $s = 3$. Hence, the general solution of the homogeneous equation is

$$k_1 e^{-2t} + k_2 e^{3t}.$$

To find a particular solution of the forced equation, we could find solutions for the forcing terms $6t$ and $3e^{4t}$ separately, but we can do both cases at once by guessing $y_p(t) = ae^{4t} + bt + c$. Substituting this guess into the left-hand side of the differential equation gives

$$\begin{aligned}\frac{d^2y_p}{dt^2} - \frac{dy_p}{dt} - 6y_p &= (16ae^{4t}) - (4ae^{4t} + b) - 6(ae^{4t} + bt + c) \\ &= 6ae^{4t} - 6bt - (b + 6c),\end{aligned}$$

so $a = 1/2$, $b = -1$, and $c = 1/6$ if $y_p(t)$ is a solution.

Hence, the general solution is

$$y(t) = k_1e^{-2t} + k_2e^{3t} + \tfrac{1}{2}e^{4t} - t + \tfrac{1}{6}.$$

21. To compute the general solution of the unforced equation, we use the method of Section 3.6. The characteristic polynomial is $s^2 - 4s + 13$, so the eigenvalues are $s = 2 \pm 3i$. Hence, the general solution of the homogeneous equation is

$$k_1e^{2t}\cos 3t + k_2e^{2t}\sin 3t.$$

To find one particular solution of the forced equation, we complexify the equation and obtain

$$\frac{d^2y}{dt^2} - 4\frac{dy}{dt} + 13y = 5e^{4it},$$

and we guess that there is a solution of the form $y_c(t) = ae^{4it}$. Substituting this guess into the differential equation yields

$$a(-16 - 16i + 13)e^{4it} = 5e^{4it},$$

which can be simplified to $a(-3 - 16i)e^{4it} = 5e^{4it}$. Thus, $y_c(t)$ is a solution if $a = 5/(-3 - 16i)$. We have

$$y_c(t) = \frac{5}{-3 - 16i}e^{4it} = \left(-\frac{3}{53} + \frac{16}{53}i\right)\left(\cos 4t + i\sin 4t\right).$$

We take the real part to obtain a solution

$$y(t) = -\tfrac{3}{53}\cos 4t - \tfrac{16}{53}\sin 4t$$

of the original nonhomogeneous equation.

The general solution of the forced equation is

$$y(t) = k_1e^{2t}\cos 3t + k_2e^{2t}\sin 3t - \tfrac{3}{53}\cos 4t - \tfrac{16}{53}\sin 4t.$$

22. To compute the general solution of the unforced equation, we use the method of Section 3.6. The characteristic polynomial is $s^2 + 3$, so the eigenvalues are $s = \pm i\sqrt{3}$. Hence, the general solution of the unforced equation is

$$k_1\cos\sqrt{3}\,t + k_2\sin\sqrt{3}\,t.$$

To find a particular solution of the forced equation, we have several choices. We could deal with the forcing terms individually, but we don't have to write as much if we deal with both at once. Also, we could complexify the equation to handle the $\cos 4t$ forcing term. However, since dy/dt does not appear in the equation, we do not expect to have a $\sin 4t$ term in the particular solution. Hence, we guess

$$y_p(t) = a \cos 4t + bt + c.$$

Substituting $y_p(t)$ into the left-hand side of the differential equation gives

$$\begin{aligned} \frac{d^2 y_p}{dt^2} + 3y_p &= -16a \cos 4t + 3(a \cos 4t + bt + c) \\ &= -13a \cos 4t + 3bt + 3c. \end{aligned}$$

In order for $y_p(t)$ to be a solution of the forced equation, we must have $a = -1/13$, $b = 2/3$, and $c = 0$. The general solution of the forced equation is

$$y(t) = k_1 \cos \sqrt{3}\,t + k_2 \sin \sqrt{3}\,t + \tfrac{2}{3}t - \tfrac{1}{13} \cos 4t.$$

23. We begin with some observations about the various equations:

(i) Solutions are periodic with period $\pi/2$.

(ii) The period of the steady-state solution is π.

(iii) The period of the steady-state solution is $\pi/2$.

(iv) The equilibrium point at the origin is a spiral source.

(v) The equilibrium point at the origin is a spiral sink.

(vi) The period of the beats is approximately 1.37.

(vii) Solutions are periodic with period $2\pi/3$.

(viii) The period of the beats is approximately 39.27.

Now we match the graphs with the equations.

(a) This graph is either a solution of an undamped homogeneous equation, or it is the steady-state solution of a damped homogeneous equation. So we consider equations (i), (ii), (iii), and (vii). The period of the solution is $2\pi/3$. Consequently, this solution comes from equation (vii).

(b) This graph corresponds to a forced equation with damping. So we consider equations (ii) and (iii). The period of the steady-state solution is π. Consequently, this solution comes from equation (ii).

(c) This graph corresponds to a homogeneous equation with damping. So we consider equations (iv) and (v). Equation (iv) has a spiral source at the origin, and equation (v) has a spiral sink at the origin. Consequently, this solution comes from equation (v).

(d) This graph is the solution of an undamped sinusoidally forced equation. So we consider equations (vi) and (viii). The period of the beat is approximately 40. Consequently, this solution comes from equation (viii).

24. **(a)** To compute the general solution of the unforced equation, we use the method of Section 3.6. The characteristic polynomial is

$$s^2 + 6s + 13,$$

so the eigenvalues are $s = -3 \pm 2i$. Hence, the general solution of the homogeneous equation is

$$k_1 e^{-3t} \cos 2t + k_2 e^{-3t} \sin 2t.$$

To find one particular solution of the nonhomogeneous equation, we complexify the equation and obtain

$$\frac{d^2y}{dt^2} + 6\frac{dy}{dt} + 13y = 2e^{3it},$$

and we guess that there is a solution of the form $y_c(t) = ae^{3it}$. Substituting this guess into the differential equation yields

$$a(-9 + 18i + 13)e^{3it} = 2e^{3it},$$

which can be simplified to $a(4 + 18i)e^{3it} = 2e^{3it}$. Thus, $y_c(t)$ is a solution if $a = 2/(4 + 18i)$. We have

$$y_c(t) = \frac{1}{2 + 9i}\, e^{3it} = \left(\frac{2}{85} - \frac{9}{85}\, i\right)\left(\cos 3t + i \sin 3t\right).$$

We take the real part to obtain a solution

$$y(t) = \tfrac{2}{85} \cos 3t + \tfrac{9}{85} \sin 3t$$

of the original nonhomogeneous equation.

The general solution of the forced equation is

$$y(t) = k_1 e^{-3t} \cos 2t + k_2 e^{-3t} \sin 2t + \tfrac{2}{85} \cos 3t + \tfrac{9}{85} \sin 3t.$$

(b) The amplitude of the steady-state solution is

$$|a| = \frac{1}{|2 + 9i|} = \frac{1}{\sqrt{85}} \approx 0.11.$$

The phase angle ϕ is the polar angle of a. In this case, it is the angle in the fourth quadrant such that $\tan \phi = -9/2$. We get $\phi \approx -77.5°$. (Note that this angle is expressed in terms of degrees.)

25. **(a)** To compute the general solution of the unforced equation, we use the method of Section 3.6. The characteristic polynomial is

$$s^2 + 2s + 3,$$

so the eigenvalues are $s = -1 \pm i\sqrt{2}$. Hence, the general solution of the homogeneous equation is

$$k_1 e^{-t} \cos \sqrt{2}\, t + k_2 e^{-t} \sin \sqrt{2}\, t.$$

To find one particular solution of the nonhomogeneous equation, we complexify the equation and obtain

$$\frac{d^2y}{dt^2} + 2\frac{dy}{dt} + 3y = e^{2it},$$

and we guess that there is a solution of the form $y_c(t) = ae^{2it}$. Substituting this guess into the differential equation yields

$$a(-4 + 4i + 3)e^{2it} = e^{2it},$$

which can be simplified to $a(-1 + 4i)e^{2it} = e^{2it}$. Thus, $y_c(t)$ is a solution if $a = 1/(-1 + 4i)$. We have

$$y_c(t) = \frac{1}{-1 + 4i}\, e^{2it} = \left(-\frac{1}{17} - \frac{4}{17}\, i\right)\left(\cos 2t + i \sin 2t\right).$$

We take the real part to obtain a solution

$$y(t) = -\tfrac{1}{17}\cos 2t + \tfrac{4}{17}\sin 2t$$

of the original nonhomogeneous equation.

The general solution of the forced equation is

$$y(t) = k_1 e^{-t}\cos\sqrt{2}\,t + k_2 e^{-t}\sin\sqrt{2}\,t - \tfrac{1}{17}\cos 2t + \tfrac{4}{17}\sin 2t.$$

(b) The amplitude of the steady-state solution is

$$|a| = \frac{1}{|-1+4i|} = \frac{1}{\sqrt{17}} \approx 0.24.$$

The phase angle ϕ is the polar angle of a. In this case, it is the angle in the third quadrant such that $\tan\phi = (-4)/(-1)$. We get $\phi \approx -104°$. (Note that this angle is expressed in terms of degrees.)

26. **(a)** To compute the general solution of the unforced equation, we use the method of Section 3.6. The characteristic polynomial is

$$s^2 + 4s + 4,$$

which has $s = -2$ as a double root. Hence, the general solution of the homogeneous equation is

$$k_1 e^{-2t} + k_2 t e^{-2t}.$$

To find one particular solution of the nonhomogeneous equation, we complexify the equation and obtain

$$\frac{d^2y}{dt^2} + 4\frac{dy}{dt} + 4y = 2e^{3it},$$

and we guess that there is a solution of the form $y_c(t) = ae^{3it}$. Substituting this guess into the differential equation yields

$$-9ae^{3it} + 12iae^{3it} + 4ae^{3it} = 2e^{3it},$$

which can be simplified to

$$(-5 + 12i)ae^{3it} = 2e^{3it}.$$

Thus, $y_c(t)$ is a solution if $a = 2/(-5 + 12i)$. We have

$$y_c(t) = \frac{2}{-5+12i}e^{3it} = -\left(\frac{10}{169} + \frac{24}{169}i\right)\left(\cos 3t + i\sin 3t\right).$$

We take the real part to obtain a solution

$$y(t) = -\tfrac{10}{169}\cos 3t + \tfrac{24}{169}\sin 3t$$

of the original nonhomogeneous equation.

The general solution of the forced equation is

$$y(t) = k_1 e^{-2t} + k_2 t e^{-2t} - \tfrac{10}{169}\cos 3t + \tfrac{24}{169}\sin 3t.$$

(b) The amplitude of the steady-state solution is

$$|a| = \frac{2}{|-5+12i|} = \frac{2}{13} \approx 0.15.$$

The phase angle ϕ is the polar angle of a. In this case, it is the angle in the third quadrant such that $\tan\phi = (-12)/(-5)$. We get $\phi \approx -113°$. (Note that this angle is expressed in terms of degrees.)

27. **(a)** To compute the general solution of the unforced equation, we use the method of Section 3.6. The characteristic polynomial is

$$s^2 + 4s + 3,$$

so the eigenvalues are $s = -3$ and $s = -1$. Hence, the general solution of the homogeneous equation is

$$k_1 e^{-3t} + k_2 e^{-t}.$$

To find one particular solution of the nonhomogeneous equation, we complexify the equation and obtain

$$\frac{d^2y}{dt^2} + 4\frac{dy}{dt} + 3y = 5e^{2it},$$

and we guess that there is a solution of the form $y_c(t) = ae^{2it}$. Substituting this guess into the differential equation yields

$$a(-4 + 8i + 3)e^{2it} = 5e^{2it},$$

which can be simplified to $a(-1+8i)e^{2it} = 5e^{2it}$. Thus, $y_c(t)$ is a solution if $a = 5/(-1+8i)$. We have

$$y_c(t) = \frac{5}{-1+8i}e^{2it} = \left(-\frac{1}{13} - \frac{8}{13}i\right)\left(\cos 2t + i\sin 2t\right).$$

We take the imaginary part to obtain a solution

$$y(t) = -\tfrac{8}{13}\cos 2t - \tfrac{1}{13}\sin 2t$$

of the original nonhomogeneous equation.

The general solution of the forced equation is

$$y(t) = k_1 e^{-3t} + k_2 e^{-t} - \tfrac{8}{13}\cos 2t - \tfrac{1}{13}\sin 2t.$$

(b) The amplitude of the steady-state solution is

$$|a| = \frac{5}{|-1+8i|} = \frac{5}{\sqrt{65}} = \frac{\sqrt{65}}{13} \approx 0.62.$$

Since we got the steady-state solution by taking the imaginary part of $y_c(t)$, the phase angle is the polar angle of the complex number

$$-\tfrac{8}{13} + \tfrac{1}{13}i.$$

In this case, it is the angle in the second quadrant such that $\tan\phi = 1/(-8)$. We get $\phi \approx -187°$. (Note that this angle is expressed in terms of degrees. It is not between $-180°$ and $0°$ due to the fact that the forcing is in terms of sine rather than cosine.)

CHAPTER 5

Nonlinear Systems

EXERCISES FOR SECTION 5.1

1. The linearizations of systems (i) and (iii) are both

$$\frac{dx}{dt} = 2x + y$$
$$\frac{dy}{dt} = -y,$$

so these two systems have the same "local picture" near (0, 0). This system has eigenvalues 2 and −1; hence, (0, 0) is a saddle for these systems. System (ii) has linearization

$$\frac{dx}{dt} = 2x + y$$
$$\frac{dy}{dt} = y,$$

which has eigenvalues 2 and 1, hence, (0, 0) is a source for this system.

2. The linearizations of systems (ii) and (iii) are both equal to

$$\frac{dx}{dt} = -3x + y$$
$$\frac{dy}{dt} = 4x$$

so these two systems have the same "local picture" near (0, 0). These systems have eigenvalues −4 and 1, hence, (0, 0) is a saddle for these systems. System (i) has linearization

$$\frac{dx}{dt} = 3x + y$$
$$\frac{dy}{dt} = 4x$$

which has eigenvalues 4 and −1 so that (0, 0) is also a saddle for this system. However, the eigenvector corresponding to the eigenvalue −4 in systems (ii) and (iii) lie on the line $y = -x$, whereas the eigenvectors corresponding to the eigenvalue −1 for system (i) lie along the line $y = -4x$.

3. **(a)** The linearized system is

$$\frac{dx}{dt} = -2x + y$$
$$\frac{dy}{dt} = -y.$$

We can see this either by "dropping higher-order terms" or by computing the Jacobian matrix

$$\begin{pmatrix} -2 & 1 \\ 2x & -1 \end{pmatrix}$$

and evaluating it at (0, 0).

(b) The eigenvalues of the linearized system are -2 and -1, so $(0, 0)$ is a sink.

(c) The vector $(1, 0)$ is an eigenvector for eigenvalue -2 and $(1, 1)$ is an eigenvector for the eigenvalue -1.

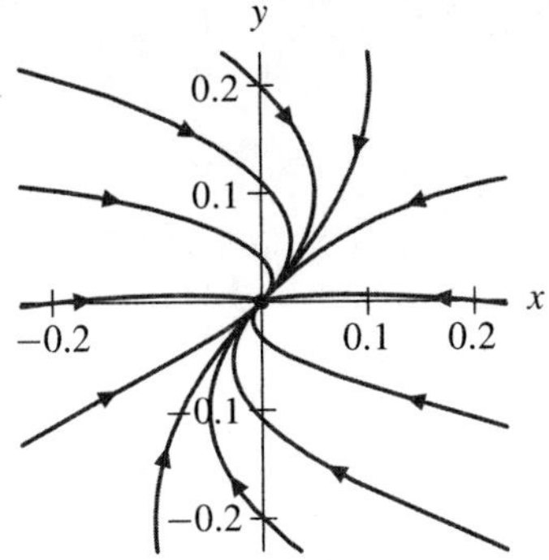

(d) By computing the Jacobian matrix

$$\begin{pmatrix} -2 & 1 \\ 2x & -1 \end{pmatrix}$$

and evaluating at $(2, 4)$, we see that linearized system at $(2, 4)$ is

$$\frac{dx}{dt} = -2x + y$$
$$\frac{dy}{dt} = 4x - y.$$

Its eigenvalues are $(-3 \pm \sqrt{17})/2$, so $(2, 4)$ is a saddle.

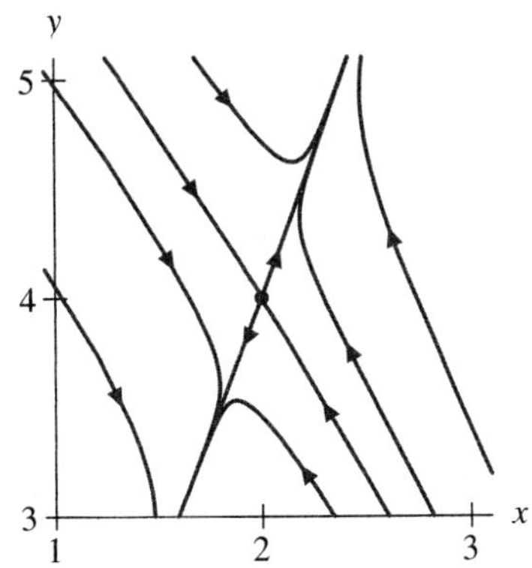

4. **(a)** The equilibrium points occur where the vector field is zero, that is, at solutions of

$$\begin{cases} -x = 0 \\ -4x^3 + y = 0. \end{cases}$$

So, $x = y = 0$ is the only equilibrium point.

(b) The Jacobian matrix of this system is

$$\begin{pmatrix} -1 & 0 \\ -12x^2 & 1 \end{pmatrix},$$

which at (0, 0) is equal to

$$\begin{pmatrix} -1 & 0 \\ 0 & 1 \end{pmatrix}.$$

So the linearized system at (0, 0) is

$$\frac{dx}{dt} = -x$$
$$\frac{dy}{dt} = y$$

(we could also see this by "dropping the higher order terms").

(c) The eigenvalues of the linearized system at the origin are -1 and 1, so the origin is a saddle. The linearized system decouples, so solutions approach the origin along the x-axis and tend away form the origin along the y-axis.

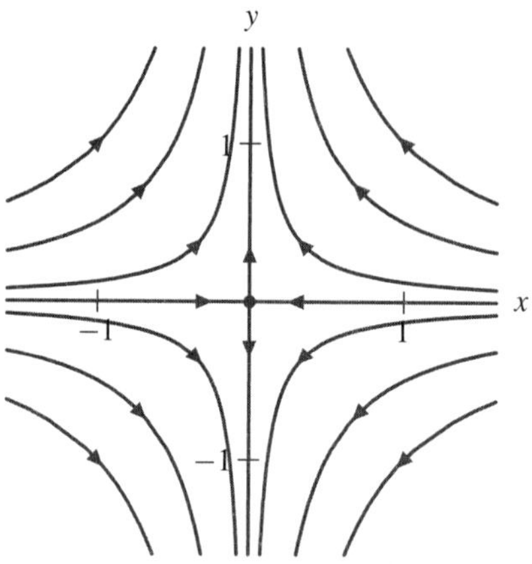

5. **(a)** Using separation of variables (or simple guessing), we have $x(t) = x_0 e^{-t}$.

(b) Using the result in part (a), we can rewrite the equation for dy/dt as

$$\frac{dy}{dt} = y - 4x_0^3 e^{-3t}.$$

This first-order equation is a nonhomogeneous linear equation.

The general solution of its associated homogeneous equation is ke^t. To find a particular solution to the nonhomogeneous equation, we rewrite it as

$$\frac{dy}{dt} - y = -4x_0^3 e^{-3t},$$

and we guess a solution of the form $y_p = \alpha e^{-3t}$. Substituting this guess into the left-hand side of the equation yields

$$\frac{dy_p}{dt} - y_p = -4\alpha e^{-3t}.$$

Therefore, y_p is a solution if $\alpha = x_0^3$. The general solution of the original equation is

$$y(t) = x_0^3 e^{-3t} + ke^t.$$

To express this result in terms of the initial condition $y(0) = y_0$, we evaluate at $t = 0$ and note that $k = y_0 - x_0^3$. We conclude that

$$y(t) = x_0^3 e^{-3t} + (y_0 - x_0^3)e^t.$$

(c) The general solution of the system is

$$x(t) = x_0 e^{-t}$$
$$y(t) = x_0^3 e^{-3t} + (y_0 - x_0^3)e^t.$$

(d) For all solutions, $x(t) \to 0$ as $t \to \infty$. For a solution to tend to the origin as $t \to \infty$, we must have $y(t) \to 0$, and this can happen only if $y_0 - x_0^3 = 0$.

(e) Since $x = x_0 e^{-t}$, we see that a solution will tend toward the origin as $t \to -\infty$ only if $x_0 = 0$. In that case, $y(t) = y_0 e^t$, and $y(t) \to 0$ as $t \to -\infty$.

(f)

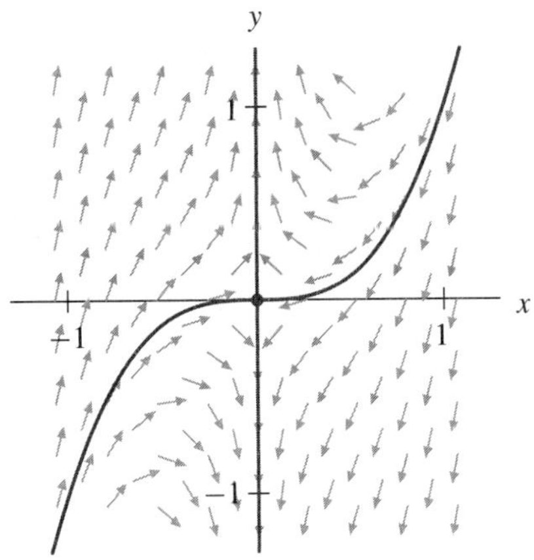

(g) Solutions tend away from the origin along the y-axis in both systems. In the nonlinear system, solutions approach the origin along the curve $y = x^3$ which is tangent to the x-axis. For the linearized system, solutions tend to the origin along the x-axis. Near the origin, the phase portraits are almost the same.

6. **(a)** The Jacobian is

$$\begin{pmatrix} 2-2x-y & -x \\ -2y & 3-2y-2x \end{pmatrix}.$$

Evaluating at (0, 0), we get the linearized system

$$\frac{dx}{dt} = 2x$$
$$\frac{dy}{dt} = 3y.$$

Evaluating at (0, 3), we get

$$\frac{dx}{dt} = -x$$
$$\frac{dy}{dt} = -6x - 3y,$$

and evaluating at (2, 0), we get

$$\frac{dx}{dt} = -2x - 2y$$
$$\frac{dy}{dt} = -y.$$

(b) At (0, 0), the eigenvalues are 2 and 3, so (0, 0) is a source. At (0, 3), eigenvalues are -1 and -3, so (0, 3) is a sink. At (2, 0), the eigenvalues are -2 and -1, so (2, 0) is a sink.

(c)

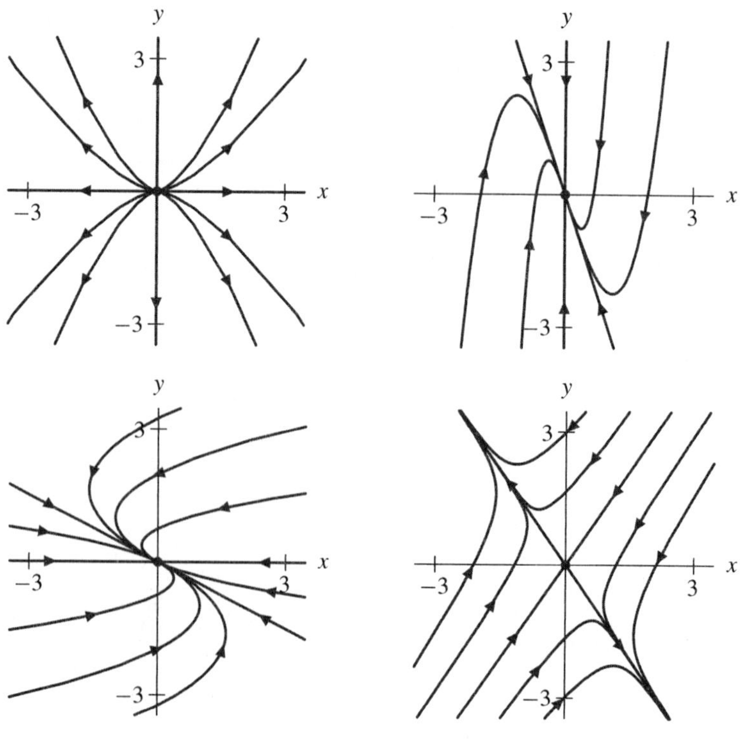

(d) The equilibrium $(0, 0)$ is a source. The y-axis is the line of eigenvectors for eigenvalue 3, and the x-axis is the line of eigenvectors for the eigenvalue 2. So solutions move away from the origin faster in the y-direction than in the x-direction.

The equilibrium $(0, 3)$ is a sink. The eigenvectors for the linearized system near $(0, 3)$ associated to the eigenvalue -3 satisfy $-3x = 2y$. The eigenvectors associated to the eigenvalue -1 form the y-axis. For the nonlinear system, $x(t) \to 0$ very quickly while the $y(t) \to 3$.

The equilibrium $(2, 0)$ is also a sink. The eigenvalues for the linearization at $(2, 0)$ are -2 and -1. The eigenvectors associated to the eigenvalue -2 for the x-axis, and the eigenvectors associated to the eigenvalue -1 satisfy $-2y = x$. For the nonlinear system, most solutions tend to the equilibrium in the direction determined by the eigenvectors for the eigenvalue -1.

The behavior of the system near the equilibrium at $(1, 1)$ is described in the text.

7. **(a)** The equilibrium points are $(0, 0)$, $(0, 100)$, $(150, 0)$, and $(30, 40)$. To determine the type of an equilibrium point, we compute the Jacobian matrix, which is

$$\begin{pmatrix} -2x - 3y + 150 & -3x \\ -2y & -2x - 2y + 100 \end{pmatrix},$$

and evaluate at the point. At $(0, 0)$, the Jacobian is

$$\begin{pmatrix} 150 & 0 \\ 0 & 100 \end{pmatrix},$$

and the eigenvalues are 150 and 100. Hence, the origin is a source. At $(0, 100)$, the Jacobian matrix is

$$\begin{pmatrix} -150 & 0 \\ -200 & -100 \end{pmatrix},$$

and the eigenvalues are -150 and -100. So $(0, 100)$ is a sink. The Jacobian at $(150, 0)$ is

$$\begin{pmatrix} -150 & -450 \\ 0 & -200 \end{pmatrix},$$

and the eigenvalues are -150 and -200. Therefore, $(150, 0)$ is a sink. Finally, the Jacobian matrix at $(30, 40)$ is

$$\begin{pmatrix} -30 & -90 \\ -80 & -40 \end{pmatrix},$$

and the eigenvalues are approximately -120 and 50. So $(30, 40)$ is a saddle.

(b)

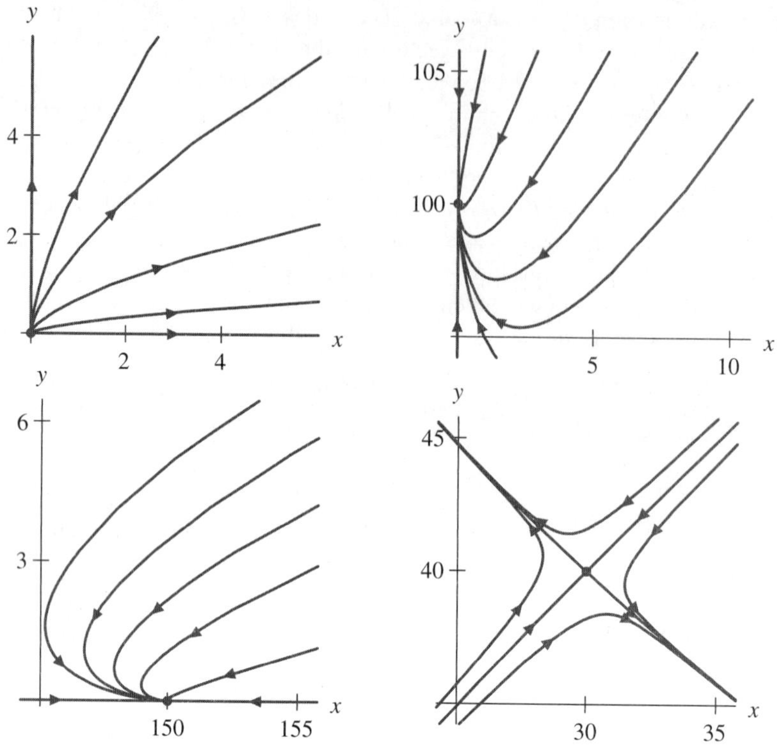

8. **(a)** The equilibrium points are (0, 0), (0, 30), and (10, 0). To determine the type of each equilibrium point, we compute the Jacobian matrix, which is

$$\begin{pmatrix} -2x - y + 10 & -x \\ -2y & -2x - 2y + 30 \end{pmatrix},$$

and evaluate it at the point. At (0, 0), the Jacobian is

$$\begin{pmatrix} 10 & 0 \\ 0 & 30 \end{pmatrix},$$

and the eigenvalues are 10 and 30. Thus, the origin is a source. At (0, 30), the Jacobian matrix is

$$\begin{pmatrix} -20 & 0 \\ -60 & -30 \end{pmatrix},$$

and the eigenvalues are -20 and -30. So (0, 30) is a sink. The Jacobian at (10, 0) is

$$\begin{pmatrix} -10 & -10 \\ 0 & 10 \end{pmatrix},$$

and the eigenvalues are -10 and 10. Therefore, (10, 0) is a saddle.

(b)

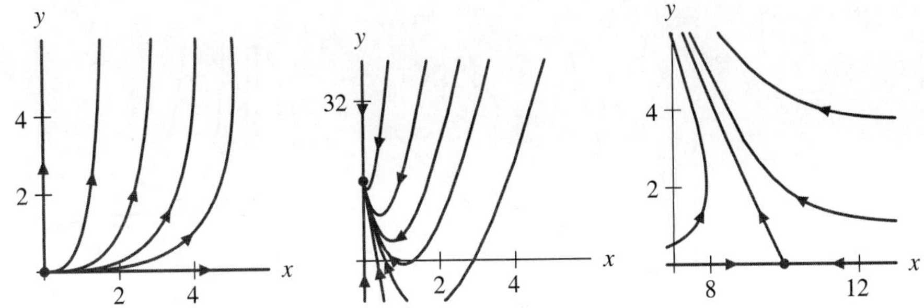

9. **(a)** The equilibrium points are (0, 0), (0, 25), (100, 0) and (75, 12.5). We classify these equilibrium points by computing the Jacobian matrix, which is

$$\begin{pmatrix} 100 - 2x - 2y & -2x \\ -y & 150 - x - 12y \end{pmatrix},$$

and evaluating it at each of the equilibrium points. At (0, 0), the Jacobian matrix is

$$\begin{pmatrix} 100 & 0 \\ 0 & 150 \end{pmatrix},$$

and the eigenvalues are 100 and 150. So this point is a source. At (0, 25), the Jacobian matrix is

$$\begin{pmatrix} 50 & 0 \\ -25 & -150 \end{pmatrix},$$

and the eigenvalues are 50 and -150. Hence, this point is a saddle. At (100, 0), the Jacobian matrix is

$$\begin{pmatrix} -100 & -200 \\ 0 & 50 \end{pmatrix},$$

and the eigenvalues are -100 and 50. Therefore, this point is a saddle. Finally, at (75, 12.5), the Jacobian matrix is

$$\begin{pmatrix} -75 & -150 \\ -12.5 & -75 \end{pmatrix},$$

and the eigenvalues are approximately -32 and -118. So this point is a sink.

(b)

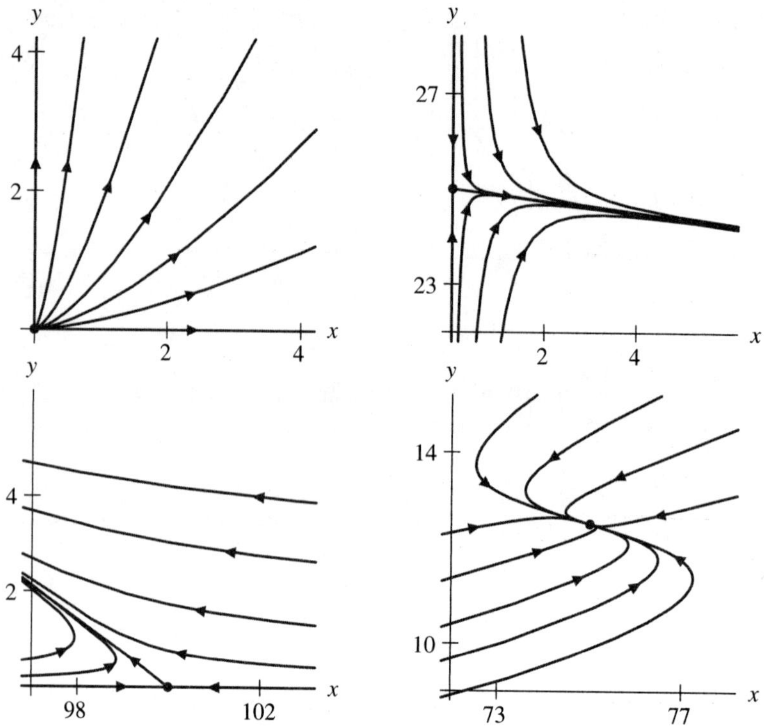

10. **(a)** The equilibrium points in the first quadrant are (0, 0), (0, 50) and (100, 0). To classify these equilibrium points, we compute the Jacobian matrix, which is

$$\begin{pmatrix} -2x - y + 100 & -x \\ -2xy & -x^2 - 3y^2 + 2500 \end{pmatrix},$$

and evaluate it at each point. At (0, 0), the Jacobian matrix is

$$\begin{pmatrix} 100 & 0 \\ 0 & 2500 \end{pmatrix},$$

which has eigenvalues 100 and 2500. So (0, 0) is a source. At (0, 50), the Jacobian matrix is

$$\begin{pmatrix} 50 & 0 \\ 0 & -5000 \end{pmatrix},$$

which has eigenvalues -10 and -5000. Hence, (0, 50) is a saddle. At (100, 0), the Jacobian matrix is

$$\begin{pmatrix} -100 & -100 \\ 0 & 900 \end{pmatrix},$$

which has eigenvalues -40 and -7500. Thus, (100, 0) is a sink.

(b)

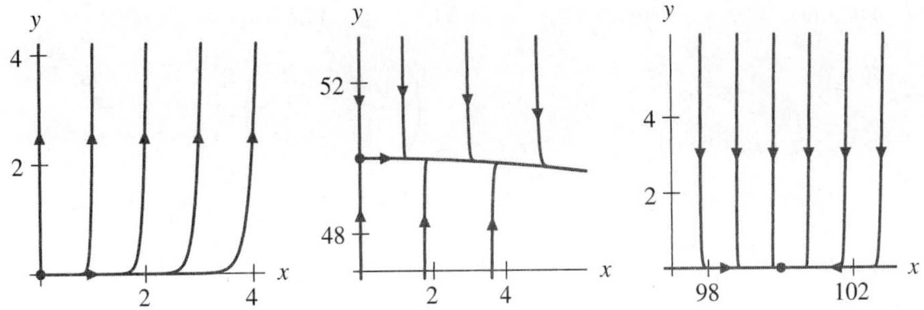

11. **(a)** The equilibrium points in the first quadrant are (0, 0), (0, 50) and (40, 0). To classify these equilibrium points, we compute the Jacobian matrix, which is

$$\begin{pmatrix} -2x - y + 40 & -x \\ -2xy & -x^2 - 3y^2 + 2500 \end{pmatrix},$$

and we evaluate it at each of the points. At (0, 0), the Jacobian matrix is

$$\begin{pmatrix} 40 & 0 \\ 0 & 2500 \end{pmatrix},$$

which has eigenvalues 40 and 2500. Therefore, (0, 0) is a source. At (0, 50), the Jacobian matrix is

$$\begin{pmatrix} -10 & 0 \\ 0 & -5000 \end{pmatrix},$$

which has eigenvalues −10 and −5000. So (0, 50) is a sink. At (40, 0), the Jacobian matrix is

$$\begin{pmatrix} -40 & -40 \\ 0 & 900 \end{pmatrix},$$

which has eigenvalues −40 and 900. Hence, (40, 0) is a saddle.

(b)

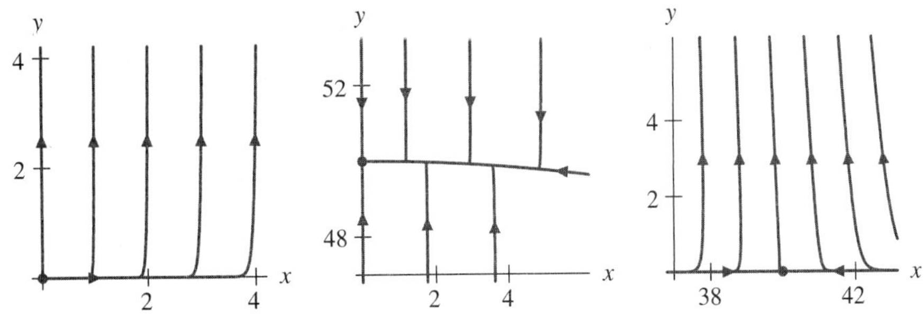

12. **(a)** The equilibrium points (in the first quadrant) are (0, 0), (0, 50), (40, 0), and (30, 40). To determine the types of the equilibria, we compute the Jacobian matrix, which is

$$\begin{pmatrix} -8x - y + 160 & -x \\ -2xy & -x^2 - 3y^2 + 2500 \end{pmatrix},$$

and evaluate it at the points. At (0, 0), the Jacobian is

$$\begin{pmatrix} 160 & 0 \\ 0 & 2500 \end{pmatrix},$$

which has eigenvalues 160 and 2500. Hence, (0, 0) is a source. At (0, 50), the Jacobian matrix is

$$\begin{pmatrix} 110 & 0 \\ 0 & -5000 \end{pmatrix},$$

which has eigenvalues 110 and -5000. So (0, 50) is a saddle. At (40, 0), the Jacobian matrix is

$$\begin{pmatrix} -320 & -40 \\ 0 & 900 \end{pmatrix},$$

which has eigenvalues 900 and -320. Therefore, (40, 0) is a saddle. At (30, 40), the Jacobian matrix is

$$\begin{pmatrix} -120 & -30 \\ -2400 & -3200 \end{pmatrix},$$

which has eigenvalues approximately equal to -3223 and -97. So (30, 40) is a sink.

(b)

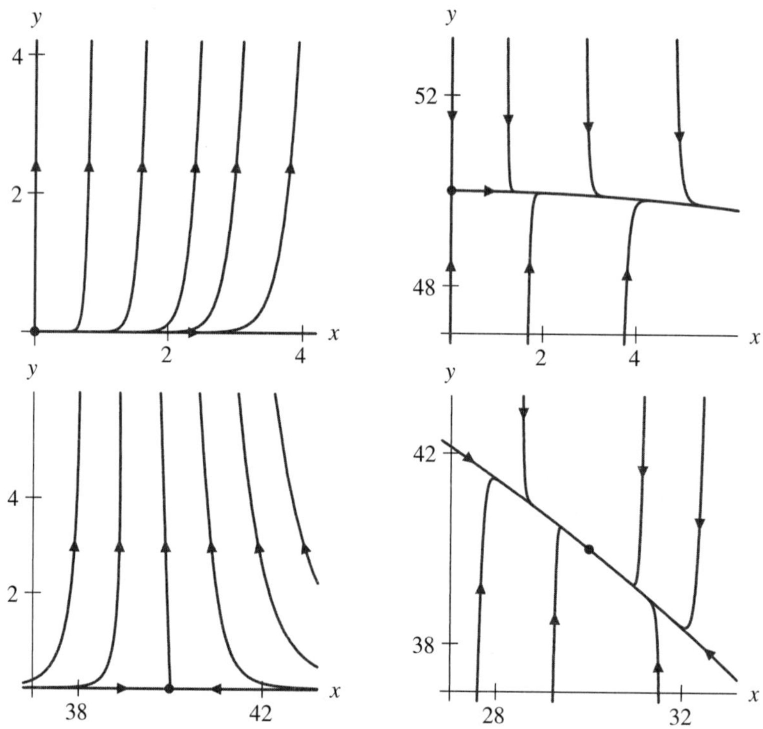

13. **(a)** The equilibrium points (in the first quadrant) are $(0, 0)$, $(0, 50)$, $(60, 0)$, $(30, 40)$ and $(234/5, 88/5)$. To classify these points, we compute the Jacobian matrix, which is,

$$\begin{pmatrix} -16x - 6y + 480 & -6x \\ -2xy & -x^2 - 3y^2 + 2500 \end{pmatrix},$$

and evaluate it at each point. At $(0, 0)$, the Jacobian is

$$\begin{pmatrix} 480 & 0 \\ 0 & 2500 \end{pmatrix},$$

which has eigenvalues 480 and 2500. Thus, $(0, 0)$ is a source. At $(0, 50)$, the Jacobian matrix is

$$\begin{pmatrix} 180 & 0 \\ 0 & -5000 \end{pmatrix},$$

which has eigenvalues 180 and -5000. So $(0, 50)$ is a saddle. At $(60, 0)$, the Jacobian matrix is

$$\begin{pmatrix} -480 & -360 \\ 0 & -1100 \end{pmatrix},$$

which has eigenvalues -480 and -1100. Hence, $(60, 0)$ is a sink. At $(30, 40)$, the Jacobian matrix is

$$\begin{pmatrix} -240 & -180 \\ -2400 & -3200 \end{pmatrix},$$

which has eigenvalues approximately equal to -3339 and -101. So, $(30, 40)$ is a sink. At $(234/5, 88/5)$, the Jacobian matrix is

$$\begin{pmatrix} -1872/5 & -1404/5 \\ -41184/25 & -15488/25 \end{pmatrix},$$

which has eigenvalues approximately equal to -1188 and 194. Therefore, $(234/5, 88/5)$ is a saddle.

(b)

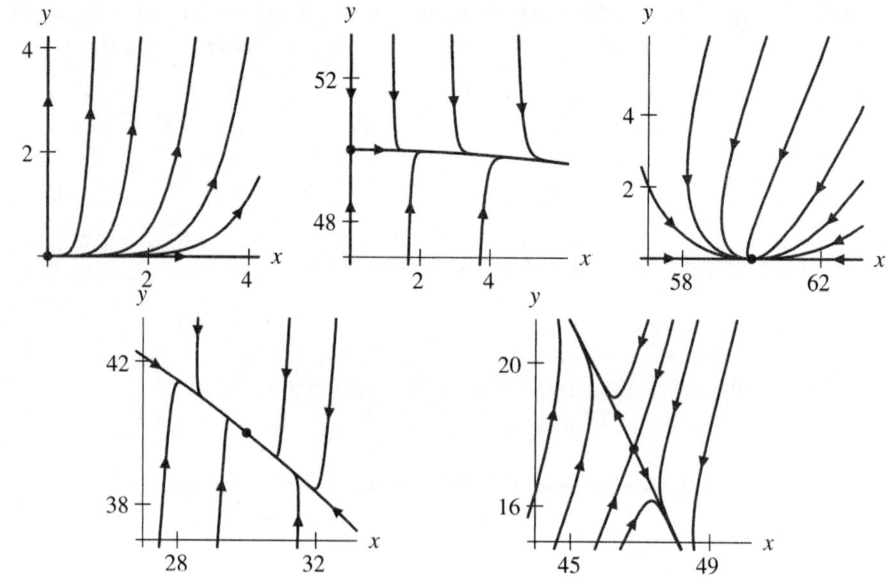

14. **(a)** The equilibrium points are $(0, 0)$, $(1, 1)$ and $(2, 0)$. We classify these points by calculating the Jacobian matrix, which is,

$$\begin{pmatrix} 2 - 2x - y & -x \\ -2xy & 2y - x^2 \end{pmatrix},$$

and evaluating it at the points. At $(0, 0)$, the Jacobian is

$$\begin{pmatrix} 2 & 0 \\ 0 & 0 \end{pmatrix},$$

which has eigenvalues 2 and 0. An eigenvector for the eigenvalue 2 is $(1, 0)$, so solutions move away from the origin parallel to the x-axis. On the line $x = 0$, we have $dy/dt = y^2$ so solutions move upwards when $y \neq 0$. Hence, $(0, 0)$ is a node. However, solutions near the origin in the first quadrant move away from the origin as t increases. At $(1, 1)$, the Jacobian is

$$\begin{pmatrix} -1 & -1 \\ -2 & 1 \end{pmatrix},$$

which has eigenvalues $\pm\sqrt{3}$. So $(1, 1)$ is a saddle. At $(2, 0)$, the Jacobian is

$$\begin{pmatrix} -2 & -2 \\ 0 & -4 \end{pmatrix},$$

which has eigenvalues -2 and -4. Thus, $(2, 0)$ is a sink.

(b)

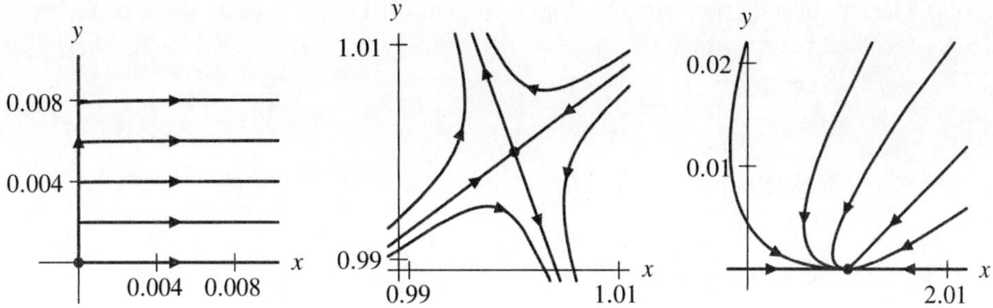

15. **(a)** The equilibrium points are (0, 0), (1, 1) and (2, 0). We determine the type of each of these points by computing the Jacobian, which is

$$\begin{pmatrix} 2-2x-y & -x \\ -y & 2y-x \end{pmatrix},$$

and evaluating it at the points. At (0, 0), the Jacobian is

$$\begin{pmatrix} 2 & 0 \\ 0 & 0 \end{pmatrix},$$

which has eigenvalues 2 and 0. An eigenvector for the eigenvalue 2 is (1, 0), so solutions move away from the origin parallel to the x-axis. On the line $x = 0$, we have $dy/dt = y^2$ so solutions move upwards when $y \neq 0$. Hence, (0, 0) is a node. However, solutions near the origin in the first quadrant move away from the origin as t increases. At (1, 1), the Jacobian is

$$\begin{pmatrix} -1 & -1 \\ -1 & 1 \end{pmatrix},$$

which has eigenvalues $\pm\sqrt{2}$. So (1, 1) is a saddle. At (2, 0), the Jacobian is

$$\begin{pmatrix} -2 & -2 \\ 0 & -2 \end{pmatrix},$$

which has a double eigenvalue of -2. Therefore, (2, 0) is a sink.

(b)

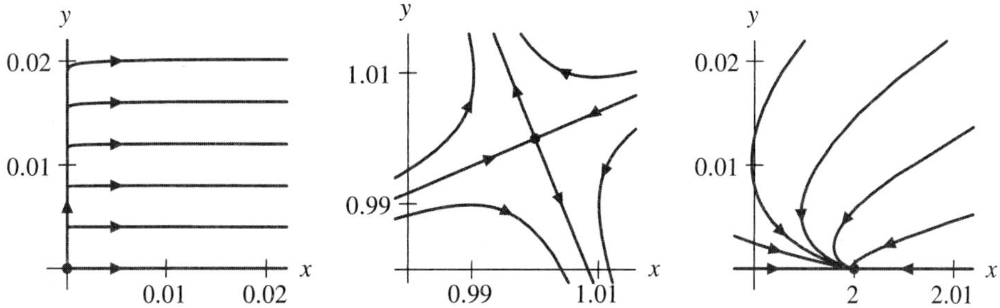

16. **(a)** The equilibrium points are $(0, 0)$, $(1, 0)$, and $(1, 1)$. We classify these points by computing the Jacobian matrix which is,

$$\begin{pmatrix} 2x - 1 & 0 \\ 2xy & x^2 - 2y \end{pmatrix},$$

and evaluating it at each equilibrium point. At $(0, 0)$, the Jacobian matrix is

$$\begin{pmatrix} -1 & 0 \\ 0 & 0 \end{pmatrix},$$

for which the eigenvalues are -1 and 0. An eigenvector for the eigenvalue -1 is $(1, 0)$, so solutions move toward the origin parallel to the x-axis. On the line $x = 0$, we have $dy/dt = -y^2$, so solutions move downwards when $y \neq 0$. Hence, $(0, 0)$ is a node. All solutions in the first quadrant near $(0, 0)$ tend toward the origin. At $(1, 0)$, the Jacobian matrix is

$$\begin{pmatrix} 1 & 0 \\ 0 & 1 \end{pmatrix},$$

which has a double eigenvalue of 1. Thus, $(1, 0)$ is a source. At $(1, 1)$, the Jacobian matrix is

$$\begin{pmatrix} 1 & 0 \\ 2 & -1 \end{pmatrix},$$

which has eigenvalues 1 and -1. So $(1, 1)$ is a saddle.

(b)

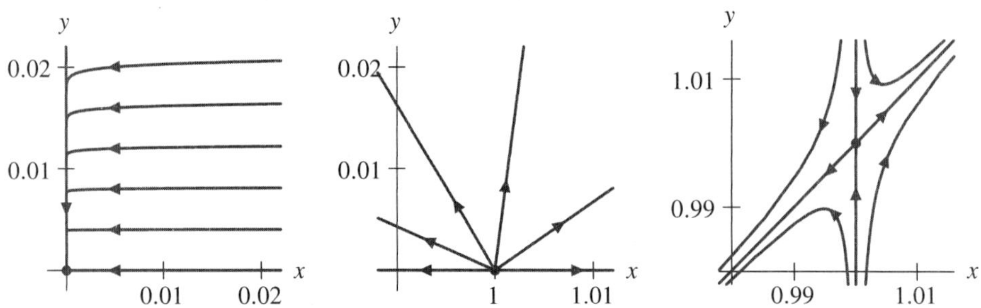

17. **(a)** The Jacobian matrix is

$$\begin{pmatrix} -3x^2 & 0 \\ 0 & -1 + 2y \end{pmatrix}$$

so the linearized system at $(0, 0)$ is

$$\frac{dx}{dt} = 0$$
$$\frac{dy}{dt} = -y.$$

(b) The eigenvalues are 0 and -1. Any vector on the x-axis is an eigenvector for the eigenvalue 0, and any vector on the y-axis is an eigenvector for the eigenvalue -1. Hence, the linearized system has a line of equilibrium points along the x-axis. Every other solution moves vertically toward the x-axis.

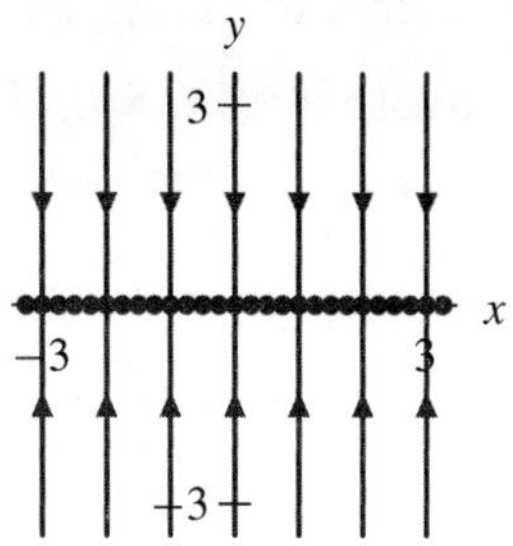

(c) The linearized system at (0, 1) is

$$\frac{dx}{dt} = 0$$
$$\frac{dy}{dt} = y.$$

(d) The eigenvalues of this system are 0 and 1. Any vector on the x-axis is an eigenvector for the eigenvalue 0, and any vector on the y-axis is an eigenvector for the eigenvalue 1. Hence, the linearized system has a line of equilibrium points along the x-axis. Every other solution moves vertically away from the x-axis. It is important to remember that the origin for the linearized system corresponds to the equilibrium point (0, 1) for the nonlinear system.

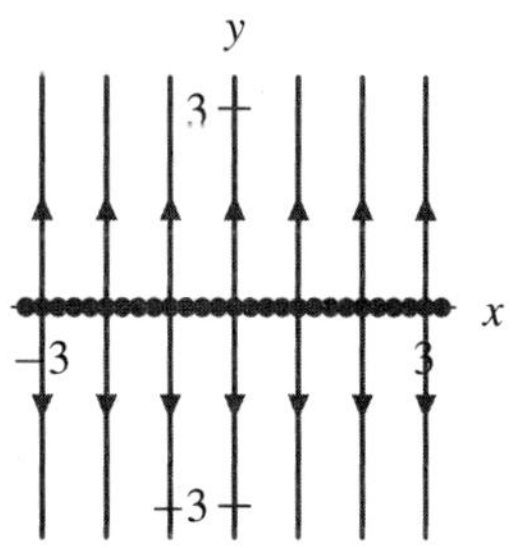

(e) The phase portrait is essentially a "combination" of two phase lines. The x-phase line has a sink at the origin. The y-phase line has a sink at the origin and a source at $y = 1$. Hence, the full phase portrait has a sink at (0, 0), and the equilibrium point at (0, 1) looks like a saddle.

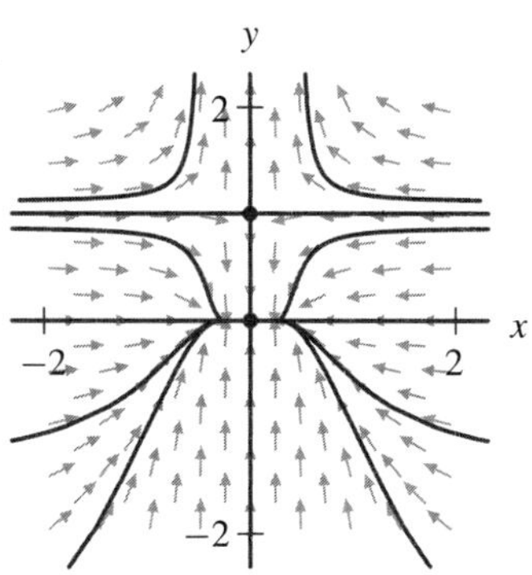

(f) The reason the linearizations and the nonlinear system look so different is that the equation for dx/dt contains only higher-order terms (just x^3 in this case). Since the equilibrium points occur along the y-axis ($x = 0$), the linearization has an entire line of equilibria in the x-direction.

18. **(a)** The equation $x^2 - a = 0$ has no solutions if $a < 0$.

(b) The equilibrium points are $(\pm\sqrt{a}, 0)$.

(c) When $a = 0$, the only equilibrium point is $(0, 0)$.

(d) The Jacobian matrix is

$$\begin{pmatrix} 2x & 0 \\ -2xy & -x^2 - 1 \end{pmatrix}.$$

At $(0, 0)$, the Jacobian matrix is

$$\begin{pmatrix} 0 & 0 \\ 0 & -1 \end{pmatrix},$$

which has eigenvalues -1 and 0. So $(0, 0)$ is a node.

19. **(a)**

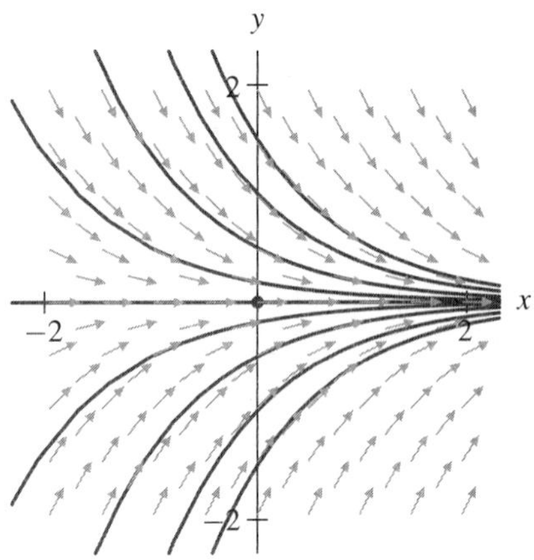

(b) The linearization of the equilibrium point at the origin has the coefficient matrix

$$\begin{pmatrix} 0 & 0 \\ 0 & -1 \end{pmatrix},$$

which has eigenvalues -1 and 0. So for the linearized system, the x-axis is a line of equilibria and solutions tend to zero in the y-direction. The nonlinear terms make solutions tend to zero in the x-direction for initial conditions with $x < 0$ and away from zero in the x-direction for initial conditions with $x > 0$.

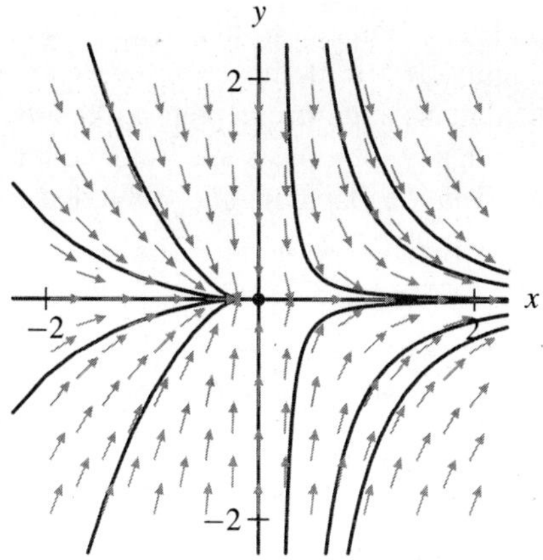

(c) The equilibria are $(\pm 1, 0)$. The coefficient matrix of the linearization at $(1, 0)$ is

$$\begin{pmatrix} 2 & 0 \\ 0 & -2 \end{pmatrix}.$$

The eigenvalues are 2 and -2, thus $(1, 0)$ is a saddle. The coefficient matrix of the linearization at $(-1, 0)$ is

$$\begin{pmatrix} -2 & 0 \\ 0 & -2 \end{pmatrix},$$

which has -2 as a repeated eigenvalue. So, $(-1, 0)$ is a sink.

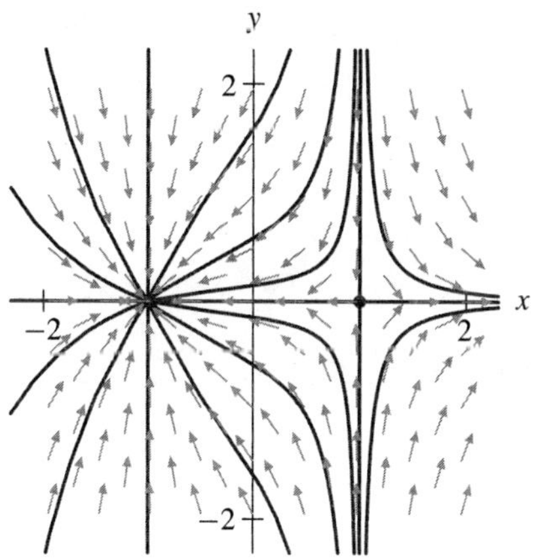

20. **(a)** The equilibrium points are $(\pm\sqrt{a}, a)$, so there are no equilibrium points if $a < 0$, one equilibrium point if $a = 0$, and two equilibrium points if $a > 0$

(b) If $a = 0$, the equilibrium point at the origin has eigenvalues 0 and 1 and is a node. If $a > 0$, the system has two equilibrium points, a saddle at $(\sqrt{a}, a)$ with eigenvalues $-2\sqrt{a}$ and 1 and a source at $(-\sqrt{a}, a)$ with eigenvalues $2\sqrt{a}$ and 1. A bifurcation occurs at $a = 0$ because the number of equilibrium points changes. It also reasonable to say that there is a bifurcation at $a = 1/4$ because the source at $(-\sqrt{a}, a)$ has repeated eigenvalues. For all other positive values of a, these eigenvalues are real and distinct.

(c) Note that for all values of the parameter a, the line $y = a$ is invariant. If $a < 0$, all solutions come from and go to infinity. If $a = 0$, most solutions come from and go to infinity, but there are separatrices associated to the equilibrium point at the origin. If $a > 0$, some solutions come from and go to infinity, but many solutions come from the source at $(-\sqrt{a}, a)$ and go to infinity. There is also a separating solution along the line $y = a$ that comes from the source at $(-\sqrt{a}, a)$ and goes to the saddle at $(\sqrt{a}, a)$.

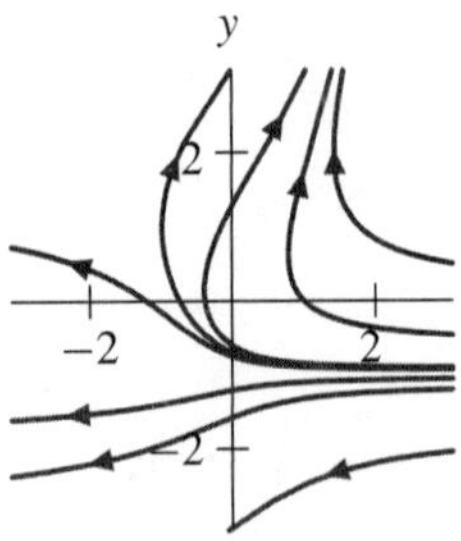

Phase portrait for $a < 0$

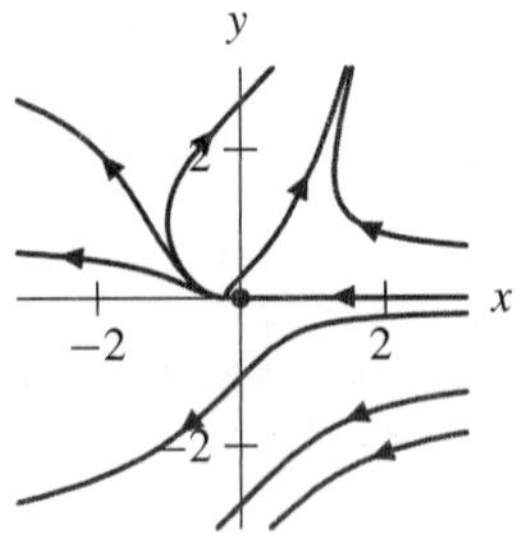

Phase portrait for $a = 0$

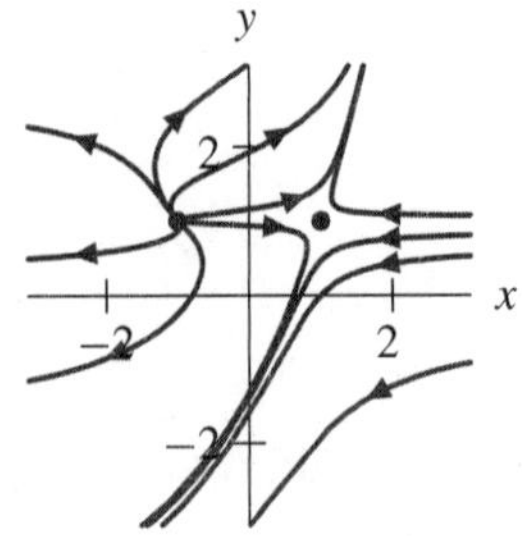

Phase portrait for $a > 0$

21. **(a)** The only equilibrium points occur if $a = 0$. Then all points on the curve $y = x^2$ are equilibrium points.

(b) The bifurcation occurs at $a = 0$.

(c) If $a < 0$, all solutions decrease in the y-direction since $dy/dt < 0$. If $a > 0$, all solutions increase in the y-direction since $dy/dt > 0$. If $a = 0$, there is a curve of equilibrium points located along $y = x^2$, and all solutions move horizontally.

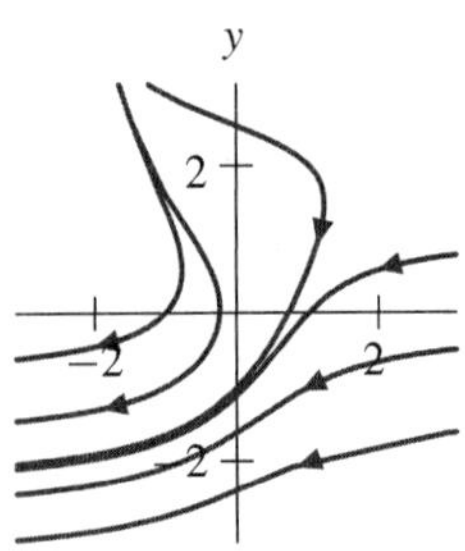

Phase portrait for $a < 0$

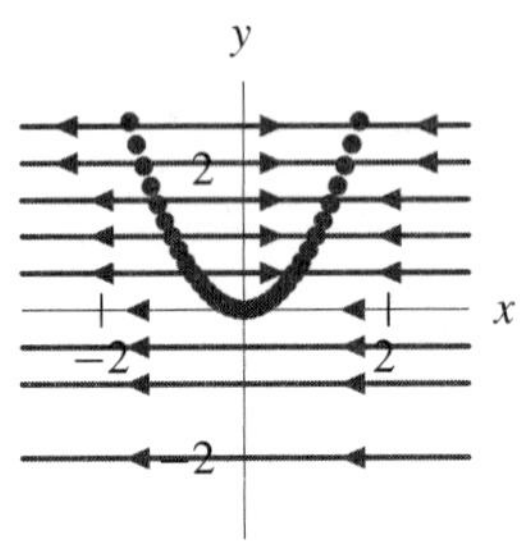

Phase portrait for $a = 0$

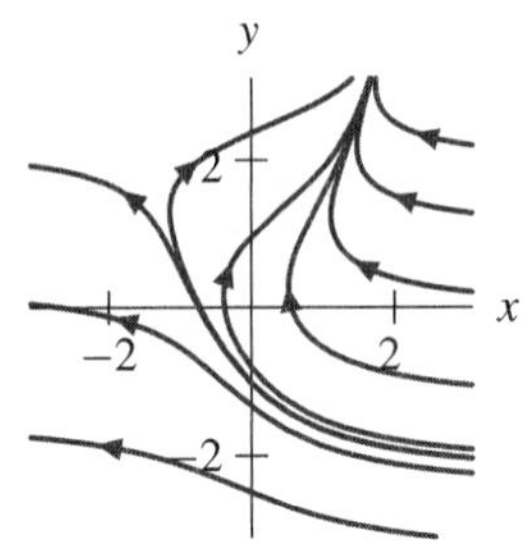

Phase portrait for $a > 0$

22. **(a)** The equilibrium points are $((1 \pm \sqrt{1+4a})/2, -(1 \pm \sqrt{1+4a})/2 - a)$, so there are no equilibrium points if $a < -1/4$, one equilibrium if $a = -1/4$, and two equilibrium points if $a > -1/4$.

(b) A bifurcation occurs at $a = -1/4$.

(c) If $a < -1/4$, there are no equilibrium points and all solutions come from and go to infinity. If $a = -1/4$, an equilibrium point appears at $(1/2, 1/4)$. This equilibrium point has both eigenvalues 0 and is a node. If $a > -1/4$, the system has two equilibrium points, at $((1 \pm \sqrt{1+4a})/2, -(1 \pm \sqrt{1+4a})/2 - a)$.

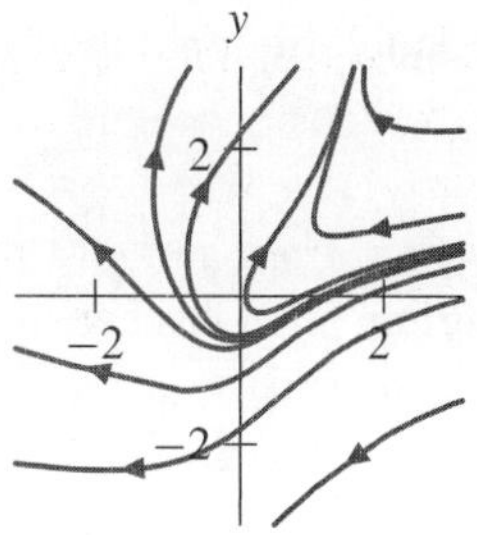
Phase portrait for $a < -1/4$

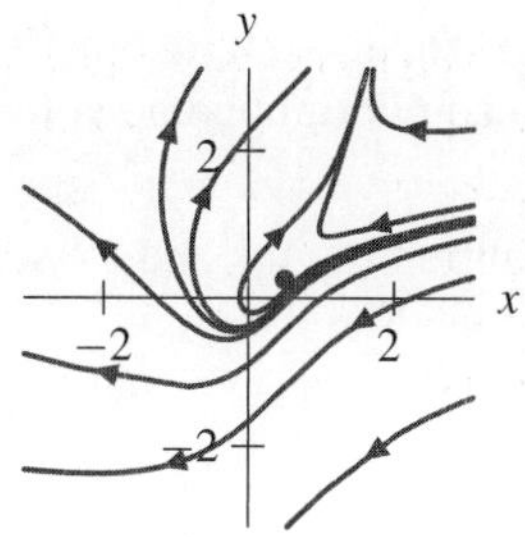
Phase portrait for $a = -1/4$

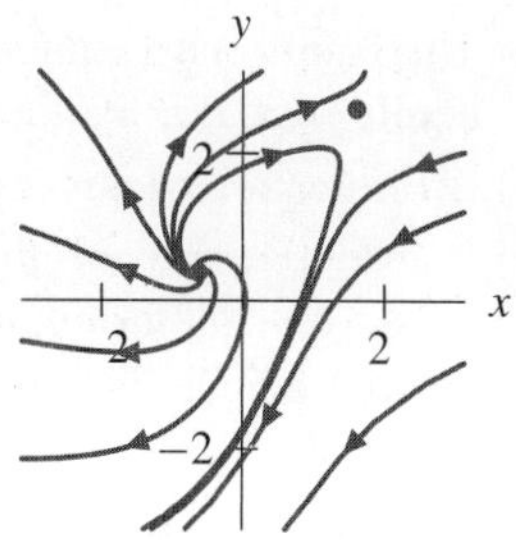
Phase portrait for $a > -1/4$

23. **(a)** The equilibrium points are $(0, 0)$, $(\pm 1/\sqrt{a}, \pm 1/\sqrt{a})$, so there is only one equilibrium point if $a \leq 0$, and three equilibrium points if $a > 0$.

(b) A bifurcation occurs at $a = 0$.

(c) If $a < 0$, there are is only one equilibrium point at the origin, and this equilibrium point is a spiral source. If $a > 0$, the system has two additional equilibrium points, at $(\pm 1/\sqrt{a}, \pm 1/\sqrt{a})$. These equilibrium points come from infinity as a increases through 0.

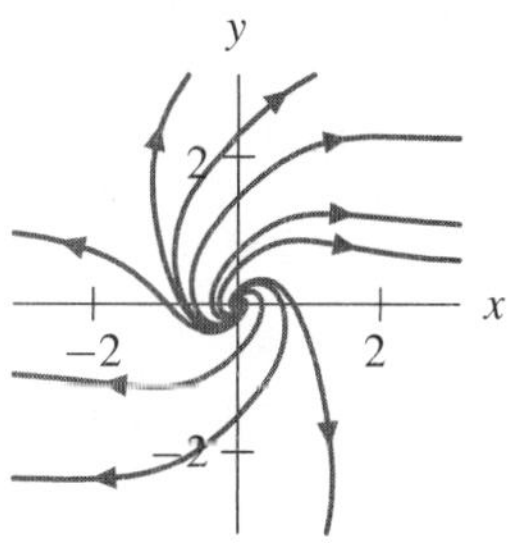
Phase portrait for $a < 0$

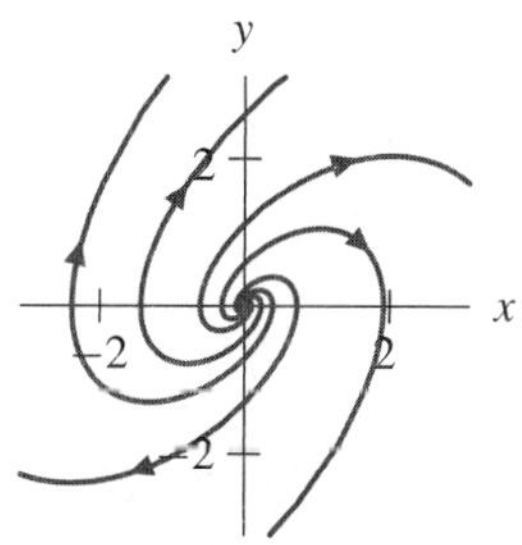
Phase portrait for $a = 0$

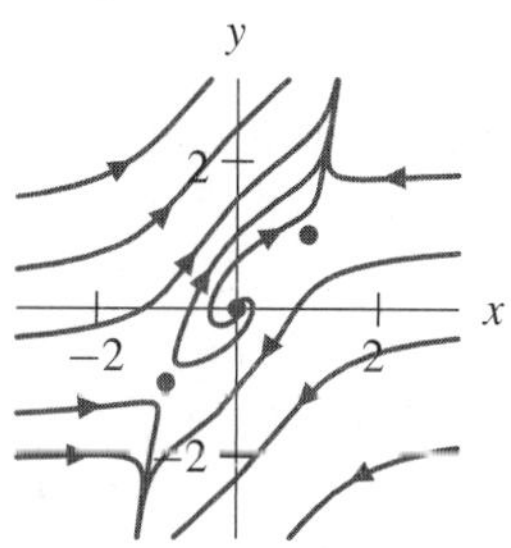
Phase portrait for $a > 0$

24. **(a)** The equilibrium points are $(\pm\sqrt{a}, 0)$, so there are no equilibrium points if $a < 0$, one equilibrium if $a = 0$, and two equilibrium points if $a > 0$

(b) A bifurcation occurs at $a = 0$.

(c) If $a < 0$, there are no equilibrium points and all solutions come from and go to infinity. If $a = 0$, an equilibrium point appears at the origin. This equilibrium point has eigenvalues 0 and 1 and is a node. If $a > 0$, the system has two equilibrium points, at $(\pm\sqrt{a}, 0)$.

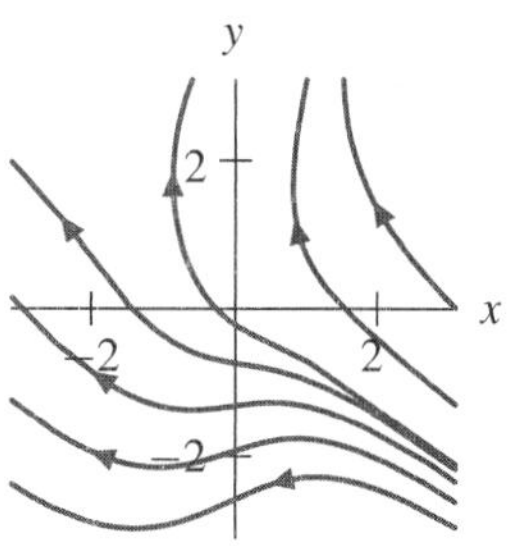
Phase portrait for $a < 0$

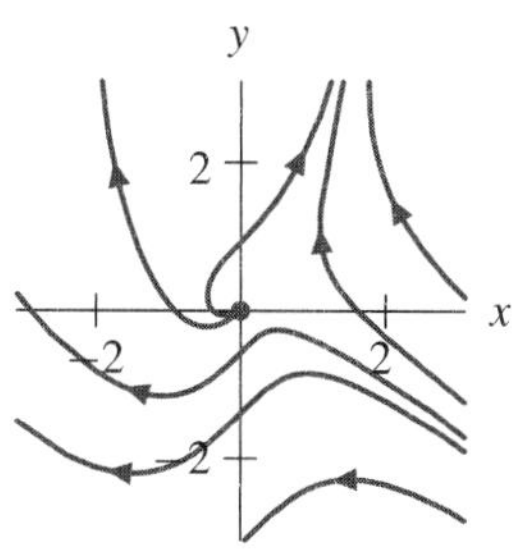
Phase portrait for $a = 0$

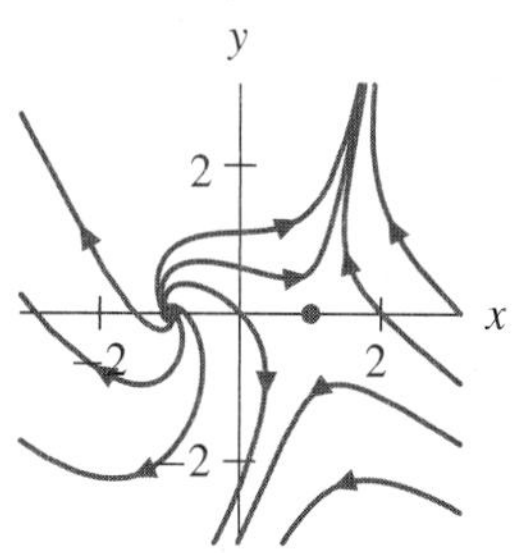
Phase portrait for $a > 0$

25. **(a)** The equilibrium points are $(\pm\sqrt{a/2}, -a/2)$, so there are no equilibrium points if $a < 0$, one equilibrium if $a = 0$, and two equilibrium points if $a > 0$

(b) A bifurcation occurs at $a = 0$.

(c) If $a < 0$, there are no equilibrium points and all solutions come from and go to infinity. If $a = 0$, an equilibrium point appears at the origin. This equilibrium point has eigenvalues 0 and 1 and is a node. If $a > 0$, the system has two equilibrium points, at $(\pm\sqrt{a/2}, -a/2)$.

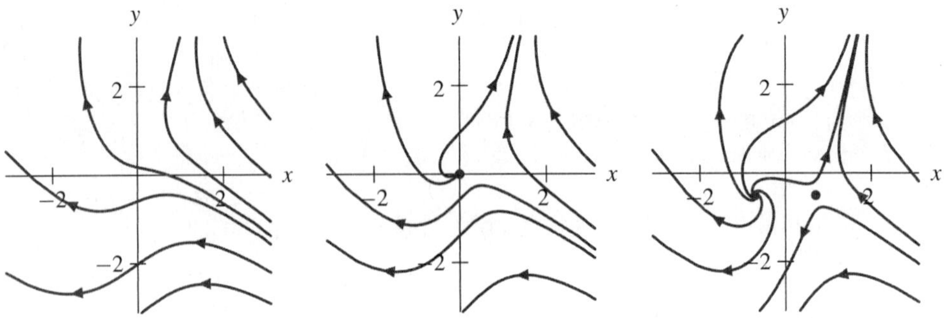

Phase portrait for $a < 0$ Phase portrait for $a = 0$ Phase portrait for $a > 0$

26. Since this is a competing species model, $a > 0$. The equilibrium points are $(0, 0)$, $(0, a)$, $(70, 0)$, and $(a - 70, 140 - a)$. If $a = 70$, the second and fourth of these points coincide. If $a = 140$, the third and fourth coincide. Hence bifurcations occur at these two a-values.

For $70 < a < 140$ there is an equilibrium point that does not lie on the axes. This equilibrium point is a saddle whose separatrices divide the first quadrant into two regions. In one region, all solutions tend to $(0, a)$ and in the other, to $(70, 0)$. If $0 < a < 70$, all solutions (not on the axes) tend to the equilibrium point at $(70, 0)$; that is, the y-species dies out. If $a > 140$, all solutions (not on the axes) tend to the equilibrium point at $(0, a)$; that is, the x-species dies out.

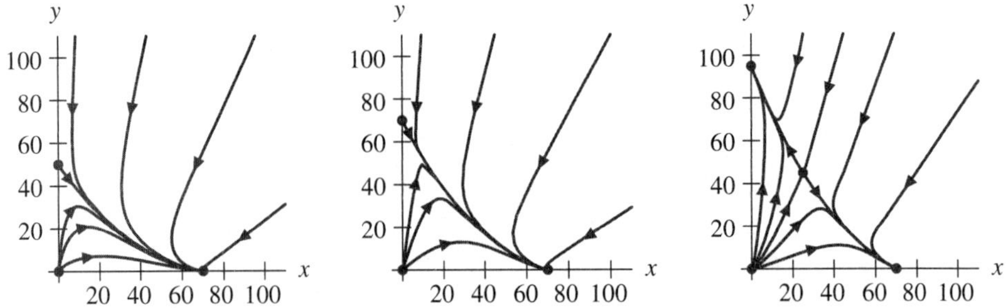

Phase portrait for $a < 70$ Phase portrait for $a = 70$ Phase portrait for $70 < a < 140$

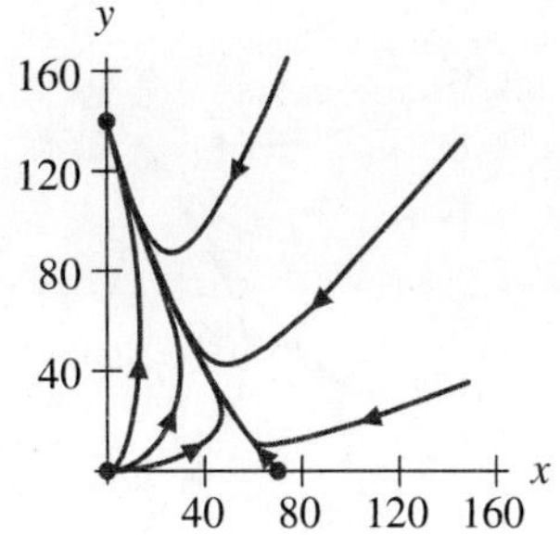

Phase portrait for $a = 140$

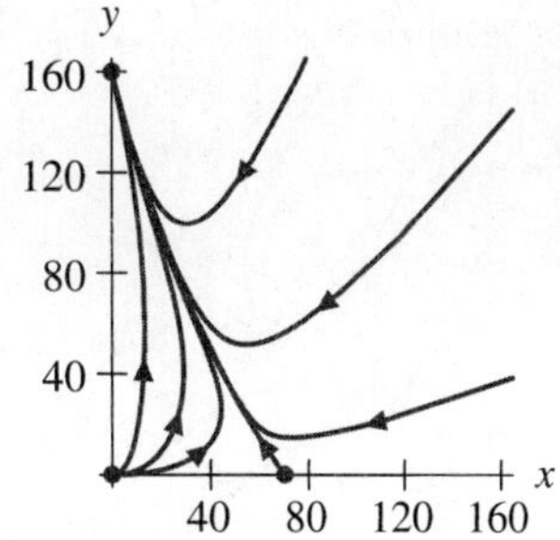

Phase portrait for $a > 140$

27. **(a)** The fact that $(0, 0)$ is an equilibrium point says that, if both X and Y are absent from the island, then neither will ever migrate to the island. However, it may be possible for one species to migrate if the other is already on the island.

(b) If a small population consisting solely of one of the species reproduces rapidly, then we expect both $\partial f/\partial x$ and $\partial g/\partial y$ to be positive and large at $(0, 0)$. We expect this because these partials are the coefficients of x and y in the linearization at $(0, 0)$.

(c) Since the species compete, an increase in y decreases dx/dt and an increase in x decreases dy/dt. Hence, both $\partial f/\partial y$ and $\partial g/\partial x$ are negative at $(0, 0)$ since $\partial f/\partial y$ is the coefficient of y in the dx/dt equation and $\partial g/\partial x$ is the coefficient for x in the dy/dt equation for the linearization at the origin.

(d) Suppose the coefficient matrix of the linearized system is

$$\begin{pmatrix} a & b \\ c & d \end{pmatrix},$$

with a and d positive and large and b and c negative. The eigenvalues are

$$\frac{(a+d) \pm \sqrt{(a-d)^2 + 4bc}}{2}.$$

If b and c are near zero, then $(0, 0)$ is a source. If b and c are very negative, then $(0, 0)$ is a saddle.

It is also possible to have 0 as an eigenvalue of the linearized system in which case the linearization fails to determine the behavior of the nonlinear system near $(0, 0)$.

(e) For the linearized system, note that $dx/dt < 0$ along the positive y-axis and $dy/dt < 0$ along the positive x-axis. If the origin is a saddle, the eigenvectors for the negative eigenvalue must be in the first and third quadrants, and a typical solution near the origin starting in the first quadrant has one of the species going extinct. If the origin is a source, then a typical solution near the origin has one or the other of the species going extinct except for one curve of solutions in the first quadrant.

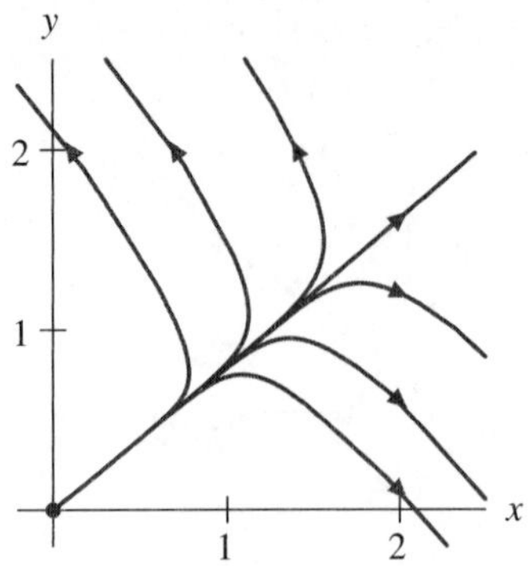

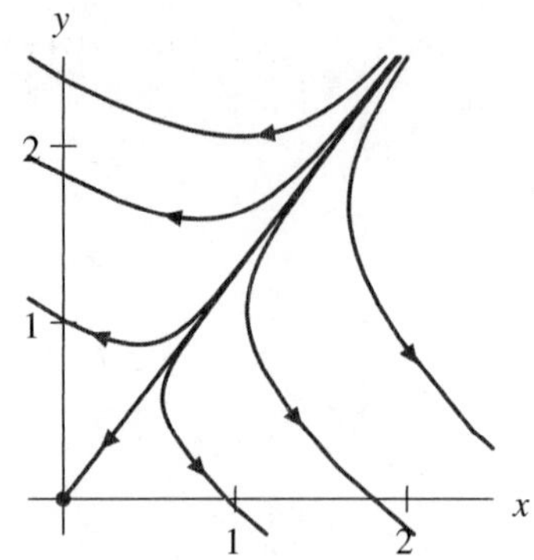

28. **(a)** At (0, 0), $\partial f/\partial x$ and $\partial g/\partial y$ are positive and small.

(b) At (0, 0), $\partial f/\partial y$ and $\partial g/\partial x$ are negative and large in absolute value.

(c) With these assumptions, the Jacobian matrix is

$$\begin{pmatrix} a & b \\ c & d \end{pmatrix}$$

where a and d are small and positive, but b and c are negative with much larger absolute value. Since the eigenvalues of this matrix are given by

$$\frac{a + d \pm \sqrt{(a+d)^2 - 4(ad - bc)}}{2}$$

and since $(a + d)^2 > 0$ and $ad - bc < 0$, the term inside the square root is positive. Thus both eigenvalues are real.

The term $a + d$ is very small and positive, but the term inside the square root is large and positive. So one of the eigenvalues is positive, and the other is negative. Thus (0, 0) is a saddle.

(d) Note that $dx/dt < 0$ on the positive y-axis and $dy/dt < 0$ on the positive x-axis. The signs are reversed on the negative axes. Hence, the eigenvectors for the negative eigenvalue are in the first and third quadrants and those for the poistive eigenvalue are in the second and fourth quadrants. Solutions starting near the origin in the first quadrant have either one or both species going extinct.

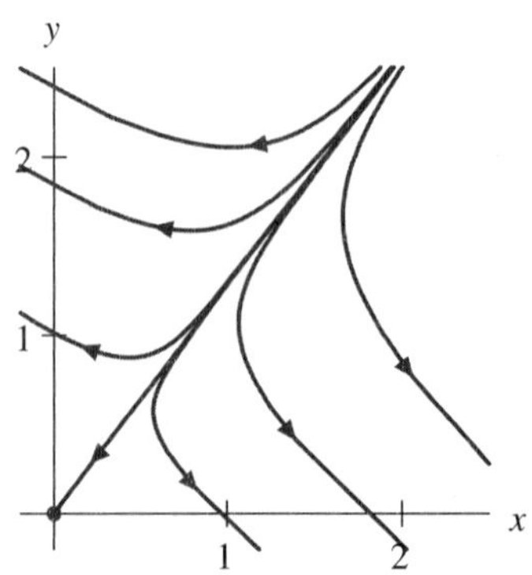

29. **(a)** At (0, 0), $\partial f/\partial x$ is positive and large, and $\partial g/\partial y$ is positive and small.
(b) At (0, 0), $\partial f/\partial y$ is negative with a large absolute value and $\partial g/\partial x = 0$.
(c) With these assumptions, the Jacobian matrix is

$$\begin{pmatrix} a & b \\ 0 & d \end{pmatrix},$$

where $a > 0$, $b < 0$, and $d > 0$ is much smaller than a. The eigenvalues of this matrix are a and d, so (0, 0) is a source.
(d) Note that for $y = 0$, $dy/dt = 0$, and the eigenvector for a is in the x-direction.

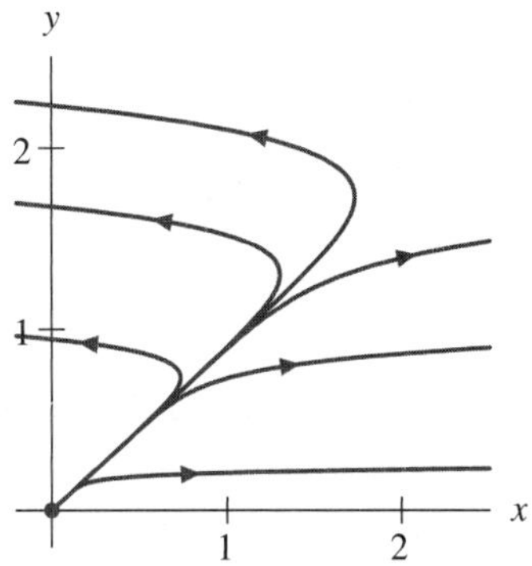

30. **(a)** If z is fixed and y increases, then our assumption is that dy/dt decreases. That is, $\partial h/\partial y < 0$. Similarly, $\partial k/\partial z < 0$.
(b) Similarly, $\partial h/\partial z$ and $\partial k/\partial y$ are both positive.
(c) With these assumptions, the Jacobian matrix is

$$\begin{pmatrix} a & b \\ c & d \end{pmatrix}$$

where $a < 0$, $b > 0$, $c > 0$, and $d < 0$. The eigenvalues of this matrix are

$$\frac{a + d \pm \sqrt{(a-d)^2 + 4bc}}{2}.$$

These eigenvalues are always real, since the term inside the square root is positive. One eigenvalue is always negative (choose the negative square root). The other may be positive or negative. Thus, we only have saddles or sinks for equilibrium points.

EXERCISES FOR SECTION 5.2

1. For x- and y-nullclines, $dx/dt = 0$, and $dy/dt = 0$ respectively. Then, we obtain $y = -x + 2$ for the x-nullcline and $y = x^2$ for the y-nullcline. To find intersections, we set $-x + 2 = x^2$, or $(x + 2)(x - 1) = 0$. Solving this for x yields $x = 1, -2$. For $x = 1$, $y = 1$, and for $x = -2$, $y = 4$. So the equilibrium points are $(1, 1)$ and $(-2, 4)$.

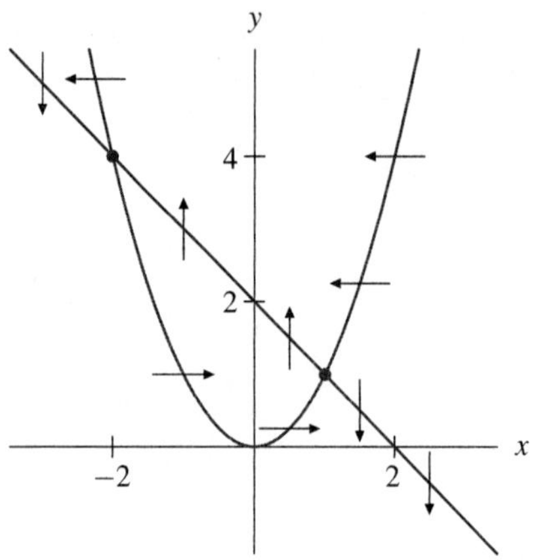

The solution for (a) is in the left-down region, and therefore, it eventually enters the region where $y < -x + 2$ and $y < x^2$. Once the solution enters this region, it stays there because the vector field on the boundaries never points out. Solutions for (b) and (c) start in this same region. Hence, all three solutions will go down and to the right without bound.

2. For x- and y-nullclines, $dx/dt = 0$, and $dy/dt = 0$ respectively. So, we have $y = -x + 2$ for the x-nullcline and $y = |x|$ for the y-nullcline. To find intersections, we set $-x + 2 = |x|$. Solving this for x yields $x = 1$, so the only equilibrium point is $(x, y) = (1, 1)$.

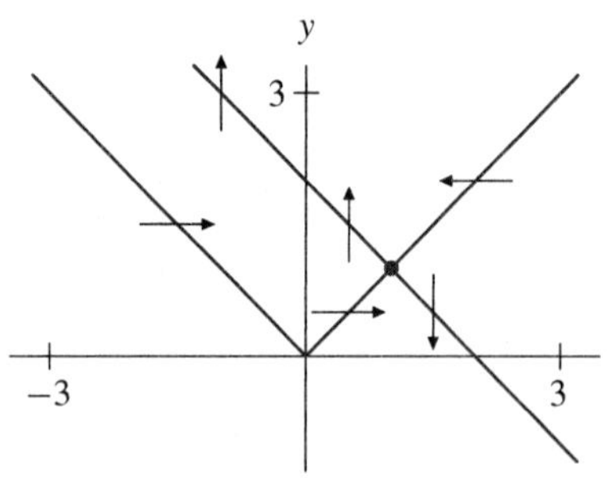

The solution for (a) begins on the y-nullcline, heads into the right-up region, eventually crosses the x-nullcline, and then tends to infinity in the left-up region.

The solution for (b) starts in the left-down position, crosses the x-nullcline, then tends to infinity in the right-down region.

The solution corresponding to (c) starts on the y-nullcline, immediately enters the left-up region, and then tends to infinity in this region.

3. For the x-nullcline, $x(x-1)=0$, or $x=0$ and $x=1$, and for the y-nullcline, $y=x^2$. The equilibrium points are the intersection points of the x- and y-nullclines. They are $(0,0)$ and $(1,1)$.

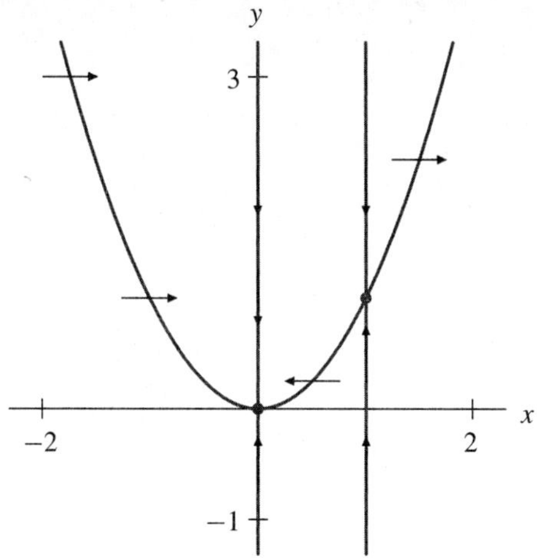

The initial conditions (a) and (b) are in right-up and left-up region respectively. Therefore, their solution curves eventually enter the region where $y > x^2$ and $x \le 1$, and tend toward the equilibrium point at $(0,0)$. The initial condition (c) is on the x-nullcline $x=1$, and therefore its solution curve tends to the equilibrium point at $(1,1)$.

4. **(a)** Equilibria are located where x-nullclines and y-nullclines intersect, so those equilibria with both $x > 0$ and $y > 0$ are located on the intersection of the lines

$$Ax + By = C \quad \text{and} \quad Dx + Ey = F.$$

However, the only way that two lines can intersect at more than one point is if they are really the same line. This happens if

$$A/D = B/E = C/F.$$

(b) To guarantee that there is exactly one equilibrium point at which the species coexist, we can stipulate that the x- and y-intercepts of the x- and y-nullclines are positioned so that these two lines are forced to intersect in the first quadrant. For example, we could require that the y-intercept of the x-nullcline, namely C/B, lies below the y-intercept of the y-nullcline, namely F/E, whereas the opposite happens for the x-intercepts. That is, we could require that

$$F/E > C/B \quad \text{but} \quad F/D < C/A.$$

Reversing both of these inequalities also guarantees that the species can coexist.

5. **(a)** The x-nullcline is made up of the lines $x = 0$ and $y = -x/3 + 50$. The y-nullcline is made up of the lines $y = 0$ and $y = -2x + 100$.

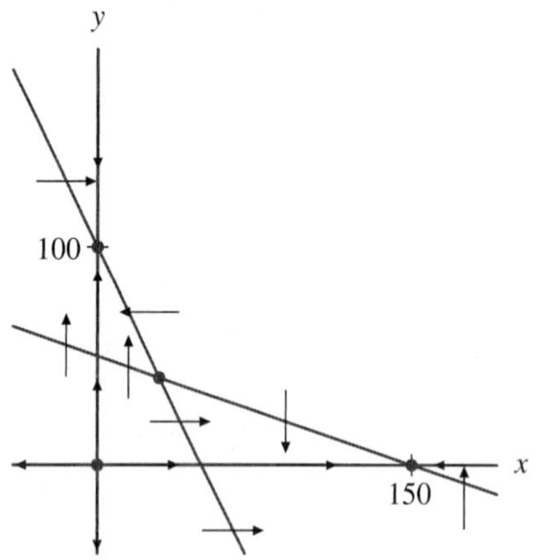

(b)

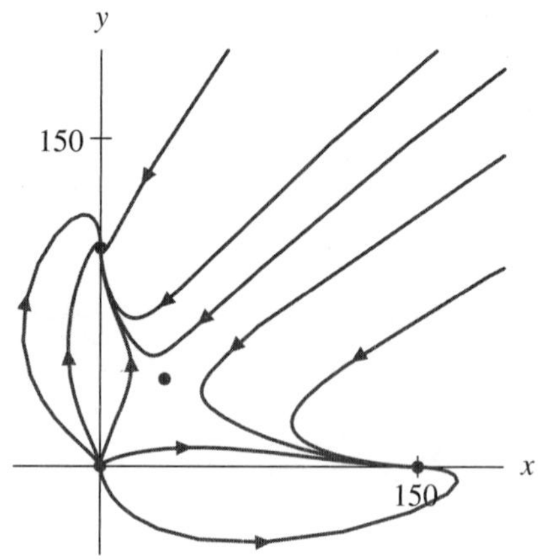

(c) Most solutions tend toward one of the equilibrium points (0, 100) or (150, 0). One curve of solutions divides these two behaviors. On this curve, solutions tend toward the saddle equilibrium at (30, 40).

6. **(a)** The x-nullcline consists of the lines $x = 0$ and $y = 10 - x$. The y-nullcline consists of the lines $y = 0$ and $y = 30 - 2x$.

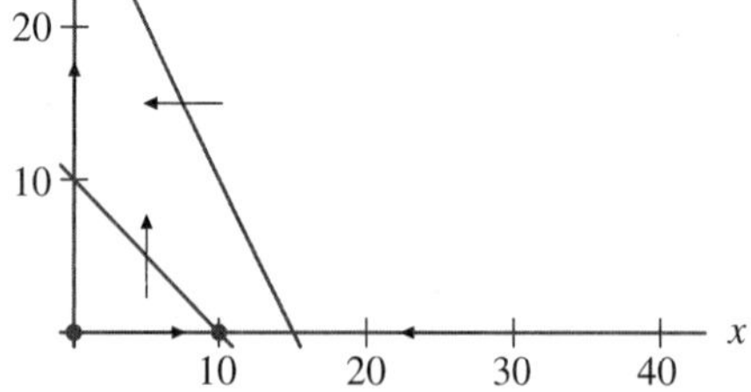

(b)

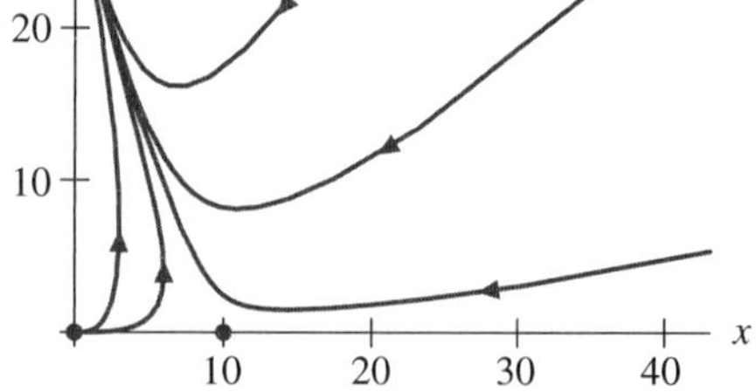

(c) All solutions (except those on the x-axis) tend to the equilibrium point at (0, 30). On the x-axis, all solutions tend to the equilibrium point at $x = 10$.

7. **(a)** The x-nullcline consists of the two lines $x = 0$ and $y = -x/2 + 50$. The y-nullcline consists of the two lines $y = 0$ and $y = -x/6 + 25$.

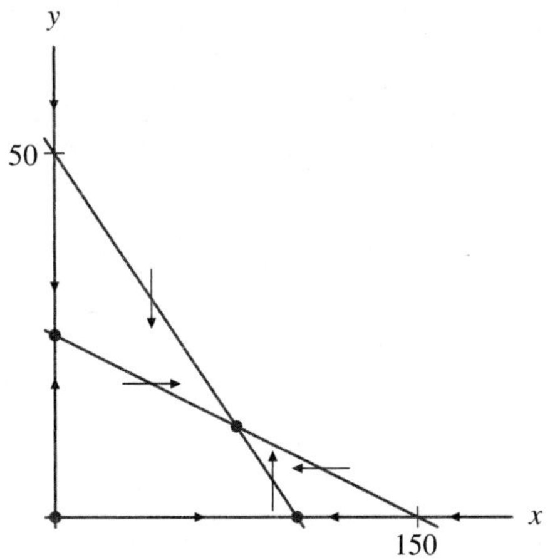

(b)

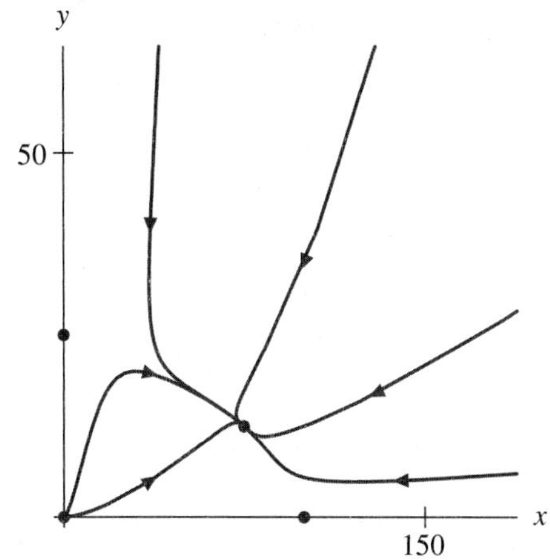

(c) All solutions off the axes tend toward the sink at $(75, 25/2)$. On the x-axis, solutions tend to the saddle at $(100, 0)$. On the y-axis, solutions tend to the saddle at $(0, 25)$.

8. **(a)** The x-nullcline is given by the two lines $x = 0$ and $y = -x + 100$. The y-nullcline is given by the line $y = 0$ and the circle $x^2 + y^2 = 50^2$.

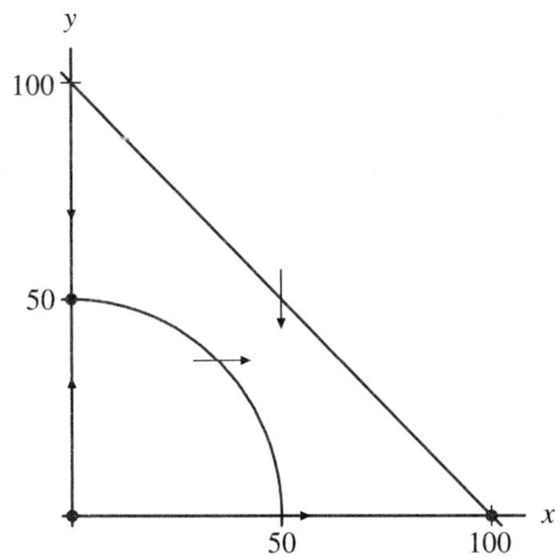

(b)

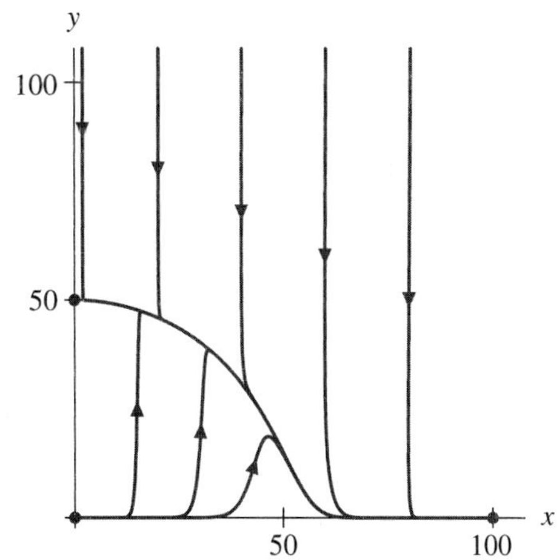

(c) All solutions (except those on the y-axis) tend to the equilibrium point at $(100, 0)$. On the y-axis, all solutions tend to the equilibrium point at $y = 50$.

9. **(a)** The x-nullcline is given by the two lines $x = 0$ and $y = -x + 40$. The y-nullcline is given by the line $y = 0$ and the circle $x^2 + y^2 = 50^2$.

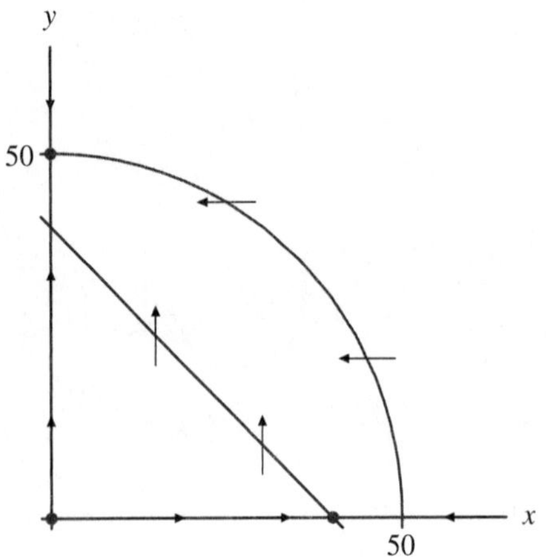

(b)

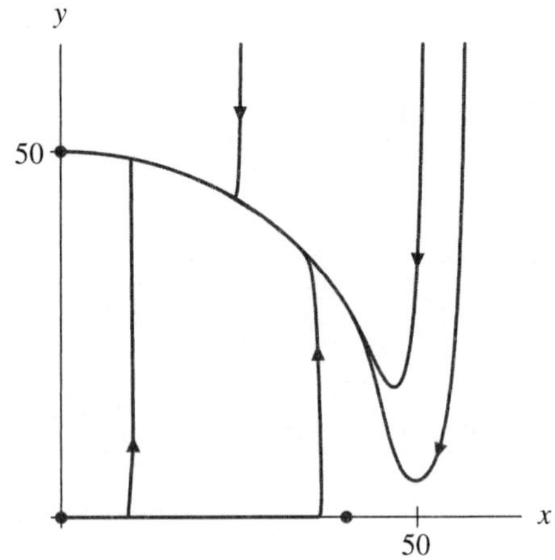

(c) Solutions off the x-axis tend toward the sink at (0, 50). Solutions on the x-axis tend toward the saddle at (40, 0).

10. **(a)** The x-nullcline is given by the lines $x = 0$ and $y = -4x + 160$. The y-nullcline is given by the line $y = 0$ and the circle $x^2 + y^2 = 50^2$. (Recall that we are only interested in the first quadrant).

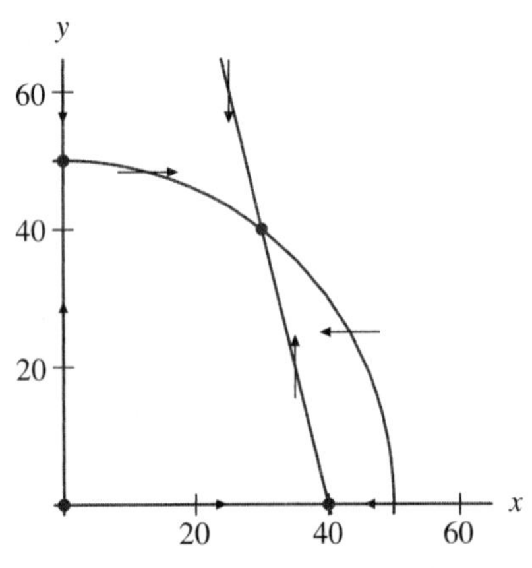

(b)

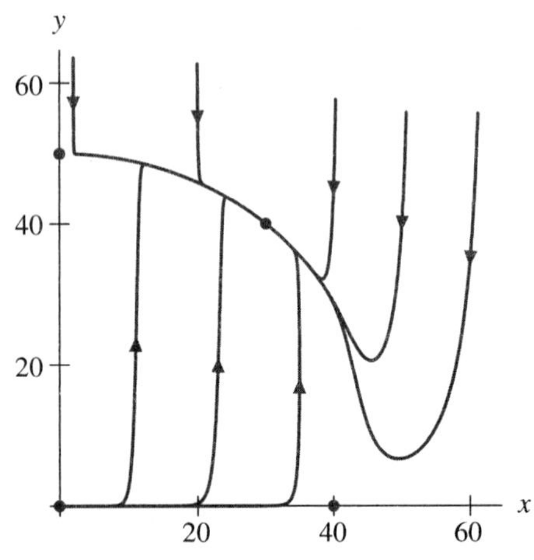

(c) All solutions (except those on the axes) tend to the equilibrium point at (40, 30). On the y-axis, all solutions tend to the equilibrium point at $y = 50$. On the x-axis, all solutions tend to the equilibrium point at $x = 40$.

11. **(a)** The x-nullcline is given by the line $x = 0$ and the line $y = -4x/3 + 80$. The y-nullcline is given by the line $y = 0$ and the circle $x^2 + y^2 = 50^2$. (Recall that we are only interested in the first quadrant).

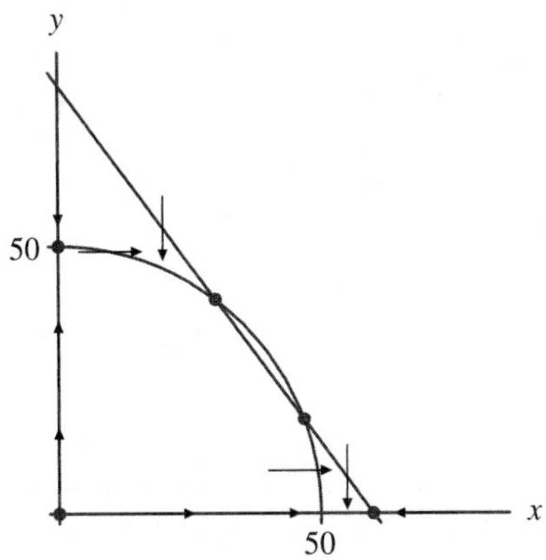

(b)

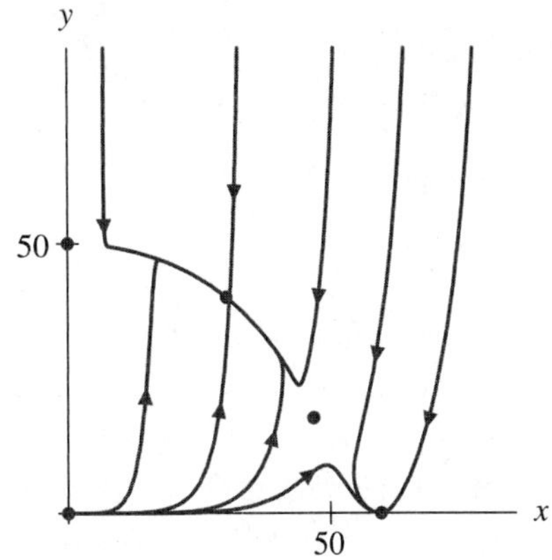

(c) Most solutions tend toward the sink at (30, 40) or the sink at (60, 0). The curve dividing these two behaviors is a curve of solutions tending toward the saddle at (234/5, 88/5). Solutions on the y-axis tend toward the saddle at (0, 50).

12. **(a)** The x-nullcline is given by the lines $x = 0$ and $y = -x + 2$. The y-nullcline is given by the line $y = 0$ and the curve $y = x^2$.

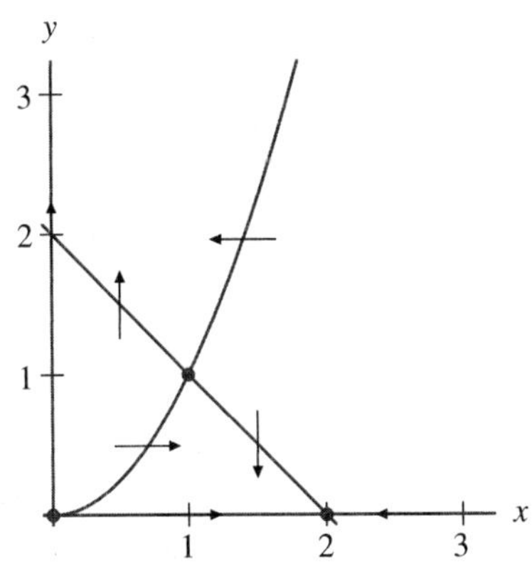

(b)

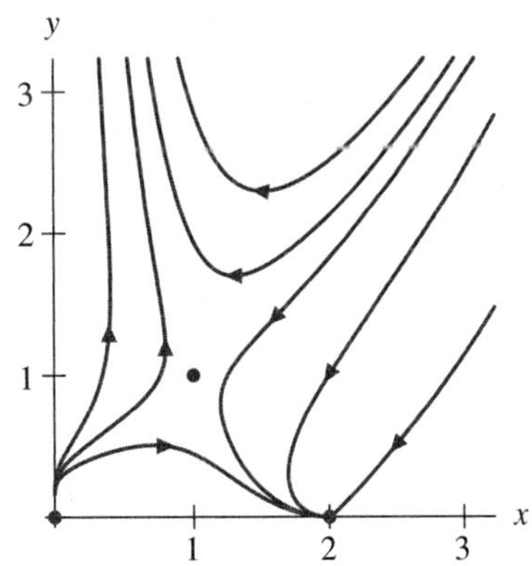

(c) There is a saddle point at (1, 1). Two solutions leave this equilibrium point, one tending to infinity, the other to the equilibrium point at (2, 0). Two solutions tend toward the saddle point, one coming from the origin, one from infinity. To the "right" of and "below" the incoming solution curve, all solutions tend to the equilibrium point at (2, 0); to the "left" all solutions tend to ∞.

13. **(a)** The x-nullcline is given by the lines $x = 0$ and $y = -x + 2$. The y-nullcline is given by the lines $y = 0$ and $y = x$.

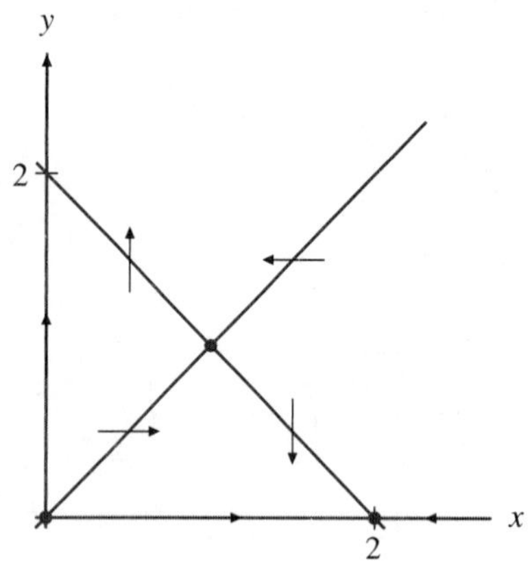

(b)

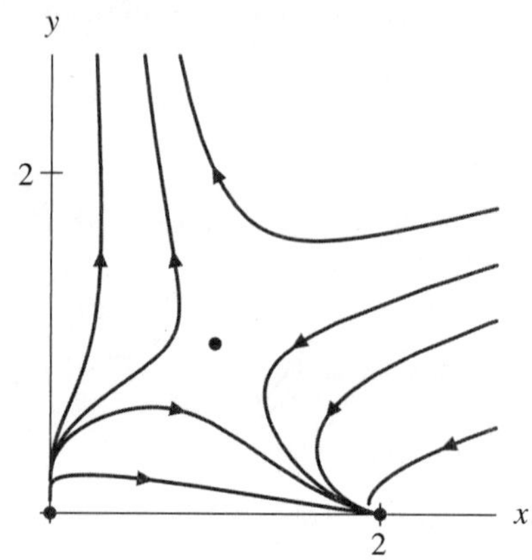

(c) Most solutions tend toward either the sink at (2, 0) or toward infinity in the y-direction (with $x < 1$). The curve separating these two behaviors is a curve of solutions that tend toward the saddle at (1, 1).

14. **(a)** The x-nullcline is given by the lines $x = 0$ and $x = 1$. The y-nullcline is given by the line $y = 0$ and the curve $y = x^2$.

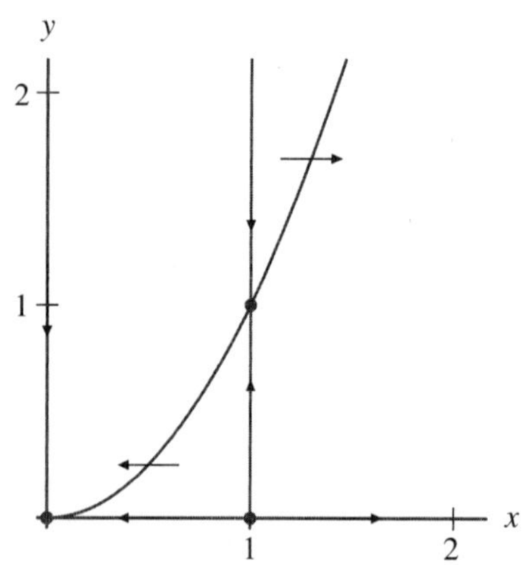

(b)

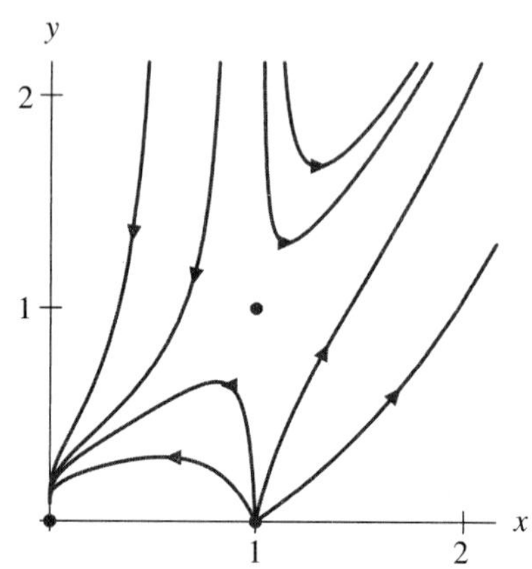

(c) There is a sink at (0, 0), a source at (1, 0), and a saddle at (1, 1). Two solution curves tend to the saddle, one coming from ∞ and one from the source. Two solutions leave the saddle, one tending to the sink and the other to ∞. Solutions leaving the source tend either to the sink or to ∞. The solution tending to the saddle divides these two regions. Solutions tending to the sink come either from ∞ or from the source. The solution originating at the saddle divides these two regions.

15. **(a)** Since the species are cooperative, an increase in y results in an increase in x and vice versa. Therefore, one needs to change the signs in front of B and D from $-$ to $+$.

(b) The x-nullcline is given by $x = 0$ or $-Ax + By + C = 0$. The y-nullcline is given by $y = 0$ or $Dx - Ey + F = 0$. The origin is always an equilibrium point. Also, $x = 0$, $y = F/E$ and $x = C/A$, $y = 0$ are equilibrium points. Equilibrium points with both x and y positive arise from solutions of

$$\begin{cases} -Ax + By + C = 0 \\ Dx - Ey + F = 0 \end{cases}$$

In matrix notation, we obtain

$$\begin{pmatrix} -A & B \\ D & -E \end{pmatrix} \begin{pmatrix} x \\ y \end{pmatrix} = \begin{pmatrix} -C \\ -F \end{pmatrix}.$$

In order for a unique solution to exist, $AE - BD \neq 0$. Then, the solution is

$$\begin{pmatrix} x \\ y \end{pmatrix} = \frac{1}{AE - BD} \begin{pmatrix} CE + BF \\ CD + AF \end{pmatrix}$$

Since A through F are all positive, we must have $AE - BD > 0$ for the solution to be in the first quadrant.

If $AE - BD = 0$, then $-Ax + By$ must be a negative multiple of $Dx - Ey$, so there are no solutions with both x and y positive.

16. **(a)** The a- and b-nullclines are identical. They both consist of both the a- and the b-axis. Hence all points on either of these axes are equilibrium points.

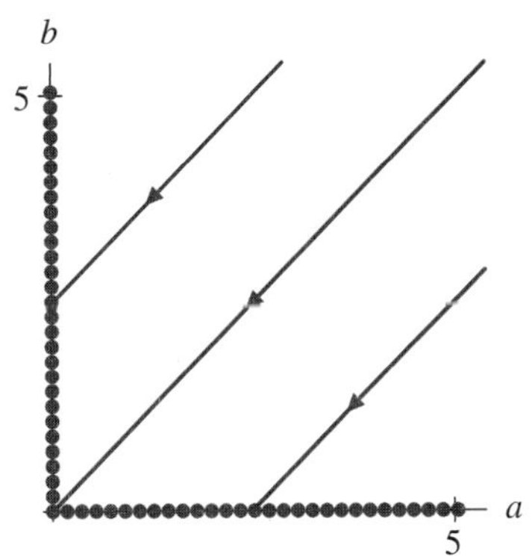

(b) Since $da/dt = db/dt$, solution curves are lines of slope 1 in the ab-plane.

(c) Along these lines, the solutions tend to the equilibrium point that is the intersection point of this line and either the positive a- or b-axis.

17. **(a)** For the a-nullcline, $da/dt = 0$, so $2 - ab/2 = 0$, or $ab = 4$. For the b-nullcline, $db/dt = 0$, so $ab = 3$. Both nullclines are hyperbolas, and the curve of $ab = 4$ is above the one of $ab = 3$. Therefore, the direction of vector field on $ab = 4$ is vertical and downward, and the one on $ab = 3$ is horizontal and to the right.

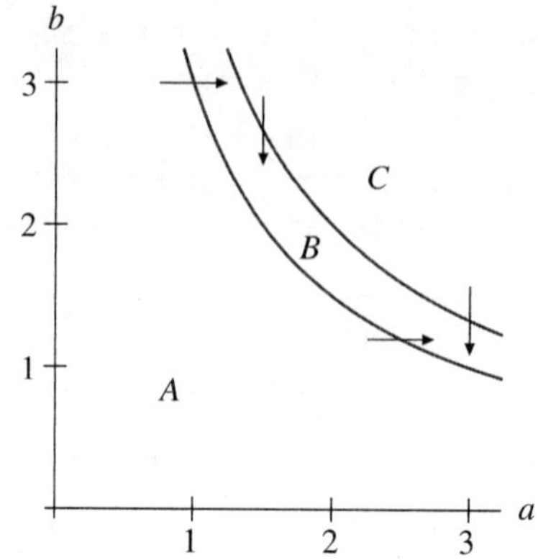

(b) Below and above $ab = 3$, $db/dt > 0$ and $db/dt < 0$ respectively. Below and above $ab = 3$, $da/dt > 0$ and $da/dt < 0$ respectively. Therefore, in region A, the vector field points up and to the right, in region B, the vector field points down and to the right, and in region C, the vector field points down and to the left.

(c) On the boundaries of B, the direction of the vector field never points out of B. Therefore, as time increases, these solutions are asymptotic to the positive x-axis from above.

18. **(a)**

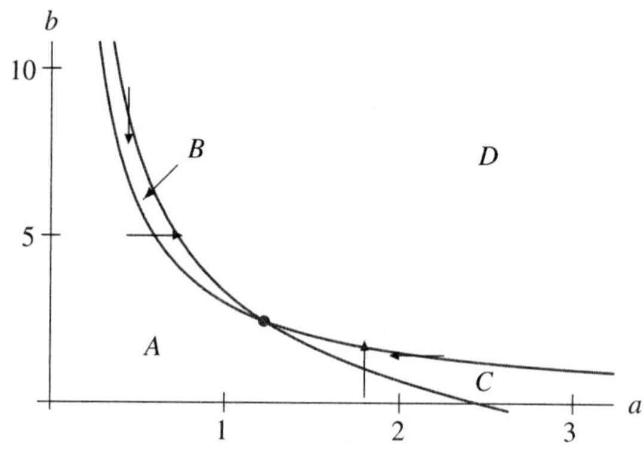

The a-nullcline is given by $b = 4/a - 2a/3$, and the b-nullcline is given by $b = 3/a$. These two nullclines meet at the equilibrium point $(a, b) = (\sqrt{3/2}, \sqrt{6})$. This is the only equilibrium point in the first quadrant.

(b) In region A the vector field points right-up. In region B, it points right-down. In region C, it points left-up, and in region D, it points left-down.

(c) All solutions in the first quadrant tend to the equilibrium point at $(\sqrt{3/2}, \sqrt{6})$.

19. **(a)** For the a-nullcline, $da/dt = 0$, so $b^2/6 - ab/2 + 2 = 0$. Multiplying by 6 yields the implicit equation $b^2 - 3ab + 12 = 0$, which can be solved using the quadratic formula.

For the b-nullcline, $db/dt = 0$, so $-b^2/3 - ab/2 + 3/2 = 0$. We multiply by -6 and obtain the implicit equation $2b^2 + 3ab - 9 = 0$, which can also be solved using the quadratic formula.

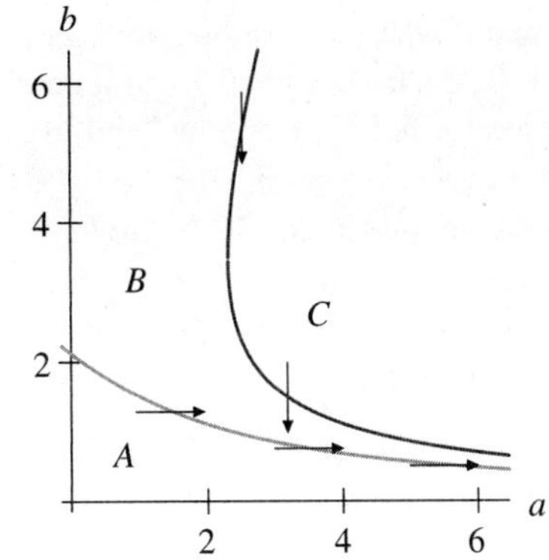

The a-nullcline is above the b-nullcline. Along the a-nullcline the vector field points down, that is, $db/dt < 0$. Along the b-nullcline the vector field points to the right, that is, $da/dt > 0$.

(b) In region A, both da/dt and db/dt are positive. In region B, $da/dt > 0$ and $db/dt < 0$. In region C, $da/dt < 0$, and $db/dt < 0$. Therefore, in region A, the vector field points up and to the right; in region B, the vector field points down and to the right; and in region C, the vector field points down and to the left.

(c) Solutions that start in region A eventually enter region B and head off to infinity in the a-direction. Solutions that start in region C head down and to the left until they enter region B and head off to infinity in the a-direction. Some solutions in region B can enter region C once. If they do, they eventually reenter region B and head off to infinity in the a-direction. Any solution in region B with $a > 2.3$ and $b < 3.2$ must stay in region B and heads off to infinity in the a direction. (The values 2.3 and 3.2 are approximations.)

20. **(a)** For the a-nullcline, $da/dt = 0$, so $2 - \frac{1}{2}ab - \frac{1}{3}ab^2 = 0$. Multiplying by 6 yields the implicit equation $12 - 3ab - 2ab^2 = 0$, which can be solved using the quadratic formula.

For the b-nullcline, $db/dt = 0$, so $\frac{3}{2} - \frac{1}{2}ab - \frac{2}{3}ab^2 = 0$. We multiply by 6 and obtain the implicit equation $9 - 3ab - 4ab^2 = 0$, which can also be solved using the quadratic formula.

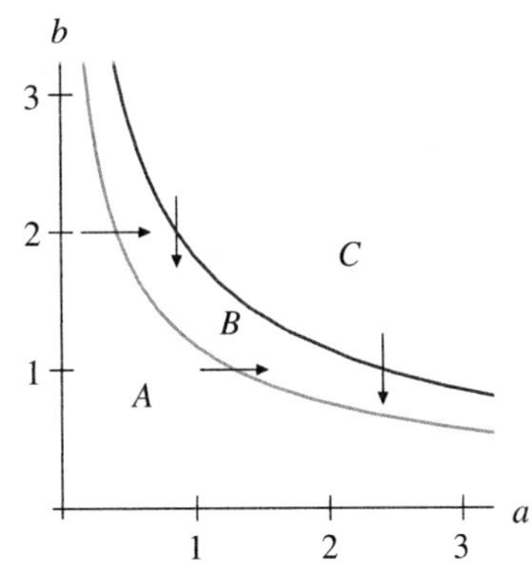

The a-nullcline is above the b-nullcline. Along the a-nullcline the vector field points down, that is, $db/dt < 0$. Along the b-nullcline the vector field points to the right, that is, $da/dt > 0$.

(b) In region A, both da/dt and db/dt are positive. In region B, $da/dt > 0$ and $db/dt < 0$. In region C, $da/dt < 0$, and $db/dt < 0$. Therefore, in region A, the vector field points up and to the right; in region B, the vector field points down and to the right; and in region C, the vector field points down and to the left.

(c) Solutions that start in region A eventually enter region B and head off to infinity in the a-direction. Solutions that start in region C head down and to the left until they enter region B and head off to infinity in the a-direction. Solutions that start in region B stay in region B and head off to infinity in the a-direction.

21. For the x-nullcline, $dx/dt = 0$; thus, $y = 0$. For the y-nullcline, $dy/dt = 0$; thus, $x(1 - x) = 0$. The line $y = 0$ is x-nullcline, and the lines $x = 0$ and $x = 1$ are y-nullclines. Since $dx/dt = y$, x increases for $y > 0$ and decreases for $y < 0$. Similarly, $dy/dt > 0$ for $0 < x < 1$ and $dy/dt < 0$ for $x < 0$ and $x > 1$. So, the function y increases for $0 < x < 1$ and decreases for $x < 0$ and $x > 1$.

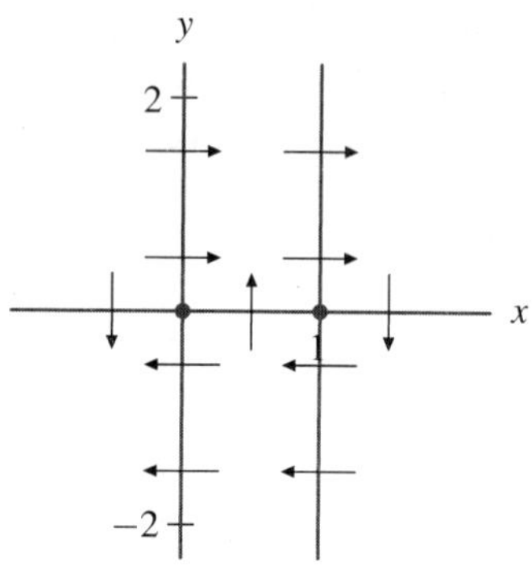

22. In the second quadrant, the vector field points right-down. Some of the solutions cross the positive y-axis and others cross the negative x-axis. The solutions that do neither must approach the equilibrium point at the origin. In the fourth quadrant, for $x < 1$, the vector field points left-up. Some solutions cross the positive x-axis and others cross the negative y-axis. Those solutions in between must approach the origin.

For solutions in the other quadrants, the same ideas work using negative time (in other words, reversing the direction of the vector field).

23. The Jacobian matrix of the vector field is

$$\begin{pmatrix} 0 & 1 \\ 1 - 2x & 0 \end{pmatrix}.$$

The coefficient matrix of the linearization at $(0, 0)$ is

$$\begin{pmatrix} 0 & 1 \\ 1 & 0 \end{pmatrix}$$

which has eigenvalues ± 1 and hence is a saddle with eigenvectors $(1, 1)$ (for 1) and $(1, -1)$ (for -1). The coefficient matrix of the linearization at $(1, 0)$ is

$$\begin{pmatrix} 0 & 1 \\ -1 & 0 \end{pmatrix}$$

which has eigenvalues $\pm i$ and is hence a center. From this information we can conclude that solutions with initial conditions near $(0, 0)$ move away from the origin in the $(1, 1)$ or $(-1, -1)$ direction while solutions near $(1, 0)$ tend to rotate around $(1, 0)$.

What we can not tell is what solutions do over the long-term. For example, do solutions spiral toward or away from $(1, 0)$? The higher order terms could cause either behavior (or all solutions could be periodic). Also, what is the behavior of the stable and unstable separatrices associated with the saddle? It turns out that this information is available for this system using techniques of Section 5.3

EXERCISES FOR SECTION 5.3

1. **(a)** We compute that

$$\frac{\partial H}{\partial x} = x - x^3$$

and so

$$\frac{dy}{dt} = -\frac{\partial H}{\partial x}.$$

Also,

$$\frac{\partial H}{\partial y} = y = \frac{dx}{dt}.$$

Hence, this is a Hamiltonian system with Hamiltonian function H.

(b) Note that $(0, 0)$ is a local minimum and $(\pm 1, 0)$ are saddle points.

(c) The equilibrium point $(0, 0)$ is a center and $(\pm 1, 0)$ are saddles. The saddles are connected by separatrix solutions.

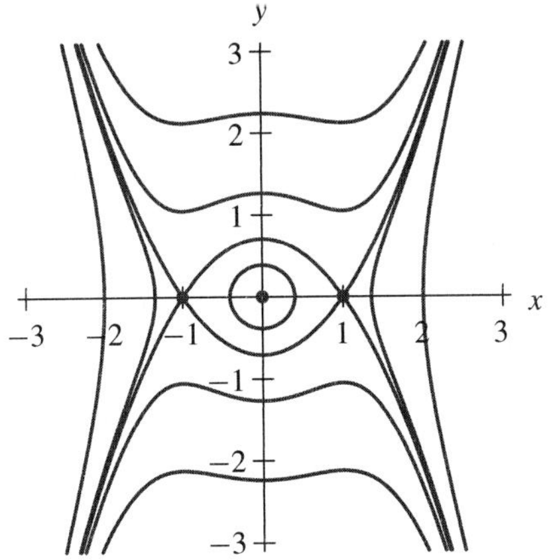

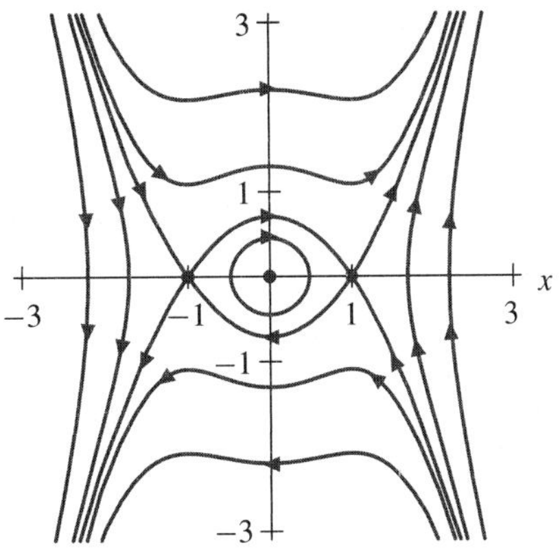

2. **(a)** If $H(x, y) = \sin(xy)$, then

$$\frac{\partial H}{\partial x} = y\cos(xy)$$

and so

$$\frac{dy}{dt} = -\frac{\partial H}{\partial x}.$$

Similarly,

$$\frac{\partial H}{\partial y} = x\cos(xy) = \frac{dx}{dt}.$$

(b) Note that the level sets of H are the same curves as those of the level sets of xy.

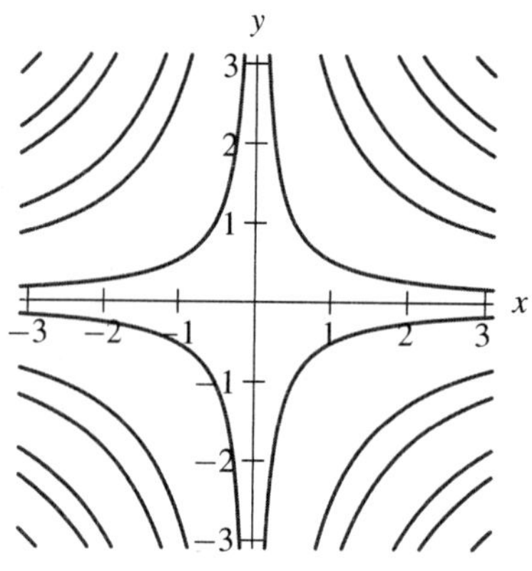

(c) Note that there are many curves of equilibrium points for this system: besides the origin, whenever $xy = n\pi + \pi/2$, the vector field vanishes.

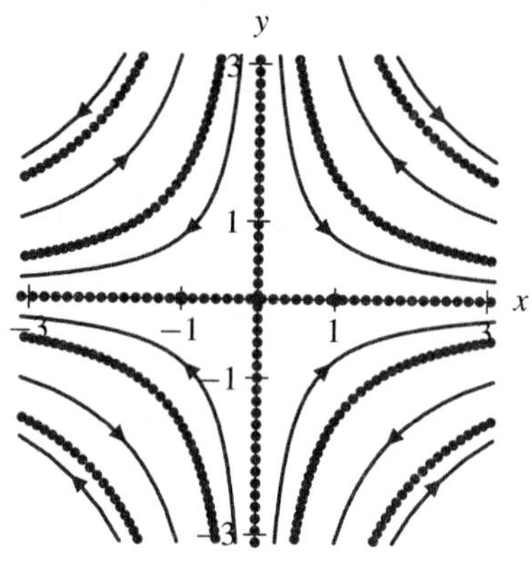

3. **(a)** If $H(x, y) = x\cos y + y^2$, then

$$\frac{\partial H}{\partial x} = \cos y$$

and so

$$\frac{dy}{dt} = -\frac{\partial H}{\partial x}.$$

Similarly,

$$\frac{\partial H}{\partial y} = -x\sin y + 2y = \frac{dx}{dt}.$$

(b)

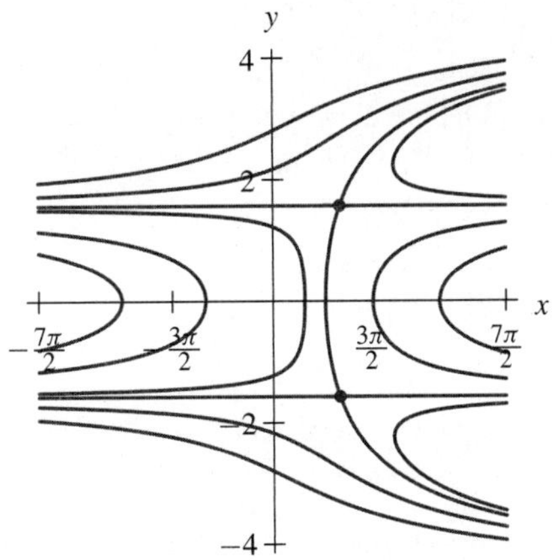

(c) The equilibrium points occur at points of the form $((1-4n)\pi, (2n-\frac{1}{2})\pi)$ and $((1+4n)\pi, (2n+\frac{1}{2})\pi)$ where n is an integer.

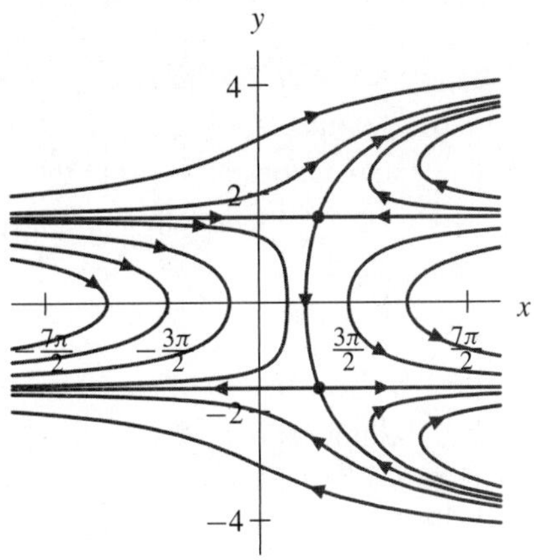

4. **(a)** The Jacobian matrix is

$$\begin{pmatrix} 0 & 1 \\ -(g/l)\cos\theta & 0 \end{pmatrix}.$$

At (0, 0), the linearization is

$$\begin{pmatrix} 0 & 1 \\ -g/l & 0 \end{pmatrix}.$$

(b) Note that the equation does not depend on m. Using $g = 9.8$, the eigenvalues for the linearization are $\pm i\sqrt{9.8/l}$ and the period of the solutions is $2\pi/\sqrt{9.8/l}$. Hence we need

$$2\pi/\sqrt{9.8/l} = 1$$

or $l = 9.8/4\pi^2$.

5. A large amplitude swing will take θ near $\pm\pi$, $v = 0$, the equilibrium point corresponding to the pendulum being balanced straight up. Near equilibrium points the vector field is very short, so solutions move very slowly. A solution passing close to $\pm\pi$, $v = 0$ must move slowly and hence, take a long time to make one complete swing. Hence, very high swings have long period. We must also be careful not to let the pendulum "swing over the top".

6. Large amplitude oscillations of an ideal pendulum have much longer period than small amplitude oscillations because they come close to the saddle points. Hence, small amplitude swings make the clock fast.

7. **(a)** The linearization at the origin is

$$\frac{d\theta}{dt} = v$$

$$\frac{dv}{dt} = -\frac{g}{l}\theta.$$

The eigenvalues of this system are $\pm i\sqrt{g/l}$, so the natural period is $2\pi/(\sqrt{g/l})$, which can also be written as $2\pi\sqrt{l}/\sqrt{g}$. Doubling the arm length corresponds to replacing l with $2l$, but the computations above stay the same. The natural period for arm length $2l$ is $2\pi\sqrt{2l}/\sqrt{g}$. Doubling the arm length multiplies the natural period by $\sqrt{2}$.

(b) Compute

$$\frac{d(2\pi\sqrt{l}/\sqrt{g})}{dl} = \frac{\pi}{\sqrt{gl}}.$$

8. Let G be the gravitational constant on the moon. Note that $G < g = 9.8$. The period of the linearization of the ideal pendulum on the moon is $2\pi/\sqrt{G/l}$. Since $G < g$, we have

$$2\pi/\sqrt{G/l} > 2\pi/\sqrt{G/l}.$$

Since the period of the pendulum is now longer, the clock runs more slowly.

9. We know that the equilibrium points of a Hamiltonian system cannot be sources or sinks. Phase portrait (b) has a spiral source, so it is not Hamiltonian. Phase portrait (c) has a sink and a source, so it is not Hamiltonian. Phase portraits (a) and (d) might come from Hamiltonian systems. (Try to imagine a function which has the solution curves as level sets.)

10. First note that

$$\frac{\partial(\sin x \cos y)}{\partial x} = \cos x \cos y = -\frac{\partial(2x - \cos x \sin y)}{\partial y}.$$

Hence, the system is Hamiltonian. Integrating dx/dt with respect to y yields

$$H(x, y) = \sin x \sin y + c(x).$$

If we differentiate $H(x, y)$ with respect to x, we get

$$\cos x \sin y + c'(x),$$

which we want to be the negative of $dy/dt = 2x - \cos x \sin y$. Hence $c'(x) = -2x$, and we pick the antiderivative $c(x) = -x^2$. A Hamiltonian function is

$$H(x, y) = -x^2 + \sin x \sin y.$$

11. First note that

$$\frac{\partial(x - 3y^2)}{\partial x} = 1 = -\frac{\partial(-y)}{\partial y}.$$

Hence, the system is Hamiltonian. Integrating dx/dt with respect to y yields

$$H(x, y) = xy - y^3 + c(x).$$

If we differentiate $H(x, y)$ with respect to x, we get

$$y + c'(x),$$

which we want to be the negative of $dy/dt = -y$. Hence $c'(x) = 0$, and we pick the antiderivative $c(x) = 0$. A Hamiltonian function is

$$H(x, y) = xy - y^3.$$

12. First we check to see if the partial derivative with respect to x of the first component of the vector field is the negative of the partial derivative with respect to y of the second component. We have

$$\frac{\partial 1}{\partial x} = 0$$

while

$$-\frac{\partial y}{\partial y} = -1.$$

Since these are not equal, the system is not Hamiltonian.

13. First we check to see if the partial derivative with respect to x of the first component of the vector field is the negative of the partial derivative with respect to y of the second component. We have

$$\frac{\partial(x \cos y)}{\partial x} = \cos y$$

while

$$-\frac{\partial(-y \cos x)}{\partial y} = \cos x.$$

Since these two are not equal, the system is not Hamiltonian.

14. First note that

$$\frac{\partial F(y)}{\partial x} = 0 = -\frac{\partial G(x)}{\partial y},$$

that is, the partial derivative of the x component of the vector field with respect to x is equal to the negative of the partial derivative of the y component with respect to y. Hence, the system is Hamiltonian. Integrating the x component of the vector field with respect to y yields

$$H(x, y) = \int F(y)\, dy + c$$

where the "constant" c could depend on x. If we differentiate this H with respect to x we get

$$-\frac{\partial H}{\partial x} = -c'(x).$$

Thus we take $c = -\int G(x)\, dx$. A Hamiltonian function is

$$H(x, y) = \int F(y)\, dy - \int G(x)\, dx.$$

15. **(a)** We note that

$$\frac{\partial(1-y^2)}{\partial x} = 0 \neq -\frac{\partial(x(1+y^2))}{\partial y} = -2xy.$$

Hence, the system is not Hamiltonian.

(b)

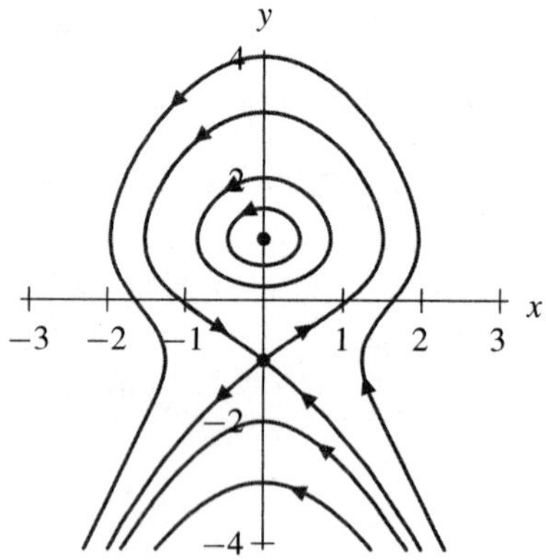

This phase portrait is consistent with the existence of a conserved quantity, but it is difficult to produce such a quantity from the figure alone.

(c) We note that

$$\frac{\partial((1-y^2)/(1+y^2))}{\partial x} = 0 = \frac{\partial(x)}{\partial y},$$

so the system is Hamiltonian. The Hamiltonian function is

$$H(x, y) = \int \frac{1-y^2}{1+y^2}\, dy - \frac{x^2}{2},$$

which is equal to

$$H(x, y) = -y + 2\arctan y - \frac{x^2}{2}.$$

(d) Multiplying the vector field by the function $f(x, y)$ changes the lengths of the vectors in the vector field, but it does not change their directions. Hence, the direction fields for the two systems agree. Consequently, the phase portraits are the same for the two systems, but solutions move at different speeds along the solution curves in the phase plane. Since H is a conserved quantity for the Hamiltonian system, the solution curves lie on level sets of H, and H is also a conserved quantity for the original system.

We can check that H is a conserved quantity by hand by computing the derivative with respect to t of $H(x(t), y(t))$, where $(x(t), y(t))$ is a solution of the original system. From the Chain Rule, we have

$$\begin{aligned}\frac{dH(x(t), y(t))}{dt} &= \frac{\partial H}{\partial x}\frac{dx}{dt} + \frac{\partial H}{\partial y}\frac{dy}{dt} \\ &= -x(1-y^2) + \frac{1-y^2}{1+y^2}(x(1+y^2)) \\ &= 0.\end{aligned}$$

(e) See the answer to part (d).

16. **(a)** We first check

$$\frac{\partial(-yx^2)}{\partial x} = -2xy \neq -\frac{\partial(x+1)}{\partial y} = 0,$$

so the system is not Hamiltonian.

(b) If we multiply the vector field by $1/x^2$, we obtain the new system

$$\begin{aligned}\frac{dx}{dt} &= -y \\ \frac{dy}{dt} &= \frac{1}{x} + \frac{1}{x^2}.\end{aligned}$$

As in Exercise 14, this system is Hamiltonian with

$$H(x, y) = \frac{1}{x} - \ln|x| - \frac{y^2}{2}.$$

17. Using the technique of Exercise 15, we we multiply the vector field by $1/(2-y)$. As in Exercise 14, the resulting system

$$\begin{aligned}\frac{dx}{dt} &= \frac{1-y^2}{2-y} \\ \frac{dy}{dt} &= x\end{aligned}$$

is Hamiltonian. The Hamiltonian is

$$\begin{aligned}H(x, y) &= -\frac{x^2}{2} + \int \frac{y^2-1}{y-2}\,dy \\ &= -\frac{x^2}{2} + \int 2 + y + \frac{3}{y-2}\,dy \\ &= -\frac{x^2}{2} + 2y + \frac{y^2}{2} + 3\ln|y-2|.\end{aligned}$$

The function

$$H(x, y) = -\frac{x^2}{2} + 2y + \frac{y^2}{2} + 3\ln|y-2|$$

is a conserved quantity for the original system. However, it is not defined on the line $y = 2$. From the system, we see that this line is a single solution curve that separates the two half-planes, $y < 2$ and $y > 2$.

18. **(a)** We have

$$\frac{\partial H}{\partial y} = y \quad \text{and} \quad -\frac{\partial H}{\partial x} = x^2 - a,$$

so this system is Hamiltonian with the given function H.

(b) Note that $dx/dt = 0$ if and only if $y = 0$ and $dy/dt = 0$ if and only if $x = \pm\sqrt{a}$. Consequently if $a < 0$, then there are no equilibrium points. If $a = 0$, there is one equilibrium point at $(0, 0)$ and if $a > 0$, there are two equilibrium points at $(\pm\sqrt{a}, 0)$.

(c) The Jacobian matrix is

$$\begin{pmatrix} 0 & 1 \\ 2x & 0 \end{pmatrix},$$

which, when evaluated at the equilibrium points, becomes

$$\begin{pmatrix} 0 & 1 \\ \pm 2\sqrt{a} & 0 \end{pmatrix}.$$

At $(\sqrt{a}, 0)$, the eigenvalues are $\pm\sqrt{2\sqrt{a}}$ so this equilibrium point is a saddle. At $(-\sqrt{a}, 0)$, the eigenvalues are $\pm i\sqrt{2\sqrt{a}}$ so this equilibrium point is a center. If $a = 0$ the eigenvalues are both 0, so this point is a node.

(d)

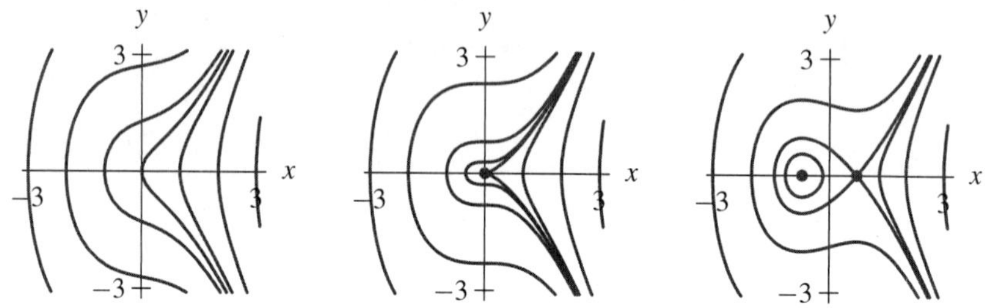

Phase portrait for $a < 0$ Phase portrait for $a = 0$ Phase portrait for $a > 0$

(e) As a increases toward 0, the phase portrait changes from having no equilibrium points to having a single equilibrium point at $a = 0$. If $a > 0$, there is a pair of equilibrium points.

19. First note that this system is Hamiltonian for every value of a. The Hamiltonian function depends on a and is given by

$$H(x, y) = x^2y + xy^2 - ax.$$

If $a > 0$, then the system has two saddle equilibrium points on the y-axis at $(0, \pm\sqrt{a})$. If $a = 0$, then system has only one equilibrium point at $(0, 0)$. If $a < 0$, the system again has two saddles, but they are now located at $(\pm 2\sqrt{-3a}/3, \mp\sqrt{-3a}/3)$. This corresponds to a change in shape of the graph of H.

20. **(a)** First note that the system is still Hamiltonian, with Hamiltonian function

$$H(x, y) = \frac{1}{2}y^2 - \frac{1}{2}x^2 + \frac{1}{3}x^3 + ax.$$

The equilibrium points are

$$\left(\frac{1 \pm \sqrt{1 - 4a}}{2}, 0\right).$$

Hence there are no equilibrium points if $a > 1/4$; one equilibrium point if $a = 1/4$; and two equilibrium points if $a < 1/4$. A bifurcation occurs at $a = 1/4$.

(b) The book would never have appeared. Wouldn't that have been awful?

EXERCISES FOR SECTION 5.4

1. **(a)** Let $(x(t), y(t))$ be a solution of the system. Then

$$\begin{aligned}\frac{d}{dt}(L(x(t), y(t))) &= \frac{\partial L}{\partial x}\frac{dx}{dt} + \frac{\partial L}{\partial y}\frac{dy}{dt} \\ &= x(-x^3) + y(-y^3) \\ &= -x^4 - y^4.\end{aligned}$$

This is negative except at $x = y = 0$, which is an equilibrium point of the system.

(b) The level sets are circles centered at the origin.

(c) Since L decreases along solutions, every solution must approach the origin as t increases.

2. **(a)** Let $(x(t), y(t))$ be a solution of the system. Then

$$\begin{aligned}\frac{d}{dt}(L(x(t), y(t))) &= \frac{\partial L}{\partial x}\frac{dx}{dt} + \frac{\partial L}{\partial y}\frac{dy}{dt} \\ &= \left(x - x^2\right)y + y\left(-x - \frac{y}{4} + x^2\right) \\ &= -\frac{y^2}{4}.\end{aligned}$$

This is negative except along $y = 0$. The origin is an equilibrium point for this system, as is the point $(1, 0)$. For all other points on the line $y = 0$ we have $dy/dt \neq 0$. Thus solutions cross this line and then L continues to decrease.

(b)

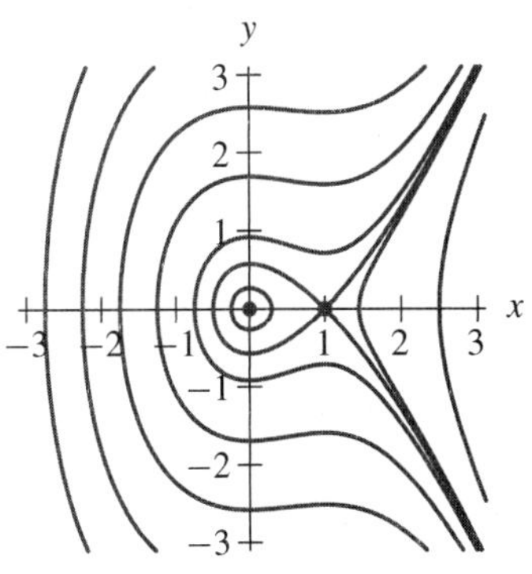

(c) Since L decreases along solutions, every solution must approach either an equilibrium point or tend toward infinity. The latter can happen since $L(x, 0)$ tends to $-\infty$ as x tends to ∞.

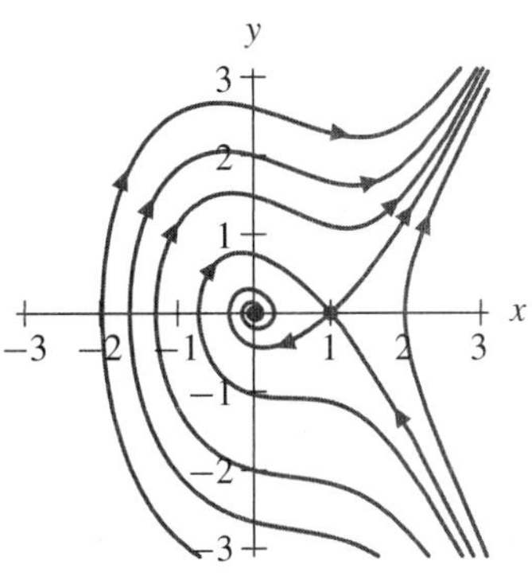

3. **(a)** We compute the eigenvalues. The characteristic polynomial is $\lambda^2 + 0.1\lambda + 4$, so the eigenvalues are

$$\lambda = \frac{-0.1 \pm i\sqrt{15.99}}{2}.$$

Hence the origin is a spiral sink.

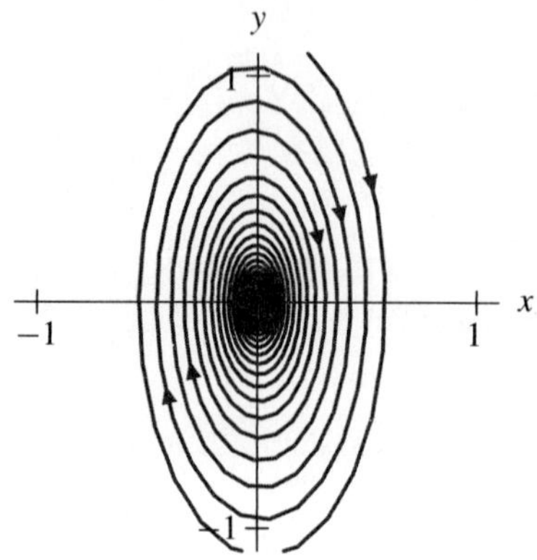

(b) Let $(x(t), y(t))$ be a solution. Then

$$\begin{aligned}\frac{d}{dt}(L(x(t), y(t))) &= 2x(y) + 2y(-4x - 0.1y) \\ &= -6xy - 0.2y^2.\end{aligned}$$

In order for L to be a Lyapunov function, this quantity must be less than or equal to zero for all (x, y). However, if we take the solution with $x(0) = -1$, $y(0) = 1$ then

$$\frac{d}{dt}(L(x(t), y(t)))|_{t=0} = 5.8$$

which is positive, so L is not a Lyapunov function.

(c) Again let $(x(t), y(t))$ be a solution and compute

$$\begin{aligned}\frac{d}{dt}(K(x(t), y(t))) &= 4x(y) + y(-4x - 0.1y) \\ &= -0.1y^2\end{aligned}$$

and this quantity is negative except at $y = 0$. (Solutions cross the $y = 0$ line at a discrete set of times. This is a technical point needed for the definition of Lyapunov functions.) So K is a Lyapunov function for the system.

4. The linearized system at the origin has Jacobian

$$\begin{pmatrix} 0 & 1 \\ -g/l & -b/m \end{pmatrix},$$

which has eigenvalues

$$\frac{-b/m \pm \sqrt{(b/m)^2 - 4g/l}}{2}.$$

For these eigenvalues to be complex, we need $(b/m)^2 - 4g/l < 0$. The period of the oscillations near the equilibrium point is

$$2\pi/\sqrt{\frac{g}{l} - \frac{b^2}{4m^2}}.$$

Hence we need

$$2\pi\left((g/l) - b^2m^{-2}/4\right)^{-1/2} = 1.$$

5. We assume that the swings are small, hence we consider solutions near $\theta = v = 0$. The linearized system at the origin has Jacobian

$$\begin{pmatrix} 0 & 1 \\ -g/l & -b/m \end{pmatrix}$$

which has eigenvalues

$$\frac{-b/m \pm \sqrt{(b/m)^2 - 4g/l}}{2}.$$

When b/m is small, the eigenvalues are complex and the natural period of the oscillations is

$$2\pi/\sqrt{\frac{g}{l} - \frac{b^2}{4m^2}}$$

or

$$2\pi\left((g/l) - b^2m^{-2}/4\right)^{-1/2}.$$

If we differentiate this expression with respect to m we obtain

$$\frac{d}{dm}\left(2\pi\left(\frac{g}{l} - \frac{b^2}{4}m^{-2}\right)^{-1/2}\right) = -\pi\left(\frac{g}{l} - \frac{b^2}{4}m^{-2}\right)^{-3/2}\left(-\frac{b^2}{4}\left(-2m^{-3}\right)\right)$$

which is less than zero. Hence, as the mass m increases, the natural period of the linearization near (0, 0) decreases and the clock runs fast.

6. **(a)** As the clock "winds down," the periods of the oscillations decreases. Hence, the ticks come closer together and the clock runs faster.

(b) The periods of very large swings are very large. Hence, the ticks are far apart, and the clock runs too slow.

7. In order to be usable as a clock, the pendulum arm must swing back and forth. That is, the equilibrium point at the origin must be a spiral sink or center so that solutions oscillate. So we study the equilibrium point at the origin. The linearized system at the origin has Jacobian

$$\begin{pmatrix} 0 & 1 \\ -g/l & -b/m \end{pmatrix},$$

which has eigenvalues

$$\frac{-b/m \pm \sqrt{(b/m)^2 - 4g/l}}{2}.$$

To have oscillating solutions we must have complex eigenvalues, that is we must have

$$\frac{b^2}{m^2} - \frac{4g}{l} < 0.$$

This holds if

$$m > \frac{b\sqrt{l}}{2\sqrt{g}}.$$

(We could also have $m < -b\sqrt{l}/(2\sqrt{g})$, but negative mass is "unphysical".)

8. **(a)** The linearized system at the origin has Jacobian

$$\begin{pmatrix} 0 & 1 \\ -g/l & -b/m \end{pmatrix},$$

which in our case yields

$$\begin{pmatrix} 0 & 1 \\ -1 & -4 \end{pmatrix}.$$

The eigenvalues are $-2 \pm \frac{1}{2}\sqrt{12}$, which are both negative. So (0, 0) is a sink.

(b) At $(\pi, 0)$ the matrix is

$$\begin{pmatrix} 0 & 1 \\ 1 & -4 \end{pmatrix}.$$

Here the eigenvalues are $-2 \pm \frac{1}{2}\sqrt{20}$, which are opposite in sign. Therefore $(\pi, 0)$ is a saddle.

(c)

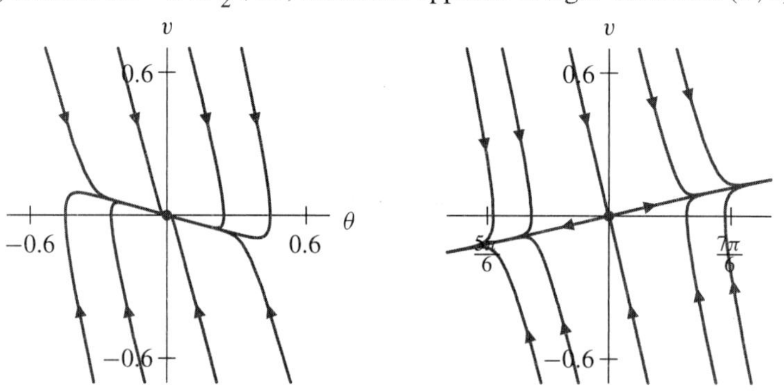

Phase portraits near (0, 0) and $(\pi, 0)$.

(d)

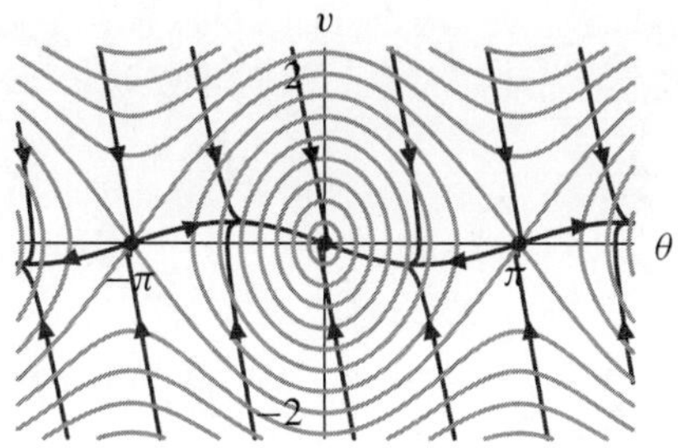

Phase portrait with level sets of undamped pendulum.

9. The linearized system at the origin has complex eigenvalues (when b/m is small) that have real part equal to $-b/(2m)$. So near $\theta = v = 0$ the amplitude of the oscillations decrease at the same rate as

$$e^{-tb/(2m)}.$$

In order for the clock to keep accurate time, we need θ and v to remain close to zero. An absolutely "worst case" for the size of θ is $|\theta| < \pi$. So we can get a rough approximation of how long it takes for the amplitude of an initial swing to decay to 0.1 by finding t such that

$$\pi e^{-tb/(2m)} = 0.1.$$

This occurs when

$$t = -\frac{2m}{b} \ln\left(\frac{0.1}{\pi}\right).$$

Pendulum clocks need to be wound because energy must be added to the system since it is dissipative.

10. **(a)**

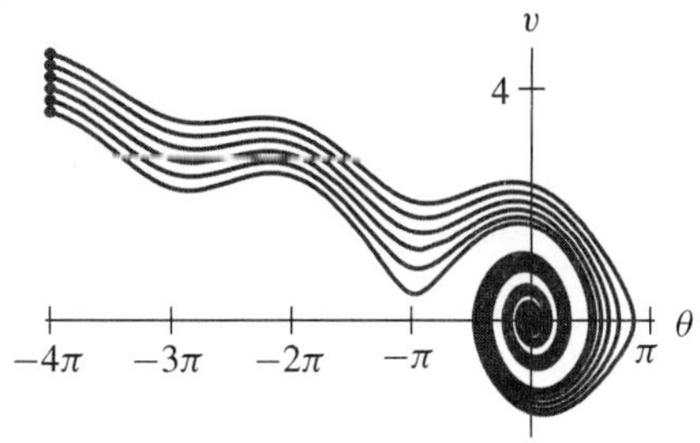

(b)

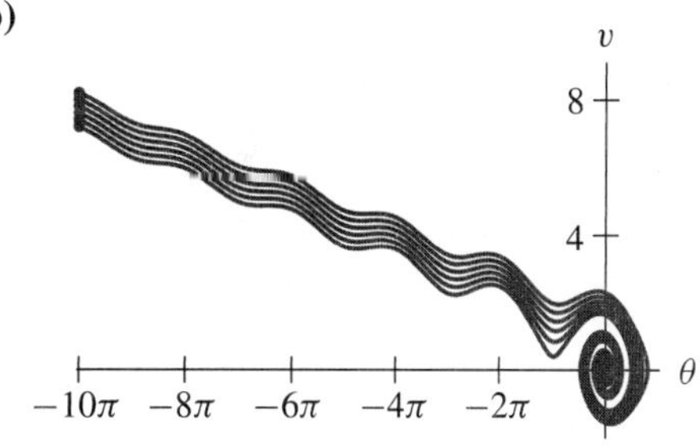

11. **(a)** The linearized system at the origin has Jacobian

$$\begin{pmatrix} 0 & 1 \\ -g/l & -b/m \end{pmatrix},$$

which has eigenvalues

$$\frac{-b/m \pm \sqrt{(b/m)^2 - 4g/l}}{2}.$$

If b/m is small and negative, these eigenvalues have a positive real part and the origin is a spiral source. The linearized system at $(\pi, 0)$ has Jacobian

$$\begin{pmatrix} 0 & 1 \\ g/l & -b/m \end{pmatrix},$$

which has eigenvalues $(-b/m \pm \sqrt{(b/m)^2 + 4g/l})/2$. If b/m is small (positive or negative), the two eigenvalues differ in sign, so this equilibrium point is a saddle. The equilibrium points $(\pm 2n\pi, 0)$ are just translates of the origin while the equilibrium points $(\pm(2n+1)\pi, 0)$ are translates of $(\pi, 0)$.

(b)

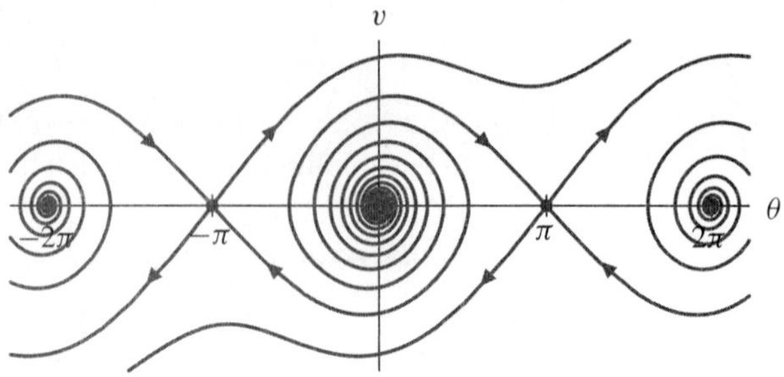

(c) A solution that has initial point near $(0, 0)$ will spiral away from the origin. Eventually, θ will become larger than π or smaller than $-\pi$ and then θ will increase or decrease monotonically toward infinity. This corresponds to the pendulum arm swinging with higher and higher amplitude until it "passes over the top" either clockwise or counter clockwise. After it passes over the top once, then it will continue to rotate in the same direction, accelerating with each rotation.

12. **(a)** We have $\nabla G(x, y) = (3x^2 - 3y^2, -6xy)$ so

$$\frac{dx}{dt} = 3x^2 - 3y^2$$
$$\frac{dy}{dt} = -6xy.$$

(b)

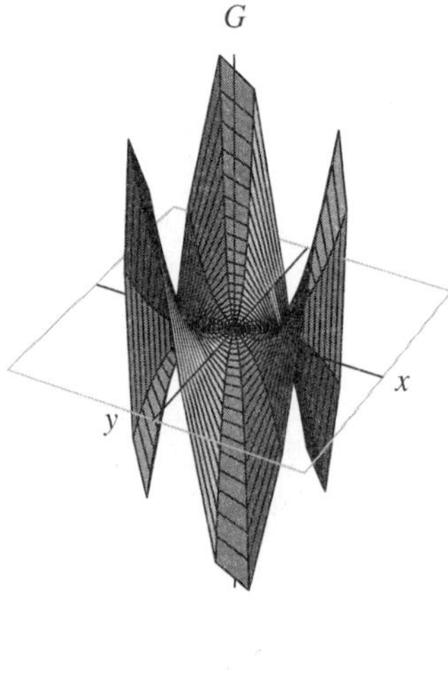

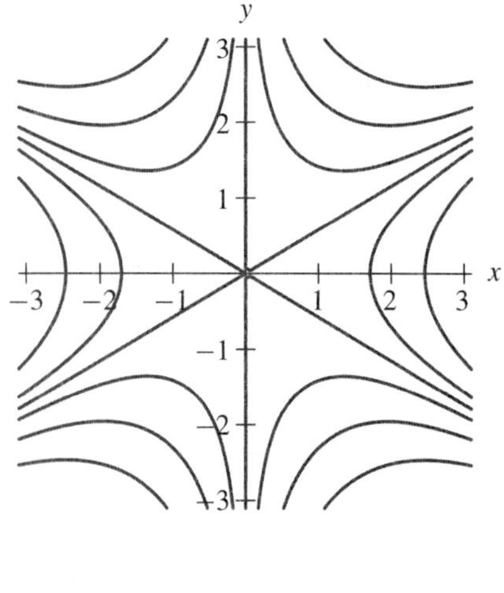

(c)

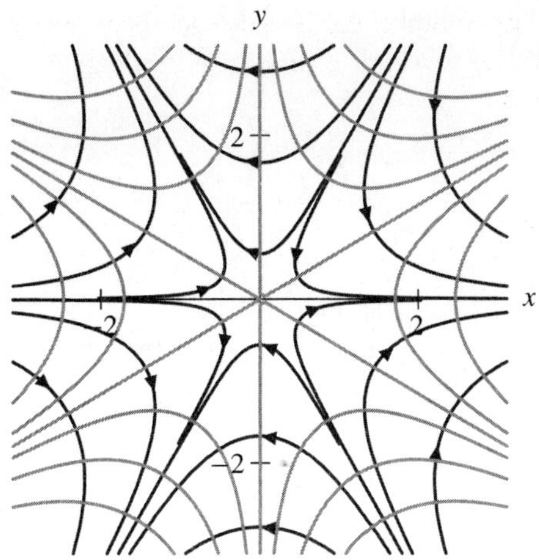

Phase portrait shown with level sets of G in gray.

13. **(a)** We have $\nabla G(x, y) = (2x, -2y)$, so

$$\frac{dx}{dt} = 2x$$
$$\frac{dy}{dt} = -2y.$$

(b) The system is linear and has eigenvalues 2 and -2. Hence the origin is a saddle.

(c) The graph of G is a saddle surface turning up in the x-direction and down in the y-direction.

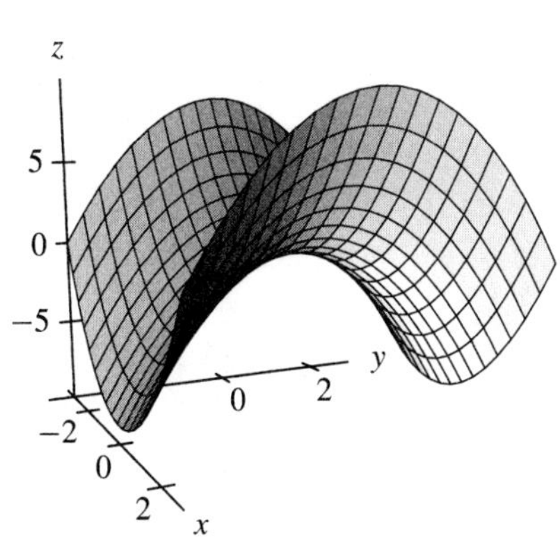

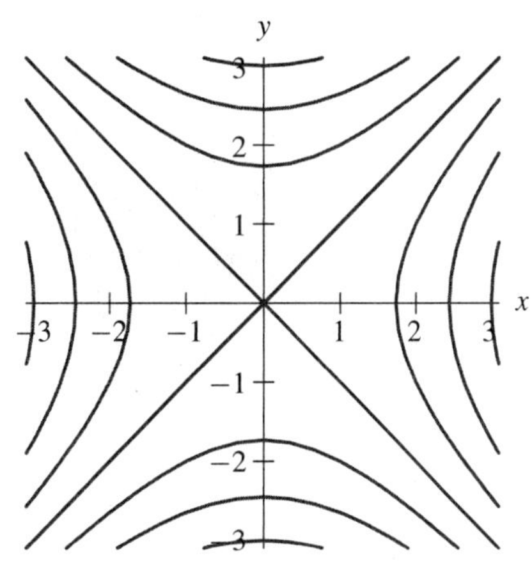

(d) The line of eigenvectors for eigenvalue 2 is the x-axis, the line of eigenvectors for eigenvalue -2 is the y-axis (see Chapter 3).

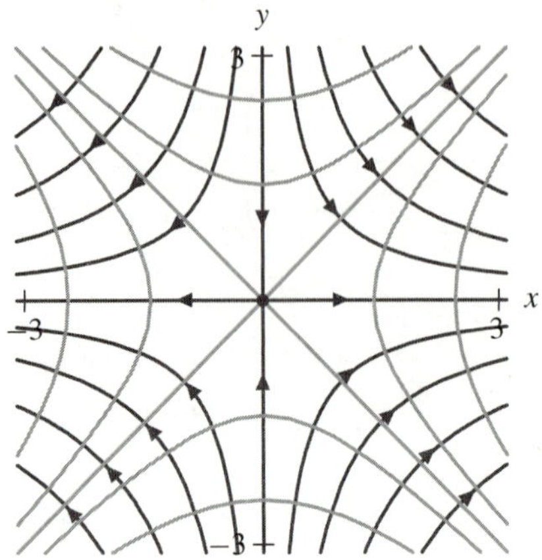

Phase portrait shown with level sets of G in gray.

14. **(a)** The gradient $\nabla G(x, y) = (2x, 2y)$ so

$$\frac{dx}{dt} = 2x$$
$$\frac{dy}{dt} = 2y.$$

(b) The system is linear and has both eigenvalues 2. Hence the origin is a source.

(c) The graph of G is a paraboloid.

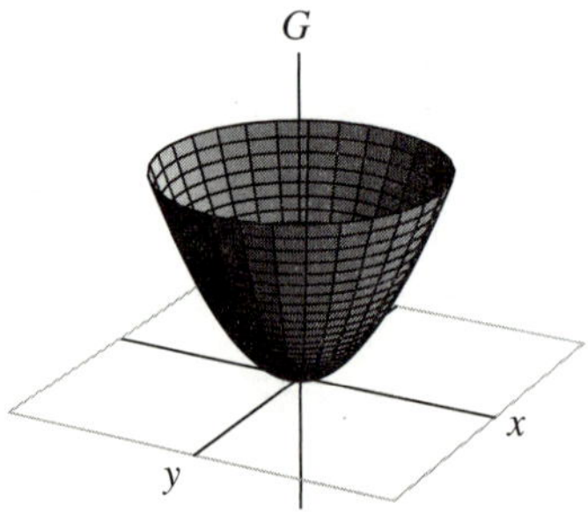

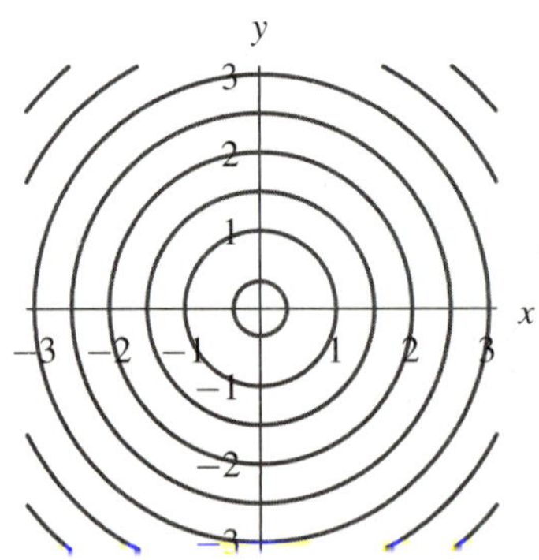

(d)

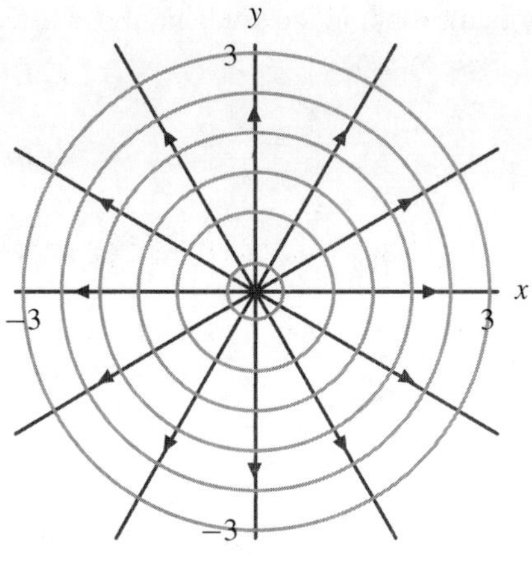

15. **(a)** The Jacobian matrix is

$$\begin{pmatrix} 1 - 3x^2 & 0 \\ 0 & -1 \end{pmatrix}$$

At the origin the coefficient matrix of the linearization is

$$\begin{pmatrix} 1 & 0 \\ 0 & -1 \end{pmatrix}$$

which has eigenvalues 1 and -1. Hence, the origin is a saddle.

(b) At $(1, 0)$ the Jacobian matrix is

$$\begin{pmatrix} -2 & 0 \\ 0 & -1 \end{pmatrix}$$

which has eigenvalues -2 and -1. Hence, the point $(1, 0)$ is a sink.

(c) Since the eigenvalue -1 has eigenvectors along the y-axis, (and $-2 < -1$), solutions which approach $(1, 0)$ as t tends to infinity do so in the y direction (the only exception being solutions on the x-axis).

(d) The Jacobian matrix at $(-1, 0)$ is the same as at $(1, 0)$.

16. **(a)**

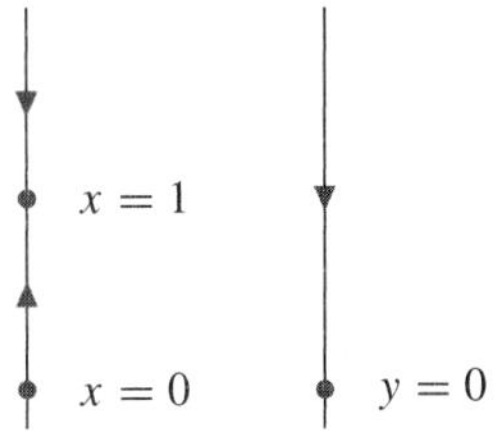

(b)

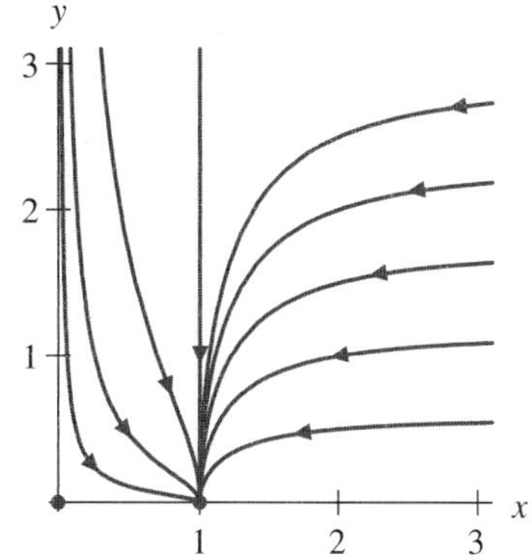

17. **(a)** The system is formed by taking the gradient of S. Hence, the system is

$$\frac{dx}{dt} = 2x - x^3 - 6xy^2$$
$$\frac{dy}{dt} = 2y - y^3 - 6x^2y.$$

(b)

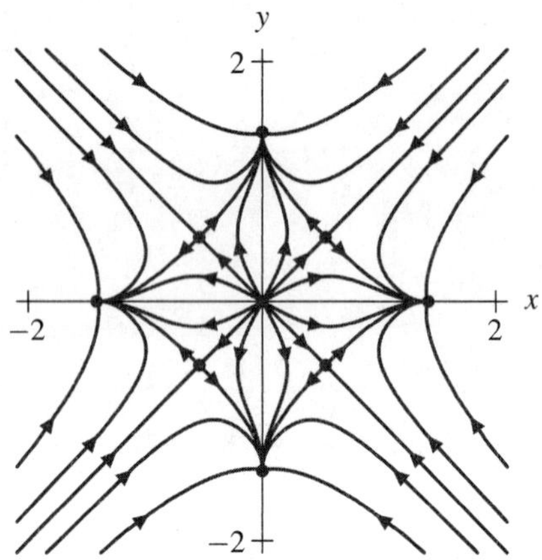

(c) From the phase portrait, we see that there are four sinks. Hence, there are four dead fish.

(d)

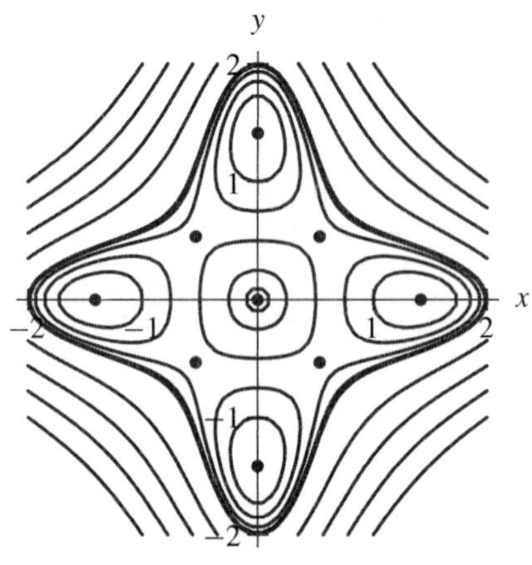

(e) For $|x|$ or $|y|$ large, S is negative, which is not physically reasonable.

18. **(a)**

$$S(x, y) = \frac{1}{(x-1)^2 + y^2 + 1} + \frac{1}{(x+1)^2 + y^2 + 1} + \frac{1}{x^2 + (y-2)^2 + 1}$$

(b)

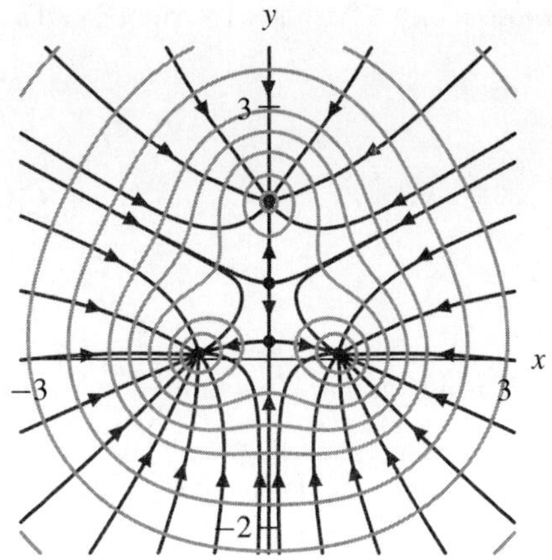

Phase portrait shown with level sets of S in gray.

(c) See above.

(d)

$$\frac{dx}{dt} = \frac{-2(x-1)}{((x-1)^2+y^2+1)^2} - \frac{2(x+1)}{((x+1)^2+y^2+1)^2} - \frac{2x}{(x^2+(y-2)^2+1)^2}$$
$$\frac{dy}{dt} = \frac{-2y}{((x-1)^2+y^2+1)^2} - \frac{2y}{((x+1)^2+y^2+1)^2} - \frac{2(y-2)}{(x^2+(y-2)^2+1)^2}$$

(e) In the first case, the smell function would tend to infinity as we near the dead fish (one smelly fish), and the differential equation would not be defined at these points. In the second case, the partial derivatives would not be defined at the locations of the smell, so the vector field would not exist there.

19. **(a)** Since $f = \partial G/\partial x$ and $g = \partial G/\partial y$,

$$\frac{\partial f}{\partial y} = \frac{\partial^2 G}{\partial y \partial x} = \frac{\partial^2 G}{\partial x \partial y} = \frac{\partial g}{\partial x}.$$

(b) We compute

$$\frac{\partial f}{\partial y} = 3x$$
$$\frac{\partial g}{\partial x} = 2,$$

and these partials are not equal. By part (a), the system is not a gradient system.

20. Note the loop consisting of a single solution curve emanating and returning to the equilibrium point at the origin. The gradient function must decrease along this solution, but note that as time goes to infinity in both directions, the solution tends to the same point. Hence the gradient must be constant on this solution, which cannot happen.

21. Let $(y(t), v(t))$ be a solution of the "damped" system (with $k > 0$). Compute

$$\begin{aligned}\frac{d}{dt}(H(y(t), v(t))) &= \frac{\partial H}{\partial y}\frac{dy}{dt} + \frac{\partial H}{\partial v}\frac{dv}{dt} \\ &= \frac{dV}{dy}v + v\left(-\frac{dV}{dy} - kv\right) \\ &= -kv^2.\end{aligned}$$

Hence, for this system, H is a Lyapunov function.

22. The solutions of the Hamiltonian system lie on the level curves of H while the solutions of the gradient system are perpendicular to the level sets of H. Therefore, the solutions of the two systems meet at right angles at all non-equilibrium points.

23. Let $(x_1(t), x_2(t), p_1(t), p_2(t))$ be a solution and differentiate

$$H(x_1(t), x_2(t), p_1(t), p_2(t))$$

with respect to t using the Chain Rule. That is,

$$\frac{dH}{dt} = \frac{\partial H}{\partial x_1}\frac{dx_1}{dt} + \frac{\partial H}{\partial x_2}\frac{dx_2}{dt} + \frac{\partial H}{\partial p_1}\frac{dp_1}{dt} + \frac{\partial H}{\partial p_2}\frac{dp_2}{dt}.$$

Note that

$$\begin{aligned}\frac{\partial H}{\partial x_1} &= k_1(x_1 - L_1) - k_2(x_2 - x_1 - L_2) \\ \frac{\partial H}{\partial x_2} &= k_2(x_2 - x_1 - L_2) \\ \frac{\partial H}{\partial p_1} &= \frac{p_1}{m_1} \\ \frac{\partial H}{\partial p_2} &= \frac{p_2}{m_2}.\end{aligned}$$

Using these partials along with the formulas for dx_1/dt, dx_2/dt, dp_1/dt, and dp_2/dt given by the system, we see that

$$\frac{dH}{dt} = \frac{\partial H}{\partial x_1}\frac{dx_1}{dt} + \frac{\partial H}{\partial x_2}\frac{dx_2}{dt} + \frac{\partial H}{\partial p_1}\frac{dp_1}{dt} + \frac{\partial H}{\partial p_2}\frac{dp_2}{dt} = 0.$$

24. The most dramatic effect should be seen when we take $\omega = 1$ so that the forcing frequency is the same as the resonant frequency of the building. Also, we assume that the building is stationary with no momentum just before the earthquake ($t = 0$).

Without damping, the amplitude of the oscillation will grow linearly without bound. The damping terms become significant when the amplitude of the oscillation is large and restricts further growth. We have

$$\frac{dH}{dt} = -b\left(\frac{p_2}{m_2} - \frac{p_1}{m_1}\right)^2 + Ap_1 \cos\omega t.$$

Graphing H as a function of t for a given solution, we see that the energy grows for a while, reaches a maximum, then oscillates toward a non-zero limit as t increases. The dependence of this limit on k_2 is shown below.

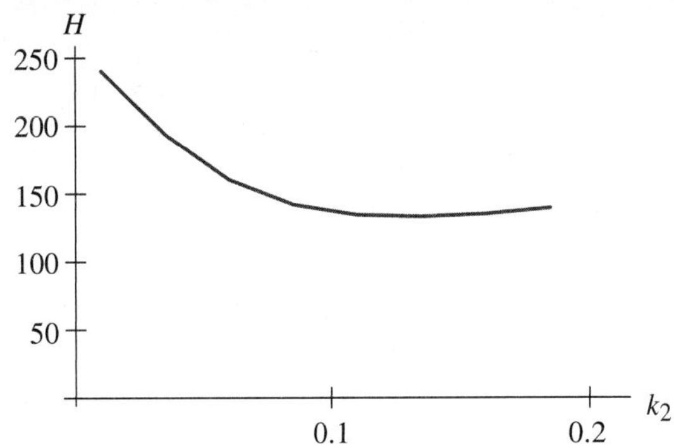

An actual earthquake would eventually stop shaking the building. After the forcing term vanishes, the analysis of the motion is the same as in the section.

EXERCISES FOR SECTION 5.5

1. **(a)** The Jacobian matrix for the system is

$$\begin{pmatrix} 1-2x-y & -x & 0 \\ y & 1-2y+x-z & -y \\ 0 & z & 1-2z+y \end{pmatrix}$$

so at $(1, 0, 0)$ the Jacobian matrix is

$$\begin{pmatrix} -1 & -1 & 0 \\ 0 & 2 & 0 \\ 0 & 0 & 1 \end{pmatrix}.$$

(b) The characteristic polynomial is $(1-\lambda)(2-\lambda)(-1-\lambda)$, and the eigenvalues are 1, 2 and -1. An eigenvector for eigenvalue 1 is $(0, 0, 1)$, an eigenvector for eigenvalue -1 is $(1, 0, 0)$ and an eigenvector for eigenvalue 2 is $(1, -3, 0)$.

(c) The equilibrium point is a saddle with two eigenvalues positive and one negative. Solutions tend toward the equilibrium point along the x-axis as t increases. Solutions in the plane formed by the two vectors $(0, 0, 1)$ and $(1, -3, 0)$ through the equilibrium point tend toward the equilibrium point as t decreases.

(d)

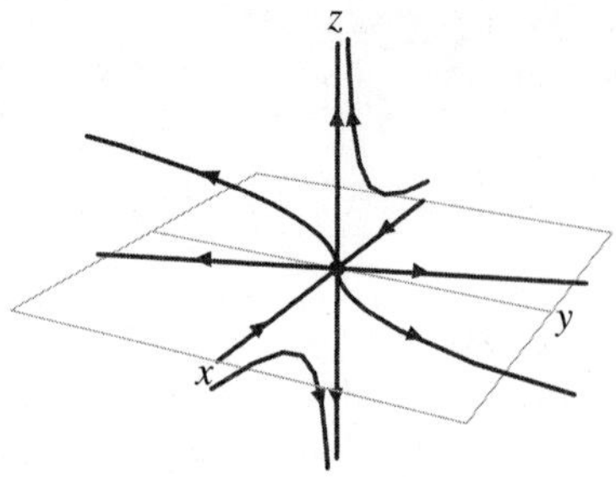

(e) If the initial condition satisfies $y = z = 0$ and x is close to 1, then the corresponding solution tends toward the equilibrium point as t increases. If y or z is small and positive and x is close to 1, then the solution may move closer to the equilibrium point if y and z are small, but eventually moves away from the equilibrium point.

2. **(a)** The Jacobian matrix for the system is

$$\begin{pmatrix} 1-2x-y & -x & 0 \\ y & 1-2y+x-z & -y \\ 0 & z & 1-2z+y \end{pmatrix}$$

so at (0, 1, 0) the Jacobian matrix is

$$\begin{pmatrix} 0 & 0 & 0 \\ 1 & -1 & -1 \\ 0 & 0 & 2 \end{pmatrix}.$$

(b) The characteristic polynomial is $(-\lambda)(2-\lambda)(-1-\lambda)$, and the eigenvalues are 0, 2 and -1. An eigenvector for eigenvalue 0 is (1, 1, 0); an eigenvector for eigenvalue -1 is (0, 1, 0); and an eigenvector for eigenvalue 2 is $(0, -1, 3)$.

(c) The equilibrium point has zero as an eigenvalue along with two other eigenvalues, one positive and one negative. Solutions tend toward the equilibrium point along the y-axis as t increases. Solutions in the xy-plane also tend toward the equilibrium point as t decreases. In the xyz-space, solutions tend away from the equilibrium, as long as $z > 0$.

(d)

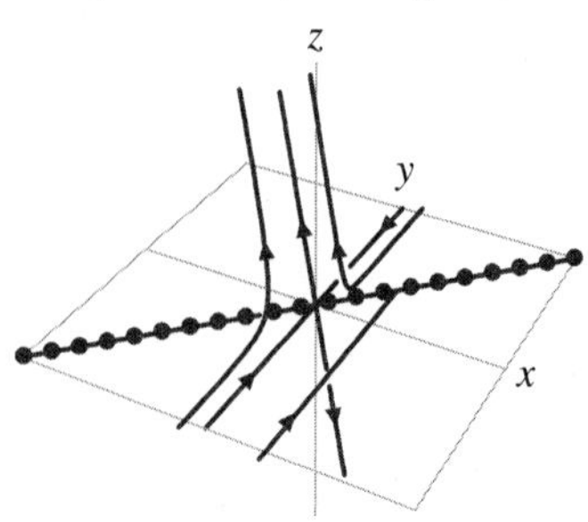

(e) Solutions in the xy-plane tend toward the equilibrium point; all others tend away from this point.

3. **(a)** The Jacobian matrix at (0, 0, 1) is

$$\begin{pmatrix} 1 & 0 & 0 \\ 0 & 0 & 0 \\ 0 & 1 & -1 \end{pmatrix}.$$

(b) The characteristic polynomial is $(1-\lambda)\lambda(-1-\lambda)$, so the eigenvalues are -1, 0 and 1. Corresponding eigenvectors are (0, 0, 1), (0, 1, 1) and (1, 0, 0) respectively. (We can see this by computation, or use the fact that x decouples from y and z in the linearized system to reduce the dimension of the problem.)

(c) The linearized system has solutions that tend toward the original equilibrium point as t increases, solutions that tend toward the original equilibrium point as t decreases, and a line of equilibrium points through the original equilibrium point.

(d)

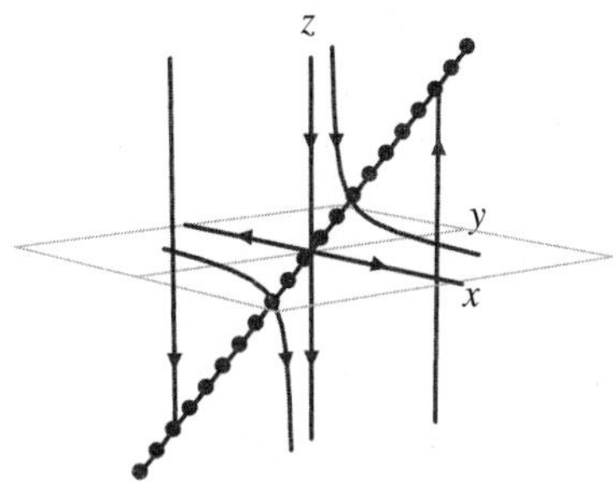

(e) For the linearized system, solutions with initial conditions with $x = 0$ tend in the z-direction toward an equilibrium point on the line of equilibrium points through the original equilibrium point. The behavior of solutions for the nonlinear system on the $x = 0$ plane is not so clear because nonlinear terms could cause solutions to tend toward or away from the equilibrium point. Off the plane $x = 0$, solutions move away from the equilibrium point in the x-direction.

4. **(a)** The Jacobian matrix for the system is

$$\begin{pmatrix} 1-2x-y & -x & 0 \\ y & 1-2y+x-z & -y \\ 0 & z & 1-2z+y \end{pmatrix}$$

so at (1, 0, 1) the Jacobian matrix is

$$\begin{pmatrix} -1 & -1 & 0 \\ 0 & 1 & 0 \\ 0 & 1 & -1 \end{pmatrix}.$$

(b) The characteristic polynomial is $(-1-\lambda)(1-\lambda)(-1-\lambda)$, so the eigenvalues are 1 and -1 (repeated). An eigenvector for eigenvalue 1 is $(-1, 2, 1)$. Two linearly-independent eigenvectors for the eigenvalue -1 are (1, 0, 0) and (0, 0, 1).

(c) The equilibrium point is a saddle with one positive and two negative eigenvalues. Solutions in the xz-plane tend toward the equilibrium point provided they do not lie on either axis. Note that $x = 1$ and $z = 1$ are invariant lines in this plane. Solutions that start off this plane ($y > 0$) tend away from the equilibrium point.

(d)

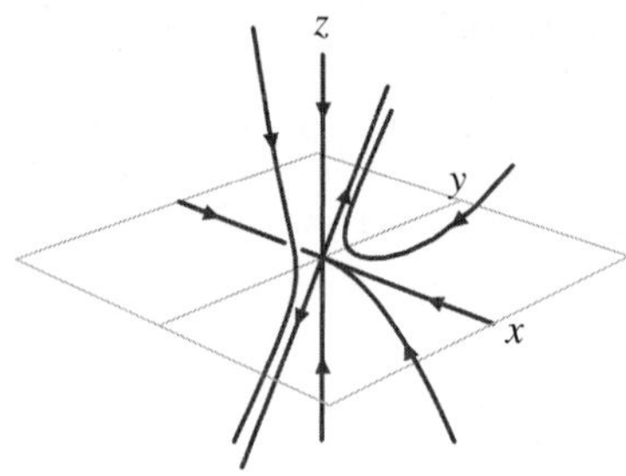

(e) Solutions in the yz-plane (not on the axes) tend toward the equilibrium point; all others tend away from this point.

5. **(a)** When $y = 0$ we have $dy/dt = 0$ and the equations for the rate of change of x and z are

$$\frac{dx}{dt} = x(1 - x)$$

$$\frac{dz}{dt} = z(1 - z).$$

These are both logistic equations with sinks at $x = 1$ and $z = 1$. Hence, $(1, 0, 1)$ is an equilibrium point. When $x = 0$ we have $dx/dt = 0$ and

$$\frac{dy}{dt} = y(1 - y) - yz$$

$$\frac{dz}{dt} = z(1 - z) + yz$$

and $y = z = 1$ is not an equilibrium point for the yz system so $(0, 1, 1)$ is not an equilibrium point for the three-dimensional system. The argument for $(1, 1, 0)$ is very similar.

(b) Having $y = 0$ corresponds to the moose being extinct. The food chain then has no connection between the wolves and the trees. Hence, both wolf and tree populations will tend toward their carrying capacity. When $x = 0$ the trees are extinct. The moose and wolfs then form a two dimensional predator-prey system. We do not expect that both moose and wolf populations to be maintained at their carrying capacitys because wolves eat moose. The argument when the wolves are extinct is similar.

6. **(a)** The new equation for moose is

$$\frac{dy}{dt} = y(1 - y) + xy - yz - \beta y.$$

(b) The equilibrium points are $(0, 0, 0)$, $(1, 0, 0)$, $(1, 0, 1)$, $(0, 0, 1)$, $(0, 1 - \beta, 0)$, and $((2 + \beta)/3, (1 - \beta)/3, (4 - \beta)/3)$.

(c) We compute the derivative with respect to β of each coordinate of this equilibrium point. We find that the derivative of the x-component is $\beta/3 > 0$, so raising β helps the trees. The derivative of the y-component is $-\beta/3$, so the moose population decreases. And the derivative with respect to the z-component is $-\beta/3$, so the wolf population also decreases.

7. **(a)** We must solve

$$
\begin{aligned}
x_0(1-x_0) - x_0y_0 &= 0\\
y_0(1-y_0) + x_0y_0 - y_0z_0 &= 0\\
\zeta z_0(1-z_0) + y_0z_0 &= 0.
\end{aligned}
$$

Since we are looking for the solution where none of the coordinates is zero, we can divide the first equation by x_0, the second by y_0 and the third by z_0 to obtain

$$
\begin{aligned}
1 - x_0 - y_0 &= 0\\
1 - y_0 + x_0 - z_0 &= 0\\
\zeta - \zeta z_0 + y_0 &= 0.
\end{aligned}
$$

The first equation gives $y_0 = 1 - x_0$ which when plugged into the second equation gives $z_0 = 2x_0$ and these in the third equation give

$$x_0 = \frac{\zeta + 1}{2\zeta + 1}.$$

Hence, $y_0 = \zeta/(2\zeta + 1)$ and $z_0 = (2\zeta + 2)/(2\zeta + 1)$.

(b) We can compute the rate of change of the position of this equilibrium point as ζ changes as follows:

$$
\begin{aligned}
\frac{dx_0}{d\zeta} &= \frac{-1}{(2\zeta+1)^2}\\
\frac{dy_0}{d\zeta} &= \frac{1}{(2\zeta+1)^2}\\
\frac{dz_0}{d\zeta} &= \frac{-2}{(2\zeta+1)^2}.
\end{aligned}
$$

Note that $dz_0/d\zeta$ is negative. Hence, a decrease in ζ, which we claimed corresponded to a decrease in the growth rate of the wolves, yields an increase in z_0. Similarly, a decrease in ζ gives a decrease in y_0 and an increase in x_0 because $dy_0/d\zeta$ is positive and $dx_0/d\zeta$ is negative (remember we are decreasing ζ).

(c) Note that the dz/dt equation is

$$\frac{dz}{dt} = \zeta z(1-z) + yz = \zeta z - \zeta z^2 + yz$$

so ζ is multiplied onto the z and the z^2 terms. If we compare this to the modification of the equations given in the section we see that the change must be causes by the fact that ζ is multiplied onto the z^2 term. It does seem "unphysical" that a decrease in the growth-rate parameter could cause an increase in the equilibrium population, but the parameter ζ gives both the growth rate for small z (the ζz term) and the proportionality constant for the effect of large populations (the ζz^2) term. This observation is interesting and deserves further study.

8. **(a)** If we let $w(t)$ represent some (normalized) measure of the wood tick population, then the equations for tree and wolf populations do not change, but the moose population becomes

$$\frac{dy}{dt} = y(1-y) + xy - yz - yw.$$

Since the wood ticks eat only moose blood, the rate of change of the tick population depends on interaction with the moose. The simplest such model is

$$\frac{dw}{dt} = wy.$$

(b) In order for $dw/dt = 0$ we must have $w = 0$ or $y = 0$. If $w = 0$ then the system becomes the original three-dimensional model with the same equilibria. If $y = 0$, then we must have $x = z = 1$ and w any value is an equilibrium. This means that the wolves reach equilibria without the moose and whatever ticks are present go into dormancy and wait for moose.

EXERCISES FOR SECTION 5.6

1. Comparing the y values at the indicated points with the y values of the $y(t)$-graphs for $t = 2n\pi$ shows this return map corresponds to graph (ii). The solution oscillates with increasing and decreasing amplitude around $y = 1$.

2. Comparing the y values at the indicated points with the y values of the $y(t)$-graphs for $t = 2n\pi$ shows this return map corresponds to graph (iii). The solution is complicated, but remains in a bounded region in y and v.

3. Comparing the y values at the indicated points with the y values of the $y(t)$-graphs for $t = 2n\pi$ shows this return map corresponds to graph (i). The solution oscillates in a regular, but fairly complicated way.

4. Comparing the y values at the indicated points with the y values of the $y(t)$-graphs for $t = 2n\pi$ shows this return map corresponds to graph (iv). The solution oscillates in a very complicated pattern, but with bounded y and v coordinates.

5. **(a)** This return map corresponds to graph (iii).

   (b) The values of the θ-coordinate of the indicated points are decreasing monotonically and are all less than one. Hence, graph (iii) is the only one that fits.

   (c) The solution should oscillate with rising and falling amplitude which corresponds to the pendulum arm swinging higher and lower, but never swinging "over the top".

6. **(a)** This return map corresponds to graph (ii).

   (b) The θ-coordinates of the indicated points appear to be fairly large positive values. Hence, this corresponds to graph (ii).

   (c) The pendulum arm frequently swings "over the top" in the direction of increasing θ.

7. **(a)** This return map corresponds to graph (iv).

(b) The θ-coordinates at the indicated points are bounded below by minus two, are all nonpositive and decrease, then increase. The graph with the corresponding behavior is graph (iv).

(c) The pendulum swings back and forth, never "going over the top" with amplitude and period oscillating in a regular way.

8. **(a)** This return map corresponds to graph (i).

(b) The θ-coordinates of the indicated points are all negative and decrease monotonically. This kind of behavior is only seen in graph (i).

(c) The pendulum arm swings "over the top" in the negative θ- direction for many turns after an initial amount of swinging back and forth (and over the top in the positive direction).

9. **(a)** The second-order differential equation is

$$\frac{d^2y}{dt^2} + 3y = 0.2 \sin t.$$

The general solution for the homogeneous equation is

$$y(t) = k_1 \sin \sqrt{3}\, t + k_2 \cos \sqrt{3}\, t.$$

Suppose $y(t) = a \sin t$. Differentiation yields

$$\frac{d^2y}{dt^2} + 3y = 2a \sin t = 0.2 \sin t.$$

Then, $a = 1/10$ and the general solution is

$$y(t) = k_1 \sin \sqrt{3}\, t + k_2 \cos \sqrt{3}\, t + \tfrac{1}{10} \sin t.$$

By the initial condition, $y(0) = 1$ and $v(0) = 0$, one obtains

$$\begin{cases} k_2 = 1 \\ \sqrt{3}\, k_1 + \frac{1}{10} = 0 \end{cases}$$

or $k_1 = -\sqrt{3}/30$ and $k_2 = 1$. The solution is

$$y(t) = -\frac{\sqrt{3}}{30} \sin \sqrt{3}\, t + \cos \sqrt{3}\, t + \tfrac{1}{10} \sin t.$$

(b) By the solution,

$$y(t) = -\frac{\sqrt{3}}{30} \sin \sqrt{3}\, t + \cos \sqrt{3}\, t + \tfrac{1}{10} \sin t.$$

After 2π in time, $0.1 \sin(t + 2\pi)$ gives the same value. Therefore, the Poincaré map is due to the first two terms and it moves on an ellipse.

(c) The change in the initial condition affects only constants, k_1 and k_2. Therefore, Poincaré map traces an ellipse and the scale of the ellipse depends on k_1 and k_2.

(d) The forced harmonic oscillator system is linear and the solution is the sum of the general solution for the homogeneous equation and the particular solution for the nonhomogeneous equation. In this case, the Poincaré map is due to the homogeneous solution as we saw in (c). On the other hand, the forced pendulum system is nonlinear and therefore, the solution becomes more complicated.

10. The natural frequency and the forcing frequency are the same, $1/\pi$. The solution will have a resonance term of the form $t \sin 2t$ and/or $t \cos 2t$. Except for the resonance term(s), all the other terms in the solution are periodic with period π; so, the Poincaré return map does not see the non-resonance terms. For every time increase of π the amplitude of the resonance term(s) increases linearly, thus one expects the Poincaré return map to be a sequence of equidistant points along a straight line.

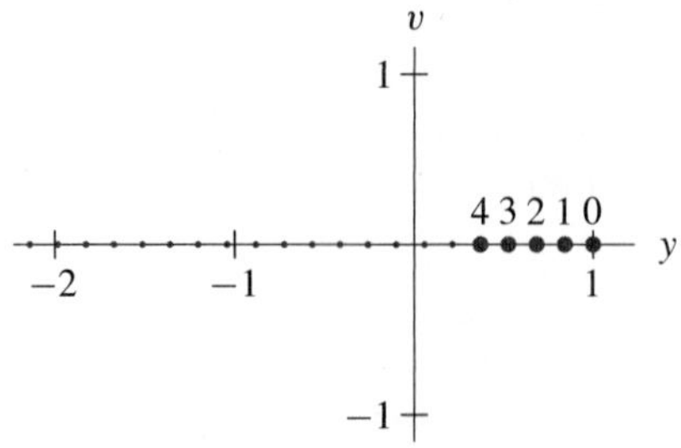

REVIEW EXERCISES FOR CHAPTER 5

1. Since the equilibrium point is at the origin and the system has only polynomial terms, the linearized system is just the linear terms in dx/dt and dy/dt, that is,

$$\frac{dx}{dt} = x$$
$$\frac{dy}{dt} = -2y.$$

2. From the linearized system in Exercise 1, we see (without any calculation) that the eigenvalues are 1 and -2. Hence, the origin is a saddle.

3. The Jacobian matrix for this system is

$$\begin{pmatrix} 2x + 3\cos 3x & 0 \\ -y\cos xy & 2 - x\cos xy \end{pmatrix},$$

and evaluating at $(0, 0)$, we get

$$\begin{pmatrix} 3 & 0 \\ 0 & 2 \end{pmatrix}.$$

So the linearized system at the origin is

$$\frac{dx}{dt} = 3x$$
$$\frac{dy}{dt} = 2y.$$

4. From the linearized system in Exercise 3, we see (without any calculation) that the eigenvalues are 3 and 2. Hence, the origin is a source.

5. The x-nullcline is where $dx/dt = 0$, that is, the line $y = x$. The y-nullcline is where $dy/dt = 0$, that is, the circle $x^2 + y^2 = 2$.

Along the x-nullcline, $dy/dt < 0$ if and only if $-\sqrt{2} < x < \sqrt{2}$. Along the y-nullcline, $dx/dt < 0$ if and only if $y > x$.

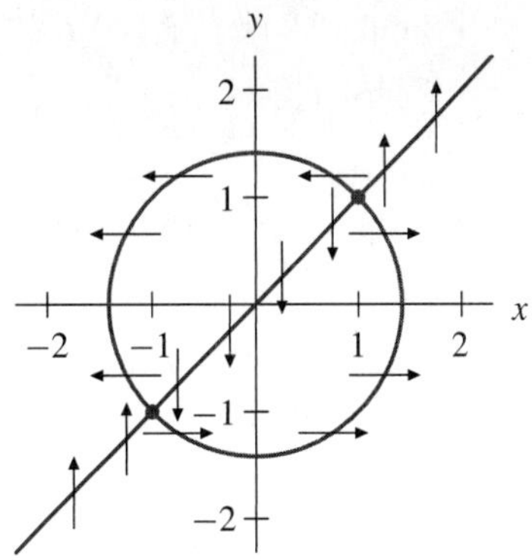

6. This system is not a Hamiltonian system. If it were, then we would have

$$\frac{\partial H}{\partial y} = \frac{dx}{dt} \quad \text{and} \quad -\frac{\partial H}{\partial x} = \frac{dy}{dt}$$

for some function $H(x, y)$. In that case, equality of mixed partials would imply that

$$\frac{\partial}{\partial x}\left(\frac{dx}{dt}\right) = -\frac{\partial}{\partial y}\left(\frac{dy}{dt}\right).$$

For this system, we have

$$\frac{\partial}{\partial x}\left(\frac{dx}{dt}\right) = 2y \quad \text{and} \quad -\frac{\partial}{\partial y}\left(\frac{dy}{dt}\right) = -2y.$$

Since these two partials do not agree, no such function $H(x, y)$ exists.

7. This system is not a gradient system. If it were, then we would have

$$\frac{\partial G}{\partial x} = \frac{dx}{dt} \quad \text{and} \quad \frac{\partial G}{\partial y} = \frac{dy}{dt}$$

for some function $G(x, y)$. In that case, equality of mixed partials would imply that

$$\frac{\partial}{\partial y}\left(\frac{dx}{dt}\right) = \frac{\partial}{\partial x}\left(\frac{dy}{dt}\right).$$

For this system, we have

$$\frac{\partial}{\partial y}\left(\frac{dx}{dt}\right) = 2x + 2y \quad \text{and} \quad \frac{\partial}{\partial x}\left(\frac{dy}{dt}\right) = 2x.$$

Since these two partials do not agree, no such function $G(x, y)$ exists.

8. Some possibilities are:

- The solution is unbounded. That is, either $|x(t)| \to \infty$ or $|y(t)| \to \infty$ (or both) as t increases.
- Similarly, $x(t)$ or $y(t)$ (or both) oscillate with increasing amplitude as t increases (similar to $t \sin t$).
- The solution tends to an equilibrium point.
- The solution tends to a periodic solution, as in the Van der Pol equation (see Section 5.1).
- The solution tends to a curve consisting of equilibrium points and solutions connecting equilibrium points.

9. If the system is a linear system, then all nonequilibrium solutions tend to infinity as t increases, that is, $|\mathbf{Y}(t)| \to \infty$ as $t \to \infty$.

If the system is not linear, it is possible for a solution to spiral toward a periodic solution. For example, consider the Van der Pol equation discussed in Section 5.1. (These two behaviors are the only possibilities.)

10. Since a solution that enters the first quadrant cannot leave, the origin cannot be a spiral sink, a spiral source, or a center.

However, a sink, a saddle, or a source are all possibilities. For example,

$$\frac{dx}{dt} = -2x + y$$
$$\frac{dy}{dt} = x - y$$

has a sink at the origin,

$$\frac{dx}{dt} = y$$
$$\frac{dy}{dt} = x$$

has a saddle at the origin, and

$$\frac{dx}{dt} = 2x + y$$
$$\frac{dy}{dt} = x + y$$

has a source at the origin.

11. True. The x-nullcline is where $dx/dt = 0$ and the y-nullcline is where $dy/dt = 0$, so any point in common must be an equilibrium point.

12. False. For example, both nullclines for the system

$$\frac{dx}{dt} = x - y$$
$$\frac{dy}{dt} = y - x$$

are the line $y = x$. Moreover, since the nullclines are identical, all points on the line are equilibrium points.

13. False. These two numbers are the diagonal entries of the Jacobian matrix. The other two entries of the Jacobian matrix also affect the eigenvalues.

14. False. The Jacobian matrix at an equilibrium point (x_0, y_0) is

$$\begin{pmatrix} f'(x_0) & 0 \\ \dfrac{\partial g}{\partial x}(x_0, y_0) & \dfrac{\partial g}{\partial y}(x_0, y_0) \end{pmatrix},$$

so its eigenvalues are $f'(x_0)$ and

$$\frac{\partial g}{\partial y}(x_0, y_0).$$

Since this partial derivative could be positive, negative, or zero, the equilibrium point could be a source, a saddle, or one of the zero eigenvalue types.

15. **(a)** Setting $dx/dt = 0$ and $dy/dt = 0$, we obtain the simultaneous equations

$$\begin{cases} x - 3y^2 = 0 \\ x - 3y - 6 = 0. \end{cases}$$

Solving for x and y yields the equilibrium points $(12, 2)$ and $(3, -1)$.

To determine the type of an equilibrium point, we compute the Jacobian matrix. We get

$$\begin{pmatrix} 1 & -6y \\ 1 & -3 \end{pmatrix}.$$

At $(12, 2)$, the Jacobian is

$$\begin{pmatrix} 1 & -12 \\ 1 & -3 \end{pmatrix},$$

and its eigenvalues are $-1 \pm 2\sqrt{2}\,i$. Hence, $(12, 2)$ is a spriral sink.

At $(3, -1)$, the Jacobian matrix is

$$\begin{pmatrix} 1 & 6 \\ 1 & -3 \end{pmatrix},$$

and the eigenvalues are $-1 \pm \sqrt{10}$. So $(3, -1)$ is a saddle.

(b) The x-nullcline is $x = 3y^2$, and the y-nullcline is $x = 3y + 6$. We compute the direction of the vector field by computing the sign of dy/dt on the x-nullcline and the sign of dx/dt on the y-nullcline.

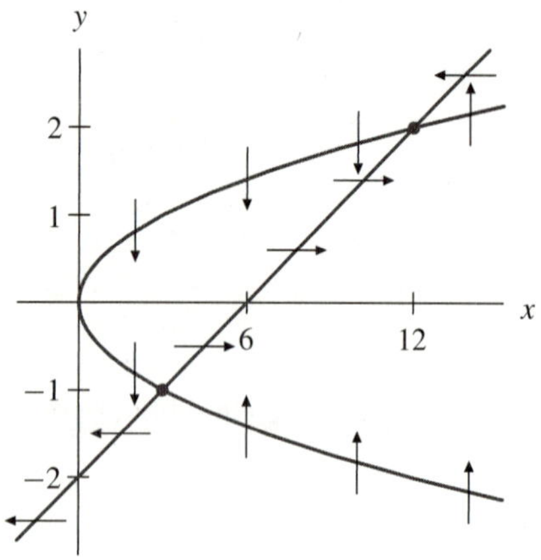

(c)

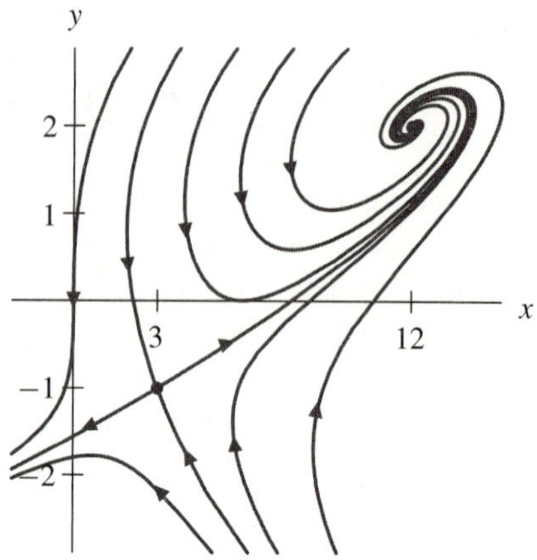

16. **(a)** Setting $dx/dt = 0$ and $dy/dt = 0$, we obtain the simultaneous equations

$$\begin{cases} 10 - x^2 - y^2 = 0 \\ 3x - y = 0. \end{cases}$$

Solving for x and y yields the equilibrium points $(1, 3)$ and $(-1, -3)$.

To determine the type of an equilibrium point, we compute the Jacobian matrix. We get

$$\begin{pmatrix} -2x & -2y \\ 3 & -1 \end{pmatrix}.$$

At $(1, 3)$, the Jacobian is

$$\begin{pmatrix} -2 & -6 \\ 3 & -1 \end{pmatrix},$$

and its eigenvalues are $(-3 \pm \sqrt{71}\, i)/2$. Hence, $(1, 3)$ is a spriral sink.

At $(-1, -3)$, the Jacobian is

$$\begin{pmatrix} 2 & 6 \\ 3 & -1 \end{pmatrix},$$

and its eigenvalues are 5 and -4. Hence, $(-1, -3)$ is a saddle.

(b) The x-nullcline is the circle $x^2 + y^2 = 10$, and the y-nullcline is the line $y = 3x$. We compute the direction of the vector field by computing the sign of dy/dt on the x-nullcline and the sign of dx/dt on the y-nullcline.

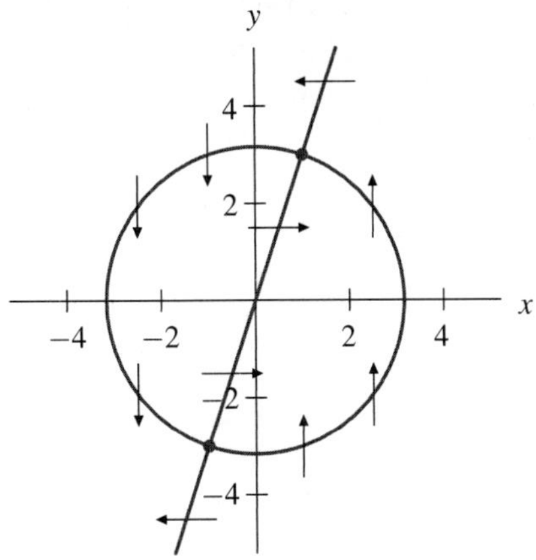

(c)

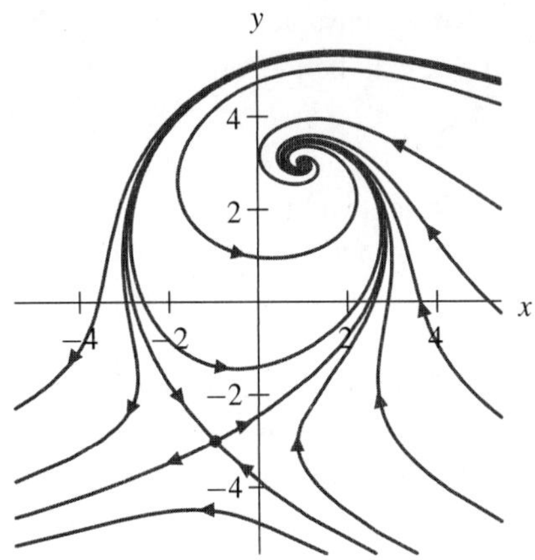

17. **(a)** To find the equilibria, we solve the system of equations

$$\begin{cases} 4x - x^2 - xy = 0 \\ 6y - 2y^2 - xy = 0. \end{cases}$$

We obtain the four equilibrium points $(0, 0)$, $(0, 3)$, $(4, 0)$, and $(2, 2)$.

To classify the equilibria, we compute the Jacobian

$$\begin{pmatrix} 4 - 2x - y & -x \\ -y & 6 - 4y - x \end{pmatrix},$$

evaluate at each equilibrium point, and compute the eigenvalues.

At $(0, 0)$, the Jacobian matrix is

$$\begin{pmatrix} 4 & 0 \\ 0 & 6 \end{pmatrix}.$$

The eigenvalues are 4 and 6, so the origin is a source.

At $(0, 3)$, the Jacobian matrix is

$$\begin{pmatrix} 1 & 0 \\ -3 & -6 \end{pmatrix}.$$

The eigenvalues are 1 and -6, so $(0, 3)$ is a saddle.

At $(4, 0)$, the Jacobian matrix is

$$\begin{pmatrix} -4 & -4 \\ 0 & 2 \end{pmatrix}.$$

The eigenvalues are -4 and 2, so (4, 0) is a saddle.

Finally, at (2, 2) the Jacobian matrix is

$$\begin{pmatrix} -6 & -2 \\ -2 & -4 \end{pmatrix}.$$

The eigenvalues are $-5 \pm \sqrt{5}$. Both are negative, so (2, 2) is a sink.

(b) The x-nullcline satisfies the equation $4x - x^2 - xy = 0$, which can be rewritten as

$$x(4 - x - y) = 0.$$

We get two lines, $x = 0$ (the y-axis) and $y = 4 - x$.

The y-nullcline satisfies the equation $6y - 2y^2 - xy = 0$, which can be rewritten as

$$y(6 - 2y - x) = 0.$$

We get two lines, $y = 0$ (the x-axis) and $x + 2y = 6$.

We compute the direction of the vector field by computing the sign of dy/dt on the x-nullcline and the sign of dx/dt on the y-nullcline.

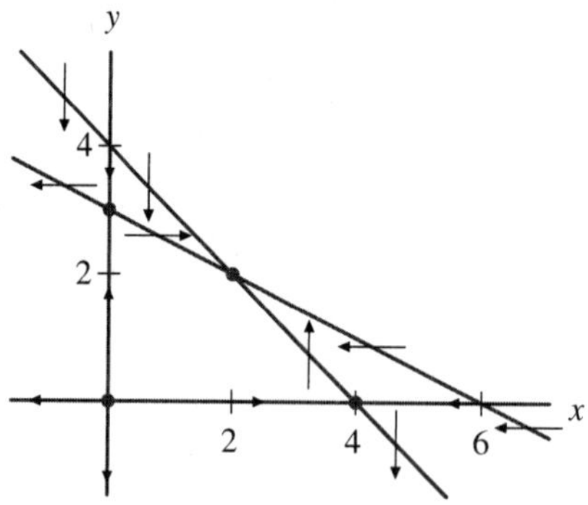

(c) The phase portrait is

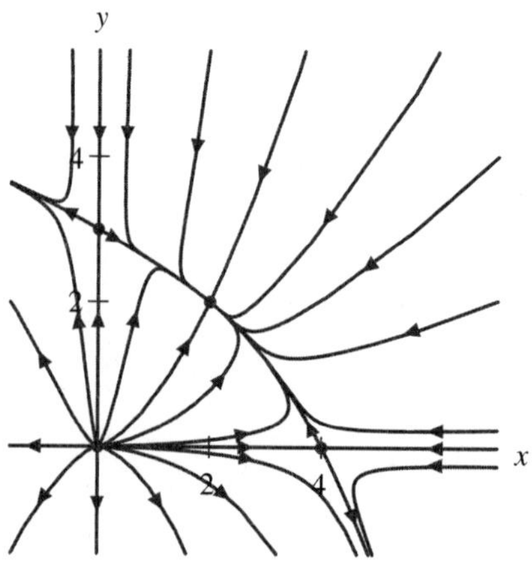

18. **(a)** Setting $dx/dt = 0$ and $dy/dt = 0$, we obtain the simultaneous equations

$$\begin{cases} xy = 0 \\ x + y - 1 = 0. \end{cases}$$

Solving for x and y yields the equilibrium points $(1, 0)$ and $(0, 1)$.

To determine the type of an equilibrium point, we compute the Jacobian matrix. We get

$$\begin{pmatrix} y & x \\ 1 & 1 \end{pmatrix}.$$

At $(1, 0)$, the Jacobian is

$$\begin{pmatrix} 0 & 1 \\ 1 & 1 \end{pmatrix},$$

and its eigenvalues are $(1 \pm \sqrt{5})/2$. Hence, $(1, 0)$ is a saddle.

At $(0, 1)$, the Jacobian is

$$\begin{pmatrix} 1 & 0 \\ 1 & 1 \end{pmatrix},$$

and 1 is a repeated eigenvalue. Hence, $(0, 1)$ is a source.

(b) The x-nullcline is given by the equation $xy = 0$, so it consists of the x- and y-axes. The y-nullcline is the line $y = 1 - x$. We compute the direction of the vector field by computing the sign of dy/dt on the x-nullcline and the sign of dx/dt on the y-nullcline.

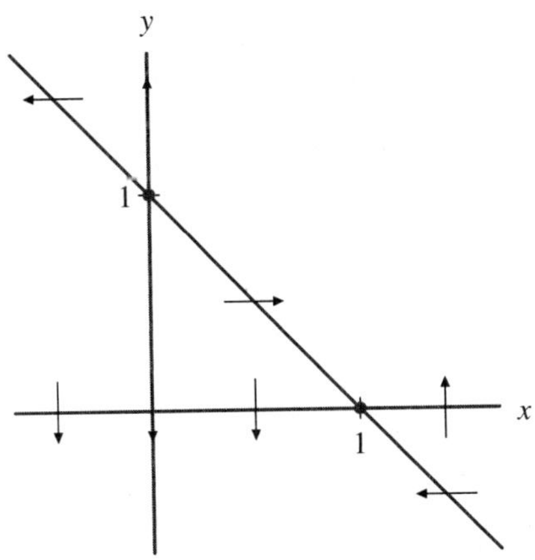

(c)

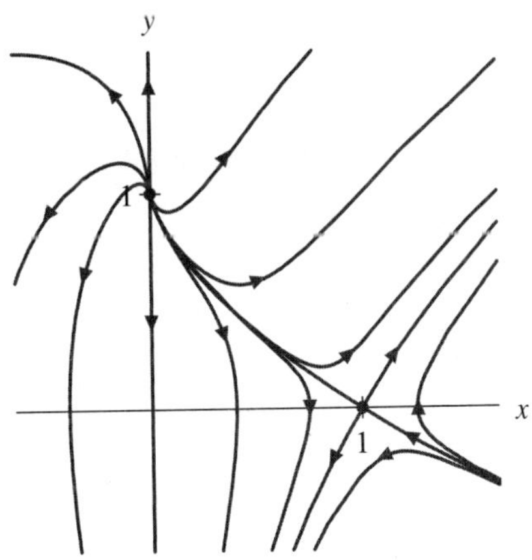

19. To find the equilibrium points, we solve the simultaneous equations

$$\begin{cases} x^2 - a = 0 \\ y^2 - b = 0 \end{cases}$$

and obtain the four points $(\sqrt{a}, \sqrt{b})$, $(-\sqrt{a}, \sqrt{b})$, $(\sqrt{a}, -\sqrt{b})$, and $(-\sqrt{a}, -\sqrt{b})$.

The Jacobian matrix for this system is

$$\begin{pmatrix} 2x & 0 \\ 0 & 2y \end{pmatrix}$$

which is diagonal. Hence, the eigenvalues are simply $2x$ and $2y$.

At $(\sqrt{a}, \sqrt{b})$, both eigenvalues are positive, so this equilibrium point is a source.

At $(-\sqrt{a}, \sqrt{b})$, one eigenvalue is negative and the other is positive. This equilibrium point is a saddle. The same is true for the equilibrium point $(\sqrt{a}, -\sqrt{b})$.

At $(-\sqrt{a}, -\sqrt{b})$, both eigenvalues are negative. This equilibrium point is a sink.

20. The x-nullclines occur where $x^2 - a = 0$, that is, $x = \pm\sqrt{a}$. The y-nullclines occur where $y^2 - b = 0$, that is, $y = \pm\sqrt{b}$. Along each x-nullcline, we have the phase line for the equation $dy/dt = y^2 - b$. In particular, solutions are increasing if $y < -\sqrt{b}$ or if $y > \sqrt{b}$ and decreasing if $-\sqrt{b} < y < \sqrt{b}$. Similar statements hold for the y-nullclines.

Using the results of Exercise 19 along with information obtained from the nullclines, we can sketch the phase portrait. In the following figures, the nullclines are on the left and the phase portrait is on the right.

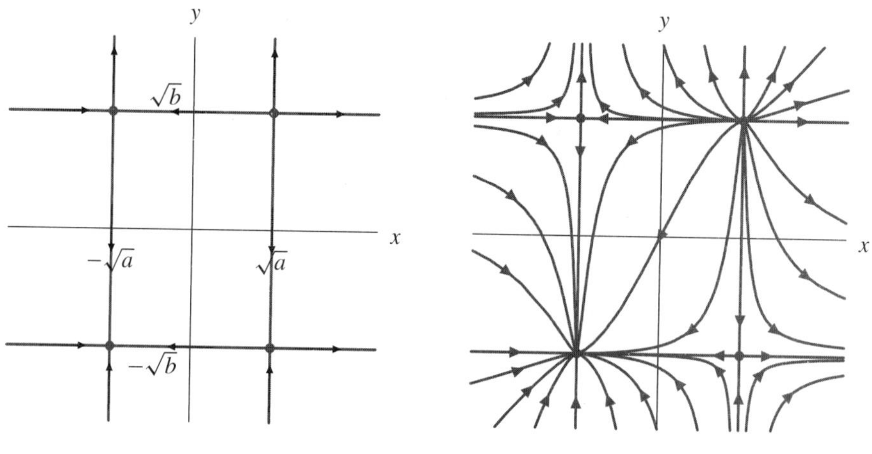

21. If $a = 0$, the equilibria are $(0, \sqrt{b})$ and $(0, -\sqrt{b})$.

The Jacobian matrix is

$$\begin{pmatrix} 2x & 0 \\ 0 & 2y \end{pmatrix},$$

and its eigenvalues are $2x$ and $2y$.

At $(0, \sqrt{b})$, the eigenvalues are 0 and $2\sqrt{b}$. Solutions of the nonlinear system approach the equilibrium along the negative x-direction, and solutions tend away from the equilibrium along the positive x-direction and along the y-axis.

At $(0, -\sqrt{b})$, the eigenvalues are 0 and $-2\sqrt{b}$. Solutions of the nonlinear system approach the equilibrium along the negative x-direction and along the y-axis and solutions tend away from the equilibrium in the positive x-direction.

The x-nullcline is the line $x = 0$, and the y-nullcline consists of the two lines $y = \pm\sqrt{b}$.
In the following figures, the nullclines are on the left and the phase portrait is on the right.

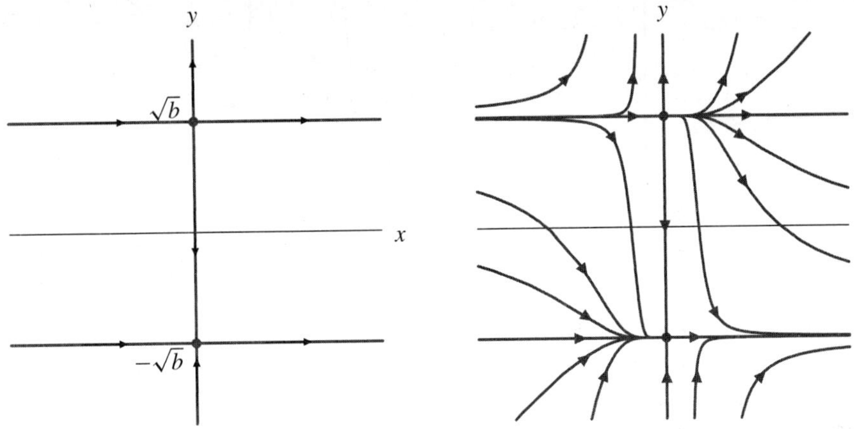

As $a \to 0$ from above, the x-nullclines merge. The pair of equilibrium points $(\sqrt{a}, \sqrt{b})$ and $(-\sqrt{a}, \sqrt{b})$ coalesce at $(0, \sqrt{b})$, and the pair of equilibrium points $(\sqrt{a}, -\sqrt{b})$ and $(-\sqrt{a}, -\sqrt{b})$ coalesce at $(0, -\sqrt{b})$.

22. This exercise is very similar to Exercise 21. We simply interchange the roles x and y.
In the following figures, the nullclines are on the left and the phase portrait is on the right.

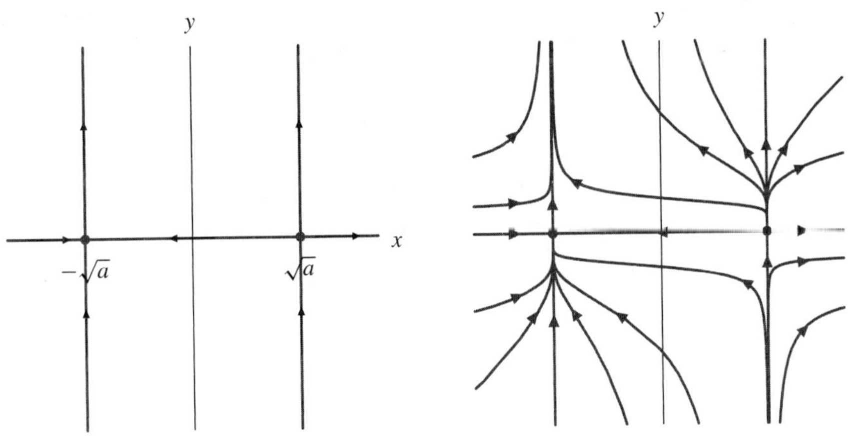

23. The equilibrium points are given by $dx/dt = x^2 = 0$ and $dy/dt = y^2 = 0$, so $(0, 0)$ is the only equilibrium point. The Jacobian at this point is the zero matrix, that is, all four entries are zero.

Solutions on the negative x- and y-axes approach the origin, and solutions on the positive x- and y-axes move away from the origin as t increases. Hence, solutions in the third quadrant approach the origin as t increases, and solutions in all other quadrants tend to infinity as t increases.

The x-nullcline is the y-axis, and the y-nullcline is the x-axis.

In the following figures, the nullclines are on the left and the phase portrait is on the right.

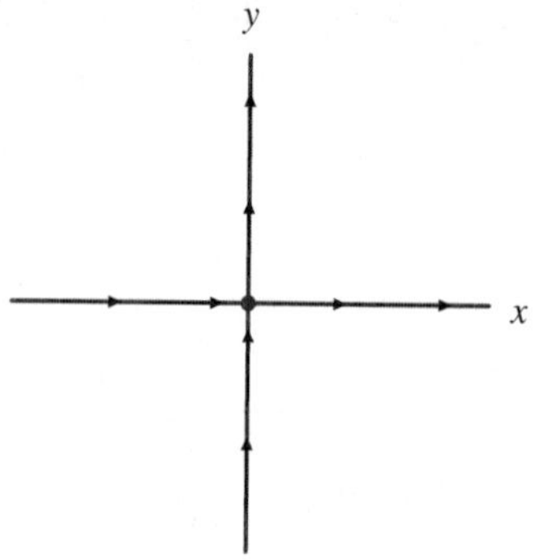

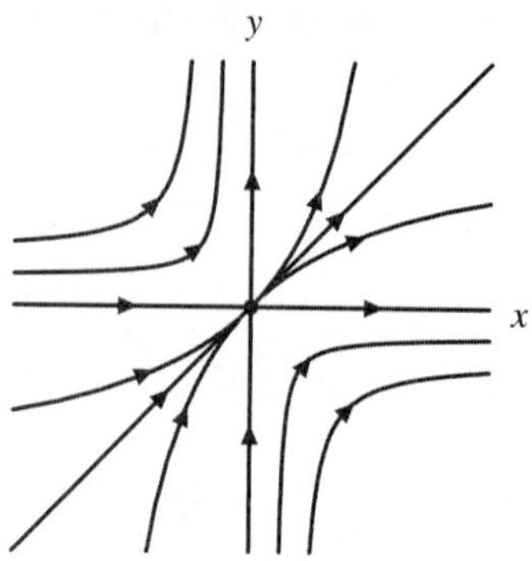

24. To understand what happens in the first quadrant of the ab-plane, we take advantage of our work in Exercises 19–23.

- If both a and b are positive, see Exercises 19 and 20.
- If $a = 0$ and $b > 0$, see Exercise 21.
- If $a > 0$ and $b = 0$, see Exercise 22.
- If both $a = 0$ and $b = 0$, see Exercise 23.

To see what happens in the rest of the ab-plane, first consider the case where $a < 0$ and $b > 0$. Then $dx/dt > 0$ at all points in the phase plane, and therefore $x(t)$ is always increasing. The two lines $y = \pm\sqrt{b}$ are y-nullclines as well as solution curves. The function $y(t)$ is decreasing for all solutions between these two nullclines, and $y(t)$ is increasing for all solutions above $y = \sqrt{b}$ or below $y = -\sqrt{b}$.

As $b \to 0$ from above, these two nullclines coalesce. If both a and b are negative, then there are no nullclines. All solutions have both $x(t)$ and $y(t)$ increasing.

The cases where $b < 0$ are essentially the same as the cases where $a < 0$ (with the roles of a and b interchanged).

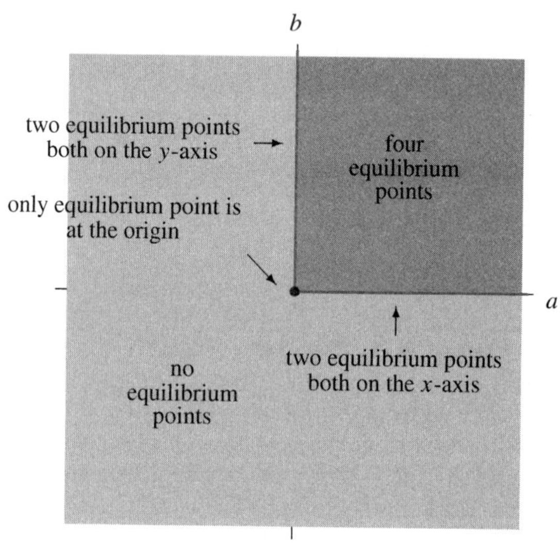

25. **(a)** The equilibrium points are the solutions of

$$\begin{cases} y^2 - x^2 - 1 = 0 \\ 2xy = 0, \end{cases}$$

that is, $(0, \pm 1)$.

The Jacobian matrix is

$$\begin{pmatrix} -2x & 2y \\ 2y & 2x \end{pmatrix}.$$

At $(0, 1)$, the Jacobian is

$$\begin{pmatrix} 0 & 2 \\ 2 & 0 \end{pmatrix}.$$

Its characteristic polynomial is $\lambda^2 - 4$, so its eigenvalues are $\lambda = \pm 2$. The equilibrium point is a saddle.

At $(0, -1)$, the Jacobian is

$$\begin{pmatrix} 0 & -2 \\ -2 & 0 \end{pmatrix}.$$

Its characteristic polynomial is $\lambda^2 - 4$, so its eigenvalues are $\lambda = \pm 2$. The equilibrium point is a saddle.

(b) The x-nullcline is the hyperbola $y^2 - x^2 = 1$, and the y-nullclines are the x- and y-axes.

In the following figures, the nullclines are on the left and the phase portrait is on the right.

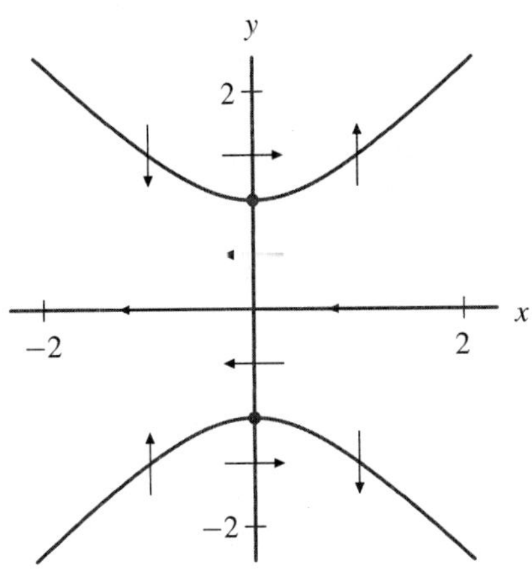

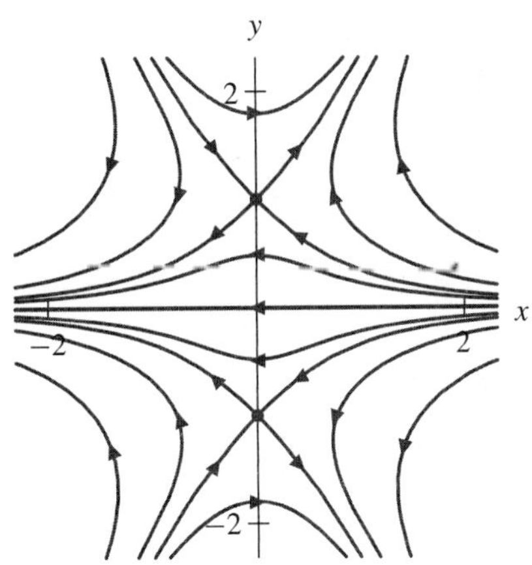

(c) To see if the system is Hamiltonian, we compute

$$\frac{\partial(y^2 - x^2 - 1)}{\partial x} = -2x \quad \text{and} \quad -\frac{\partial(2xy)}{\partial y} = -2x.$$

Since these partials agree, the system is Hamiltonian.

The Hamiltonian is a function $H(x, y)$ such that

$$\frac{\partial H}{\partial y} = \frac{dx}{dt} = y^2 - x^2 - 1 \quad \text{and} \quad \frac{\partial H}{\partial x} = -\frac{dy}{dt} = -2xy.$$

We integrate the second equation with respect to x to see that

$$H(x, y) = -x^2 y + \phi(y),$$

where $\phi(y)$ represents the terms whose derivative with respect to x are zero. Using this expression for $H(x, y)$ in the first equation, we obtain

$$-x^2 + \phi'(y) = y^2 - x^2 - 1.$$

Hence, $\phi'(y) = y^2 - 1$, and we can take $\phi(y) = \frac{1}{3}y^3 - y$. The function

$$H(x, y) = -x^2 y + \frac{y^3}{3} - y$$

is a Hamiltonian function for this system.

(d) To see if the system is a gradient system, we compute

$$\frac{\partial(y^2 - x^2 - 1)}{\partial y} = 2y \quad \text{and} \quad \frac{\partial(2xy)}{\partial x} = 2y.$$

Since these partials agree, the system is a gradient system.

We must now find a function $G(x, y)$ such that

$$\frac{\partial G}{\partial x} = \frac{dx}{dt} = y^2 - x^2 - 1 \quad \text{and} \quad \frac{\partial G}{\partial y} = \frac{dy}{dt} = 2xy.$$

Integrating the second equation with respect to y, we obtain

$$G(x, y) = xy^2 + h(x),$$

where $h(x)$ represents the terms whose derivative with respect to y are zero.

Using this expression for $G(x, y)$ in the first equation, we obtain

$$y^2 + h'(x) = y^2 - x^2 - 1.$$

Hence, $h'(x) = -x^2 - 1$, and we can take $h(x) = -\frac{1}{3}x^3 - x$. The function

$$G(x, y) = xy^2 - \frac{x^3}{3} - x$$

is the required function.

26. **(a)** Letting $y = dx/dt$, we obtain the system

$$\frac{dx}{dt} = y$$
$$\frac{dy}{dt} = 3x - x^3 - 2y.$$

From the first equation, we see that $y = 0$ for any equilibrium point. Substituting $y = 0$ in the equation $3x - x^3 - 2y = 0$ yields $x = 0$ or $x^2 = 3$. Hence, the equilibria are $(0, 0)$ and $(\pm\sqrt{3}, 0)$.

(b) The Jacobian matrix is

$$\begin{pmatrix} 0 & 1 \\ 3 - 3x^2 & -2 \end{pmatrix}.$$

Evaluating the Jacobian at (0, 0) yields

$$\begin{pmatrix} 0 & 1 \\ 3 & -2 \end{pmatrix},$$

which has eigenvalues -3 and 1. Hence, the origin is a saddle. At $(\pm\sqrt{3}, 0)$, the Jacobian matrix is

$$\begin{pmatrix} 0 & 1 \\ -6 & -2 \end{pmatrix},$$

which has eigenvalues $-1 \pm i\sqrt{5}$. Hence, these two equilibria are spiral sinks.

27. To see if the system is Hamiltonian, we compute

$$\frac{\partial(-3x + 10y)}{\partial x} = -3 \quad \text{and} \quad -\frac{\partial(-x + 3y)}{\partial y} = -3.$$

Since these partials agree, the system is Hamiltonian.

To find the Hamiltonian function, we use the fact that

$$\frac{\partial H}{\partial y} = \frac{dx}{dt} = -3x + 10y.$$

Integrating with respect to y gives

$$H(x, y) = -3xy + 5y^2 + \phi(x),$$

where $\phi(x)$ represents the terms whose derivative with respect to y are zero. Differentiating this expression for $H(x, y)$ with respect to x gives

$$-3y + \phi'(x) = -\frac{dy}{dt} = x - 3y.$$

We choose $\phi(x) = \frac{1}{2}x^2$ and obtain the Hamiltonian function

$$H(x, y) = -3xy + 5y^2 + \frac{x^2}{2}.$$

We know that the solution curves of a Hamiltonian system remain on the level sets of the Hamiltonian function. Hence, solutions of this system satisfy the equation

$$-3xy + 5y^2 + \frac{x^2}{2} = h$$

for some constant h. Multiplying through by 2 yields the equation

$$x^2 - 6xy + 10y^2 = k$$

where $k = 2h$ is a constant.

28. **(a)** To see if the system is Hamiltonian, we compute

$$\frac{\partial(ax+by)}{\partial x} = a \quad \text{and} \quad -\frac{\partial(cx+dy)}{\partial y} = -d.$$

For these partials to agree, we must have $a = -d$.

Assuming that $d = -a$, we want a function $H(x, y)$ such that

$$\frac{\partial H}{\partial y} = \frac{dx}{dt} = ax + by \quad \text{and} \quad \frac{\partial H}{\partial x} = -\frac{dy}{dt} = -cx + ay.$$

We integrate the second equation with respect to x to see that

$$H(x, y) = -\frac{c}{2}x^2 + axy + \phi(y),$$

where $\phi(y)$ represents the terms whose derivative with respect to x are zero.

Using this expression for $H(x, y)$ in the first equation, we obtain

$$ax + \phi'(y) = ax + by.$$

In other words, $\phi'(y) = by$, and we can take $\phi(y) = by^2/2$. The function

$$H(x, y) = -\frac{c}{2}x^2 + axy + \frac{b}{2}y^2$$

is a Hamiltonian function for this system if $d = -a$.

(b) To see if the system is a gradient system, we compute

$$\frac{\partial(ax+by)}{\partial y} = b \quad \text{and} \quad \frac{\partial(cx+dy)}{\partial x} = c.$$

The linear system is a gradient system if $b = c$.

Assuming that $b = c$, we want a function $G(x, y)$ such that

$$\frac{\partial G}{\partial x} = \frac{dx}{dt} = ax + by \quad \text{and} \quad \frac{\partial G}{\partial y} = \frac{dy}{dt} = bx + dy.$$

Integrating the first equation with respect to x, we obtain

$$G(x, y) = \frac{a}{2}x^2 + bxy + h(y),$$

where $h(y)$ represents the terms whose derivative with respect to y are zero.

Using this expression for $G(x, y)$ in the second equation, we obtain

$$bx + h'(y) = bx + dy.$$

Hence, $h'(y) = dy$, and we can take $h(y) = dy^2/2$. The function

$$G(x, y) = \frac{a}{2}x^2 + bxy + \frac{d}{2}y^2$$

is the required function if $c = b$.

(c) The system is Hamiltonian if $d = -a$ and gradient if $b = c$. Both conditions are satisfied if the system has the form

$$\frac{d\mathbf{Y}}{dt} = \begin{pmatrix} a & b \\ b & -a \end{pmatrix} \mathbf{Y}.$$

The eigenvalues of the coefficient matrix are $\pm\sqrt{a^2 + b^2}$, so the origin is a saddle if the system is both Hamiltonian and gradient.

(d) Any matrix

$$\begin{pmatrix} a & b \\ c & d \end{pmatrix}$$

where $d \neq -a$ and $b \neq c$ gives a system that is neither Hamiltonian nor gradient. (Recall that both gradient and Hamiltonian systems cannot have equilibrium points that are spiral sources or spiral sinks.)

29. **(a)** Since θ represents an angle in this model, we restrict θ to the interval $-\pi < \theta < \pi$.

The equilibria must satisfy the equations

$$\begin{cases} \cos\theta = s^2 \\ \sin\theta = -s^2. \end{cases}$$

Therefore,

$$\tan\theta = \frac{\sin\theta}{\cos\theta} = \frac{-s^2}{s^2} = -1,$$

and consequently, $\theta = -\arctan 1 = -\pi/4$.

To find s, we note that $s^2 = \cos(-\pi/4) = 1/\sqrt{2}$. Hence, $s = 1/\sqrt[4]{2}$, and the only equilibrium point is

$$(\theta, s) = \left(-\frac{\pi}{4}, \frac{1}{\sqrt[4]{2}}\right).$$

(b) The Jacobian matrix for this system is

$$\begin{pmatrix} \dfrac{\sin\theta}{s} & 1 + \dfrac{\cos\theta}{s^2} \\ -\cos\theta & -2s \end{pmatrix}.$$

Evaluating at the equilibrium point, we get

$$\begin{pmatrix} -2^{-3/4} & 2 \\ -2^{-1/2} & -2^{3/4} \end{pmatrix}.$$

The characteristic polynomial of this matrix is

$$\lambda^2 + \frac{1 + 2\sqrt{2}}{2^{3/4}}\lambda + (1 + \sqrt{2}).$$

Since

$$\left(\frac{1 + 2\sqrt{2}}{2^{3/4}}\right)^2 - 4(1 + \sqrt{2}) < 0,$$

the eigenvalues are complex. Their real part is

$$-\frac{1+2\sqrt{2}}{2^{7/4}},$$

which is negative. Consequently, the equilibrium point is a spiral sink.

CHAPTER 6

Laplace Transforms

EXERCISES FOR SECTION 6.1

1. We have

$$\begin{aligned}\mathcal{L}[3] &= \int_0^\infty 3e^{-st}\,dt \\ &= \lim_{b\to\infty}\int_0^b 3e^{-st}\,dt \\ &= \lim_{b\to\infty}\left(\frac{-3}{s}e^{-st}\bigg|_0^b\right) \\ &= \lim_{b\to\infty} -\frac{3}{s}\left(e^{-sb}-e^0\right) \\ &= \frac{3}{s} \quad \text{if } s>0,\end{aligned}$$

since $\lim_{b\to\infty} e^{-sb} = \lim_{b\to\infty} 1/e^{sb} = 0$ if $s > 0$.

2. We have

$$\mathcal{L}[t] = \int_0^\infty te^{-st}\,dt = \lim_{b\to\infty}\int_0^b te^{-st}\,dt.$$

To evaluate the integral we use integration by parts with $u = t$ and $dv = e^{-st}dt$. Then $du = dt$ and $v = -e^{-st}/s$. Thus

$$\begin{aligned}\lim_{b\to\infty}\int_0^b te^{-st}\,dt &= \lim_{b\to\infty}\left(-\frac{te^{-st}}{s}\bigg|_0^b - \int_0^b -\frac{e^{-st}}{s}\,dt\right) \\ &= \lim_{b\to\infty}\left(-\frac{be^{-sb}}{s} - \frac{e^{-st}}{s^2}\bigg|_0^b\right) \\ &= \lim_{b\to\infty}\left(-\frac{be^{-sb}}{s} - \frac{e^{-sb}}{s^2} + \frac{e^0}{s^2}\right) \\ &= \frac{1}{s^2}\end{aligned}$$

since

$$\lim_{b\to\infty} -\frac{be^{-sb}}{s} = \lim_{b\to\infty}\frac{-b}{se^{sb}} = \lim_{b\to\infty}\frac{-1}{s^2e^{sb}} = 0$$

by L'Hôpital's Rule if $s > 0$.

3. We use the fact that $\mathcal{L}[df/dt] = s\mathcal{L}[f] - f(0)$. Letting $f(t) = t^2$ we have $f(0) = 0$ and

$$\mathcal{L}[2t] = s\mathcal{L}[t^2] - 0$$

or

$$2\mathcal{L}[t] = s\mathcal{L}[t^2]$$

using the fact that the Laplace transform is linear. Then since $\mathcal{L}[t] = 1/s^2$ (by the previous exercise), we have

$$\mathcal{L}[-5t^2] = -5\mathcal{L}[t^2] = -5\left(\frac{2\mathcal{L}[t]}{s}\right) = -\frac{10}{s^3}.$$

4. We have shown thus far that $\mathcal{L}[t] = 1/s^2$ and $\mathcal{L}[t^2] = 2/s^3$. Let's compute $\mathcal{L}[t^3]$ and see if a pattern emerges. Using $\mathcal{L}[df/dt] = s\mathcal{L}[f] - f(0)$ with $f(t) = t^3$, we have

$$\mathcal{L}[3t^2] = 3\mathcal{L}[t^2] = s\mathcal{L}[t^3] - f(0)$$

which yields

$$\mathcal{L}[t^3] = \frac{3}{s}\mathcal{L}[t^2] = \frac{3 \cdot 2}{s^4} = \frac{3!}{s^4}.$$

If we were to continue, we would see that

$$\mathcal{L}[t^4] = \frac{4}{s}\mathcal{L}[t^3] = \frac{4!}{s^5}.$$

A clear pattern has emerged (which we could prove by induction should we be so inclined):

$$\mathcal{L}[t^n] = \frac{n!}{s^{n+1}}$$

so that $\mathcal{L}[t^5] = 5!/s^6$.

5. To show a rule by induction, we need two steps. First, we need to show the rule is true for $n = 1$. Then, we need to show that if the rule holds for n, then it holds for $n + 1$.

(a) $n = 1$. We need to show that $\mathcal{L}[t] = 1/s^2$.

$$\mathcal{L}[t] = \int_0^\infty te^{-st}\,dt.$$

Using integration by parts with $u = t$ and $dv = e^{-st}\,dt$, we find

$$\begin{aligned}
\mathcal{L}[t] &= \left.\frac{te^{-st}}{-s}\right|_0^\infty + \int_0^\infty \frac{e^{-st}}{s}\,dt \\
&= \lim_{b\to\infty}\left[\left.\frac{te^{-st}}{-s}\right|_0^b\right] + \int_0^\infty \frac{e^{-st}}{s}\,dt \\
&= \int_0^\infty \frac{e^{-st}}{s}\,dt \\
&= -\left.\frac{e^{-st}}{s^2}\right|_0^\infty \\
&= \frac{1}{s^2} \quad (s > 0).
\end{aligned}$$

(b) Now we assume that the rule holds for n, that is, that $\mathcal{L}[t^n] = n!/s^{n+1}$, and show it holds true for $n + 1$, that is, $\mathcal{L}[t^{n+1}] = (n + 1)!/s^{n+2}$. There are two different methods to do so:

(i)

$$\mathcal{L}[t^{n+1}] = \int_0^\infty t^{n+1}e^{-st}\,dt$$

Using integration by parts with $u = t^{n+1}$ and $dv = e^{-st}\,dt$, we find

$$\mathcal{L}[t^{n+1}] = -\frac{t^{n+1}e^{-st}}{s}\bigg|_0^\infty + \int_0^\infty \frac{n+1}{s} t^n\, e^{-st}\,dt.$$

Now,

$$\begin{aligned}
-\frac{t^{n+1}e^{-st}}{s}\bigg|_0^\infty &= \lim_{b\to\infty}\left[-\frac{t^{n+1}e^{-st}}{s}\bigg|_0^b\right] \\
&= \lim_{b\to\infty}\frac{-b^{n+1}e^{-sb}}{s} + 0 \\
&= 0 \quad (s > 0).
\end{aligned}$$

So

$$\begin{aligned}
\mathcal{L}[t^{n+1}] &= \int_0^\infty \frac{n+1}{s} t^n\, e^{-st}\,dt \\
&= \frac{n+1}{s}\int_0^\infty t^n\, e^{-st}\,dt \\
&= \frac{n+1}{s}\mathcal{L}[t^n].
\end{aligned}$$

Since we assumed that $\mathcal{L}[t^n] = n!/s^{n+1}$, we get that

$$\mathcal{L}[t^{n+1}] = \frac{n+1}{s}\cdot\frac{n!}{s^{n+1}} = \frac{(n+1)!}{s^{n+2}}$$

which is what we wanted to show.

(ii) We use the fact that $\mathcal{L}[df/dt] = s\mathcal{L}[f] - f(0)$. Letting $f(t) = t^{n+1}$ we have $f(0) = 0$ and

$$\mathcal{L}[(n+1)t^n] = s\mathcal{L}[t^{n+1}] - 0$$

or

$$(n+1)\mathcal{L}[t^n] = s\mathcal{L}[t^{n+1}]$$

using the fact that the Laplace transform is linear. Since we assumed $\mathcal{L}[t^n] = n!/s^{n+1}$, we have

$$\mathcal{L}[t^{n+1}] = \frac{n+1}{s}\mathcal{L}[t^n] = \frac{n+1}{s}\cdot\frac{n!}{s^{n+1}} = \frac{(n+1)!}{s^{n+2}},$$

which is what we wanted to show.

6. Using the formula that $\mathcal{L}[t^n] = n!/s^{n+1}$, we can see by linearity that

$$\mathcal{L}[a_i t^i] = a_i \frac{i!}{s^{i+1}},$$

so

$$\begin{aligned}
\mathcal{L}[a_0 + a_1 t + \cdots + a_n t^n] &= \mathcal{L}[a_0] + \mathcal{L}[a_1 t] + \cdots + \mathcal{L}[a_n t^n] \\
&= \frac{a_0}{s} + \frac{a_1}{s^2} + \cdots + \frac{a_n n!}{s^{n+1}}.
\end{aligned}$$

7. Since we know that $\mathcal{L}[e^{at}] = 1/(s-a)$, we have $\mathcal{L}[e^{3t}] = 1/(s-3)$, and therefore,

$$\mathcal{L}^{-1}\left[\frac{1}{s-3}\right] = e^{3t}.$$

8. We see that

$$\frac{5}{3s} = \frac{5}{3} \cdot \frac{1}{s},$$

so

$$\mathcal{L}^{-1}\left[\frac{5}{3s}\right] = \frac{5}{3},$$

since $\mathcal{L}^{-1}[1/s] = 1$.

9. We see that

$$\frac{2}{3s+5} = \frac{2}{3} \cdot \frac{1}{s+5/3},$$

so

$$\mathcal{L}^{-1}\left[\frac{2}{3s+5}\right] = \frac{2}{3}e^{-\frac{5}{3}t}.$$

10. Using the method of partial fractions,

$$\frac{14}{(3s+2)(s-4)} = \frac{A}{3s+2} + \frac{B}{s-4}.$$

Putting the right-hand side over a common denominator gives $A(s-4) + B(3s+2) = 14$, which can be written as $(A+3B)s + (-4A+2B) = 14$. So, $A + 3B = 0$ and $-4A + 2B = 14$. Thus $A = -3$ and $B = 1$, and

$$\mathcal{L}^{-1}\left[\frac{14}{(3s+2)(s-4)}\right] = \mathcal{L}^{-1}\left[\frac{1}{s-4} - \frac{3}{3s+2}\right].$$

Finally,

$$\mathcal{L}^{-1}\left[\frac{5}{(s-1)(s-2)}\right] = e^{4t} - e^{-2t/3}.$$

11. Using the method of partial fractions, we write

$$\frac{4}{s(s+3)} = \frac{A}{s} + \frac{B}{s+3}.$$

Putting the right-hand side over a common denominator gives $A(s+3) + B(s) = 4$, which can be written as $(A+B)s + 3A = 4$. Thus, $A + B = 0$, and $3A = 4$. This gives $A = 4/3$ and $B = -4/3$, so

$$\mathcal{L}^{-1}\left[\frac{4}{s(s+3)}\right] = \mathcal{L}^{-1}\left[\frac{4/3}{s} - \frac{4/3}{s+3}\right].$$

Hence,

$$\mathcal{L}^{-1}\left[\frac{4}{s(s+3)}\right] = \frac{4}{3} - \frac{4}{3}e^{-3t}.$$

12. Using the method of partial fractions, we write

$$\frac{5}{(s-1)(s-2)} = \frac{A}{s-1} + \frac{B}{s-2}.$$

Putting the right-hand side over a common denominator gives $A(s-2) + B(s-1) = 5$, which can be written as $(A+B)s + (-2A-B) = 5$. Thus, $A+B = 0$, and $-2A - B = 5$. This gives $A = -5$ and $B = 5$, so

$$\mathcal{L}^{-1}\left[\frac{5}{(s-1)(s-2)}\right] = \mathcal{L}^{-1}\left[\frac{5}{s-2} - \frac{5}{s-1}\right].$$

Thus,

$$\mathcal{L}^{-1}\left[\frac{5}{(s-1)(s-2)}\right] = 5e^{2t} - 5e^{t}.$$

13. Using the method of partial fractions, we have

$$\frac{2s+1}{(s-1)(s-2)} = \frac{A}{s-1} + \frac{B}{s-2}.$$

Putting the right-hand side over a common denominator gives $A(s-2) + B(s-1) = 2s+1$, which can be written as $(A+B)s + (-2A-B) = 2s+1$. So, $A+B = 2$, and $-2A - B = 1$. Thus $A = -3$ and $B = 5$, which gives

$$\mathcal{L}^{-1}\left[\frac{2s+1}{(s-1)(s-2)}\right] = \mathcal{L}^{-1}\left[\frac{5}{s-2} - \frac{3}{s-1}\right].$$

Finally,

$$\mathcal{L}^{-1}\left[\frac{2s+1}{(s-1)(s-2)}\right] = 5e^{2t} - 3e^{t}.$$

14. Using the method of partial fractions,

$$\frac{2s^2+3s-2}{s(s+1)(s-2)} = \frac{A}{s} + \frac{B}{s+1} + \frac{C}{s-2}.$$

Putting the right-hand side over a common denominator gives

$$A(s+1)(s-2) + Bs(s-2) + Cs(s+1) = 2s^2 + 3s - 2,$$

which can be written as $(A+B+C)s^2 + (-A-2B+C)s - 2A = 2s^2 + 3s - 2$. So, $A+B+C = 2$, $-A - 2B + C = 3$, and $-2A = -2$. Thus $A = 1$, $B = -1$, and $C = 2$, and

$$\mathcal{L}^{-1}\left[\frac{2s^2+3s-2}{s(s+1)(s-2)}\right] = \mathcal{L}^{-1}\left[\frac{2}{s-2} - \frac{1}{s+1} + \frac{1}{s}\right].$$

Hence,

$$\mathcal{L}^{-1}\left[\frac{2s^2+3s-2}{s(s+1)(s-2)}\right] = 2e^{2t} - e^{-t} + 1.$$

15. **(a)** We have

$$\mathcal{L}\left[\frac{dy}{dt}\right] = s\mathcal{L}[y] - y(0)$$

and

$$\mathcal{L}[-y + e^{-2t}] = \mathcal{L}[-y] + \mathcal{L}[e^{-2t}] = -\mathcal{L}[y] + \frac{1}{s+2}$$

using linearity of the Laplace transform and the formula $\mathcal{L}[e^{at}] = 1/(s-a)$ from the text.

(b) Substituting the initial condition yields

$$s\mathcal{L}[y] - 2 = -\mathcal{L}[y] + \frac{1}{s+2}$$

so that

$$(s+1)\mathcal{L}[y] = 2 + \frac{1}{s+2}$$

which gives

$$\mathcal{L}[y] = \frac{1}{(s+1)(s+2)} + \frac{2}{s+1} = \frac{2s+5}{(s+1)(s+2)}.$$

(c) Using the method of partial fractions,

$$\frac{2s+5}{(s+1)(s+2)} = \frac{A}{s+1} + \frac{B}{s+2}.$$

Putting the right-hand side over a common denominator gives $A(s+2) + B(s+1) = 2s+5$, which can be written as $(A+B)s + (2A+B) = 2s+5$. So we have $A+B = 2$, and $2A+B = 5$. Thus, $A = 3$ and $B = -1$, and

$$\mathcal{L}[y] = \frac{3}{s+1} - \frac{1}{s+2}.$$

Therefore, $y(t) = 3e^{-t} - e^{-2t}$ is the desired function.

(d) Since $y(0) = 3e^0 - e^0 = 2$, $y(t)$ satisfies the given initial condition. Also,

$$\frac{dy}{dt} = -3e^{-t} + 2e^{-2t}$$

and

$$-y + e^{-2t} = -3e^{-t} + e^{-2t} + e^{-2t} = -3e^{-t} + 2e^{-2t},$$

so our solution also satisfies the differential equation.

16. **(a)** Taking Laplace transforms of both sides of the equation and simplifying gives

$$\mathcal{L}\left[\frac{dy}{dt}\right] + 5\mathcal{L}[y] = \mathcal{L}[e^{-t}]$$

so

$$s\mathcal{L}[y] - y(0) + 5\mathcal{L}[y] = \frac{1}{s+1}$$

and $y(0) = 2$ gives

$$s\mathcal{L}[y] - 2 + 5\mathcal{L}[y] = \frac{1}{s+1}.$$

(b) Solving for $\mathcal{L}[y]$ gives

$$\mathcal{L}[y] = \frac{2}{s+5} + \frac{1}{(s+5)(s+1)} = \frac{2s+3}{(s+5)(s+1)}.$$

(c) Using the method of partial fractions,

$$\frac{2s+3}{(s+5)(s+1)} = \frac{A}{s+5} + \frac{B}{s+1}.$$

Putting the right-hand side over a common denominator gives $A(s+1) + B(s+5) = 2s+3$, which can be written as $(A+B)s + (A+5B) = 2s+3$. So $A+B=2$, and $A+5B=3$. Hence, $A = 7/4$ and $B = 1/4$ and we have

$$\mathcal{L}[y] = \frac{7/4}{s+5} + \frac{1/4}{s+1}.$$

Therefore,

$$y(t) = \frac{7}{4}e^{-5t} + \frac{1}{4}e^{-t}.$$

(d) To check, compute

$$\frac{dy}{dt} + 5y = -\frac{35}{4}e^{-5t} - \frac{1}{4}e^{-t} + 5\left(\frac{7}{4}e^{-5t} + \frac{1}{4}e^{-t}\right) = e^{-t},$$

and $y(0) = 7/4 + 1/4 = 2$.

17. **(a)** Taking Laplace transforms of both sides of the equation and simplifying gives

$$\mathcal{L}\left[\frac{dy}{dt}\right] + 7\mathcal{L}[y] = \mathcal{L}[1]$$

so

$$s\mathcal{L}[y] - y(0) + 7\mathcal{L}[y] = \frac{1}{s}$$

and $y(0) = 3$ gives

$$s\mathcal{L}[y] - 3 + 7\mathcal{L}[y] = \frac{1}{s}.$$

(b) Solving for $\mathcal{L}[y]$ gives

$$\mathcal{L}[y] = \frac{3}{s+7} + \frac{1}{s(s+7)} = \frac{3s+1}{s(s+7)}.$$

(c) Using the method of partial fractions, we get

$$\frac{3s+1}{s(s+7)} = \frac{A}{s} + \frac{B}{s+7}.$$

Putting the right-hand side over a common denominator gives $A(s+7) + Bs = 3s+1$, which can be written as $(A+B)s + 7A = 3s+1$. So $A+B=3$, and $7A=1$. Hence, $A = 1/7$ and $B = 20/7$, and we have

$$\mathcal{L}[y] = \frac{1/7}{s} + \frac{20/7}{s+7}.$$

Thus,

$$y(t) = \frac{20}{7}e^{-7t} + \frac{1}{7}.$$

(d) To check, we compute

$$\frac{dy}{dt} + 7y = -20e^{-7t} + 7\left(\frac{20}{7}e^{-7t} + \frac{1}{7}\right) = 1,$$

and $y(0) = 20/7 + 1/7 = 3$, so our solution satisfies the initial-value problem.

18. **(a)** Taking Laplace transforms of both sides of the equation and simplifying gives

$$\mathcal{L}\left[\frac{dy}{dt}\right] + 4\mathcal{L}[y] = \mathcal{L}[6]$$

so

$$s\mathcal{L}[y] - y(0) + 4\mathcal{L}[y] = \frac{6}{s}$$

and $y(0) = 0$ gives

$$s\mathcal{L}[y] + 4\mathcal{L}[y] = \frac{6}{s}.$$

(b) Solving for $\mathcal{L}[y]$ gives

$$\mathcal{L}[y] = \frac{6}{s(s+4)}.$$

(c) Using the method of partial fractions,

$$\frac{6}{s(s+4)} = \frac{A}{s} + \frac{B}{s+4}.$$

Putting the right-hand side over a common denominator gives $A(s+4) + Bs = 6$, which can be written as $(A + B)s + 4A = 6$. So, $A + B = 0$, and $4A = 6$. Hence, $A = 3/2$ and $B = -3/2$, and we have

$$\mathcal{L}[y] = \frac{3/2}{s} - \frac{3/2}{s+4}.$$

Thus,

$$y(t) = \frac{3}{2} - \frac{3}{2}e^{-4t}.$$

(d) To check, we compute

$$\frac{dy}{dt} + 4y = 6e^{-4t} + 4\left(\frac{3}{2} - \frac{3}{2}e^{-4t}\right) = 6,$$

and $y(0) = 3/2 - 3/2 = 0$, so our solution satisfies the initial-value problem.

19. **(a)** Taking Laplace transforms of both sides of the equation and simplifying gives

$$\mathcal{L}\left[\frac{dy}{dt}\right] + 9\mathcal{L}[y] = \mathcal{L}[2]$$

so

$$s\mathcal{L}[y] - y(0) + 9\mathcal{L}[y] = \frac{2}{s}$$

and $y(0) = -2$ gives

$$s\mathcal{L}[y] + 2 + 9\mathcal{L}[y] = \frac{2}{s}.$$

(b) Solving for $\mathcal{L}[y]$ gives

$$\mathcal{L}[y] = -\frac{2}{s+9} + \frac{2}{s(s+9)} = \frac{-2s+2}{s(s+9)}.$$

(c) Using the method of partial fractions,

$$\frac{-2s+2}{s(s+9)} = \frac{A}{s} + \frac{B}{s+9}.$$

Putting the right-hand side over a common denominator gives $A(s+9)+Bs = -2s+2$, which can be written as $(A+B)s+9A = -2s+2$. So $A+B = -2$ and $9A = 2$. Hence, $A = 2/9$ and $B = -20/9$, which gives us

$$\mathcal{L}[y] = \frac{2/9}{s} - \frac{20/9}{s+9}.$$

Finally,

$$y(t) = -\frac{20}{9}e^{-9t} + \frac{2}{9}.$$

(d) To check, we compute

$$\frac{dy}{dt} + 9y = 20e^{-9t} + 9\left(\frac{-20}{9}e^{-9t} + \frac{2}{9}\right) = 2,$$

and $y(0) = -20/9 + 2/9 = -2$, so our solution satisfies the initial-value problem.

20. **(a)** First we put the equation in the form

$$\frac{dy}{dt} + y = 2.$$

Taking Laplace transforms of both sides of the equation and simplifying gives

$$\mathcal{L}\left[\frac{dy}{dt}\right] + \mathcal{L}[y] = \mathcal{L}[2]$$

so

$$s\mathcal{L}[y] - y(0) + \mathcal{L}[y] = \frac{2}{s}$$

and $y(0) = 4$ gives

$$s\mathcal{L}[y] - 4 + \mathcal{L}[y] = \frac{2}{s}.$$

(b) Solving for $\mathcal{L}[y]$ gives

$$\mathcal{L}[y] = \frac{4}{s+1} + \frac{2}{s(s+1)} = \frac{4s+2}{s(s+1)}.$$

(c) Using the method of partial fractions, we write

$$\frac{4s+2}{s(s+1)} = \frac{A}{s} + \frac{B}{s+1}.$$

Putting the right-hand side over a common denominator gives $A(s+1) + Bs = 4s + 2$, which can be written as $(A + B)s + A = 4s + 2$. Thus, $A + B = 4$, and $A = 2$. Hence $B = 2$ and

$$\mathcal{L}[y] = \frac{2}{s+1} + \frac{2}{s}.$$

So,

$$y(t) = 2e^{-t} + 2.$$

(d) To check, we compute

$$\frac{dy}{dt} + y = -2e^{-t} + \left(2e^{-t} + 2\right) = 2,$$

and $y(0) = 2 + 2 = 4$, so our solution satisfies the initial-value problem.

21. **(a)** Putting the equation in the form

$$\frac{dy}{dt} + y = e^{-2t},$$

taking Laplace transforms of both sides of the equation and simplifying gives

$$\mathcal{L}\left[\frac{dy}{dt}\right] + \mathcal{L}[y] = \mathcal{L}[e^{-2t}]$$

so

$$s\mathcal{L}[y] - y(0) + \mathcal{L}[y] = \frac{1}{s+2}$$

and $y(0) = 1$ gives

$$s\mathcal{L}[y] - 1 + \mathcal{L}[y] = \frac{1}{s+2}.$$

(b) Solving for $\mathcal{L}[y]$ gives

$$\mathcal{L}[y] = \frac{1}{s+1} + \frac{1}{(s+1)(s+2)} = \frac{s+3}{(s+1)(s+2)}.$$

(c) Using partial fractions,

$$\frac{s+3}{(s+1)(s+2)} = \frac{A}{s+1} + \frac{B}{s+2}.$$

Putting the right-hand side over a common denominator gives $A(s+2) + B(s+1) = s + 3$, which can be written as $(A + B)s + (2A + B) = s + 3$. Thus, $A + B = 1$, and $2A + B = 3$. So $A = 2$ and $B = -1$, which gives us

$$\mathcal{L}[y] = \frac{2}{s+1} - \frac{1}{s+2}.$$

Hence,

$$y(t) = 2e^{-t} - e^{-2t}.$$

(d) To check, we compute

$$\frac{dy}{dt} + y = -2e^{-t} + 2e^{-2t} + \left(2e^{-t} - e^{-2t}\right) = e^{-2t},$$

and $y(0) = 2 - 1 = 1$, so our solution satisfies the initial-value problem.

22. **(a)** Putting the equation in the form

$$\frac{dy}{dt} - 2y = t,$$

taking Laplace transforms of both sides of the equation and simplifying gives

$$\mathcal{L}\left[\frac{dy}{dt}\right] - 2\mathcal{L}[y] = \mathcal{L}[t].$$

Using the formulas

$$\mathcal{L}\left[\frac{dy}{dt}\right] = s\mathcal{L}[y] - y(0),$$

and

$$\mathcal{L}[t^n] = n!/s^{n+1},$$

we have

$$s\mathcal{L}[y] - y(0) - 2\mathcal{L}[y] = \frac{1}{s^2}.$$

The initial condition $y(0) = 0$ gives

$$s\mathcal{L}[y] - 2\mathcal{L}[y] = \frac{1}{s^2}.$$

(b) Solving for $\mathcal{L}[y]$ gives

$$\mathcal{L}[y] = \frac{1}{s^2(s-2)}.$$

(c) Using partial fractions, we seek constants A, B, and C so that

$$\frac{1}{s^2(s-2)} = \frac{A}{s} + \frac{B}{s^2} + \frac{C}{s-2}.$$

Putting the right-hand side over a common denominator gives $As(s-2) + B(s-2) + Cs^2 = 1$, which can be written as $(A+C)s^2 + (-2A+B)s - 2B = 1$. This gives us $A + C = 0$, $-2A + B = 0$, and $-2B = 1$. Hence, $A = -1/4$, $B = -1/2$, and $C = 1/4$, and we get

$$\mathcal{L}[y] = \frac{1/4}{s-2} - \frac{1/2}{s^2} - \frac{1/4}{s}.$$

So,

$$y(t) = \frac{1}{4}e^{2t} - \frac{t}{2} - \frac{1}{4}.$$

(d) To check, we compute

$$\frac{dy}{dt} - 2y = \frac{1}{2}e^{2t} - \frac{1}{2} - 2\left(\frac{1}{4}e^{2t} - \frac{t}{2} - \frac{1}{4}\right) = t,$$

and $y(0) = 1/4 - 1/4 = 0$, so our solution satisfies the initial-value problem.

23. **(a)** We have

$$\mathcal{L}\left[\frac{dy}{dt}\right] = s\mathcal{L}[y] - y(0)$$

and

$$\mathcal{L}[-y + t^2] = \mathcal{L}[-y] + \mathcal{L}[t^2] = -\mathcal{L}[y] + \frac{2}{s^3}$$

using linearity of the Laplace transform and the formula $\mathcal{L}[t^n] = n!/s^{n+1}$ from Exercise 5.

(b) Substituting the initial condition yields

$$s\mathcal{L}[y] - 1 = -\mathcal{L}[y] + \frac{2}{s^3}$$

so that

$$\mathcal{L}[y] = \frac{2/s^3 + 1}{1 + s} = \frac{2 + s^3}{s^3(s+1)}.$$

(c) The best way to deal with this problem is with partial fractions. We seek constants A, B, C, and D such that

$$\frac{A}{s} + \frac{B}{s^2} + \frac{C}{s^3} + \frac{D}{s+1} = \frac{2 + s^3}{s^3(s+1)}.$$

Multiplying through by $s^3(s+1)$ and equating like terms yields the system of equations

$$\begin{cases} A + D = 1 \\ A + B = 0 \\ B + C = 0 \\ C = 2. \end{cases}$$

Solving simultaneously gives us $A = 2$, $B = -2$, $C = 2$, and $D = -1$. Therefore we seek a function $y(t)$ whose Laplace transform is

$$\frac{2}{s} - \frac{2}{s^2} + \frac{2}{s^3} - \frac{1}{s+1}.$$

We have $\mathcal{L}[e^{-t}] = 1/(s+1)$ so that

$$\mathcal{L}[-e^{-t}] = -\mathcal{L}[e^{-t}] = -\frac{1}{s+1}.$$

Also, using the formula from Exercise 5, we have

$$\mathcal{L}[t^2] = \frac{2}{s^3}, \quad \mathcal{L}[t] = \frac{1}{s^2}, \quad \text{and} \quad \mathcal{L}[1] = \frac{1}{s}$$

so that

$$\mathcal{L}[t^2 - 2t + 2] = \mathcal{L}[t^2] - 2\mathcal{L}[t] + 2\mathcal{L}[1] = \frac{2}{s^3} - \frac{2}{s^2} + \frac{2}{s}.$$

Therefore, $y(t) = t^2 - 2t + 2 - e^{-t}$ is the desired function.

(d) Since $y(0) = 2 - e^0 = 1$, $y(t)$ satisfies the given initial condition. Also,

$$\frac{dy}{dt} = 2t - 2 + e^{-t}$$

and

$$-y + t^2 = -t^2 + 2t - 2 + e^{-t} + t^2 = 2t - 2 + e^{-t}$$

so our solution also satisfies the differential equation.

24. **(a)** Taking Laplace transforms of both sides of the equation and simplifying gives

$$\mathcal{L}\left[\frac{dy}{dt}\right] + 4\mathcal{L}[y] = \mathcal{L}[2] + \mathcal{L}[3t].$$

Using the formulas

$$\mathcal{L}\left[\frac{dy}{dt}\right] = s\mathcal{L}[y] - y(0),$$

and

$$\mathcal{L}[t^n] = n!/s^{n+1},$$

we have

$$s\mathcal{L}[y] - y(0) + 4\mathcal{L}[y] = \frac{2}{s} + \frac{3}{s^2}.$$

The initial condition $y(0) = 1$ gives

$$s\mathcal{L}[y] - 1 + 4\mathcal{L}[y] = \frac{2}{s} + \frac{3}{s^2}.$$

(b) Solving for $\mathcal{L}[y]$ gives

$$\mathcal{L}[y] = \frac{1}{s+4} + \frac{2}{s(s+4)} + \frac{3}{s^2(s+4)} = \frac{s^2 + 2s + 3}{s^2(s+4)}.$$

(c) Using the method of partial fractions, we have

$$\frac{s^2 + 2s + 3}{s^2(s+4)} = \frac{A}{s} + \frac{B}{s^2} + \frac{C}{s+4}.$$

Putting the right-hand side over a common denominator gives

$$As(s+4) + B(s+4) + Cs^2 = s^2 + 2s + 3,$$

which can be written as $(A+C)s^2 + (4A+B)s + 4B = s^2 + 2s + 3$. So, $A + C = 1$, $4A + B = 2$, and $4B = 3$. Thus $A = 5/16$, $B = 3/4$, and $C = 11/16$, and we get

$$\mathcal{L}[y] = \frac{11/16}{s+4} + \frac{3/4}{s^2} + \frac{5/16}{s}.$$

Therefore,

$$y(t) = \frac{11}{16}e^{-4t} + \frac{3t}{4} + \frac{5}{16}.$$

(d) To check, we compute

$$\frac{dy}{dt} + 4y = -\frac{11}{4}e^{-4t} + \frac{3}{4} + 4\left(\frac{11}{16}e^{-4t} + \frac{3t}{4} + \frac{5}{16}\right) = 3t + 2,$$

and $y(0) = 11/16 + 5/16 = 1$, so our solution satisfies the initial-value problem.

25. First take Laplace transforms of both sides of the equation

$$\mathcal{L}\left[\frac{dy}{dt}\right] = 2\mathcal{L}[y] + 2\mathcal{L}[e^{-3t}]$$

and use the rules to simplify, obtaining

$$s\mathcal{L}[y] - y(0) = 2\mathcal{L}[y] + \frac{2}{s+3}$$

$$(s-2)\mathcal{L}[y] = y(0) + \frac{2}{s+3}$$

$$\mathcal{L}[y] = \frac{y(0)}{s-2} + \frac{2}{(s-2)(s+3)}.$$

Next note that

$$\mathcal{L}[y(0)e^{2t}] = y(0)/(s-2).$$

For the other summand, first simplify using partial fractions,

$$\frac{2}{(s-2)(s+3)} = \frac{A}{s-2} + \frac{B}{s+3}.$$

Putting the right-hand side over a common denominator gives $A(s+3) + B(s-2) = 2$, which can be written as $(A+B)s + (3A-2B) = 2$. This yields $A+B=0$ and $3A-2B=2$. Hence $B=-2/5$ and $A=2/5$, and

$$\frac{2}{(s-2)(s+3)} = \frac{2/5}{s-2} - \frac{2/5}{s+3}.$$

Now, $\mathcal{L}[e^{2t}] = 1/(s-2)$ and $\mathcal{L}[e^{-3t}] = 1/(s+3)$ so

$$\mathcal{L}[y] = \frac{y(0)}{s-2} + \frac{2}{5}\frac{1}{s-2} - \frac{2}{5}\frac{1}{s+3}.$$

Hence,

$$y(t) = y(0)e^{2t} + \frac{2}{5}e^{2t} - \frac{2}{5}e^{-3t}.$$

The first two terms can be combined into one, giving

$$y(t) = ce^{2t} - \frac{2}{5}e^{-3t},$$

where $c = y(0) + 2/5$.

26. We know that

$$\frac{dg}{dt} = f.$$

So

$$\mathcal{L}[f] = \mathcal{L}\left[\frac{dg}{dt}\right] = s\mathcal{L}[g] - g(0).$$

Hence,

$$\mathcal{L}[g] = \frac{\mathcal{L}[f] + g(0)}{s}.$$

27. As always the first step must be to take Laplace transform of both sides of the differential equation, giving

$$\mathcal{L}\left[\frac{dy}{dt}\right] = \mathcal{L}[y^2].$$

Simplifying, we obtain

$$s\mathcal{L}[y] - 1 = \mathcal{L}[y^2].$$

To solve for $\mathcal{L}[y]$ we must come up with an expression for $\mathcal{L}[y^2]$ in terms of $\mathcal{L}[y]$. This is not so easy! In particular, there is no easy way to simplify

$$\mathcal{L}[y^2] = \int_0^\infty y^2 e^{-st}\, dt$$

since we do not have a rule for the Laplace transform of a product.

EXERCISES FOR SECTION 6.2

1. **(a)** The function $g_a(t) = 1$ precisely when $u_a(t) = 0$, and $g_a(t) = 0$ precisely when $u_a(t) = 1$, so

$$g_a(t) = 1 - u_a(t).$$

(b) We can compute the Laplace transform of $g_a(t)$ from the definition

$$\mathcal{L}[g_a] = \int_0^a 1e^{-st}\, dt = -\frac{e^{-as}}{s} + \frac{e^{-0s}}{s} = \frac{1}{s} - \frac{e^{-as}}{s}.$$

Alternately, we can use the table

$$\mathcal{L}[g_a] = \mathcal{L}[1 - u_a(t)] = \frac{1}{s} - \frac{e^{-as}}{s}.$$

2. **(a)** We have $r_a(t) = u_a(t)y(t-a)$, where $y(t) = kt$. Now

$$\mathcal{L}[y(t)] = k\mathcal{L}[t] = \frac{k}{s^2},$$

so using the rules of Laplace transform,

$$\mathcal{L}[r_a(t)] = \mathcal{L}[u_a(t)y(t-a)] = \frac{k}{s^2}e^{-as}.$$

(b)

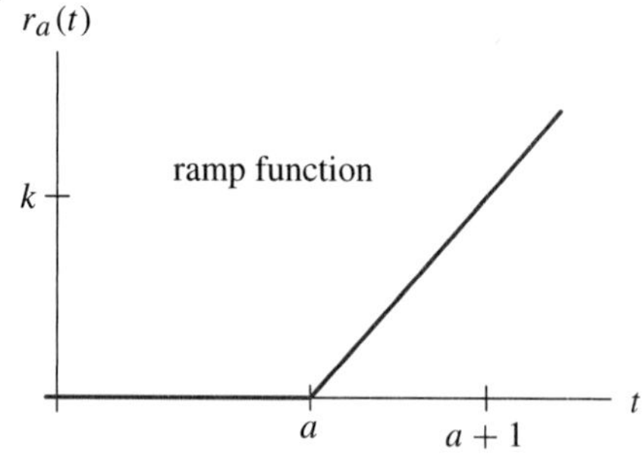

3.

$$\begin{aligned}\mathcal{L}[g_a(t)] &= \int_0^\infty g_a(t)\, e^{-st}\, dt \\ &= \int_0^a \frac{t}{a} e^{-st}\, dt + \int_a^\infty e^{-st}\, dt\end{aligned}$$

Using integration by parts with $u = t$ and $dv = e^{-st}dt$, we have $du = dt$, $v = -e^{-st}/s$ and

$$\begin{aligned}\int_0^a \frac{t}{a} e^{-st}\, dt &= \frac{1}{a}\int_0^a te^{-st}\, dt \\ &= \frac{1}{a}\left(-\frac{te^{-st}}{s}\bigg|_0^a - \int_0^a -\frac{e^{-st}}{s}\, dt\right) \\ &= \frac{1}{a}\left(-\frac{ae^{-as}}{s} - \frac{1}{s^2}e^{-st}\bigg|_0^a\right) \\ &= \frac{1}{a}\left(-\frac{ae^{-as}}{s} - \frac{1}{s^2}(e^{-as} - 1)\right) \\ &= -\frac{e^{-as}}{s} - \frac{1}{as^2}\left(e^{-as} - 1\right).\end{aligned}$$

Also,

$$\begin{aligned}\int_a^\infty e^{-st}\, dt &= \lim_{b\to\infty}\int_a^b e^{-st}\, dt \\ &= \lim_{b\to\infty} -\frac{1}{s}e^{-st}\bigg|_a^b \\ &= \lim_{b\to\infty} -\frac{1}{s}\left(e^{-sb} - e^{-as}\right) \\ &= \frac{1}{s}e^{-as}.\end{aligned}$$

Therefore,

$$\begin{aligned}\mathcal{L}[g_a(t)] &= -\frac{e^{-as}}{s} - \frac{1}{as^2}\left(e^{-as} - 1\right) + \frac{1}{s}e^{-as} \\ &= \frac{1}{as^2}\left(1 - e^{-as}\right).\end{aligned}$$

4. We have

$$\mathcal{L}[e^{3t}] = \frac{1}{s-3},$$

so using the rule

$$\mathcal{L}[u_a(t)y(t-a)] = e^{-as}\mathcal{L}[y(t)],$$

we determine that

$$\mathcal{L}[u_2(t)e^{3(t-2)}] = \frac{e^{-2s}}{s-3}.$$

The desired function is $u_2(t)e^{3(t-2)}$.

5. First use partial fractions to write

$$\frac{1}{(s-1)(s-2)} = \frac{A}{s-1} + \frac{B}{s-2}.$$

Putting the right-hand side over a common denominator yields $As - 2A + Bs - B = 1$ which can be written as $(A+B)s + (-2A - B) = 1$. Thus, $A + B = 0$, and $-2A - B = 1$. Solving for A and B yields $A = -1$ and $B = 1$, so

$$\frac{1}{(s-1)(s-2)} = \frac{-1}{s-1} + \frac{1}{s-2}.$$

Now, as above

$$\mathcal{L}[u_3(t)e^{2(t-3)}] = \frac{e^{-3s}}{s-2}$$

and

$$\mathcal{L}[u_3(t)e^{t-3}] = \frac{e^{-3s}}{s-1}$$

and the desired function is

$$u_3(t)\left(e^{2(t-3)} - e^{(t-3)}\right).$$

6. Using partial fractions, we write

$$\frac{4}{s(s+3)} = \frac{A}{s} + \frac{B}{s+3}.$$

Hence, we must have $As + 3A + Bs = 4$ which can be written as $(A+B)s + 3A = 4$. So, $A + B = 0$, and $3A = 4$. This gives us $A = 4/3$ and $B = -4/3$, so

$$\frac{4}{s(s+3)} = \frac{4/3}{s} - \frac{4/3}{s+3}.$$

Applying the rules

$$\mathcal{L}[u_2(t)] = \frac{e^{-2s}}{s}$$

and

$$\mathcal{L}[u_2(t)e^{-3(t-2)}] = \frac{e^{-2s}}{s+3},$$

the desired function is

$$y(t) = u_2(t)\left(\frac{4}{3} - \frac{4e^{-3(t-2)}}{3}\right)$$

or

$$y(t) = \frac{4}{3}u_2(t)\left(1 - e^{-3(t-2)}\right).$$

7. Using partial fractions, we get

$$\frac{14}{(3s+2)(s-4)} = \frac{A}{3s+2} + \frac{B}{s-4}.$$

Hence, we must have $As - 4A + 3Bs + 2B = 14$, which can be written as

$$(A + 3B)s + (-4A + 2B) = 14.$$

Therefore, $A + 3B = 0$, and $-4A + 2B = 14$. Solving for A and B yields $A = -3$ and $B = 1$, so

$$\frac{14}{(3s+2)(s-4)} = \frac{1}{s-4} - \frac{3}{3s+2} = \frac{1}{s-4} - \frac{1}{s+2/3}.$$

Applying the rules

$$\mathcal{L}[u_1(t)e^{4(t-1)}] = \frac{e^{-s}}{s-4}$$

and

$$\mathcal{L}[u_1(t)e^{-\frac{2}{3}(t-1)}] = \frac{e^{-s}}{s+2/3},$$

the desired function is

$$y(t) = u_1(t)\left(e^{4(t-1)} - e^{-\frac{2}{3}(t-1)}\right).$$

8. Taking the Laplace transform of both sides of the equation gives us

$$\mathcal{L}\left[\frac{dy}{dt}\right] = \mathcal{L}[u_2(t)],$$

so

$$s\mathcal{L}[y] - y(0) = \frac{e^{-2s}}{s}.$$

Substituting the initial condition yields

$$s\mathcal{L}[y] - 3 = \frac{e^{-2s}}{s},$$

so that

$$\mathcal{L}[y] = \frac{e^{-2s}}{s^2} + \frac{3}{s}.$$

By taking the inverse of the Laplace transform, we get

$$y(t) = u_2(t)(t-2) + 3.$$

To check our answer, we compute

$$\frac{dy}{dt} = \frac{du_2}{dt}(t-2) + u_2(t),$$

and since $du_2/dt = 0$ except at $t = 2$ (where it is undefined),

$$\frac{dy}{dt} = u_2(t).$$

Hence, our $y(t)$ satisfies the differential equation except when $t = 2$. (We cannot expect $y(t)$ to satisfy the differential equation at $t = 2$ because the differential equation is not continuous there.) Note that $y(t)$ also satisfies the initial condition $y(0) = 3$.

9. Taking the Laplace transform of both sides of the equation, we have

$$\mathcal{L}\left[\frac{dy}{dt}\right] + 9\mathcal{L}[y] = \mathcal{L}[u_5(t)],$$

which is equivalent to

$$s\mathcal{L}[y] - y(0) + 9\mathcal{L}[y] = \frac{e^{-5s}}{s}.$$

Since $y(0) = -2$, we have

$$s\mathcal{L}[y] + 2 + 9\mathcal{L}[y] = \frac{e^{-5s}}{s},$$

which yields

$$\mathcal{L}[y] = \frac{-2}{s+9} + \frac{e^{-5s}}{s(s+9)}.$$

Using the partial fractions decomposition

$$\frac{1}{s(s+9)} = \frac{1/9}{s} - \frac{1/9}{s+9},$$

we see that

$$\mathcal{L}[y] = \frac{-2}{s+9} + \frac{1}{9}\left(\frac{e^{-5s}}{s}\right) - \frac{1}{9}\left(\frac{e^{-5s}}{s+9}\right).$$

Taking the inverse of the Laplace transform, we obtain

$$\begin{aligned} y(t) &= -2e^{-9t} + \frac{1}{9}u_5(t) - \frac{1}{9}u_5(t)e^{-9(t-5)} \\ &= -2e^{-9t} + \frac{1}{9}u_5(t)\left(1 - e^{-9(t-5)}\right). \end{aligned}$$

To check our answer, we compute

$$\frac{dy}{dt} = 18e^{-9t} + \frac{1}{9}\frac{du_5}{dt}\left(1 - e^{-9(t-5)}\right) + \frac{1}{9}u_5(t)\left(9e^{-9(t-5)}\right),$$

and since $du_5/dt = 0$ except at $t = 5$ (where it is undefined),

$$\begin{aligned} \frac{dy}{dt} + 9y &= 18e^{-9t} + u_5(t)e^{-9(t-5)} + 9\left(-2e^{-9t} + \frac{1}{9}u_5(t)\left(1 - e^{-9(t-5)}\right)\right) \\ &= u_5(t). \end{aligned}$$

Hence, our $y(t)$ satisfies the differential equation except when $t = 5$. (We cannot expect $y(t)$ to satisfy the differential equation at $t = 5$ because the differential equation is not continuous there.) Note that $y(t)$ also satisfies the initial condition $y(0) = -2$.

10. Taking the Laplace transform of both sides of the equation, we have

$$\mathcal{L}\left[\frac{dy}{dt}\right] + \mathcal{L}[7y] = \mathcal{L}[u_2(t)],$$

which is equivalent to

$$s\mathcal{L}[y] - y(0) + 7\mathcal{L}[y] = \frac{e^{-2s}}{s},$$

using linearity of the Laplace transform and the formula $\mathcal{L}[u_a(t)] = e^{-as}/s$ from the text. Substituting the initial condition yields

$$s\mathcal{L}[y] - 3 + 7\mathcal{L}[y] = \frac{e^{-2s}}{s},$$

which gives us

$$\mathcal{L}[y] = \frac{3}{s+7} + \frac{e^{-2s}}{s(s+7)}.$$

We have $\mathcal{L}[e^{-7t}] = 1/(s+7)$ so that

$$\mathcal{L}[3e^{-7t}] = 3\mathcal{L}[e^{-7t}] = \frac{3}{s+7}.$$

Also, using partial fractions, we see that

$$\frac{1/7}{s} - \frac{1/7}{s+7} = \frac{1}{s(s+7)},$$

so

$$\mathcal{L}[y] = \frac{3}{s+7} + \frac{1}{7}\left(\frac{e^{-2s}}{s}\right) - \frac{1}{7}\left(\frac{e^{-2s}}{s+7}\right).$$

We know that $\mathcal{L}[u_2(t)] = e^{-2s}/s$. To find the function whose Laplace transform is $e^{-2s}/(s+7)$ we use the rule

If $\mathcal{L}[f] = F(s)$ then $\mathcal{L}[u_a(t)f(t-a)] = e^{-as}F(s)$

from the text. Since we have $F(s) = 1/(s+7)$ and we know that $\mathcal{L}[e^{-7t}] = 1/(s+7)$, we let $f(t) = e^{-7t}$. Then according to the formula,

$$\mathcal{L}[u_2(t)e^{-7(t-2)}] = \mathcal{L}[u_2(t)f(t-2)] = \frac{e^{-2s}}{s+7}.$$

Therefore,

$$\begin{aligned} y(t) &= 3e^{-7t} + \frac{1}{7}u_2(t) - \frac{1}{7}u_2(t)e^{-7(t-2)} \\ &= 3e^{-7t} + \frac{1}{7}u_2(t)\left(1 - e^{-7(t-2)}\right) \end{aligned}$$

is the desired function.

To check our answer, we compute

$$\frac{dy}{dt} = -21e^{-7t} + \frac{1}{7}\frac{du_2}{dt}\left(1 - e^{-7(t-2)}\right) + \frac{1}{7}u_2(t)\left(7e^{-7(t-2)}\right),$$

and since $du_2/dt = 0$ except at $t = 2$ (where it is undefined),

$$\begin{aligned} \frac{dy}{dt} + 7y &= -21e^{-7t} + u_2(t)e^{-7(t-2)} + 7\left(3e^{-7t} + \frac{1}{7}u_2(t)\left(1 - e^{-7(t-2)}\right)\right) \\ &= u_2(t). \end{aligned}$$

Hence, our $y(t)$ satisfies the differential equation except when $t = 2$. (We cannot expect $y(t)$ to satisfy the differential equation at $t = 2$ because the differential equation is not continuous there.) Note that $y(t)$ also satisfies the initial condition $y(0) = 3$.

11. Taking the Laplace transform of both sides of the equation, we obtain

$$\mathcal{L}\left[\frac{dy}{dt}\right] = \mathcal{L}[-y] + \mathcal{L}[u_2(t)e^{-2(t-2)}],$$

which is equivalent to

$$s\mathcal{L}[y] - y(0) = -\mathcal{L}[y] + \frac{e^{-2s}}{s+2}$$

(using linearity of the Laplace transform and the formula

$$\text{If } \mathcal{L}[f] = F(s) \text{ then } \mathcal{L}[u_a(t)f(t-a)] = e^{-as}F(s)$$

where $f(t) = e^{-2t}$ and $a = 2$.)
Substituting the initial condition yields

$$s\mathcal{L}[y] - 1 = -\mathcal{L}[y] + \frac{e^{-2s}}{s+2}$$

so that

$$\mathcal{L}[y] = \frac{1}{s+1} + \frac{e^{-2s}}{(s+1)(s+2)}.$$

By partial fractions, we know that

$$\frac{1}{s+1} - \frac{1}{s+2} = \frac{1}{(s+1)(s+2)},$$

so we have

$$\frac{e^{-2s}}{(s+1)(s+2)} = e^{-2s}\left(\frac{1}{s+1} - \frac{1}{s+2}\right) = \frac{e^{-2s}}{s+1} - \frac{e^{-2s}}{s+2}.$$

Taking the inverse of the Laplace transform yields

$$\begin{aligned} y(t) &= e^{-t} + u_2(t)e^{-(t-2)} - u_2(t)e^{-2(t-2)} \\ &= e^{-t} + u_2(t)\left(e^{-(t-2)} - e^{-2(t-2)}\right). \end{aligned}$$

To check our answer, we compute

$$\frac{dy}{dt} = -e^{-t} + \frac{du_2}{dt}\left(e^{-(t-2)} - e^{-2(t-2)}\right) + u_2(t)\left(-e^{-(t-2)} + 2e^{-2(t-2)}\right),$$

and since $du_2/dt = 0$ except at $t = 2$ (where it is undefined),

$$\begin{aligned} \frac{dy}{dt} + y &= -e^{-t} + u_2(t)\left(-e^{-(t-2)} + 2e^{-2(t-2)}\right) + e^{-t} + u_2(t)\left(e^{-(t-2)} - e^{-2(t-2)}\right) \\ &= u_2(t)e^{-2(t-2)}. \end{aligned}$$

Hence, our $y(t)$ satisfies the differential equation except when $t = 2$. (We cannot expect $y(t)$ to satisfy the differential equation at $t = 2$ because the differential equation is not continuous there.) Note that $y(t)$ also satisfies the initial condition $y(0) = 1$.

12. Taking the Laplace transform of both sides yields

$$\mathcal{L}\left[\frac{dy}{dt}\right] = -\mathcal{L}[y] + 2\mathcal{L}[u_3(t)],$$

which is equivalent to

$$s\mathcal{L}[y] - y(0) = -\mathcal{L}[y] + 2\frac{e^{-3s}}{s}.$$

Using the initial condition $y(0) = 4$ gives

$$s\mathcal{L}[y] - 4 = -\mathcal{L}[y] + 2\frac{e^{-3s}}{s}.$$

Solving for $\mathcal{L}[y]$, we get

$$\mathcal{L}[y] = \frac{4}{s+1} + \frac{2e^{-3s}}{s(s+1)}.$$

Using partial fractions, we see that

$$\frac{2}{s(s+1)} = \frac{2}{s} - \frac{2}{s+1},$$

so

$$\mathcal{L}[y] = \frac{4}{s+1} + \frac{2e^{-3s}}{s} - \frac{2e^{-3s}}{s+1}.$$

Taking the inverse of the Laplace transform, we obtain

$$\begin{aligned} y(t) &= 4e^{-t} + 2u_3(t) - 2u_3(t)e^{-(t-3)} \\ &= 4e^{-t} + 2u_3(t)\left(1 - e^{-(t-3)}\right). \end{aligned}$$

To check our answer, we compute

$$\frac{dy}{dt} = -4e^{-t} + 2\frac{du_3}{dt}\left(1 - e^{-(t-3)}\right) + 2u_3(t)\left(e^{-(t-3)}\right),$$

and since $du_3/dt = 0$ except at $t = 3$ (where it is undefined),

$$\begin{aligned} \frac{dy}{dt} + y &= -4e^{-t} + 2u_3(t)\left(e^{-(t-3)}\right) + \left(4e^{-t} + 2u_3(t)\left(1 - e^{-(t-3)}\right)\right) \\ &= 2u_3(t). \end{aligned}$$

Hence, our $y(t)$ satisfies the differential equation except when $t = 3$. (We cannot expect $y(t)$ to satisfy the differential equation at $t = 3$ because the differential equation is not continuous there.) Note that $y(t)$ also satisfies the initial condition $y(0) = 4$.

13. Taking the Laplace transform of both sides of the equation, we obtain

$$\mathcal{L}\left[\frac{dy}{dt}\right] = -\mathcal{L}[y] + \mathcal{L}[u_1(t)(t-1)],$$

which is equivalent to

$$s\mathcal{L}[y] - y(0) = -\mathcal{L}[y] + \frac{e^{-s}}{s^2}.$$

Substituting the initial condition yields

$$s\mathcal{L}[y] - 2 = -\mathcal{L}[y] + \frac{e^{-s}}{s^2}$$

so that

$$\mathcal{L}[y] = \frac{e^{-s}}{s^2(s+1)} + \frac{2}{s+1}.$$

Using the technique of partial fractions, we write

$$\frac{1}{s^2(s+1)} = \frac{A}{s} + \frac{B}{s^2} + \frac{C}{s+1}.$$

Putting the right-hand side over a common denominator gives us $As(s+1) + B(s+1) + Cs^2 = 1$ which can be written as $(A+C)s^2 + (A+B)s + B = 1$. So $A + C = 0$, $A + B = 0$, and $B = 1$. Thus $A = -1$ and $C = 1$, and

$$\frac{1}{s^2(s+1)} = \frac{-1}{s} + \frac{1}{s^2} + \frac{1}{s+1}.$$

Taking the inverse of the Laplace transform gives us

$$\begin{aligned}
y(t) &= \mathcal{L}^{-1}\left[\frac{e^{-s}}{s^2(s+1)}\right] + \mathcal{L}^{-1}\left[\frac{2}{s+1}\right] \\
&= -\mathcal{L}^{-1}\left[\frac{e^{-s}}{s}\right] + \mathcal{L}^{-1}\left[\frac{e^{-s}}{s^2}\right] + \mathcal{L}^{-1}\left[\frac{e^{-s}}{s+1}\right] + \mathcal{L}^{-1}\left[\frac{2}{s+1}\right] \\
&= -u_1(t) + u_1(t)\,(t-1) + u_1(t)e^{-(t-1)} + 2e^{-t} \\
&= u_1(t)\left((t-2) + e^{-(t-1)}\right) + 2e^{-t}.
\end{aligned}$$

To check our answer, we compute

$$\frac{dy}{dt} = \frac{du_1}{dt}\left((t-2) + e^{-(t-1)}\right) + u_1(t)\left(1 - e^{-(t-1)}\right) - 2e^{-t},$$

and since $du_1/dt = 0$ except at $t = 1$ (where it is undefined),

$$\begin{aligned}
\frac{dy}{dt} + y &= u_1(t)\left(1 - e^{-(t-1)}\right) - 2e^{-t} + u_1(t)\left((t-2) + e^{-(t-1)}\right) + 2e^{-t} \\
&= u_1(t) + u_1(t)\,(t-2) \\
&= u_1(t)\,(t-1)\,.
\end{aligned}$$

Hence, our $y(t)$ satisfies the differential equation except when $t = 1$. (We cannot expect $y(t)$ to satisfy the differential equation at $t = 1$ because the differential equation is not continuous there.) Note that $y(t)$ also satisfies the initial condition $y(0) = 2$.

14. First, rewrite the term $u_2(t)e^{-t}$ as

$$u_2(t)e^{-(t-2)-2} = e^{-2}u_2(t)e^{-(t-2)},$$

so we can apply the rule

$$\text{If } \mathcal{L}[f] = F(s) \text{ then } \mathcal{L}[u_a(t)f(t-a)] = e^{-as}F(s)$$

where $f(t) = e^{-t}$ and $a = 2$.
Then take the Laplace transform of both sides of the equation to get

$$\mathcal{L}\left[\frac{dy}{dt}\right] = -2\mathcal{L}[y] + e^{-2}\mathcal{L}[u_2(t)e^{-(t-2)}],$$

which gives us

$$s\mathcal{L}[y] - y(0) = -2\mathcal{L}[y] + \frac{e^{-2s}}{e^2(s+1)}.$$

Substituting the initial condition yields

$$s\mathcal{L}[y] - 3 = -2\mathcal{L}[y] + \frac{e^{-2s}}{e^2(s+1)}$$

so that

$$\mathcal{L}[y] = \frac{e^{-2s}}{e^2(s+1)(s+2)} + \frac{3}{s+2}.$$

Using the technique of partial fractions, we see that

$$\frac{1}{(s+2)(s+1)} = \frac{1}{s+1} - \frac{1}{s+2}.$$

Therefore, taking the inverse of the Laplace transform gives us

$$\begin{aligned} y(t) &= \mathcal{L}^{-1}\left[e^{-2}\frac{e^{-2s}}{(s+2)(s+1)}\right] + \mathcal{L}^{-1}\left[\frac{3}{s+2}\right] \\ &= e^{-2}\mathcal{L}^{-1}\left[\frac{e^{-2s}}{s+1} - \frac{e^{-2s}}{s+2}\right] + \mathcal{L}^{-1}\left[\frac{3}{s+2}\right] \\ &= e^{-2}u_2(t)\left(e^{-(t-2)} - e^{-2(t-2)}\right) + 3e^{-2t}. \end{aligned}$$

To check our answer, we compute

$$\frac{dy}{dt} = \frac{1}{e^2}\frac{du_2}{dt}\left(e^{-(t-2)} - e^{-2(t-2)}\right) + \frac{1}{e^2}u_2(t)\left(-e^{-(t-2)} + 2e^{-2(t-2)}\right) - 6e^{-2t},$$

and since $du_2/dt = 0$ except at $t = 2$ (where it is undefined),

$$\begin{aligned} \frac{dy}{dt} + 2y &= \frac{1}{e^2}u_2(t)\left(-e^{-(t-2)} + 2e^{-2(t-2)}\right) - 6e^{-2t} \\ &\quad +2\left(e^{-2}u_2(t)\left(e^{-(t-2)} - e^{-2(t-2)}\right) + 3e^{-2t}\right) \\ &= e^{-2}u_2(t)e^{-(t-2)} \\ &= u_2(t)e^{-t}. \end{aligned}$$

Hence, our $y(t)$ satisfies the differential equation except when $t = 2$. (We cannot expect $y(t)$ to satisfy the differential equation at $t = 2$ because the differential equation is not continuous there.) Note that $y(t)$ also satisfies the initial condition $y(0) = 3$.

15. Taking the Laplace transform of both sides of the equation, we have

$$\mathcal{L}\left[\frac{dy}{dt}\right] = -\mathcal{L}[y] + \mathcal{L}[u_a(t)],$$

which is equivalent to

$$s\mathcal{L}[y] - y(0) = -\mathcal{L}[y] + \frac{e^{-as}}{s}.$$

Solving for $\mathcal{L}[y]$ yields

$$\mathcal{L}[y] = \frac{e^{-as}}{s(s+1)} + \frac{y(0)}{s+1}.$$

Using the partial fractions decomposition

$$\frac{1}{s(s+1)} = \frac{1}{s} - \frac{1}{s+1},$$

we get

$$\mathcal{L}[y] = \frac{e^{-as}}{s} - \frac{e^{-as}}{s+1} + \frac{y(0)}{s+1}.$$

Taking the inverse Laplace transform, we obtain

$$\begin{aligned} y(t) &= u_a(t) - u_a(t)e^{-(t-a)} + y(0)e^{-t} \\ &= u_a(t)\left(1 - e^{-(t-a)}\right) + y(0)e^{-t}. \end{aligned}$$

To check our answer, we compute

$$\frac{dy}{dt} = \frac{du_a}{dt}\left(1 - e^{-(t-a)}\right) + u_a(t)e^{-(t-a)} - y(0)e^{-t}$$

and since $du_a/dt = 0$ except at $t = a$ (where it is undefined),

$$\begin{aligned} \frac{dy}{dt} + y &= u_a(t)e^{-(t-a)} - y(0)e^{-t} + u_a(t)\left(1 - e^{-(t-a)}\right) + y(0)e^{-t} \\ &= u_a(t). \end{aligned}$$

Hence, our $y(t)$ satisfies the differential equation except when $t = a$. (We cannot expect $y(t)$ to satisfy the differential equation at $t = a$ because the differential equation is not continuous there.)

16. We can write $\mathcal{L}[f]$ as the sum of two integrals, that is,

$$\mathcal{L}[f] = \int_0^\infty f(t)\,e^{-st}\,dt = \int_0^T f(t)\,e^{-st}\,dt + \int_T^\infty f(t)\,e^{-st}\,dt.$$

Next, we use the substitution $u = t - T$ on the second integral. Note that $t = u + T$. We get

$$\int_T^\infty f(t)\,e^{-st}\,dt = \int_0^\infty f(u+T)\,e^{-s(u+T)}\,du.$$

Since f is periodic with period T, we can rewrite the last integral as

$$e^{-Ts}\int_0^{\infty} f(u)\,e^{-su}\,du,$$

which is just $e^{-Ts}\mathcal{L}[f]$. Hence,

$$\mathcal{L}[f] = \int_0^T f(t)\,e^{-st}\,dt + e^{-Ts}\mathcal{L}[f].$$

We have

$$(1 - e^{-Ts})\mathcal{L}[f] = \int_0^T f(t)\,e^{-st}\,dt.$$

Consequently,

$$\mathcal{L}[f] = \frac{1}{1 - e^{-Ts}}\int_0^T f(t)\,e^{-st}\,dt.$$

17. From the formula in Exercise 16, we see that we need only compute the integral $\int_0^2 w(t)\,e^{-st}\,dt$. We write

$$\begin{aligned}
\int_0^2 w(t)\,e^{-st}\,dt &= \int_0^1 e^{-st}\,dt - \int_1^2 e^{-st}\,dt \\
&= \left.\frac{e^{-st}}{-s}\right|_0^1 - \left.\frac{e^{-st}}{-s}\right|_1^2 \\
&= \left(-\frac{e^{-s}}{s} + \frac{1}{s}\right) - \left(-\frac{e^{-2s}}{s} + \frac{e^{-s}}{s}\right) \\
&= \frac{1}{s} - 2\frac{e^{-s}}{s} + \frac{e^{-2s}}{s} \\
&= \frac{1}{s}(1 - 2e^{-s} + e^{-2s}) \\
&= \frac{1}{s}(1 - e^{-s})^2.
\end{aligned}$$

From the formula of Exercise 16, we get

$$\begin{aligned}
\mathcal{L}[w] &= \frac{1}{1 - e^{-2s}}\left(\frac{1}{s}(1 - e^{-s})^2\right) \\
&= \frac{(1 - e^{-s})^2}{s(1 - e^{-s})(1 + e^{-s})} \\
&= \frac{1 - e^{-s}}{s(1 + e^{-s})}.
\end{aligned}$$

18. From the formula in Exercise 16, we see that we need only compute the integral $\int_0^1 te^{-st}\,dt$. Using integration by parts (as in Exercise 2 of Section 6.1), we get

$$\begin{aligned}\mathcal{L}[z] &= \frac{1}{1-e^{-s}}\left(\frac{1}{s^2}-\frac{e^{-s}}{s^2}-\frac{e^{-s}}{s}\right)\\ &= \frac{1}{s^2}-\frac{e^{-s}}{s(1-e^{-s})}.\end{aligned}$$

19. **(a)** Transforming both sides of the equation, we have $\mathcal{L}\,[dy/dt] = -\mathcal{L}[y] + \mathcal{L}[w(t)]$, and using the result of Exercise 17, we get

$$s\mathcal{L}[y] - y(0) = -\mathcal{L}[y] + \frac{1-e^{-s}}{s(1+e^{-s})}.$$

Solving for $\mathcal{L}[y]$ using the fact that $y(0) = 0$, we obtain

$$\mathcal{L}[y] = \frac{1-e^{-s}}{s(s+1)(1+e^{-s})}.$$

(b) The function $w(t)$ is alternatively 1 and -1. While $w(t) = 1$, the solution decays exponentially toward $y = 1$. When $w(t)$ changes to -1, the solution then decays toward $y = -1$.

In fact, there is a periodic solution with initial condition $y(0) = (1-e)/(1+e) \approx -0.462$, and our solution tends toward this periodic solution as $t \to \infty$.

20. **(a)** Transforming both sides of the equation, we have $\mathcal{L}\,[dy/dt] = -\mathcal{L}[y] + \mathcal{L}[z(t)]$, and using the result of Exercise 18, we get

$$s\mathcal{L}[y] - y(0) = -\mathcal{L}[y] + \frac{1}{s^2} - \frac{e^{-s}}{s(1-e^{-s})}.$$

Solving for $\mathcal{L}[y]$ using the fact that $y(0) = 0$, we obtain

$$\mathcal{L}[y] = \frac{1}{s^2(s+1)} - \frac{e^{-s}}{s(s+1)(1-e^{-s})}.$$

(b) We know that the solution to the unforced equation decays exponentially to 0, so the solution to the forced equation decays exponentially toward the forcing term. For $0 \le t < 1$, the forcing term is simply t, and the solution is $t - 1 + e^{-t}$. At $t = 1$, the forcing function $z(t)$ jumps back to 0, yet $y(1) = 1/e$. Thus the solution starts to decrease. However, for some value of t in the interval $1 \le t \le 2$, $z(t) = y(t)$, so the solution begins to increase again. At $t = 2$, the forcing function $z(t)$ jumps back to zero again, and the solution begins to decrease again. Once again, for some value of t in the interval $2 \le t \le 3$, $z(t) = y(t)$, so the solution begins to increase again. This "oscillating" phenomenon repeats.

In fact, there is a periodic solution with initial condition $y(0) = 1/(e-1) \approx 0.582$, and our solution tends toward this periodic solution as $t \to \infty$.

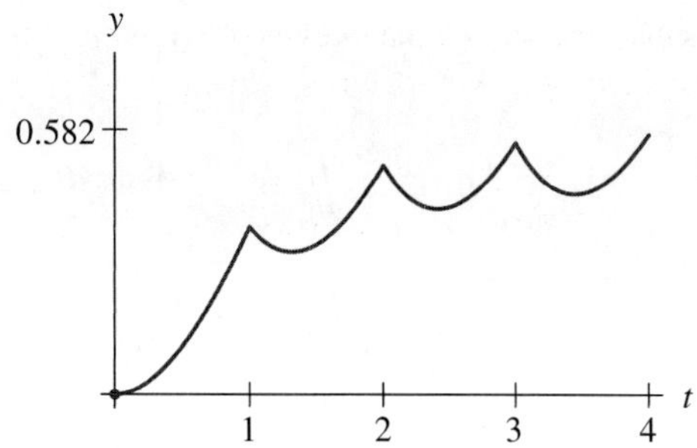

EXERCISES FOR SECTION 6.3

1. We use integration by parts twice to compute

$$\mathcal{L}[\sin \omega t] = \int_0^\infty \sin \omega t \, e^{-st} \, dt.$$

First, letting $u = \sin \omega t$ and $dv = e^{-st}\, dt$, we get

$$\begin{aligned}
\mathcal{L}[\sin \omega t] &= \sin \omega t \left. \frac{e^{-st}}{-s} \right|_0^\infty - \int_0^\infty \frac{e^{-st}}{-s} \omega \cos \omega t \, dt \\
&= \lim_{b\to\infty} \left[\frac{e^{-st}}{-s} \sin \omega t \Big|_0^b \right] + \frac{\omega}{s} \int_0^\infty e^{-st} \cos \omega t \, dt \\
&= \frac{\omega}{s} \int_0^\infty e^{-st} \cos \omega t \, dt,
\end{aligned}$$

since the limit of $e^{-sb} \sin \omega b$ is 0 as $b \to \infty$ and $s > 0$.

Using integration by parts on

$$\int_0^\infty e^{-st} \cos \omega t \, dt,$$

with $u = \cos \omega t$ and $dv = e^{-st}\, dt$, we get

$$\begin{aligned}
\int_0^\infty e^{-st} \cos \omega t \, dt &= \frac{e^{-st}}{-s} \cos \omega t \Big|_0^\infty - \int_0^\infty \frac{e^{-st}}{-s} (-\omega \sin \omega t) \, dt \\
&= \lim_{b\to\infty} \left[\frac{e^{-st}}{-s} \cos \omega t \Big|_0^b \right] - \frac{\omega}{s} \int_0^\infty e^{-st} \sin \omega t \, dt \\
&= \lim_{b\to\infty} \left[\frac{e^{-sb}}{-s} \cos \omega b \right] + \frac{1}{s} - \frac{\omega}{s} \int_0^\infty e^{-st} \sin \omega t \, dt \\
&= \frac{1}{s} - \frac{\omega}{s} \int_0^\infty e^{-st} \sin \omega t \, dt,
\end{aligned}$$

since the limit of $e^{-sb}\cos\omega b$ is 0 as $b \to \infty$ and $s > 0$.

Thus,

$$\int_0^\infty \sin\omega t\, e^{-st}\,dt = \frac{\omega}{s}\int_0^\infty e^{-st}\cos\omega t\,dt$$

$$= \frac{\omega}{s}\left(\frac{1}{s} - \frac{\omega}{s}\int_0^\infty \sin\omega t\, e^{-st}\,dt\right)$$

$$= \frac{\omega}{s^2} - \frac{\omega^2}{s^2}\int_0^\infty \sin\omega t\, e^{-st}\,dt.$$

So

$$\frac{s^2+\omega^2}{s^2}\int_0^\infty \sin\omega t\, e^{-st}\,dt = \frac{\omega}{s^2},$$

and

$$\int_0^\infty \sin\omega t\, e^{-st}\,dt = \frac{\omega}{s^2+\omega^2}.$$

2. We use integration by parts twice to compute

$$\mathcal{L}[\cos\omega t] = \int_0^\infty \cos\omega t\, e^{-st}\,dt.$$

First, letting $u = \cos\omega t$ and $dv = e^{-st}\,dt$, we get

$$\mathcal{L}[\sin\omega t] = \frac{e^{-st}}{-s}\cos\omega t\bigg|_0^\infty - \int_0^\infty \frac{e^{-st}}{-s}(-\omega\sin\omega t)\,dt$$

$$= \lim_{b\to\infty}\left[\frac{e^{-st}}{-s}\cos\omega t\bigg|_0^b\right] - \frac{\omega}{s}\int_0^\infty e^{-st}\sin\omega t\,dt$$

$$= \frac{1}{s} - \int_0^\infty \frac{\omega}{s}e^{-st}\sin\omega t\,dt,$$

since the limit of $e^{-sb}\cos\omega b$ is 0 as $b \to \infty$ and $s > 0$.

Using integration by parts on

$$\int_0^\infty e^{-st}\sin\omega t\,dt,$$

with $u = \sin\omega t$ and $dv = e^{-st}\,dt$, we get that

$$\int_0^\infty e^{-st}\sin\omega t\,dt = \frac{e^{-st}}{-s}\sin\omega t\bigg|_0^\infty - \int_0^\infty \frac{e^{-st}}{-s}\omega\cos\omega t\,dt$$

$$= \lim_{b\to\infty}\left[\frac{e^{-st}}{-s}\sin\omega t\bigg|_0^b\right] + \frac{\omega}{s}\int_0^\infty e^{-st}\cos\omega t\,dt$$

$$= \frac{\omega}{s}\int_0^\infty e^{-st}\cos\omega t\,dt,$$

since the limit of $e^{-sb}\sin\omega b$ is 0 as $b \to \infty$ and $s > 0$.

Thus

$$\int_0^\infty \cos\omega t\, e^{-st}\, dt = \frac{1}{s} - \frac{\omega}{s}\int_0^\infty e^{-st}\sin\omega t\, dt$$

$$= \frac{1}{s} - \frac{\omega^2}{s^2}\int_0^\infty e^{-st}\cos\omega t\, dt.$$

So

$$\frac{s^2+\omega^2}{s^2}\int_0^\infty \cos\omega t\, e^{-st}\, dt = \frac{1}{s},$$

and

$$\int_0^\infty \cos\omega t\, e^{-st}\, dt = \frac{s}{s^2+\omega^2}.$$

3. We need to compute

$$\mathcal{L}[e^{at}\sin\omega t] = \int_0^\infty e^{at}\sin\omega t e^{-st}\, dt = \int_0^\infty \sin\omega t e^{-(s-a)t}\, dt.$$

We can do this using integration by parts twice and ending up with $\mathcal{L}[e^{at}\sin\omega t]$ on both sides of the equation. Alternately, if we let $r = s - a$, then

$$\int_0^\infty \sin\omega t\, e^{-(s-a)t}\, dt = \int_0^\infty \sin\omega t\, e^{-rt}\, dt$$

The integral on the right is the Laplace transform of $\sin\omega t$ with r as the new independent variable. From Exercise 1, we know

$$\int_0^\infty \sin\omega t\, e^{-rt}\, dt = \frac{\omega}{r^2+\omega^2}.$$

Substituting back we have

$$\mathcal{L}[e^{at}\sin\omega t] = \frac{\omega}{(s-a)^2+\omega^2}.$$

4. We need to compute

$$\mathcal{L}[e^{at}\cos\omega t] = \int_0^\infty e^{at}\cos\omega t\, e^{-st}\, dt = \int_0^\infty \cos\omega t\, e^{-(s-a)t}\, dt.$$

We can do this using integration by parts twice to end up with $\mathcal{L}[e^{at}\cos\omega t]$ on both sides of the equation. Alternately, if we let $r = s - a$, then

$$\int_0^\infty \cos\omega t\, e^{-(s-a)t}\, dt = \int_0^\infty \cos\omega t\, e^{-rt}\, dt.$$

The integral on the right is the Laplace transform of $\cos\omega t$ with r as the new independent variable. From the table, we know

$$\int_0^\infty \cos\omega t\, e^{-rt}\, dt = \frac{r}{r^2+\omega^2}.$$

Then substituting back we have

$$\mathcal{L}[e^{at}\cos\omega t] = \frac{s-a}{(s-a)^2+\omega^2}.$$

5. Using the formula

$$\mathcal{L}\left[\frac{d^2y}{dt^2}\right] = s^2\mathcal{L}[y] - y'(0) - sy(0),$$

and the linearity of the Laplace transform, we get that

$$s^2\mathcal{L}[y] - y'(0) - sy(0) + \omega^2\mathcal{L}[y] = 0.$$

Substituting the initial conditions and solving for $\mathcal{L}[y]$ gives

$$\mathcal{L}[y] = \frac{s}{s^2+\omega^2}.$$

6. Since

$$\mathcal{L}[\cos\omega t] = \frac{s}{s^2+\omega^2},$$

we can compute that

$$\frac{d}{d\omega}\mathcal{L}[\cos\omega t] = \frac{-s(2\omega)}{(s^2+\omega^2)^2} = \frac{-2\omega s}{(s^2+\omega^2)^2},$$

but

$$\frac{d}{d\omega}\mathcal{L}[\cos\omega t] = \mathcal{L}\left[\frac{d}{d\omega}\cos\omega t\right] = \mathcal{L}[-t\sin\omega t].$$

We can bring the derivative with respect to ω inside the Laplace transform because the Laplace transform is an integral with respect to t, that is,

$$\frac{d}{d\omega}\mathcal{L}[\cos\omega t] = \frac{d}{d\omega}\int_0^\infty \cos\omega t\, e^{-st}\,dt = \int_0^\infty \frac{d}{d\omega}\left(\cos\omega t\, e^{-st}\right)\,dt.$$

Canceling the minus signs on left and right gives

$$\mathcal{L}[t\sin\omega t] = \frac{2\omega s}{(s^2+\omega^2)^2}.$$

7. Since

$$\mathcal{L}[\sin\omega t] = \frac{\omega}{s^2+\omega^2},$$

we can compute that

$$\frac{d}{d\omega}\mathcal{L}[\sin\omega t] = \frac{s^2-\omega^2}{(s^2+\omega^2)^2},$$

but

$$\frac{d}{d\omega}\mathcal{L}[\sin\omega t] = \mathcal{L}\left[\frac{d}{d\omega}\sin\omega t\right] = \mathcal{L}[t\cos\omega t].$$

So

$$\mathcal{L}[t\cos\omega t] = \frac{s^2-\omega^2}{(s^2+\omega^2)^2}.$$

8. We need to compute

$$\mathcal{L}[te^{at}] = \int_0^\infty te^{at}e^{-st}\,dt.$$

We can do this using the hint, by differentiating $\mathcal{L}[e^{at}]$ with respect to a. Another method is to write

$$\mathcal{L}[te^{at}] = \int_0^\infty te^{at}e^{-st}\,dt = \int_0^\infty te^{-(s-a)t}\,dt = \int_0^\infty te^{-rt}\,dt$$

where $r = s - a$. The last integral is the Laplace transform of t using r as the new independent variable. Hence, from the table we have

$$\int_0^\infty te^{-rt}\,dt = \frac{1}{r^2}.$$

Substituting back $r = s - a$ we have

$$\mathcal{L}[te^{at}] = \frac{1}{(s-a)^2}.$$

9. From Exercise 10, we know that

$$\mathcal{L}[te^{at}] = \frac{1}{(s-a)^2}.$$

Differentiating both sides of this formula with respect to a gives

$$\frac{d}{da}\mathcal{L}[te^{at}] = \mathcal{L}\left[\frac{d}{da}te^{ta}\right] = \mathcal{L}[t^2e^{at}]$$

while

$$\frac{d}{da}\frac{1}{(s-a)^2} = \frac{2}{(s-a)^3}.$$

Hence,

$$\mathcal{L}[t^2e^{at}] = \frac{2}{(s-a)^3}.$$

10. Using the results of Exercise 9, we can work by induction on n, with induction hypothesis

$$\mathcal{L}[t^ne^{at}] = \frac{n!}{(s-a)^{n+1}}.$$

Alternately, we can compute

$$\mathcal{L}[t^ne^{at}] = \int_0^\infty t^ne^{at}e^{-st}\,dt = \int_0^\infty t^ne^{-(s-a)t}\,dt = \int_0^\infty t^ne^{-rt}\,dt$$

where $r = s - a$. Now the last integral is the Laplace transform of t^n using r as the independent variable, so

$$\int_0^\infty t^ne^{-rt}\,dt = \frac{n!}{r^{n+1}}$$

from the table. Hence, substituting $r = s - a$ we have

$$\mathcal{L}[t^ne^{at}] = \frac{n!}{(s-a)^{n+1}}.$$

11. In this case, $b = 2$, and $(s + b/2)^2 = (s + 1)^2 = s^2 + 2s + 1$, so $s^2 + 2s + 10 = (s + 1)^2 + 3^2$.

12. In this case, $b = -4$, and $(s + b/2)^2 = (s - 2)^2 = s^2 - 4s + 4$, so $s^2 - 4s + 5 = (s - 2)^2 + 1^2$.

13. In this case, $b = 1$, and $(s+b/2)^2 = (s+1/2)^2 = s^2+s+1/4$, so $s^2+s+1 = (s+1/2)^2+3/4 = (s + 1/2)^2 + (\sqrt{3}/2)^2$.

14. In this case, $b = 6$, and $(s + b/2)^2 = (s + 3)^2 = s^2 + 6s + 9$, so $s^2 + 6s + 10 = (s + 3)^2 + 1^2$.

15. In Exercise 11, we completed the square and obtained $s^2 + 2s + 10 = (s + 1)^2 + 3^2$, so

$$\begin{aligned}\mathcal{L}^{-1}\left[\frac{1}{s^2+2s+10}\right] &= \mathcal{L}^{-1}\left[\frac{1}{(s+1)^2+3^2}\right]\\ &= \frac{1}{3}\mathcal{L}^{-1}\left[\frac{3}{(s+1)^2+3^2}\right]\\ &= \frac{1}{3}e^{-t}\sin 3t.\end{aligned}$$

16. In Exercise 12, we completed the square and obtained $s^2 - 4s + 5 = (s - 2)^2 + 1^2$, so

$$\begin{aligned}\mathcal{L}^{-1}\left[\frac{s}{s^2-4s+5}\right] &= \mathcal{L}^{-1}\left[\frac{s}{(s-2)^2+1^2}\right]\\ &= \mathcal{L}^{-1}\left[\frac{s-2}{(s-2)^2+1^2}\right] + \mathcal{L}^{-1}\left[\frac{2}{(s-2)^2+1^2}\right]\\ &= e^{2t}\cos t + e^{2t}(2\sin t) = e^{2t}(\cos t + 2\sin t).\end{aligned}$$

17. In Exercise 13, we completed the square and obtained $s^2 + s + 1 = (s + 1/2)^2 + (\sqrt{3}/2)^2$, so

$$\frac{2s+3}{s^2+s+1} = \frac{2s+3}{(s+1/2)^2+(\sqrt{3}/2)^2}.$$

We want to put this fraction in the right form so that we can use the formulas for $\mathcal{L}[e^{at}\cos\omega t]$ and $\mathcal{L}[e^{at}\sin\omega t]$. We see that

$$\begin{aligned}\frac{2s+3}{(s+1/2)^2+(\sqrt{3}/2)^2} &= \frac{2s+1}{(s+1/2)^2+(\sqrt{3}/2)^2} + \frac{2}{(s+1/2)^2+(\sqrt{3}/2)^2}\\ &= \frac{2(s+1/2)}{(s+1/2)^2+(\sqrt{3}/2)^2} + \frac{(4/\sqrt{3})(\sqrt{3}/2)}{(s+1/2)^2+(\sqrt{3}/2)^2}.\end{aligned}$$

So

$$\begin{aligned}\mathcal{L}^{-1}\left[\frac{2s+3}{s^2+s+1}\right] &= 2\mathcal{L}^{-1}\left[\frac{(s+1/2)}{(s+1/2)^2+(\sqrt{3}/2)^2}\right] + \frac{4}{\sqrt{3}}\mathcal{L}^{-1}\left[\frac{\sqrt{3}/2}{(s+1/2)^2+(\sqrt{3}/2)^2}\right]\\ &= 2e^{-t/2}\cos\left(\frac{\sqrt{3}}{2}t\right) + \frac{4}{\sqrt{3}}e^{-t/2}\sin\left(\frac{\sqrt{3}}{2}t\right).\end{aligned}$$

18. In Exercise 14, we completed the square and obtained $s^2 + 6s + 10 = (s+3)^2 + 1^2$, so

$$\frac{s+1}{s^2+6s+10} = \frac{s+1}{(s+3)^2+1^2}.$$

We want to put this fraction in the right form so that we can use the formulas for $\mathcal{L}[e^{at}\cos\omega t]$ and $\mathcal{L}[e^{at}\sin\omega t]$. We see that

$$\frac{s+1}{(s+3)^2+1^2} = \frac{s+3}{(s+3)^2+1^2} - \frac{2}{(s+3)^2+1^2}.$$

So

$$\mathcal{L}^{-1}\left[\frac{s+1}{s^2+6s+10}\right] = e^{-3t}\cos t - 2e^{-3t}\sin t.$$

19. We compute

$$\begin{aligned}
\mathcal{L}\left[e^{(a+ib)t}\right] &= \int_0^\infty e^{(a+ib)t}e^{-st}\,dt \\
&= \int_0^\infty e^{-(s-(a+ib))t}\,dt \\
&= -\frac{1}{s-(a+ib)}\left(\lim_{u\to\infty}\left[e^{-(s-a)u}e^{-ibu}\right]-1\right).
\end{aligned}$$

The limit is zero as long as $s > a$. Hence,

$$\mathcal{L}\left[e^{(a+ib)t}\right] = \frac{1}{s-(a+ib)}$$

if $s > a$ and undefined otherwise. This is the same formula as for real exponentials. It can also be written

$$\mathcal{L}\left[e^{(a+ib)t}\right] = \frac{s-a+ib}{(s-a)^2+b^2}.$$

20. This follows from linearity:

$$\begin{aligned}
\mathcal{L}[y] &= \mathcal{L}[y_{\text{re}} + i y_{\text{im}}] \\
&= \int_0^\infty (y_{\text{re}} + i y_{\text{im}})\,e^{-st}\,dt \\
&= \int_0^\infty y_{\text{re}}(t)\,e^{-st}\,dt + i\int_0^\infty y_{\text{im}}(t)\,e^{-st}\,dt \\
&= \mathcal{L}[y_{\text{re}}] + i\mathcal{L}[y_{\text{im}}].
\end{aligned}$$

21. We recall that

$$e^{at}\cos\omega t = \mathrm{Re}(e^{(a+ib)t}).$$

So

$$\begin{aligned}\mathcal{L}[e^{at}\cos\omega t] &= \mathrm{Re}(\mathcal{L}[e^{(a+ib)t}]) \\ &= \mathrm{Re}\left(\frac{s-a+i\omega}{(s-a)^2+\omega^2}\right) \\ &= \frac{s-a}{(s-a)^2+\omega^2}.\end{aligned}$$

Similarly,

$$\begin{aligned}\mathcal{L}[e^{at}\sin\omega t] &= \mathrm{Im}(\mathcal{L}[e^{(a+ib)t}]) \\ &= \mathrm{Im}\left(\frac{s-a+i\omega}{(s-a)^2+\omega^2}\right) \\ &= \frac{\omega}{(s-a)^2+\omega^2}.\end{aligned}$$

22. **(a)** The roots of s^2+2s+5 are $-1\pm 2i$, so the quadratic factors into

$$(s-(-1+2i))(s-(-1-2i)).$$

(b) We write

$$\frac{1}{s^2+2s+5} = \frac{A}{s+1+2i} + \frac{B}{s+1-2i}.$$

So, finding common denominators (that is, usual partial fractions but with complex numbers) gives

$$\begin{cases} A+B=0 \\ A+B+2i(-A+B)=1. \end{cases}$$

Solving, we get $A = i/4$ and $B = -i/4$, so

$$\frac{1}{s^2+2s+5} = \frac{i/4}{s+1+2i} + \frac{-i/4}{s+1-2i}.$$

(c) We know that

$$\begin{aligned}\mathcal{L}^{-1}\left[\frac{i/4}{s+1+2i}\right] &= \frac{i}{4}e^{(-1-2i)t} \\ &= \frac{i}{4}\left(e^{-t}\cos(-2t) + ie^{-t}\sin(-2t)\right) \\ &= \frac{1}{4}\left(e^{-t}\sin 2t + ie^{-t}\cos 2t\right)\end{aligned}$$

and

$$\mathcal{L}^{-1}\left[\frac{-i/4}{s+1-2i}\right] = -\frac{i}{4}e^{(-1+2i)t}$$

$$= -\frac{i}{4}\left(e^{-t}\cos 2t + ie^{-t}\sin 2t\right)$$

$$= \frac{1}{4}\left(e^{-t}\sin 2t - ie^{-t}\cos 2t\right).$$

(d) Adding, we get

$$\mathcal{L}^{-1}\left[\frac{1}{s^2+2s+5}\right] = \frac{1}{2}e^{-t}\sin 2t.$$

23. Using the quadratic formula, we see that the roots of $s^2 + 2s + 10 = 0$ are $s = -1 \pm 3i$. Thus $s^2 + 2s + 10 = (s + 1 + 3i)(s + 1 - 3i)$. So we want to find A and B so that

$$\frac{1}{s^2+2s+10} = \frac{A}{s+1+3i} + \frac{B}{s+1-3i}.$$

So, finding common denominators (that is, usual partial fractions only with complex numbers) gives

$$\begin{cases} A + B = 0 \\ A + B + 3i(-A + B) = 1. \end{cases}$$

Solving, we get $A = i/6$ and $B = -i/6$, so

$$\frac{1}{s^2+2s+10} = \frac{i/6}{s+1+3i} + \frac{-i/6}{s+1-3i}.$$

Thus

$$\mathcal{L}^{-1}\left[\frac{1}{s^2+2s+10}\right] = \mathcal{L}^{-1}\left[\frac{i/6}{s+1+3i} + \frac{-i/6}{s+1-3i}\right]$$

$$= \frac{i}{6}e^{(-1-3i)t} - \frac{i}{6}e^{(-1+3i)t}$$

$$= \frac{i}{6}\left(e^{-t}\cos(-3t) + ie^{-t}\sin(-3t)\right) - \frac{i}{6}\left(e^{-t}\cos 3t + ie^{-t}\sin 3t\right)$$

$$= -\frac{i}{6}\left(2ie^{-t}\sin 3t\right)$$

$$= \frac{1}{3}e^{-t}\sin 3t.$$

24. Using the quadratic formula, we find that the roots of the denominator are $2 \pm i$. Hence, we can factor the denominator into

$$s^2 - 4s + 5 = (s - (2 + i))(s - (2 - i)).$$

Using partial fraction decomposition, we get

$$\frac{s}{s - 4s + 5} = \frac{A}{s - (2 + i)} + \frac{B}{s - (2 - i)},$$

which gives the equations

$$\begin{cases} A + B = 1 \\ -(2 - i)A - (2 + i)B = 0. \end{cases}$$

Solving for A and B gives $A = 1/2 - i$ and $B = 1/2 + i$ so

$$\frac{s}{s - 4s + 5} = \frac{\frac{1}{2} - i}{s - (2 + i)} + \frac{\frac{1}{2} + i}{s - (2 - i)}.$$

Taking the inverse Laplace transform of the terms on the right gives

$$\left(\tfrac{1}{2} - i\right) e^{(2+i)t} + \left(\tfrac{1}{2} + i\right) e^{(2-i)t}.$$

Using Euler's formula to expand the exponentials and simplifying gives

$$e^{2t}(\cos t + 2 \sin t).$$

25. Using the quadratic formula, the roots of the denominator are $(-1 \pm i\sqrt{3})/2$. Hence, we can factor the denominator into

$$\left(s - \left(\tfrac{-1+i\sqrt{3}}{2}\right)\right)\left(s - \left(\tfrac{-1-i\sqrt{3}}{2}\right)\right).$$

We then do the partial fractions decomposition

$$\frac{2s + 3}{s^2 + s + 1} = \frac{A}{s - \left(\frac{-1+i\sqrt{3}}{2}\right)} + \frac{B}{s - \left(\frac{-1-i\sqrt{3}}{2}\right)},$$

which gives rise to the equations

$$\begin{cases} A + B = 2 \\ \left(\tfrac{1+i\sqrt{3}}{2}\right) A + \left(\tfrac{1-i\sqrt{3}}{2}\right) B = 3. \end{cases}$$

Solving yields $A = 1 - \frac{2}{\sqrt{3}}i$ and $B = 1 + \frac{2}{\sqrt{3}}i$. So

$$\frac{2s + 3}{s^2 + s + 1} = \frac{1 - \frac{2}{\sqrt{3}}i}{s - \left(\frac{-1+i\sqrt{3}}{2}\right)} + \frac{1 + \frac{2}{\sqrt{3}}i}{s - \left(\frac{-1-i\sqrt{3}}{2}\right)}.$$

Taking inverse Laplace transforms of the right-hand side gives

$$\left(1-\tfrac{2}{\sqrt{3}}i\right)e^{(-1+i\sqrt{3})t/2}+\left(1+\tfrac{2}{\sqrt{3}}i\right)e^{(-1-i\sqrt{3})t/2}.$$

Using Euler's formula to replace the complex exponentials and simplifying yields

$$2e^{-t/2}\cos\left(\tfrac{\sqrt{3}}{2}t\right)+\tfrac{4}{\sqrt{3}}e^{-t/2}\sin\left(\tfrac{\sqrt{3}}{2}t\right).$$

26. Using the quadratic formula, we find the roots of the denominator are $-3\pm i$ so the denominator can be factored

$$s^2+6s+10=(s-(-3+i))(s-(-3-i)).$$

The partial fractions decomposition is

$$\frac{s+1}{s^2+6s+10}=\frac{A}{s-(-3-i)}+\frac{B}{s-(-3+i)},$$

which leads to the equations

$$\begin{cases} A+B=1 \\ (3-i)A+(3+i)B=1. \end{cases}$$

Solving, we find $A=\frac{1}{2}-i$ and $B=\frac{1}{2}+i$, so

$$\frac{s+1}{s^2+6s+10}=\frac{\frac{1}{2}-i}{s-(-3-i)}+\frac{\frac{1}{2}+i}{s-(-3+i)}.$$

Taking inverse Laplace transform of the right-hand side gives

$$\left(\tfrac{1}{2}-i\right)e^{(-3-i)t}+\left(\tfrac{1}{2}+i\right)e^{(-3+i)t}$$

and using Euler's formula and simplifying gives

$$e^{-3t}\cos t-2e^{-3t}\sin t.$$

27. **(a)** Taking the Laplace transform of both sides of the equation, we obtain

$$\mathcal{L}\left[\frac{d^2y}{dt^2}\right]+4\mathcal{L}[y]=\frac{8}{s},$$

and using the fact that $\mathcal{L}[d^2y/dt^2]=s^2\mathcal{L}[y]-sy(0)-y'(0)$, we have

$$(s^2+4)\mathcal{L}[y]-sy(0)-y'(0)=\frac{8}{s}.$$

(b) Substituting the initial conditions yields

$$(s^2+4)\mathcal{L}[y] - 11s - 5 = \frac{8}{s},$$

and solving for $\mathcal{L}[y]$ we get

$$\mathcal{L}[y] = \frac{11s+5}{s^2+4} + \frac{8}{s(s^2+4)}.$$

The partial fractions decomposition of $8/(s(s^2+4))$ is

$$\frac{8}{s(s^2+4)} = \frac{A}{s} + \frac{Bs+C}{s^2+4}.$$

Putting the right-hand side over a common denominator gives us

$$(A+B)s^2 + Cs + 4A = 8,$$

and consequently, $A = 2$, $B = -2$, and $C = 0$. In other words,

$$\frac{8}{s(s^2+4)} = \frac{2}{s} + \frac{-2s}{s^2+4}.$$

We obtain

$$\mathcal{L}[y] = \frac{2}{s} + \frac{9s+5}{s^2+4}.$$

(c) To take the inverse Laplace transform, we rewrite $\mathcal{L}[y]$ in the form

$$\mathcal{L}[y] = \frac{2}{s} + 9\left(\frac{s}{s^2+4}\right) + \frac{5}{2}\left(\frac{2}{s^2+4}\right).$$

Therefore, $y(t) = 2 + 9\cos 2t + \frac{5}{2}\sin 2t$.

28. **(a)** Taking the Laplace transform of both sides of the equation, we obtain

$$\mathcal{L}\left[\frac{d^2y}{dt^2}\right] - \mathcal{L}[y] = \frac{1}{s-2},$$

and using the fact that $\mathcal{L}[d^2y/dt^2] = s^2\mathcal{L}[y] - sy(0) - y'(0)$, we have

$$(s^2-1)\mathcal{L}[y] - sy(0) - y'(0) = \frac{1}{s-2}.$$

(b) Substituting the initial conditions yields

$$(s^2-1)\mathcal{L}[y] - s + 1 = \frac{1}{s-2},$$

and solving for $\mathcal{L}[y]$ we get

$$\mathcal{L}[y] = \frac{1}{s+1} + \frac{1}{(s-2)(s^2-1)}.$$

Using the partial fractions decomposition

$$\frac{1}{(s-2)(s^2-1)} = \frac{\frac{1}{3}}{s-2} + \frac{-\frac{1}{2}}{s-1} + \frac{\frac{1}{6}}{s+1},$$

we obtain

$$\mathcal{L}[y] = \frac{\frac{1}{3}}{s-2} + \frac{-\frac{1}{2}}{s-1} + \frac{\frac{7}{6}}{s+1}.$$

(c) Taking the inverse Laplace transform, we have

$$y(t) = \tfrac{1}{3}e^{2t} - \tfrac{1}{2}e^{t} + \tfrac{7}{6}e^{-t}.$$

29. **(a)** Taking the Laplace transform of both sides of the equation, we obtain

$$\mathcal{L}\left[\frac{d^2y}{dt^2}\right] - 4\mathcal{L}\left[\frac{dy}{dt}\right] + 5\mathcal{L}[y] = \frac{2}{s-1},$$

and using the formulas for $\mathcal{L}[dy/dt]$ and $\mathcal{L}[d^2y/dt^2]$ in terms of $\mathcal{L}[y]$, we have

$$(s^2 - 4s + 5)\mathcal{L}[y] - sy(0) - y'(0) + 4y(0) = \frac{2}{s-1}.$$

(b) Substituting the initial conditions yields

$$(s^2 - 4s + 5)\mathcal{L}[y] - 3s + 11 = \frac{2}{s \quad 2},$$

and solving for $\mathcal{L}[y]$ we get

$$\mathcal{L}[y] = \frac{3s - 11}{s^2 - 4s + 5} + \frac{2}{(s-1)(s^2 - 4s + 5)}.$$

Using the partial fractions decomposition

$$\frac{2}{(s-1)(s^2-4s+5)} = \frac{1}{s-1} + \frac{-s+3}{s^2-4s+5},$$

we obtain

$$\mathcal{L}[y] = \frac{1}{s-1} + \frac{2s-8}{s^2-4s+5}.$$

(c) In order to compute the inverse Laplace transform, we first write

$$s^2 - 4s + 5 = (s-2)^2 + 1$$

by completing the square, and then we write

$$\frac{2s-8}{s^2-4s+5} = \frac{2(s-2)}{(s-2)^2+1} - \frac{4}{(s-2)^2+1}.$$

Taking the inverse Laplace transform, we have

$$y(t) = e^t + 2e^{2t}\cos t - 4e^{2t}\sin t.$$

30. **(a)** Taking the Laplace transform of both sides of the equation, we obtain

$$\mathcal{L}\left[\frac{d^2y}{dt^2}\right] + 6\mathcal{L}\left[\frac{dy}{dt}\right] + 13\mathcal{L}[y] = 13\frac{e^{-4s}}{s},$$

and using the formulas for $\mathcal{L}[dy/dt]$ and $\mathcal{L}[d^2y/dt^2]$ in terms of $\mathcal{L}[y]$, we have

$$(s^2 + 6s + 13)\mathcal{L}[y] - sy(0) - y'(0) - 6y(0) = 13\frac{e^{-4s}}{s}.$$

(b) Substituting the initial conditions yields

$$(s^2 + 6s + 13)\mathcal{L}[y] - 3s - 19 = 13\frac{e^{-4s}}{s},$$

and solving for $\mathcal{L}[y]$ we get

$$\mathcal{L}[y] = \frac{3s + 19}{s^2 + 6s + 13} + \left(\frac{13}{s(s^2 + 6s + 13)}\right) e^{-4s}.$$

Using the partial fractions decomposition

$$\frac{13}{s(s^2 + 6s + 13)} = \frac{1}{s} - \frac{s + 6}{s^2 + 6s + 13},$$

we obtain

$$\mathcal{L}[y] = \frac{3s + 19}{s^2 + 6s + 13} + \left(\frac{1}{s} - \frac{s + 6}{s^2 + 6s + 13}\right) e^{-4s}.$$

(c) In order to compute the inverse Laplace transform, we first write

$$s^2 + 6s + 13 = (s + 3)^2 + 4$$

by completing the square, and then we write

$$\frac{3s + 19}{s^2 + 6s + 13} = 3\left(\frac{s + 3}{(s + 3)^2 + 4}\right) + 5\left(\frac{2}{(s + 3)^2 + 4}\right)$$

and

$$\frac{s + 6}{s^2 + 6s + 13} = \left(\frac{s + 3}{(s + 3)^2 + 4}\right) + \frac{3}{2}\left(\frac{2}{(s + 3)^2 + 4}\right).$$

Taking the inverse Laplace transform, we have

$$y(t) = 3e^{-3t}\cos 2t + 5e^{-3t}\sin 2t + u_4(t)\left(1 - e^{-3(t-4)}\cos 2(t - 4) - \tfrac{3}{2}e^{-3(t-4)}\sin 2(t - 4)\right).$$

31. **(a)** Note that this is resonant forcing of an undamped oscillator. We take the Laplace transform of both sides

$$\mathcal{L}\left[\frac{d^2y}{dt^2}\right] + 4\mathcal{L}[y] = \mathcal{L}[\cos 2t]$$

and obtain

$$s^2\mathcal{L}[y] + 2s + 4\mathcal{L}[y] = \frac{s}{s^2 + 4}.$$

(b) Solving for $\mathcal{L}[y]$, we get

$$\mathcal{L}[y] = -\frac{2s}{s^2+4} + \frac{s}{(s^2+4)^2}.$$

(c) To take the inverse Laplace transform, we note that

$$\mathcal{L}^{-1}\left[-\frac{2s}{s^2+4}\right] = -2\cos 2t$$

and

$$\mathcal{L}^{-1}\left[\frac{s}{(s^2+4)^2}\right] = \frac{1}{4}\mathcal{L}^{-1}\left[\frac{4s}{(s^2+4)^2}\right] = \frac{t}{4}\sin 2t.$$

So

$$y(t) = -2\cos 2t + \frac{t}{4}\sin 2t,$$

which is of the form we would expect for a resonant response.

32. **(a)** We take Laplace transform of both sides to obtain

$$\mathcal{L}\left[\frac{d^2y}{dt^2}\right] + 3\mathcal{L}[y] = \mathcal{L}[u_4(t)\cos(5(t-4))],$$

which is equivalent to

$$s^2\mathcal{L}[y] - sy(0) - y'(0) + 3\mathcal{L}[y] = \frac{e^{-4s}s}{s^2+25}.$$

Using the given initial conditions, we have

$$(s^2+3)\mathcal{L}[y] + 2 = \frac{e^{-4s}s}{s^2+25}.$$

(b) Solving for $\mathcal{L}[y]$ gives

$$\mathcal{L}[y] = \frac{-2}{s^2+3} + \frac{e^{-4s}s}{(s^2+3)(s^2+25)}.$$

(c) Now to find the inverse Laplace transform, we first note that

$$\mathcal{L}^{-1}\left[\frac{-2}{s^2+3}\right] = \frac{-2}{\sqrt{3}}\mathcal{L}^{-1}\left[\frac{\sqrt{3}}{s^2+3}\right] = \frac{-2}{\sqrt{3}}\sin\sqrt{3}\,t.$$

For the second term, we first use partial fractions to write

$$\frac{s}{(s^2+3)(s^2+25)} = \frac{1}{22}\left(\frac{s}{s^2+3} - \frac{s}{s^2+25}\right).$$

Hence

$$\mathcal{L}^{-1}\left[\frac{e^{-4s}s}{(s^2+3)(s^2+25)}\right] = \frac{1}{22}u_4(t)\left(\cos(\sqrt{3}\,(t-4)) - \cos(5(t-4))\right).$$

Combining the two results, we obtain the solution of the initial-value problem

$$y(t) = -\frac{2}{\sqrt{3}}\sin\sqrt{3}\,t + \frac{1}{22}u_4(t)\left(\cos(\sqrt{3}\,(t-4)) - \cos(5(t-4))\right).$$

33. **(a)** We take Laplace transform of both sides to obtain

$$\mathcal{L}\left[\frac{d^2y}{dt^2}\right] + 4\mathcal{L}\left[\frac{dy}{dt}\right] + 9\mathcal{L}[y] = \mathcal{L}[20u_2(t)\sin(t-2)],$$

which is equivalent to

$$s^2\mathcal{L}[y] - y(0)s - y'(0) + 4\left(s\mathcal{L}[y] - y(0)\right) + 9\mathcal{L}[y] = 20e^{-2s}\left(\frac{1}{s^2+1}\right).$$

Using the given initial conditions, we have

$$(s^2+4s+9)\mathcal{L}[y] - s - 6 = e^{-2s}\left(\frac{20}{s^2+1}\right).$$

(b) Solving for $\mathcal{L}[y]$ gives

$$\mathcal{L}[y] = \frac{s+6}{s^2+4s+9} + e^{-2s}\left(\frac{20}{(s^2+4s+9)(s^2+1)}\right).$$

(c) Now to find the inverse Laplace transform, we first note that

$$\frac{s+6}{s^2+4s+9} = \frac{s+2}{(s+2)^2+5} + \frac{4}{(s+2)^2+5}.$$

Therefore,

$$\mathcal{L}^{-1}\left[\frac{s+6}{s^2+4s+9}\right] = \mathcal{L}^{-1}\left[\frac{s+2}{(s+2)^2+5}\right] + \frac{4}{\sqrt{5}}\mathcal{L}^{-1}\left[\frac{\sqrt{5}}{(s+2)^2+5}\right]$$

$$= e^{-2t}\cos\sqrt{5}\,t + \frac{4}{\sqrt{5}}e^{-2t}\sin\sqrt{5}\,t.$$

For the second term, we use partial fractions to write

$$\frac{20}{(s^2+4s+9)(s^2+1)} = \frac{s+2}{s^2+4s+9} - \frac{s-2}{s^2+1}.$$

We have already seen that

$$\mathcal{L}^{-1}\left[\frac{s+2}{s^2+4s+9}\right] = e^{-2t}\cos\sqrt{5}\,t.$$

Moreover,

$$\mathcal{L}^{-1}\left[\frac{s-2}{s^2+1}\right] = \mathcal{L}^{-1}\left[\frac{s}{s^2+1}\right] - 2\mathcal{L}^{-1}\left[\frac{1}{s^2+1}\right] = \cos t - 2\sin t.$$

Therefore, the inverse Laplace transform

$$\mathcal{L}^{-1}\left[e^{-2s}\left(\frac{s+2}{s^2+4s+9} - \frac{s-2}{s^2+1}\right)\right]$$

is

$$u_2(t)\left(e^{-2(t-2)}\cos(\sqrt{5}(t-2)) - \cos(t-2) + 2\sin(t-2)\right).$$

Combining the two results, we obtain the solution of the initial-value problem as

$$y(t) = e^{-2t}\cos\sqrt{5}\,t + \frac{4}{\sqrt{5}}e^{-2t}\sin\sqrt{5}\,t + u_2(t)\left(e^{-2(t-2)}\cos(\sqrt{5}\,(t-2)) - \cos(t-2) + 2\sin(t-2)\right).$$

34. **(a)** First take the Laplace transform of both sides of the equation

$$\mathcal{L}\left[\frac{d^2y}{dt^2}\right] + 3\mathcal{L}[y] = \mathcal{L}[w(t)].$$

We need to compute $\mathcal{L}[w(t)]$. One way to do this is to use the definition

$$\mathcal{L}[w(t)] = \int_0^\infty w(t)e^{-st}\,dt = \int_0^1 te^{-st}\,dt + \int_1^\infty e^{-st}\,dt$$

Evaluating the first integral by parts and the second integral directly yields

$$\mathcal{L}[w(t)] = -\frac{e^{-s}}{s} - \frac{e^{-s}}{s^2} + \frac{1}{s^2} + \frac{e^{-s}}{s} = \frac{1-e^{-s}}{s^2}.$$

(We could also write $w(t) = t - (t-1)u_1(t)$ and use the table.)
Hence, after we transform the equation, we get

$$s^2\mathcal{L}[y] - 2s + 3\mathcal{L}[y] = \frac{1-e^{-s}}{s^2}.$$

(b) Solving for $\mathcal{L}[y]$, we obtain

$$\mathcal{L}[y] = \frac{2s}{s^2+3} + \frac{1-e^{-s}}{s^2(s^2+3)}.$$

(c) To compute the inverse Laplace transform, we note that

$$\mathcal{L}^{-1}\left[\frac{2s}{s^2+3}\right] = 2\cos\sqrt{3}\,t.$$

Next, we note by partial fractions that

$$\frac{1}{s^2(s^2+3)} = \frac{1}{3}\left(\frac{1}{s^2} - \frac{1}{s^2+3}\right)$$

so

$$\mathcal{L}^{-1}\left[\frac{1-e^{-s}}{s^2(s^2+3)}\right] = \frac{1}{3}\left(t - \frac{1}{\sqrt{3}}\sin(\sqrt{3}\,t)\right) - \frac{1}{3}u_1(t)\left((t-1) + \frac{1}{\sqrt{3}}\sin(\sqrt{3}\,(t-1))\right).$$

Combining these two inverses, we obtain the solution

$$y(t) = 2\cos\sqrt{3}\,t + \frac{1}{3}\left(t - \frac{1}{\sqrt{3}}\sin(\sqrt{3}\,t)\right) - \frac{1}{3}u_1(t)\left((t-1) + \frac{1}{\sqrt{3}}\sin(\sqrt{3}\,(t-1))\right).$$

35. **(a)** Consider

$$\mathcal{L}[f] = F(s) = \int_0^\infty f(t)\,e^{-st}\,dt.$$

We can calculate dF/ds by differentiating under the integral sign. That is,

$$\begin{aligned}\frac{dF}{ds} &= \int_0^\infty \frac{\partial}{\partial s}\left(f(t)\,e^{-st}\right)\,dt \\ &= \int_0^\infty f(t)(-t)e^{-st}\,dt \\ &= -\mathcal{L}[tf(t)].\end{aligned}$$

(b) If we apply this result to

$$\mathcal{L}[\sin \omega t] = \frac{\omega}{s^2+\omega^2} = \omega(s^2+\omega^2)^{-1},$$

we obtain

$$\begin{aligned}\mathcal{L}[t\sin \omega t] &= -\omega(-1)(s^2+\omega^2)^{-2}(2s) \\ &= \frac{2\omega s}{(s^2+\omega^2)^2}.\end{aligned}$$

Compare this result with the result of Exercise 6.

EXERCISES FOR SECTION 6.4

1. This is the $\frac{0}{0}$ case of L'Hôpital's Rule. Differentiating numerator and denominator with respect to Δt, we obtain

$$\frac{se^{s\Delta t} - (-s)e^{-s\Delta t}}{2},$$

which simplifies to

$$\frac{s(e^{s\Delta t} + e^{-s\Delta t})}{2}.$$

Since both $e^{s\Delta t}$ and $e^{-s\Delta t}$ tend to 1 as $\Delta t \to 0$, the desired limit is s.

2. Taking Laplace transforms of both sides and applying the rules yields

$$s^2\mathcal{L}[y] - sy(0) - y'(0) + 3\mathcal{L}[y] = 5\mathcal{L}[\delta_2].$$

Simplifying, using the initial conditions, and the fact that $\mathcal{L}[\delta_2] = e^{-2s}$, we get

$$(s^2+3)\mathcal{L}[y] = 5e^{-2s}.$$

Hence,

$$\mathcal{L}[y] = 5\frac{e^{-2s}}{s^2+3}.$$

This can be written as

$$\mathcal{L}[y] = \frac{5}{\sqrt{3}}e^{-2s}\frac{\sqrt{3}}{s^2+3},$$

which yields

$$y(t) = \frac{5}{\sqrt{3}}u_2(t)\sin\left(\sqrt{3}(t-2)\right).$$

3. Applying the Laplace transform to both sides, using the rules, and the fact that $\mathcal{L}[\delta_3] = e^{-3s}$, we get

$$s^2\mathcal{L}[y] - sy(0) - y'(0) + 2s\mathcal{L}[y] - 2y(0) + 5\mathcal{L}[y] = e^{-3s}.$$

Substituting the given initial conditions, we have

$$\mathcal{L}[y] = \frac{s+3}{s^2+2s+5} + \frac{e^{-3s}}{s^2+2s+5}.$$

Using the fact that $s^2+2s+5 = (s+1)^2+4$, we obtain

$$\mathcal{L}[y] = \frac{s+1}{(s+1)^2+4} + \frac{2}{(s+1)^2+4} + \frac{1}{2}e^{-3s}\frac{2}{(s+1)^2+4}.$$

Therefore,

$$y(t) = e^{-t}\cos 2t + e^{-t}\sin 2t + \frac{1}{2}u_3(t)e^{-(t-3)}\sin(2(t-3)).$$

4. Taking the Laplace transform of both sides, using the rules, and the fact that $\mathcal{L}[\delta_2] = e^{-2s}$, we get

$$s^2\mathcal{L}[y] - sy(0) - y'(0) + 2s\mathcal{L}[y] - 2y(0) + 2\mathcal{L}[y] = -2e^{-2s}.$$

Substituting the given initial conditions, we obtain

$$\mathcal{L}[y] = \frac{2s+4}{s^2+2s+2} - \frac{2e^{-2s}}{s^2+2s+2}.$$

Using $s^2+2s+2 = (s+1)^2+1$ in the denominator gives us

$$\mathcal{L}[y] = 2\frac{s+1}{(s+1)^2+1} + 2\frac{1}{(s+1)^2+1} - 2e^{-2s}\frac{1}{(s+1)^2+1}.$$

Taking the inverse Laplace transform, we have

$$y(t) = 2e^{-t}\cos t + 2e^{-t}\sin t - 2u_2(t)e^{-(t-2)}\sin(t-2).$$

5. Applying Laplace transform to both sides, using the rules, and the fact that $\mathcal{L}[\delta_a] = e^{-as}$, we get

$$s^2\mathcal{L}[y] - sy(0) - y'(0) + 2s\mathcal{L}[y] - 2y(0) + 3\mathcal{L}[y] = e^{-s} - 3e^{-4s}.$$

Substituting the initial conditions gives us

$$\mathcal{L}[y] = \frac{e^{-s}}{s^2 + 2s + 3} - \frac{3e^{-4s}}{s^2 + 2s + 3}.$$

Now, using that $s^2 + 2s + 3 = (s + 1)^2 + 2$, we have

$$\mathcal{L}[y] = \frac{1}{\sqrt{2}} e^{-s} \frac{\sqrt{2}}{(s + 1)^2 + 2} + \frac{3}{\sqrt{2}} e^{-4s} \frac{\sqrt{2}}{(s + 1)^2 + 2}.$$

So,

$$y(t) = \frac{1}{\sqrt{2}} u_1(t) e^{-(t-1)} \sin(\sqrt{2}(t - 1)) - \frac{3}{\sqrt{2}} u_4(t) e^{-(t-4)} \sin(\sqrt{2}(t - 4)).$$

6. **(a)** The characteristic polynomial of the unforced oscillator is $\lambda^2 + 2\lambda + 3$, and the eigenvalues are $\lambda = -1 \pm \sqrt{2}i$. Hence, the natural period is $\sqrt{2}\pi$ and the damping causes the solutions of the unforced equation to tend to zero like e^{-t}. At $t = 4$, the system is given a jolt, so the solution rises. After $t = 4$, the equation is unforced, so the solution again tends to zero as e^{-t}.

(b) Taking Laplace transforms of both sides of the equation, we have

$$s^2\mathcal{L}[y] - sy(0) - y'(0) + 2s\mathcal{L}[y] - 2y(0) + 3\mathcal{L}[y] = \mathcal{L}[\delta_4].$$

Plugging in the initial conditions and solving for $\mathcal{L}[y]$ gives us

$$\mathcal{L}[y] = \frac{s + 2}{s^2 + 2s + 3} + \frac{e^{-4s}}{s^2 + 2s + 3}.$$

If we complete the square for the polynomial $s^2 + 2s + 3$, we get $s^2 + 2s + 3 = (s + 1)^2 + 2$, so

$$\mathcal{L}[y] = \frac{s + 1}{(s + 1)^2 + 2} + \frac{1}{\sqrt{2}} \frac{\sqrt{2}}{(s + 1)^2 + 2} + \frac{1}{\sqrt{2}} e^{-4s} \frac{\sqrt{2}}{(s + 1)^2 + 2}.$$

Therefore,

$$y(t) = e^{-t} \cos \sqrt{2}t + \frac{1}{\sqrt{2}} e^{-t} \sin \sqrt{2}t + \frac{1}{\sqrt{2}} u_4(t) e^{-(t-4)} \sin(\sqrt{2}(t - 4)).$$

(c)

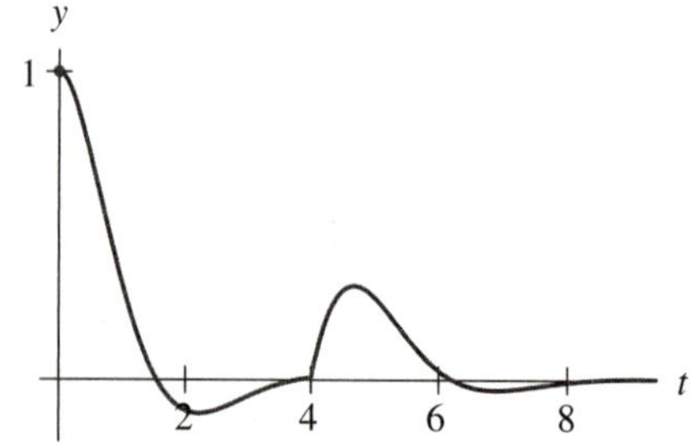

Note that the solution goes through about 3/4 of a natural period before the application of the delta function. The delta function forcing causes the second maximum of the solution to be much higher than it would have been without the forcing, but the long term effect is small because the damping is fairly large.

7. **(a)** From the table

$$\mathcal{L}[\delta_a] = e^{-as}$$

$$s\mathcal{L}[u_a] - u_a(0) = s\frac{e^{-as}}{s} - 0 = e^{-as}.$$

(b) The formula for the Laplace transform of a derivative is

$$\mathcal{L}\left[\frac{dy}{dt}\right] = s\mathcal{L}[y] - y(0)$$

and this is exactly the relationship between the Laplace transforms of $u_a(t)$ and $\delta_a(t)$. Hence, it is tempting to think of the Dirac delta function as the derivative of the Heaviside function.

(c) We can think of the Heaviside function $u_a(t)$ as a limit of piecewise linear functions equal to zero for t less than $a - \Delta t$, equal to one for t greater than $a + \Delta t$ and a straight line for t between $a - \Delta t$ and $a + \Delta t$. The derivative of this function is precisely the function $g_{\Delta t}$ used to define the Dirac delta function. This is still just an informal relationship until we specify in what sense we are taking the limit.

8. Actually, this exercise is a little more complicated than it seems at first. We can think of g as a periodic function with period a and apply Exercise 16 in Section 6.2, but to do so, we must decide how to integrate $\delta_a(t)$ over the interval $0 \le t \le a$. In other words, is the impulse inside or outside the interval?

To avoid this issue, we consider the function

$$f(t) = \sum_{n-0}^{\infty} \delta_{na+u/2}(t).$$

We can apply the periodicity formula from Exercise 16 in Section 6.2 to this function to get

$$\mathcal{L}[f] = \frac{1}{1 - e^{-as}} \int_0^a f(t)\, e^{-st}\, dt = \frac{1}{1 - e^{-as}} \int_0^a \delta_{a/2}(t)\, e^{-st}\, dt,$$

because $\delta_{na+a/2}(t) = 0$ for all $n > 0$ on the interval $[0, a]$. Moreover,

$$\int_0^a \delta_{a/2}(t)\, e^{-st}\, dt = \mathcal{L}[\delta_{a/2}]$$

because $\delta_{a/2}(t) = 0$ for all $t > a/2$. Therefore, we have

$$\mathcal{L}[f] = \frac{e^{-as/2}}{1 - e^{-as}}.$$

To obtain $\mathcal{L}[g]$, we use the relation $g(t) = u_{a/2}(t) f(t - a/2)$ to obtain

$$\mathcal{L}[g] = e^{-as/2}\frac{e^{-as/2}}{1 - e^{-as}} = \frac{e^{-as}}{1 - e^{-as}}.$$

Note that this is the same answer we get if we apply the periodicity formula directly to $g(t)$ assuming that the entire impulse takes place inside the interval $0 \le t \le a$. In other words, if we assume that

$$\int_0^a \delta_a(t)\, e^{-st}\, dt = e^{-as},$$

then we get

$$\begin{aligned}\mathcal{L}[g] &= \frac{1}{1-e^{-as}} \int_0^a g(t)\, e^{-st}\, dt \\ &= \frac{1}{1-e^{-as}} \int_0^a \delta_a(t)\, e^{-st}\, dt \\ &= \frac{e^{-as}}{1-e^{-as}}.\end{aligned}$$

9. **(a)** To compute the Laplace transform of the infinite sum on the right-hand side of the equation, we can either sum the geometric series that results from the fact that $\mathcal{L}[\delta_n] = e^{-ns}$ or use Exercise 16 in Section 6.2. Either way, we get

$$\mathcal{L}\left[\sum_{n=1}^{\infty} \delta_n(t)\right] = \frac{e^{-s}}{1-e^{-s}} = \frac{1}{e^s - 1}.$$

For our purposes, it is actually better to leave the Laplace transform of the right-hand side as

$$\mathcal{L}\left[\sum_{n=1}^{\infty} \delta_n(t)\right] = \sum_{n=1}^{\infty} e^{-ns}.$$

Since $y(0) = 0$ and $y'(0) = 0$, the transformed equation is

$$s^2\mathcal{L}[y] + 2\mathcal{L}[y] = \sum_{n=1}^{\infty} e^{-ns},$$

which simplifies to

$$\mathcal{L}[y] = \frac{1}{s^2+2} \sum_{n=1}^{\infty} e^{-ns} = \sum_{n=1}^{\infty} \frac{e^{-ns}}{s^2+2}.$$

(b) Since

$$\mathcal{L}^{-1}\left[\frac{e^{-ns}}{s^2+2}\right] = \frac{1}{\sqrt{2}} u_n(t) \sin(\sqrt{2}\,(t-n)),$$

we have

$$y(t) = \frac{1}{\sqrt{2}} \sum_{n=1}^{\infty} u_n(t) \sin(\sqrt{2}\,(t-n)).$$

(c) The period of the forcing is different from the natural period of the unforced oscillator. Hence, the solution oscillates but not periodically.

10. **(a)** To compute the Laplace transform of the infinite sum on the right-hand side of the equation, we can either sum the geometric series or use Exercise 16 in Section 6.2 (see Exercise 9 as well). We get

$$\mathcal{L}\left[\sum_{n=1}^{\infty} \delta_{2n\pi}(t)\right] = \frac{e^{-2\pi}}{1-e^{-2\pi}}.$$

Since $y(0) = 0$ and $y'(0) = 0$, the transform of the left-hand side of the equation is

$$(s^2 + 1)\mathcal{L}[y].$$

Therefore,

$$\mathcal{L}[y] = \frac{e^{-2\pi}}{(1 - e^{-2\pi})(s^2 + 1)}.$$

For our purposes, we are better off leaving the transform of the right-hand side of the equation as

$$\sum_{n=1}^{\infty} e^{-2n\pi s}.$$

(b) To take the inverse Laplace transform, we note that

$$\begin{aligned}
y(t) &= \mathcal{L}^{-1}\left[\frac{e^{-2\pi}}{s^2 + 1}\right] + \mathcal{L}^{-1}\left[\frac{e^{-4\pi}}{s^2 + 1}\right] + \dots \\
&= u_{2\pi}(t)\sin(t - 2\pi) + u_{4\pi}(t)\sin(t - 4\pi) + \dots \\
&= \sum_{n=1}^{\infty} u_{2n\pi}(t)\sin(t - 2n\pi) \\
&= (\sin t)\sum_{n-1}^{\infty} u_{2n\pi}(t).
\end{aligned}$$

(c) The period of the forcing is the same as the natural period of the unforced oscillator. Thus, we have resonant forcing. We expect the solution to oscillate periodically and the amplitudes of the oscillations to grow linearly. These expectations agree with the formula for $y(t)$ derived in part (b).

EXERCISES FOR SECTION 6.5

1. Using the definition of the convolution with f and g, we see that

$$\begin{aligned}
(f * g)(t) &= \int_0^t 1 \cdot e^{-u}\,du \\
&= \int_0^t e^{-u}\,du \\
&= -e^{-u}\Big|_0^t \\
&= 1 - e^{-t}.
\end{aligned}$$

Checking the convolution property ($\mathcal{L}[f * g] = \mathcal{L}[f] \cdot \mathcal{L}[g]$) for Laplace transforms, we have

$$\mathcal{L}[f] = \frac{1}{s}, \quad \mathcal{L}[g] = \frac{1}{s+1},$$

and

$$\begin{aligned}
\mathcal{L}[f * g] &= \frac{1}{s} - \frac{1}{s+1} \\
&= \frac{s+1-s}{s(s+1)} \\
&= \frac{1}{s(s+1)}.
\end{aligned}$$

So, $\mathcal{L}[f] \cdot \mathcal{L}[g] = \mathcal{L}[f * g]$.

2. Using the definition of the convolution with f and g, we see that

$$\begin{aligned}
(f * g)(t) &= \int_0^t e^{-a(t-u)} e^{-bu}\, du \\
&= \int_0^t e^{-at} e^{(a-b)u}\, du \\
&= -e^{-at} \left. \frac{e^{(a-b)u}}{a-b} \right|_0^t \\
&= e^{-at} \left(\frac{e^{(a-b)t}}{a-b} - \frac{1}{a-b} \right) \\
&= \frac{e^{-bt}}{a-b} - \frac{e^{-at}}{a-b}.
\end{aligned}$$

Checking the convolution property ($\mathcal{L}[f * g] = \mathcal{L}[f] \cdot \mathcal{L}[g]$) for Laplace transforms, we have

$$\mathcal{L}[f] = \frac{1}{s+a}, \quad \mathcal{L}[g] = \frac{1}{s+b},$$

and

$$\begin{aligned}
\mathcal{L}[f * g] &= \frac{1}{(s+b)(a-b)} - \frac{1}{(s+a)(a-b)} \\
&= \frac{s+a}{(s+a)(s+b)(a-b)} - \frac{s+b}{(s+a)(s+b)(a-b)} \\
&= \frac{a-b}{(s+a)(s+b)(a-b)} \\
&= \frac{1}{(s+a)(s+b)}.
\end{aligned}$$

Therefore, $\mathcal{L}[f] \cdot \mathcal{L}[g] = \mathcal{L}[f * g]$.

3. Using the definition of the convolution with f and g, we see that

$$(f * g)(t) = \int_0^t \cos(t - v)u_2(v)\,dv.$$

(We're using v as the integrating variable instead of u so as not to confuse it with the Heaviside function.) First, notice that if $0 < v < 2$, then $u_2(v) = 0$. Thus, if $t < 2$, then the function $u_2(v)$ is always 0, which means the integral is 0. Now, if $t \geq 2$, then

$$\begin{aligned}\int_0^t \cos(t - v)u_2(v)\,dv &= \int_0^2 \cos(t - v)\,0\,dv + \int_2^t \cos(t - v)\,1\,dv \\ &= \int_2^t \cos(t - v)\,dv.\end{aligned}$$

So,

$$\int_0^t \cos(t - v)u_2(v)\,dv = \begin{cases} 0, & \text{if } t < 2, \\ \displaystyle\int_2^t \cos(t - v)\,dv, & \text{if } t \geq 2. \end{cases}$$

Evaluating the second integral, we get

$$\int_2^t \cos(t - v)\,dv = -\sin(t - v)\Big|_2^t = \sin(t - 2).$$

We have a function that is 0 for $t < 2$ and equal to $\sin(t - 2)$ for $t \geq 2$, so our function is

$$u_2(t)\sin(t - 2).$$

Checking the convolution property ($\mathcal{L}[f * g] = \mathcal{L}[f] \cdot \mathcal{L}[g]$) for Laplace transforms, we have

$$\mathcal{L}[f] = \frac{s}{s^2 + 1}, \qquad \mathcal{L}[g] = \frac{e^{-2s}}{s},$$

and

$$\mathcal{L}[f * g] = \frac{e^{-2s}}{s^2 + 1}.$$

Hence, $\mathcal{L}[f] \cdot \mathcal{L}[g] = \mathcal{L}[f * g]$.

4. Using the definition of the convolution with f and g, we see that

$$(f * g)(t) = \int_0^t u_2(t - v)u_3(v)\,dv.$$

(We're using v as the integrating variable instead of u so as not to confuse it with the Heaviside function.) First, notice that if $0 < v < 3$, then $u_3(v) = 0$. Also, if $t - 2 < v < t$, then $u_2(t - v) = 0$. So if $t < 5$, then one or the other of those functions is equal to 0 for all $0 < v < t$, so the integral is 0. But if $t \geq 5$, then

$$\begin{aligned}\int_0^t u_2(t - v)u_3(v)\,dv &= \int_0^3 u_2(t - v)\,0\,dv + \int_3^{t-2} u_2(t - v)u_3(v)\,dv + \int_{t-2}^t 0\,u_3(v)\,dv \\ &= \int_3^{t-2} 1\,dv.\end{aligned}$$

So

$$\int_0^t u_2(t-v)u_3(v)\,dv = \begin{cases} 0, & \text{if } t < 5, \\ \displaystyle\int_3^{t-2} 1\,dv, & \text{if } t \geq 5. \end{cases}$$

Evaluating the second integral, we get

$$\int_3^{t-2} 1\,dv = t\Big|_3^{t-2} = t - 5.$$

We have a function that is 0 for $t < 5$ and equal to $t-5$ for $t \geq 5$, so our function is $u_5(t)(t-5)$.

Checking the convolution property ($\mathcal{L}[f * g] = \mathcal{L}[f] \cdot \mathcal{L}[g]$) for Laplace transforms, we have

$$\mathcal{L}[f] = \frac{e^{-2s}}{s}, \quad \mathcal{L}[g] = \frac{e^{-3s}}{s},$$

and

$$\mathcal{L}[f * g] = \frac{e^{-5s}}{s^2}.$$

So, $\mathcal{L}[f] \cdot \mathcal{L}[g] = \mathcal{L}[f * g]$.

5. Using the definition of the convolution with f and g, we see that

$$(f * g)(t) = \int_0^t 3\sin(t-u)\cos(2u)\,du.$$

We will use four trigonometric identities to evaluate this integral:

$$\sin(t-u) = \sin t \cos u - \cos t \sin u$$

$$\sin(mt)\sin(nt) = \tfrac{1}{2}\left[\cos((m-n)t) - \cos((m+n)t)\right]$$

$$\cos(mt)\cos(nt) = \tfrac{1}{2}\left[\cos((m+n)t) + \cos((m-n)t)\right]$$

$$\sin(mt)\cos(nt) = \tfrac{1}{2}\left[\sin((m+n)t) + \sin((m-n)t)\right].$$

So

$$\int_0^t 3\sin(t-u)\cos(2u)\,du$$

$$= \int_0^t \left[3\cos 2u \cos u \sin t - \cos 2u \sin u \cos t\right] du$$

$$= \int_0^t \left[\tfrac{3}{2}(\cos 3u + \cos u)\sin t - \tfrac{3}{2}(\sin 3u - \sin u)\cos t\right] du$$

$$= \sin t\left[\tfrac{1}{2}\sin 3u + \tfrac{3}{2}\sin u\right]_0^t + \cos t\left[\tfrac{1}{2}\cos 3u - \tfrac{3}{2}\cos u\right]_0^t$$

$$= \sin t\left(\tfrac{1}{2}\sin 3t + \tfrac{3}{2}\sin t\right) + \cos t\left(\tfrac{1}{2}\cos 3t - \tfrac{3}{2}\cos t + 1\right)$$

$$= \tfrac{1}{2}\sin 3t \sin t + \tfrac{3}{2}\sin^2 t + \tfrac{1}{2}\cos 3t \cos t - \tfrac{3}{2}\cos^2 t + \cos t$$

$$= \tfrac{1}{4}(\cos 2t - \cos 4t) + \tfrac{3}{2}\sin^2 t + \tfrac{1}{4}(\cos 4t + \cos 2t) - \tfrac{3}{2}\cos^2 t + \cos t$$

$$= \tfrac{1}{2}\cos 2t + \tfrac{3}{2}\left(\sin^2 t - \cos^2 t\right) + \cos t$$

$$= \tfrac{1}{2}\cos 2t - \tfrac{3}{2}\cos 2t + \cos t$$

$$= \cos t - \cos 2t,$$

which is the same answer obtained in the text using the technique of Laplace transforms.

6. We will use the substitution $v = t - u$, so that $u = t - v$, and $du = -dv$. Also, as u goes from 0 to t, v goes from t to 0, so we have

$$\begin{aligned}(f * g)(t) &= \int_0^t f(t-u)g(u)\,du \\ &= -\int_t^0 f(v)g(t-v)\,dv \\ &= \int_0^t f(v)g(t-v)\,dv \\ &= (g * f)(t).\end{aligned}$$

7. Taking Laplace transform of both sides of the equation and solving for $\mathcal{L}[\zeta]$ (see page 607), we obtain

$$\mathcal{L}[\zeta] = \frac{1}{s^2 + ps + q}.$$

Hence, if we let

$$z(s) = s^2 + ps + q,$$

we have that $z(0) = 5$ and $z(2) = 17$. Now $z(0) = 5$ implies $q = 5$. Using $z(2) = 17 = 2^2 + 2p + 5$, we see that $p = 4$.

8. Since $\eta(t)$ solves the first equation, we know that

$$\frac{d\eta}{dt} + a\eta = f(t), \quad \eta(0) = 0.$$

Taking the Laplace transform of both sides of the equation, we get

$$s\mathcal{L}[\eta] - \eta(0) + a\mathcal{L}[\eta] = \mathcal{L}[f].$$

Substituting the initial condition and solving for $\mathcal{L}[\eta]$, we have

$$\mathcal{L}[\eta] = \frac{\mathcal{L}[f]}{s + a}.$$

Now, since $\zeta(t)$ solves the second equation, we know that

$$\frac{d\zeta}{dt} + a\zeta = \delta_0.$$

So

$$s\mathcal{L}[\zeta] + a\mathcal{L}[\zeta] = \mathcal{L}[\delta_0],$$

and

$$\mathcal{L}[\zeta] = \frac{1}{s+a}.$$

Hence,

$$\mathcal{L}[\zeta] \cdot \mathcal{L}[f] = \mathcal{L}[\eta].$$

9. **(a)** Since ζ solves the initial-value problem above, we know that

$$\frac{d\zeta^2}{dt^2} + p\frac{d\zeta}{dt} + q\zeta = \delta_0(t), \quad \zeta(0) = \zeta'(0) = 0^-.$$

Taking Laplace transforms of both sides, and substituting initial conditions gives us

$$s^2\mathcal{L}[\zeta] + ps\mathcal{L}[\zeta] + q\mathcal{L}[\zeta] = 1,$$

which yields

$$\mathcal{L}[\zeta] = \frac{1}{s^2 + ps + q}.$$

Now, taking Laplace transforms of both sides of

$$\frac{d^2y}{dt^2} + p\frac{dy}{dt} + qy = 0, \quad y(0) = a, \, y'(0) = 0$$

gives us

$$s^2\mathcal{L}[y] - sa + ps\mathcal{L}[y] - pa + q\mathcal{L}[y] = 0.$$

Solving for $\mathcal{L}[y]$ gives

$$\mathcal{L}[y] = \frac{a(s+p)}{s^2 + ps + q},$$

so

$$\mathcal{L}[y] = a(s+p)\mathcal{L}[\zeta].$$

(b) Taking Laplace transforms of both sides of

$$\frac{d^2y}{dt^2} + p\frac{dy}{dt} + qy = 0, \quad y(0) = 0, \, y'(0) = b$$

gives us

$$s^2\mathcal{L}[y] - b + ps\mathcal{L}[y] + q\mathcal{L}[y] = 0.$$

Solving for $\mathcal{L}[y]$ gives

$$\mathcal{L}[y] = \frac{b}{s^2 + ps + q},$$

so

$$\mathcal{L}[y] = b\mathcal{L}[\zeta].$$

(c) Taking Laplace transforms of both sides of

$$\frac{d^2y}{dt^2} + p\frac{dy}{dt} + qy = f(t), \quad y(0) = a,\, y'(0) = b$$

gives us

$$s^2\mathcal{L}[y] - sa - b + ps\mathcal{L}[y] - pa + q\mathcal{L}[y] = \mathcal{L}[f].$$

Solving for $\mathcal{L}[y]$ gives

$$\mathcal{L}[y] = \frac{\mathcal{L}[f] + a(s+p) + b}{s^2 + ps + q},$$

so

$$\mathcal{L}[y] = (\mathcal{L}[f] + a(s+p) + b)\,\mathcal{L}[\zeta].$$

10. Since η solves the first initial-value problem, we know that

$$\frac{d\eta^2}{dt^2} + p\frac{d\eta}{dt} + q\eta = u_0(t), \quad \eta(0) = \eta'(0) = 0^-.$$

Taking Laplace transforms of both sides and replacing the initial conditions gives us

$$s^2\mathcal{L}[\eta] + ps\mathcal{L}[\eta] + q\mathcal{L}[\eta] = \frac{1}{s}.$$

Solving for $\mathcal{L}[\eta]$ gives

$$\mathcal{L}[\eta] = \frac{1}{s(s^2 + ps + q)}.$$

If we take the Laplace transform of both sides of the second initial-value problem, and solve for $\mathcal{L}[y]$, we have

$$\mathcal{L}[y] = \frac{\mathcal{L}[f]}{s^2 + ps + q}.$$

Using the convolution property for Laplace Transforms, we get

$$\begin{aligned}\mathcal{L}[y] &= s\,(\mathcal{L}[f] \cdot \mathcal{L}[\eta]) \\ &= s\,(\mathcal{L}[f * \eta]).\end{aligned}$$

Now,

$$(f * \eta)(t) = \int_0^t f(t-u)\eta(u)\,du,$$

so

$$(f * \eta)(0) = \int_0^0 f(t-u)\eta(u)\,du = 0.$$

Using the rule that

$$\mathcal{L}\left[\frac{dy}{dt}\right] = s\mathcal{L}[y] - y(0),$$

we have that

$$\begin{aligned}\mathcal{L}\left[\frac{d}{dt}(f * \eta)\right] &= s\mathcal{L}[(f * \eta)] - (f * \eta)(0) \\ &= s\mathcal{L}[(f * \eta)] \\ &= \mathcal{L}[y].\end{aligned}$$

Therefore

$$y(t) = \frac{d}{dt} \int_0^t f(t-u)\eta(u)\,du.$$

This could be evaluated further, but the resulting expression is much more complicated.

11. **(a)** We know that

$$\frac{d^2y_1}{dt^2} + p\frac{dy_1}{dt} + qy_1 = f_1(t).$$

Taking Laplace transforms of both sides and solving for $\mathcal{L}[y_1]$ gives

$$\mathcal{L}[y_1] = \frac{\mathcal{L}[f_1]}{s^2 + ps + q}.$$

(b) As above, we know that

$$\frac{d^2y_2}{dt^2} + p\frac{dy_2}{dt} + qy_2 = f_2(t),$$

so

$$\mathcal{L}[y_2] = \frac{\mathcal{L}[f_2]}{s^2 + ps + q}.$$

Now we see that

$$\frac{\mathcal{L}[f_1]}{\mathcal{L}[y_1]} = s^2 + ps + q,$$

and

$$\frac{\mathcal{L}[f_2]}{\mathcal{L}[y_2]} = s^2 + ps + q,$$

so

$$\frac{\mathcal{L}[f_1]}{\mathcal{L}[y_1]} = \frac{\mathcal{L}[f_2]}{\mathcal{L}[y_2]}.$$

(c) Solving for $\mathcal{L}[y_2]$ gives us

$$\mathcal{L}[y_2] = \mathcal{L}[f_2]\frac{\mathcal{L}[y_1]}{\mathcal{L}[f_1]}.$$

EXERCISES FOR SECTION 6.6

1. **(a)** Taking Laplace transform of both sides and substituting the initial conditions yields

$$s^2\mathcal{L}[y] - 2s + 2 + 2s\mathcal{L}[y] - 4 + 2\mathcal{L}[y] = \frac{4}{(s+2)^2 + 16}.$$

Hence,

$$\mathcal{L}[y] = \frac{2s+2}{s^2+2s+2} + \frac{4}{(s^2+4s+20)(s^2+2s+2)}.$$

(b) The poles are the roots of $s^2 + 2s + 2$ and $s^2 + 4s + 20$, or $-1 \pm i$ and $-2 \pm 4i$.

(c) Since all poles have negative real parts, the solution tends to zero at an exponential rate. The real part closest to 0 is -1 so solutions tend to zero at a rate of e^{-t}. Since the poles are complex, the solutions oscillate. The oscillations with period $2\pi/4 = \pi/2$ decay at the rate e^{-2t} while the oscillations with period 2π decay at the rate e^{-t}.

2. **(a)** Taking Laplace transform of both sides and substituting the initial conditions gives

$$s^2\mathcal{L}[y] + 2s + s\mathcal{L}[y] + 2 + 5\mathcal{L}[y] = \mathcal{L}[u_2(t)\sin(4(t-2))].$$

So,

$$(s^2 + s + 5)\mathcal{L}[y] + 2s + 2 = \frac{4e^{-2s}}{s^2+16},$$

and

$$\mathcal{L}[y] = -\frac{2s+2}{s^2+s+5} + \frac{4e^{-2s}}{(s^2+s+5)(s^2+16)}.$$

(b) The poles are the roots of $s^2 + 16$ and $s^2 + s + 5$ or $s = \pm 4i$ and $s = -1/2 \pm i\sqrt{19}/2$.

(c) The poles $s = \pm 4i$ indicate that the long-term behavior is oscillation with constant amplitude and period $2\pi/4$, the forcing period. The forcing term does not "turn on" until time $t = 2$. The natural response corresponds to the poles $s = -1/2 \pm i\sqrt{19}/2$. This decays like $e^{-t/2}$ and oscillates with period $4\pi/\sqrt{19}$.

3. **(a)** Taking the Laplace transform of both sides and plugging in the initial conditions gives

$$s^2\mathcal{L}[y] + s\mathcal{L}[y] + 8\mathcal{L}[y] = \mathcal{L}[\cos(t-4)] - \mathcal{L}[u_4(t)\cos(t-4)].$$

To take the Laplace transform of the first term on the right we recall that

$$\cos(t-4) = \cos(4)\cos(t) + \sin(4)\sin(t),$$

so

$$(s^2 + s + 8)\mathcal{L}[y] = \cos(4)\frac{s}{s^2+1} + \sin(4)\frac{1}{s^2+1} - \frac{se^{-4s}}{s^2+1}.$$

Therefore,

$$\mathcal{L}[y] = \frac{\cos(4)s + \sin(4) - se^{-4s}}{(s^2+s+8)(s^2+1)}.$$

(b) The poles are given by $s^2 + 1 = 0$ and $s^2 + s + 8 = 0$ or $s = \pm i$ and $s = -1/2 \pm i\sqrt{31}/2$.

(c) Note that the forcing "turns off" at time $t = 4$. Hence up until time $t = 4$ there is a forced response with period 2π. Since $t = 4$ is only about half a period for the forced response, this is not very significant. The natural response is an oscillation with period $4\pi/\sqrt{31}$ which decays like $e^{-t/2}$, and this is the long-term behavior of the solution.

4. **(a)** It is easy to see that the main problem will be to compute the Laplace transform of the forcing function. We can break it into two parts. The easier part to calculate is

$$\mathcal{L}\left[-u_2(t)e^{-(t-2)/10}\sin(t-2)\right] = \frac{-e^{-2s}}{(s+1/10)^2+1}.$$

Now, to calculate

$$\mathcal{L}\left[e^{-(t-2)/10}\sin(t-2)\right],$$

we need to do some precalculation first. Since we don't have the $u_2(t)$ in front, we need to make the variable look like t instead of $t - 2$. So

$$\begin{aligned} e^{-(t-2)/10}\sin(t-2) &= e^{2/10}e^{-t/10}\sin(t-2) \\ &= e^{1/5}e^{-t/10}\left(\cos(2)\sin t - \sin(2)\cos t\right). \end{aligned}$$

So

$$\mathcal{L}\left[e^{-(t-2)/10}\sin(t-2)\right] = e^{1/5}\cos(2)\frac{1}{(s+1/10)^2+1} - e^{1/5}\sin(2)\frac{s+1/10}{(s+1/10)^2+1}.$$

Putting all of this together, we see that

$$\mathcal{L}\left[(1-u_2(t))e^{-(t-2)/10}\sin(t-2)\right] = \frac{e^{1/5}\cos(2) - e^{1/5}\sin(2)(s+1/10) - e^{-2s}}{(s+1/10)^2+1}.$$

Putting this into the original equation, we have that

$$\begin{aligned} s^2\mathcal{L}[y] - s - 2 + s\mathcal{L}[y] - 1 + 3\mathcal{L}[y] \\ = \frac{e^{1/5}\cos(2) - e^{1/5}\sin(2)(s+1/10) - e^{-2s}}{(s+1/10)^2+1}. \end{aligned}$$

Solving for $\mathcal{L}[y]$, we get

$$\mathcal{L}[y] = \frac{e^{1/5}\cos(2) - e^{1/5}\sin(2)(s+1/10) - e^{-2s}}{(s^2+s+3)((s+1/10)^2+1)} + \frac{s+3}{s^2+s+3}.$$

(b) The poles are $s = -1/2 \pm i\sqrt{11}/2$ and $s = -1/10 \pm i/2$.

(c) Note that the forcing "turns off" at time $t = 2$. Hence up until time $t = 2$ there is a forced response with period 2π. Since $t = 2$ is less than half a period for the forced response, this is insignificant. The natural response is an oscillation with period $4\pi/\sqrt{11}$ which decays like $e^{-t/2}$, and this is the long-term behavior of the solution.

5. **(a)** Computing the Laplace transform of both sides of the equation and using the initial conditions gives

$$s^2\mathcal{L}[y] - s - 1 + 16\mathcal{L}[y] = 0,$$

so

$$\mathcal{L}[y] = \frac{s+1}{s^2+16}.$$

(b) The poles are the solutions of $s^2 + 16 = 0$ or $s = \pm 4i$.

(c) If we find the inverse Laplace transform, we see that

$$y(t) = \cos(4t) + (1/4)\sin(4t).$$

So a reasonable conjecture would be that when the poles of the Laplace transform are on the imaginary axis, the solution is periodic and does not decay.

6. **(a)** Computing the Laplace transform of both sides of the equation and substituting the initial conditions gives

$$s^2\mathcal{L}[y] + 4\mathcal{L}[y] = \frac{2}{s^2+4}.$$

Hence,

$$\mathcal{L}[y] = \frac{2}{(s^2+4)^2}.$$

(b) The poles are $s = \pm 2i$, however, these are both "double roots" of $(s^2+4)^2$.

(c) Since this is an undamped harmonic oscillator with resonant forcing, we expect that the solution will contain terms of the form $t\sin 2t$ and $t\cos 2t$. Hence, we conjecture that double roots on the imaginary axis correspond to terms of this form (that is, oscillation with linearly increasing amplitude).

7. **(a)** As usual, we take the Laplace transform of both sides of the equation and substitute the initial conditions to get

$$s^2\mathcal{L}[y] - s - 2 + 2s\mathcal{L}[y] - 2 + \mathcal{L}[y] = 0.$$

Hence,

$$\mathcal{L}[y] = \frac{s+4}{s^2+2s+1}.$$

(b) The poles are the roots of $s^2 + 2s + 1 = (s+1)^2$. Hence, there is a "double pole" at $s = -1$.

(c) For homogeneous second-order equations, double poles play the same role as double eigenvalues. In the case of a double pole on the real axis, the solution is a critically damped oscillator.

8. **(a)** Taking the Laplace transform of both sides and substituting the initial conditions gives

$$s^2\mathcal{L}[y] - s - 1 + 16\mathcal{L}[y] = \frac{1}{s^2},$$

so

$$\mathcal{L}[y] = \frac{s+1}{s^2+16} + \frac{1}{s^2(s^2+16)}.$$

(b) The poles are $s = \pm 4i$ and $s = 0$.

(c) If we find the inverse Laplace transform, we see that

$$y(t) = \frac{t}{16} + \cos(4t) + \frac{15}{64}\sin(4t).$$

So a reasonable conjecture would be that when there is a "double pole" at zero, the solution grows linearly with t, much like the case of resonance, where you also have a double pole. If we look at how the double pole arose in the algebraic manipulations, the double pole is not caused by a resonant forcing term but by a linearly growing forcing term, but the resulting behavior is very similar.

9. **(a)** As usual, we compute

$$s^2\mathcal{L}[y] - sy(0) - y'(0) + 20s\mathcal{L}[y] - 20y(0) + 200\mathcal{L}[y] = \mathcal{L}[w(t)]$$

and using the initial conditions and Exercise 17 of Section 6.2, we have that

$$(s^2 + 20s + 200)\mathcal{L}[y] - s - 20 = \frac{1-e^{-s}}{s(1+e^{-s})}.$$

Hence,

$$\mathcal{L}[y] = \frac{s+20}{s^2+20s+200} + \frac{1-e^{-s}}{s(1+e^{-s})(s^2+20s+200)}.$$

(b) The poles are the roots of $s^2+20s+200 = 0$ ($s = -10 \pm 10i$) and the zeros of $s(1+e^{-s}) = 0$. One zero is $s = 0$, and using Euler's formula, we obtain the other zeros $s = (2n+1)i\pi$ for $n = 0, \pm 1, \pm 2, \ldots$.

(c) The natural response corresponds to the poles $s = -10 \pm 10i$, so it decays like e^{-10t} (very rapidly), while oscillating with a period of $\pi/5$. The remainder of the poles correspond to the forcing and indicate a forced response, which is an oscillation between $\pm 1/200$. Because the forcing is discontinuous, solutions settle (quickly) toward $1/200$ for $0 < t < 1$. When the forcing switches, they tend (quickly) to $-1/200$. (The function $y(t) = 1/200$ is a particular solution of the equation

$$\frac{d^2y}{dt^2} + 20\frac{dy}{dt} + 200y = 1,$$

and $y(t) = -1/200$ is a solution for the same equation if the forcing is -1 rather than 1.)

10. **(a)** Taking Laplace transforms of both sides, using the initial conditions, and the result of Exercise 18 of Section 6.2, we get

$$s^2\mathcal{L}[y] - s + 20s\mathcal{L}[y] - 20 + 200\mathcal{L}[y] = \frac{1}{s^2} - \frac{e^{-s}}{s(1-e^{-s})}.$$

Hence,

$$\mathcal{L}[y] = \frac{s+20}{s^2+20s+200} + \frac{1}{s^2(s^2+20s+200)} - \frac{e^{-s}}{s(1-e^{-s})(s^2+20s+200)}.$$

(b) The poles are $s = -10 \pm 10i$ and $s = 0$.

(c) Each successive push from the forcing results in a quickly decaying oscillation. In one unit of time the solutions decrease by a factor of e^{-10}, hence the total amount of all the small pushes is bounded by the sum of a geometric series and is finite.

REVIEW EXERCISES FOR CHAPTER 6

1. Note that

$$\frac{4}{s^2-1} = \frac{2}{s-1} - \frac{2}{s+1}.$$

Since

$$\mathcal{L}^{-1}\left[\frac{1}{s-1}\right] = e^t \quad \text{and} \quad \mathcal{L}^{-1}\left[\frac{1}{s+1}\right] = e^{-t},$$

we have

$$\mathcal{L}^{-1}\left[\frac{4}{s^2-1}\right] = 2(e^t - e^{-t}).$$

2. We know that $\mathcal{L}[dy/dt] = s\mathcal{L}[y] - y(0)$ and $\mathcal{L}[d^2y/dt^2] = s^2\mathcal{L}[y] - sy(0) - y'(0)$. If we apply the second formula to dy/dt rather than $y(t)$, we get

$$\mathcal{L}\left[\frac{d^3y}{dt^3}\right] = s^2\mathcal{L}\left[\frac{dy}{dt}\right] - sy'(0) - y''(0),$$

and using the first formula, we obtain

$$\begin{aligned}\mathcal{L}\left[\frac{d^3y}{dt^3}\right] &= s^2\left(s\mathcal{L}[y] - y(0)\right) - sy'(0) - y''(0)\\ &= s^3\mathcal{L}[y] - s^2y(0) - sy'(0) - y''(0).\end{aligned}$$

3. We know that the general solution to this differential equation is $y(t) = \alpha \cos \sqrt{5}\,t + \beta \sin \sqrt{5}\,t$. Therefore, the solution to the given initial-value problem is $y(t) = \cos \sqrt{5}\,t$. Its Laplace transform is

$$\frac{s}{s^2+5}.$$

4. Using the definition of $u_4(t)$, we get

$$1 - u_4(t) = \begin{cases} 1, & \text{if } t < 4; \\ 0, & \text{if } t \geq 4. \end{cases}$$

Hence,

$$\int_0^\infty (1 - u_4(t))t\,dt = \int_0^4 t\,dt = 8.$$

5. Note that $y(t) = 1 - u_6(t)$. Therefore,

$$\mathcal{L}[y(t)] = \mathcal{L}[1 - u_6(t)] = \mathcal{L}[1] - \mathcal{L}[u_6(t)] = \frac{1}{s} - \frac{e^{-6s}}{s} = \frac{1 - e^{-6s}}{s}.$$

6. The function that is 1 for $t < 1$ and 0 for $t \geq 1$ is $1 - u_1(t)$. The function that is $\sin t$ for $1 \leq t < \pi$ and 0 elsewhere is

$$(u_1(t) - u_\pi(t)) \sin t.$$

The function that is 0 for $t < \pi$ and 2 for $t \geq \pi$ is $2u_\pi(t)$. Hence,

$$\begin{aligned} y(t) &= (1 - u_1(t)) + (u_1(t) - u_\pi(t)) \sin t + 2u_\pi(t) \\ &= 1 + u_1(t)(-1 + \sin t) + u_\pi(t)(2 - \sin t). \end{aligned}$$

7. This improper integral is the Laplace transform of $\delta_2(t)$ evaluated at $s = 2$. Since $\mathcal{L}[\delta_2(t)] = e^{-2s}$,

$$\int_0^\infty \delta_2(t)e^{-2t}\,dt = e^{-4}.$$

8. Taking the Laplace transform of both sides, we get

$$\begin{aligned} \mathcal{L}\left[\frac{dy}{dt}\right] &= \mathcal{L}[\delta_1(t)] \\ s\mathcal{L}[y] - y(0) &= e^{-s} \\ s\mathcal{L}[y] &= e^{-s} \\ \mathcal{L}[y] &= \frac{e^{-s}}{s}. \end{aligned}$$

Therefore,

$$y(t) = \mathcal{L}^{-1}\left[\frac{e^{-s}}{s}\right] = u_1(t).$$

This computation justifies the statement "the derivative of a Heaviside function is a delta function."

9. We can use the formula in Exercise 16 from Section 6.2. However, the result of Review Exercise 8 can be generalized to yield the fact that $u_n(t)$ is the solution to the initial-value problem

$$\frac{dy}{dt} = \delta_n(t), \quad y(0) = 0.$$

If we do not worry about the convergence of any of the infinite sums involved, we can use the linearity of the differential equation to produce the solution

$$y(t) = u_1(t) + u_2(t) + u_3(t) + \dots .$$

(See the result of Exercise 19 in Section 1.8).

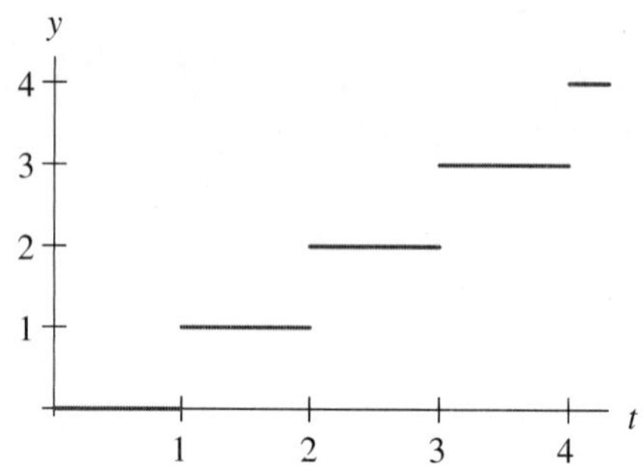

10. Following the procedure of Exercise 9 exactly, we obtain

$$y(t) = u_1(t) - u_2(t) + u_3(t) - u_4(t) + \dots .$$

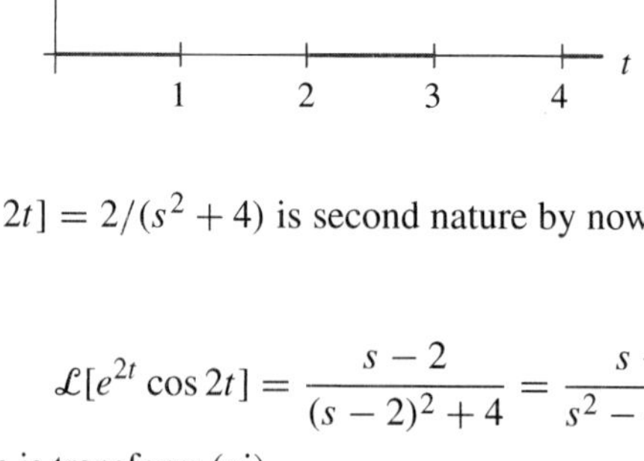

11. **(a)** The transform $\mathcal{L}[\sin 2t] = 2/(s^2 + 4)$ is second nature by now. Hence, this transform is transform (iii).

(b) We know that

$$\mathcal{L}[e^{2t}\cos 2t] = \frac{s-2}{(s-2)^2+4} = \frac{s-2}{s^2-4s+8}.$$

Hence, this transform is transform (xi).

(c) We have

$$\mathcal{L}[e^{2t} - e^{-2t}] = \frac{1}{s-2} - \frac{1}{s+2} = \frac{4}{s^2-4}.$$

This transform is transform (i).

(d) We know that $\sin(t-\pi) = -\sin t$. Therefore,

$$\mathcal{L}[\sin(t-\pi)] = \mathcal{L}[-\sin t] = \frac{-1}{s^2+1},$$

and this transform is transform (vi).

(e) The transform $\mathcal{L}[\cos 2t] = s/(s^2+4)$ is also second nature. Hence, this transform is transform (vii).

(f) We have

$$\mathcal{L}[u_2(t)e^{3(t-2)}] = e^{-2s}\mathcal{L}[e^{3t}] = \frac{e^{-2s}}{s-3}.$$

Hence, this transform is transform (ix).

12. Strictly speaking, this statement is false because the formula $\mathcal{L}\,[dy/dt] = s\mathcal{L}[y] - y(0)$ replaces the operation of differentiation with multiplication by s and subtraction of a constant. However, we consider the statement to be "essentially" true. Perhaps a better true/false exercise would be the statement: The Laplace transform replaces operations of calculus with operations of algebra.

13. True. The Laplace transform of $\delta_1(t)$ is e^{-s}, which is an infinitely differentiable function in s. (The Laplace transform is a "smoothing" operator. The transformed function $Y(s) = \mathcal{L}[y(t)]$ tends to be more differentiable than the original function $y(t)$.)

14. False. Since the differential equation is linear, the function $2y_1(t)$ is a solution to

$$\frac{d^2y}{dt^2} + 2\frac{dy}{dt} + 6y = 2\delta_3(t).$$

Consequently, $2y_1(t) = 0$ for $t < 3$, but $2y_1(t)$ is nonzero for (most) $t > 3$. The solution of the second initial-value problem is zero for $t < 6$.

15. True. We use the linearity of the equation and solve it for the forcing functions $\delta_\pi(t)$, $\delta_{2\pi}(t)$, $\delta_{4\pi}(t)$, ... individually. For example, for the initial-value problem

$$\frac{d^2y}{dt^2} + 4y = \delta_\pi(t), \quad y(0) = 0, \quad y'(0) = 0,$$

we get

$$\mathcal{L}[y] = \frac{e^{-\pi s}}{s^2+4},$$

and therefore, $y(t) = u_\pi(t)\sin 2(t-\pi)$.

Next we solve

$$\frac{d^2y}{dt^2} + 4y = \delta_{2\pi}(t), \quad y(0) = 0, \quad y'(0) = 0,$$

and we obtain $y(t) = u_{2\pi}(t)\sin 2(t-2\pi)$. Continuing in this manner, we obtain

$$\begin{aligned} y(t) &= u_\pi(t)\sin 2(t-\pi) + u_{2\pi}(t)\sin 2(t-2\pi) + u_{4\pi}(t)\sin 2(t-4\pi) + \dots \\ &= u_\pi(t)\sin 2t + u_{2\pi}(t)\sin 2t + u_{4\pi}(t)\sin 2t + \dots \\ &= (u_\pi(t) + u_{2\pi}(t) + u_{4\pi}(t) + \dots)\sin 2t, \end{aligned}$$

as the solution to the original initial-value problem. This solution oscillates with increasing and unbounded amplitude.

16. False. The partial fractions decomposition of this rational function has the form

$$\frac{7s}{(s^2+1)(s^2+3)(s^2+5)} = \frac{As+B}{s^2+1} + \frac{Cs+D}{s^2+3} + \frac{Es+F}{s^2+5},$$

where A, B, C, D, E, and F are constants. Therefore, its inverse Laplace transform is a linear combination of the functions $\sin t$, $\cos t$, $\sin\sqrt{3}\,t$, $\cos\sqrt{3}\,t$, $\sin\sqrt{5}\,t$, and $\cos\sqrt{5}\,t$. Such a function is bounded for all t.

17. False. Since $s^4+6s^2+9=(s^2+3)^2$, the fraction

$$\frac{3s+5}{s^4+6s^2+9}$$

is already in its partial fractions form. Furthermore, the denominator of $(s^2+3)^2$ indicates that the inverse Laplace transform contains terms that involve $t\sin\sqrt{3}\,t$ and $t\cos\sqrt{3}\,t$ (see Exercise 35 in Section 6.3). These functions oscillate with amplitudes that grow linearly. If t is large enough, $|y(t)| > 15$.

18. First we note that $s^2+5s+6=(s+2)(s+3)$. Hence, the partial fractions decomposition is

$$\frac{3}{s^2+5s+6} = \frac{A}{s+2} + \frac{B}{s+3} = \frac{A(s+3)+B(s+2)}{s^2+2s+8}.$$

We solve the equations

$$\begin{cases} A+B=0 \\ 3A+2B=3 \end{cases}$$

and obtain $A=3$ and $B=-3$. Therefore,

$$\frac{3}{s^2+5s+6} = \frac{3}{s+2} - \frac{3}{s+3}.$$

The inverse Laplace transform is

$$\mathcal{L}^{-1}\left[\frac{3}{s^2+5s+6}\right] = 3\mathcal{L}^{-1}\left[\frac{1}{s+2}\right] - 3\mathcal{L}^{-1}\left[\frac{1}{s+3}\right] = 3e^{-2t} - 3e^{-3t}.$$

19. First we note that $s^2+2s-8=(s-2)(s+4)$. Hence, the partial fractions decomposition is

$$\frac{s+16}{s^2+2s-8} = \frac{A}{s-2} + \frac{B}{s+4} = \frac{A(s+4)+B(s-2)}{s^2+2s+8}.$$

We solve the equations

$$\begin{cases} A+B=1 \\ 4A-2B=16 \end{cases}$$

and obtain $A=3$ and $B=-2$. Therefore,

$$\frac{s+16}{s^2+2s-8} = \frac{3}{s-2} - \frac{2}{s+4}.$$

The inverse Laplace transform is

$$\mathcal{L}^{-1}\left[\frac{s+16}{s^2+2s-8}\right] = 3\mathcal{L}^{-1}\left[\frac{1}{s-2}\right] - 2\mathcal{L}^{-1}\left[\frac{1}{s+4}\right] = 3e^{2t} - 2e^{-4t}.$$

20. We note that the roots of $s^2 - 2s + 4$ are complex, so we complete the square. We get $s^2 - 2s + 4 = (s-1)^2 + 3$, and we write

$$\begin{aligned}\frac{2s+3}{s^2-2s+4} &= \frac{2s+3}{(s-1)^2+3} \\ &= 2\left(\frac{s-1}{(s-1)^2+3}\right) + \frac{5}{\sqrt{3}}\left(\frac{\sqrt{3}}{(s-1)^2+3}\right).\end{aligned}$$

So

$$\mathcal{L}^{-1}\left[\frac{2s+3}{s^2-2s+4}\right] = 2e^t \cos\sqrt{3}\,t + \frac{5}{\sqrt{3}} e^t \sin\sqrt{3}\,t.$$

21. First we compute the partial fractions decomposition of

$$\frac{5s-12}{s^2-5s+6}.$$

We note that $s^2 - 5s + 6 = (s-2)(s-3)$. Hence, the partial fractions decomposition is

$$\frac{5s-12}{s^2-5s+6} = \frac{A}{s-2} + \frac{B}{s-3} = \frac{A(s-3)+B(s-2)}{s^2-5s+6}.$$

We solve the equations

$$\begin{cases} A + B = 5 \\ 3A + 2B = 12 \end{cases}$$

and obtain $A = 2$ and $B = 3$. Therefore,

$$\frac{5s-12}{s^2-5s+6} = \frac{2}{s-2} + \frac{3}{s-3}.$$

We conclude that

$$\mathcal{L}^{-1}\left[\frac{5s-12}{s^2-5s+6}\right] = 2e^{2t} + 3e^{3t},$$

and consequently,

$$\mathcal{L}^{-1}\left[\frac{(5s-12)e^{-3s}}{s^2-5s+6}\right] = u_3(t)\left(2e^{2(t-3)} + 3e^{3(t-3)}\right).$$

22. The partial fractions decomposition of this rational function has the form

$$\frac{5s^2-27s+49}{(s-2)(s^2-6s+13)} = \frac{A}{s-2} + \frac{Bs+C}{s^2-6s+13}.$$

To find the values of A, B, and C, we put the fractions on the right-hand side over a common denominator. We get

$$\begin{aligned}\frac{5s^2-27s+49}{(s-2)(s^2-6s+13)} &= \frac{As^2-6As+13A+Bs^2+Cs-2Bs-2C}{(s-2)(s^2-27s+13)} \\ &= \frac{(A+B)s^2+(-6A-2B+C)s+(13A-2C)}{(s-2)(s^2-27s+13)}.\end{aligned}$$

So we must have

$$\begin{cases} A + B = 5 \\ -6A - 2B + C = -27 \\ 13A - 2C = 49. \end{cases}$$

Solving this simultaneous system of equations, we get $A = 3$, $B = 2$, and $C = -5$. Therefore,

$$\frac{5s^2 - 27s + 49}{(s-2)(s^2 - 6s + 13)} = \frac{3}{s-2} + \frac{2s - 5}{s^2 - 6s + 13}.$$

Then

$$\mathcal{L}^{-1}\left[\frac{3}{s-2}\right] = 3e^{2t},$$

and

$$\mathcal{L}^{-1}\left[\frac{2s-5}{s^2 - 6s + 13}\right] = \mathcal{L}^{-1}\left[2\left(\frac{s-3}{(s-3)^2 + 4}\right) + \frac{1}{2}\left(\frac{2}{(s-3)^2 + 4}\right)\right]$$

$$= 2e^{3t}\cos 2t + \tfrac{1}{2}e^{3t}\sin 2t.$$

We obtain

$$\mathcal{L}^{-1}\left[\frac{5s^2 - 27s + 49}{(s-2)(s^2 - 6s + 13)}\right] = 3e^{2t} + 2e^{3t}\cos 2t + \tfrac{1}{2}e^{3t}\sin 2t.$$

23. Note that $s^2 - 4s + 4 = (s-2)^2$, so $s = 2$ is a repeated root of the denominator. The fraction

$$\frac{1}{s^2 - 4s + 4}$$

cannot be further decomposed, and we must use the result of Exercise 35 in Section 6.3.

We know that

$$\mathcal{L}\left[e^{2t}\right] = \frac{1}{s-2},$$

so

$$\mathcal{L}\left[te^{2t}\right] = -\frac{d}{ds}\left(\frac{1}{s-2}\right) = \frac{1}{(s-2)^2}.$$

Consequently,

$$\mathcal{L}^{-1}\left[\frac{1}{(s-2)^2}\right] = te^{2t}.$$

24. **(a)** The characteristic polynomial of the unforced equation is

$$s^2 - 2s,$$

which has $s = 0$ and $s = 2$ as roots. Therefore, the general solution of the associated homogeneous equation is

$$y_h(t) = k_1 + k_2 e^{2t}.$$

The natural guess of $y_p(t) = a$, where a is a constant, fails to be a solution of the forced equation because it is a solution of the homogeneous equation. Hence, we guess $y_p(t) = at$. Substituting this guess into the equation gives $-2a = 4$, and $y_p(t) = -2t$ is one solution of the forced equation. The general solution of the forced equation is

$$y(t) = k_1 + k_2 e^{2t} - 2t.$$

To satisfy the initial conditions, we note that $y'(t) = 2k_2 e^{2t} - 2$, so we must have

$$\begin{cases} y(0) = k_1 + k_2 = -1 \\ y'(0) = 2k_2 - 2 = 2. \end{cases}$$

Therefore, $k_2 = 2$ and $k_1 = -3$. Hence, the solution of the initial-value problem is

$$y(t) = -3 + 2e^{2t} - 2t.$$

(b) Transforming both sides of the differential equation, we have

$$s^2\mathcal{L}[y] + s - 2 - 2(s\mathcal{L}[y] + 1) = \frac{4}{s}.$$

Solving for $\mathcal{L}[y]$ gives

$$(s^2 - 2s)\mathcal{L}[y] = \frac{4}{s} - s + 4,$$

and therefore,

$$\mathcal{L}[y] = \frac{4}{s(s^2 - 2s)} + \frac{-s + 4}{s^2 - 2s} = \frac{-s^2 + 4s + 4}{s^2(s - 2)}.$$

The partial fractions decomposition of this rational function is

$$\frac{-s^2 + 4s + 4}{s^2(s - 2)} = \frac{A}{s} + \frac{B}{s^2} + \frac{C}{s - 2}.$$

Putting the right-hand side over a common denominator, we get

$$As(s - 2) + B(s - 2) + Cs^2 = (A + C)s^2 + (B - 2A)s - 2B = -s^2 + 4s + 4.$$

Solving for A, B, and C, we obtain $B = -2$, $A = -3$, and $C = 2$.

To obtain the solution to the initial-value problem, we take the inverse Laplace transform

$$\begin{aligned} y(t) &= \mathcal{L}^{-1}\left[\frac{-3}{s} + \frac{-2}{s^2} + \frac{2}{s - 2}\right] \\ &= -3 - 2t + 2e^{2t}. \end{aligned}$$

(c) The solutions in parts (a) and (b) were more or less equally complicated. One advantage of the solution in part (a) is that we obtained the general solution to the equation as part of the process.

25. **(a)** The characteristic polynomial of the unforced equation is

$$s^2 - 4s + 4,$$

which has $s = 2$ as a double root. Therefore, the natural guesses of $y_p(t) = ke^{2t}$ and $y_p(t) = kte^{2t}$ fail to be solutions of the forced equation because they are both solutions of the unforced equation.

So we guess $y_p(t) = kt^2e^{2t}$. Substituting this guess into the left-hand side of the differential equation gives

$$\begin{aligned}\frac{d^2y_p}{dt^2} - 4\frac{dy_p}{dt} + 4y_p &= (2ke^{2t} + 8kte^{2t} + 4kt^2e^{2t}) - 4(2kte^{2t} + 2kt^2e^{2t}) + 4kt^2e^{2t} \\ &= 2ke^{2t}.\end{aligned}$$

So $k = 1/2$ yields the solution

$$y_p(t) = \tfrac{1}{2}t^2e^{2t}.$$

From the characteristic polynomial, we know that the general solution of the unforced equation is

$$k_1e^{2t} + k_2te^{2t}.$$

Consequently, the general solution of the forced equation is

$$y(t) = k_1e^{2t} + k_2te^{2t} + \tfrac{1}{2}t^2e^{2t}.$$

To solve the initial-value problem, we note that $y(0) = 1$ implies that $k_1 = 1$. We find k_2 by differentiation. We get

$$y'(t) = 2e^{2t} + k_2(1 + 2t)e^{2t} + (t^2 + t)e^{2t}.$$

Since $y'(0) = 5$, we have $k_2 = 3$ and

$$y(t) = (1 + 3t + \tfrac{1}{2}t^2)e^{2t}.$$

(b) Transforming both sides of the differential equation, we have

$$\begin{aligned}s^2\mathcal{L}[y] - s - 5 - 4(s\mathcal{L}[y] - 1) + 4\mathcal{L}[y] &= \frac{1}{s-2} \\ (s^2 - 4s + 4)\mathcal{L}[y] - s - 1 &= \frac{1}{s-2} \\ (s-2)^2\mathcal{L}[y] &= s + 1 + \frac{1}{s-2}.\end{aligned}$$

So

$$\mathcal{L}[y] = \frac{s+1}{(s-2)^2} + \frac{1}{(s-2)^3}.$$

Note that

$$\frac{s+1}{(s-2)^2} = \frac{s-2}{(s-2)^2} + \frac{3}{(s-2)^2} = \frac{1}{s-2} + \frac{3}{(s-2)^2}.$$

Writing $\mathcal{L}[y]$ in terms of its partial fractions decomposition, we have

$$\frac{1}{s-2} + \frac{3}{(s-2)^2} + \frac{1}{(s-2)^3}.$$

The inverse Laplace transform of the first term is routine. The inverse transform of the second term can be deduced from the result of Exercise 23, and the inverse transform of the third term can also be obtained using the method of solution for Exercise 23. That is, we use the result of Exercise 35 in Section 6.3. In particular, we know that

$$\mathcal{L}[te^{2t}] = \frac{1}{(s-2)^2}$$

from Exercise 23. Consequently,

$$\mathcal{L}[t^2e^{2t}] = -\frac{d}{ds}\left(\frac{1}{(s-2)^2}\right) = \frac{2}{(s-2)^3}.$$

Therefore,

$$\mathcal{L}^{-1}\left[\frac{1}{(s-2)^3}\right] = \tfrac{1}{2}t^2e^{2t},$$

and the solution to the initial-value problem is

$$y(t) = (1 + 3t + \tfrac{1}{2}t^2)e^{2t}.$$

(c) The solutions in parts (a) and (b) were more or less equally complicated. One advantage of the solution in part (a) is that we obtained the general solution to the equation as part of the process.

26. Note that the equation can be expressed as

$$\frac{dy}{dt} - 3y = 6(1 - u_3(t)).$$

Taking the Laplace transform of both sides of the equation and using the fact that $y(0) = 0$, we have

$$\mathcal{L}\left[\frac{dy}{dt}\right] - 3\mathcal{L}[y] = 6\mathcal{L}[1 \quad u_3(t)]$$

$$s\mathcal{L}[y] - y(0) - 3\mathcal{L}[y] = \frac{6}{s}\left(1 - e^{-3s}\right)$$

$$\mathcal{L}[y] = \frac{6}{s(s-3)}\left(1 - e^{-3s}\right).$$

Using partial fractions, we see that

$$\frac{6}{s(s-3)} = \frac{2}{s-3} - \frac{2}{s},$$

so

$$\mathcal{L}^{-1}\left[\frac{6}{s(s-3)}\right] = 2\left(e^{3t} - 1\right).$$

We also obtain

$$\mathcal{L}^{-1}\left[\frac{6e^{-3s}}{s(s-3)}\right] = 2u_3(t)\left(e^{3(t-3)} - 1\right).$$

The solution of the original initial-value problem is

$$y(t) = 2\left(e^{3t} - 1\right) - 2u_3(t)\left(e^{3(t-3)} - 1\right).$$

27. We transform both sides of the differential equation and obtain

$$\mathcal{L}\left[\frac{dy}{dt}\right] - 4\mathcal{L}[y] = \mathcal{L}[50u_2(t)\sin(3(t-2))]$$

$$s\mathcal{L}[y] - y(0) - 4\mathcal{L}[y] = 50e^{-2s}\left(\frac{3}{s^2+9}\right)$$

$$(s-4)\mathcal{L}[y] - 5 = \frac{150e^{-2s}}{s^2+9}.$$

Solving for $\mathcal{L}[y]$, we get

$$\mathcal{L}[y] = \frac{5}{s-4} + \frac{150e^{-2s}}{(s^2+9)(s-4)}.$$

To find the inverse Laplace transform, we first note that

$$\mathcal{L}^{-1}\left[\frac{5}{s-4}\right] = 5e^{4t}.$$

To invert the second term, we need the partial fractions decomposition of

$$\frac{150}{(s^2+9)(s-4)}.$$

We set

$$\frac{150}{(s^2+9)(s-4)} = \frac{As+B}{s^2+9} + \frac{C}{s-4} = \frac{(As+B)(s-4) + C(s^2+9)}{(s^2+9)(s-4)},$$

and we get $A = -6$, $B = -24$, and $C = 6$. Therefore,

$$\mathcal{L}^{-1}\left[\frac{150}{(s^2+9)(s-4)}\right] = \mathcal{L}^{-1}\left[\frac{-6s}{s^2+9}\right] + \mathcal{L}^{-1}\left[\frac{-24}{s^2+9}\right] + \mathcal{L}^{-1}\left[\frac{6}{s-4}\right]$$

$$= -6\cos 3t - 8\sin 3t + 6e^{4t}.$$

Hence,

$$\mathcal{L}^{-1}\left[\frac{150e^{-2s}}{(s^2+9)(s-4)}\right] = u_2(t)\left(-6\cos(3(t-2)) - 8\sin(3(t-2)) + 6e^{4(t-2)}\right).$$

Combining the two results, we obtain the solution of the initial-value problem

$$y(t) = 5e^{4t} + u_2(t)\left(6e^{4(t-2)} - 6\cos(3(t-2)) - 8\sin(3(t-2))\right)$$

28. First, we rewrite the equation as

$$\frac{d^2y}{dt^2} + 4\frac{dy}{dt} + 7y = 1 - u_3(t).$$

Next, we transform both sides and obtain

$$s^2\mathcal{L}[y] - 3s + 4(s\mathcal{L}[y] - 3) + 7\mathcal{L}[y] = \frac{1}{s} - \frac{e^{-3s}}{s}.$$

Solving for $\mathcal{L}[y]$, we get

$$\begin{aligned}(s^2 + 4s + 7)\mathcal{L}[y] &= 3s + 12 + \frac{1}{s} - \frac{e^{-3s}}{s}\\ \mathcal{L}[y] &= \frac{3s + 12}{s^2 + 4s + 7} + \frac{1}{s(s^2 + 4s + 7)} - \frac{e^{-3s}}{s(s^2 + 4s + 7)}.\end{aligned}$$

To compute the inverse Laplace transform, we first note that $s^2 + 4s + 7 = (s + 2)^2 + 3$. To invert the first term, we write it as

$$\frac{3s + 12}{s^2 + 4s + 7} = \frac{3(s + 2)}{(s + 2)^2 + 3} + 2\sqrt{3}\,\frac{\sqrt{3}}{(s + 2)^2 + 3}.$$

We see that

$$\mathcal{L}^{-1}\left[\frac{3s + 12}{s^2 + 4s + 7}\right] = 3e^{-2t}\cos\sqrt{3}\,t + 2\sqrt{3}\,e^{-2t}\sin\sqrt{3}\,t.$$

To invert the second term, we compute

$$\begin{aligned}\frac{1}{s(s^2 + 4s + 7)} &= \frac{A}{s} + \frac{Bs + C}{s^2 + 4s + 7}\\ &= \frac{As^2 + 4As + 7A + Bs^2 + Cs}{s(s^2 + 4s + 7)}\\ &= \frac{(A + B)s^2 + (4A + C)s + 7A}{s(s^2 + 4s + 7)}\end{aligned}$$

and obtain the system of equations

$$\begin{cases} A + B = 0\\ 4A + C = 0\\ 7A = 1.\end{cases}$$

We get $A = 1/7$, $B = -1/7$, and $C = -4/7$. Hence,

$$\begin{aligned}\frac{1}{s(s^2 + 4s + 7)} &= \frac{1}{7}\left(\frac{1}{s} - \frac{s + 4}{s^2 + 4s + 7}\right)\\ &= \frac{1}{7}\left(\frac{1}{s} - \frac{s + 2}{(s + 2)^2 + 3} - \frac{2}{(s + 2)^2 + 3}\right).\end{aligned}$$

We conclude that

$$\mathcal{L}^{-1}\left[\frac{1}{s(s^2+4s+7)}\right] = \frac{1}{7}\left(1 - e^{-2t}\cos\sqrt{3}\,t - \frac{2}{\sqrt{3}}e^{-2t}\sin\sqrt{3}\,t\right)$$

and

$$\mathcal{L}^{-1}\left[\frac{e^{-3s}}{s(s^2+4s+7)}\right] = \frac{u_3(t)}{7}\left(1 - e^{-2(t-3)}\cos\sqrt{3}\,(t-3) - \frac{2}{\sqrt{3}}e^{-2(t-3)}\sin\sqrt{3}\,(t-3)\right).$$

So the solution of the initial-value problem is

$$\begin{aligned} y(t) &= 3e^{-2t}\cos\sqrt{3}\,t + 2\sqrt{3}\,e^{-2t}\sin\sqrt{3}\,t \\ &\quad + \frac{1}{7}\left(1 - e^{-2t}\cos\sqrt{3}\,t - \frac{2}{\sqrt{3}}e^{-2t}\sin\sqrt{3}\,t\right) \\ &\quad + \frac{u_3(t)}{7}\left(1 - e^{-2(t-3)}\cos\sqrt{3}\,(t-3) - \frac{2}{\sqrt{3}}e^{-2(t-3)}\sin\sqrt{3}\,(t-3)\right). \end{aligned}$$

29. We transform both sides of the differential equation and obtain

$$s^2\mathcal{L}[y] - 2s - 1 + 2(s\mathcal{L}[y] - 2) + 3\mathcal{L}[y] = e^{-4s}.$$

Solving for $\mathcal{L}[y]$, we get

$$(s^2+2s+3)\mathcal{L}[y] = 2s + 5 + e^{-4s}$$

$$\mathcal{L}[y] = \frac{2s+5}{s^2+2s+3} + \frac{e^{-4s}}{s^2+2s+3}.$$

To compute the inverse Laplace transform, we first note that $s^2+2s+3 = (s+1)^2+2$. To invert the first term, we write it as

$$\frac{2s+5}{s^2+2s+3} = \frac{2(s+1)}{(s+1)^2+2} + \frac{3}{(s+1)^2+2}.$$

We see that

$$\mathcal{L}^{-1}\left[\frac{2s+5}{s^2+2s+3}\right] = 2e^{-t}\cos\sqrt{2}\,t + \frac{3}{\sqrt{2}}e^{-t}\sin\sqrt{2}\,t.$$

To invert the second term, we compute

$$\mathcal{L}^{-1}\left[\frac{1}{s^2+2s+3}\right] = \frac{1}{\sqrt{2}}e^{-t}\sin\sqrt{2}\,t.$$

Therefore,

$$\mathcal{L}^{-1}\left[\frac{e^{-4s}}{s^2+2s+3}\right] = \frac{u_4(t)}{\sqrt{2}}e^{-(t-4)}\sin\sqrt{2}\,(t-4).$$

Combining both results, we get the solution to the original initial-value problem

$$y(t) = 2e^{-t}\cos\sqrt{2}\,t + \frac{3}{\sqrt{2}}e^{-t}\sin\sqrt{2}\,t + \frac{u_4(t)}{\sqrt{2}}e^{-(t-4)}\sin\sqrt{2}\,(t-4).$$

30. We transform both sides of the differential equation and obtain

$$s^2\mathcal{L}[y] - s - 2 + 2(s\mathcal{L}[y] - 1) + 5\mathcal{L}[y] = e^{-3s} + \frac{e^{-6s}}{s}.$$

Solving for $\mathcal{L}[y]$, we get

$$\begin{aligned}(s^2 + 2s + 5)\mathcal{L}[y] &= s + 4 + e^{-3s} + \frac{e^{-6s}}{s}\\ \mathcal{L}[y] &= \frac{s+4}{s^2+2s+5} + \frac{e^{-3s}}{s^2+2s+5} + \frac{e^{-6s}}{s(s^2+2s+5)}.\end{aligned}$$

To compute the inverse Laplace transform, we note that $s^2 + 2s + 5 = (s+1)^2 + 4$. To invert the first term, we write it as

$$\frac{s+4}{s^2+2s+5} = \frac{s+1}{(s+1)^2+4} + \frac{3}{(s+1)^2+4}.$$

We see that

$$\mathcal{L}^{-1}\left[\frac{s+4}{s^2+2s+5}\right] = e^{-t}\cos 2t + \tfrac{3}{2}e^{-t}\sin 2t.$$

To invert the second term, we note that

$$\mathcal{L}^{-1}\left[\frac{1}{s^2+2s+5}\right] = \mathcal{L}^{-1}\left[\frac{1}{(s+1)^2+4}\right] = \tfrac{1}{2}e^{-t}\sin 2t.$$

Consequently,

$$\mathcal{L}^{-1}\left[\frac{e^{-3s}}{s^2+2s+5}\right] = \frac{u_3(t)}{2}e^{-(t-3)}\sin 2(t-3).$$

Finally, to invert the third term, we write

$$\begin{aligned}\frac{1}{s(s^2+2s+5)} &= \frac{A}{s} + \frac{Bs+C}{s^2+2s+5}\\ &= \frac{As^2 + 2As + 5A + Bs^2 + Cs}{s(s^2+2s+5)}\\ &= \frac{(A+B)s^2 + (2A+C)s + 5A}{s(s^2+2s+5)}.\end{aligned}$$

We solve

$$\begin{cases} A + B = 0\\ 2A + C = 0\\ 5A = 1\end{cases}$$

and obtain $A = 1/5$, $B = -1/5$, and $C = -2/5$. So

$$\frac{1}{s(s^2+2s+5)} = \frac{1}{5}\left(\frac{1}{s} - \frac{s+2}{s^2+2s+5}\right) = \frac{1}{5}\left(\frac{1}{s} - \frac{s+1}{(s+1)^2+4} - \frac{1}{(s+1)^2+4}\right),$$

and

$$\mathcal{L}^{-1}\left[\frac{e^{-6s}}{s(s^2+2s+5)}\right] = \frac{u_6(t)}{5}\left(1 - e^{-(t-6)}\cos 2(t-6) - \frac{1}{2}e^{-(t-6)}\sin 2(t-6)\right).$$

Combining all three results, we obtain the solution to the original initial-value problem

$$y(t) = e^{-t}\cos 2t + \frac{3}{2}e^{-t}\sin 2t + \frac{u_3(t)}{2}e^{-(t-3)}\sin 2(t-3)$$

$$+\frac{u_6(t)}{5}\left(1 - e^{-(t-6)}\cos 2(t-6) - \frac{1}{2}e^{-(t-6)}\sin 2(t-6)\right).$$

CHAPTER 7

Numerical Methods

EXERCISES FOR SECTION 7.1

1. **(a)** The differential equation is separable. Therefore, one way to obtain the solution to the initial-value problem is to integrate

$$\int y^{-2}\,dy = \int -2t\,dt.$$

We obtain

$$\frac{y^{-1}}{-1} = -t^2 + c$$

$$y^{-1} = t^2 + k$$

$$y = \frac{1}{t^2 + k}.$$

We determine the value of k using the initial condition $y(0) = 1$. Hence, $k = 1$, and the solution to the given initial-value problem is $y(t) = 1/(t^2 + 1)$.

(b) To calculate y_{20}, we must apply Euler's method 20 times. Table 7.1 contains the results of a number of intermediate calculations.

(c) The total error e_{20} is the difference between the actual value $y(2) = 0.2$ and the approximate value $y_{20} = 0.193342$. Therefore, $e_{20} = 0.0066581$.

(d) Table 7.2 contains the results of Euler's method and the corresponding total errors for $n = 1000$, 2000, …, 6000.

Table 7.1
Results of Euler's method

k	t_k	y_k
0	0	1.0
1	0.1	1.0
2	0.2	0.98
3	0.3	0.941584
4	0.4	0.888389
5	0.5	0.82525
⋮	⋮	⋮
10	1.0	0.503642
⋮	⋮	⋮
19	1.9	0.210119
20	2.0	0.193342

Table 7.2
Results of Euler's method and the corresponding total errors

n	y_n	e_n
1000	0.199874	0.000125848
2000	0.199937	0.0000628901
3000	0.199958	0.0000419192
4000	0.199969	0.0000314366
5000	0.199975	0.0000251479
6000	0.199979	0.0000209558

(e) Table 7.3 gives values of e_n for some n that are intermediate to the ones above in case that you want to double check the ones you have computed. Also, the graph of e_n as a function of n for $100 \le n \le 6000$ is given.

Table 7.3
Selected total errors

n	e_n
100	0.00127095
200	0.000631985
1400	0.0000898639
3700	0.0000339862
5600	0.000022453

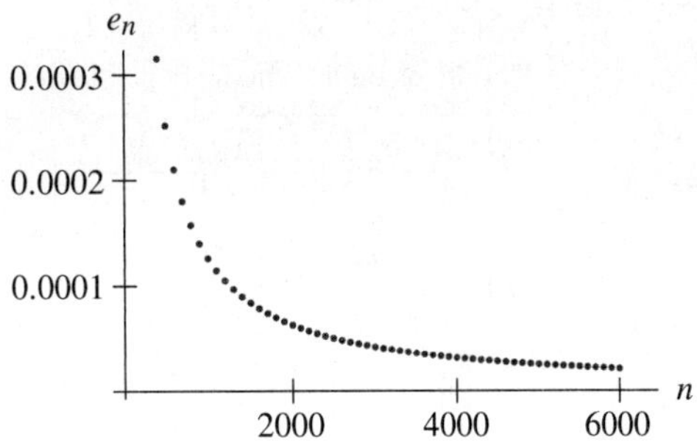

(f) Our computer math system fits the data to the function $0.126731/n$. The following figure includes both the data and the graph of this function.

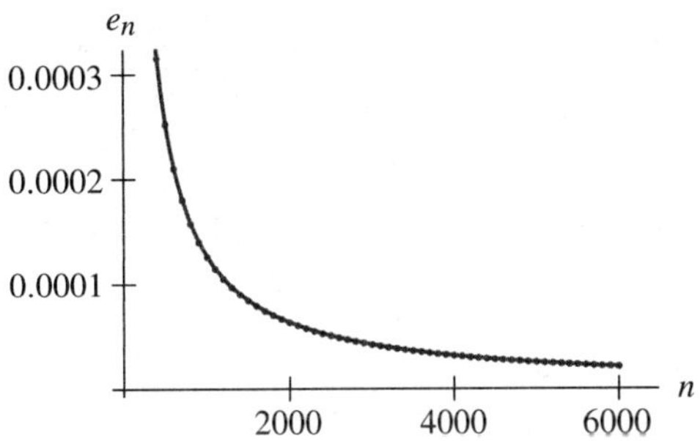

2. **(a)** The differential equation is both separable and linear. Therefore, one way to obtain the solution to the initial-value problem is to integrate

$$\int \frac{1}{1-y}\,dy = \int 1\,dt.$$

We obtain

$$-\ln|1-y| = t + c$$

$$y = 1 - ke^{-t}.$$

We determine the value of k using the initial condition $y(0) = 0$. Hence, $k = 1$, and the solution to the given initial-value problem is $y(t) = 1 - e^{-t}$.

(b) To calculate y_{20}, we must apply Euler's method 20 times. Table 7.4 contains the results of a number of intermediate calculations.

(c) The total error e_{20} is the difference between the actual value $y(1) = 1 - e^{-1}$ and the approximate value $y_{20} = 0.641514$. Therefore, $e_{20} = 0.00939352$.

(d) Table 7.5 contains the results of Euler's method and the corresponding total errors for $n = 1000$, 2000, ..., 6000.

Table 7.4
Results of Euler's method

k	t_k	y_k
0	0.0	0.0
1	0.05	0.05
2	0.10	0.0975
3	0.15	0.142625
4	0.20	0.185494
5	0.25	0.226219
⋮	⋮	⋮
10	0.5	0.401263
⋮	⋮	⋮
19	0.95	0.622646
20	1.00	0.641514

Table 7.5
Results of Euler's method and the corresponding total errors

n	y_n	e_n
1000	0.632305	0.000184016
2000	0.632213	0.000091989
3000	0.632182	0.0000613218
4000	0.632167	0.0000459897
5000	0.632157	0.000036791
6000	0.632151	0.0000306587

(e) Table 7.6 gives values of e_n for some n that are intermediate to the ones above in case that you want to double check the ones you have computed. Also, the graph of e_n as a function of n for $100 \leq n \leq 6000$ is given.

Table 7.6
Selected total errors

n	e_n
100	0.0018471
200	0.000921619
1400	0.000131425
3700	0.000049719
5600	0.0000328488

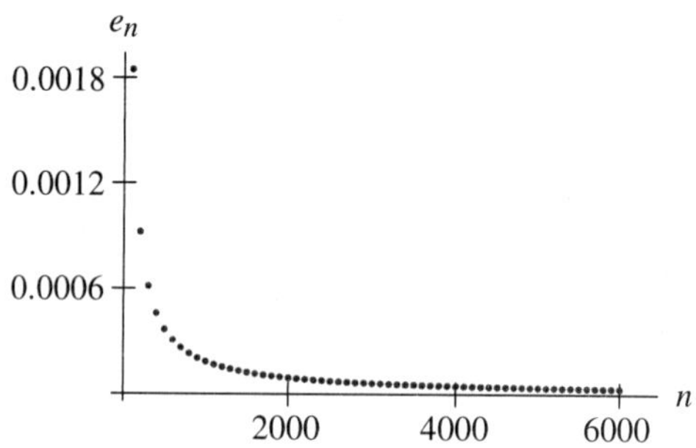

(f) Our computer math system fits the data to the function $0.184508/n$. The following figure includes both the data and the graph of this function.

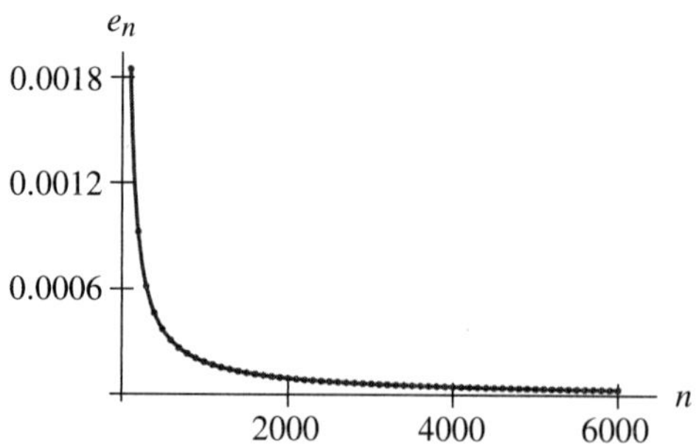

3. **(a)** The differential equation is both separable and linear. Therefore, one way to obtain the solution to the initial-value problem is to integrate

$$\int \frac{1}{y}\,dy = \int t\,dt.$$

We obtain

$$\ln|y| = \frac{t^2}{2} + c$$

$$y = ke^{t^2/2}.$$

We determine the value of k using the initial condition $y(0) = 1$. Hence, $k = 1$, and the solution to the given initial-value problem is $y(t) = e^{t^2/2}$.

(b) To calculate y_{20}, we must apply Euler's method 20 times. Table 7.7 contains the results of a number of intermediate calculations.

(c) The total error e_{20} is the difference between the actual value $y(\sqrt{2}) = e$ and the approximate value $y_{20} = 2.51066$. Therefore, $e_{20} = 0.20762$.

(d) Table 7.8 contains the results of Euler's method and the corresponding total errors for $n = 1000$, $2000, \ldots, 6000$.

Table 7.7
Results of Euler's method

k	t_k	y_k
0	0	1
1	0.0707107	1
2	0.141421	1.005
3	0.212132	1.01505
4	0.282843	1.03028
5	0.353553	1.05088
⋮	⋮	⋮
10	0.707107	1.24797
⋮	⋮	⋮
19	1.3435	2.29284
20	1.41421	2.51066

Table 7.8
Results of Euler's method and the corresponding total errors

n	y_n	e_n
1000	2.71376	0.00452218
2000	2.71602	0.00226316
3000	2.71677	0.00150923
4000	2.71715	0.0011321
5000	2.71738	0.000905762
6000	2.71753	0.000754848

(e) Table 7.9 gives values of e_n for some n that are intermediate to the ones above in case that you want to double check the ones you have computed. Also, the graph of e_n as a function of n for $100 \le n \le 6000$ is given.

Table 7.9
Selected total errors

n	e_n
100	0.0444901
200	0.0224467
1400	0.00323182
3700	0.00122384
5600	0.000808748

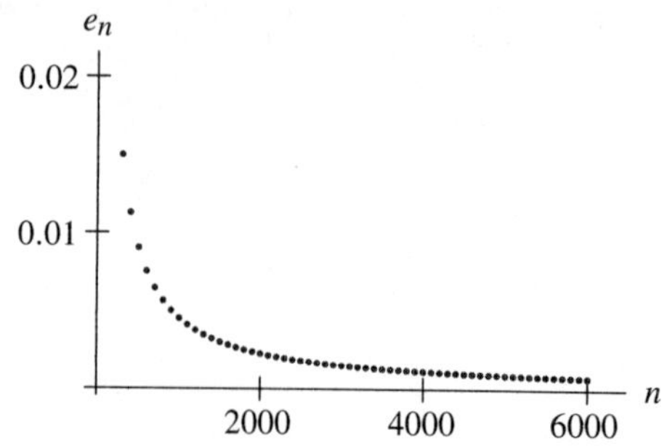

(f) Our computer math system fits the data to the function $4.47023/n$. The following figure includes both the data and the graph of this function.

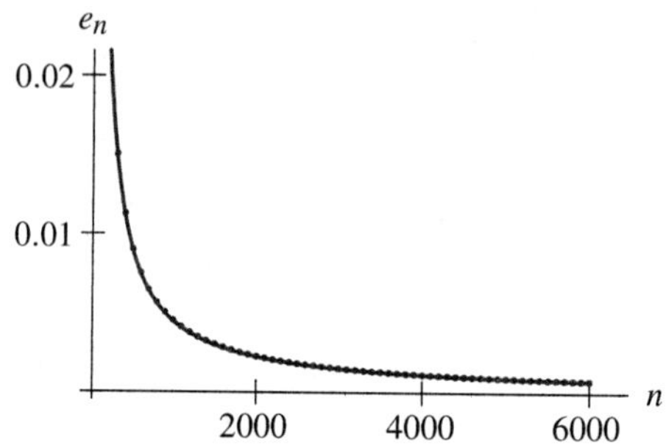

4. **(a)** The differential equation is separable. Therefore, we integrate

$$\int \frac{1}{y^2}\,dy = \int -1\,dt.$$

We obtain

$$-\frac{1}{y} = -t + c$$

$$y = \frac{1}{t+k}.$$

We determine the value of k using the initial condition $y(0) = 1/2$. Hence, $k = 2$, and the solution to the given initial-value problem is $y(t) = 1/(t+2)$.

(b) To calculate y_{20}, we must apply Euler's method 20 times. Table 7.10 contains the results of a number of intermediate calculations.

(c) The total error e_{20} is the difference between the actual value $y(2) = e$ and the approximate value $y_{20} = 2.51066$. Therefore, $e_{20} = 0.00444754$.

(d) Table 7.11 contains the results of Euler's method and the corresponding total errors for $n = 1000, 2000, \ldots, 6000$.

Table 7.10
Results of Euler's method

k	t_k	y_k
0	0	0.5
1	0.1	0.475
2	0.2	0.452437
3	0.3	0.431968
4	0.4	0.413308
5	0.5	0.396226
⋮	⋮	⋮
10	1.	0.328637
⋮	⋮	⋮
19	1.9	0.251898
20	2.	0.245552

Table 7.11
Results of Euler's method and the corresponding total errors

n	y_n	e_n
1000	0.249913	0.000086688
2000	0.249957	0.0000433328
3000	0.249971	0.0000288861
4000	0.249978	0.0000216636
5000	0.249983	0.0000173305
6000	0.249986	0.0000144418

(e) Table 7.12 gives values of e_n for some n that are intermediate to the ones above in case that you want to double check the ones you have computed. Also, the graph of e_n as a function of n for $100 \le n \le 6000$ is given.

Table 7.12
Selected total errors

n	e_n
100	0.000870919
200	0.000434334
1400	0.0000619109
3700	0.0000234204
5600	0.0000154735

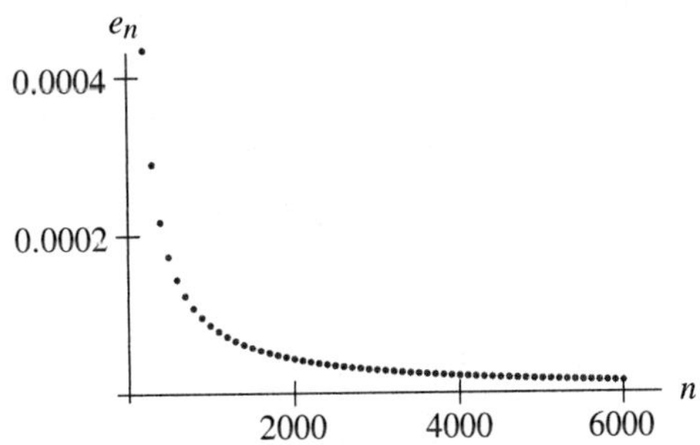

(f) Our computer math system fits the data to the function $0.0869742/n$. The following figure includes both the data and the graph of this function.

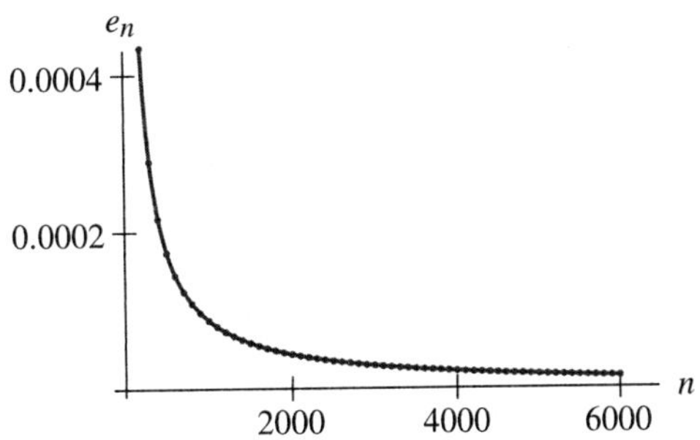

5. **(a)** The differential equation is linear, and we can use a guessing technique to find an analytic solution (see Section 1.8). First, we rewrite the equation as

$$\frac{dy}{dt} - 3y = 1 - t.$$

This form of the equation suggests the guess $y_p(t) = at + b$. Substituting this guess into the differential equation yields

$$a - 3(at + b) = 1 - t.$$

We obtain the equations $-3a = -1$ and $a - 3b = 1$. Then $y_p(t)$ is a solution if and only if $a = 1/3$ and $b = -2/9$.

The general solution of the associated homogeneous equation is ce^{3t}. Therefore, the general solution to this linear equation is

$$y(t) = \tfrac{1}{3}t - \tfrac{2}{9} + ce^{3t}.$$

We determine the value of c using the initial condition $y(0) = 1$. We get $c = 11/9$, and the solution to the given initial-value problem is

$$y(t) = \tfrac{11}{9}e^{3t} + \tfrac{1}{3}t - \tfrac{2}{9}.$$

(b) To calculate y_{20}, we must apply Euler's method 20 times. Table 7.13 contains the results of a number of intermediate calculations.

(c) The total error e_{20} is the difference between the actual value $y(1) = (11e^3 + 1)/9$ and the approximate value $y_{20} = 20.1147$. Therefore, $e_{20} = 4.54544$.

(d) Table 7.14 contains the results of Euler's method and the corresponding total errors for $n = 1000, 2000, \ldots, 6000$.

Table 7.13
Results of Euler's method

k	t_k	y_k
0	0	1.
1	0.05	1.2
2	0.1	1.4275
3	0.15	1.68663
4	0.2	1.98212
5	0.25	2.31944
⋮	⋮	⋮
10	0.5	4.88902
⋮	⋮	⋮
19	0.95	17.4888
20	1.	20.1147

Table 7.14
Results of Euler's method and the corresponding total errors

n	y_n	e_n
1000	24.5501	0.110003
2000	24.605	0.0551181
3000	24.6233	0.0367714
4000	24.6325	0.0275883
5000	24.638	0.0220753
6000	24.6417	0.0183987

(e) Table 7.3 gives values of e_n for some n that are intermediate to the ones above in case that you want to double check the ones you have computed. Also, the graph of e_n as a function of n for $100 \le n \le 6000$ is given.

Table 7.15
Selected total errors

n	e_n
100	1.05955
200	0.540843
1400	0.0786686
3700	0.0298226
5600	0.0197119

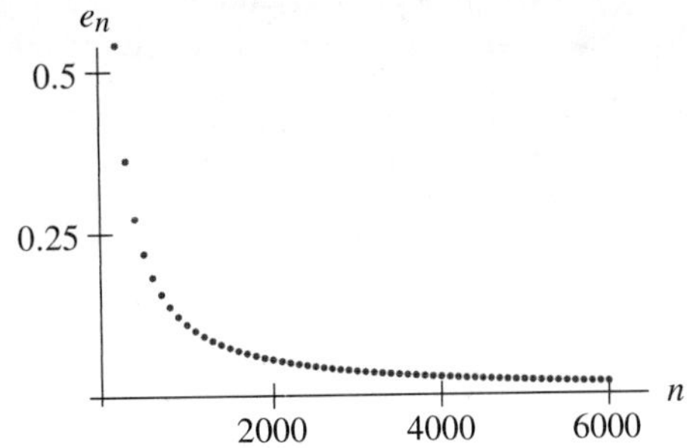

(f) Our computer math system fits the data to the function $107.125/n$. The following figure includes both the data and the graph of this function.

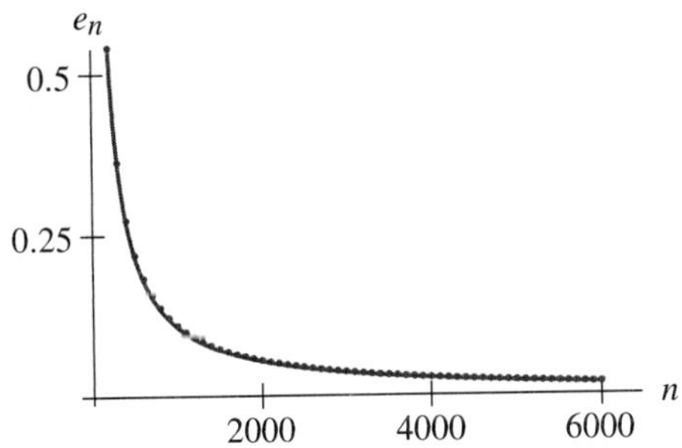

6. For Euler's method, we assume that the error using n steps is of the form K/n for some constant K. Therefore, we assume that $e_{2000} = 0.000063 \approx K/2000$. We could use this observation to obtain an approximation for K, but unless we need to know K for some other reason, we can skip that step. Basically, we need only observe that, since Euler's method is a first order numerical method, we need to increase the number of steps by a factor of 63 in order to lower the error by a factor of 63. Since $n = 2000$ yields an error of 0.000063, we must use $n = 63 \times 2000 = 126,000$ steps to obtain an error of 0.000001.

7. **(a)** The partial derivative $\partial f/\partial y$ of $f(t, y) = -2ty^2$ is $-4ty$. Thus, $M_2 = -4ty$.
The partial derivative $\partial f/\partial t$ of $f(t, y) = -2ty^2$ is $-2y^2$. Therefore,

$$M_1 = \frac{\partial f}{\partial t} + \frac{\partial f}{\partial y} f(t, y) = -2y^2 + (-4ty)(-2ty^2) = -2y^2 + 8t^2y^3.$$

(b) Since $f(t_0, y_0) = 0$, the result of the first step of Euler's method is the point $(t_1, y_1) = (0.02, 1)$. Consequently, to estimate the error, we compute the quantity M_1 at the point $(0.01, 1)$. We obtain $M_1 \approx -1.9992$.

Once we have M_1 at this point, we can estimate the error e_1 (which is the same as the truncation error) by computing $|M_1(\Delta t)^2/2|$. In this case, we obtain 0.00039984.

To calculate the actual error, we can compare the result y_1 of Euler's method with the value of the solution $y(0.2)$. Since we know that the solution is $1/(1+t^2)$, we have $y(0.2) = 0.9996$, and thus the actual error is 0.00039984. For this computation, note that the estimated error and the actual error essentially agree. (In order to see a difference in these two quantities, we had to do the calculations to 11 decimal places.)

(c) The second point (t_2, y_2) obtained from Euler's method is the point $(0.04, 0.9992)$. For the second step, the estimated error is no longer simply the truncation error, so we must compute both M_1 and M_2. Evaluating M_1 at the point $(0.03, 0.9996)$, we obtain $M_1 = -1.99121$. Evaluating M_2 at the point $(0.02, 0.9996)$, we obtain $M_2 = -0.079968$.

Now to estimate the error in the second step, we use the approximation

$$e_k \approx (1 + M_2 \Delta t)\, e_{k-1} + M_1 \frac{(\Delta t)^2}{2}$$

and obtain $e_2 \approx 0.000797442$.

To compare this estimate to the actual error, we compute the value of the solution $y(0.04) = 0.998403$. Since $y_2 = 0.9992$, we see that the actual error e_2 is 0.000797444. Note that our estimate of e_2 and the true value of e_2 are very close.

(d) Table 7.16 gives the values of e_k and our estimates of e_k for $k = 10, 20, 30, \ldots, 100$ in case that you want to double check your computations.

Table 7.16
Selected total errors

k	t_k	e_k	estimated e_k
10	0.2	-0.00354477	-0.00354363
20	0.4	-0.00487796	-0.00486646
30	0.6	-0.00399684	-0.00396879
40	0.8	-0.00226733	-0.00223171
50	1.	-0.000714495	-0.000683793
60	1.2	0.000335713	0.000355439
70	1.4	0.000929043	0.000937796
80	1.6	0.00120617	0.00120657
90	1.8	0.00129161	0.0012866
100	2.	0.00127095	0.00126289

(e) Compare your plots with the Figures 7.5 and 7.6 in Section 7.1.

8. **(a)** The partial derivative $\partial f/\partial y$ of $f(t, y) = t - y^3$ is $-3y^2$. Thus, $M_2 = -3y^2$.

The partial derivative $\partial f/\partial t$ of $f(t, y) = t - y^3$ is 1. Therefore,

$$M_1 = \frac{\partial f}{\partial t} + \frac{\partial f}{\partial y} f(t, y) = 1 - 3y^2(t - y^3).$$

(b) The result of the first step of Euler's method is the point $(t_1, y_1) = (0.01, 0.99)$. Consequently, to estimate the error, we compute the quantity M_1 at the point $(0.005, 0.995)$. We obtain $M_1 \approx 3.9109$.

Once we have M_1 at this point, we can estimate the error e_1 (which is the same as the truncation error) by computing $|M_1(\Delta t)^2/2|$. In this case, we obtain 0.000195545.

We cannot compare our estimate to the true error since we do not know how to calculate the true error for this differential equation. (We cannot find a closed-form solution.)

(c) The second point (t_2, y_2) obtained from Euler's method is the point $(0.02, 0.980397)$. For the second step, the estimated error is no longer simply the truncation error, so we must compute both M_1 and M_2. Evaluating M_1 at the point $(0.015, 0.985199)$, we obtain $M_1 = 3.74078$. Evaluating M_2 at the point $(0.01, 0.985199)$, we obtain $M_2 = -2.91185$.

Now to estimate the error in the second step, we use the approximation

$$e_k \approx (1 + M_2 \Delta t)\, e_{k-1} + M_1 \frac{(\Delta t)^2}{2}$$

and obtain $e_2 \approx 0.00037689$.

(d) Table 7.17 gives the values of our estimates of e_k for $k = 10, 20, 30, \ldots, 100$ in case that you want to double check your computations. We also plot our estimates of the error as a function of k.

Table 7.17
Selected error estimates

k	t_k	estimated e_k
10	0.1	0.00143941
20	0.2	0.00216289
30	0.3	0.00253222
40	0.4	0.00270004
50	0.5	0.00273695
60	0.6	0.00267832
70	0.7	0.00254412
80	0.8	0.00234837
90	0.9	0.00210387
100	1.	0.00182453

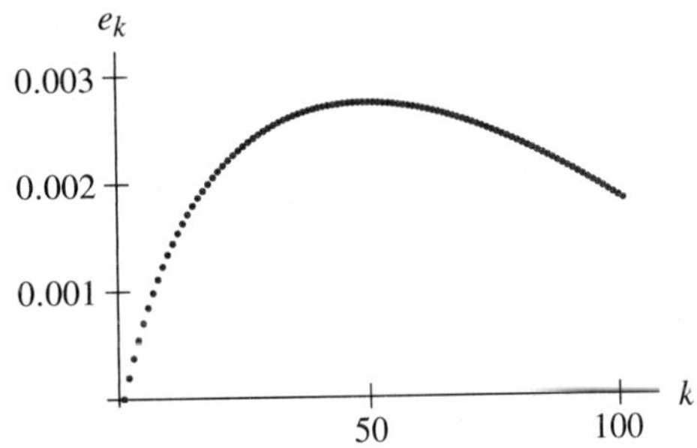

9. **(a)** The partial derivative $\partial f/\partial y$ of $f(t, y) = \sin ty$ is $t \cos ty$. Thus, $M_2 = t \cos ty$.
The partial derivative $\partial f/\partial t$ of $f(t, y) = \sin ty$ is $y \cos ty$. Therefore,

$$M_1 = \frac{\partial f}{\partial t} + \frac{\partial f}{\partial y} f(t, y) = (y + t \sin ty) \cos ty.$$

(b) The result of the first step of Euler's method is the point $(t_1, y_1) = (0.03, 3.0)$. Consequently, to estimate the error, we compute the quantity M_1 at the point $(0.015, 3)$. We obtain $M_1 \approx 2.99764$.

Once we have M_1 at this point, we can estimate the error e_1 (which is the same as the truncation error) by computing $|M_1(\Delta t)^2/2|$. In this case, we obtain 0.00134894.

We cannot compare our estimate to the true error since we do not know how to calculate the true error for this differential equation. (We cannot find a closed-form solution.)

(c) The second point (t_2, y_2) obtained from Euler's method is the point $(0.06, 3.0027)$. For the second step, the estimated error is no longer simply the truncation error, so we must compute both M_1 and M_2. Evaluating M_1 at the point $(0.045, 3.00135)$, we obtain $M_1 = 2.98002$. Evaluating M_2 at the point $(0.03, 3.00135)$, we obtain $M_2 = 0.0298785$.

Now to estimate the error in the second step, we use the approximation

$$e_k \approx (1 + M_2 \Delta t)\, e_{k-1} + M_1 \frac{(\Delta t)^2}{2}$$

and obtain $e_2 \approx 0.00269115$.

(d) The following table gives the values of our estimates of e_k for $k = 10, 20, 30, \ldots, 100$ in case that you want to double check your computations. We also plot our estimates of the error as a function of k.

Table 7.18
Selected error estimates

k	t_k	estimated e_k
10	0.3	0.0123521
20	0.6	0.0138193
30	0.9	0.00135629
40	1.2	0.0109331
50	1.5	0.0121639
60	1.8	0.0119345
70	2.1	0.0129518
80	2.4	0.0164855
90	2.7	0.0240115
100	3.	0.0343504

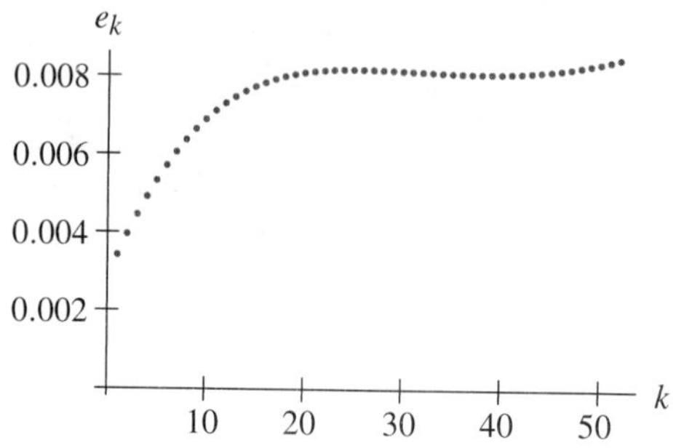

10. The Taylor series for e^{α} is

$$1 + \alpha + \frac{\alpha^2}{2!} + \frac{\alpha^3}{3!} + \ldots.$$

Since all of the terms in this series are positive if $\alpha > 0$, we can truncate the series anywhere and obtain quantity that is less than e^{α}. In this case, we truncate the series after the first two terms and obtain

$$1 + \alpha < e^{\alpha}.$$

11. **(a)** The argument that justifies the inequality

$$e_1 \le M_1 \frac{(\Delta t)^2}{2}$$

is given on pages 631 and 632. In particular, the truncation error in the first step is given by Taylor's Theorem.

(b) The total error e_2 involved in the second step is discussed on pages 633 and 634. On the right-hand side of the inequality

$$e_2 \le e_1 + M_2 e_1 \Delta t + M_1 \frac{(\Delta t)^2}{2},$$

the first term is the error involved in the previous step. The second term measures the contribution to the error that arises from evaluating the right-hand side of the differential equation at the point (t_1, y_1) rather than at $(t_1, y(t_1))$. The third term measures the truncation error associated this step of the approximation (see Figure 7.4).

(c) The analysis of the error e_{k+1} is essentially identical to the analysis of the error e_2. That is,

$$e_{k+1} = |y(t_{k+1}) - y_{k+1}|$$

where $y(t_{k+1})$ is the actual value of the function and y_{k+1} is the value given by Euler's method. In other words,

$$y_{k+1} = y_k + f(t_k, y_k)\,\Delta t.$$

Applying Taylor's Theorem to $y(t)$ at the point $(t_k, y(t_k))$, we have

$$y(t_{k+1}) = y(t_k) + f(t_k, y(t_k)) + y''(\xi_k)\frac{(\Delta t)^2}{2}.$$

Thus,

$$\begin{aligned}
e_{k+1} &= |y(t_{k+1}) - y_{k+1}| \\
&\le |y(t_k) + f(t_k, y(t_k)) + y''(\xi_k)\frac{(\Delta t)^2}{2} - (y_k + f(t_k, y_k)\,\Delta t)| \\
&\le |y(t_k) - y_k| + |f(t_k, y(t_k)) - f(t_k, y_k)|\,\Delta t + |y''(\xi_k)|\frac{(\Delta t)^2}{2} \\
&\le e_k + |f(t_k, y(t_k)) - f(t_k, y_k)|\,\Delta t + |y''(\xi_k)|\frac{(\Delta t)^2}{2}.
\end{aligned}$$

The term $|f(t_k, y(t_k)) - f(t_k, y_k)|$ is bounded by the product of M_2 and e_k, and the third term (the truncation error) is bounded by $M_1(\Delta t)^2/2$. Hence, we have

$$\begin{aligned}
e_{k+1} &\le e_k + (M_2)(e_k)(\Delta t) + M_1\frac{(\Delta t)^2}{2} \\
&= (1 + M_2\Delta t)e_k + M_1\frac{(\Delta t)^2}{2}.
\end{aligned}$$

(d) We know that $e_1 \le K_2$ because e_1 is the same as the truncation error. Using the result of part (b) (or part (c) with $k = 1$), we know that

$$e_2 \le (1 + M_2\Delta t)e_1 + M_1\frac{(\Delta t)^2}{2} = (K_1)e_1 + K_2.$$

However, given that $e_1 \le K_2$, we have

$$e_2 \le (K_1 + 1)K_2.$$

Finally, from part (c), we know that

$$e_3 \le (K_1)e_2 + K_2 \le (K_1)(K_1 + 1)K_2 + K_2 = (K_1^2 + K_1 + 1)K_2.$$

(e) We can verify this assertion by induction. In fact, we can use the result of part (d) as the first step. Then we assume the inductive hypothesis that

$$e_{n-1} \leq \left(K_1^{n-2} + K_1^{n-3} + \cdots + K_1 + 1\right) K_2.$$

Since part (c) says that $e_n \leq (K_1)(e_{n-1}) + K_2$, we can use the inductive hypothesis to obtain

$$\begin{aligned} e_n &\leq (K_1)\left(K_1^{n-2} + K_1^{n-3} + \cdots + K_1 + 1\right) K_2 + K_2 \\ &= \left(K_1^{n-1} + K_1^{n-2} + \cdots + K_1 + 1\right) K_2. \end{aligned}$$

(f) We can verify the formula

$$K_1^{n-1} + K_1^{n-2} + \cdots + K_1 + 1 = \frac{K_1^n - 1}{K_1 - 1}$$

using induction. However, it is probably easier to see why it holds if we multiply both sides by the factor $(K_1 - 1)$ and cancel on the left-hand side.

Applying this formula to the result of part (e), we get

$$e_n \leq \left(\frac{K_1^n - 1}{K_1 - 1}\right) K_2.$$

(g) Since $K_1 = 1 + M_2\Delta t$ and $K_2 = M_2(\Delta t)^2/2$, we have

$$\begin{aligned} e_n &\leq \left(\frac{(1 + M_2\Delta t)^n - 1}{M_2\Delta t}\right)\left(M_1\frac{(\Delta t)^2}{2}\right) \\ &= \frac{M_1}{2M_2}\left((1 + M_2\Delta t)^n - 1\right)\Delta t. \end{aligned}$$

(h) If we let $\alpha = M_2\Delta t$ in Exercise 10, then

$$\begin{aligned} \left((1 + M_2\Delta t)^n - 1\right) &\leq \left(\left(e^{(M_2\Delta t)}\right)^n - 1\right) \\ &= \left(e^{(M_2\Delta t)n} - 1\right). \end{aligned}$$

Consequently, we can conclude that

$$e_n \leq \frac{M_1}{2M_2}\left(e^{(M_2\Delta t)n} - 1\right)\Delta t.$$

(i) Since $\Delta t = (t_n - t_0)/n$, the product $(M_2\Delta t)n$ is equal to $M_2(t_n - t_0)$. Therefore, we can conclude that

$$e_n \leq \frac{M_1}{2M_2}\left(e^{M_2(t_n - t_0)} - 1\right)\Delta t.$$

(j) Note that the quantities M_1 and M_2 are determined by the right-hand side of the differential equation and the rectangle R in the ty-plane. Also, the quantity $(t_n - t_0)$ is precisely the length of the interval over which we are approximating the solution. Therefore, all of the terms in the expression

$$\frac{M_1}{2M_2}\left(e^{M_2(t_n-t_0)}-1\right)$$

do not depend on the number of steps involved in the application of Euler's method. In other words, when considering the effectiveness of Euler's method, we can treat this expression as a constant C determined only by the right-hand side of the differential equation and the rectangle R under consideration.

This expression includes all of the right-hand side of part (i) with the exception of the Δt factor. Consequently, we can consider this long and involved derivation as a justification of the simple inequality

$$e_n \le C \cdot \Delta t.$$

(k) This estimate is a rigorous one. In other words, given the hypotheses stated at the beginning of the exercise, we can be certain that the error is indeed bounded by the quantities specified. The "estimates" calculated in Exercises 7–9 are not as certain. The logic that justifies their calculation is valid, but we cannot be certain that these quantities always give us a true indication of the accuracy of our approximations.

12. **(a)** Since

$$\left|\frac{\partial f}{\partial t}+\frac{\partial f}{\partial y}f(t,y)\right| = \left|-2y^2+4ty(2ty^2)\right|,$$

its maximum value M_1 is assumed at the point $(t, y) = (2, 1)$. At this point, we get $M_1 = 30$. Similarly,

$$\left|\frac{\partial f}{\partial y}\right| = |-4ty|,$$

so $M_2 = 8$ over this rectangle.

(b) To calculate the constant C as given in part (j) of Exercise 11, we need to evaluate

$$\frac{M_1}{2M_2}\left(e^{M_2(t_n-t_0)}-1\right).$$

This quantity is just slightly less than 16,661,456.

(c) Since the interval is $0 \le t \le 2$, we know that

$$C \cdot \Delta t = C \cdot \frac{(2-0)}{n}.$$

Hence, $K = 2C$ in this example. Note that K is roughly 34 million.

(d) These estimates are so conservative because we estimated the quantities M_1 and M_2 over the entire rectangle R. Without additional information, the theory requires that make these estimates over such a large region. However, we know that using these values for M_1 and M_2 is severe overkill. That is why the method of estimating the error as outlined on page 637 is more indicative of the actual errors involved in the calculation.

EXERCISES FOR SECTION 7.2

1. Table 7.19 includes the approximate values y_k obtained using improved Euler's method. In addition to the results of improved Euler's method, we graph the results of Euler's method and the results obtained when we used a built-in numerical solver.

Table 7.19
Results of improved Euler's method

k	t_k	y_k
0	0.0	3.0000
1	0.5	8.2500
2	1.0	21.3750
3	1.5	54.1875
4	2.0	136.2187

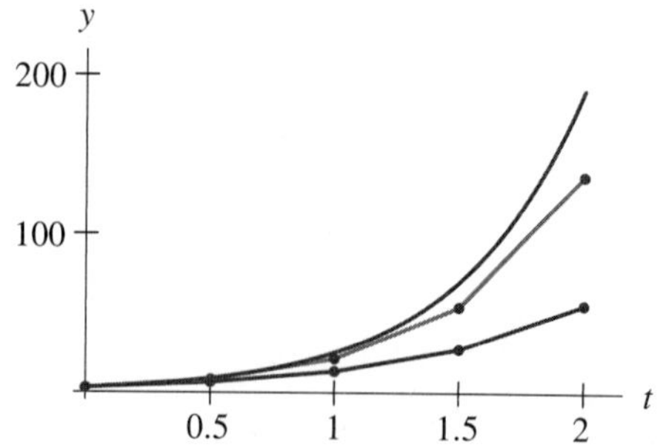

2. Table 7.20 includes the approximate values y_k obtained using improved Euler's method. In addition to the results of improved Euler's method, we graph the results of Euler's method and the results obtained when we used a built-in numerical solver.

Table 7.20
Results of improved Euler's method

k	t_k	y_k
0	0.	1.
1	0.25	0.835937
2	0.5	0.776864
3	0.75	0.787177
4	1.	0.844469

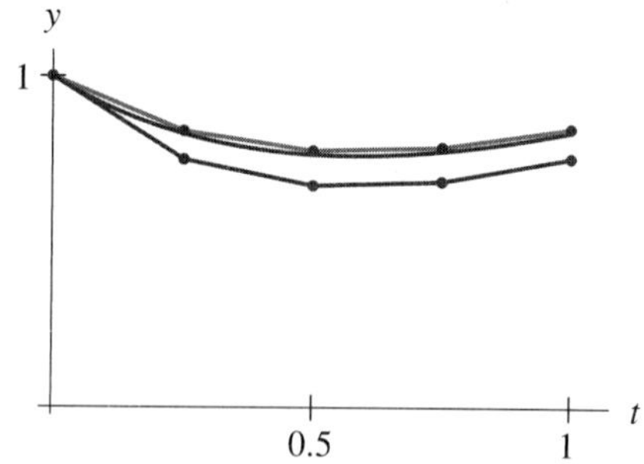

3. Table 7.21 includes the approximate values y_k obtained using improved Euler's method. In addition to the results of improved Euler's method, we graph the results of Euler's method and the results obtained when we used a built-in numerical solver.

Table 7.21
Results of improved Euler's method

k	t_k	y_k	k	t_k	y_k
0	0.00	0.5	5	1.25	−1.70535
1	0.25	0.445801	6	1.50	−2.09616
2	0.50	0.103176	7	1.75	−2.39212
3	0.75	−0.501073	8	2.00	−2.63277
4	1.00	−1.16818			

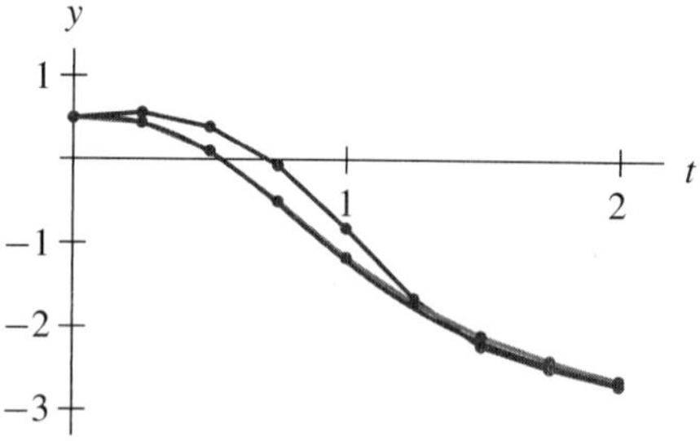

4. Table 7.22 includes the approximate values y_k obtained using improved Euler's method. In addition to the results of improved Euler's method, we graph the results of Euler's method and the results obtained when we used a built-in numerical solver.

Table 7.22
Results of improved Euler's method

k	t_k	y_k
0	0.	1.
1	0.5	1.45756
2	1.	1.93779
3	1.5	2.33918
4	2.	2.62608
5	2.5	2.81577
6	3.	2.93705

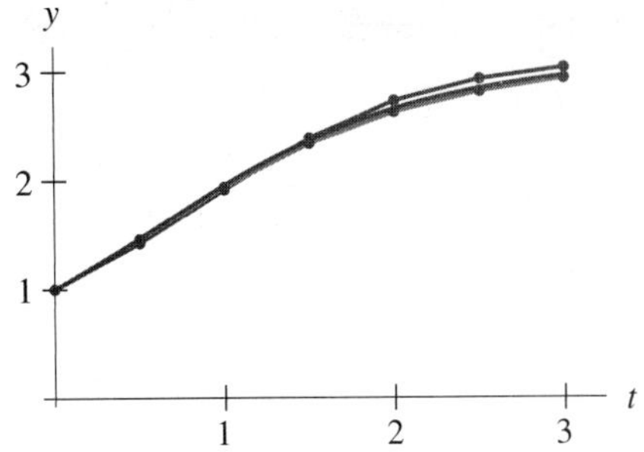

5. Table 7.23 includes the approximate values w_k obtained using improved Euler's method. In addition to the results of improved Euler's method, we graph the results of Euler's method and the results obtained when we used a built-in numerical solver.

Table 7.23
Results of improved Euler's method

k	t_k	y_k
0	0.	0.
1	0.5	1.6875
2	1.	2.06728
3	1.5	2.22283
4	2.	2.31738
5	2.5	2.38333
6	3.	2.43292
7	3.5	2.47205
8	4.	2.50398
9	4.5	2.53071
10	5.	2.55351

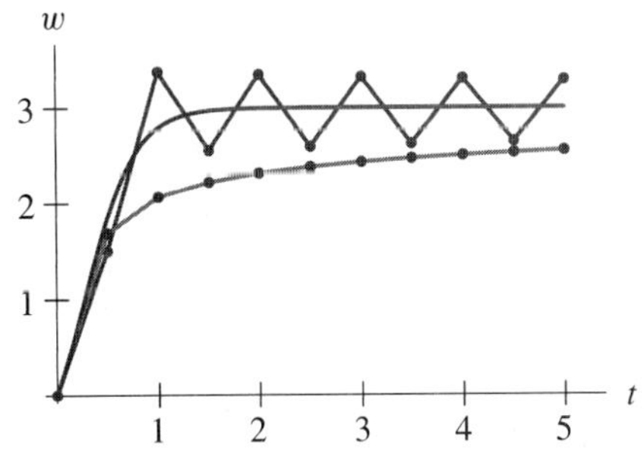

6. Table 7.24 includes the approximate values y_k obtained using improved Euler's method. In addition to the results of improved Euler's method, we graph the results of Euler's method and the results obtained when we used a built-in numerical solver.

Table 7.24
Results of improved Euler's method

k	t_k	y_k
0	0.	2.
1	0.5	3.13301
2	1.	4.01452
3	1.5	4.80396
4	2.	5.54124

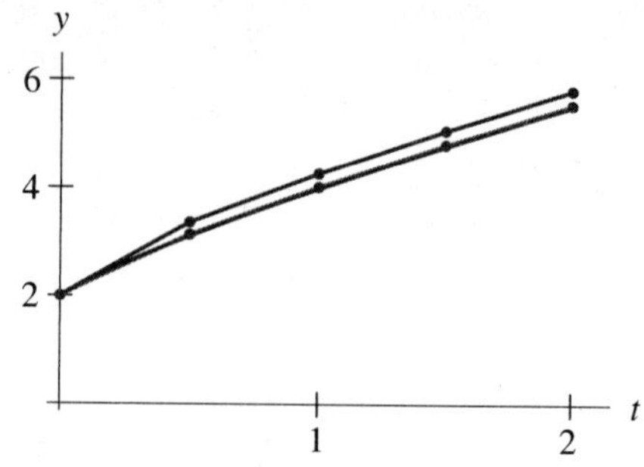

7. Table 7.25 includes the approximate values y_k obtained using improved Euler's method. Compare the results of this calculation with the results obtained in Exercise 6. In addition to the results of improved Euler's method, we graph the results of Euler's method and the results obtained when we used a built-in numerical solver.

Table 7.25
Results of improved Euler's method

k	t_k	y_k
0	1.	2.
1	1.5	3.13301
2	2.	4.01452
3	2.5	4.80396
4	3.	5.54124

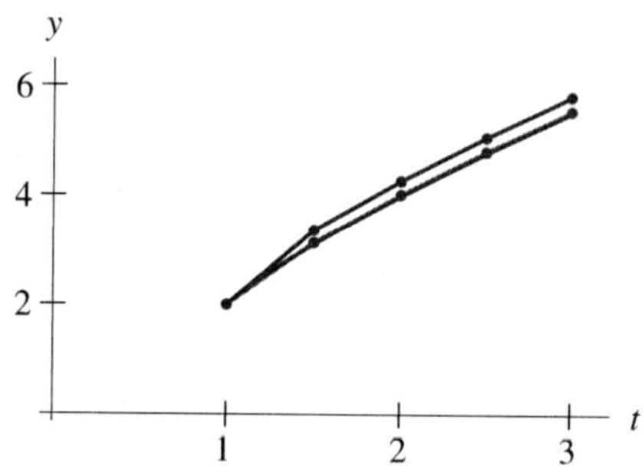

8. Table 7.26 includes the approximate values y_k obtained using improved Euler's method. In addition to the results of improved Euler's method, we graph the results of Euler's method and the results obtained when we used a built-in numerical solver. Note that it is basically impossible to distinguish the three graphs.

Table 7.26
Results of improved Euler's method

k	t_k	y_k
0	0.0	0.2
1	0.1	0.203
2	0.2	0.207
3	0.3	0.210
⋮	⋮	
99	9.9	0.989
100	10.0	0.990

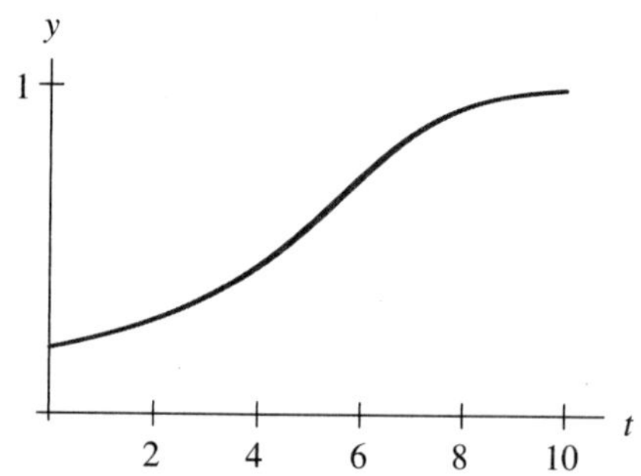

9. **(a)** The differential equation is both separable and linear. Therefore, one way to obtain the solution to the initial-value problem is to integrate

$$\int \frac{1}{1-y}\,dy = \int 1\,dt.$$

We obtain

$$-\ln|1-y| = t + c$$
$$y = 1 - ke^{-t}.$$

We determine the value of k using the initial condition $y(0) = 0$. Hence, $k = 1$, and the solution to the given initial-value problem is $y(t) = 1 - e^{-t}$.

(b) Table 7.27 contains the steps involved in applying improved Euler's method to this initial-value problem.

Table 7.27
Results of improved Euler's method

t_k	y_k
0.	0.
0.25	0.21875
0.5	0.389648
0.75	0.523163
1.	0.627471

Using the analytic solution, we know that the actual value of $y(1)$ is $1 - 1/e$. Therefore, we can compute the error

$$e_4 = |y(1) - y_4| = 0.00464959.$$

(c) If we want an approximation that is accurate to 0.0001, we need an improvement by a factor of

$$\frac{0.00464959}{0.0001} = 46.4959.$$

Since improved Euler's method is a second-order numerical scheme, we expect to get that kind of improvement if we increase the number of steps by a factor of $\sqrt{46.4959}$. In other words, we compute the smallest integer larger than $4\sqrt{46.4959} = 27.2752$. Using $n = 28$ steps, we get the approximate value $y_{28} = 0.63204$ and, consequently, an error $e_{28} \approx 0.00008$.

10. **(a)** The differential equation is both separable and linear. Therefore, one way to obtain the solution to the initial-value problem is to integrate

$$\int \frac{1}{y}\,dy = \int t\,dt.$$

We obtain

$$\ln |y| = \frac{t^2}{2} + c$$

$$y = ke^{t^2/2}.$$

We determine the value of k using the initial condition $y(0) = 1$. Hence, $k = 1$, and the solution to the given initial-value problem is $y(t) = e^{t^2/2}$.

(b) Table 7.28 contains the steps involved in applying improved Euler's method to this initial-value problem.

Table 7.28
Results of improved Euler's method

t_k	y_k
0.	1.
0.353553	1.0625
0.707107	1.27832
1.06066	1.73772
1.41421	2.66088

Using the analytic solution, we know that the actual value of $y(\sqrt{2})$ is e. Therefore, we can compute the error

$$e_4 = |y(\sqrt{2}) - y_4| = 0.0574032.$$

(c) If we want an approximation that is accurate to 0.0001, we need an improvement by a factor of

$$\frac{0.0574032}{0.0001} = 574.032.$$

Since improved Euler's method is a second-order numerical scheme, we expect to get that kind of improvement if we increase the number of steps by a factor of $\sqrt{574.032}$. In other words, we compute the smallest integer larger than $4\sqrt{574.032} = 95.8358$. Using $n = 96$ steps, we get the approximate value $y_{96} = 2.71818$ and, consequently, an error $e_{96} \approx 0.00009$.

11. **(a)** The differential equation is separable. Therefore, we integrate

$$\int \frac{1}{y^2}\, dy = \int -1\, dt.$$

We obtain

$$-\frac{1}{y} = -t + c$$

$$y = \frac{1}{t+k}.$$

We determine the value of k using the initial condition $y(0) = 1/2$. Hence, $k = 2$, and the solution to the given initial-value problem is $y(t) = 1/(t+2)$.

(b) Table 7.29 contains the steps involved in applying improved Euler's method to this initial-value problem.

Table 7.29
Results of improved Euler's method

t_k	y_k
0.	0.5
0.5	0.402344
1.	0.336049
1.5	0.288275
2.	0.252281

Using the analytic solution, we know that the actual value of $y(2)$ is $1/4$. Therefore, we can compute the error

$$e_4 = |y(2) - y_4| = 0.002281.$$

(c) If we want an approximation that is accurate to 0.0001, we need an improvement by a factor of

$$\frac{0.002281}{0.0001} = 22.81.$$

Since improved Euler's method is a second-order numerical scheme, we expect to get that kind of improvement if we increase the number of steps by a factor of $\sqrt{22.81}$. In other words, we compute the smallest integer larger than $4\sqrt{22.81} = 19.103$. Using $n = 20$ steps, we get the approximate value $y_{20} = 0.250081$ and, consequently, an error $e_{20} \approx 0.00008$.

12. **(a)** The differential equation is linear, and we use the Extended Linearity Principle to find an analytic solution (see Section 1.8).

The general solution of the associated homogeneous equation is ce^{3t}. We rewrite the nonhomogeneous equation as

$$\frac{dy}{dt} - 3y = 1 - t$$

and guess a solution of the form $y_p(t) = at + b$. Substituting this guess into the left-hand side of the differential equation, we get

$$\begin{aligned}\frac{dy_p}{dt} - 3y &= a - 3(at + b)\\ &= (-3a)t + (a - 3b).\end{aligned}$$

Therefore, $y_p(t)$ is a solution if $-3a = -1$ and $a - 3b = 1$. In other words, $a = 1/3$ and $b = -2/9$. The general solution of the original nonhomogeneous equation is

$$y(t) = ce^{3t} + \tfrac{1}{3}t - \tfrac{2}{9}.$$

We determine the value of c using the initial condition $y(0) = 1$. We get $c = 11/9$, and the solution to the given initial-value problem is

$$y(t) = \tfrac{11}{9}e^{3t} + \tfrac{1}{3}t - \tfrac{2}{9}.$$

(b) Table 7.30 contains the steps involved in applying improved Euler's method to this initial-value problem.

Table 7.30
Results of improved Euler's method

t_k	y_k
0.	1.
0.25	2.34375
0.5	4.9873
0.75	10.2711
1.	20.9178

Using the analytic solution, we know that the actual value of $y(1)$ is $(11e^3 + 1)/9$. Therefore, we can compute the error

$$e_4 = |y(1) - y_4| = 3.74227.$$

(c) If we want an approximation that is accurate to 0.0001, we need an improvement by a factor of

$$\frac{3.74227}{0.0001} = 37422.7.$$

Since improved Euler's method is a second-order numerical scheme, we expect to get that kind of improvement if we increase the number of steps by a factor of $\sqrt{37422.7}$. In other words, we compute the smallest integer larger than $4\sqrt{37422.7} = 773.798$. Using $n = 774$ steps, we get the approximate value $y_{774} = 24.6599$ and, consequently, an error $e_{774} \approx 0.000184$.

Unfortunately, this result is not within the tolerance specified in the statement of the exercise, and we must increase the number of steps once more. We want an additional improvement of a factor of

$$\frac{0.000184}{0.0001} = 1.84.$$

To determine our second choice for the number of steps, we compute $774\sqrt{1.84} = 1049.9$ Therefore, rather than 774 steps, we use 1050 steps. In this case, we get an error $e_{1050} \approx 0.0000999$.

13. **(a)** If we want an approximation that is accurate to 0.0001, we need an improvement by a factor of

$$\frac{0.000695}{0.0001} = 6.95.$$

Since improved Euler's method is a second-order numerical scheme, we expect to get that kind of improvement if we increase the number of steps by a factor of $\sqrt{6.95}$. In other words, we compute the smallest integer larger than $20\sqrt{6.95} = 52.7257$.

(b) Using $n = 53$ steps, we get the approximate value $y_{53} = 0.200095$.

(c) Consequently, the error e_{53} is the difference between the actual value $y(2) = 0.2$ and y_{53}. We get $e_{53} = 0.000095$.

14.

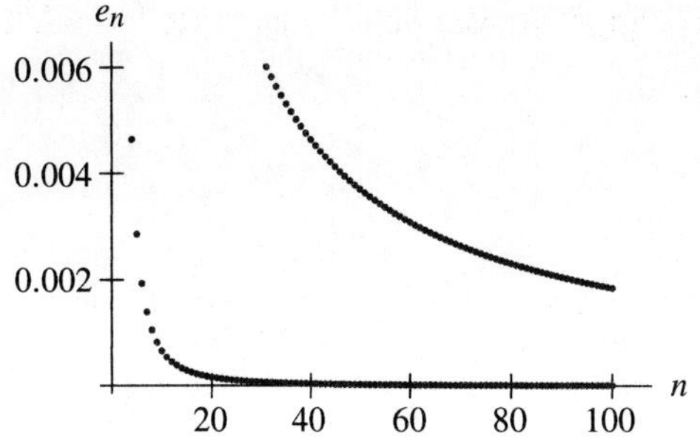

15.

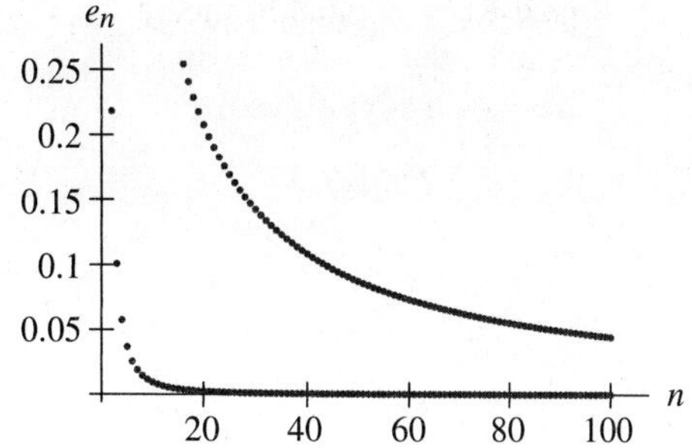

16.

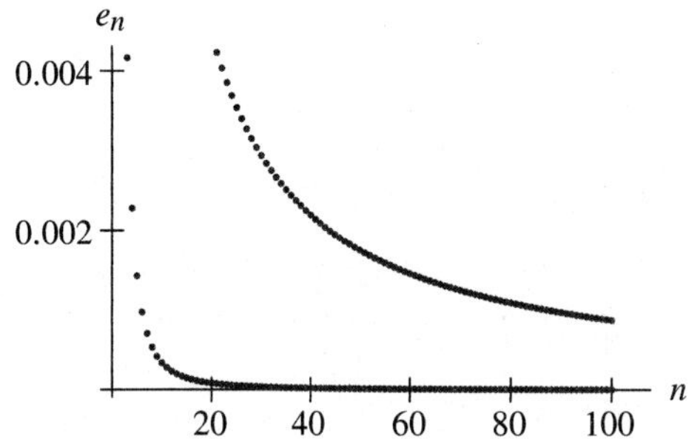

17.

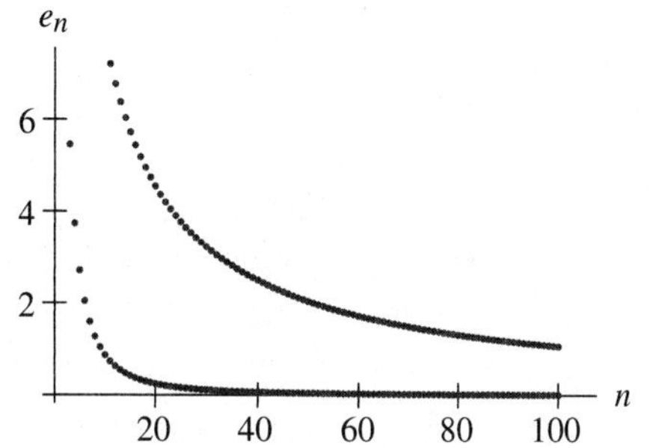

EXERCISES FOR SECTION 7.3

1. Runge-Kutta applied to this initial-value problem yields the points given in Table 7.31. The graph illustrates the results of Runge-Kutta as compared to those of Euler's method, improved Euler's method, and a built-in solver.

Table 7.31
Results of Runge-Kutta

t_k	y_k
0.0	3.000
0.5	8.979
1.0	25.173
1.5	69.030
2.0	187.811

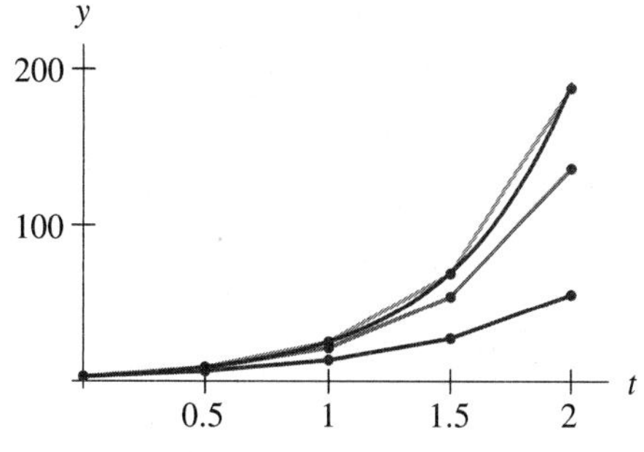

2. Runge-Kutta applied to this initial-value problem yields the points given in Table 7.32. The graph illustrates the results of Runge-Kutta as compared to those of Euler's method, improved Euler's method, and a built-in solver.

Table 7.32
Results of Runge-Kutta

t_k	y_k
0.	1.
0.25	0.827265
0.5	0.765304
0.75	0.775139
1.	0.83343

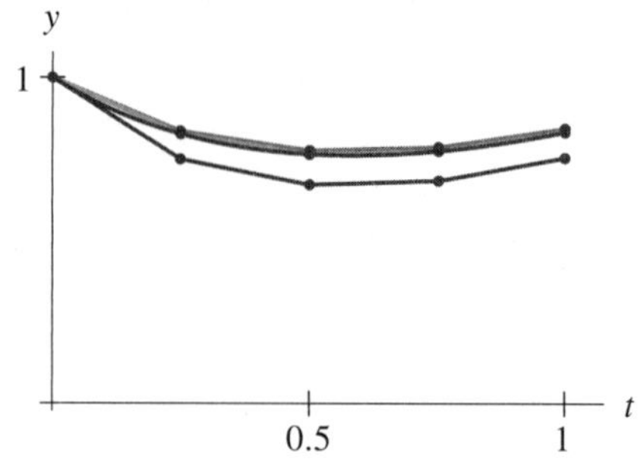

3. Runge-Kutta applied to this initial-value problem yields the points given in Table 7.33. The graph illustrates the results of Runge-Kutta as compared to those of Euler's method, improved Euler's method, and a built-in solver.

Table 7.33
Results of Runge-Kutta

t_k	y_k	t_k	y_k
0.0	0.00000	3.0	2.99645
0.5	1.82290	3.5	2.99882
1.0	2.70058	4.0	2.99961
1.5	2.90368	4.5	2.99987
2.0	2.96803	5.0	2.99996
2.5	2.98935		

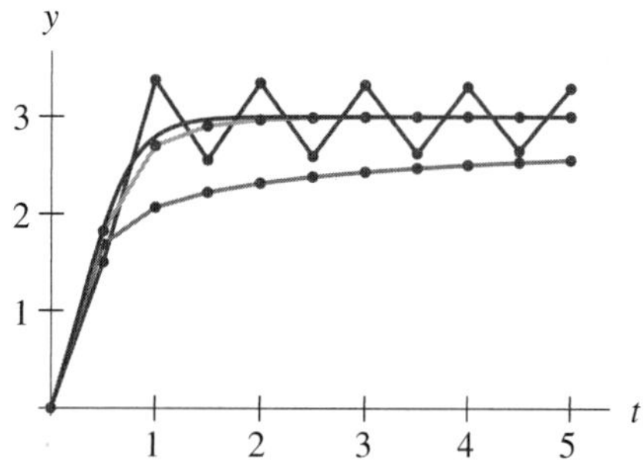

4. Runge-Kutta applied to this initial-value problem yields the points given in Table 7.34. The graph illustrates the results of Runge-Kutta as compared to those of Euler's method, improved Euler's method, and a built-in solver.

Table 7.34
Results of Runge-Kutta

t_k	y_k
0.	2.
0.5	3.10456
1.	3.98546
1.5	4.77554
2.	5.51352

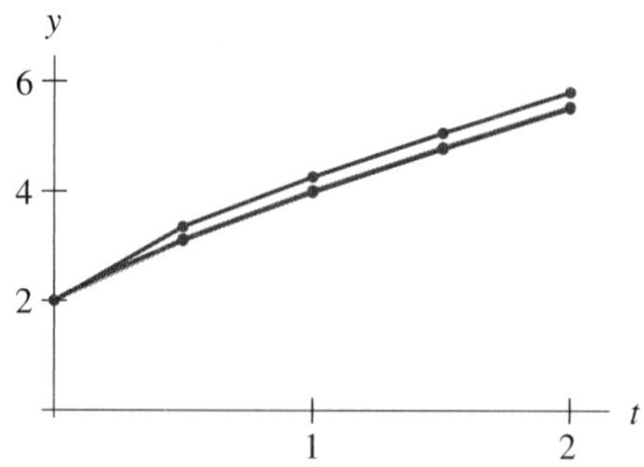

5. Note the relationship between this exercise and Exercise 4.

Runge-Kutta applied to this initial-value problem yields the points given in Table 7.35. The graph illustrates the results of Runge-Kutta as compared to those of Euler's method, improved Euler's method, and a built-in solver.

Table 7.35
Results of Runge-Kutta

t_k	y_k
1.0	2.00000
1.5	3.10456
2.0	3.98546
2.5	4.77554
3.0	5.51352

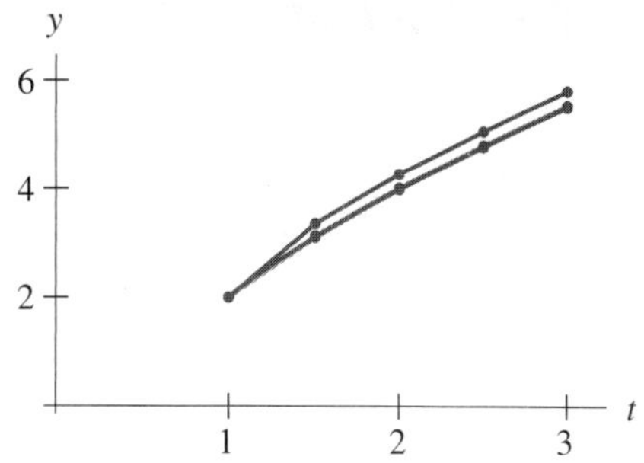

6. **(a)** The steps involved in Runge-Kutta applied to this initial-value problem are shown in Table 7.36.

Table 7.36
Results of Runge-Kutta

t_k	y_k
0.	1.
0.5	0.798379
1.	0.499702
1.5	0.308167
2.	0.200406

(b) From the text, we know that the solution to this initial-value problem is the function $y(t) = 1/(1+t^2)$ and, therefore, $y(2) = 0.2$. Hence, the total error $e_4 = 0.000406$.

(c) We would like an error e_n of 0.0001. Therefore, we need to improve our estimate by a factor of $0.000406/0.0001 = 4.06$. Since Runge-Kutta is a fourth-order method, we expect to have to increase the number of steps by a factor of $\sqrt[4]{4.06} = 1.41949$. In other words, a good guess for the appropriate number of steps is the lowest integer greater than $4\sqrt[4]{4.06}$.

As a result, we try 6 steps. In that case, we get a total error $e_6 \approx 0.000087$.

7. **(a)** The results of Runge-Kutta applied to the predator-prey system are given in Table 7.37 and the figure illustrates this computation in the phase plane.

Table 7.37
Results of Runge-Kutta

t_k	R_k	F_k
0	1.	1.
1	1.50412	1.91806
2	0.641301	2.48192
3	0.416812	1.62154
4	0.636774	1.08434

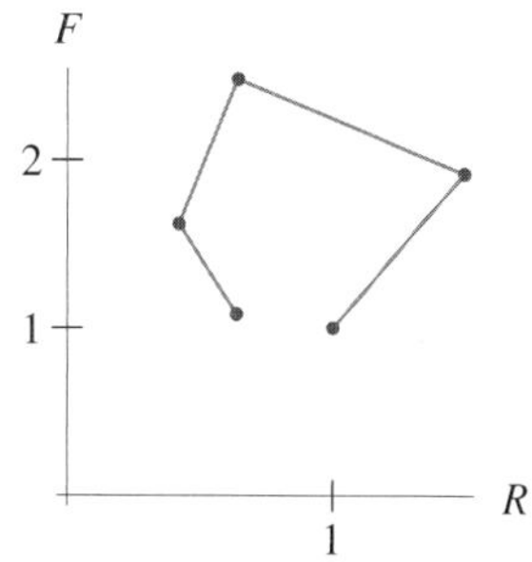

(b)

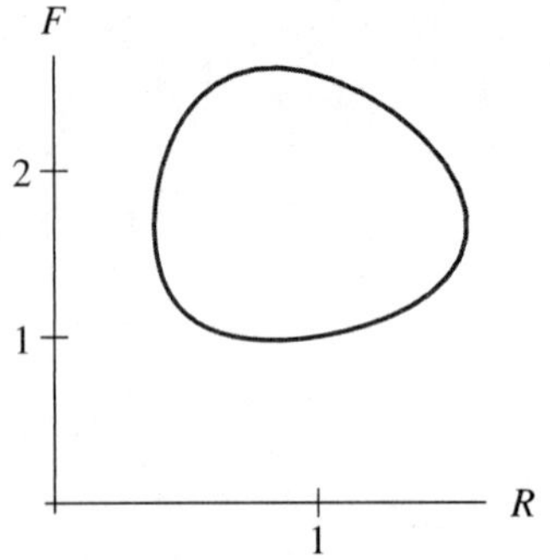

(c)

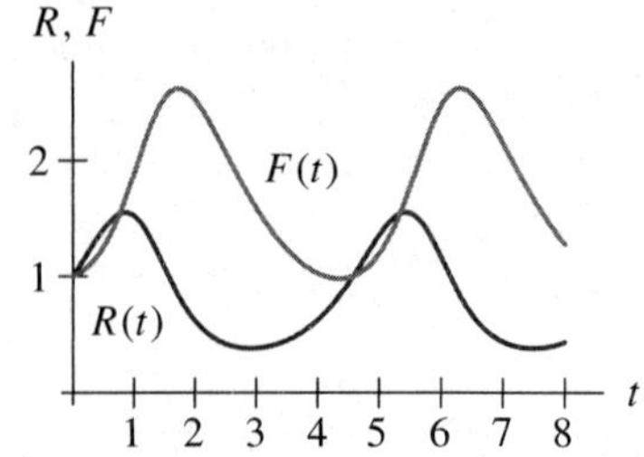

8. We introduce a new variable $v = dy/dt$ and convert the second-order equation into a first-order system where

$$\frac{dy}{dt} = v$$
$$\frac{dv}{dt} = -5y + (3 - y^2)v.$$

Table 7.38 contains some of the results of Runge-Kutta applied to the initial condition $(y(0), v(0)) = (1, 0)$, and the figures plot the results in the phase plane (the yv-plane) and as a $y(t)$-graph.

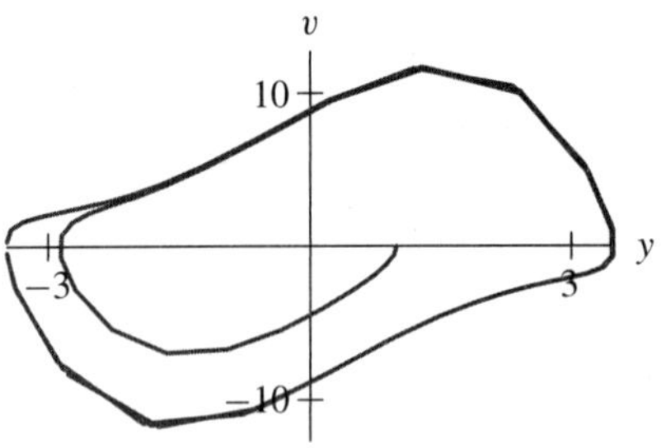

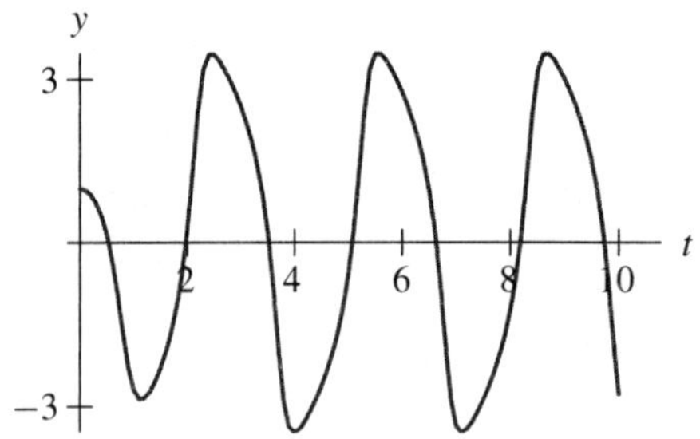

Table 7.38
Results of Runge-Kutta for the second-order equation

t_k	y_k	v_k	t_k	y_k	v_k
0	1.	0	⋮	⋮	⋮
0.1	0.973343	-0.549664	5.	-0.608744	7.0434
0.2	0.886652	-1.20244	⋮	⋮	⋮
0.3	0.728727	-1.97995	7.5	-2.89422	2.19975
0.4	0.485011	-2.92749	⋮	⋮	⋮
0.5	0.135717	-4.09813	10.	-2.77529	-8.30128

EXERCISES FOR SECTION 7.4

1. Table 7.39 contains the results of approximating the value $y(1)$ of the solution using y_k, where y_k is obtained by applying Runge-Kutta over the interval $0 \le t \le 1$ using k steps with single precision arithmetic. Note that we obtain the approximation $y(1) \approx 0.941274$, but we cannot be confident about the next digit.

Table 7.39
Runge-Kutta approximations

k	y_k	k	y_k
2	0.94188476	64	0.94127458
4	0.92484933	128	0.94127458
8	0.94101000	256	0.94127434
16	0.94127089	512	0.94127452
32	0.94127452	1024	0.94127500

2. Table 7.40 contains the results of approximating the value $y(3)$ of the solution using y_k, where y_k is obtained by applying Runge-Kutta over the interval $0 \le t \le 3$ using k steps with single precision arithmetic. Note that we obtain the approximation $y(3) \approx 3.52803$, but we cannot be confident about the next digit.

Table 7.40
Runge-Kutta approximations

k	y_k	k	y_k
2	3.52767634	64	3.52803159
4	3.52797222	128	3.52803111
8	3.52802801	256	3.52803183
16	3.52803159	512	3.52803278
32	3.52803206		

3. Table 7.41 contains the results of approximating the value $y(2)$ of the solution using y_k, where y_k is obtained by applying Runge-Kutta over the interval $0 \le t \le 2$ using k steps with single precision arithmetic. Note that we obtain the approximation $y(2) \approx 1.25938$, but we cannot be confident about the next digit.

Table 7.41
Runge-Kutta approximations

k	y_k	k	y_k
2	1.33185744	64	1.25938189
4	1.25684679	128	1.25938177
8	1.25911093	256	1.25938165
16	1.25936496	512	1.25938165
32	1.25938094	1024	1.25938201

REVIEW EXERCISES FOR CHAPTER 7

1. True. In a first-order method, the error is, at worst, proportional to the first power of the step size. In other words, if e_n is the error at the nth step, there is a constant C such that

$$e_n \le C \cdot (\Delta t)^1.$$

Assuming that this inequality is basically an equality for large values of n, we see that halving Δt would halve e_n.

2. False. In a second-order method, the error is, at worst, proportional to the second power of the step size. In other words, if e_n is the error at the nth step, there is a constant C such that

$$e_n \le C \cdot (\Delta t)^2.$$

Doubling the number of steps halves the step size. Assuming that the inequality for e_n is basically an equality for large values of n, we see that doubling the number of steps would lower e_n by a factor of $(1/2)^2 = 1/4$.

3. False. In a fourth-order method, the error is, at worst, proportional to the fourth power of the step size. In other words, if e_n is the error at the nth step, there is a constant C such that

$$e_n \le C \cdot (\Delta t)^4.$$

Assuming that this inequality is basically an equality for large values of n, we see that halving Δt would lower e_n by a factor of $(1/2)^4 = 1/16$.

4. True. Improved Euler is a second-order method. See the discussion in the text on page 646.

5. False. Runge-Kutta is a fourth-order method. See the discussion in the text on page 653.

6. **(a)** The following table has the results of Euler's method for $n = 4$. The graph is shown in part (c).

k	t_k	y_k
0	0.00	1.000
1	0.25	0.750
2	0.50	0.707
3	0.75	0.744
4	1.00	0.828

(b) The following table has the results of improved Euler's method for $n = 4$. The graph is shown in part (c).

k	t_k	y_k
0	0.00	1.000
1	0.25	0.854
2	0.50	0.815
3	0.75	0.838
4	1.00	0.899

(c) The following table has the results of Runge-Kutta for $n = 4$. The graphs correspond to Euler's method, improved Euler's method, and Runge-Kutta. See parts (a) and (b).

k	t_k	y_k
0	0.00	1.000
1	0.25	0.843
2	0.50	0.801
3	0.75	0.825
4	1.00	0.888

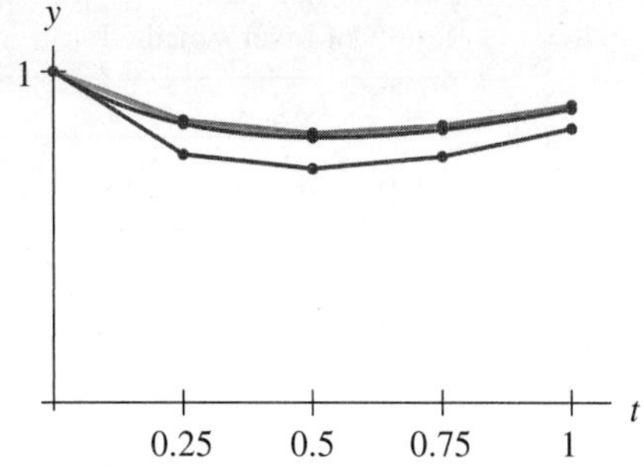

(d) Results of Euler's method

t_k	y_k	t_k	y_k
0.00	1.	⋮	⋮
0.01	0.99	⋮	⋮
0.02	0.980397	0.97	0.87697
0.03	0.971174	0.98	0.879925
0.04	0.962314	0.99	0.882912
0.05	0.953802	1.00	0.885929

Results of improved Euler's method

t_k	y_k	t_k	y_k
0.00	1.	⋮	⋮
0.01	0.990199	⋮	⋮
0.02	0.98078	0.97	0.878917
0.03	0.971727	0.98	0.881843
0.04	0.963026	0.99	0.884801
0.05	0.954663	1.00	0.887789

Results of Runge-Kutta

t_k	y_k	t_k	y_k
0.00	1.	⋮	⋮
0.01	0.990197	⋮	⋮
0.02	0.980777	0.97	0.878917
0.03	0.971723	0.98	0.881843
0.04	0.963021	0.99	0.884801
0.05	0.954656	1.00	0.887789

The graphs of Euler, improved Euler, and Runge-Kutta with $n = 100$ steps. Note that it is basically impossible to distinguish the three graphs.

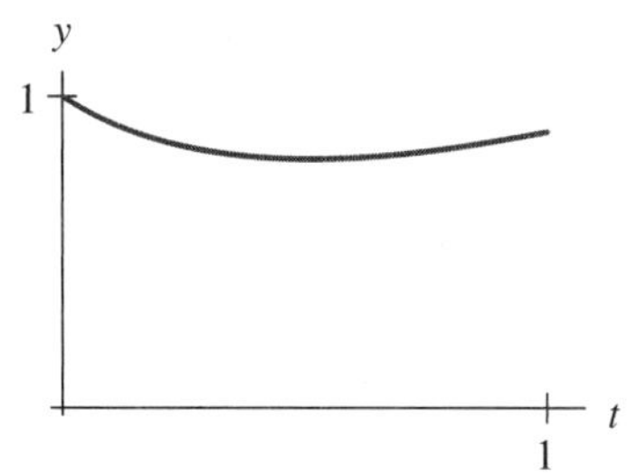

7. **(a)** Euler's method:

Table 7.42
Results of Euler's method

t_k	x_k	y_k	t_k	x_k	y_k
0.0	3.	3.	⋮	⋮	⋮
0.1	1.8	1.2	⋮	⋮	⋮
0.2	1.62	0.984	4.7	1.87797	0.0737875
0.3	1.52215	0.863558	4.8	1.88703	0.0676651
0.4	1.46344	0.785159	4.9	1.89558	0.0619696
0.5	1.42706	0.729253	5.0	1.90363	0.0566828

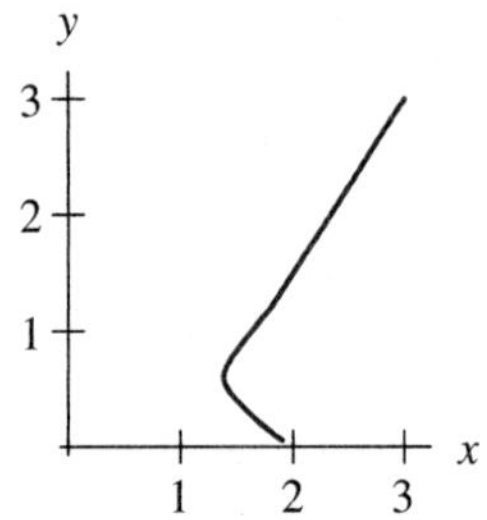

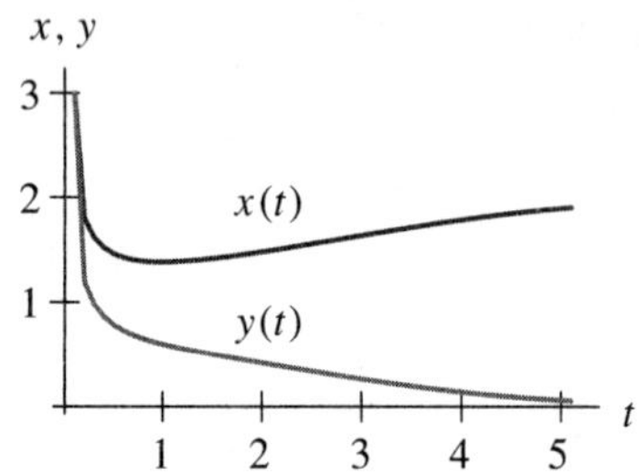

(b) improved Euler's method:

Table 7.43
Results of improved Euler's method

t_k	x_k	y_k	t_k	x_k	y_k
0.0	3.	3.	⋮	⋮	⋮
0.1	2.31	1.992	⋮	⋮	⋮
0.2	1.9507	1.51588	4.7	1.80751	0.124714
0.3	1.7399	1.24768	4.8	1.81953	0.115741
0.4	1.60693	1.07962	4.9	1.83107	0.107255
0.5	1.5191	0.96594	5.0	1.84212	0.0992498

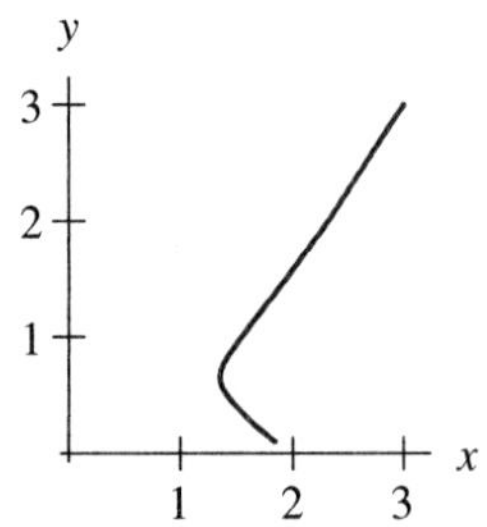

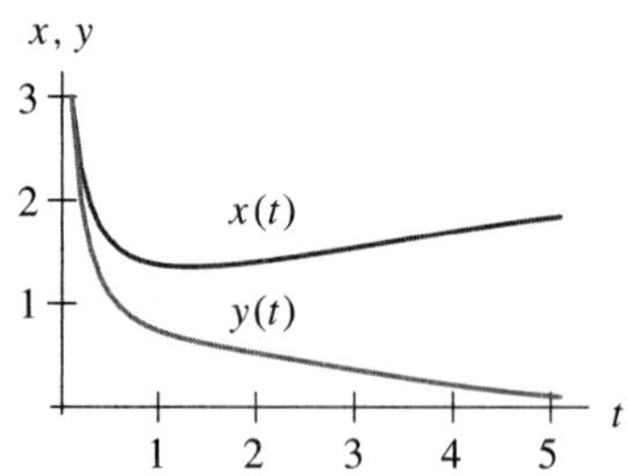

(c) Runge Kutta:

Table 7.44
Results of Runge Kutta

t_k	x_k	y_k	t_k	x_k	y_k
0.0	3.	3.	⋮	⋮	⋮
0.1	2.23985	1.91269	⋮	⋮	⋮
0.2	1.88951	1.45316	4.7	1.80448	0.126944
0.3	1.69272	1.2038	4.8	1.81662	0.117848
0.4	1.57072	1.04912	4.9	1.82828	0.10924
0.5	1.49087	0.944565	5.0	1.83946	0.101114

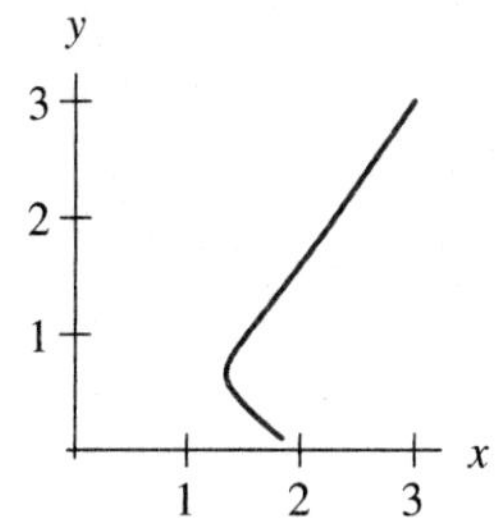

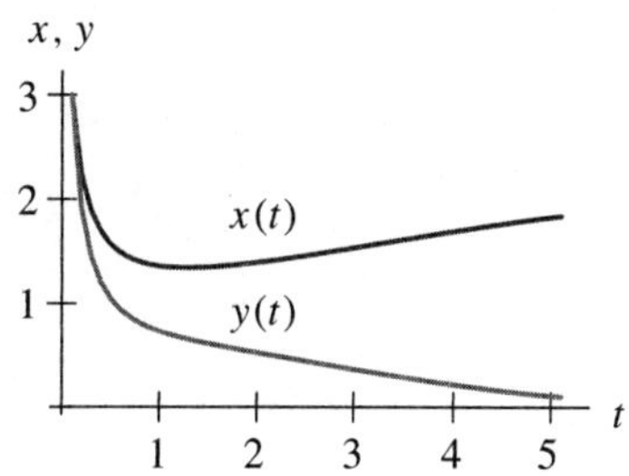

CHAPTER 8

Discrete Dynamical Systems

EXERCISES FOR SECTION 8.1

1. $x_0 = 0$, $x_1 = -2$, $x_2 = 2$, and $x_3 = 2$. The orbit is eventually fixed.

2. $x_0 = 0$, and $x_1 = 0$. The orbit is a fixed point.

3. $x_0 = 0$, $x_1 = 1$, $x_2 = e$, and $x_3 = e^e$. The orbit tends to infinity.

4. $x_0 = 0$, $x_1 = 1$, $x_2 = 2$, $x_3 = 5$, and $x_4 = 26$. The orbit tends to infinity.

5. $x_0 = 0$, $x_1 = 2$, $x_2 = 0$, and $x_3 = 2$. The orbit is periodic of period 2.

6. $x_0 = 0$, $x_1 = 1$, $x_2 = 0.54$, $x_3 = 0.858$, $x_4 = 0.65$, $x_5 = 0.79$, $x_6 = 0.70$, $x_7 = 0.76$, $x_8 = 0.72$, $x_9 = 0.75$, and $x_{10} = 0.73$. The orbit tends to a fixed point, and therefore the answer is none of the above.

7. $x_0 = 0$, $x_1 = 1$, $x_2 = 1$, and $x_3 = 1$. The orbit is eventually fixed.

8. $x_0 = 0$, $x_1 = 1$, $x_2 = 1/2$, $x_3 = 7/8$, and $x_4 = 79/128$. The orbit tends to a fixed point, and, therefore, the answer is none of the above.

9. To find the fixed points, we solve $F(x) = x$ or

$$-x + 2 = x.$$

The solution is $x = 1$. For periodic points of period 2, we solve $F^2(x) = x$ or

$$-(-x + 2) + 2 = x.$$

This equation reduces to the identity $x = x$. Therefore, all real numbers (except the fixed point $x = 1$) are periodic points of period 2.

10. To find the fixed points, we solve $F(x) = x$ or

$$x^4 = x.$$

This equation reduces to

$$x(x - 1)(x^2 + x + 1) = 0.$$

Since the roots of the quadratic factor are imaginary, $x = 0, 1$ are the only fixed points. For periodic points of period 2, we first compute that $F^2(x) = (x^4)^4 = x^{16}$. Then we solve $F^2(x) = x$ or

$$x^{16} = x.$$

Since the graph of $F^2(x)$ meets the diagonal only at $x = 0, 1$ and these points are fixed points, there are no periodic points of period 2.

11. To find the fixed points, we solve $F(x) = x$ or

$$x^2 + 1 = x.$$

This equation can be written as $x^2 - x + 1 = 0$ which has imaginary solutions. Alternatively, since

$$F(x) - x = x^2 - x + 1 = (x - \frac{1}{2})^2 + \frac{3}{4} > 0,$$

we see that $F(x) > x$. Thus, the graph of $F(x)$ does not intersect the diagonal and there are no fixed points. For periodic points of period 2, we first compute that

$$F^2(x) = (x^2+1)^2 + 1 = x^4 + 2x^2 + 2.$$

Then we solve $F^2(x) = x$ or

$$x^4 + 2x^2 - x + 2 = 0.$$

We already know two solutions of this equation. They are the two imaginary solutions to

$$x^2 - x + 1 = 0.$$

(Any x-value satisfying $F(x) = x$ also satisfies $F^2(x) = x$.) Using long division, we have

$$x^4 + 2x^2 - x + 2 = (x^2 - x + 1)(x^2 + x + 2).$$

Since the roots of the quadratic factor $x^2 + x + 2$ are imaginary also, we have no period 2 points. We also could have used the fact that

$$F^2(x) - x = (x^2+1)^2 + 1 - x = x^4 + 2(x - \frac{1}{4})^2 + \frac{15}{8} > 0.$$

Therefore, $F^2(x) > x$ and the graph of $F^2(x)$ does not intersect the diagonal.

12. To find the fixed points, we solve $F(x) = x$ or

$$x^2 - 3 = x.$$

The quadratic formula yields $x = (1 \pm \sqrt{13})/2$ as the solution. For periodic points of period 2, we first obtain

$$F^2(x) = (x^2 - 3)^2 - 3 = x^4 - 6x^2 + 6.$$

Then we solve $F^2(x) = x$ or

$$x^4 - 6x^2 - x + 6 = 0.$$

We already know two solutions of this equation. They are the fixed points, the two solutions to $x^2 - x - 3 = 0$. (Any x-value satisfying $F(x) = x$ also satisfies $F^2(x) = x$.) Using long division, we have

$$x^4 - 6x^2 - x + 6 = (x^2 - x - 3)(x^2 + x - 2).$$

The roots of the quadratic factor $x^2 + x - 2$ give us the period two points $x = -2, 1$.

13. To find the fixed points, we solve $F(x) = x$ or

$$\sin x = x.$$

Since the graph of $F(x) = \sin x$ only crosses the diagonal at the origin, $x = 0$ is the only fixed point. For periodic points of period 2, consider the function

$$G(x) = F^2(x) - x = \sin(\sin x) - x.$$

Any period 2 point will be a root of $G(x)$. Differentiation yields

$$G'(x) = (\cos(\sin x)) \cos x - 1 < 0 \text{ for } x \neq 0,$$

so that $G(x)$ is strictly decreasing. Since $G(0) = 0$, we have $G(x) > 0$ for $x < 0$ and $G(x) < 0$ for $x > 0$. This means that $x = 0$ is the only root of $G(x)$ and thus there are no periodic points of period 2. Another method would be to use the fact that $|\sin x| < |x|$ for $x \neq 0$. This implies that

$$|\sin(\sin x)| < |\sin x| < |x| \text{ for } x \neq 0.$$

14. To find the fixed points, we solve $F(x) = x$ or

$$\frac{1}{x} = x.$$

This equation reduces to $x^2 - 1 = 0$, which has solutions $x = \pm 1$. For periodic points of period 2, we solve $F^2(x) = x$ or

$$\frac{1}{1/x} = x.$$

This equation reduces to the identity $x = x$. Therefore, all real numbers except 0 and the fixed points ± 1 are periodic points of period 2.

15. For fixed points,

$$F(x) = -2x - x^2 = x,$$

or $x = 0, -3$. For periodic points of period 2,

$$F^2(x) = -x^4 - 4x^3 - 2x^2 + 4x = x,$$

or

$$x(x+3)(-x^2 - x + 1) = 0.$$

Therefore, the periodic points of period 2 are $x = (-1 \pm \sqrt{5})/2$.

16. For fixed points, consider

$$f(x) = e^x - x.$$

Differentiation yields

$$f'(x) = e^x - 1,$$

and

$$f''(x) = e^x > 0.$$

Therefore, $f(0) = 1$ is the minimum and $f(x) > 0$. Therefore, $e^x > x$. This means that there are no fixed points.

Since $e^x > x$, we have

$$e^{e^x} > e^x > x$$

and so there are no periodic points of period 2.

17. From the graph of $F(x) = -e^x$ and $f(x) = x$, there is one intersection and, therefore, there is one fixed point. For periodic points of period 2, consider

$$g(x) = -e^{-e^x} - x.$$

By the graph of g shown to the right (or by computing g'), $g(x)$ is zero only once, so this point must be the fixed point and there are no periodic points of period 2.

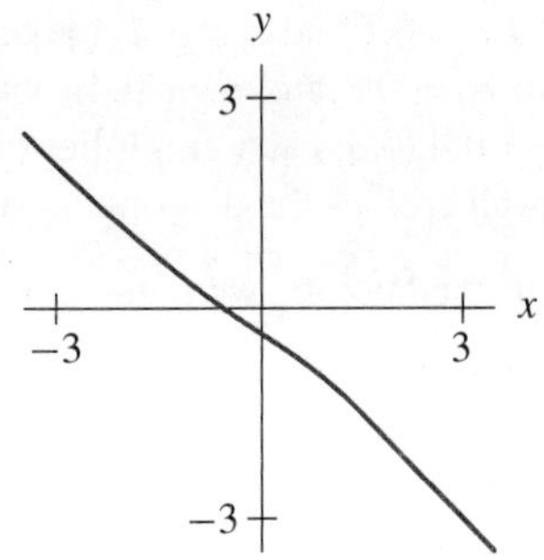

18. For fixed points, we must have $x^3 - x = 0$, or $x = 0, \pm 1$. The graph of $F^2(x) = x^9$ meets the diagonal at precisely these points. Therefore there are no periodic points of period 2.

19. The graph of $y = -x$ meets $y = x$ only at $x = 0$, so 0 is a fixed point. Since $F^2(x) = x$, it follows that all other real numbers lie on periodic orbits of period 2.

20. For fixed points, we solve $-2x + 1 = x$, or $x = 1/3$. For periodic points of period 2, we must have $F^2(x) = -2(-2x + 1) + 1 = x$, or $4x - 1 = x$. This yields only the fixed point $x = 1/3$, so there are no periodic points of period 2.

21. There is a fixed point at $x = 2$; all other points are sent to $x = 2$ by F so they are eventually fixed. Therefore there are no other periodic points.

22. $x_0 = x, x_1 = x^3, x_2 = x^9$ and, therefore, $x_n = x^{3^n}$. Given real number, x,

$$F^n(x) = x^{3^n}.$$

Therefore, $x = \pm 1$ and $x = 0$ are fixed points, $|x| < 1$ tends to zero, and $|x| > 1$ tends to plus or minus infinity.

23. $x_0 = x, x_1 = -x + 4, x_2 = x, x_3 = -x + 4$, and $x_4 = x$. The orbit of any real number is a periodic orbit of period 2, except for $x = 2$, which is fixed.

24. The orbit is $2/3, 2/3, 2/3, \ldots$, so $2/3$ is a fixed point.

25. The orbit is $1/6, 1/3, 2/3, 2/3, \ldots$, so $1/6$ is eventually fixed.

26. The orbit is $2/5, 4/5, 2/5, 4/5, \ldots$ so $2/5$ is periodic with period 2.

27. The orbit is $2/7, 4/7, 6/7, 2/7, \ldots$, so $2/7$ is periodic with period 3.

28. The orbit is $3/14, 3/7, 6/7, 2/7, 4/7, 6/7, \ldots$, so $3/14$ is eventually periodic.

29. The orbit is $1/8, 1/4, 1/2, 1, 0, 0, 0, \ldots$, so $1/8$ is eventually fixed.

30. The orbit is $1/9, 2/9, 4/9, 8/9, 2/9, \ldots$, so $1/9$ is eventually periodic.

31. The orbit of $6/11$ is $6/11, 10/11, 2/11, 4/11, 8/11, 6/11, \ldots$, so $6/11$ is periodic with period 5.

32. The orbit is $2/9, 4/9, 8/9, 2/9, \ldots$, so it is periodic with period 3.

33. The orbit is $0, 0, 0, \ldots$, so it is a fixed point.

34. The orbit is $1/4, 1/2, 1, 0, 0, \ldots$, so it is eventually fixed.

35. The orbit is $1/2, 1, 0, 0, \ldots$, so it is eventually fixed.

36. For $c = 0.1$ and $c = 0.2$, the orbits tend to the fixed points, $x = 0.1127$ and $x = 0.2764$ respectively. For $c = 0.3$ and $c = 0.4$, both orbits tend to infinity. From the graphs of $x^2 + c$, we see that for $c = 0.4, 0.3$, each graph lies above the diagonal so there are no fixed points. For $c = 0.2, 0.1$, each graph crosses the diagonal at two points (fixed points), one of which "attracts" the orbit of 0.

37. For fixed points, we must solve

$$F_c(x) = x^2 + c = x$$

or

$$x^2 - x + c = 0.$$

From the quadratic formula we find

$$x = (1 \pm \sqrt{1 - 4c})/2.$$

Thus we need $1 - 4c \geq 0$ or $c \leq \frac{1}{4}$. For $c = 1/4$, we have $x = 1/2$ and this is the only fixed point. For $c < 1/4$, there are two roots and thus there are two fixed points. For $c > 1/4$, there are no fixed points.

38. For $c = -0.6$ and $c = -0.7$, the orbits oscillate and tend to the fixed points, $x = -0.42$ and $x = -0.47$. For $c = -0.8$ and $c = -0.9$, the orbits tend to cycles of period 2. For $c = -0.8$, the orbit oscillates between -0.7236 and -0.2764. For $c = -0.9$, the orbit oscillates between -0.8873 and -0.1127. From the graphs of F^2, we see that these graphs cross the diagonal line $y = x$ at only two points (the fixed points) when $c = -0.6, -0.7$. When $c = -0.8, -0.9$, the graphs of F^2 cross the diagonal at four points, the fixed points and a cycle of period 2.

39. For periodic points of period 2, we have

$$F_c^2(x) = (x^2 + c)^2 + c = x,$$

or

$$x^4 + 2cx^2 - x + c^2 + c = 0.$$

Since we have fixed points when $x^2 + c = x$, the left-hand side can be factored and one obtains

$$(x^2 - x + c)(x^2 + x + c + 1) = 0.$$

This is zero when $x^2 - x + c = 0$ or when $x^2 + x + c + 1 = 0$. The first factor corresponds to the fixed points, which are present when $c \leq 1/4$. The second factor corresponds to the period two points and has two solutions when $c < -3/4$. At $c = -3/4$ there are two fixed points (roots of the second factor coincide with one of the fixed points).

EXERCISES FOR SECTION 8.2

1. For the fixed points,

$$F(x) = x^2 - 2x = x,$$

or

$$x^2 - 3x = 0.$$

Then, $x = 0, 3$ are fixed points. Differentiation yields

$$F'(x) = 2x - 2.$$

Then, $F'(0) = -2$ and $F'(3) = 4$. Therefore, both $x = 0$ and $x = 3$ are repelling fixed points.

2. For the fixed points,

$$F(x) = x^5 = x,$$

or

$$x(x+1)(x-1)(x^2+1) = 0.$$

Then, $x = 0, \pm 1$ are fixed points. Differentiation yields

$$F'(x) = 5x^4.$$

Then, $F'(0) = 0$ and $F'(\pm 1) = 5$. Therefore, $x = 0$ is attracting and $x = \pm 1$ are repelling.

3. For the fixed points,

$$F(x) = \sin x = x,$$

or $x = 0$. Differentiation yields $F'(0) = 1$. For $x > 0$, $x > \sin x$ and for $x < 0$, $x < \sin x$. Therefore, $x = 0$ is attracting by graphical analysis.

4. For the fixed points,

$$F(x) = x^3 - x = x,$$

or

$$x^3 - 2x = 0.$$

Then, $x = 0, \pm\sqrt{2}$ are fixed points. Differentiation yields

$$F'(x) = 3x^2 - 1.$$

Then, $F'(0) = -1$ and $F'(\pm\sqrt{2}) = 5$. Since $F(x) - x = x^3 - 2x$, for $-\sqrt{2} < x < 0$, $F(x) - x > 0$ and $F(x) - x < 0$ for $0 < x < \sqrt{2}$. Therefore, $x = 0$ is attracting. $x = \pm\sqrt{2}$ is repelling.

5. For the fixed points,

$$F(x) = \arctan x = x.$$

Suppose $f(x) = \arctan x - x$ and differentiation yields

$$f'(x) = \frac{-x^2}{1+x^2} - 1.$$

Then, $f'(x) < 0$ for $x \neq 0$ and $f'(x) = 0$ only for $x = 0$. $f(x)$ is decreasing for $x \neq 0$. Since $f(0) = 0$, $f(x)$ is tangent to x-axis at $x = 0$. Therefore, $x = 0$ is the only fixed point. Also, $f(x) = F(x) - x > 0$ for $x < 0$ and $f(x) = F(x) - x < 0$ for $x > 0$. Therefore, $x = 0$ is attracting.

6. For the fixed points,

$$F(x) = 3x(1-x) = x,$$

or

$$x(2 - 3x) = 0.$$

Then, $x = 0, 2/3$ are fixed points. Differentiation yields

$$F'(x) = -6x + 3.$$

Then, $F'(0) = 3$ and $F'(2/3) = -1$. By careful graphical analysis, $x = 2/3$ is attracting, and $x = 0$ is repelling.

7. For the fixed points,

$$F(x) = \frac{\pi}{2} \sin x = x.$$

Then, $x = 0, \pm\pi/2$ are fixed points. Differentiation yields

$$F'(x) = \frac{\pi}{2} \cos x.$$

Then, $F'(0) = \pi/2$ and $F'(\pm\pi/2) = 0$. Therefore, $x = 0$ is repelling, and $x = \pm\pi/2$ are attracting.

8. For the fixed points,

$$F(x) = x^2 - 3 = x,$$

or

$$x^2 - x - 3 = 0.$$

Then, $x = (1 \pm \sqrt{13})/2$ are fixed points. Differentiation yields

$$F'(x) = 2x.$$

Then, $F'((1 \pm \sqrt{13})/2) = 1 \pm \sqrt{13}$ and $|1 \pm \sqrt{13}| > 1$. Therefore, $x = (1 \pm \sqrt{13})/2$ are repelling.

9. For the fixed points,

$$F(x) = \frac{1}{x} = x,$$

or

$$x^2 = 1.$$

Then, $x = \pm 1$ are fixed points. Differentiation yields

$$F'(x) = -\frac{1}{x^2}.$$

Then, $F'(\pm 1) = -1$. Since $F^2(x) = x$ for any non-zero x, $x = \pm 1$ are neutral.

10. For the fixed points,

$$F(x) = \frac{1}{x^2} = x,$$

or

$$x^3 = 1.$$

Then, $x = 1$ is a fixed point. Differentiation yields

$$F'(x) = -\frac{2}{x^3}.$$

Then, $F'(1) = -2$ and $x = 1$ is repelling.

11. For the fixed points,

$$F(x) = e^x = x.$$

Suppose $f(x) = e^x - x$ and differentiation yields

$$f'(x) = e^x - 1.$$

Therefore, $f(x)$ is increasing for $x > 0$, decreasing for $x < 0$ and equal to 1 for $x = 0$. $f(x) \neq 0$ for any x and there are no fixed points.

12. $F(0) = 1$, and $F^2(0) = 0$. The period is 2 and since $(F^2)'(x) = 0$, it is attracting.

13. $F(0) = 1$, and $F^2(0) = 0$. The period is 2 and since $(F^2)'(0) = 0$, it is attracting.

14. $F(0) = \pi/2$, and $F^2(0) = 0$. The period is 2 and since $(F^2)'(0) = 0$, the orbit is attracting.

15. $F(0) = 1$, $F(1) = 2$, $F(2) = 3$, $F(3) = 4$, and $F(4) = 0$. The period is 5 and since $(F^5)'(0) = 2$, the cycle is repelling.

16. $F(0) = 1$, and $F^2(0) = 0$. The period is 2 and since $(F^2)'(0) = 0$, the cycle is attracting.

17. $F(0) = 1$ and $F^2(0) = 0$. The period is 2. For $x < 2$ close to zero, $F(x) = 1 - x$. Therefore, any x close to zero is periodic point and the cycle of $x = 0$ is neutral.

18. For fixed points,

$$F(x) = x - x^2 = x,$$

or $x = 0$. By graphical analysis, the fixed point is neutral.

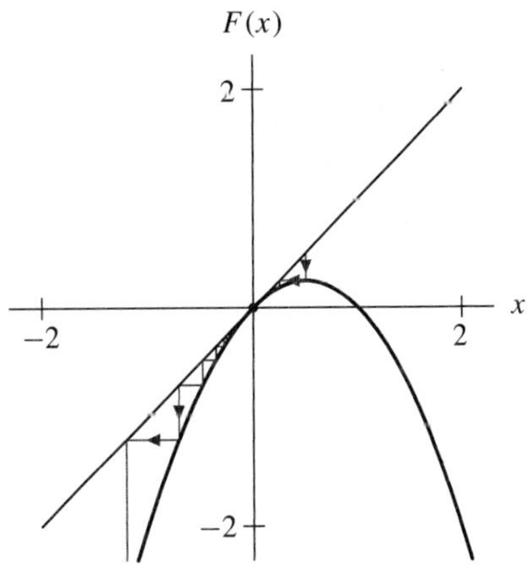

19. For fixed points,

$$F(x) = \frac{1}{x} = x,$$

or $x = \pm 1$. Since $F^2(x) = x$ for all other x, these fixed points are neutral. We can also see this by graphical analysis.

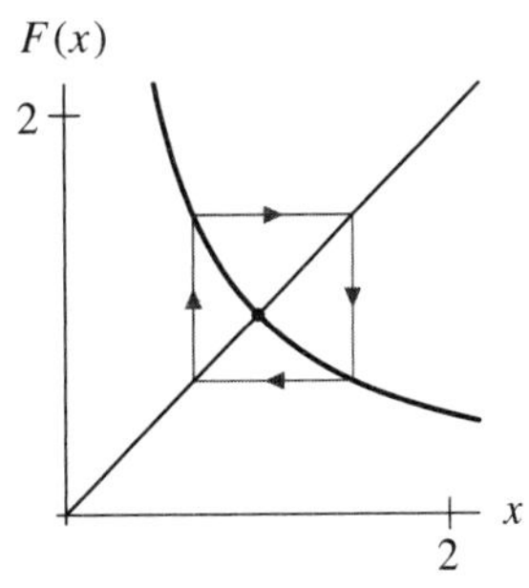

20. For fixed points,

$$F(x) = \sin x = x,$$

or $x = 0$. By graphical analysis, the fixed point is attracting.

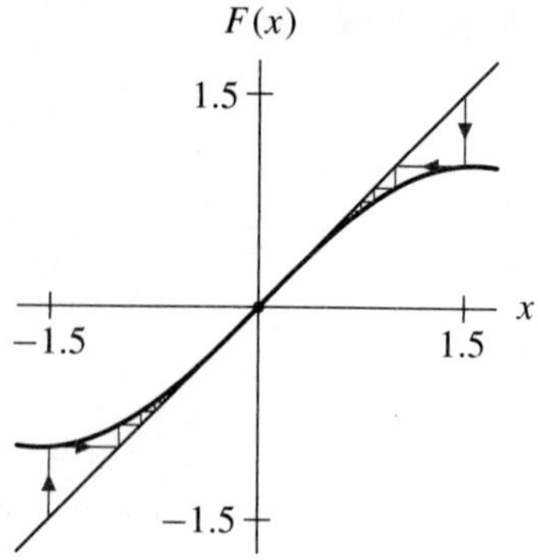

21. For fixed points,

$$F(x) = \tan x = x,$$

or $x = 0$. Since $(\tan x)' = 1/\cos^2 x = 1$ if and only if $x = 0, \pi$, $x = 0$ is the only fixed point with $F'(0) = 1$. By graphical analysis, the fixed point is repelling.

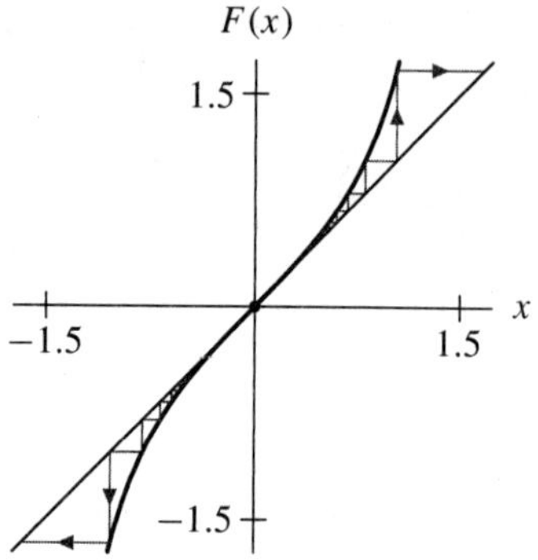

22. For fixed points,

$$F(x) = -x + x^3 = x,$$

or $x = 0, \pm\sqrt{2}$. The fixed points at $\pm\sqrt{2}$ are repelling. By graphical analysis, the fixed point at $x = 0$ is attracting.

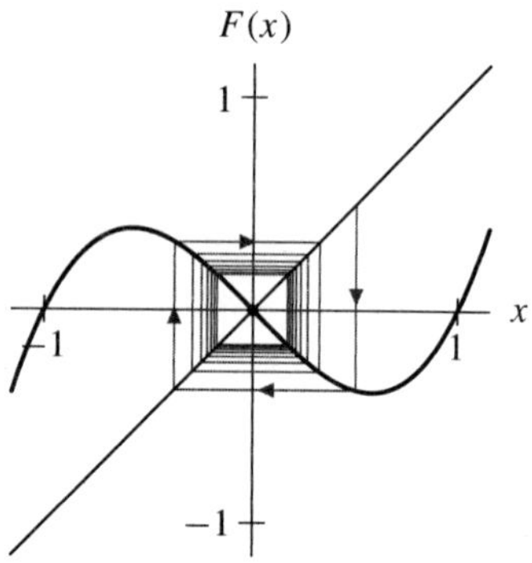

23. For fixed points,

$$F(x) = -x - x^3 = x,$$

or $x = 0$. By graphical analysis, the fixed point is repelling.

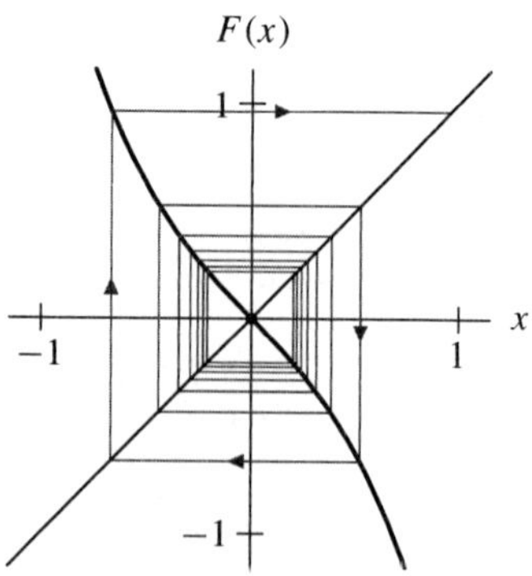

24. For fixed points,

$$F(x) = e^{x-1} = x,$$

or $x = 1$. By graphical analysis, the fixed point is neutral.

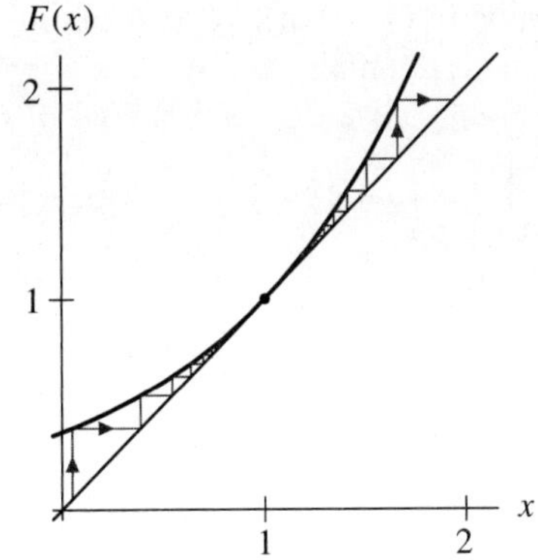

25. For fixed points,

$$F(x) = -e \cdot e^x = x,$$

or $x = -1$. By graphical analysis, the fixed point is attracting.

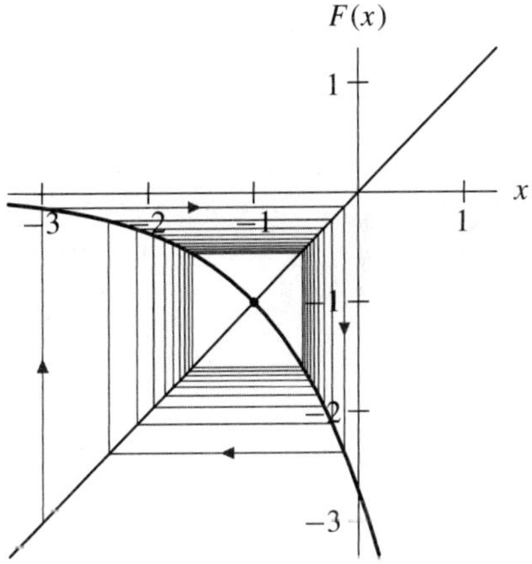

26. For fixed points, $F_c(x) = x^2 + c = x$, or

$$x = \frac{1 \pm \sqrt{1-4c}}{2}.$$

For $1 - 4c < 0$ or $c > 1/4$, there are no fixed points. For $1 - 4c = 0$ or $c = 1/4$, $x = 1/2$ is the only fixed point. For $1 - 4c > 0$ or $c < 1/4$, $x = (1 \pm \sqrt{1-4c})/2$ are fixed points. For $c = 1/4$, the graph of F_c is tangent to $y = x$ from the above, and, therefore, the fixed point is neutral. For $c < 1/4$,

$$F_c'\left(\frac{1 \pm \sqrt{1-4c}}{2}\right) = 1 \pm \sqrt{1-4c}.$$

Since $1 + \sqrt{1-4c} > 1$, $x = (1 + \sqrt{1-4c})/2$ is repelling. For $-1 < 1 - \sqrt{1-4c} < 1$ or $-3/4 < c < 1/4$, $x = (1 - \sqrt{1-4c})/2$ is attracting, and for $-1 > 1 - \sqrt{1-4c}$ or $c < -3/4$, $x = (1 - \sqrt{1-4c})/2$ is repelling. At $c = -3/4$,

$$F_{-3/4}^2(x) - x = (x^2 - x - \frac{3}{4})(x + \frac{1}{2})^2$$

and, therefore, $F_{-3/4}^2(x) > x$ for x slightly less than $-1/2$ and $F_{-3/4}^2(x) < x$ for x slightly larger than $-1/2$. The fixed point, $x = -1/2$, for $c = -3/4$ is attracting.

27. For $F(x) = \tan x = x$, there are infinitely many fixed points because $\tan x$ is periodic, and $|\tan x|$ tends to infinity as x tends to $n\pi/2$ for odd integers n. As we saw in Exercise 21, $x = 0$ is repelling. First, note that for $x = n\pi$ for $n \neq 0$,

$$F(n\pi) = \tan n\pi = 0 \neq n\pi.$$

Therefore, $x = n\pi$ for $n \neq 0$ are not fixed points. At the fixed points,

$$F'(x) = \frac{1}{\cos^2 x} > 1$$

because $|\cos x| < 1$. Therefore, all other fixed points are repelling.

28. Let $f(x) = F_c(x) - x = ce^x - x$. Differentiation yields $f'(x) = ce^x - 1$ and $f''(x) = ce^x$. For $f'(x) = 0$, $x = -\ln c$ by $c > 0$ and $f''(-\ln c) = 1$. Therefore, at $x = -\ln c$, $f(x)$ is minimum, and

$$f(-\ln c) = 1 + \ln c.$$

If $f(-\ln c) > 0$, $f(x) = F_c(x) - x > 0$ or F_c does not have any fixed points. Then, $c > 1/e$. If $f(-\ln c) = 0$, $F_c(x) - x \geq 0$, where equality is true only at $x = -\ln c$. Therefore, there is only one fixed point where the graph of F_c is tangent to $y = x$ from the above. The fixed point is neutral. If $f(-\ln c) < 0$, $F_c(x) - x = 0$ at two different x. Therefore, there are two fixed points and the graph of F_c is below $y = x$ between the two fixed points. Therefore, one fixed point is attracting and the other is repelling.

29. There is a cycle for T. Try $x = 16/21$. Suppose T has n-cycle, $x_0, x_1, \cdots, x_n = x_0$. Then,

$$\begin{aligned}|(T^n)'(x_0)| &= |T'(x_0) \cdot T'(x_1) \cdots T'(x_{n-1})| \\ &= 4^n,\end{aligned}$$

where n is integer larger than zero. Therefore, the cycle is repelling.

30. **(a)** Suppose y is some root for P. Then, $P(y) = 0$ and

$$y - \frac{P(y)}{P'(y)} = y.$$

Therefore, y is a fixed point of the Newton iteration.

(b) First, we find the root of $P(x)$.

$$x^3 - x = 0$$

or $x = 0, \pm 1$. These are the fixed points of Newton's method for P. The derivative of the Newton's method function is

$$1 - \frac{(P')^2 - PP''}{(P')^2}.$$

Hence, at the roots of P this is zero and the points are attracting fixed points of the Newton's method iteration.

EXERCISES FOR SECTION 8.3

1. For fixed points, we must have

$$F_\alpha(x) = x + x^2 + \alpha = x$$

or $x^2 + \alpha = 0$. Therefore, for $\alpha > 0$, there are no fixed points; for $\alpha = 0$, there is one fixed point; and for $\alpha < 0$, there are two fixed points at $x = \pm\sqrt{-\alpha}$. Differentiation yields

$$F'_\alpha(x) = 1 + 2x$$

and

$$F'_\alpha(\pm\sqrt{-\alpha}) = 1 \pm 2\sqrt{-\alpha}.$$

Therefore, for $-\alpha$ small, $0 < 1 - 2\sqrt{-\alpha} < 1$ and $x = -\sqrt{-\alpha}$ is attracting. Since $1 + 2\sqrt{-\alpha} > 1$, $x = \sqrt{-\alpha}$ is repelling. For $\alpha = 0$, $F_\alpha(x)$ is tangent to $y = x$ from above and, therefore, $x = 0$ is neutral. The bifurcation is a tangent bifurcation.

2. Since $F'_\alpha = \alpha$, for $0 > \alpha > -1$, the fixed point, the origin, is attracting. For $\alpha = -1$, $x = 0$ is neutral since all nonzero orbits are periodic with period 2. For $\alpha < -1$, the fixed point is repelling. Therefore, the bifurcation is none of the above.

3. For α slightly smaller than 1, the origin is the only fixed point and it is attracting. For $\alpha = 1$, F_1 is tangent to $y = x$ and 0 is attracting. For $\alpha > 1$, two more fixed points appear and they are attracting for α slightly larger than 1. The origin becomes a repelling fixed point. This is a pitchfork bifurcation.

4. For α slightly larger than -1, $|F'_\alpha(0)| = |\alpha \cos 0| = |\alpha| < 1$, therefore, when $\alpha > -1$, the origin is attracting. For $\alpha < -1$, $|F'_\alpha(0)| = |\alpha| > 1$, therefore, the origin is repelling. Also, two new period 2 points appear. They are the intersections of $y = x$ and F^2_α. The cycle is attracting by careful graphical analysis. For $\alpha = -1$, by the graphical analysis, the origin is attracting. The bifurcation is a period-doubling bifurcation.

5. For fixed points,

$$\alpha - x^2 = x,$$

and $x = (-1 \pm \sqrt{1 + 4\alpha})/2$. Therefore, there are no fixed point for $\alpha < -1/4$, there is one fixed point for $\alpha = -1/4$, and there are two fixed points for α slightly larger than $-1/4$. This is a tangent bifurcation. Also, for α slightly larger than $-1/4$,

$$\left| F'_\alpha\left(\frac{-1 + \sqrt{1 + 4\alpha}}{2}\right)\right| = |-1 + \sqrt{1 + 4\alpha}| < 1$$

and

$$\left| F'_\alpha\left(\frac{-1 - \sqrt{1 + 4\alpha}}{2}\right)\right| = |-1 - \sqrt{1 + 4\alpha}| > 1.$$

Therefore, $x = (-1 + \sqrt{1 + 4\alpha})/2$ is attracting and $x = (-1 - \sqrt{1 + 4\alpha})/2$ is repelling. For $\alpha = -1/4$, $F'_{-1/4}(-1/2) = 1$ and, therefore, $F_{-1/4}$ is tangent to $y = x$ from below at $x = -1/2$. By graphical analysis, $x = -1/2$ is neutral. Nearby orbits on the right of the fixed point tend to the fixed point and nearby orbits on the left tend away from it.

6. This bifurcation is a pitchfork bifurcation. For $\alpha \leq 1$, the origin is an attracting fixed point since $F'_\alpha(0) = \alpha$. When $\alpha = 1$, the graph is tangent to the diagonal, but the graph shows that 0 is still attracting. For $\alpha > 1$, two attracting fixed points emerge and the origin becomes a repelling fixed point.

7. This bifurcation is none of the above. When $\alpha = 0$, we have $F_\alpha(x) = 0$ for all x. So all nonzero points are eventually fixed. For $0 < |\alpha| < 1$, the origin is an attracting fixed point and there is a second (repelling) fixed point at $x = (1-\alpha)/\alpha$.

8. The addition of β simply lowers or raises the graph of $F_0(x) = x^3$. Hence we expect that there will be two β-values for which the graph of F_β is tangent to the diagonal. We can find them by solving the equations

$$x^3 + \beta = x, \quad F'_\beta(x) = 3x^2 = 1$$

simultaneously. This yields $x = \pm 1/\sqrt{3}$, and so $\beta = \pm 2/3\sqrt{3}$. Graphical analysis shows that, if $\beta > 2/3\sqrt{3}$ or $\beta < -2/3\sqrt{3}$, then F_β has a single repelling fixed point. For β-values between these two values, F_β has a single attracting fixed point and a pair of repelling fixed points. Therefore there are tangent bifurcations at $\beta = \pm 2/3\sqrt{3}$.

9. The function T_μ has a single fixed point at the origin when $0 < \mu < 1$. This fixed point is attracting. If $\mu = 1$, all $x \leq 1/2$ are fixed points, and each is neutral. If $\mu > 1$, T_μ has two fixed points, at $x = 0$ and $x = \mu/(\mu+1)$, and both are repelling.

10. We must solve $F_k^2(x) = x$. This equation is

$$-k^3x^4 + 2k^3x^3 - (k^3 + k^2)x + (k^2 - 1) = 0.$$

Now we know that $x = 0$ and $x = k/(k-1)$ are fixed points and so solve this equation. Therefore we may use long division to obtain

$$\frac{-k^3x^4 + 2k^3x^3 - (k^3 + k^2)x + (k^2 - 1)}{-kx^2 + (k-1)x} = k^2x^2 - (k^2 + k)x + (k+1).$$

The roots of this quadratic expression are real only if $k \geq 3$ (or $k \leq -1$).

11. The graph of the function $F_c(x)$ is a parabola with minimum at $x = 0$ and $F_c(0) = c$. The function $F_c^2(x)$ is a quartic with local maximum at $x = 0$. Consider the small portion of the graph of F_c^2 defined over the interval $[p_c, -p_c]$, where p_c is the leftmost fixed point of F_c (not F_c^2). This piece of the graph of F_c^2 resembles that of F_c, only upside down. Note that, as c decreases, this piece of graph behaves similarly to that of F_c relative to the diagonal. Thus we expect F_c^2 to undergo a period-doubling bifurcation in this interval, much the same as F_c did.

EXERCISES FOR SECTION 8.4

1.

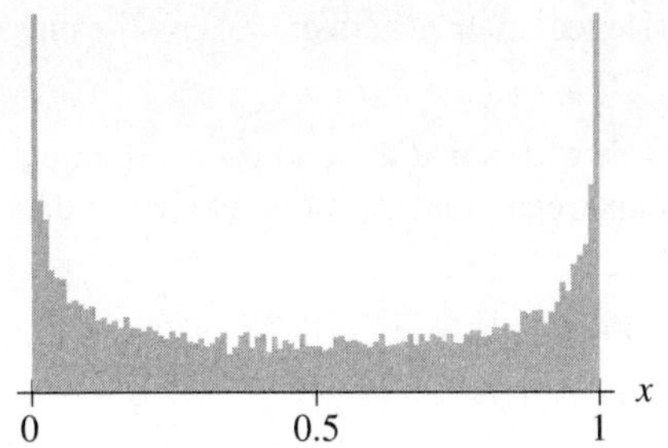

Histogram for seed .3.

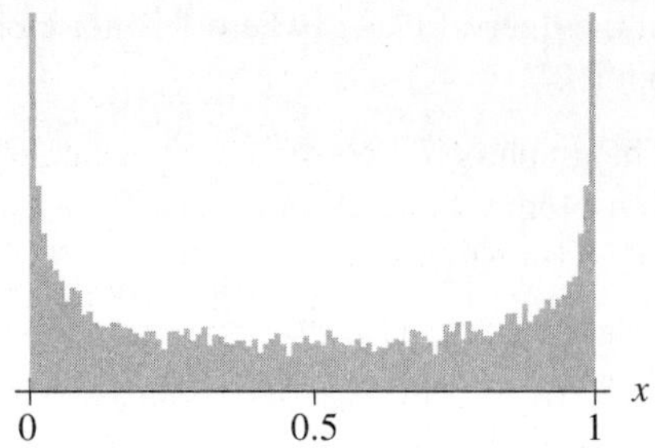

Histogram for seed .3001.

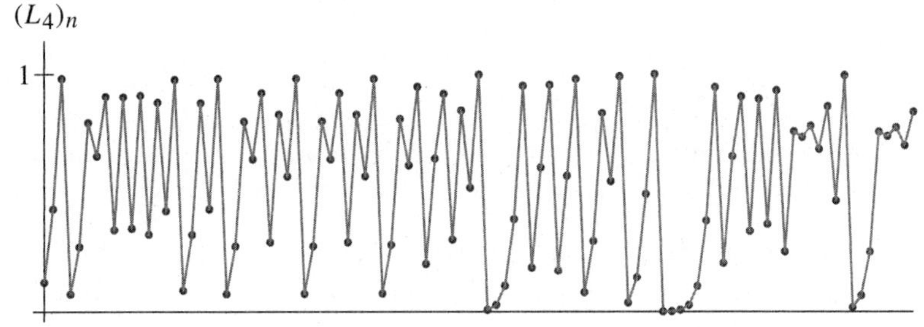

Time series for seed .3.

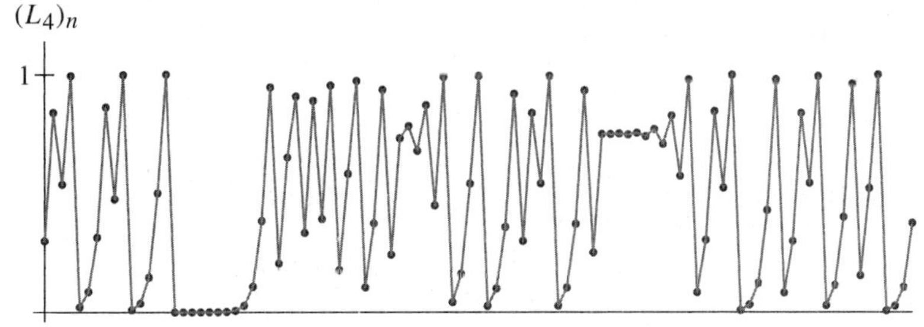

Time series for seed .3001.

2. For $x_0 = 1/5$, the orbit is periodic with period 4: $1/5, 2/5, 4/5, 3/5, 1/5, \dots$
For $x_0 = 2/7$, the orbit is periodic with period 3: $2/7, 4/7, 1/7, 2/7, \dots$
For $x_0 = 3/11$, the orbit is periodic with period 10:

$$3/11, 6/11, 1/11, 2/11, 4/11, 8/11, 5/11, 10/11, 9/11, 7/11, 3/11, \dots$$

For $x_0 = 1/10$, the orbit is eventually periodic: $1/10, 1/5, 2/5, 4/5, 3/5, 1/5, \dots$
For $x_0 = 1/6$, the orbit is eventually periodic: $1/6, 1/3, 2/3, 1/3, \dots$
For $x_0 = 4/14$, the orbit is eventually periodic: $4/14, 4/7, 1/7, 2/7, 4/7, \dots$
For $x_0 = 4/15$, the orbit is periodic with period 5: $4/15, 8/15, 1/15, 2/15, 4/15, \dots$

3. The image under T of a point of the form $p/2^n$ where p is an integer is of the form $q/2^{n-1}$ where q is an integer. This is because T multiplies by 2 and then subtracts one if necessary to keep the image in the interval $[0, 1)$ (where 1 is not included). Hence, after n iterates, $T^n(p/2^n)$ must be the fixed point 0.

4. The graph of T^n consists of 2^n straight lines, each with slope 2^n, and each extending from 0 to 1. Therefore we expect T^n to have $2^n - 1$ fixed points (recall that 1 is not included in the domain).

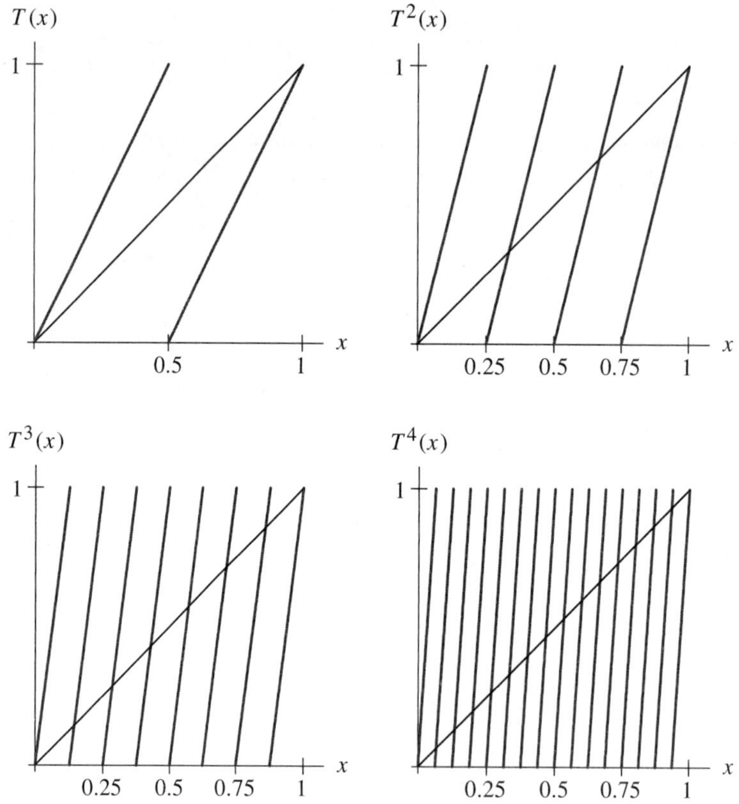

5. The graph of T^n crosses the diagonal in each interval $[i/2^n, (i+1)/2^n)$ for $i = 0, 1, \ldots, 2^n - 2$. Hence, each of these intervals contains a point of period n (n may not be the least period). We may take n as large as we like, so periodic points of T are dense.

6. If $x_0 < 1/2$ then $a_1 = 0$ so that the binary expansion of x_0 is

$$x_0 = \frac{a_2}{2^2} + \frac{a_3}{2^3} + \cdots.$$

Hence

$$T(x_0) = 2x_0 = \frac{a_2}{2} + \frac{a_3}{2^2} + \frac{a_4}{2^3} + \cdots.$$

So the binary representation of $T(x_0)$ is $.a_2a_3a_4\ldots$.

If $1/2 \leq x_0 < 1$, then $a_1 = 1$, so the binary expansion of x_0 is

$$x_0 = \frac{1}{2} + \frac{a_2}{2^2} + \frac{a_3}{2^3} + \cdots.$$

Thus we have

$$T(x_0) = 2x_0 - 1 = \frac{a_2}{2} + \frac{a_3}{2^2} + \frac{a_4}{2^3} + \cdots.$$

Again, the binary representation of $T(x_0)$ is $.a_2a_3a_4\ldots$. Note that T simply knocks off the first binary digit in each case.

In similar fashion, T^n knocks off the first n binary digits of x_0, so the binary representation of $T^n(x_0)$ is $a_{n+1}a_{n+2}a_{n+3}\ldots$.

7. The eventually fixed points of T are those whose orbits hit 0. These are precisely the points which have finite binary expansion, that is, points of the form

$$\frac{a_1}{2} + \frac{a_2}{2^2} + \frac{a_3}{2^3} + \ldots + \frac{a_n}{2^n}$$

where $a_1, a_2, \ldots, a_n$ equal zero or one. Using 2^n as a common denominator, we can express such a point in the form $p/2^n$ for an integer p between zero and $2^n - 1$.

8. By exercise 6, any cycle for T corresponds to a repeating binary sequence. That is, the periodic points of period 2 must have binary representations either $.\overline{01} = .010101\ldots$ or $.\overline{10}$. Similarly, the 3-cycles have orbits (in binary form)

$$.\overline{001} \to .\overline{010} \to .\overline{100} \to .\overline{001}\cdots$$

or

$$.\overline{011} \to .\overline{110} \to .\overline{101} \to .\overline{011}\cdots.$$

As a fraction, one can check for instance that $.\overline{100}$ corresponds to

$$\frac{1}{2} + \frac{1}{2^4} + \frac{1}{2^7} + \cdots = \frac{1}{2}\sum_{i=0}^{\infty}\left(\frac{1}{8}\right)^i = 4/7.$$

9. The points which are fixed point T^n are those whose binary expansion repeats every n summands, that is, those of the form

$$x = \frac{a_1}{2} + \frac{a_2}{2^2} + \ldots + \frac{a_n}{2^n} + \frac{a_1}{2^{n+1}} + \frac{a_2}{2^{n+2}} + \ldots + \frac{a_n}{2^{2n}} + \frac{a_1}{2^{2n+1}} + \ldots.$$

We may choose $a_1, a_2, \ldots$, however we like provided they are not all one (since this corresponds to $x = 1$ which is not in the domain). Hence, T^n has $2^n - 1$ fixed points (and T has $2^n - 1$ points of period n). This can also be seen by looking at the graph of T^n.

10. T exhibits sensitive dependence. One way to see this is to note that the distance between two nearby seeds is effectively doubled by each iteration of T (unless the 2 points lie on opposite sides of $1/2$, in which case T takes them very far apart). Equivalently, from exercise 4, the graph of T^n takes intervals of the form $[i/2^n, (i+1)/2^n)$ onto $[0, 1)$, so nearby points are sent far away by T^n.

11. First form a list of all the possible finite strings of zeros and ones,

$$0, 1, 00, 01, 10, 11, 000, 001, 010, 011, 100, 101, 110, 111, 0000, \ldots$$

then concatenate this list into the binary expansion of a number x_0,

$$x_0 = .0100011011000001010011100101110111\ldots.$$

Since every finite string appears in this list and application of T shifts the decimal expansion to the left by one place, the orbit of this point will eventually come close to every point in $[0, 1)$.

12. The graph of T^n consists of 2^n line segments extending from 0 to 1, each of which has slopes that are alternately 2^n and -2^n. Hence there are 2^n fixed points for T^n.

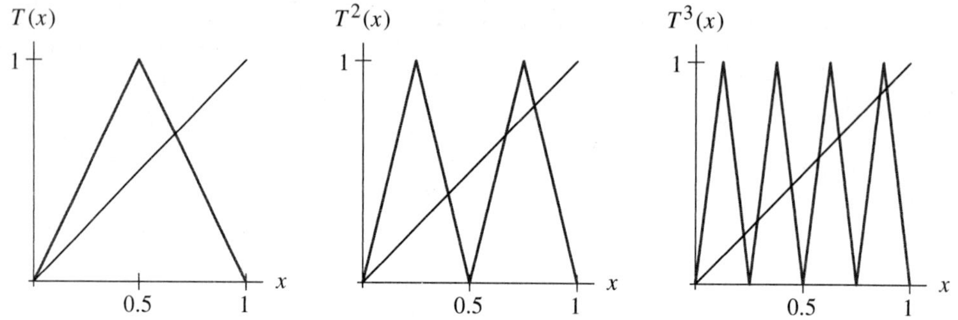

13. The graph of T^n on the segment $[i/2^n, (i+1)/2^n]$ is a straight line from zero to one or one to zero depending on whether i is even or odd. In either case, the graph of T^n crosses the diagonal in each such interval. Hence, T^n has a fixed point and T has a periodic point in each such interval. Since n can be taken as large as we like, periodic points are dense.

14. From exercise 13, T takes segments of the form $[i/2^n, (i+1)/2^n]$ onto $[0, 1]$, so nearby orbits separate under T^n. Therefore T exhibits sensitive dependence.

REVIEW EXERCISES FOR CHAPTER 8

1. We solve $x^3 - x/2 = x$ and obtain $x = 0$ or $x = \pm\sqrt{3/2}$.

To determine the type of each fixed point, we compute $F'(x) = 3x^2 - 1/2$. So $F'(0) = -1/2$, and $x = 0$ is attracting. On the other hand, $F'(\pm\sqrt{3/2}) = 4$, so both of these fixed points are repelling.

2. We solve $3.5x(1-x) = x$, which is the same as $\frac{7}{2}x(1-x) = x$, and we obtain $x = 0$ and $x = 5/7$. To determine the type of each fixed point, we compute

$$L'_{3.5}(x) = \tfrac{7}{2} - 7x,$$

so $L'_{3.5}(0) = 7/2$ and $L'_{3.5}(5/7) = -3/2$. Both fixed points are repelling.

3. The graph of the second iterate of T consists of four line segments.

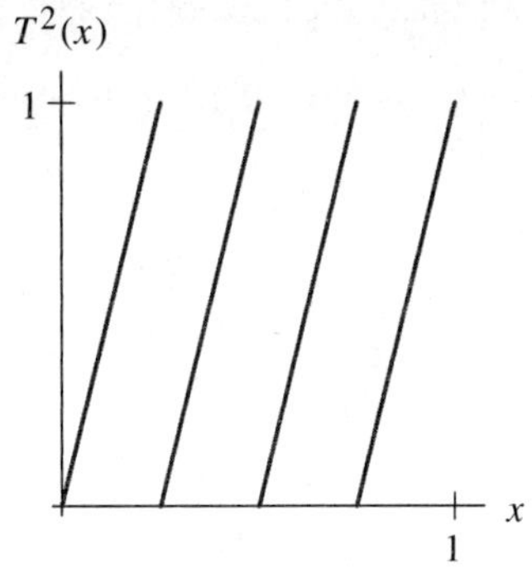

The points $x = 0$ and $x = 1$ are fixed points. To find the cycle of period two, we note that $T^2(x) = 4x - 1$ for $1/4 \le x < 1/2$. Solving $4x - 1 = x$ produces the period-two point $x = 1/3$. Note that $T(1/3) = 2/3$ and $T(2/3) = 1/3$.

4. The derivative of T at every point except $x = 1/2$ is 2. Since $T(1/2) = 1$ and $T(1) = 1$, the point $x = 1/2$ is not periodic. Therefore, the derivative is 2 at every point of any periodic cycle, and consequently, the derivative along the periodic cycle is a power of 2. Hence, there are no attracting cycles.

5. The bifurcation at $c = 1$ is a pitchfork bifurcation. The attracting fixed point at $x = 0$ for $-1 < c < 1$ becomes repelling for $c > 1$, and two new attracting fixed points appear, one on each side of $x = 0$.

6. If a function $F(x)$ has a fixed point at x_0, then its graph touches the diagonal at (x_0, x_0). If x_0 is an attracting fixed point, then the graph of $F(x)$ must be above the diagonal for all x slightly less than x_0 and below the diagonal for all x slightly greater than x_0. It is impossible for the graph of a continuous function to satisfy these conditions for exactly two fixed points.

7. For every x, $-1 \le F(x) \le 1$. So the first iterate of every point lies in this interval. Moreover, in this interval, it is easy to see that $|F(x)| < |x|$ for $x \ne 0$. Since the only fixed point is $x = 0$, all orbits tend to this fixed point.

8. The equation $\cos x = x$ has exactly one solution between 0 and $\pi/2$ because $\cos x$ decreases from 1 to 0 monotonically on this interval. Since $-1 \le \cos x \le 1$ for all x and cosine is an even function, all fixed points must lie in the interval $0 \le x \le 1$. Hence, cosine has exactly one fixed point.

Since $F'(x) = -\sin x$, the derivative at the fixed point is strictly between -1 and 0, so the fixed point is attracting.

9. For every x, $-1 \le F(x) \le 1$. For all x between -1 and 1, the orbit of x tends to the attracting fixed point at $0.73908\ldots$ (see Exercise 8). Hence, every orbit tends to this attracting fixed point.

10. The function $\tan x$ is periodic of period π. Moreover, in each interval of the form

$$(n\pi - \pi/2, n\pi + \pi/2)$$

where n is an integer, $\tan x$ is monotonically increasing. It tends to $+\infty$ as $x \to n\pi + \pi/2$ from below, and it tends to $-\infty$ as $x \to n\pi - \pi/2$ from above. Hence, each such interval contains exactly one fixed point.

11. False. If $c > 1/4$, the equation $x^2 + c = x$ has no solutions.

12. False. Note that $F(\sqrt{2}) = 0 \neq \sqrt{2}$, so $x = \sqrt{2}$ is not a fixed point.

13. False. Note that $F(1) = -1$ and $F(-1) = 1$, so $x = 1$ does lie on a cycle of period two. However, using the fact that $F'(x) = -3x^2$, we see that $F'(\pm 1) = -3$, so $(F^2)'(\pm 1) = 9$. Therefore, $x = 1$ lies on a repelling cycle of period two.

14. False. The 99th iterate of $1/2^{100}$ is $1/2$, so its 100th iterate is zero, a fixed point.

15. False. The bifurcation that occurs at $a = 1$ is a pitchfork bifurcation. For $a < 1$, there is one fixed point (at $x = 0$). For $a > 1$, there are three fixed points, $x = 0$ and one on each side of $x = 0$.

16. False. If $k \neq 0$, the function $L_k(x) = kx(1 - x)$ has two fixed points, $x = 0$ and $x = (k - 1)/k$. So there are two distinct fixed points for any $k \neq 0$, and no tangent bifurcation occurs at $k = 4$.

17. All points of the form $x = .abababab\ldots$ where $a \neq b$ are digits from 0 to 9 are periodic points of period two. Note that the points of the form $.aaaa\ldots$ are fixed points, not points of period two.

18. For $a \neq 1$, the function has exactly one fixed point, $x = b/(1 - a)$.

- If $a > 1$, the fixed point is repelling, and all other orbits tend to either $+\infty$ or $-\infty$ depending on whether they are larger or smaller than the fixed point.
- If $a = 1$ and $b > 0$, then all orbits tend to $+\infty$. If $a = 1$ and $b = 0$, then the function is the identity function, and all points are fixed. If $a = 1$ and $b < 0$, then all orbits tend to $-\infty$.
- If $0 < a < 1$, the fixed point attracts all other orbits.
- If $a = 0$, every orbit lands on the fixed point $x = b$ after one iterate.
- If $-1 < a < 0$, the fixed point attracts all other orbits. Each orbit alternates above and below the fixed point as it converges.
- If $a = -1$, all orbits except the fixed point are periodic of period two.
- If $a < -1$, all orbits except the fixed point tend to $\pm\infty$. Each orbit alternates above and below the fixed point as it becomes unbounded.

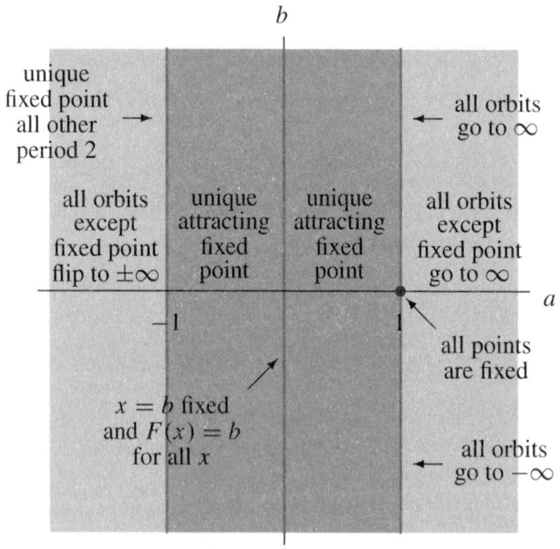

19. The graph of $F(x)$ shows that each point in the interval $-2 < x \leq 2$ is the image of exactly two points in -2 to 2. The same observation holds for the logistic map $L_4(x)$ on the interval $[0, 1]$, so there are infinitely many periodic points for F.

20. **(a)** Its graph looks like a tent.

(b) For $a < 1$, the only fixed point is $x = 0$, and it is attracting. For $a = 1$, every point in the interval $0 \leq x \leq 1/2$ is a fixed point. For $a > 1$ and close to 1, there are two fixed points, a repelling fixed point at $x = 0$ and an attracting fixed point in the interval $1/2 < x < 1$.

(c) The graph of F_2^n crosses the diagonal 2^n times in the interval $0 \leq x \leq 1$, so F_2^n has 2^n fixed points in this interval.

APPENDIXES

EXERCISES FOR APPENDIX A

1. We rewrite the equation as

$$\frac{dy}{dt} = (y - 4t) + (y - 4t)^2 + 4,$$

use that

$$\frac{dy}{dt} = \frac{du}{dt} + 4,$$

and substitute to obtain

$$\frac{du}{dt} + 4 = u + u^2 + 4.$$

This equation simplifies to

$$\frac{du}{dt} = u^2 + u,$$

which is nonlinear, autonomous, and separable.

2. Note that $y = tu$, so

$$\frac{dy}{dt} = t\frac{du}{dt} + u.$$

Substituting for y and dy/dt, we obtain

$$t\frac{du}{dt} + u = \frac{(tu)^2 + t(tu)}{(tu)^2 + 3t^2} = \frac{t^2(u^2 + u)}{t^2(u^2 + 3)}.$$

Simplifying we obtain

$$\frac{du}{dt} = \frac{1}{t}\left(\frac{u^2 + u}{u^2 + 3} - u\right).$$

This equation is nonlinear and nonautonomous.

3. Rewrite the equation as

$$\frac{dy}{dt} = ty + (ty)^2 + \cos(ty).$$

Using that $y = u/t$, we have

$$\frac{dy}{dt} = \frac{1}{t}\frac{du}{dt} - \frac{1}{t^2}u.$$

We substitute to obtain

$$\frac{1}{t}\frac{du}{dt} - \frac{1}{t^2}u = u + u^2 + \cos u,$$

which simplifies to

$$\frac{du}{dt} = \frac{u}{t} + t(u + u^2 + \cos u).$$

This equation is nonlinear and nonautonomous.

This is a good example of a change of variables that looks like it is going to greatly simplify the equation but does not because the term that replaces dy/dt is complicated.

4. We have $y = \ln u$, so
$$\frac{dy}{dt} = \frac{1}{u}\frac{du}{dt}.$$
Substituting we obtain
$$\frac{1}{u}\frac{du}{dt} = u + \frac{t^2}{u},$$
which simplifies to
$$\frac{du}{dt} = u^2 + t^2.$$
This equation is nonlinear and nonautonomous.

5. Let $u = y - t$. Then
$$\frac{du}{dt} = \frac{dy}{dt} - 1,$$
and the differential equation becomes
$$\frac{du}{dt} + 1 = u^2 - u - 1,$$
which simplifies to
$$\frac{du}{dt} = u^2 - u - 2 = (u-2)(u+1).$$
The equilibrium points are $u = 2$ (a source) and $u = -1$ (a sink). These equilibria correspond to the solutions $y_1(t) = 2 + t$ and $y_2(t) = -1 + t$.

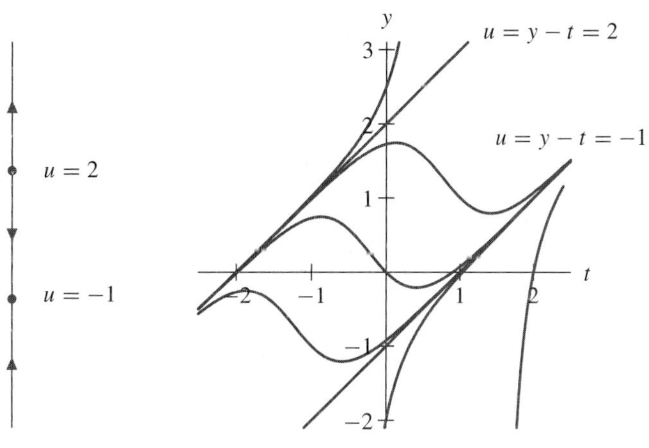

6. We let $u = y/t$. Then $y = tu$, so
$$\frac{dy}{dt} = u + t\frac{du}{dt}.$$
Replacing y by tu, we obtain
$$u + t\frac{du}{dt} = \frac{(tu)^2}{t} + 2(tu) - 4t + u,$$

which simplifies to

$$\frac{du}{dt} = u^2 + 2u - 4.$$

The equilibrium points of this equation are $u = -1 \pm \sqrt{5}$. The equilibrium $u = -1 - \sqrt{5}$ is a sink, and the equilibrium $u = -1+\sqrt{5}$ is a source. Note that the equilibrium solutions for $u(t)$ correspond to the solutions $y(t) = (-1 \pm \sqrt{5})t$ for $y(t)$.

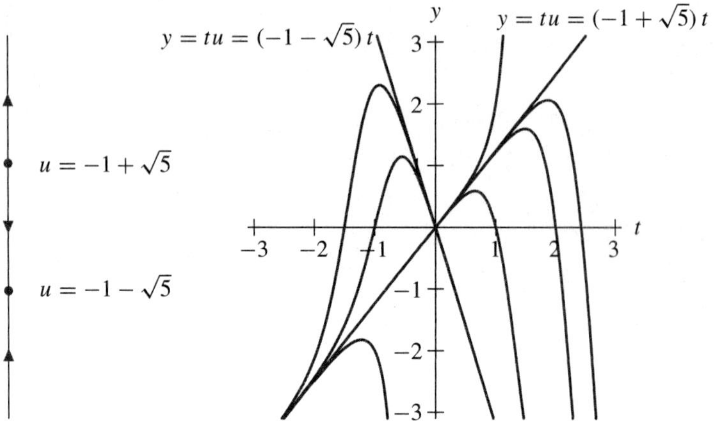

7. First we clear the denominator, so the equation becomes

$$t\frac{dy}{dt} = ty(\cos ty) - y.$$

Then we change variables using $u = ty$. We have

$$\frac{du}{dt} = t\frac{dy}{dt} + y,$$

and the differential equation becomes

$$\frac{du}{dt} - \frac{u}{t} = u(\cos u) - \frac{u}{t}.$$

Thus we get

$$\frac{du}{dt} = u \cos u.$$

The equilibrium point $u = 0$ corresponds to the solution $y(t)$ that is constantly zero for all t, and the equilibrium points $u = \pi/2 \pm n\pi$ for $n = 0, \pm 1, \pm 2, \ldots$ correspond to solutions $y(t) = (\pi/2 \pm n\pi)/t$, for $n = 0, \pm 1, \pm 2, \ldots$. Note that the change of variables "blows up" at $t = 0$.

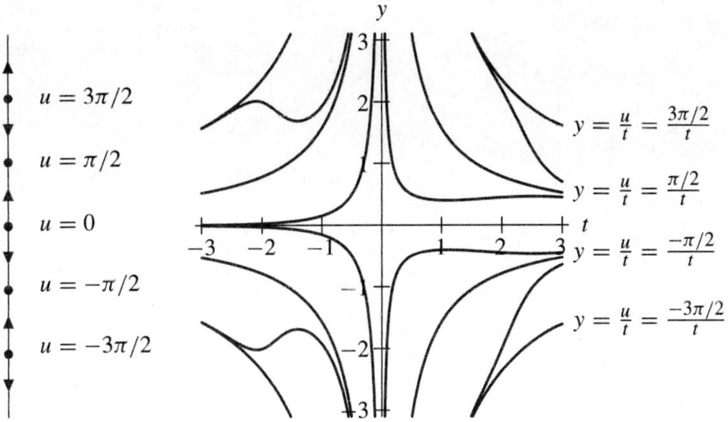

8. If $y = \sqrt{u}$, then

$$\frac{dy}{dt} = \frac{1}{2\sqrt{u}}\frac{du}{dt},$$

so replacing y with u, we have

$$\frac{1}{2\sqrt{u}}\frac{du}{dt} = \frac{t\sqrt{u}}{2} + \frac{e^{t^2/2}}{2\sqrt{u}},$$

which simplifies to

$$\frac{du}{dt} = tu + e^{t^2/2}.$$

This equation is linear. We rewrite it as

$$\frac{du}{dt} - tu = e^{t^2/2},$$

and the integrating factor is $\mu(t) = e^{\int -t\,dt} = e^{-t^2/2}$. Multiplying both sides by $\mu(t)$, we obtain

$$e^{-t^2/2}\frac{du}{dt} - te^{-t^2/2}u = 1,$$

which is equivalent to

$$\frac{d(e^{-t^2/2}u)}{dt} = 1.$$

Integrating both sides, we have

$$e^{-t^2/2}u = t + c,$$

where c is an arbitrary constant. The general solution is $u = (t + c)e^{t^2/2}$. In terms of the original dependent variable y, we have

$$y(t) = \pm\sqrt{(t + c)e^{t^2/2}}.$$

The choice of sign depends on the initial condition.

9. If $u = y/(1+t)$, we have

$$\frac{du}{dt} = \frac{1}{1+t}\frac{dy}{dt} - \frac{y}{(1+t)^2}.$$

Then

$$\frac{dy}{dt} = (1+t)\frac{du}{dt} + u,$$

and the differential equation becomes

$$(1+t)\frac{du}{dt} + u = u - \frac{u(1+t)}{t} + t^2(t+1),$$

which reduces to

$$\frac{du}{dt} = -\frac{u}{t} + t^2.$$

This differential equation is linear, and we rewrite it as

$$\frac{du}{dt} + \frac{u}{t} = t^2.$$

Its integrating factor is

$$\mu(t) = e^{\int (1/t)\,dt} = e^{\ln t} = t.$$

Multiplying both sides of the differential equation by $\mu(t)$, we obtain

$$t\frac{du}{dt} + u = t^3,$$

which is equivalent to

$$\frac{d(tu)}{dt} = t^3.$$

Integrating both sides, we have

$$tu = \frac{t^4}{4} + c,$$

where c is an arbitrary constant. Therefore,

$$u(t) = \frac{t^3}{4} + \frac{c}{t}.$$

To determine the general solution for $y(t)$, we have $y = u(1+t)$, and therefore

$$\begin{aligned} y(t) &= \frac{t^3(1+t)}{4} + \frac{c(1+t)}{t} \\ &= \frac{t^3(1+t)}{4} + \frac{c}{t} + c. \end{aligned}$$

10. If $u = y - t$, then

$$\frac{du}{dt} = \frac{dy}{dt} - 1.$$

We have

$$\frac{du}{dt} = \frac{dy}{dt} - 1 = (y^2 - 2yt + t^2 + y - t + 1) - 1,$$

and substituting $u + t$ for y gives

$$\frac{du}{dt} = (u + t)^2 - 2(u + t)t + t^2 + (u + t) - t,$$

which simplifies to

$$\frac{du}{dt} = u^2 + u.$$

This differential equation is autonomous and consequently separable. Separating variables we have

$$\int \frac{du}{u^2 + u} = \int dt,$$

and integrating we obtain

$$\ln \left| \frac{u}{1 + u} \right| = t + c.$$

Solving for u gives

$$u(t) = \frac{ke^t}{1 - ke^t},$$

where k is an arbitrary constant. This expression omits one solution—the equilibrium solution $u = -1$. The general solution of the original differential equation is

$$y(t) = \frac{ke^t}{1 - ke^t} + t$$

along with the solution $y(t) = t - 1$.

11. We know that

$$\frac{du}{dt} = \frac{dy}{dt} - 1.$$

Substituting y and dy/dt into the equation $du/dt = (1 - u)u$, we get

$$\frac{dy}{dt} - 1 = (1 - (y - t))(y - t),$$

which simplifies to

$$\frac{dy}{dt} = -y^2 + 2yt + y - t^2 - t + 1.$$

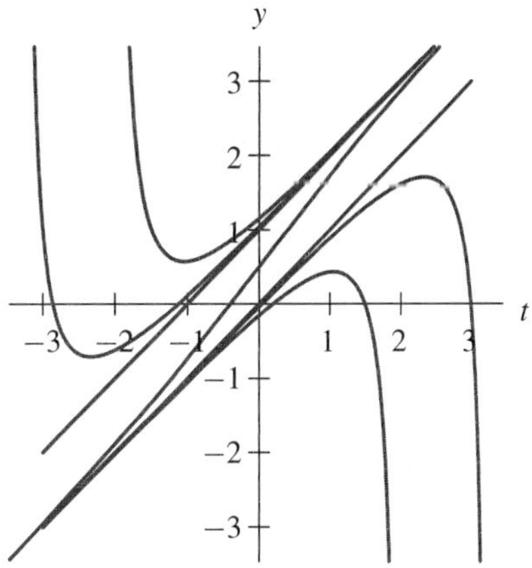

Graphs of solutions $y(t)$ are obtained from the graphs of the solutions to $du/dt = (1 - u)u$ by "tilting up" the plane so that horizontal lines in the tu-plane become lines of slope 1 in the ty-plane. As a result, the graphs of the equilibrium solutions in the tu-plane become lines with slope 1 in the ty-plane.

12. We use $y^2 = u$. Differentiating we obtain

$$2y\frac{dy}{dt} = \frac{du}{dt} = (1-u)u = (1-y^2)y^2,$$

so

$$\frac{dy}{dt} = \tfrac{1}{2}(y - y^3).$$

This is an autonomous equation. Note that the original change of variables is defined (real valued) only if $u \geq 0$ and results in $y \geq 0$. However, the change of variables $y^2 = u$ applies to any y.

This change of variables fixes the equilibrium points in the sense that $u = 0$ and $u = 1$ correspond to $y = 0$ and $y = 1$. If we consider $y < 0$, we obtain a third equilibrium point, $y = -1$. Note that the equation in y is symmetric about $y = 0$.

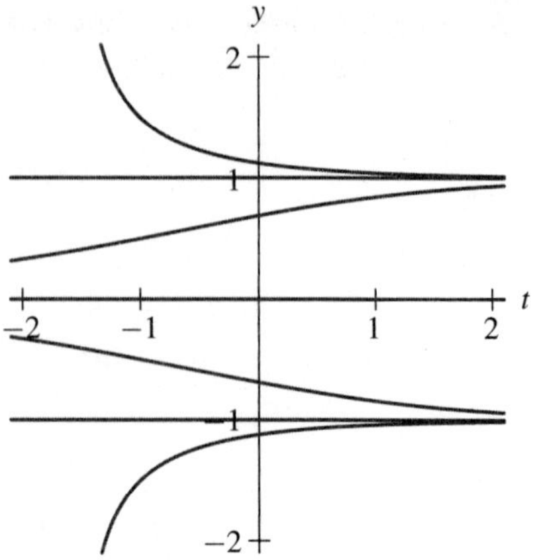

13. If $y = u^2$, then $u = \sqrt{y}$ and

$$\frac{du}{dt} = \frac{1}{2\sqrt{y}}\frac{dy}{dt}.$$

Note that the change of variables $u = \sqrt{y}$ is defined (real valued) only if $y \geq 0$ and results in $u \geq 0$. Substituting y and dy/dt into $du/dt = (1-u)u$, we have

$$\frac{1}{2\sqrt{y}}\frac{dy}{dt} = (1-\sqrt{y})\sqrt{y},$$

which simplifies to

$$\frac{dy}{dt} = 2y(1-\sqrt{y}).$$

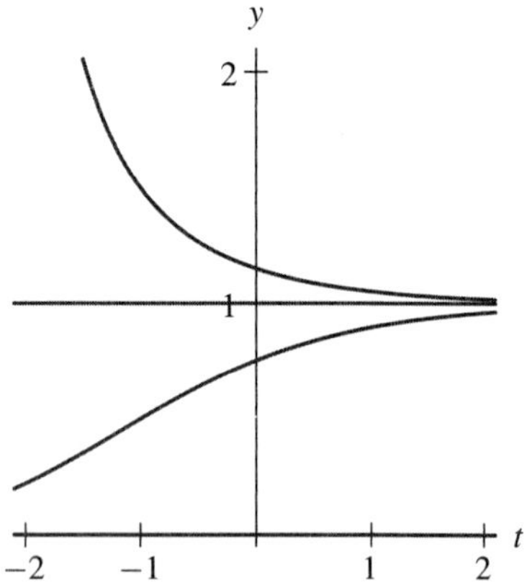

Note that the change of variables fixes the equilibrium points $u = 0$ and $u = 1$. In other words, $y = 0$ and $y = 1$ are also equilibrium points.

14. **(a)** We let $S(t)$ denote the amount of salt in the vat at time t. The volume $V(t)$ of salt water in the tank at time t is

$$V(t) = 5 + 5t$$

because 5 gallons are in the tank at $t = 0$ and 7 gallons per minute are added while only 2 gallons per minute are removed. From the first pipe, 6 pounds per minute of salt enter the vat, and from the second pipe, 2 pounds per minute of salt enter the vat. Well mixed salt water leaves the vat at the rate of 2 gallons per minute, so the amount of salt leaving the vat is twice the concentration. Hence,

$$\begin{aligned}\frac{dS}{dt} &= 6 + 2 - 2\left(\frac{S}{5+5t}\right)\\ &= 8 - \frac{2S}{5+5t}.\end{aligned}$$

(b) To convert this differential equation into one that involves the concentration of salt in the salt water, we let $C(t)$ be the concentration (in pounds per gallon) at time t. Then

$$C(t) = \frac{S(t)}{V(t)} = \frac{S(t)}{5+5t},$$

and

$$\begin{aligned}\frac{dC}{dt} &= \frac{1}{5+5t}\frac{dS}{dt} - \frac{5}{(5+5t)^2}S \\ &= \frac{1}{5+5t}\frac{dS}{dt} - \frac{5C}{5+5t}.\end{aligned}$$

Note that the differential equation for dS/dt can be rewritten as

$$\frac{dS}{dt} = 8 - 2C.$$

We have

$$\begin{aligned}\frac{dC}{dt} &= \frac{1}{5+5t}(8-2C) - \frac{5C}{5+5t} \\ &= \frac{8-7C}{5+5t}.\end{aligned}$$

(c) To find the concentration at the time that the tank overflows ($t = 3$), we can solve either differential equation. The differential equation for concentration is separable, and we separate variables to obtain

$$\frac{dC}{8-7C} = \frac{dt}{5+5t}.$$

After integrating and solving for $C(t)$, we have

$$C(t) = \tfrac{8}{7} + k(5+5t)^{-7/5},$$

where k is a constant determined by the initial condition. Solving for k using $C(0) = 0$ gives

$$k = -\tfrac{8}{7} \cdot 5^{7/5}.$$

With this value of k, we obtain $C(3) \approx 0.979$ pounds per gallon.

15. **(a)** Since 1 gallon per minute of salt water containing 2 pounds of salt per gallon and 5 gallons per minute of salt water containing 0.2 pounds of salt per gallon enter the vat, 3 pounds of salt per minute enters the vat. The concentration of the salt water at time t is $S(t)/(10+3t)$. Since 3 gallons per minute of salt water leaves the vat, $3S(t)/(10+3t)$ pounds of salt is removed. The rate of change of the amount of salt in the vat is

$$\frac{dS}{dt} = 3 - \frac{3S}{10+3t}.$$

(b) The concentration of the salt in the vat at time t is $C(t) = S(t)/(10 + 3t)$. Differentiating, we obtain

$$\frac{dC}{dt} = \frac{1}{10 + 3t}\frac{dS}{dt} - \frac{3C}{10 + 3t},$$

and thus

$$\frac{dS}{dt} = (10 + 3t)\frac{dC}{dt} + 3C.$$

Since $dS/dt = 3 - 3(S/(10 + 3t)) = 3 - 3C$, we obtain

$$(10 + 3t)\frac{dC}{dt} + 3C = 3 - 3C,$$

which yields

$$\frac{dC}{dt} = \frac{3 - 6C}{10 + 3t}.$$

(c) From the right-hand side of the differential equation, note that $C(t) = 1/2$ is an equilibrium solution. Moreover, for solutions whose initial conditions satisfy $C(0) < 1/2$, we see that $dC/dt > 0$. In our case, $C(0) = 0$, and we know that $C(t) \to 1/2$ as $t \to \infty$.

(d) From the differential equation in part (b), we separate variables and obtain

$$\frac{dC}{3 - 6C} = \frac{dt}{10 + 3t}.$$

Integration yields

$$C = \frac{1}{2} - \frac{k}{(10 + 3t)^2}.$$

At $t = 0$, $C(t) = 0$. Therefore, we have $0 = 1/2 - k/100$. Hence $k = 50$, and

$$C(t) = \frac{1}{2} - \frac{50}{(10 + 3t)^2}.$$

Evaluating this expression at $t = 5$ yields $C(5) = 0.42$.

16. If $u = y/t$, then the Product Rule yields

$$\begin{aligned}\frac{du}{dt} &= \frac{1}{t}\frac{dy}{dt} - \frac{1}{t^2}y \\ &= \frac{1}{t}\left(\frac{dy}{dt} - \frac{y}{t}\right) \\ &= \frac{1}{t}\Big(g(u) - u\Big),\end{aligned}$$

which is a separable differential equation.

17. **(a)** To find the equilibrium points, we solve $dy/dt = 0$, which is equivalent to

$$10y^3 = 1.$$

We get $y = \sqrt[-3]{10}$.

(b) Let $u = y - \sqrt[3]{10}$. Then, $du/dt = dy/dt$, and

$$\begin{aligned} 10y^3 - 1 &= 10\left(u + \sqrt[3]{10}\right)^3 - 1 \\ &= 10\left(u^3 + 3\sqrt[3]{10}\,u^2 + 3\sqrt[3]{10^2}\,u + \frac{1}{10}\right) - 1 \\ &= 10u^3 + 3 \cdot 10^{2/3}u^2 + 3\sqrt[3]{10}\,u. \end{aligned}$$

The linear approximation is

$$\frac{du}{dt} = 3\sqrt[3]{10}\,u.$$

Since $3\sqrt[3]{10} > 0$, the origin is a source.

(c) Solving the linear system, we get

$$u(t) = u_0 e^{3\sqrt[3]{10}\,t}$$

near the origin. In other words, solutions move away from the equilibrium solution exponentially fast with exponent $3\sqrt[3]{10}\,t$. In particular, to derive the amount of time that it takes a solution to double its distance from the equilibrium, we solve

$$2u_0 = u_0 e^{3\sqrt[3]{10}\,t}$$

and obtain $t = (\ln 2)/(3\sqrt[3]{10})$.

18. **(a)** To find the equilibria, we solve $dy/dt = 0$ and obtain $y = -1$ and $y = 3$.

(b) For $y = -1$, we let $u = y + 1$. To change variables, we write

$$\frac{du}{dt} = \frac{dy}{dt} = (y + 1)(y - 3) = u(u - 4),$$

which expands to

$$\frac{du}{dt} = -4u + u^2.$$

The linearization is

$$\frac{du}{dt} = -4u,$$

so $u = 0$ is a sink.

Similarly, for $y = 3$, we let $v = y - 3$ and obtain

$$\frac{dv}{dt} = (v + 4)v = 4v + v^2.$$

The linearization is

$$\frac{dv}{dt} = 4v,$$

so $v = 0$ is a source.

(c) The linearizations at $y = -1$ and $y = 3$ have general solutions $u = u_0 e^{-4t}$ and $v = v_0 e^{4t}$, respectively. Hence, the time necessary to halve or double the distance to the equilibrium point starting nearby is $t = (\ln 2)/4$.

19. **(a)** To find the equilibria, we solve $dy/dt = 0$ and obtain $y = -1$ and $y = 3$.

(b) For $y = -1$, let $u = y + 1$. Then, $du/dt = dy/dt$ and

$$\frac{du}{dt} = \frac{dy}{dt} = (y+1)(3-y) = u(4-u).$$

The linear approximation is

$$\frac{du}{dt} = 4u,$$

so $u = 0$ is a source.

Similarly, for $y = 3$, we let $v = y - 3$ and obtain

$$\frac{dv}{dt} = -v(v+4) = -4v - v^2.$$

The linear approximation is

$$\frac{dv}{dt} = -4v.$$

so $v = 0$ is a sink.

(c) The linearizations at $y = -1$ and $y = 3$ have general solutions $u = u_0 e^{4t}$ and $v = v_0 e^{-4t}$, respectively. Hence, the time necessary to halve or double the distance to the equilibrium point starting nearby is $t = (\ln 2)/4$.

20. **(a)** To find the equilibria, we solve $dy/dt = y^3 - 3y^2 + y = 0$ by factoring

$$y^3 - 3y^2 + y = y(y^2 - 3y + 1).$$

We obtain $y = 0$ and $y = (3 \pm \sqrt{5})/2$.

(b) For $y = 0$ there is no need for a change of variables. The linearization is

$$\frac{dy}{dt} = y,$$

so $y = 0$ is a source.

For $y = (3 + \sqrt{5})/2$, we let $u = y - (3 + \sqrt{5})/2$. Then $du/dt = dy/dt$ and

$$\begin{aligned}\frac{du}{dt} &= y\left(y - \frac{3+\sqrt{5}}{2}\right)\left(y - \frac{3-\sqrt{5}}{2}\right)\\ &= \left(u + \frac{3+\sqrt{5}}{2}\right)u\left(u + \sqrt{5}\right).\end{aligned}$$

The linearization is

$$\begin{aligned}\frac{du}{dt} &= \sqrt{5}\left(\frac{3+\sqrt{5}}{2}\right)u\\ &= \left(\frac{5+3\sqrt{5}}{2}\right)u,\end{aligned}$$

so $u = 0$ is a source.

For $y = (3 - \sqrt{5})/2$, we let $v = y - (3 - \sqrt{5})/2$. After a similar computation, we determine that the linearization at $v = 0$ is

$$\frac{dv}{dt} = \left(\frac{5 - 3\sqrt{5}}{2}\right) v,$$

so $v = 0$ is a sink.

(c) The times to halve or double the distance between the equilibria and nearby initial points are given by finding the general solution of the linearization and solving. For example, the linearization at $y = 0$ has $y_0 e^t$ as its general solution. We solve $2y_0 = y_0 e^t$ and determine that the doubling time is $t = \ln 2$. For $u = 0$, the doubling time is $t = (2\ln 2)/(5 + 3\sqrt{5})$, and for $v = 0$, the halving time is $t = (2\ln 2)/(3\sqrt{5} - 5)$.

21. **(a)** Since $y = u + y_0$, $du/dt = dy/dt$, and we can replace y by $u + y_0$ in the right-hand side of the differential equation. We get

$$\frac{du}{dt} = f(u + y_0).$$

(b) Using the Taylor series

$$f(y) = f(y_0) + f'(y_0)(y - y_0) + \frac{f''(y_0)}{2!}(y - y_0)^2 + \dots$$

about $y = y_0$, we have

$$\frac{du}{dt} = f(y_0) + f'(y_0)\, u + \frac{f''(y_0)}{2!} u^2 + \dots$$

because $u = y - y_0$. Since y_0 is an equilibrium point, $f(y_0) = 0$. Thus, when we truncate higher-order terms (order ≥ 2) in u, we get the linearized equation

$$\frac{du}{dt} = f'(y_0)\, u.$$

22. **(a)** Using the Chain Rule and the Product Rule, we have

$$\frac{dy}{ds} = \frac{dy}{dt}\frac{dt}{ds} = (3t^2 y + 3t^5)\frac{s^{-2/3}}{3} = y + s$$

because $s^{-2/3} = (t^3)^{-2/3} = t^{-2}$.

(b) This equation for dy/ds is linear, so we can derive the general solution using an integrating factor. We rewrite it as

$$\frac{dy}{ds} - y = s$$

and determine that an integrating factor is $\mu(s) = e^{-s}$. Multiplying both sides by $\mu(s)$, we obtain

$$\frac{d(\mu \cdot y)}{ds} = se^{-s}.$$

Integration yields

$$e^{-s} y = -se^{-s} - e^{-s} + c,$$

where c is a constant of integration. Therefore,

$$y(s) = -s - 1 + ce^s.$$

(c) Recalling that $s = t^3$, we have

$$y(t) = -t^3 - 1 + ce^{t^3}$$

as the general solution of the original equation.

23. This equation is separable, and we can solve it by separating variables. However, treating it as a Bernoulli equation and changing variables is easier.

First, note that the equilibrium solution $y(t) = 0$ for all t is a solution to this equation. So we assume $y \neq 0$ when we change variables.

We change variables using $z = y^{-2}$. We get

$$\begin{aligned}\frac{dz}{dt} &= -2y^{-3}\left(\frac{dy}{dt}\right)\\ &= -2y^{-3}(y + y^3)\\ &= -2y^{-2} - 2\\ &= -2z - 2 = -2(z + 1).\end{aligned}$$

The differential equation for $z(t)$ is autonomous and linear. One particular solution is the equilibrium solution $z(t) = -1$. Also, the general solution of the associated homogeneous equation is ke^{-2t}. Hence, the general solution of the differential equation for $z(t)$ is $z(t) = ke^{-2t} - 1$.

In our case, $z = y^{-2}$. Therefore, $z(t)$ must be positive, and we discard those solutions for which $k \leq 0$. Converting back to y, using $y = z^{-1/2}$, we have

$$y(t) = \frac{\pm 1}{\sqrt{ke^{-2t} - 1}},$$

where $k > 0$.

24. This equation is a Bernoulli equation, and we solve it by changing variables.

First, note that the equilibrium solution $y(t) = 0$ for all t is a solution to this equation. So we assume $y \neq 0$ when we change variables.

We change variables using $z = y^{-2}$. We get

$$\begin{aligned}\frac{dz}{dt} &= -2y^{-3}\left(\frac{dy}{dt}\right)\\ &= -2y^{-3}(y + ty^3)\\ &= -2y^{-2} - 2t\\ &= -2z - 2t.\end{aligned}$$

The differential equation for $z(t)$ is linear, and the general solution of the associated homogeneous equation is ke^{-2t}. Guessing $z_p(t) = at + b$ for a particular solution of the nonhomogeneous equation, we get

$$\frac{dz_p}{dt} = a \quad \text{and} \quad -2z_p - 2t = -2(a+1)t - 2b.$$

Then $z_p(t)$ is a solution if $a = -1$ and $b = 1/2$. Hence, the general solution of the differential equation for $z(t)$ is $z(t) = ke^{-2t} - t + 1/2$.

In our case, $z = y^{-2}$. Therefore, $z(t)$ must be positive, and we discard those solutions for which $k \leq -1/2$. Converting back to y, using $y = z^{-1/2}$, we have

$$y(t) = \frac{\pm 1}{\sqrt{ke^{-2t} - t + 1/2}},$$

where $k > -1/2$.

25. Note that this differential equation is not defined at $t = 0$. Therefore, we assume that $t \neq 0$ when we derive the general solution. Also, the equilibrium solution $y(t) = 0$ for all $t \neq 0$ is a solution. So we also assume that $y \neq 0$.

This equation is a Bernoulli equation, so we change variables using $z = y^{-3}$. We obtain

$$\begin{aligned}\frac{dz}{dt} &= -3y^{-4}\left(\frac{dy}{dt}\right) \\ &= 3y^{-4}\left(\frac{1}{t}y - y^4\right) \\ &= -\frac{3}{t}z + 3.\end{aligned}$$

This equation is linear, but it does not have constant coefficients. Therefore, we solve it using an integrating factor (see Section 1.9). We rewrite the equation as

$$\frac{dz}{dt} + \frac{3}{t}z = 3,$$

and observe that the integrating factor is $e^{\int (3/t)\,dt} = e^{3\ln t} = t^3$. Multiplying both sides of the differential equation for z by t^3, we have

$$\begin{aligned}t^3\frac{dz}{dt} + 3t^2z &= 3t^3 \\ t^3z &= \tfrac{3}{4}t^4 + c \\ z &= \tfrac{3}{4}t + ct^{-3}.\end{aligned}$$

So

$$y(t) = \left(\tfrac{3}{4}t + ct^{-3}\right)^{-1/3}.$$

26. This equation is separable, and we can solve it by separating variables. However, treating it as a Bernoulli equation and changing variables is easier.

First, note that the equilibrium solution $y(t) = 0$ for all t is a solution to this equation. So we assume $y \neq 0$ when we change variables.

We change variables using $z = y^{-49}$. We get

$$\begin{aligned}\frac{dz}{dt} &= -49y^{-50}\left(\frac{dy}{dt}\right) \\ &= -49y^{-50}(y + y^{50}) \\ &= -49y^{-49} - 49 \\ &= -49z - 49 = -49(z+1).\end{aligned}$$

The differential equation for $z(t)$ is autonomous and linear. One particular solution is the equilibrium solution $z(t) = -1$. Also, the general solution of the associated homogeneous equation is ke^{-49t}. Hence, the general solution of the differential equation for $z(t)$ is $z(t) = ke^{-49t} - 1$. Converting back to y, using $y = z^{-1/49}$, we have

$$y(t) = \frac{1}{\sqrt[49]{ke^{-49t} - 1}}.$$

27. **(a)** Substituting y_1 into the right-hand side of the equation, we get

$$\left(2t + \frac{1}{t}\right)t - t^2 - t^2 = 1,$$

which agrees with dy_1/dt. Hence, $y_1(t) = t$ is a solution.

(b) This differential equation is a Riccati equation, and we know one solution. Therefore, we change variables using $w = y - t$. We get

$$\begin{aligned}\frac{dw}{dt} &= \frac{dy}{dt} - 1 \\ &= \left(2t + \frac{1}{t}\right)y - y^2 - t^2 - 1 \\ &= \left(2t + \frac{1}{t}\right)(w + t) - (w + t)^2 - t^2 - 1 \\ &= \left(2t + \frac{1}{t}\right)w + 2t^2 + 1 - \left(w^2 + 2tw + t^2\right) - t^2 - 1 \\ &= \frac{1}{t}w - w^2.\end{aligned}$$

This differential equation for w is a Bernoulli equation. So change variables once more using $z = w^{-1}$. Hence,

$$\frac{dz}{dt} = -w^{-2}\left(\frac{dw}{dt}\right)$$

$$= -w^{-2}\left(\frac{1}{t}\,w - w^2\right)$$

$$= -\frac{1}{t}\,w^{-1} + 1$$

$$= -\frac{1}{t}\,z + 1.$$

This differential equation for z is linear, but it does not have constant coefficients. Therefore, we use an integrating factor to solve it (see Section 1.9). We rewrite the equation as

$$\frac{dz}{dt} + \frac{1}{t}\,z = 1,$$

and derive the integrating factor $\mu = e^{\int (1/t)dt} = e^{\ln t} = t$.

Multiplying both sides by t and integrating, we obtain

$$t\frac{dz}{dt} + z = t$$

$$\frac{d(tz)}{dt} = t$$

$$tz = \frac{t^2}{2} + c,$$

where c is the constant of integration. Therefore, the solution is

$$z(t) = \frac{t}{2} + \frac{c}{t}.$$

Hence, the solution to the equation in w is

$$w(t) = \frac{2t}{t^2 + k},$$

where $k = 2c$ is an arbitrary constant. The solution of the original differential equation is

$$y(t) = w(t) + t = \frac{(2+k)t + t^3}{t^2 + k}.$$

28. **(a)** Substituting y_1 into the right-hand side of the equation, we get

$$(t^4 - t^2 + 2t) + (1 - 2t^2)t^2 + (t^2)^2 = 2t,$$

which agrees with dy_1/dt. Hence, $y_1(t) = t^2$ is a solution.

(b) This differential equation is a Riccati equation, and we know one solution. Therefore, we change variables using $w = y - t^2$. We get

$$\begin{aligned}\frac{dw}{dt} &= \frac{dy}{dt} - 2t \\ &= (t^4 - t^2 + 2t) + (1 - 2t^2)y + y^2 - 2t \\ &= (t^4 - t^2 + 2t) + (1 - 2t^2)(w + t^2) + (w + t^2)^2 - 2t \\ &= (t^4 - t^2 + 2t) + (w - 2t^2 w + t^2 - 2t^4) + (w^2 + 2t^2 w + t^4) - 2t \\ &= w + w^2.\end{aligned}$$

This differential equation for w is a Bernoulli equation. So we change variables once more using $z = w^{-1}$. Hence,

$$\begin{aligned}\frac{dz}{dt} &= -w^{-2}\left(\frac{dw}{dt}\right) \\ &= -w^{-2}\left(w + w^2\right) \\ &= -w^{-1} - 1 \\ &= -z - 1 = -(z + 1).\end{aligned}$$

The differential equation for $z(t)$ is autonomous and linear. One particular solution is the equilibrium solution $z(t) = -1$. Also, the general solution of the associated homogeneous equation is ke^{-t}. Hence, the general solution of the differential equation for $z(t)$ is $z(t) = ke^{-t} - 1$. Therefore, the solution to the equation in w is

$$w(t) = \frac{1}{ke^{-t} - 1}$$

along with the equilibrium solution $w(t) = 0$ for all t. The solution of the original differential equation is

$$y(t) = w(t) + t^2 = t^2 + \frac{1}{ke^{-t} - 1}$$

along with the solution $y_1(t) = t^2$.

29. **(a)** Substituting y_1 into the right-hand side of the equation, we get

$$2\sin t + \cos t + t^2 \sin^2 t - 2(1 + t^2 \sin t)\sin t + t^2 \sin^2 t = \cos t,$$

which equals dy_1/dt. Hence, $y_1(t) = \sin t$ is a solution.

(b) This differential equation is a Riccati equation, and we know one solution. Therefore, we change variables using $w = y - \sin t$. We get

$$\begin{aligned}\frac{dw}{dt} &= \frac{dy}{dt} - \cos t \\ &= 2\sin t + t^2 \sin^2 t - 2(1 + t^2 \sin t)y + t^2 y^2 \\ &= 2\sin t + t^2 \sin^2 t - 2(1 + t^2 \sin t)(w + \sin t) + t^2(w + \sin t)^2 \\ &= -2w + t^2 w^2.\end{aligned}$$

This differential equation for w is a Bernoulli equation. So change variables once more using $z = w^{-1}$. Hence,

$$\frac{dz}{dt} = -w^{-2}\left(\frac{dw}{dt}\right)$$

$$= -w^{-2}\left(-2w + t^2 w^2\right)$$

$$= 2z - t^2.$$

This differential equation for z is linear. We solve it using the Extended Linearity Principle and a guessing technique. The general solution of the associated homogeneous equation is ke^{2t}. To find one particular solution of the nonhomogeneous equation, we rewrite the equation in the form

$$\frac{dz}{dt} - 2z = -t^2,$$

and we quess $z_p(t) = at^2 + bt + c$. Substituting this guess into the equation yields

$$(2at + b) - 2(at^2 + bt + c) = -t^2,$$

which simplifies to $(-2a)t^2 + (2a - 2b)t + (b - 2c) = -t^2$. The function $z_p(t)$ is a solution if $a = 1/2$, $b = 1/2$, and $c = 1/4$. Therefore,

$$z(t) = \tfrac{1}{2}t^2 + \tfrac{1}{2}t + \tfrac{1}{4} + ke^{2t}$$

is the general solution to the linear equation for z.

Hence, the functions

$$w(t) = \frac{4}{2t^2 + 2t + 1 + Ke^{2t}}$$

where K is an arbitrary constant are solutions to the the Bernoulli equation for w. Note also that the equilibrium solution $w(t) = 0$ for all t is a solution.

Solutions to the original Riccati equation consist of function of the form

$$y(t) = \frac{4}{2t^2 + 2t + 1 + Ke^{2t}} + \sin t$$

as well as the solution $y(t) = \sin t$.

30. We assume that $n \neq 1$. Otherwise, the equation is already a homogeneous linear equation, and we can solve it by separating variables.

If $z = y^\alpha$, then

$$\frac{dz}{dt} = \alpha y^{\alpha-1}\frac{dy}{dt}$$

$$= \alpha y^{\alpha-1}\left(r(t)y + a(t)y^n\right)$$

$$= \alpha r(t)y^\alpha + a(t)y^{\alpha-1+n}$$

$$= \alpha r(t)z + a(t)y^{\alpha-1+n}.$$

To complete the change of variables, we must replace $y^{\alpha-1+n}$ with an expression in z. If the resulting equation is to be linear, we must have $\alpha = 1 - n$ because we have already ruled out the case where $n = 1$.

31. Let $u = (y - y_1)^{-1}$. Then

$$\begin{aligned}\frac{du}{dt} &= -(y - y_1)^{-2}\left(\frac{dy}{dt} - \frac{dy_1}{dt}\right)\\ &= -(y - y_1)^{-2}\left(r(t) + a(t)y + b(t)y^2 - r(t) - a(t)y_1 - b(t)y_1^2\right)\\ &= -(y - y_1)^{-2}\left(a(t)(y - y_1) + b(t)(y^2 - y_1^2)\right)\\ &= -(y - y_1)^{-2}\left(a(t)(y - y_1) + b(t)(y - y_1)(y + y_1)\right)\\ &= -(y - y_1)^{-2}\left(a(t)(y - y_1) + b(t)(y - y_1)(y - y_1 + 2y_1)\right)\\ &= -u^2\left(a(t)u^{-1} + b(t)u^{-1}\left(u^{-1} + 2y_1\right)\right)\\ &= -a(t)u - b(t) - 2b(t)y_1(t)u\\ &= -(a(t) + 2b(t)y_1(t))u - b(t),\end{aligned}$$

which is a linear differential equation for $u(t)$.

32. Since $n = 5$, we have $z = y^{-4}$. Then

$$\frac{dz}{dt} = -4y^{-5}\frac{dy}{dt},$$

and consequently,

$$\begin{aligned}\frac{dy}{dt} &= -\frac{y^5}{4}\frac{dz}{dt}\\ &= -\frac{y^5}{4}\left(z + e^{2t}\right)\\ &= -\frac{y^5}{4}\left(y^{-4} + e^{2t}\right)\\ &= -\frac{1}{4}y - \frac{e^{2t}}{4}y^5,\end{aligned}$$

which is the desired Bernoulli equation.

33. We begin with $w = y - y_1$, so $w = y - (t^2 + 2)$. Equivalently, $y = w + t^2 + 2$. We have

$$\begin{aligned}\frac{dy}{dt} &= \frac{dw}{dt} + 2t\\ &= w + t^2w^2 + 2t\\ &= (y - (t^2 + 2)) + t^2(y - (t^2 + 2))^2 + 2t\\ &= y - t^2 - 2 + t^2y^2 - 2t^4y - 4t^2y + t^6 + 4t^4 + 4t^2 + 2t\\ &= (t^6 + 4t^4 + 3t^2 + 2t - 2) + (1 - 4t^2 - 2t^4)y + t^2y^2.\end{aligned}$$

Hence, our Riccati equation is

$$\frac{dy}{dt} = (t^6 + 4t^4 + 3t^2 + 2t - 2) + (1 - 4t^2 - 2t^4)y + t^2y^2.$$

EXERCISES FOR APPENDIX B

1. We guess the power series $y(t) = a_0 + a_1 t + a_2 t^2 + a_3 t^3 + a_4 t^4 + \dots$. Since the exercise only requests the coefficients $a_0, a_1, \dots, a_4$, we stop at the $a_4 t^4$ term. Substituting the power series for $y(t)$ and dy/dt into the differential equation, we get

$$a_1 + 2a_2 t + 3a_3 t^2 + 4a_4 t^3 + \dots = a_0 + a_1 t + a_2 t^2 + a_3 t^3 + a_4 t^4 + \dots.$$

Equating coefficients for each power of t yields

$$\begin{aligned} a_1 &= a_0 \\ 2a_2 &= a_1 \\ 3a_3 &= a_2 \\ 4a_4 &= a_3 \\ &\vdots \end{aligned}$$

Expressing all of the coefficients in terms of a_0, we get

$$a_1 = a_0, \quad a_2 = \frac{a_0}{2}, \quad a_3 = \frac{a_0}{2 \cdot 3}, \quad a_4 = \frac{a_0}{2 \cdot 3 \cdot 4}, \quad \dots.$$

The power series solution is

$$y(t) = a_0 \left(1 + t + \frac{t^2}{2!} + \frac{t^3}{3!} + \frac{t^4}{4!} + \dots\right).$$

As expected, this series is the power series for $a_0 e^t$, the general solution of the differential equation $dy/dt = y$.

2. We guess the power series $y(t) = a_0 + a_1 t + a_2 t^2 + a_3 t^3 + a_4 t^4 + \dots$. Since the exercise only requests the coefficients $a_0, a_1, \dots, a_4$, we stop at the $a_4 t^4$ term. Substituting the power series for $y(t)$ and dy/dt into the differential equation, we get

$$a_1 + 2a_2 t + 3a_3 t^2 + 4a_4 t^3 + \dots = -a_0 - a_1 t - a_2 t^2 - a_3 t^3 - a_4 t^4 - \dots + 1.$$

Equating coefficients for each power of t yields

$$\begin{aligned} a_1 &= -a_0 + 1 \\ 2a_2 &= -a_1 \\ 3a_3 &= -a_2 \\ 4a_4 &= -a_3 \\ &\vdots \end{aligned}$$

Expressing all of the coefficients in terms of a_0, we get

$$a_1 = -a_0 + 1, \quad a_2 = \frac{a_0 - 1}{2}, \quad a_3 = \frac{-a_0 + 1}{2 \cdot 3}, \quad a_4 = \frac{a_0 - 1}{2 \cdot 3 \cdot 4}, \dots.$$

The power series solution is

$$y(t) = a_0 + (-a_0 + 1)t + \frac{a_0 - 1}{2}t^2 + \frac{-a_0 + 1}{2 \cdot 3}t^3 + \frac{a_0 - 1}{2 \cdot 3 \cdot 4}t^4 + \dots.$$

Note that this solution can be written as

$$y(t) = a_0\left(1 - t + \frac{t^2}{2!} - \frac{t^3}{3!} + \frac{t^4}{4!} - \dots\right) + \left(t - \frac{t^2}{2!} + \frac{t^3}{3!} - \frac{t^4}{4!} + \dots\right).$$

The first term is the power series for $a_0 e^{-t}$, and the second term is the power series for $-e^{-t} + 1$.

The equation $dy/dt = -y + 1$ is linear and separable with the equilibrium solution $y(t) = 1$ for all t. Therefore, from the Extended Linearity Principle, we know that the general solution is

$$y(t) = ke^{-t} + 1.$$

This solution agrees with the power series solution obtained above if $k = a_0 - 1$.

3. We guess the power series $y(t) = a_0 + a_1 t + a_2 t^2 + a_3 t^3 + a_4 t^4 + \dots$. Since the exercise only requests the coefficients $a_0, a_1, \dots, a_4$, we stop at the $a_4 t^4$ term. Substituting the power series for $y(t)$ and dy/dt into the differential equation, we get

$$a_1 + 2a_2 t + 3a_3 t^2 + 4a_4 t^3 + \dots = -2t(a_0 + a_1 t + a_2 t^2 + a_3 t^3 + a_4 t^4 + \dots).$$

Equating coefficients for each power of t yields

$$\begin{aligned} a_1 &= 0 \\ 2a_2 &= -2a_0 \\ 3a_3 &= -2a_1 \\ 4a_4 &= -2a_2 \\ &\vdots \end{aligned}$$

Expressing all of the coefficients in terms of a_0, we get

$$a_1 = 0, \quad a_2 = -a_0, \quad a_3 = 0, \quad a_4 = \frac{a_0}{2}, \quad \dots.$$

The power series solution is

$$y(t) = a_0\left(1 - t^2 + \frac{t^4}{2} \mp \dots\right).$$

As expected, this series is the power series for $a_0 e^{-t^2}$, the general solution of the differential equation $dy/dt = -2ty$.

4. We guess the power series $y(t) = a_0 + a_1 t + a_2 t^2 + a_3 t^3 + a_4 t^4 + \dots$. Since the exercise only requests the coefficients $a_0, a_1, \dots, a_4$, we stop at the $a_4 t^4$ term. Substituting the power series for $y(t)$ and dy/dt into the differential equation, we get

$$a_1 + 2a_2 t + 3a_3 t^2 + 4a_4 t^3 + \dots = t^2(a_0 + a_1 t + a_2 t^2 + a_3 t^3 + a_4 t^4 + \dots) + 1.$$

Equating coefficients for each power of t yields

$$\begin{aligned} a_1 &= 1 \\ 2a_2 &= 0 \\ 3a_3 &= a_0 \\ 4a_4 &= a_1 \\ &\vdots \end{aligned}$$

Expressing all of the coefficients in terms of a_0, we get

$$a_1 = 1, \quad a_2 = 0, \quad a_3 = \frac{1}{3}a_0, \quad a_4 = \frac{1}{4}, \quad a_5 = 0, \quad a_6 = \frac{1}{3 \cdot 6}a_0, \ldots .$$

The power series solution is

$$y(t) = a_0 \left(1 + \frac{t^3}{3} + \frac{t^6}{3 \cdot 6} + \ldots\right) + \left(t + \frac{t^4}{4} + \ldots\right).$$

We can attempt to find the general solution to this linear equation using integrating factors, but the integrals involved cannot be explicitly computed.

5. We guess the power series $y(t) = a_0 + a_1 t + a_2 t^2 + a_3 t^3 + a_4 t^4 + \ldots$. Since the exercise only requests the coefficients $a_0, a_1, \ldots, a_4$, we stop at the $a_4 t^4$ term. Using the power series for e^u with $u = 2t$ (see Appendix C), we have

$$e^{2t} = 1 + 2t + \frac{4t^2}{2!} + \frac{8t^3}{3!} + \frac{16t^4}{4!} + \ldots .$$

Substituting the power series for $y(t)$, dy/dt, and e^{2t} into the differential equation, we get

$$a_1 + 2a_2 t + 3a_3 t^2 + 4a_4 t^3 + \ldots = -\left(a_0 + a_1 t + a_2 t^2 + a_3 t^3 + a_4 t^4 + \ldots\right) + \left(1 + 2t + \frac{4t^2}{2!} + \frac{8t^3}{3!} + \frac{16t^4}{4!} + \ldots\right).$$

Equating coefficients for each power of t yields

$$\begin{aligned} a_1 &= -a_0 + 1 \\ 2a_2 &= -a_1 + 2 \\ 3a_3 &= -a_2 + \frac{4}{2!} \\ 4a_4 &= -a_3 + \frac{8}{3!} \\ &\vdots \end{aligned}$$

Expressing all of the coefficients in terms of a_0, we get

$$a_1 = -a_0 + 1, \quad a_2 = \frac{a_0 + 1}{2!}, \quad a_3 = \frac{-a_0 + 3}{3!}, \quad a_4 = \frac{a_0 + 5}{4!}, \quad \ldots .$$

The power series solution is

$$y(t) = a_0 + (-a_0 + 1)\,t + \left(\frac{a_0 + 1}{2!}\right)t^2 + \left(\frac{-a_0 + 3}{3!}\right)t^3 + \left(\frac{a_0 + 5}{4!}\right)t^4 + \dots.$$

6. We guess the power series $y(t) = a_0 + a_1 t + a_2 t^2 + a_3 t^3 + a_4 t^4 + \dots$. Since the exercise only requests the coefficients $a_0, a_1, \dots, a_4$, we stop at the $a_4 t^4$ term. Substituting the power series for $y(t)$, dy/dt, and

$$\sin t = t - \frac{t^3}{3!} + \frac{t^5}{5!} \mp \dots$$

into the differential equation, we get

$$a_1 + 2a_2 t + 3a_3 t^2 + 4a_4 t^3 + \dots = 2(a_0 + a_1 t + a_2 t^2 + a_3 t^3 + a_4 t^4 + \dots)$$
$$+ \left(t - \frac{t^3}{3!} + \frac{t^5}{5!} + \dots\right).$$

Equating coefficients for each power of t yields

$$\begin{aligned} a_1 &= 2a_0 \\ 2a_2 &= 2a_1 + 1 \\ 3a_3 &= 2a_2 \\ 4a_4 &= 2a_3 - \frac{1}{3!} \\ &\vdots \end{aligned}$$

Expressing all of the coefficients in terms of a_0, we get

$$a_1 = 2a_0, \quad a_2 = 2a_0 + \tfrac{1}{2}, \quad a_3 = \tfrac{4}{3}a_0 + \tfrac{1}{3}, \quad a_4 = \tfrac{2}{3}a_0 + \tfrac{1}{8}, \dots.$$

The power series solution is

$$y(t) = a_0 + 2a_0 t + \left(2a_0 + \tfrac{1}{2}\right)t^2 + \left(\tfrac{4}{3}a_0 + \tfrac{1}{3}\right)t^3 + \left(\tfrac{2}{3}a_0 + \tfrac{1}{8}\right)t^4 + \dots.$$

7. We guess the power series $y(t) = a_0 + a_1 t + a_2 t^2 + a_3 t^3 + a_4 t^4 + \dots$. Since the exercise only requests the coefficients $a_0, a_1, \dots, a_4$, we stop at the $a_4 t^4$ term. The power series for $\cos t$ is

$$1 - \frac{t^2}{2!} + \frac{t^4}{4!} - \frac{t^6}{6!} \pm \dots$$

(see Appendix C). Substituting the power series for $y(t)$, d^2y/dt^2, and $\cos t$ into the differential equation, we get

$$(2a_2 + (2 \cdot 3)a_3 t + (3 \cdot 4)a_4 t^2 + \dots) + 2(a_0 + a_1 t + a_2 t^2 + a_3 t^3 + a_4 t^4 + \dots)$$
$$= 1 - \frac{t^2}{2!} + \frac{t^4}{4!} \mp \dots.$$

Equating coefficients for each power of t yields

$$\begin{aligned} 2a_2 + 2a_0 &= 1 \\ 6a_3 + 2a_1 &= 0 \\ 12a_4 + 2a_2 &= -\tfrac{1}{2} \\ &\vdots \end{aligned}$$

Expressing all of the coefficients in terms of a_0 and a_1, we get

$$a_2 = \tfrac{1}{2} - a_0, \quad a_3 = -\tfrac{1}{3}a_1, \quad a_4 = \tfrac{1}{6}a_0 - \tfrac{1}{8}, \quad \ldots.$$

The power series solution is

$$y(t) = a_0 + a_1 t + \left(\tfrac{1}{2} - a_0\right)t^2 - \tfrac{1}{3}a_1 t^3 + \left(\tfrac{1}{6}a_0 - \tfrac{1}{8}\right)t^4 + \ldots.$$

8. We guess the power series $y(t) = a_0 + a_1 t + a_2 t^2 + a_3 t^3 + a_4 t^4 + \ldots$. Since the exercise only requests the coefficients $a_0, a_1, \ldots, a_4$, we stop at the $a_4 t^4$ term. Using the power series for $\sin u$ where $u = 2t$ (see Appendix C), we have

$$\sin 2t = 2t - \frac{8t^2}{3!} + \frac{32t^5}{5!} \mp \ldots.$$

Substituting the power series for $y(t)$, dy/dt, d^2y/dt^2, and $\sin 2t$ into the differential equation, we get

$$\begin{aligned} &(2a_2 + 6a_3 t + 12a_4 t^2 + \ldots) + 5(a_1 + 2a_2 t + 3a_3 t^2 + 4a_4 t^3 + \ldots) + \\ &\qquad (a_0 + a_1 t + a_2 t^2 + a_3 t^3 + a_4 t^4 + \ldots) = 2t - \frac{8t^2}{3!} + \frac{32t^5}{5!} \mp \ldots. \end{aligned}$$

Equating coefficients for each power of t yields

$$\begin{aligned} 2a_2 + 5a_1 + a_0 &= 0 \\ 6a_3 + 10a_2 + a_1 &= 2 \\ 12a_4 + 15a_3 + a_2 &= 0 \\ &\vdots \end{aligned}$$

Expressing all of the coefficients in terms of a_0 and a_1, we get

$$a_2 = -\tfrac{1}{2}a_0 - \tfrac{5}{2}a_1, \quad a_3 = \tfrac{2}{3} + \tfrac{5}{6}a_0 + 4a_1, \quad a_4 = -\tfrac{5}{6} - a_0 - \tfrac{115}{24}a_1, \ldots.$$

The power series solution is

$$y(t) = a_0 + a_1 t - \left(\tfrac{1}{2}a_0 + \tfrac{5}{2}a_1\right)t^2 + \left(\tfrac{2}{3} + \tfrac{5}{6}a_0 + 4a_1\right)t^3 - \left(\tfrac{5}{6} + a_0 + \tfrac{115}{24}a_1\right)t^4 + \ldots.$$

9. First we verify that $y(t) = \tan t$ is a solution of the differential equation. We note that

$$\begin{aligned}\frac{dy}{dt} &= \sec^2 t \\ &= 1 + \tan^2 t \\ &= 1 + y^2.\end{aligned}$$

We guess the power series $y(t) = a_0 + a_1 t + a_2 t^2 + a_3 t^3 + a_4 t^4 + \ldots$. Since the exercise only requests the coefficients $a_0, a_1, \ldots, a_6$, we stop at the $a_6 t^6$ term. We also need the coefficients of the power series for y^2 up to degree five. We compute

$$\begin{aligned}\left(a_0 + a_1 t + a_2 t^2 + a_3 t^3 + a_4 t^4 + a_5 t^5 + \ldots\right)^2 = a_0^2 + (2a_0 a_1)\, t + (2a_0 a_2 + a_1^2)\, t^2 +& \\ (2a_0 a_3 + 2a_1 a_2)\, t^3 +& \\ (2a_0 a_4 + 2a_1 a_3 + a_2^2)\, t^4 +& \\ (2a_0 a_5 + 2a_1 a_4 + 2a_2 a_3)\, t^5 +& \ldots.\end{aligned}$$

Substituting the power series for dy/dt and y^2 into the differential equation, we get

$$\begin{aligned}&a_1 + 2a_2 t + 3a_3 t^2 + 4a_4 t^3 + 5a_5 t^4 + 6a_6 t^5 + \ldots = \\ &\quad 1 + a_0^2 + (2a_0 a_1)\, t + (2a_0 a_2 + a_1^2)\, t^2 + (2a_0 a_3 + 2a_1 a_2)\, t^3 + \\ &\quad (2a_0 a_4 + 2a_1 a_3 + a_2^2)\, t^4 + (2a_0 a_5 + 2a_1 a_4 + 2a_2 a_3)\, t^5 + \ldots.\end{aligned}$$

Equating coefficients for each power of t, we obtain the equations

$$\begin{aligned}a_1 &= 1 + a_0^2 \\ 2a_2 &= 2a_0 a_1 \\ 3a_3 &= 2a_0 a_2 + a_1^2 \\ 4a_4 &= 2a_0 a_3 + 2a_1 a_2 \\ 5a_5 &= 2a_0 a_4 + 2a_1 a_3 + a_2^2 \\ 6a_6 &= 2a_0 a_5 + 2a_1 a_4 + 2a_2 a_3 \\ &\ \vdots\end{aligned}$$

The function $y(t) = \tan t$ satisfies the initial condition $y(0) = 0$, so we know that $a_0 = 0$. We have

$$a_1 = 1, \quad a_2 = 0, \quad a_3 = \tfrac{1}{3}, \quad a_4 = 0, \quad a_5 = \tfrac{2}{15}, \quad a_6 = 0, \quad \ldots.$$

The power series for $\tan t$ is

$$t + \tfrac{1}{3} t^3 + \tfrac{2}{15} t^5 + \ldots.$$

10. We guess the power series $y(t) = a_0 + a_1 t + a_2 t^2 + a_3 t^3 + a_4 t^4 + \ldots$. Since the exercise only requests the coefficients $a_0, a_1, \ldots, a_6$, we stop at the $a_6 t^6$ term. Substituting the power series for $y(t)$ and d^2y/dt^2 into the differential equation, we get

$$\begin{aligned}&(2a_2 + 6a_3 t + 12a_4 t^2 + 20a_5 t^3 + 30a_6 t^4 + \ldots) \\ &\quad + 2(a_0 + a_1 t + a_2 t^2 + a_3 t^3 + a_4 t^4 + a_5 t^5 + a_6 t^6 + \ldots) = 0.\end{aligned}$$

Equating coefficients for each power of t yields

$$\begin{aligned} 2a_2 + 2a_0 &= 0 \\ 6a_3 + 2a_1 &= 0 \\ 12a_4 + 2a_2 &= 0 \\ 20a_5 + 2a_3 &= 0 \\ 30a_6 + 2a_4 &= 0 \\ &\vdots \end{aligned}$$

Expressing all of the coefficients in terms of a_0 and a_1, we get

$$a_2 = -a_0, \quad a_3 = -\tfrac{1}{3}a_1, \quad a_4 = \tfrac{1}{6}a_0, \quad a_5 = \tfrac{1}{30}a_1, \quad a_6 = -\tfrac{1}{90}a_0, \ldots .$$

The power series solution is

$$y(t) = a_0 + a_1 t - a_0 t^2 - \tfrac{1}{3}a_1 t^3 + \tfrac{1}{6}a_0 t^4 + \tfrac{1}{30}a_1 t^5 - \tfrac{1}{90}a_0 t^6 + \ldots .$$

11. We guess the power series $y(t) = a_0 + a_1 t + a_2 t^2 + a_3 t^3 + a_4 t^4 + \ldots$. Since the exercise only requests the coefficients $a_0, a_1, \ldots, a_6$, we stop at the $a_6 t^6$ term. Substituting the power series for $y(t)$, dy/dt, and d^2y/dt^2 into the differential equation, we get

$$\begin{aligned} &(2a_2 + 6a_3 t + 12a_4 t^2 + 20a_5 t^3 + 30a_6 t^4 + \ldots) \\ &\quad + (a_1 + 2a_2 t + 3a_3 t^2 + 4a_4 t^3 + 5a_5 t^4 + 6a_6 t^5 + \ldots) \\ &\quad\quad + (a_0 + a_1 t + a_2 t^2 + a_3 t^3 + a_4 t^4 + a_5 t^5 + a_6 t^6 + \ldots) = 0. \end{aligned}$$

Equating coefficients for each power of t yields

$$\begin{aligned} 2a_2 + a_1 + a_0 &= 0 \\ 6a_3 + 2a_2 + a_1 &= 0 \\ 12a_4 + 3a_3 + a_2 &= 0 \\ 20a_5 + 4a_4 + a_3 &= 0 \\ 30a_6 + 5a_5 + a_4 &= 0 \\ &\vdots \end{aligned}$$

Expressing all of the coefficients in terms of a_0 and a_1, we get

$$a_2 = -\tfrac{1}{2}(a_0 + a_1), \quad a_3 = \tfrac{1}{6}a_0, \quad a_4 = \tfrac{1}{24}a_1, \quad a_5 = -\tfrac{1}{120}(a_0 + a_1), \quad a_6 = \tfrac{1}{720}a_0, \quad \ldots .$$

The power series solution is

$$y(t) = a_0 + a_1 t - \tfrac{1}{2}(a_0 + a_1)t^2 + \tfrac{1}{6}a_0 t^3 + \tfrac{1}{24}a_1 t^4 - \tfrac{1}{120}(a_0 + a_1)t^5 + \tfrac{1}{720}a_0 t^6 + \ldots .$$

12. We guess the power series

$$y(t) = a_0 + a_1 t + a_2 t^2 + a_3 t^3 + a_4 t^4 + a_5 t^5 + a_6 t^6 + \dots .$$

Since the exercise only requests the coefficients $a_0, a_1, \dots, a_6$, we stop at the $a_6 t^6$ term. Substituting the power series for $y(t)$, dy/dt, d^2y/dt^2, and $\cos t$ (see Appendix C) into the differential equation, we get

$$\begin{aligned}(2a_2 + 6a_3 t + 12a_4 t^2 + 20a_5 t^3 + 30a_6 t^4 + \dots)\\ + (a_1 + 2a_2 t + 3a_3 t^2 + 4a_4 t^3 + 5a_5 t^4 + \dots)\\ + t^2(a_0 + a_1 t + a_2 t^2 + a_3 t^3 + a_4 t^4 + \dots)\\ = 1 - \frac{t^2}{2!} + \frac{t^4}{4!} \mp \dots .\end{aligned}$$

Equating coefficients for each power of t yields

$$\begin{aligned} 2a_2 + a_1 &= 1\\ 6a_3 + 2a_2 &= 0\\ 12a_4 + 3a_3 + a_0 &= -\tfrac{1}{2}\\ 20a_5 + 4a_4 + a_1 &= 0\\ 30a_6 + 5a_5 + a_2 &= \tfrac{1}{24}\\ &\vdots \end{aligned}$$

Expressing all of the coefficients in terms of a_0 and a_1, we get

$$a_2 = \tfrac{1}{2}(1 - a_1), \quad a_3 = \tfrac{1}{6}(a_1 - 1), \quad a_4 = -\tfrac{1}{24}(a_0 + a_1), \quad a_5 = \tfrac{1}{60}a_0 - \tfrac{1}{24}a_1,$$

and

$$a_6 = -\tfrac{1}{360}a_0 + \tfrac{17}{720}a_1 - \tfrac{11}{720}.$$

The power series solution is

$$\begin{aligned} y(t) = a_0 + a_1 t + \tfrac{1}{2}(1 - a_1)t^2 + \tfrac{1}{6}(a_1 - 1)t^3 - \tfrac{1}{24}(a_0 + a_1)t^4\\ + (\tfrac{1}{60}a_0 - \tfrac{1}{24}a_1)t^5 - (\tfrac{1}{360}a_0 - \tfrac{17}{720}a_1 + \tfrac{11}{720})t^6 + \dots . \end{aligned}$$

13. We guess the power series $y(t) = a_0 + a_1 t + a_2 t^2 + a_3 t^3 + a_4 t^4 + \dots$. Since the exercise only requests the coefficients $a_0, a_1, \dots, a_6$, we stop at the $a_6 t^6$ term. Using the power series for e^u with $u = -2t$ (see Appendix C), we have

$$e^{-2t} = 1 - 2t + \frac{4t^2}{2!} - \frac{8t^3}{3!} + \frac{16t^4}{4!} - \frac{32t^5}{5!} \pm \dots .$$

Substituting the power series for $y(t)$, dy/dt, d^2y/dt^2, and e^{-2t} into the differential equation, we get

$$\begin{aligned}(2a_2 + 6a_3 t + 12a_4 t^2 + 20a_5 t^3 + 30a_6 t^4 + \dots)\\ + t\,(a_1 + 2a_2 t + 3a_3 t^2 + 4a_4 t^3 + 5a_5 t^4 + 6a_6 t^5 + \dots)\\ + (a_0 + a_1 t + a_2 t^2 + a_3 t^3 + a_4 t^4 + a_5 t^5 + a_6 t^6 + \dots) =\\ 1 - 2t + 2t^2 - \tfrac{4}{3}t^3 + \tfrac{2}{3}t^4 + \dots .\end{aligned}$$

Equating coefficients for each power of t yields

$$\begin{aligned} 2a_2 + a_0 &= 1 \\ 6a_3 + 2a_1 &= -2 \\ 12a_4 + 3a_2 &= 2 \\ 20a_5 + 4a_3 &= -\tfrac{4}{3} \\ 30a_6 + 5a_4 &= \tfrac{2}{3} \\ &\vdots \end{aligned}$$

Expressing all of the coefficients in terms of a_0 and a_1, we get

$$a_2 = \tfrac{1}{2}(1 - a_0),\ a_3 = -\tfrac{1}{3}(a_1 + 1),\ a_4 = \tfrac{1}{24}(3a_0 + 1),\ a_5 = \tfrac{1}{15}a_1,\ a_6 = \tfrac{1}{720}(11 - 15a_0),\ \ldots.$$

The power series solution is

$$y(t) = a_0 + a_1 t + \tfrac{1}{2}(1 - a_0)t^2 - \tfrac{1}{3}(a_1 + 1)t^3 + \tfrac{1}{24}(3a_0 + 1)t^4 + \tfrac{1}{15}a_1 t^5 + \tfrac{1}{720}(11 - 15a_0)t^6 + \ldots.$$

14. Letting $p = 3$, $a_0 = 0$, and $a_1 = 1$, we get

$$H_3(t) = t - \tfrac{2}{3}t^3.$$

To check this result, we calculate $H_3'(t) = 1 - 2t^2$ and $H_3''(t) = -4t$, and we substitute these results into the equation

$$\begin{aligned} \frac{d^2H_3}{dt^2} - 2t\frac{dH_3}{dt} + 6H_3 &= (-4t) - 2t(1 - 2t^2) + 6(t - \tfrac{2}{3}t^3) \\ &= 0. \end{aligned}$$

Hence, $H_3(t) = t - \frac{2}{3}t^3$ is the desired solution with $p = 3$.

Letting $p = 4$, $a_0 = 1$, and $a_1 = 0$, we get

$$H_4(t) = 1 - 4t^2 + \tfrac{4}{3}t^4.$$

To check this result, we calculate $H_4'(t) = -8t + \frac{16}{3}t^3$ and $H_4''(t) = -8 + 16t^2$, and we substitute these results into the equation

$$\begin{aligned} \frac{d^2H_4}{dt^2} - 2t\frac{dH_4}{dt} + 8H_4 &= (-8 + 16t^2) - 2t(-8t + \tfrac{16}{3}t^3) + 8(1 - 4t^2 + \tfrac{4}{3}t^4) \\ &= 0. \end{aligned}$$

Hence, $H_4(t) = 1 - 4t^2 + \frac{4}{3}t^4$ is the desired solution with $p = 4$.

Letting $p = 5$, $a_0 = 0$, and $a_1 = 1$, we get

$$H_5(t) = t - \tfrac{4}{3}t^3 + \tfrac{4}{15}t^5.$$

To check this result, we calculate $H_5'(t) = 1 - 4t^2 + \frac{4}{3}t^4$ and $H_5''(t) = -8t + \frac{16}{3}t^3$, and we substitute these results into the equation

$$\begin{aligned} \frac{d^2H_5}{dt^2} - 2t\frac{dH_5}{dt} + 10H_5 &= (-8t + \tfrac{16}{3}t^3) - 2t(1 - 4t^2 + \tfrac{4}{3}t^4) + 10(t - \tfrac{4}{3}t^3 + \tfrac{4}{15}t^5) \\ &= 0. \end{aligned}$$

Hence, $H_5(t) = t - \frac{4}{3}t^3 + \frac{4}{15}t^5$ is the desired solution with $p = 5$.

15. **(a)** We guess the power series $y(t) = a_0 + a_1 t + a_2 t^2 + a_3 t^3 + a_4 t^4 + \ldots$. To do this part of the exercise, we need only consider terms up to degree four. Substituting the power series for $y(t)$, dy/dt, and d^2y/dt^2 into the differential equation, we get

$$\begin{aligned}(1-t^2)&(2a_2 + 6a_3 t + 12a_4 t^2 + \ldots)\\ &-2t(a_1 + 2a_2 t + 3a_3 t^2 + 4a_4 t^3 + \ldots)\\ &\qquad + \nu(\nu+1)(a_0 + a_1 t + a_2 t^2 + a_3 t^3 + a_4 t^4 + \ldots) = 0.\end{aligned}$$

Collecting the coefficients of the constant terms, linear terms, and quadratic terms, we see that

$$\begin{aligned}2a_2 + \nu(\nu+1)a_0 &= 0\\ 6a_3 - 2a_1 + \nu(\nu+1)a_1 &= 0\\ 12a_4 - 6a_2 + \nu(\nu+1)a_2 &= 0.\end{aligned}$$

Therefore,

$$\begin{aligned}a_2 &= \frac{-\nu(\nu+1)}{2}a_0\\ a_3 &= \frac{2-\nu(\nu+1)}{6}a_1\\ a_4 &= \left(\frac{6-\nu(\nu+1)}{12}\right)\left(\frac{-\nu(\nu+1)}{2}\right)a_0.\end{aligned}$$

(b) If we substitute the power series for $y(t)$, dy/dt, and d^2y/dt^2 into the differential equation and collect all of the t^n terms ($n \geq 2$), we get

$$(n+1)(n+2)a_{n+2}t^n - n(n-1)a_n t^n - 2na_n t^n + \nu(\nu+1)a_n t^n.$$

Consequently, $y(t)$ is a solution only if

$$(n+1)(n+2)a_{n+2} - n(n-1)a_n - 2na_n + \nu(\nu+1)a_n = 0$$

for all $n \geq 2$. This equality is satisfied if

$$a_{n+2} = \frac{n(n+1) - \nu(\nu+1)}{(n+1)(n+2)}a_n.$$

Note that this relationship between a_n and a_{n+2} generalizes the results obtained in part (a).

An important consequence of this relationship is that, if any coefficient, $a_1, a_3, a_5, \ldots$, of an odd power of t is zero, then all coefficients of higher odd powers are also zero. For example, if $a_7 = 0$, then $a_9 = 0$, $a_{11} = 0, \ldots$. Moreover, the same can be said for coefficients of even powers of t.

To produce polynomial solutions if ν is a positive integer, we consider two cases. If ν is even, we use the initial condition $(y(0), y'(0)) = (1, 0)$. Then $a_1 = 0$, and all of the coefficients of odd powers are zero. Moreover, the formula that relates a_{n+2} to a_n implies that $a_{\nu+2} = 0$. Therefore, all of the coefficients of higher even powers are also zero, and the solution is a polynomial (with no odd powers of t).

If ν is odd, we use the initial condition $(y(0), y'(0)) = (0, 1)$ and obtain a polynomial solution with no even powers of t.

(c) To compute $P_0(t)$, take $\nu = 0$ and $(y(0), y'(0)) = (1, 0)$. Then $a_0 = 1$ and $a_1 = 0$. From part (b), $a_3 = a_5 = a_7 = \ldots = 0$. Also, from the formula in part (a), we get $a_2 = 0$, and consequently, $a_4 = a_6 = \ldots = 0$ as well. Hence, the solution is

$$P_0(t) = 1.$$

Similarly, taking $\nu = 1$ and $(y(0), y'(0)) = (0, 1)$, we obtain $a_0 = a_2 = \ldots = 0$, and since $a_3 = 0$, $a_5 = a_7 = \ldots = 0$. Hence, the solution is

$$P_1(t) = t.$$

To compute $P_2(t)$, take $\nu = 2$ and $(y(0), y'(0)) = (1, 0)$. We have $a_1 = a_3 = \ldots = 0$. Also, $a_0 = 1$ implies that $a_2 = -3$ and $a_4 = a_6 = a_8 = \ldots = 0$. We have the solution

$$P_2(t) = 1 - 3t^2.$$

(d) To compute $P_3(t)$, take $\nu = 3$ and $(y(0), y'(0)) = (0, 1)$. Then $a_3 = -5/3$ and

$$P_3(t) = t - \tfrac{5}{3}t^3.$$

To compute $P_4(t)$, take $\nu = 4$ and $(y(0), y'(0)) = (1, 0)$. Then $a_2 = -10$, $a_4 = 35/3$, and

$$P_4(t) = 1 - 10t^2 + \tfrac{35}{3}t^4.$$

To compute $P_5(t)$, take $\nu = 5$ and $(y(0), y'(0)) = (0, 1)$. Then $a_3 = -14/3$, $a_5 = 21/5$, and

$$P_5(t) = t - \tfrac{14}{3}t^3 + \tfrac{21}{5}t^5.$$

To compute $P_6(t)$, take $\nu = 6$ and $(y(0), y'(0)) = (1, 0)$. Then $a_2 = -21$, $a_4 = 63$, $a_6 = -231/5$, and

$$P_6(t) = 1 - 21t^2 + 63t^4 - \tfrac{231}{5}t^6.$$

(e) Legendre's equation is a homogeneous linear equation. The Linearity Principle implies that $kP_\nu(t)$ is a solution for any constant k if $P_\nu(t)$ is a solution. (However, the initial conditions change by a factor of k.)

16. Throughout this solution, we assume that the two series converge for all t near zero (including $t = 0$).

(a) If $f(t) = g(t)$ for all t near zero, then $f(0) = g(0)$. Moreover, $f(0) = a_0$ and $g(0) = b_0$. So $a_0 = b_0$.

(b) Using the fact that $f'(t) = a_1 + 2a_2t + 3a_3t^2 + \ldots$, we see that $f'(0) = a_1$. Similarly, $g'(0) = b_1$. If $f(t) = g(t)$ for all t near zero, then $f'(t) = g'(t)$ for all t near zero. In particular, $f'(0) = g'(0)$, and therefore, $a_1 = b_1$.

(c) In general, the nth derivative of f is given by a power series of the form

$$n!\,a_n + \text{ terms having } t \text{ as a factor.}$$

Consequently, $n!\,a_n = f^{(n)}(0)$. Similarly, $g^{(n)}(0) = n!\,b_n$. If $f(t) = g(t)$ for all t near zero, then $f^{(n)}(t) = g^{(n)}(t)$ for all t near zero. In particular, $f^{(n)}(0) = g^{(n)}(0)$, and therefore, $a_n = b_n$ for all n.

17. We guess the power series $y(t) = a_0 + a_1 t + a_2 t^2 + a_3 t^3 + a_4 t^4 + \ldots$. Using the power series for e^u with $u = -t$ (see Appendix C), we have

$$e^{-t} = 1 - t + \frac{t^2}{2!} - \frac{t^3}{3!} \pm \ldots.$$

Substituting the power series for $y(t)$ and dy/dt into the differential equation, we get

$$a_1 + 2a_2 t + 3a_3 t^2 + 4a_4 t^3 + \ldots = \\ -\left(a_0 + a_1 t + a_2 t^2 + a_3 t^3 + \ldots\right) + \left(1 - t + \frac{t^2}{2!} - \frac{t^3}{3!} \pm \ldots\right).$$

Equating coefficients for each power of t yields

$$\begin{aligned} a_1 &= -a_0 + 1 \\ 2a_2 &= -a_1 - 1 \\ 3a_3 &= -a_2 + \tfrac{1}{2} \\ 4a_4 &= -a_3 - \tfrac{1}{6} \\ &\vdots \end{aligned}$$

In fact, by comparing coefficients of t^n, we get

$$(n+1)a_{n+1} = -a_n + (-1)^n \frac{1}{n!}.$$

Using the initial condition $y(0) = a_0 = 0$, we observe that

$$a_0 = 0, \quad a_1 = 1, \quad a_2 = -1, \quad a_3 = \tfrac{1}{2}, \quad a_4 = -\tfrac{1}{6}, \quad \ldots.$$

Using induction, we can show that

$$a_n = (-1)^{n-1} \frac{1}{(n-1)!}$$

for $n \geq 2$. The power series solution is

$$\begin{aligned} y(t) &= t - t^2 + \frac{t^3}{2!} - \frac{t^4}{3!} \pm \ldots \\ &= t\left(1 - t + \frac{t^2}{2!} - \frac{t^3}{3!} \pm \ldots\right), \end{aligned}$$

which is the power series for te^{-t}.

18. We guess the power series $y(t) = a_0 + a_1 t + a_2 t^2 + a_3 t^3 + a_4 t^4 + \ldots$. Using the power series for $\cos u$ where $u = 2t$ (see Appendix C), we have

$$\cos 2t = 1 - \frac{(2t)^2}{2!} + \frac{(2t)^4}{4!} \mp \ldots.$$

Substituting the power series for $y(t)$, d^2y/dt^2, and $\cos 2t$ into the differential equation, we get

$$(2a_2 + (2\cdot 3)a_3t + (3\cdot 4)a_4t^2 + \ldots) + 4(a_0 + a_1t + a_2t^2 + a_3t^3 + a_4t^4 + \ldots)$$
$$= 1 - \frac{(2t)^2}{2!} + \frac{(2t)^4}{4!} \mp \ldots.$$

Note that the coefficient of t^n on the left-hand side is

$$(n+2)(n+1)a_{n+2} + 4a_n,$$

and on the right-hand side, it is 0 if n is odd and

$$\frac{(-1)^{n/2}2^n}{n!}$$

if n is even. Therefore, the value of a_{n+2} is determined by the value of a_n.

We are assuming that $a_0 = 0$ and $a_1 = 1$. To determine the coefficients of the odd powers of t, we use the relation

$$(n+2)(n+1)a_{n+2} + 4a_n = 0.$$

In other words,

$$a_{n+2} = \frac{-4}{(n+2)(n+1)}a_n.$$

To see the pattern, it helps to write a few of these coefficients in terms of a_1. That is,

$$a_3 = \frac{-2^2}{3\cdot 2}a_1, \quad a_5 = \frac{2^4}{5\cdot 4\cdot 3\cdot 2}a_1, \quad a_7 = \frac{-2^6}{7\cdot 6\cdot 5\cdot 4\cdot 3\cdot 2}a_1, \ldots.$$

Writing these coefficients using the notation $n = 2k+1$ and recalling that $a_1 = 1$, we get

$$a_{2k+1} = \frac{(-1)^k 2^{2k}}{(2k+1)!}.$$

The odd powers of t form the series

$$t - \frac{2^2t^3}{3!} + \frac{2^4t^5}{5!} - \frac{2^6t^7}{7!} \pm \ldots$$

which can be rewritten as

$$\frac{1}{2}\left((2t) - \frac{(2t)^3}{3!} + \frac{(2t)^5}{5!} - \frac{(2t)^7}{7!} \pm \ldots\right).$$

This series is the power series of the function $y_1(t) = \frac{1}{2}\sin 2t$, which is a solution of the associated homogeneous equation.

To determine the coefficients of the even powers of t, we use the relation

$$(n+2)(n+1)a_{n+2} + 4a_n = \frac{(-1)^{n/2}2^n}{n!}.$$

In other words,

$$a_{n+2} = \frac{-4}{(n+2)(n+1)} a_n + \frac{(-1)^{n/2} 2^n}{(n+2)!}.$$

Since $a_0 = 0$, we see that $a_2 = 1/2$. Similarly, we get

$$a_4 = -\frac{1}{3}, \quad a_6 = \frac{1}{15}, \quad a_8 = -\frac{2}{312}, \quad \ldots.$$

The pattern in this case is harder to observe. Using the notation $n = 2k$, we see that

$$a_{2k} = \frac{(-1)^{k+1} 2^{2k-3}}{(2k-1)!}.$$

Using this formula, we note that the even powers of t form the series

$$\frac{t^2}{2} - \frac{2^1 t^4}{3!} + \frac{2^3 t^6}{5!} - \frac{2^5 t^8}{7!} \pm \ldots,$$

which can be rewritten as

$$\frac{t}{4}\left((2t) - \frac{(2t)^3}{3!} + \frac{(2t)^5}{5!} - \frac{(2t)^7}{7!} \pm \ldots \right).$$

This series is the power series of the function $y_2(t) = \frac{1}{4} t \sin 2t$, which is the particular solution of the nonhomogeneous equation obtained by using the guessing technique discussed in Chapter 4.